Cloud research is a rapidly developing branch of climate science that is vital to climate modelling. With new observational and simulation technologies, our knowledge of clouds and their role in the warming climate is increasing. This book provides a comprehensive overview of research on clouds and their role in our present and future climate, covering theoretical, observational and modelling perspectives. Part I discusses clouds from three different viewpoints: as particles, light and fluid. Part II describes our capability to model clouds, ranging from theoretical conceptual models to applied parameterised representations. Part III describes the interaction of clouds with the large-scale circulation in the tropics, mid latitudes and polar regions. Part IV describes how clouds are perturbed by aerosols, the land surface and global warming. Each chapter contains end-of-chapter exercises and further-reading sections, making this an ideal resource for advanced students and researchers in climatology, atmospheric science, meteorology and climate change.

A. Pier Siebesma is a professor of atmospheric science at the Delft University of Technology, and is also affiliated with the Royal Netherlands Meteorological Institute (KNMI), where he has been researching the role of clouds and convection in our climate system since 1990. During a consultancy at the European Centre for Medium-Range Weather Forecasts (ECMWF), he co-designed the eddy-diffusivity mass-flux (EDMF) approach to parameterise convective transport by cumulus convection in numerical weather prediction and climate models. He has been the coordinator of the FP7 European Project Euclipse (European Union Cloud Intercomparison, Process Study and Evaluation Project).

Sandrine Bony is a director of research at CNRS (the French National Centre for Scientific Research), working in the Laboratory of Dynamical Meteorology of Sorbonne University, the École Normale Supérieure and the École Polytechnique. Her research aims at understanding the role of clouds in climate and climate change through the analysis of model simulations, satellite observations and in-situ data from field experiments. She was a lead author of the Fourth Assessment Report of the Intergovernmental Panel on Climate Change and has been coordinating several activities of the World Climate Research Programme (WCRP) related to cloud and climate research. Currently, she co-leads (with B. Stevens) the WCRP Grand Challenge on Clouds, Circulation and Climate Sensitivity.

Christian Jakob is a professor at the School of Earth, Atmosphere and Environment at Monash University. He is interested in climate models and how scientists around the world can work together to improve them. He has worked for organisations as varied as the Monash Univeristy, the European Centre for Medium-Range Weather Forecasts and the Australian Bureau of Meteorology. He has led several weather and climate model development activities of the World Meteorological Organization, and was a lead author of the chapter on model evaluation for the Fifth Assessment Report of the Intergovernmental Panel on Climate Change.

Bjorn Stevens is a director at the Max-Planck-Institute for Meteorology, and an honorary professor at the University of Hamburg. His research aims to understand how water in the atmosphere, especially in the form of clouds, shapes climate. He has played a leading role in developing and executing major international field programmes, modelling studies and scientific assessments related to clouds and climate. He served as a lead-author of Chapter 7, 'Cloud and Aerosols', for the Fifth Assessment Report of the Intergovernmental Panel on Climate Change and co-leads (with S. Bony) the World Climate Research Programme (WCRP) Grand Challenge on Clouds, Circulation and Climate Sensitivity.

Clouds and Climate

Climate Science's Greatest Challenge

Edited by

A. Pier Siebesma
Delft University of Technology

Sandrine Bony
Laboratoire de Meteorologie Dynamique

Christian Jakob
Monash University

Bjorn Stevens
Max-Planck-Institut für Meteorologie

CAMBRIDGE
UNIVERSITY PRESS

CAMBRIDGE
UNIVERSITY PRESS

Shaftesbury Road, Cambridge CB2 8EA, United Kingdom

One Liberty Plaza, 20th Floor, New York, NY 10006, USA

477 Williamstown Road, Port Melbourne, VIC 3207, Australia

314–321, 3rd Floor, Plot 3, Splendor Forum, Jasola District Centre, New Delhi – 110025, India

103 Penang Road, #05–06/07, Visioncrest Commercial, Singapore 238467

Cambridge University Press is part of Cambridge University Press & Assessment,
a department of the University of Cambridge.

We share the University's mission to contribute to society through the pursuit of
education, learning and research at the highest international levels of excellence.

www.cambridge.org
Information on this title: www.cambridge.org/9781107061071

DOI: 10.1017/9781107447738

First published 2020

A catalogue record for this publication is available from the British Library

Library of Congress Cataloging-in-Publication data
Names: Siebesma, A. Pier, editor. | Bony, Sandrine, editor. |
    Jakob, Christian, 1967– editor. | Stevens, Bjorn, editor.
Title: Clouds and climate : climate science's greatest challenge / edited
    by A. Pier Siebesma, Sandrine Bony, Christian Jakob, Bjorn Stevens.
Description: Cambridge ; New York, NY : Cambridge University Press, 2020. |
    Includes bibliographical references and index.
Identifiers: LCCN 2019055984 (print) | LCCN 2019055985 (ebook) |
    ISBN 9781107061071 (hardback) | ISBN 9781107637559 (paperback) |
    ISBN 9781107447738 (epub)
Subjects: LCSH: Clouds.
Classification: LCC QC921 .C566 2020 (print) | LCC QC921 (ebook) |
    DDC 551.57/6–dc23
LC record available at https://lccn.loc.gov/2019055984
LC ebook record available at https://lccn.loc.gov/2019055985

ISBN   978-1-107-06107-1   Hardback

Additional resources for this publication at www.cambridge.org/siebesma

Contents

v

Contributors

Gilles Bellon
Department of Physics, University of Auckland

Sandrine Bony
Laboratoire de Météorologie Dynamique / Institut Pierre Simon Laplace, Centre National de la Recherche Scientifique and Sorbonne Université

Hélène Chepfer
Laboratoire de Météorologie Dynamique / Institut Pierre Simon Laplace, Centre National de la Recherche Scientifique and Sorbonne Université

Stephan de Roode
Department of Geoscience & Remote Sensing, Delft University of Technology

Jean-Louis Dufresne
Laboratoire de Météorologie Dynamique / Institut Pierre Simon Laplace, Centre National de la Recherche Scientifique and Sorbonne Université

Kevin Grise
Department of Environmental Sciences, University of Virginia

Cathy Hohenegger
Max Planck Institute for Meteorology

Christian Jakob
School of Earth, Atmosphere and Environment, Monash University

Ulrike Lohmann
Institute for Atmospheric and Climate Science ETH Zürich

Thorsten Mauritsen
Department of Meteorology, Stockholm University

Roel Neggers
Institut für Geophysik und Meteorologie, Universität zu Köln

Louise Nuijens
Department of Geoscience & Remote Sensing, Delft University of Technology

Hanna Pawlowska
Institute of Geophysics, Faculty of Physics, University of Warsaw

Robert Pincus
Cooperative Institute for Research in Environmental Sciences, University of Colorado, and Physical Sciences Division, NOAA/Earth System Research Lab

Johannes Quaas
Institute for Meteorology, Universität Leipzig

Christoph Schär
Institute for Atmospheric and Climate Science, ETH Zürich

Axel Seifert
Deuscher Wetterdienst

Ben Shipway
UK Met Office

A. Pier Siebesma
Department of Geoscience & Remote Sensing, Delft University of Technology and Royal Netherlands Institute for Meteorology

Bjorn Stevens
Max Planck Institute for Meteorology and Center for Earth System Research and Sustainability, University of Hamburg

Gunilla Svensson
Department of Meteorology, Stockholm University

George Tselioudis
NASA Goddard Institute for Space Studies

Preface

Clouds have always fascinated humans, but never has the need to understand them been so vital. As global surface temperature increases, and human activities influence particulate matter in the atmosphere and the properties of the land surface, clouds are expected to change, with manifold consequences for the climate. Clouds influence climate through their regulation of radiant energy transfer, through their role in convective energy transport, and in mediating the water cycle.

Cloud research has never been so exciting. It is a topic with many new opportunities. New satellite observations from space with active instruments such as lidar and radar allow for the first time to reconstruct the three-dimensional distribution of clouds across Earth. Likewise, numerical simulations are beginning to globally resolve the three spatial dimensions of clouds as well as their dynamic evolution. This is enabling researchers, for the first time, to link the small-scale cloud processes and the basic laws that govern them to the general circulation of the atmosphere. These new types of observation, and simulation, are enabling researchers to distinguish how different cloud types influence the climate, and are guiding conceptual representations of their collective behaviour.

The rich variety of approaches used to study clouds demands a great deal from the modern researcher. The great majority of books approach clouds from a particular perspective, such as cloud microphysics, cloud dynamics or clouds and radiation. A few books mention how clouds influence the climate system, but usually in quite an elementary fashion. No book brings these different aspects together in a way designed to bring the reader to the forefront of research on clouds and climate. This book aims to fill that gap. The main objective is to provide researchers, students and teachers a modern overview of the theoretical, observational and modelling perspectives that contribute to our scientific understanding of clouds.

This book grew out of the European Union Cloud Intercomparison, Process Study and Evaluation Project (EUCLIPSE). This project brought together a collection of world-class scientists and made it possible to write a book that aspired to illuminate clouds from so many different directions. Despite being a multi-authored enterprise, care has been taken to create a coherent textbook, both in content and notation, rather than a collection of independent chapters.

Clouds and Climate is organised into four parts. Fundamentals are presented in Part I and discuss clouds from three different viewing points: a fluid-dynamical description of clouds (Chapter 2), clouds seen as a collection of particles (Chapter 3) and clouds interacting with radiation (Chapter 4). Part II describes our understanding of clouds from a modelling perspective through conceptual models (Chapter 5) and parameterisations (Chapter 6), which represent the effect of clouds in weather and climate models. Chapter 7 presents evaluation methods that aim to assess how successful these parameterisations end up being. Part III (Chapters 8–10) describes the interaction of clouds with the large-scale circulation in the tropics, the mid-latitudes and the polar regions. Part IV concludes by describing how clouds are perturbed by: aerosols (Chapter 11), the land surface (Chapter 12), and global warming (Chapter 13).

Most chapters outline material that could, and in some cases do, concern an entire textbook. Rather than reviewing the literature, chapters emphasise the essentials, with the aim of outlining the frontiers of our knowledge, and providing an entry point to the broader literature. Sections on Further Reading are included for those desiring additional background or preparation. Each chapter also ends with exercises that extend and complement the material presented in the body of the chapter.

Clouds and Climate benefitted greatly from the critical feedback and insights of colleagues who kindly reviewed each chapter, as well as from informal comments and feedback provided by many others. For the reviews we thank Bruce A. Albrecht, Peter Bechtold, Alan K. Betts, Christopher S. Bretherton, Anthony D. Del Genio, Hervé Douville, Annica Ekman, Kerry A. Emanuel, Graham Feingold, Richard Forbes, Piers M. Forster, Pierre Gentine, Andrew Gettelman, Jean-Christophe Golaz, Jerry Y. Harrington, Dennis L. Hartmann, Richard H. Johnson, Jennifer E. Kay, Axel Kleidon, Stephen A. Klein, Pascal Marquet, Juan Pedro Mellado, John F. B. Mitchell, J. David Neelin, Paul O'Gorman, Olivier Pauluis, David A. Randall, Mark A. Ringer, Brian J. Soden and David W. Thompson. We would also, and especially, like to thank the many students and post docs who, in addition to providing feedback and corrections, also served as motivation for this book. Among these, we are particularly indebted to the participants of the 2013 Clouds and Climate Summer School, which was held at the beautiful Les Houches School of Physics, nestled in the French Alps, near Chamonix. Lectures from the summer school formed the outline of the material presented in this book and the enthusiasm of the participants gave impetus to our efforts. We hope that this, the resulting book, helps transport some of the spirit of that school to stimulate future generations of researchers to study cloud processes and their role in our changing climate.

1 Cloudy Perspectives

Louise Nuijens and Christian Jakob

A collection of vortices with an eddying nature ... full of water, being necessarily precipitated when full of rain, then they fall heavily upon each other, and burst, and clap.

(Socrates in Aristophanes' play *The Clouds*, 417 BC)

What is a cloud?

Everyone can picture a cloud or describe a cloud. Yet there is no singular description of a cloud, even with physical laws in mind, such as Socrates' speculations in Aristophanes' play. Formal descriptions of clouds may read, for instance: clouds are turbulent flows; clouds are multiphase flows; clouds are a collection of hydrometeors; clouds are a means to modulate electromagnetic radiation; clouds are a source of heat; or clouds are a crucial component to circulations.

Long before clouds were scientifically recognised, they had different meanings to society. Clouds brought rain and therefore fertility, but clouds also brought hail, thunder and gusts, and along with that devastation. In mythology, clouds were closely associated with the expressions of gods: gods would send rain when in good spirits, but on bad days, skies would be threatening and filled with thunderbolts. This fascination for and fear of clouds inspired painters and poets for many centuries.

This chapter describes different views on clouds and their meaning throughout history, leading to where cloud research stands today. In the absence of a unique description, the chapter relies on everyone's own picture of a cloud. But one thing will become clear: clouds are key to understanding and predicting how Earth's climate is changing, and, therefore, they are even more fascinating than painters and poets could have thought. Clouds have become one of the wildcards of climate change, a fact that has changed cloud research.

In today's research, clouds are associated with phenomena that have scales much larger than that of a cloud, and different views on clouds have become intertwined. That clouds are connected to large-scale

phenomena through the flows in which they are embedded is especially clear when viewing clouds from space. A collection of photographs taken from space exemplify the perspective we have gained (Fig. 1.1). The long-stretched thick cloud decks, the streets of clouds over land and over ocean, and the neighbouring cells of buoyant convective towers all reveal structures that are far larger than that of individual clouds. Clouds align with the atmospheric flow and they can be influenced by the underlying surface, such as in Fig. 1.1a. In Fig. 1.1b clouds are absent only over the river and its banks. Clouds also organise themselves within the mean flow, as Fig. 1.1c suggests, or clump together to form much bigger clusters, as in Fig. 1.1d. Different types of clouds can also occur in the same region, such as the cellular convection underneath the thick cloud decks in the mid-latitude cyclone in Fig. 1.1a, or the shallow cumulus clouds surrounding the deep thunderstorms in Fig. 1.1d. In other words, clouds are not just blown around by the winds as passive tracers, but are actively arranging themselves within the mean flow.

An emerging idea is that clouds do not just influence climate through their reflection of sunlight or by trapping infrared radiation but that they help shape the flow and circulations in which they are embedded (Bony et al., 2015). They could do so in several ways, many of which are not understood, and many of which are not easy to study because of the wide range of scales that are involved. Clouds may change heating rates of the atmosphere, which they can do through precipitation or through their interaction with radiant energy transfer through the atmosphere (not just at the surface or at the top of the atmosphere [TOA]). Precipitation and cloud–radiation interactions depend on how small-scale turbulence, large-scale dynamics and microphysics help shape the structure of clouds. Differential heating rates, and thus temperature and pressure gradients, can help drive atmospheric winds. More directly, convective clouds transport air across different height levels within the atmosphere, through

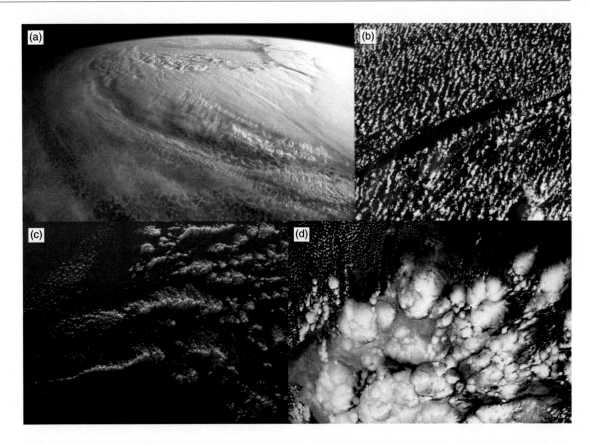

FIGURE 1.1: Photographs taken by the astronaut Alexander Gerst from the International Space Station (Gerst, 2017). (a) Cloud fields in an extratropical cyclone; (b) cumulus streets over land; (c) field of cumulus over the ocean; (d) organised convective cells. Copyright © [2014] ESA/NASA

their updrafts and downdrafts. Those air masses have a different temperature, humidity and momentum, so that clouds effectively mix the atmosphere, and influence large-scale momentum and energy budgets.

It is not hard to imagine that cloud-radiative effects and precipitation over a large area depend on the structure of clouds and of water vapour, which is the most abundant greenhouse gas. For instance, the convective cells that are clumped together in Fig. 1.1d are full of water vapour and very bright. But the areas around it reveal the dark ocean, and are likely much drier. Such contrasts in water vapour and clouds lead to contrasts in atmospheric heating rates, which are a function of vertical gradients in the fluxes of short-wave and long-wave radiation. In another example, Fig. 1.1a shows cellular-like structures below the thick cloud decks, which are thought to be caused by precipitation. A strong rain shaft can trigger a convective downdraft, which brings cold and dry air towards the surface. The resulting cold pool spreads out like a density current, and triggers new convective clouds when it collides with other currents.

Such examples demonstrate that all facets of cloud research are important, including microphysics, turbulence, radiation, the large-scale flow and more. The purpose of this book is to bring together these different meanings of clouds and help a new generation of scientists understand how clouds may influence climate in many different ways.

Part I of the book (Chapters 2–4) includes the necessary theory. It will explain basic theories of clouds as radiative entities, clouds as turbulent and multiphase flows, and clouds as a collection of hydrometeors. None of these are designed to provide an ultimate background in radiation, thermodynamics, convection or microphysics. Excellent books have been written about these subjects, and each chapter will guide the interested reader to such books. Instead, these chapters are meant to provide the basic knowledge necessary to understand Parts II–IV of this book.

Part II of the book (Chapters 5–7) explains current techniques that are used to study clouds, including theoretical models, global models and comparing models to observations. Part III of the book (Chapters 8–10) covers cloud systems that are found in three broad but distinct regions on Earth (the subtropics and tropics, the extra-tropics and the polar regions) and how these cloud systems matter to climate. Part IV of the book (Chapters 11–13) describes interactions between clouds and other aspects of the Earth system, including aerosols, the Earth's surface and climate (change).

The current understanding of clouds, and the ideas and questions that are emerging today, are the result of more than a century of cloud research. Being mindful of lessons learned in the past can help understand why important realisations were made, why cloud research is where it is today and why this is an exciting time. This introductory chapter will review how humankind first started to observe and document clouds in a coordinated way, which has become increasingly important as trends in cloudiness can provide insights into climate change. The chapter will also review how the scientific view on clouds has shifted throughout the century. Which questions interested researchers at the beginning of last century and why? Have they been answered? What discoveries spurred technologies? The chapter will describe important observing systems and modelling capabilities, and it will discuss the challenges and opportunities in cloud research that lie ahead.

1.1 Observing and Documenting Clouds

Already before clouds were scientifically recognised, they had a strong imprint on society, but any knowledge on clouds remained largely unorganised and undocumented. Only with the introduction of a cloud nomenclature in the early nineteenth century did clouds become scientifically meaningful. The nomenclature allowed humankind to observe clouds in similar ways all around the globe and to classify them, which marked the change from an ancient to a modern mind frame, when imagination and fear made place for the sense that measuring and describing clouds would lead to a better understanding of nature, and with that an ability to conquer nature.

1.1.1 The International Cloud Atlas

The *Essay on the Modification of Clouds* by Luke Howard is the most widely used cloud nomenclature. First published in 1803, his essay uses the Latin system of Linneaus to categorise clouds in similar ways as plants and animals were categorised many years before. Howard defines ten basic categories of clouds based on their

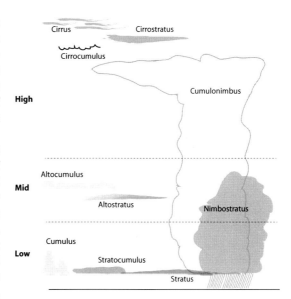

FIGURE 1.2: A schematic of Luke Howard's cloud classification.

conformation and height level as seen by the eye (Table 1.1 and Fig. 1.2) divided into low-level, mid-level and high-level clouds. For each 'genus' he also specifies different species and varieties that are not listed here. Howard specifically chose to name his essay the *Modification of Clouds,* because he recognised that clouds were not necessarily steady in time and space. He noted that clouds could transition from one type into another, even though at that time the mechanisms driving cloud transitions were largely unknown to him. He defined low-level clouds as **cumulus**, which in Latin means 'heaps' and which are those clouds with relatively flat bases and a puffy appearance. He also defined **stratus**, which refers to a flat layered type of cloud, and the intermediate form: **stratocumulus**. **Altocumulus** and **altostratus** belong to the mid-level category. As high-level clouds, Howard defined **cirrocumulus** as the intermediate form of **cirrus** and **cirrostratus**, where cirrus is Latin for 'a curly lock of hair', thus referring to thin and wispy clouds. The last two cloud types span multiple levels. **Cumulonimbus** stretches from low to high levels and is often associated with heavy rain. It usually has a well-defined cloud base and therefore belongs to the low-level cloud category. **Nimbostratus** refers to a raining grey deck of cloud, which stretches from low to mid levels and has a much less well-defined cloud base. It therefore belongs to the mid-level cloud category.

An influential man at that time named Ralph Abercromby promoted Howard's work by noting that it provided a guideline for how to observe clouds in a similar manner all around the world. He pushed the notion that similar clouds can be observed irrespective of one's location, and even travelled around the world for a few years to convince himself that this was true at least to a first approximation. He advocated that understanding clouds through observing them would allow humankind to use that knowledge to utilise nature for purposes poets and painters had never dreamt of. These goals were perhaps overly ambitious, but the 'cloud modification' experiments that took place many years later (Section 1.2.3) were not far from what Abercromby might have had in mind. Foremost, however, Abercromby's efforts played a major role in leading people to associate clouds with recently discovered physical laws on multiphase dynamical flows and turbulence, rather than with imaginary terrors. In a lecture he gave at the Philosophical Institute of Edinburgh in the year 1887, he stated:

> Contrast ancient and modern thought. Our ancestors saw in a thunderstorm the conflict between a many-headed, hairy monster with a being of superhuman strength throwing lightning and thunderbolts about. Such an attitude of mind can only induce terror. Now, when we see a thunderstorm we might observe a squall from the southwest with a velocity of sixty miles an hour just as the rain commences. Then we might measure the height of the lower base of the clouds and find it not more than five thousand feet above the Earth, while the rocky summits rise no less than fifteen thousand feet above the ground, and the rain-gauge might show that water to the depth of three inches fell out of these ten thousand feet of cloud . . . the moral effect of weighing and measuring is so great. (Abercromby, 1888, p. 114)

Following up on Howard's cloud classification and Abercromby's efforts, the first *International Cloud Atlas* was published in 1896, written by members of the cloud commission of the International Meteorological Committee (IMC). The Atlas included cloud naming schemes and photographs, but also instructions for how to observe clouds. For instance, an observer should not only determine whether a cloud was located at low, mid or high levels and whether it belonged to the cumulus or stratus type, he or she should also register the direction from which the cloud comes, the direction of parallel and equidistant stripes in the cloud deck – which we nowadays attribute to gravity waves – or the direction in which a bank of cirrus was most dense. In the authors' opinion, one of the most important applications of the *Cloud Atlas* was to document the direction of the wind at different altitudes, thereby revealing information about underlying (changing) weather patterns.

Detailed surface observations of the clouds that accompanied certain weather led to the first models of the extratropical cyclone in the early 1900s. The Norwegian frontal cyclone model was developed during and after the First World War within the Bergen School of Meteorology, and was solely based on surface-based weather observations. Descriptions of clouds found near frontal boundaries played a crucial role. Meteorologists also became increasingly interested in clouds themselves, and not just in what they revealed about the underlying wind patterns. The interest in more detailed characterisations of cloud features led to requests for more instructions on how to observe them and the *Atlas* underwent several revisions. Its latest digital version was published in 2017 by the World Meteorological Organisation (WMO) (WMO, 2017), which replaced the IMC in 1951. The ten cloud types as defined by Howard still form the core of the cloud classification (Table 1.1). The *Atlas* has also maintained the three height levels: low, mid and high, whose exact heights vary between the tropics and the polar regions approximately, as listed in Table 1.2. Because clouds can have different appearances depending on the location from which they are observed, the *Atlas* now lists detailed instructions on how to observe clouds not only from the surface, but also from aircraft and mountain stations. It also asks the observer to estimate the height of the cloud base above the point of observations and if possible the vertical extent of clouds and the direction and speed of horizontal movement. The *Atlas* also defines several parameters that reveal information about the radiative effects of clouds:

(1) **Total cloud cover**: the fraction of the celestial dome covered by all clouds visible.
(2) **Cloud amount**: the fraction of sky covered by clouds of a particular genus, species, variety, layer or combination of clouds.
(3) **Optical thickness**: the degree to which the cloud prevents sunlight from passing through it.

Instructions describe how to estimate these parameters. They are fairly specific, but it cannot be avoided that estimates remain ambiguous or subjective to a certain degree. The estimation of cloud cover and cloud amount from the surface is done using the octa scale whereby '0' corresponds to a completely clear sky (0 %) and '8' to a completely cloudy sky (100 %). Estimating cloud amount and cloud cover in the 1–7 octa bins will be much harder for a surface observer than to estimate situations that fall in the 0 and 8 octa bins. The estimation of cloud amount is also more difficult than the estimation of cloud cover. For instance, when clouds occur in superposed layers, the

TABLE 1.1: The original cloud classification defined by Luke Howard, whereby clouds belong to one of ten gender types ('*genus*') based on their aesthetic appearance and their altitude at three different 'Étages' (levels). Also shown are the official abbreviations for cloud types as identified by the World Meteorological Organisation, along with the additional cloud type *fog* and types of precipitation. The annual mean cloud amount is shown, which is derived from surface observations over land (1971–97) and from ships (1954–2008), collected by Hahn and Warren (2007). The cloud amount is the product of the frequency of occurrence and the amount-when-present.

| Étages | Genus | Abbreviation | Cloud amount (%) | | Frequency (%) | | Amount-when-present (%) | |
			Land	Ocean	Land	Ocean	Land	Ocean
High	Cirrus	Ci	—	—	—	—	—	—
	Cirrocumulus	Cc	22	12	45	34	49	35
	Cirrostratus	Cs	—	—	—	—	—	—
Mid	Altocumulus	Ac	17	17	32	37	52	47
	Altostratus	As	4	6	5	10	82	59
	Nimbostratus	Ns	5	5	5	5	98	99
Low	Cumulus	Cu	5	13	14	33	34	40
	Cumulonimbus	Cb	4	6	7	11	59	58
	Stratus	St	5	12	6	14	72	84
	Stratocumulus	Sc	12	22	21	31	34	40
	Fog	*Fo*	*1*	*2*	*1*	*2*	*100*	*100*
	Total cloud cover	**Tc**	**54**	**68**	—	—	—	—
	Clear sky	Cr	—	—	22	3	—	—
	Precipitation	Pt	—	—	9	10	—	—

TABLE 1.2: Approximate height ranges of the three 'Étages' (levels) for polar, temperature and tropical regions, from the WMO *International Cloud Atlas* (WMO, 2017).

Étages	Polar	Temperate	Tropical
High	3–8 km	5–13 km	6–18 km
Mid	2–4 km	2–7 km	2–8 km
Low	0–2 km	0–2 km	0–2 km

topmost layer may be concealed for an observer on the ground. Surface observations of low and mid cloud are therefore more accurate than that of high cloud.[1] For the optical thickness, the *Atlas* defines a scale that provides a qualitative estimate: ranging from '1', when the blue sky is still discernible through the cloud, to '5', when the cloud is dark except for its edges, which should be exposed to the sun and brilliantly white. The *Atlas* also commemorates early views on clouds by noting that a cloud with an optical thickness of 5 is well recognised by its threatening appearance.

[1] Boers et al. (2010) contains a translation from octa's to percent cloud cover.

By 1925, the observation of clouds around the world had greatly expanded, which allowed scientists to produce geographical distributions of cloud types. As Abercromby advocated, these revealed that the same cloud type can be found all over the world. However, they also revealed that specific cloud types favour specific places – the beginning of cloud climatologies and linking clouds to weather patterns.

1.1.2 Geographical Distribution of Cloud Types: Cloud Climatologies

A climatology is constructed by spatially and temporally sampling observations, and then averaging and interpolating them. How an interpolation or averaging is performed can lead to important differences in climatologies, and for some of the early climatologies such details are not well known. Nevertheless, zonal distributions of cloud cover stemming from the earlier part of the twentieth century appear remarkably similar to climatologies made with longer records of surface observations collected later in the century.

A map of the global distribution of total cloud cover is shown in Fig. 1.3. The map is constructed using what

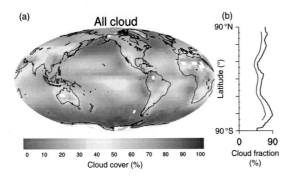

FIGURE 1.3: (a) Mean global and (b) zonal distribution of total cloud cover, derived from the longest record of surface observations over land and ocean in Hahn and Warren (2007). The red line in the zonal distribution represents one of the first (hand-drawn) climatologies of cloudiness by Brooks (1927). The maximum value on the x-axis of the zonal distribution is the maximum value found.

is now the longest record of surface-based observations of clouds, which contains twenty-five years of observations over land (1971–96) and forty-three years of observations over oceans (1954–97) (Hahn and Warren, 2007). The cloud information was collected every three hours, leading to 185 million quality-checked reports from 5,388 weather stations on continents and islands and 50 million reports from ships or land stations. The white pixels correspond to insufficient data. The map reveals that the Earth has remarkably little clear sky: the mean zonal distribution shows that no latitude on average experiences less than 45 % cloud cover. The least cloudy places on Earth, with less than 20 % of cloud cover, are found over the Sahara, the Arabian peninsula, southern Africa, Australia and the south-west coast of the United States. These are the continental desert areas, which are areas where large-scale subsidence prevails, which brings warm and dry air towards the surface, suppressing cloud formation. Over oceans, the central parts of tropical oceans just a few degrees off the equator tend to have the lowest cloud amount. Here, mostly smaller cumulus clouds are found, formed by convection driven by large surface latent heat fluxes. Their vertical extent is limited by large-scale subsidence (see also the cumulus map in Fig. 1.4). The least cloudy place over the ocean, which has also been called the great oceanic desert, is over the east-central part of the tropical Pacific just south of the equator.

Deeper cumulonimbus are more prevalent in narrow bands closer to the equator (Fig. 1.4), and towards the western boundary of the tropical Pacific (Chapter 8). In comparison, the far eastern ocean boundaries are cloudier, which is a result from the more frequent occurrence of stratocumulus clouds. At these locations, the upwelling

of cold water, which promotes large relative humidities above the surface, coincides with the subsiding branch of the Hadley circulation, which leads to favourable conditions for stratocumulus clouds (Chapter 5). That oceans are on average cloudier than land with a cloud cover of 68 % versus 54 % (Table 1.1) is mainly due to stratocumulus, along with the other two low-level cloud types: shallow cumulus and stratus.

Cloud cover tends to exceed 30 % in tropical regions, and even 70 % in temperate regions and the poles. Cloudiness is largest in the mid latitudes between 40° and 70°, especially in the southern hemisphere, where both boundary-layer clouds and frontal cloud systems or storms prevail (Chapter 9). These storms form when baroclinic waves develop due to strong upper-level westerlies that become unstable, and they are manifested at lower levels by cyclonic disturbances. These frontal cyclones consist of strong low-level pressure centres and warm and cold fronts. Another example are polar lows, which are smaller cyclones that form when cold polar air moves over warmer waters and triggers convection (Chapter 10). The cloud fields in these storms are diverse, including low and mid-level clouds that form in cold air outbreaks following the storm passage, as well as deep clouds and cirrus bands that curl outward of the storm centres. Especially the amount of low- and mid-level cloud is large here (Fig. 1.4).

Fig. 1.3 also shows the zonal distribution of one of the first climatologies, as a red line. This climatology was presented by Brooks (1927). He used more than 1,000 land stations distributed as evenly as possible over the Earth to derive monthly averages of mean cloud cover in 10 × 10° boxes. Over some regions such as the Pacific he did not have any data at all. Nevertheless, Brooks' early climatology fairly accurately estimated the zonal distribution of cloud, at least qualitatively. Several other climatologies that were made at the time, following Brooks, also show that same zonal distribution, even when their data was constructed for different years (Hughes, 1984). The same is true for the global and zonal distribution of clouds that are observed with current state-of-the-art remote sensing instruments aboard satellites (similar maps as in Figs. 1.3 and 1.4, but using satellite data, are shown in Chapter 4). This finding implies that clouds are to a first approximation persistent in time and space: locations that are cloudy one year are very likely just as cloudy the following year. The explanation is that clouds are tied to weather or circulation patterns on Earth, and if scientists had visible satellite images at the time (such as in Fig. 1.1), they would have seen the influence of weather systems on the global distribution of cloud.

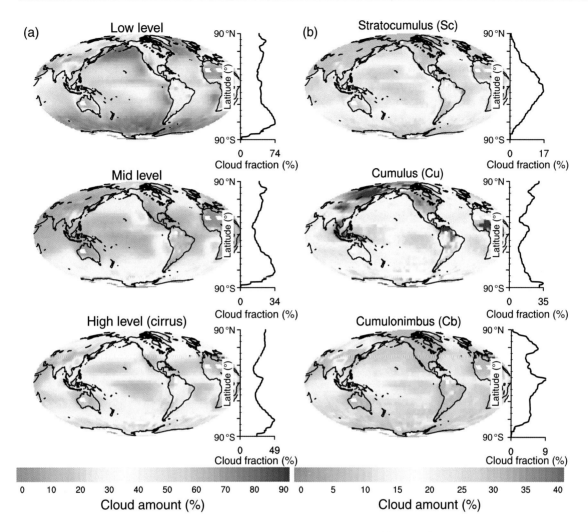

FIGURE 1.4: (a) Mean global and (b) zonal distribution of cloud amount between 90°S and 90°N for different cloud layers (low, mid and high) and for three cloud types (stratocumulus, cumulus and cumulonimbus). Data from Hahn and Warren (2007). The high-level cloud category in this figure only includes the cirrus class of clouds. The maximum value on the x-axis of the zonal distribution is the maximum value found.

Whereas the different surface-based and space-borne cloud climatologies are qualitatively similar, they differ quantitatively. One of the difficulties with estimating quantities such as cloud cover and cloud amount, as defined in the Cloud Atlas (Section 1.1.1), is that they are subjective. Two observers are not exactly alike. The issue is not only the human observer, but rather that cloudiness itself is poorly defined. How many cloud droplets need to be within a given volume to still qualify as a cloud? Do raindrops qualify as cloud? If not visible by eye, is it a cloud? Such issues have remained as human observers have been replaced by remote sensing instruments, which also include subjective choices in the retrievals they use to derive cloud parameters. For instance, they use a threshold

on the signal-to-noise ratio to decide whether a scanned volume is cloudy. As the next section describes, different remote sensing instruments view cloudiness differently, and the previous questions may seem odd, but are questions that cloud research deals with today, and they have become particularly important in the hope to derive trends in long-term records of cloudiness.

1.1.3 Remote Sensing Replacing Human Observers

Because many parts of the globe are unoccupied by humans, human observers have been mostly replaced by ground-based and space-borne remote sensing instruments. In remote sensing, a cloud or precipitation is not measured directly. Instead, cloud and precipitation properties are inferred from the way they interact with electromagnetic radiation. Generally, there are three ways that this interaction is used to infer information. These include the attenuation of a defined source of radiation through clouds and precipitation, for instance the degree at which a light beam is transmitted through a cloud, which provides information on the depth or optical thickness of that cloud. Furthermore, as a rule, matter emits radiation, and at Earth-like temperatures this radiation is predominantly emitted in the thermal infrared. By measuring the amount of emitted infrared radiation, the temperature of a cloud can be inferred. Cloud droplets and raindrops also scatter or absorb radiation depending on their size, shape and composition. For example, radars use the scattering and absorbing properties of cloud droplets and rain drops in the microwave part of the electromagnetic spectrum.

All these methods rely on what is called a retrieval method, which is a model that inverts the measured effect of clouds on radiation into inferences about cloud properties themselves. This means that different instrument sensors, or even different channels of the same instrument, can be used to retrieve the same physical parameter based on a somewhat different retrieval method. Often the retrieved parameters differ from one other, which introduces an uncertainty that needs careful attention. Each instrument carries its own source of errors. Different instruments also have a different detection sensitivity and field of view, or footprint. The larger the footprint, the larger the heterogeneity within that footprint. For instance, the measured area can include both cloudy and clear sky, both large and small clouds, and those that may or may not precipitate. Separating the signals from clear skies from the signals from clouds or precipitation is a challenging task.

At most airports the human observer has been replaced by instruments that measure cloudiness and their height level. Meteorological stations of national weather centres also perform broadband radiation measurements, which measure the downwelling and upwelling fluxes of short wave and long wave radiation. More detailed cloud measurements are performed with radars, lidars and radiometers at dedicated measurement sites, under the umbrella of national or international projects and institutions. Such sites are strategically located at places with different types of clouds. For instance, one of the largest programmes, the Atmospheric Radiation Measurements (ARM) programme from the Department

of Energy in the United States (Ackerman and Stokes, 2003) performs continuous measurements over land in Barrow (Alaska), the Southern Great Plains in Oklahoma and over subtropical oceans on the Azores. ARM sites were also located on Darwin, Manus and Nauru, but these three have been phased out. The only (sub)tropical measurement site currently present is the Barbados Cloud Observatory, initiated and maintained by the Max Planck Institute of Meteorology and the Caribbean Institute for Meteorology and Hydrology. In Europe, cloud measurements are made under the auspices of CloudNet, which includes sites in Great Britain, Germany, the Netherlands, Italy and France.

To measure clouds over oceans researchers rely on infrequent measurements aboard ships or aircraft, and of course from global measurements by satellites. The first long-term programmes to collate observations of clouds and precipitation from space, and which still exist today, are the International Satellite Cloud Climatology Project (ISCCP) and the Global Precipitation Climatology Project (GPCP), which are coordinated by the World Climate Research Programme. Both projects combine passive remote sensing information aboard polar orbiting satellites to construct a long-term global data set of cloud and precipitation. The ISCCP uses infrared and visible radiances obtained from imaging radiometers to derive cloud amount, cloud optical thickness and cloud top pressures (CTPs), whereas the GPCP merges infrared and microwave satellite estimates of precipitation with rain gauge data from more than 6,000 stations on the ground. The biggest achievement of these projects lies in the fact that they combine and blend different sources of information from different sensors and instruments to arrive at the best possible estimates of cloud and precipitation properties. The projects have by now collected records that are (almost) thirty-years long and have become the standard reference data sets to evaluate modelled global distributions of clouds and precipitation.

Many of the most recent remote sensing instruments that probe the atmosphere are aligned in a satellite constellation that is called the (Afternoon) A-Train, because it crosses the equator during the afternoon. An example of the different views of active and remote sensors aboard the A-Train is shown in Fig. 1.5. The advantage of flying the satellites in this constellation is that measurements are almost simultaneous, and therefore probe exactly the same atmosphere, albeit with a different footprint (indicated by the shaded areas or vertical lines). Passive remote sensing instruments only provide an integrated measure of the effect of clouds through the entire atmosphere.

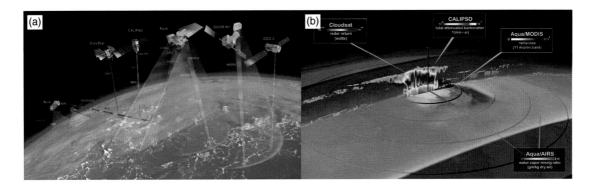

FIGURE 1.5: (a) The A-Train of satellites is a constellation of international satellites that includes OCO-2, GCOM-W1, Aqua, CALIPSO, CloudSat, PARASOL and Aura (but PARASOL ceased operation at the end of 2013). The instruments on these satellites make almost simultaneous measurements of clouds, aerosols, atmospheric chemistry and more. The figure illustrates the footprint of each of the A-Train's instruments: active instruments aboard CALIPSO/CALIOP and CloudSat/CPR are indicated with dashed lines, whereas swaths indicate the area of the Earth's surface, or the surface of its atmosphere, over which data are collected. Microwaves (observed by both AMSRs, AMSU-A, CPR, and MLS) are represented by red-purple to deep purple colours; solar wavelengths (POLDER, OMI, and OCO-2) by yellow; solar and infrared wavelengths (MODIS and CERES) by grey; and other infrared wavelengths (IIR, AIRS, TES, and HIRDLS) are represented by reds. (b) Different A-Train borne instruments provide different views on cloud systems. In this example, provided on NASA's website, tropical storm Debby crossed the central Atlantic on 24 August 2006 and was observed by four instruments: radiances measured by Aqua/MODIS are projected onto the surface plane and give an overview of the storm; Aqua/AIRS water vapour mixing ratio's are superimposed in colour onto the same surface plane; returns from CloudSat's CPR and CALIOP are shown in the colour in the y-z slice through the storm's centre. Whereas the CPR can penetrate through the cloud system towards the surface, CALIOP tends to observe only the upper parts of cloud layers and the aerosol layers closer to the surface (when no cloud is overhead). Copyright © 2012 NASA, http://atrain.nasa.gov/

For instance, the Moderate-Resolution Imaging Spectro-radiometer (MODIS) aboard the Aqua satellite measures radiances at different frequencies, from which the amount and thickness of cloud can be derived. In Fig. 1.5b, clouds are deeper as the MODIS signal turns from grey to white. Since 2006, the A-Train constellation also includes satellites carrying active remote sensing instruments. These are cloud profiling radar (CPR) aboard the CloudSat satellite and the Cloud-Aerosol Lidar with Orthogonal Polarisation (CALIOP) aboard the Cloud-Aerosol Lidar and Infrared Pathfinder Satellite Observation (CALIPSO) satellite. An important advantage of active remote sensing is that the signal is range-resolved, which means that the vertical structure of cloud and precipitation properties can be derived. For instance, the CPR aboard CloudSat shows the strongest returns in brownish-red colours where the cloud has the largest rain drops or ice crystals. The cloud in this figure is about 12 km deep and is accompanied by substantial precipitation that reaches all the way down to the surface. The change from red to green that returns at about 5 km indicates the melting level, where ice crystals turn into rain drops. The very thin layers of outflow and cirrus that accompany the cloud are best seen by CALIOP, which is better at detecting thin and small cloud, such as those farther north and south of the deeper raining cloud.

The remote sensing of clouds forms a tremendous source of information and has made automatic collection of cloud data much simpler. But some aspects of clouds are far harder to observe with remote sensing compared to the human eye. For instance, classifying cloud types based on their characteristics – as first proposed by Luke Howard (Section 1.1.1) – is a challenge. However, what matters is the way that a cloud interacts with the climate system, regardless of its genus, species or variety. Although the cloud classification is still invaluable as a way of communicating different cloud types with similar properties, cloud research relies far less on the distinction of cloud types. Throughout the years different ways of categorising clouds have been developed, largely driven by an interest in the processes underlying clouds, or by the desire to compare observations with models. In the following sections we take a step back in time to describe which interests drove cloud research, how clouds gained different meanings, and what tools and classifications have been developed to infer the role of clouds in climate.

1.2   Through Science, Clouds Gain New Meanings

This dive into history starts with ideas on how clouds modify the radiation budget of the Earth, around the time that the first *Cloud Atlas* was published. It is followed by studies that viewed clouds as turbulent entities, as producers of global rain and as crucial component of circulation patterns. We will also present new ways of classifying observations of clouds. This will make it clear that some aspects of clouds and their interaction with climate cannot be observed. These gaps are filled in by models, which have become essential in cloud research.

1.2.1 Clouds as Radiative Entities

1.2.1.1 Earth's Energy Balance

During the nineteenth century, physicists were intrigued with the question of whether the atmosphere plays a large role in determining the surface temperature on Earth. It was suspected that atmospheric gases absorbed sunlight and infrared radiation from the surface, so that heat is retained in the atmosphere, which prevents the surface from getting rather cold. Of all the gases that the atmosphere contains, water vapour and carbon dioxide were considered the most important players, but their spectroscopic details were poorly understood. In 1896, Svante Arrhenius was the first to build a simple model of the Earth and its atmosphere to assess the influence of water vapour and carbon dioxide on the surface temperature (Arrhenius, 1896). His model assumed a simple equilibrium between the Earth's surface and the absorbing atmosphere, whereby the atmosphere consisted of just a single uniform layer (see also the pure radiative equilibrium model in Chapter 5). In his calculation he took into account that atmospheric gases absorb both in the solar and in the infrared part of the electromagnetic spectrum, and that the Earth's surface reflects sunlight by a factor A, which we know as the *surface albedo*. To determine this albedo, Arrhenius first considered the Earth's surface, covered by either ground, water or snow. Following that, he shortly considered how to deal with clouds in his calculations, which he rightfully noted prevent a large part of the sunlight reaching the Earth surface. He decided to assume both a constant number for the cloud cover and a constant number for the albedo of clouds. To estimate the first, he used the latest work of his colleagues, who were putting together the first *Cloud Atlas*, and ended up with a mean value of 52.5 %. To estimate the second, he noted that clouds likely have an albedo that lies somewhere between that of fresh and old snow and used $A = 0.5$.

Arrhenius mentioned the influence that clouds may have on radiation emitted by the underlying surface. He also realised that multiple atmospheric layers would be necessary to accurately predict the radiative transfer between the surface and outer space. He also took note

of the fact that cloudiness differs with latitude, and that his final calculations might change if those variations were taken into account. Nevertheless, although he knew that clouds are abundant, covering at least 50 % of the Earth's surface, and that different cloud types occur at different height levels in the atmosphere, clouds and their variability were not given serious consideration. And, like Arrhenius, most studies in the earlier half of the twentieth century assumed clouds as given. Instead, Arrhenius' paper focused on the speculation that by burning all fuel at the current rate, humankind could raise atmospheric gas concentrations and cause a global warming – a speculation which remained largely controversial until Guy Stewart Callendar took up a hobby of recovering a carbon dioxide time series in 1938.

At the beginning of the twentieth century the Smithsonian Institute in the United States started a programme to routinely measure radiation from the sun and its variability. These measurements were done using pyrheliometry, in which the heating from absorption of solar radiation is compared with the heating from the same detector by the dissipation of a given amount of electrical power. These instruments were deployed at mountain observatories in the United States, and estimated the solar irradiance to be within 2 % of its currently accepted value ($1360.8(5)\,\mathrm{W\,m^{-2}}$). When contrasting solar irradiances for clear skies and cloudy skies at these locations, the albedo of clouds could be estimated. Additionally, pyrheliometers were deployed aboard balloons which could fly over low-lying cloud. One of the first reported cloud albedo's was that from the stratocumulus decks and fog layers that prevail over the coastal plains of Southern California. In the middle of the twentieth century, these few 'golden' numbers estimated by researchers at the Smithsonian Institute, as well as Arrhenius' estimated cloud cover of 52.5 %, were used by others to estimate the albedo of the combined Earth-atmosphere system, the *planetary albedo*. They derived the mean effect of clouds on the solar irradiance just for the United States, using the Smithsonian network, and extrapolated these effects to other latitudes and longitudes where such measurements were not standardised. These simple calculations provided a planetary albedo of around 0.43. This number was soon challenged by astronomers, who used a very different measurement to estimate the planetary albedo, something they called 'earthshine observations'. They observed the side of the moon that is not lit by sunlight directly, but only indirectly by the sunlight reflected off the Earth. The range of albedos derived from the degree of illumination on the dark side of the moon suggested that an albedo larger than 0.35 would be unreasonable.

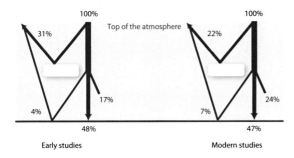

Early studies   Modern studies

FIGURE 1.6: The contributions of different components of the short-wave radiative fluxes to the total reflected radiation, as derived from early studies on the Earth energy budget and modern satellite observations. The white boxes indicate that most, but not all, atmospheric reflection of solar radiation is associated with clouds.

Indeed, using space-borne observations designed specifically for measuring Earth's radiation budget, the planetary albedo is currently estimated at 0.29. The fluxes of short-wave radiation through the atmosphere in early and modern views of the energy budget are illustrated in Fig. 1.6. Early estimates of the cloud albedo turned out to be too large (31 % compared to 23 %), whereas early estimates of the surface albedo were too small (4% compared to 6%). Estimates of the amount of solar absorption by atmospheric gases were also too small (17 % compared to 24 %). In other words: more sunlight is reflected by the surface than what was thought, but this larger surface albedo is largely compensated for by less reflection of sunlight by clouds and more absorption by gases. Because we now know that Earth has a cloud cover that is larger than the estimated 52.5 % used by Arrhenius, the optical thickness of clouds turns out to be smaller than initially thought. One of the cloud types responsible for this discrepancy is optically thin high cloud, which is ubiquitous on Earth, but can be difficult to spot by an observer at the surface. Nevertheless, it is fair to say that these studies in the pre-satellite era were quite successful, because they obtained a reasonable estimate of the planetary albedo with very few observations of clouds.

Modern estimates of all major components of the Earth's global mean energy budget are depicted in Fig. 1.7. The energy budget is constructed by combining a large number of observations with modelling of energy flows where necessary, following the pioneering work of Trenberth et al. (2009). Despite their fundamental importance to the climate system, and indeed to life on Earth, quantifying the energy flows through the system, even in a global mean sense, remains challenging. Fig. 1.7 expresses the uncertainties in each of the

components based on estimates by Wild et al. (2015). Thanks to advances in our ability to observe the Earth from space, the values at the TOA are, perhaps surprisingly, the most certain. The energy received from the sun, on average 340 W m^{-2} to 341 W m^{-2} is balanced by reflected solar radiation (29 %) and the emission of long-wave radiation (71 %). Uncertainties in our estimates of the energy flows at the surface of the Earth are larger. Here, satellite observations cannot easily be used to estimate all components (e.g., downward fluxes of radiation) and surface observations do not cover the globe sufficiently. As a result, estimates must necessarily include model results, such as those from radiative transfer models or reanalyses. This leads to uncertainties in the surface energy budget components that are an order of magnitude larger than those at the TOA. Considering radiation alone, the surface warms through solar radiation and long-wave radiation from the atmosphere (the greenhouse effect) and cools through the emission of long-wave radiation. An imbalance of O (100 W m^{-2}) exists between these three fluxes. This imbalance is accounted for by the turbulent transport of heat – the sensible heat flux – and water vapour – the latent heat flux – from the surface into the atmosphere. The evaporation of water from the surface, in a global average sense, is the sole source of water in the atmosphere, and ultimately gives rise to the clouds and associated precipitation that are the subject of this book.

1.2.1.2   Long-Wave in Addition to Short-Wave Effects of Clouds

The next step towards making more and better observations of clouds and their optical properties did not take place until the end of the 1960s, when Arrhenius' ideas about the influence of the atmosphere on surface temperature were more seriously picked up by climate scientists. In one of the first breakthrough studies, Manabe and Wetherald (1967) turned Arrhenius' one layer of cloud into three layers and found that an atmosphere with more low-level cloud has a colder Earth surface, because more solar radiation is reflected. However, an atmosphere with more high-level cloud, due to its interference with outgoing infrared radiation, had the ability to warm the surface. The researchers probably did not anticipate the impact of these results at the time, because they only included them as a side comment in the conclusions. Nevertheless, they were one of the first to show that the vertical location of clouds helps determine whether they heat or cool the atmosphere relative to clear skies.

The effect of clouds on radiation was given a more formal footing a few years later by Schneider (1972). He

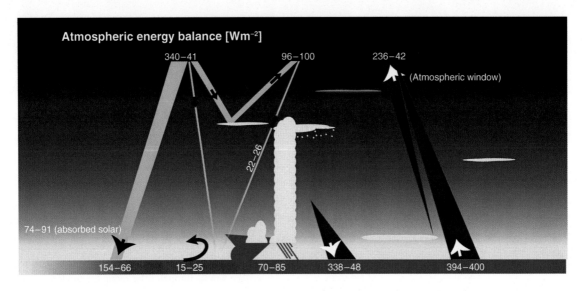

FIGURE 1.7: The major components of Earth's global mean energy balance. Figure adapted from Stevens and Schwartz (2012) with numbers from Wild et al. (2015). Fluxes associated with solar radiation are depicted in yellow, while those in the long-wave part of the spectrum are shown in red. The surface fluxes of sensible and latent heat are shown as red and blue arrows at the surface.

introduced what is currently known as the 'background' cloud radiative effect, which is the difference between the radiative fluxes in an atmosphere with clouds, the all-sky component, and the radiative fluxes in an atmosphere with zero cloud amount, the clear-sky component. Schneider asked if it is possible that clouds can globally amplify or dampen the changes induced by changing other components of the climate system, but, although he talked about clouds as a feedback mechanism, his calculations did not actually estimate the cloud feedback as we know it (Chapter 13).

More explicitly than Manabe and Wetherald, Schneider advocated for a more detailed treatment of clouds, because clouds with different cloud top heights, thicknesses and absorptivity can have different effects on the surface temperature. Hereby, both short-wave and long-wave effects need to be considered. Thin cirrus can lead to a heating, because long-wave effects dominate over short-wave effects, but thick cirrus can lead to a cooling. These two components are not independent, but coupled with other components of the atmosphere, such as atmospheric water vapour or temperature. This coupled and non-linear nature of feedbacks was not considered in such early studies, but since then methods have been developed that account for this coupling (Chapter 13).

The need for more detailed knowledge of clouds pushed the launch of a couple of large observational efforts: the start of the satellite-observing era. This included ISCCP (Section 1.1.3), which focused on

mapping the amount, height and optical thickness of clouds globally, as well as their long-term, seasonal, synoptic and diurnal variability (Rossow and Schiffer, 1999). Furthermore, the Earth Radiation Budget Experiment (ERBE) was designed to measure both the short-wave radiation and the long-wave radiation fluxes at the TOA using broadband radiation measurements. It was launched in 1984 and decommissioned in 1993. The goal of these missions was to determine the global net effect of clouds, and also whether and how this effect could change with an increase in surface temperature. Some researchers speculated that even when cloud amount changes with surface temperature, it would not influence the net radiative budget, because changes in the short-wave and long-wave optical properties of clouds would compensate each other (Cess, 1976). ERBE provided definite proof that short-wave and long-wave effects of clouds do not cancel in the global mean (Ramanathan et al., 1989), and its successor, the Clouds and Earth's Radiant Energy System (CERES) mission, launched in 1997, further confirmed that even in regions such as the deep tropics, where short-wave and long-wave effects are in a subtle balance, a change in surface temperature can disrupt this balance.

1.2.1.3   The ISCCP's Optical Thickness: CTP Histograms

The ISCCP led to the first space-based climatology of clouds by collecting different measurements of radiative

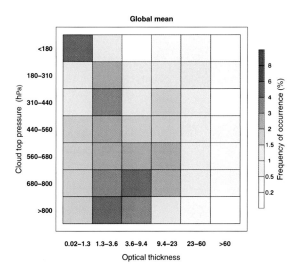

FIGURE 1.8: The global cloud top pressure-optical thickness joint histogram from the ISCCP in frequency of occurrence (%). Data from Tselioudis et al. (2013)

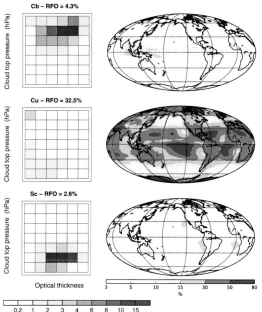

FIGURE 1.9: The cloud top pressure optical thickness histograms of three cloud states, stratocumulus (Sc), cumulus (Cu) and cumulonimbus (Cb), and their geographical distributions. Data from Tselioudis et al. (2013). RFO denotes the relative frequency of occurrence

properties of clouds (Section 1.1.3). This led to a widely used new method of classifying clouds using optical thickness – CTP diagrams. These diagrams help categorise clouds using their CTP, which holds information about their long-wave effects, and the optical thickness of clouds (τ), which holds information about their short-wave effects. The global frequency of occurrence of clouds in CTP-τ space using ISCCP data is shown in Fig. 1.8. Inspired by Howards' classification, ISCCP adopted the notion of three main layers of clouds. Low-level clouds are defined as having a CTP >680 hPa, mid-level clouds have a CTP >440 hPa and <680 hPa and high-level clouds have a CTP <440 hPa. In addition, ISCCP allows clouds to be categorised by their optical thickness. Thin clouds are those clouds with τ <3.6. Examples of such clouds are cumulus, altocumulus and cirrus. Optically thick clouds on the other hand have a τ >23, and include cumulonimbus, nimbostratus and sometimes stratus. In accordance with Table 1.1, Fig. 1.8 reveals that the most common clouds on Earth are those that are located at low levels, which tend to have an τ <10. Also frequent are very high and thin clouds, and to a lesser extent mid-level clouds with moderate optical thickness.

A natural question is whether Howard's idea of classifying clouds can be reimagined and applied to satellite-observing systems. In the early 2000s, researchers at the Bureau of Meteorology in Australia and at NASA GISS in New York developed a simple method to use the ISCCP CTP-τ joint histograms as the foundation of an entirely

objective satellite-based cloud classification. They made use of the simple idea that in mathematical terms, each histogram represents a pattern in CTP-τ space. This means that simple pattern recognition algorithms can be applied to the millions of observed CTP-τ histograms available every three hours. Each of these histograms contains a mixture of cloud types over an area of 280 × 280 km, slightly larger than the region a surface observer would see, but roughly the size of climate model grid boxes at the time. By applying a simple clustering algorithm to daily-average histograms all over the globe, eleven recurring cloud states emerge, as well as an additional – very rare – state of completely clear sky.

Fig. 1.9 shows the average CTP-τ for three cloud states, stratocumulus (Sc), cumulus (Cu) and cumulonimbus (Cb), and their corresponding geographical distribution. The most common type of cloud on Earth – occurring more then 32 % of the time – is shallow cumulus. Closer inspection reveals that this type is very similar to the cumulus cloud type defined by Howard, with a very large frequency of occurrence over the trade-wind regions of the subtropical oceans (Fig. 1.4), but also regular occurrences in the polar regions. Exactly because these clouds are so frequent, they can effectively modulate short-wave radiation, even when an individual cloud is much thinner than a deep convective cumulus (Chapter 13). The stratocumulus clouds occur very close to the coasts of the subtropical western continental boundaries, whereas the largest deep convective systems are confined to the equator.

The ability to objectively classifying clouds from satellite observations, such as using ISCCP cloud states, has also enabled a decomposition of the Earth's radiative budget by cloud type, and opened the door for evaluating clouds in climate models, which is further discussed in Chapter 7.

1.2.1.4   Cloud-Radiative Effects: Advances and Uncertainties

Through ERBE, CERES and ground-based broadband radiation measurements the energy balance of the Earth and the planetary albedo are well documented, and the energy fluxes at the TOA and at the surface are known to within a few Watts per square metre. The influence of different cloud types on the energy budget has also been studied using the instruments carried aboard the A-Train such as MODIS, CloudSat and CALIPSO, in addition to ISCCP data. This whole new space-borne view on cloudiness and the radiation budget will be described in Chapter 4.

Much of what has been described so far deals with the radiation that is either leaving to space at the TOA, or that is received at the surface. But, as pointed out in the review paper by Stephens (2005), the vertical distribution of the heating (or cooling) is just as important. This distribution can influence atmospheric dynamics by leading to horizontal pressure gradients, and therefore drive atmospheric circulations. This topic will come up again in this chapter in Section 1.2.4. This profile of radiative heating is not actually measured by any instrument, but has to be derived using an offline model that solves radiative transfer equations (Chapter 4). Such a model needs the measured clouds, water vapour and temperature profiles as input. Profiles of clouds can only be measured with active remote sensing instruments, such as the CPR and CALIOP aboard the A-Train (Section 1.1.3) or radars and lidars deployed at ground-based sites. Because such measurement records now provide over a decade of data, research on this topic is gaining ground, for example, the use of ARM and A-Train data in Thorsen et al. (2013).

Despite the advances, a number of components of the energy budget still carry considerable uncertainty (Fig. 1.7). This is especially true for the energy fluxes over oceans, which include the downwelling short-wave

and long-wave radiation, the sensible heat flux and the latent heat flux (evaporation). These fluxes are harder to observe with instruments from space, and near-surface measurements over open ocean are limited. Several journal publications review the present state of the energy budget and the adjustments to the budget that have been made using the latest measurements (Stephens et al., 2012a; Wild et al., 2015). But even at the TOA the accuracy that is needed to detect cloud trends or imbalances in the Earth system is not sufficient. One problem is that data records are interrupted every time a measurement platform in space is decommissioned, so that real trends in cloudiness need to be separated from trends that arise from the exchange of instruments, or even drifts in the orbits or altitudes of the satellite. So far, data records from satellites appear too short and ambiguous to provide solid proof that clouds and their radiative effect have changed during the twentieth and twenty-first centuries (Norris and Slingo, 2009).

1.2.2   Clouds as Turbulent Flows

From the previous section it is clear that the interaction of clouds with radiation is a critical component of climate, and how clouds first mattered to climate. Most of our instrumentation used to measure clouds is also based on principles of radiative transfer. But, throughout much of the earlier part of the twentieth century, a different line of cloud research developed around the view of clouds as multiphase flows, in which turbulence occurs on a variety of scales. As this section will describe, the turbulent motions associated with clouds are critical to the development and appearance of clouds, and hence as important as interactions with radiation.

1.2.2.1   The Growth of Cumulus Clouds

By the 1950s, linear perturbation methods and theoretical models were readily explored to study disturbances in the atmosphere. Convection and clouds were a nice prototype of such disturbances. In the bubble theory or thermal theory, developed by Scorer and Ludlam, convection was described in terms of isolated buoyant bubbles rising through an ambient fluid. In one of their journal publications they drew analogies to cumulus clouds whose tops look much like their bubble caps. In their theory, the bubble's ascent rate was determined by competing buoyancy and drag forces (Fig. 1.10). Upon ascent, the surface layer of the spherical cap of the bubble would continuously dilute as a function of density gradients between the bubble and its environment. It would sink

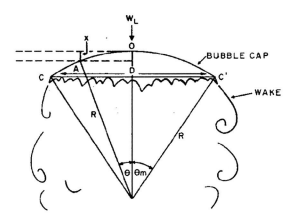

FIGURE 1.10: Schematic diagram of a buoyant bubble with a spherical surface (bubble cap) and a turbulent wake area. Figure from Malkus and Scorer (1955). Copyright © 1955 American Meteorological Society (AMS)

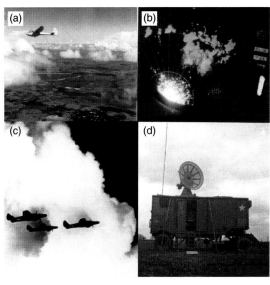

FIGURE 1.11: Photographs made during the Thunderstorm Project, whose first phase started in 1946 near Orlando, Florida. (a) Glider in a cumulus field; (b) radar storm depiction; (c) squadron F pilots of the P 61s; (d) SCR 584 radar tracking van. Copyright © 2001 NOAA

relatively to the motion of the rising cap and be shed into its underlying wake region, until the whole bubble would be eroded.

Joanne Simpson, one of the pioneers of cumulus research,[2] who at the time worked at Woods Hole Oceanographic Institution, expressed her opinion on this applicability of this 'bubble' theory to cumulus clouds in a published correspondence (Malkus et al., 1953, p. 288).

It is in their attempt to apply their bubble analogy to cumulus cloud processes that these authors wander farthest afield, and must shut their eyes to the accumulation of observational evidence. In contradistinction to the bubble model, I should like to contend that a cumulus cloud is more fruitfully regarded as a process or a field of motion rather than an entity or a group of entities. In illustration, it can be demonstrated that a single cumulus may exchange its entire air content with its environment once or even twice during its life cycle.

Malkus's concern was one of their assumptions, namely, that no heat or mass is transferred through the boundaries of the separate bubble entities and that pressure disturbances are ignored. In other words, the bubbles did not mix with their environment. To support her scepticism of this assumption, Malkus made use of measurements that were made inside of clouds during recent field studies

whereby airplanes equipped with in situ sensors flew through clouds, and newly developed cloud radars were deployed from the ground. Such studies had picked up since the first of its kind, the Thunderstorm Project in 1946 (Byers and Braham Jr., 1949), which marked the start of more focused field work on convective clouds. The Thunderstorm Project was a pretty adventurous undertaking, and would by no means pass security tests today. A total of 1,363 penetrations were made through thunderstorms at various altitudes near Orlando, Florida, where thunderstorms occurred most frequently (Fig. 1.11). Such field experiments were initially driven by the need for a better understanding of deeper convective clouds, which during the Second World War had posed some serious risks to aircraft operations.

The measurements used by Malkus consisted of cross-sections made by aeroplanes through active cloud 'bubbles', which showed that maximum temperatures fell short of the temperature that an unmixed bubble should have acquired upon ascent (*moist adiabatic* ascent, see also Chapter 2). Malkus explained that the bubble must have mixed with drier environmental air. Upon mixing, part of the bubble's condensate would evaporate and introduce a cooling, which would reduce the bubble's buoyancy and thus its updraft speed. Indeed, the updraft speeds that were measured were only 4–5 m s^{-1}, instead of the 13 m s^{-1} that moist adiabatic ascent would have predicted. In her letter

[2] Joanne Simpson was born Joanne Gerould and, by virtue of different husbands and customs at the time, she was variously named also Joanne Starr and Joanne Malkus. Here she is often referred to as Malkus as it is this name that is associated with her many pioneering studies of clouds.

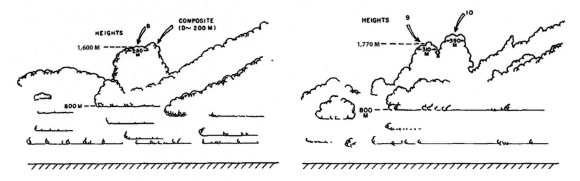

FIGURE 1.12: Drawings derived from time-lapse photography made on 8 August 1950 of clouds formed over Nantucket Island in a set of field programmes of the marine meteorology group at the Woods Hole Oceanographic Institution. The two frames are separated in time by 2.5 min. Large 'bubbles' or turrets are indicated by arrows. Figure from Malkus and Scorer (1955). Copyright © 1955 American Meteorological Society (AMS)

to Scorer and Ludlam, she strongly advocated that their theory makes place for an analytical formulation that can be fitted to measurements.

Their formal exchange turned out to be rewarding. Scorer visited the Woods Hole Oceanographic Institute to work with Malkus on observations, including time-lapse photography to estimate the growth velocities of individual turrets in cumulus clouds and link them to temperature and humidity profiles from radiosondes launched in noncloudy regions (Malkus and Scorer, 1955). Some of those photographs were turned into hand-drawn illustrations (Fig. 1.12). These time-lapse photos of clouds alluded to many important aspects about the motions that underlie clouds. For instance, even though clouds continuously change their appearance, they can also be remarkably steady in time and persist for periods of well over an hour. The fact that turrets always reappear at similar locations during the cloud's lifetime, such as the ones indicated with arrows in Fig. 1.12, also show that cloud tops are clearly connected to a source below cloud base and that circulations might take place on scales much larger than that of individual 'bubbles'. Furthermore, smaller bumpy nodules at the edges of cumuli were observed to continuously erode, whereas the 'bubbles' in the centres of cumuli would turn into bigger turrets, seemingly protected from the drier environment. These insights were beautifully summarised a few decades later by Malkus herself:

Like people, cumulus clouds have a life cycle. They are born, they grow up and eventually age and die, but unlike people, the fatter they are the longer and more vigorously they live and the taller they grow. (Simpson, 1973)

The observations evidently helped to indicate why different aspects of theories were not applicable to

cumulus clouds. In plume models, for instance, the cloud is modelled as a steady turbulent plume that continuously engulfs environmental air at its sides, a process called **entrainment** – the missing process to which Malkus referred. However, in nature, clouds obviously do not have such a well-defined source region (at the start of the plume). Moreover, the model did not consider a temporal development of the depth and width of the plume. On the other hand, the complete absence of a continuous source of heat and moisture also made bubble theories inapplicable, because the existence of sharp flat bases of cumulus clouds clearly implies that a continuous renewal of cloud by condensation is taking place.

Based on their insights from observations, Malkus and Scorer envisaged a model that would model a cumulus cloud as an agglomerate of bubbles that temporally and spatially interact and together form the observed larger turrets. One of the promising features of the agglomerate model they envisaged would be to study the life cycle of a cloud. This was also not possible with a steady-state plume model, which somewhat circumvented the entrainment mechanism by modelling a homogeneous plume of fluid whose properties differ from the surrounding fluid by a continuous jump.

Despite Malkus's initial criticism of the bubble theory's lack of entrainment, they based their agglomerate model on Scorer and Ludlam's previous theory, but discussed that entrainment would more naturally arise from the rate at which bubbles succeeded to exist in each other's wakes and protect each other or agglomerate.

More generally, the discussion between Malkus, Scorer and Ludlam is a great example of how theory, models and observations are all needed to advance knowledge.

1.2.2.2 Beyond Cloudy Updrafts: Downdrafts and Mesoscale Air Motions

Warfare had proven to be rather beneficial for cloud science by advancing observing techniques. Not only did the Second World War rapidly expand the existing air fleet, making the first airborne field experiments such as the Thunderstorm Project possible, it also spurred the development of computer and radar technology. The latter was discovered by chance: radars were used during the Second World War to detect and follow aircraft, but, first, the images had to be cleared from clutter. This clutter turned out to be precipitating clouds within the radar field of view. Meteorologists realised that radar could thus be used to track and observe the evolution and growth of clouds. Especially Doppler radars, which measure the velocity with which objects move towards or away from radar, were crucial for revealing the three-dimensional structure and dimensions of cloud.

The use of aircraft and radar led to another great insight, namely, that deep cumulus clouds are associated with strong downdrafts, which gain substantial speeds and induce gusts that are more devastating to aircraft than the upward thermals within clouds. During two more important field campaigns, the Line Islands Experiment in 1969 and the Global Atmospheric Research Program Atlantic Tropical Experiment (GATE) in 1974, it was discovered that these downdrafts could be confined to scales of individual clouds, but also reach mesoscales far larger than a single cloud (Zipser, 1969). The downdrafts would transport very dry air from further aloft and fill the boundary layer with denser air, effectively stabilising the boundary layer and preventing further convection from developing. GATE also led to the discovery that cumulus clouds tend to occur in 'clusters', a clumping together of cloudy updrafts whose individual structure cannot be clearly separated from another (Houze and Betts, 1981). These cloud clusters, later called mesoscale convective systems (MCSs), could either be slowly moving or rapidly moving clusters called squall lines. More generally, this led to questions about how convective clouds impact the motions on scales far larger than that of a single cloud, some of which we return to later (Section 1.2.4).

Numerical integrations of the equations of motion using computers has been a great leap forward for this line of cloud research. The rapid development of three-dimensional numerical experiments of convection was one of the reasons that Malkus and Scorer never developed their theoretical model of an agglomerate of bubbles.

Nowadays, high-resolution numerical models are the key tool to viewing clouds as turbulent flows (see also Section 1.2.5). Such flows are simply very difficult to measure, even using radar. The full evolution and path travelled by large cloud systems cannot be followed and observed, unless by an extensive network of radars, but these are employed only over land, and not over open ocean. Furthermore, intensive field campaigns such as GATE, where airplanes help measure in-cloud turbulence, ship-based radars help measure the structure of clouds, and soundings reveal the thermodynamic and wind structure of the atmosphere, are sporadic.

Present-day research also asks how important the clustering or organisation of clouds is for climate change (Bony et al., 2015). Model development still struggles with how to represent mesoscale motions in global models that do not have fine enough grid spacing to resolve such motions (Sections 1.2.5 and 1.3). But before we continue with an even larger-scale view of clouds, namely, clouds as drivers of atmospheric circulations, we will review another elementary view of clouds: a collection of droplets.

1.2.3 Clouds as a Collection of Droplets

Until the late 1940s, studies on cloud microphysics and precipitation had received remarkably little attention, given that it is so important for agriculture and society. But the Second World War also proved beneficial for this branch of cloud research. Warfare triggered a desire to forecast and regulate precipitation, which led to controversial cloud modification programmes that stimulated research on cloud microphysics, even if those were ultimately driven by academic interests.

1.2.3.1 What Makes Rain?

Since the 1870s, it was understood that the cooling induced by adiabatic expansion of air leads to saturation, but also that the phase transition from vapour to liquid droplets is associated with a large energy barrier. Unrealistically large supersaturations of a few hundred per cent would be necessary to overcome that barrier and form cloud droplets from pure water alone. It was also understood that the atmosphere could overcome that problem, because it contains particles with specific chemical compositions that serve as nucleation cores (*cloud condensation nuclei [CCN]*), which require much lower supersaturations. Finally, from theories on diffusion of water vapour onto a sphere, it was understood that smaller spheres (droplets) would grow faster and therefore catch up with larger spheres. However, a mathematical synthesis of these different processes was only established in 1949 by Wallace Howell (Howell, 1949), who put them together in a model. He showed that the supersaturations near cloud base reach only a few tenths of a per cent and that the diffusional growth of cloud droplets yields only a very small range of cloud droplet sizes. At the same time, airborne field studies provided more evidence on the concentrations of droplets (and ice crystals) in air and their sizes and compositions. Glass slides coated with some material, for example, soot or oil, would be exposed to the air stream for a short time interval and examined and photographed afterwards. Cloud droplets would leave size-proportional pits in the soot or would be captured as water bubbles in the oil on these so-called impactors.

As one of the active scientists in the field of cloud and weather modification at the time, James McDonald nicely laid out the apparent discrepancy between what would be necessary for a cloud to rain in existing theories and which clouds were actually observed to precipitate (McDonald, 1958). In an extensive review, he wrote that clouds that rain should have droplets large enough to gain a fall speed that exceeds that of the updraft speed. Condensation processes alone would yield the largest droplets when: (1) the fewest possible CCN are available; (2) cloud bases are low; and (3) warm and rising air parcels in the cloud updraft are carried to the greatest possible altitudes. Such a hypothetically large cloud droplet could, according to Howell's model, attain a radius of about 40 μm. However, such a droplet would require about a day to travel from cloud top to cloud base, even if it avoided all updrafts, and then it would likely be completely evaporated before reaching the ground. In other words, condensation alone will never be capable of producing drops large enough to precipitate. Hence, the biggest question to tackle was: how can precipitation-sized drops form so quickly?

The theoretical foundation for one of the two processes that prove essential for rain formation had already been laid out a long time ago. Alfred Wegener noted in 1911 that the (water) vapour pressure at which a liquid droplet is in equilibrium with the surrounding vapour, which means that condensation and evaporation are in balance, is larger than the vapour pressure needed for an equilibrium between ice crystals and the surrounding vapour. Tor Bergeron picked up on this in 1933 (and independently so did Walter Findeisen in 1938) and explained that when ice crystals and liquid droplets coexist at a given temperature below 0°C, water vapour will be deposited onto the ice crystal and the decrease in vapour will be compensated for by evaporation of the liquid droplets. In other words, the ice crystals will grow at the expense of liquid droplets (Chapter 3). Only in 1946 was this *Wegener–Bergeron–Findeisen* process actually demonstrated in a laboratory.

Vincent Schaefer created a cloud of supercooled water droplets in a cloud chamber in which he dropped crushed dry ice made from solid carbon dioxide with a similar structure as ice, and observed the cloud quickly turning into ice crystals.

Later that same year, he repeated the experiment in real clouds by dropping dry ice from an aeroplane into a cloud and observed snow falling from its base. These cloud-seeding experiments were closely linked to cloud modification programmes, which raised moral questions for cloud scientists. For instance, programmes to get rid of fog as a nuisance to air traffic could also be used as a tool of warfare. But it was soon recognised that the actual effect of cloud seeding would be very difficult to weigh against natural mechanisms – how would the cloud have behaved and rained had it not been seeded? Not only would it be very costly to seed clouds worldwide at such a rate that it would lead to a measurable effect, it would also be very difficult to verify that the seeding was the primary process responsible for the observed rainfall. Therefore, one of the reasons that cloud modification programmes did not become a large success was that laboratory findings were not realisable in seeding experiments.

Another reason was that observations demonstrated that ice is not needed to let clouds rain. In 1943, Herbert Riehl, followed by the Thunderstorm Project and other field campaigns, observed that heavy rain fell from clouds whose tops were much warmer than 0°C. Apparently, nature has another mechanism that can efficiently produce larger drops. In this process, *coalescence*, two cloud droplets collide and form a single larger drop (Chapter 3). The process is especially effective when a few large drops or particles are present along with many smaller droplets. This happens for example when cloudy air mixes with clear air that contains only a few CCN, which shows that microphysical processes need to be viewed in light of the turbulent flows that accompany clouds, and this recognition helped stimulate studies on cloud dynamics and the numerical modelling of clouds. In turn, such modelling studies recognised that microphysical processes are important for understanding cloud dynamics. For instance, the first numerical simulations of clouds were largely unrealistic, because they did not capture the maturing and dying of clouds due to precipitation, or the influence of drag from falling rain drops on the updraft speed, or the contribution of latent heating to the buoyancy of the updraft.

Several decades later, a similar lesson was learned. Namely, that the influence of aerosol (and thus CCN) on clouds needs to be viewed in light of other (meteorological) factors that influence clouds.

1.2.3.2   Aerosol–Cloud Interactions

The ideas that initially motivated cloud-seeding experiments also motivated a better understanding of the relationships between nucleating particles in the air (*aerosols*) and the concentration of droplets in clouds. The droplet concentration does not just influence the processes that form rain, but also influences the albedo of clouds. A study with a large impact was that of Sean Twomey, who asked how pollution, or anthropogenic aerosols, would impact cloud droplet number concentrations and thereby the cloud albedo (Twomey, 1977). Twomey responded to a report from the Study of Critical Environmental Problems (SCEP) (SCEP Study of Critical Environmental Problems, 1970), who suggested that enhanced pollution (more aerosols) would lead to dirtier and therefore darker clouds. Twomey instead argued that two competing effects are at play when more aerosols are introduced in a cloud. Aerosols that absorb sunlight can make a cloud darker. But a larger number of aerosols also implies more cloud droplets, which scatter sunlight. The latter would make the cloud brighter and whiter. Using approximate computations, Twomey showed that for thinner clouds, the brightening effect outweighs the darkening effect.

This effect can be seen in Fig. 1.13. Take, for instance, the curves that start at point C, which is for a cloud that is 4 km thick. For this cloud, any further increase in the

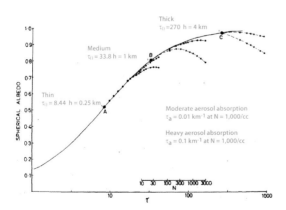

optical thickness, τ, a measure of how many absorbing aerosols are present, leads to a decrease in the spherical albedo (a proxy for the planetary albedo). This is true for both moderately and heavy absorbing types of aerosols. However, for thinner clouds, such as clouds at point A (250 m thick) and point B (1 km thick), a further increase in τ can increase the albedo, as long as aerosols are only moderately absorbing.

Twomey was also one of the first to postulate that by increasing the brightness or the albedo of clouds by increasing pollution, the atmosphere could be cooled, and global warming might be offset. He postulated that this could be done by increasing cloud droplet concentrations and the cloud optical thickness. But other studies subsequently speculated that changes in cloud droplet concentrations can also influence cloud amount or cloud depth. While the effects of CCN on cloud microphysical and radiative properties is well established, the impact on cloud macrophysical properties remains controversial.

Since then, much work has been inspired by such ideas, but a clear consensus on the global impact of aerosol–cloud interactions has not been reached, plagued by a number of challenges. An important one is that effects of aerosols on clouds are hard to disentangle from other effects on clouds, especially in observational records. For example, a review paper by Chen et al. (2014) illustrates the potential cooling of stratocumulus decks: only a 6 % increase in the stratocumulus albedo would be needed to offset global warming. But using a long record of collocated aerosol and cloud measurements from the A-Train satellites, they also find that the stability and humidity of the lower troposphere, and precipitation, strongly regulate how liquid water contents in stratocumulus clouds responds to aerosol. High-resolution modelling studies also suggest that changes in clouds from aerosols may be buffered by the turbulent flows associated with clouds. So far, such mechanisms are not well represented in global climate models, which makes interpretations of aerosol effects in global models ambiguous (Chapter 11).

FIGURE 1.13: The spherical albedo is plotted as a function of the optical thickness, τ. The spherical albedo is the fraction of incident radiation reflected by a sphere covered by a layer of cloud with the prescribed properties. At $\tau \approx 10$, the sun's disc can no longer be observed through a cloud when viewed from the ground. The curves are for three clouds (A, B, C) that have different thickness h and different τs starting at a CCN concentration $N = 25$ cm^{-3}. The top and bottom dashed trajectories belong to moderately and heavily absorbing aerosols, whose τ at $N = 1,000$ per cc is shown with text in the graph. Each dot in the trajectory corresponds to a larger N, from 10 to 3,000 per cc. Figure from Twomey (1977). Copyright © 1977 American Meteorological Society (AMS)

1.2.3.3   A Classification Based on Hydrometeor Type

How much precipitation reaches the surface or how clouds modulate the radiation budget is determined not only by the concentration of cloud droplets and rain drops, but also by their phase and shapes. Even though the Wegener–Bergeron–Findeisen process is not responsible for all precipitation that falls to the ground, it still has an important influence on clouds in which both liquid water and ice coexist. The wide variety of shapes of ice crystals and the presence of supercooled liquid is also important for

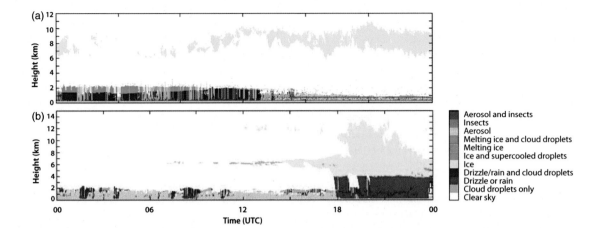

FIGURE 1.14: Time-height cross sections of the cloud field observed by remote sensing instrumentation at the (a) Chilbolton Facility for Atmospheric and Radio Research, part of the Rutherford Appleton Laboratory and (b) at the Barbados Cloud Observatory operated by the Max Planck Institute for Meteorology. Colours represent the different hydrometeor classifications derived using CloudNet algorithms, available as quicklooks online, www.cloud-net.org/index.html. Copyright © 2015 The Cloudnet project (European Union contract EVK2-2000-00611)

how a cloud interacts with radiation. Looking at clouds in terms of their microphysical structure is based on the type of hydrometeor in a cloud (a liquid cloud droplet, ice crystal, rain drop, hail, graupel, and many more), which provides yet another way of categorising clouds: the distinction between warm-phased, cold-phased, and mixed-phased clouds, and a detailed hydrometeor target classification.

Deriving the distributions of hydrometeors and water phase inside clouds requires the use of both millimetre-wavelength radar and lidar (see also Chapter 4). Passive remote sensing instrumentation lacks information on the vertical structure of clouds, and can only be used to infer the vertically integrated amount of liquid or ice in the atmosphere. Global estimates of these are still subject to high uncertainty. Fig. 1.14 shows different hydrometeor categories, derived from algorithms applied to air volumes viewed by both radar and lidar. Essentially these algorithms make use of the fact that radars are sensitive to large particles, such as drizzle, rain drops and ice crystals, whereas lidars are sensitive to high concentrations of particles, such as cloud droplets and aerosol layers. Because lidar is so sensitive to cloud droplets, it enables the detection of supercooled liquid layers even when embedded inside ice clouds. Information of the 0°C temperature line also helps distinguish between ice and melting ice.

The classifications in Fig. 1.14 are derived from ground-based remote sensing sites in the mid latitudes and in the subtropics, and equal classifications have been developed for the merged CloudSat and CALIPSO

observations from space. The quicklooks reveal the presence of multiple layers of warm liquid clouds ($T > 0\,°C$) and cold clouds consisting of ice. Clouds are rarely composed of only cloud droplets, but often contain both cloud droplets and larger drizzle or rain drops. It is also evident that clouds consisting of a mixture of ice and supercooled droplets, which are often referred to as **mixed-phase** clouds, can exist well above the freezing level. Such mixed-phase clouds are common at lower latitudes, but especially persist over large areas in the Arctic.

The persistence of mixed-phase layers is somewhat of a puzzle, because, according to the Wegener–Bergeron–Findeisen process (Section 1.2.3 and Chapter 3), this mixture would be unstable: ice crystals would grow at the expense of the liquid droplets. A subtle balance between different processes inside such clouds, including the formation and growth of ice and cloud droplets, radiative cooling, turbulence and entrainment as well as surface fluxes of heat and moisture, appear responsible for creating such a resilient system. Such clouds are especially of interest in the Arctic regions, where they can significantly alter the radiation received at the snow- and ice-covered surface, and are thus important for understanding changes in snow and ice cover under global warming (Chapter 10).

A great advancement made with CloudSat is the ability to look at light rain, drizzle, produced by warm-phased clouds, in particular over oceans. The low rain-rates associated with drizzle are more challenging to observe and

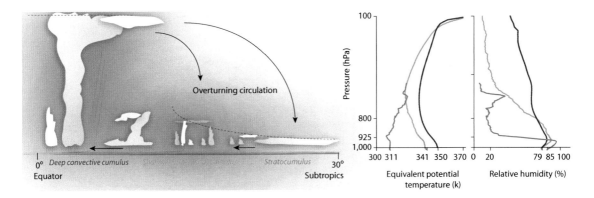

FIGURE 1.15: The zonal-mean large-scale overturning circulation between the subtropics and tropics (the Hadley circulation), along with representative profiles of (saturation) equivalent potential temperature and relative humidity that are typical of the three cloud regimes found throughout the Hadley circulation: in blue, low-level stratocumulus over the eastern ocean boundaries; in orange, shallow (trade-wind) cumulus regime over warmer sea surface temperatures in the trades and, in red, deep cumulus clusters in the inner tropics.

the recent CloudSat measurements have confirmed what earlier measurements from the Tropical Rainfall Measurement Mission (TRMM) already indicated: that light rain from warm clouds comprises a significant fraction of the total rain over oceans. With TRMM and CloudSat we have also gained more insight into the frequency and intensity of precipitation. Even when the mean precipitation is close to observations, constrained by the surface evaporation in global models, the frequency and intensity are often biased, and compensate each other (Stephens et al., 2010). Moreover, the spatial patterns of precipitation can be biased, which has an influence on atmospheric circulations, through latent heating released with precipitation. This brings us to the last historical review and view of clouds which we consider in this chapter: the view of clouds as crucial components of the general circulation of Earth's atmosphere.

This view has long been established for deeper convection, as the next section will discuss, but has recently emerged as a factor that is crucial in understanding how clouds change in response to warming. As discussed earlier, any change in clouds will lead to changes in radiative heating or cooling rates, as well as latent heating rates, turbulent mixing and mesoscale flow patterns. All of these factors may change the atmospheric circulation, which in turn can further adjust clouds, making the interaction of clouds with climate a closed loop.

1.2.4 Clouds as Crucial Components of Circulations

In the late 1950s, it was suggested that the deep convective cumulus clouds found near the equator are necessary components of the zonal mean overturning circulation, the Hadley circulation, illustrated in Fig. 1.15. Herbert Riehl and Joanne Malkus applied a novel theoretical analysis on a set of observations which included surface measurements and radiosondes all around the world (Riehl and Malkus, 1958). They hypothesised that the buoyant updrafts within the cores of these clouds, which have substantial horizontal dimensions, would be protected from mixing with the environment, so that they carry air with very large heat contents from near the surface all the way to the tropopause. This was an important hypothesis, because it would help explain how the upper troposphere in the tropics gains the heat that is transported poleward, and which is necessary to warm other parts on Earth that receive less energy from the sun.

The thermal structure of the tropical atmosphere sketched by the solid red profiles in Fig. 1.15 is represented by θ_e, the equivalent potential temperature, which is a form of reference temperature which measures the moist entropy of the air (see Chapter 2 for a more complete description). At altitudes where temperatures are potentially high and the water vapour content large, θ_e is large. In the inner tropics (10°S–10°N), θ_e is high near the surface, because large surface fluxes supply abundant heat and moisture. θ_e minimises at mid levels near 700–600 hPa where both water vapour content and potential temperature are low, but increases again towards 100 hPa, which is mostly due to a larger potential heat content, because the atmosphere there is rather dry. Riehl and Malkus showed that exactly because of the mid-level minimum in θ_e, neither a gradual transport of mass to this altitude nor a simple diffusion mechanism through cumulus convection could account for the large heat contents near 100 hPa. Instead, they postulated that a relatively small sample of about 1,500–5,000 undiluted cloud towers (cores) is responsible for the transport of heat upwards. The importance of such cumulus towers was also pointed out in several studies on the formation and growth of tropical cyclones. It became clear that for models of large-scale circulations, cumulus convection needs to be included as a heating mechanism. The first theoretical formulations of the coupling of convection to the large-scale flow appeared in the 1960s (Arakawa, 2004). However, despite these theoretical advancements and hypotheses, there was still a lack of knowledge on how this apparent heat was utilised to warm the large-scale environment, as well as a lack of observations of the properties of cumulus ensembles.

A huge leap forward was made by Michio Yanai, then a professor at University of California, Los Angeles, who at the time worked together with Akio Arakawa on improving cumulus convection in a general circulation model (GCM). With his colleagues, Yanai demonstrated that by using observed heat and moisture budgets over an area covering a cumulus ensemble, combined with a model of a cumulus ensemble which exchanges mass, heat and moisture with the environment through entrainment and detrainment, it was possible to determine the bulk properties of that cloud cluster, such as the vertical mass flux, the excess temperature and moisture of the cluster and the liquid water content (Yanai et al., 1973).

Yanai used observations made at five stations in the Marshall Islands region, which enclosed an area of about $62 \times 10^4 \mathrm{km}^2$. At these stations, soundings were made four times a day, which Yanai used to derive the (changes in the) vertical profiles of temperature and humidity, as well the horizontal divergence or large-scale vertical motion in the enclosed region. Deploying the equations for mass continuity, heat and moisture conservation, he derived that:

$$\frac{\partial \bar{\eta}_d}{\partial t} + \underbrace{\overline{\nabla \cdot sV} + \frac{\partial \bar{\eta}_d \bar{\omega}}{\partial p}}_{\text{large-scale flow}}$$

$$\equiv Q_R + \underbrace{\ell_v(c-e) - \frac{\partial}{\partial p}\overline{\eta'_d \omega'}}_{\text{convection}} \equiv Q_1 \qquad (1.1)$$

and

$$-\ell_v\left(\frac{\partial \bar{q}_v}{\partial t} + \underbrace{\overline{\nabla \cdot q_v V} + \frac{\partial \bar{q}_v \bar{\omega}}{\partial p}}_{\text{large-scale flow}}\right)$$

$$\equiv \underbrace{\ell_v(c-e) + \ell_v \frac{\partial}{\partial p}\overline{q'_v \omega'}}_{\text{convection}} \equiv Q_2 \qquad (1.2)$$

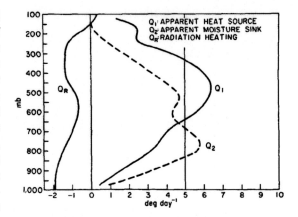

FIGURE 1.16: Profiles of the *apparent heat source* Q_1, *apparent moisture sink* Q_2 and radiative heating Q_R derived from observations made over the Marshall Islands in 1956. Figure from Yanai et al. (1973). Copyright © 1973 American Meteorological Society (AMS)

The terms on the left-hand side of the equations show the time derivative of the dry static energy $\eta_d = c_{p_d}T + gz$ and specific humidity q_v, as well as their horizontal advection by a wind vector V and vertical advection by large-scale vertical motion ω (for a full explanation of the static energy and other terms in these equations, see Chapter 2). The terms on the right-hand side of the equations show the radiative heating or cooling rate Q_R, condensational or evaporational effects proportional to c and e, where the vaporisation enthalpy is denoted by ℓ_v, as well as the vertical divergence of the vertical eddy fluxes of heat and humidity. Yanai called the terms Q_1 and Q_2 the *apparent heat source* and the *apparent moisture sink*. The observations provided estimates of all terms on the left-hand side of the equations, allowing Yanai and his team to derive Q_1 and Q_2.

Yanai showed that the apparent heating Q_1 is positive everywhere (Fig. 1.16) with a maximum in the upper troposphere near 400–500 hPa. The moisture sink Q_2 of the atmosphere is also positive everywhere, but its maximum is located lower down, at 800 hPa. He explained that the decrease of Q_1 towards the surface could be caused by the larger occurrence of cooling from evaporation of detrained liquid. We now interpret the decrease in Q_1 due to evaporation of stratiform rain often associated with organised mesoscale convective cloud systems, which at the time were just on the verge of being discovered. The peak in Q_2 near 800 hPa was explained by the condensation of large amounts of liquid low down in the atmosphere, where the water vapour content is largest.

To further interpret Q_1 and Q_2 in terms of the interaction of convection with the environment, Yanai employed

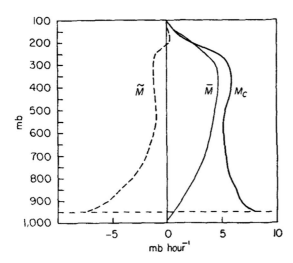

FIGURE 1.17: Profiles of the mean vertical motion or mass flux of the atmosphere $\bar{M} = -\bar{\omega}$, as well as its two components $\bar{M} = M_c + \tilde{M}$, whereby M_c is the mass flux from cumulus clouds and $\tilde{M}$ is the residual mass flux. These are derived from the apparent heat source Q_1 and moisture sink Q_2. Figure from Yanai et al. (1973). Copyright © 1973 American Meteorological Society (AMS)

a model of a population of shallow and deep cumulus clouds. His model assumed that the average mean ascent in cumulus clouds $\bar{M}$ would balance the slow descent in the clear sky environment $\bar{\omega}$ such that $\bar{M} = -\bar{\omega}$. He decomposed $\bar{M}$ into $M_c + \tilde{M}$, the upward mass flux of M_c and a residual mass flux in the surrounding environment $\tilde{M}$, whereby the vertical change of M_c depends on the entrainment of environmental air into the cloudy updraft minus the detrainment of cloudy air into the environment. In linking these terms to the observed Q_1 and Q_2, one of the most important results that Yanai arrived at was that the cloud mass flux M_c exceeds ω, leaving a considerable residual $\tilde{M}$ (Fig. 1.17). Hence, there must be a considerable compensating sinking motion in between active clouds: a downward mass flux. The warming from the adiabatic compression of this downward mass flux, and the dry air transported from upper to lower levels, are the heating source and the moisture sink that effectively change the environment (as discussed further in Chapters 2 and 6).

With the establishment of this picture, clouds were no longer just visual expressions of the state of the atmosphere, or a collection of small droplets that could produce considerable rainfall. Instead, they were a crucial component of the dynamics of the atmosphere as a whole. Deep convective towers produce the largest amount of heating through compensating motion, but other cloud types within circulations were also gaining attention through

these studies, such as the shallower clouds underneath the subsiding branches of the Hadley circulation, also called the trades.

Riehl and Malkus showed that the trades, the region in between 5° and 20° in both hemispheres (Fig. 1.15), produce a net export of latent and sensible heat (Riehl and Malkus, 1957). As such, the trades themselves play a role in establishing the temperature and pressure gradients that drive the trade winds. Ernst Augstein, together with Herbert Riehl, showed further evidence of the heat and moisture export from the trades using a rich set of observations collected aboard three research vessels, which were positioned in a triangle and drifted downstream in the North Atlantic trades from 13.5°N, 34.7°W to 9.5°N, 40.6°W during a period of three weeks during the North Atlantic Trade-Wind Experiment (Augstein et al., 1973). With those data, the mass, heat and moisture budgets were calculated in a similar way as Yanai did for the inner tropics. They showed that between the sea surface and the trade-wind inversion, which is located at roughly 800 hPa (see the orange profiles in Fig. 1.15), a large vertical eddy flux of moisture from shallow cumulus convection counteracts the drying effect of the mean downward motion, which maintains the strong humidity gradient in the inversion layer.

The fact that shallow cumulus and stratocumulus are so abundant over the oceans (Fig. 1.4) also gives them a strong cooling potential (via their interaction with radiation) and a susceptibility to aerosol perturbations. This shifted the research focus back to low-level clouds, which had received somewhat less attention compared with deep cumulus, perhaps because they were not as spectacular. Stratocumulus decks started to interest scientists because of the extreme sharpness of the inversion near their top (blue profiles in Fig. 1.15). It was understood that this inversion, like the trade-wind inversion, is produced by turbulent motions which transport moisture towards the inversion, where it meets the very dry and warm subsiding air. They asked how air could be exchanged across such a sharp inversion, which was an important question, because it connects turbulence in the boundary layer with large-scale mean vertical motion.

Scientists were particularly intrigued by the idea that radiative cooling near the stratocumulus tops plays a major role in setting this inversion. The strong cooling near the cloud top was shown to give rise to negatively buoyant air parcels that sink and drive turbulence, which in turn can help break up the stratocumulus deck into fields of cumuli. Because shallow cumulus has smaller cloud cover than stratocumulus, the transition from stratocumulus to cumulus matters for the planetary albedo. These clouds have been in the spotlight of climate studies precisely for that reason. More recently, another idea is emerging, namely, that low clouds can help drive larger-scale circulations due to their radiative effects, which may also matter for the aggregation of deeper convection (Muller and Held, 2012). Evidently, research on cloud–radiation interactions, on the organisation and clustering of deep clouds, and on large-scale circulations and climate have become increasingly intertwined.

To test such ideas, missions such as CloudSat, and the planned EarthCare mission by the European and Japanese Space Agencies have been motivated and designed. Already before these missions, precipitation radar (PR) measurements from TRMM did the same for heating rates associated with deep convective precipitation, which revealed large variations in the latent heating contribution to Q_1. Viewing clouds by their precipitation (or latent heating profiles) has been yet another way of classifying clouds, and is discussed in the next section.

1.2.4.1   Latent Heating Profiles

Two types of precipitation are generally distinguished: convective precipitation and stratiform precipitation. Stratiform here means cloudiness that accompanies deep convective clouds, such as nimbostratus associated with cold fronts, hurricanes or MCSs (i.e., we do not mean low-level stratocumulus cloud). The difference in convective and stratiform precipitation lies in the fall velocity of rain-sized drops and ice crystals relative to the vertical motion of the air. With convective precipitation, the fall velocity of ice crystals may only marginally exceed or even be smaller than the strong vertical updrafts that exist in deep cumulonimbus clouds (thunderstorms). Stratiform precipitation instead falls out of widespread stratiform cloud attached to deep cumulus towers, and the ice fall velocity exceeds the vertical air motion. In the tropics, a large fraction of the precipitation results from the convective precipitation and stratiform precipitation associated with MCSs.

Idealised profiles of the latent heating part of Q_1 that belong to these different cloud types are shown in Fig. 1.18, and are based on measurements of the TRMM PR between 20°N and 20°S (Schumacher et al., 2007). In reality, the profiles will have different magnitudes depending on the exact amount of precipitation, but the shape of the profiles is representative. For deep convection, the profile is positive throughout the atmosphere as condensation dominates phase changes throughout the depth of the atmosphere, introducing latent heating everywhere. The maximum heating takes place in the

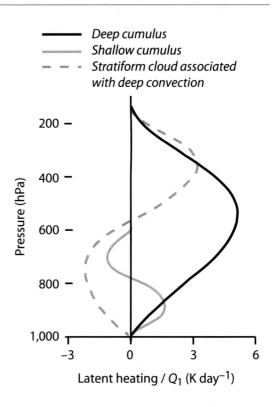

FIGURE 1.18: Idealised latent heating profiles for different cloud types, loosely based on TRMM PR heating profiles, adapted from fig. 1 in Schumacher et al. (2007). Copyright © 2007 American Meteorological Society (AMS)

mid troposphere (500 hPa) and decreases towards the surface. This decrease is caused by the evaporation of precipitation, which starts to play a role as precipitation falls through drier air outside the cloud. The heating profile of stratiform cloud that is associated with such deep convective systems is dominated by heating in the mid and upper troposphere (from condensation in the stratiform anvil clouds). But at heights below 600 hPa, where ice particles start to melt and evaporate, only a cooling is present.

A latent heating profile of a shallower cumulus cloud that detrains moisture at its top is illustrated with the solid grey line, which is positive in the lower troposphere but negative at mid levels. Larger detrainment takes place when the atmosphere is stable with respect to cumulus updrafts, for instance, near the trade-wind inversion or the freezing level. When cumulus updrafts encounter such stable layers they loose their buoyancy, which leads them to shed cloud water into the environment. This water will subsequently evaporate in the drier environment surrounding the cloud, where it will lead to a cooling instead of heating.

Although the shape of the latent heating profiles are believed to be reasonably accurate, the actual heating rates carry large uncertainties. One uncertainty is that a model is required to turn profiles of reflectivity and precipitation rates into heating rates. The model simulates an ensemble of convective clouds and provides relationships that can be used to convert the observed reflectivity to rain, but even this model relies on empirical relationships to model the small scales associated with cloud microphysics and rain formation. Hence, it is fair to assume that 'observed' heating rates might not be representative. In the next section we describe more generally which models have been developed over the years, and how they are used to view clouds in different ways.

1.2.5 A Hierarchy of Models to View Clouds and Their Effects

Along with a century of cloud research and an increasing number of observations, our theoretical understanding of processes that govern clouds, whether those relate to radiation, large scale dynamics, turbulence or microphysics, has greatly advanced. This understanding has led to the development of a hierarchy of modelling tools to study clouds from the smallest scales to the largest scales (see also Chapter 5). These models have been built on physical laws and equations as well as empiricism. Generally, models can be distinguished into two types: those that are built around simple concepts and used for testing ideas and understanding; and those that attempt to simulate a system as realistically as possible. The latter especially have been gaining ground with the ability to numerically integrate equations on computers. Here we give a first overview of these models, many of which are defined more precisely in subsequent chapters.

The first category of more conceptual models already existed before computers emerged. For instance, these include the first models of the Earth energy budget in which clouds intercepted incoming short-wave radiative fluxes (Section 1.2.1), or the first descriptions of buoyant air parcels in an ambient fluid (Section 1.2.2). The conceptual models have in common that clouds are described in simplified or idealised ways. Often these models are used to understand the effect of clouds in larger systems, such as the boundary layer, the atmosphere at a certain location or even the entire Earth system. Often these systems are represented by (a few) single layers, columns or boxes. Because they tend to be simplified descriptions, they are cheaper to run on computers and hence a great tool for doing a large number of sensitivity experiments. Because these models are inspired by the latest understanding and empiricism, they tend to be more straightforward to interpret.

The second category represents more realistic models which rely more strongly on computational power, because they solve equations on numerical grids. The rule of thumb is that the smaller the system that is modelled, the smaller the grid spacing, and the closer to nature the simulated clouds will be. A model of the entire atmosphere or Earth system is currently run at a grid spacing on the order of 100 km, whereas a model of a boundary layer of 50×50 km^2 is run at a grid spacing on the order of 100 m. In the entire atmosphere model, larger atmospheric motions and the global distribution of clouds are resolved on the model grid and therefore close to reality, but the convection, turbulent flows and microphysical processes associated with clouds are not explicitly simulated and may thus differ from nature. In the smaller area models, the convection and larger turbulent eddies associated with clouds are resolved on the model grid and appear more realistic, but smaller-scale turbulence and the formation of precipitation will still be simplified.

A schematic of the different models in use to study clouds is shown in Fig. 1.19 (Bony et al. 2013). The

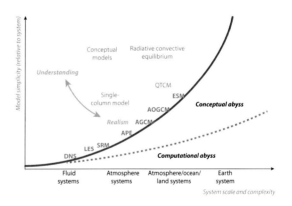

FIGURE 1.19: Schematic of models used to study clouds and their role in climate, with the scale and complexity of the system considered on the horizontal axis and the simplified nature of models on the vertical axis. The red line represents the boundary of current modelling capabilities, limited by computational power and conceptual understanding to its right. Models in blue close to the red line are more realistic at the scale of the system considered, whereas models in grey away from the red line have a more simplified nature, but represent our current understanding of the systems and are good tools for testing conceptual ideas. AGCM, atmosphere-only GCM; AOGCM, atmosphere–ocean GCM; APE, aquaplanet experiment; DNS, direct numerical stimulation; ESM, Earth system model; LES, large eddy simulation; QTCM, quasi-equilibrium tropical circulation model; SRM, storm-resolving model. Adapted from fig. 4 in Bony et al.(2013). Copyright © 2013 Springer

horizontal axis represents the scale of the system considered: from small particle systems to larger atmosphere and Earth systems, which become increasingly complex due to the number of processes considered. The vertical axis represents the simplified nature of models, and by definition a model of a system cannot lie on the zero axis, or it would not be a model. The red line represents the boundary of our current modelling capabilities. Models on the red line are those that best simulate the processes at the respective scale of the system shown on the horizontal axis. On the smallest scale, direct numerical simulation (DNS) explicitly resolves the Navier–Stokes equations on very fine grids down to Kolmogorov millimetre scales, but (for realistic values of the molecular properties of air) this implies using domains of just a few metres on a side. DNS can thus be used to study turbulent flows in clouds, such as the mixing that takes place at cloud boundaries influenced by evaporation and radiative cooling. A visualisation of the turbulent structure of the top of a stratocumulus cloud is shown in Fig. 1.20a.

Moving up in scale, large eddy simulation (LES) models resolve the largest turbulent eddies $O(100\,\text{m})$, and a storm-resolving model (SRM, sometimes called a cloud-resolving model) resolves the vertical convective overturning associated with convective cloud systems on grids $O(1\,\text{km})$. LES makes use of the fact that, within the boundary layer, the largest energy-containing eddies are on the scale of the boundary-layer depth and carry most of the turbulent fluxes and variances. Unlike DNS, LES relies on a model of turbulent motions on scales smaller than the model grid. LES is often used for simulations of shallow clouds which are driven by turbulence, such as shallow cumulus and stratocumulus, but, nowadays, LES on larger domains are also used for deep convection. SRMs have been traditionally used for deeper convective systems, but do not resolve turbulent motions.

An aquaplanet experiment (APE) or atmosphere-only GCM (AGCM) resolve atmospheric flows on grids $O(100\,\text{km})$. When coupled to an ocean model, these are sometimes referred to as an atmosphere–ocean GCM (AOGCM) , or just as a global climate model. When coupled to a representation of the biosphere, these models are often referred to as Earth system models (ESMs). This increases the complexity of the model. Because these interactions are represented in ways more simple than in nature, the models move away from reality (following the upward curve of the red line in Fig. 1.19) towards a more simplistic view of the entire system. However, this should not be mistaken for the model being more simple to interpret. On the contrary, it becomes increasingly

difficult to interpret the effect of individual processes in such models.

Moving up and leftwards from the red line are models which become increasingly more conceptual and less complex. Some of these models have been built specifically for testing the performance of their complex counterparts. The single-column model (SCM), for example, is analogous to a grid column of a GCM and can be used for less expensive sensitivity experiments that lack the complex interactions with the rest of the model. The radiative convective equilibrium (RCE) framework on the other hand represents the entire global atmosphere, like an APE or AGCM, but imposes an equilibrium between the heating due to (deep) convection and the cooling due to outgoing infrared radiation. Another model of the atmosphere with intermediate complexity is the quasi-equilibrium tropical circulation model (QTCM). In both the RCE and QTCM frameworks the focus is on deep convective clouds, because they play a central role in driving the atmospheric circulations that are needed to obtain an accurate (tropical) climate (Section 1.2.4).

That no model lies in the area to the right of the solid red line is related to limited computational power and limited conceptual understanding. But, increasingly computational power is available that allows models with a fine grid mesh (LES, SRMs) to simulate larger domains, or, vice versa, allows APE or AGCMs to run simulations at finer grid meshes. Pioneering global simulations of an aquaplanet with a 3.5 and 7 km grid spacing were performed with the non-hydrostatic icosahedral atmospheric model by Hirofumi Tomita and Masaki Satoh in 2005 (Tomita et al., 2005). Simulations by other models have followed, and a recent example of the global cloud field at a grid spacing of 2.5 km using the German icosahedral non-hydrostatic model is shown in Fig. 1.20b. This simulation produces many of the cloud systems that exist in nature, such as the comma-shaped cold fronts, the clusters of deep convection in the Inter-Tropical Convergence Zone and the widespread fields of shallow cumuli in the trades. Such capabilities are now moving SRMs to the right and bottom of the red line in Fig. 1.19.

Until it is computationally affordable to run such models for long periods of time, coarser grid model are needed to predict climate and climate change. Under the umbrella of the Coupled Model Intercomparison Project (CMIP) simulations by AOGCMs and ESMs are used to inform policy regarding climate change. These feature prominently in the reports of the Intergovernmental Panel on Climate Change. These models also have the capability to become more realistic (and move into the grey shaded area) when we better understand how to represent

(a)

z

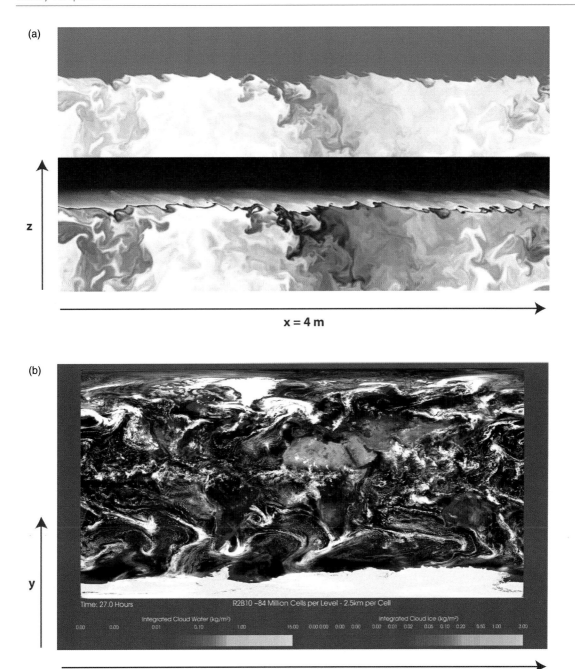

x = 4 m

(b)

y

x ~ 40,000 km

FIGURE 1.20: (a) Visualisation of the liquid water and buoyancy field of a DNS of a stratocumulus top (cropped domain spanning about 4 m in the horizontal direction) with a grid spacing of about 4 mm. The red region denotes the region where a strong capping inversion limits stratocumulus growth. Figure courtesy: Juan-Pedro Mellado. (b) Visualisation of the global integrated cloud water and cloud ice field of an SRM simulation with the icosahedral non-hydrostatic GCM conducted at the Deutsches Klima Rechenzentrum (German Climate Computing Centre) as part of the EU Excellence in Simulation of Weather and Climate in Europe project. Horizontal grid spacing is 2.5 km. Figure courtesy: Niklas Röber the coupling between ocean and atmosphere, the coupling between small- and large-scale processes, and the representation of the biosphere. Here a number of challenges lie ahead, many of which are related to clouds in particular, as described in the final section of this chapter, and throughout this book.

1.3 Uncertainties and Challenges in Climate Modelling Related to Clouds

To make climate projections we rely on climate models, such as AOGCMs. The AOGCMs that are part of CMIP are capable of reproducing our past and our current climate. For instance, they reproduce the observed planetary albedo and the energy fluxes at the TOA within reasonable bounds. They also reproduce annual variations in the planetary albedo, related to seasonal variations in the ice and snow coverage of high latitudes.

However, experience tells us that if these models are not tuned first, their energy budget at the TOA will substantially deviate from that observed. Tuning means that radiative properties, or parameters that control cloudiness, are adjusted in order to reproduce the observed energy budget. Getting the energy budget right is important, because other features of our climate strongly depend on the net energy received from the sun and on the global mean surface temperature.

But even with tuning, or perhaps because of tuning, the clouds in climate models differ from observations in a number of important ways.

Observations suggest that the Earth system prefers a state in which the energy balance is as symmetric across the equator as possible. Thirteen years of CERES data show that the difference in the planetary albedo between the northern and southern hemispheres (NH, SH) is small on average (Fig. 1.21a and 1.21b). Whereas the NH has on average a larger surface albedo than the SH, related to the larger land coverage in the NH, the SH has a larger albedo coming from the atmosphere, which is caused by a larger cloud coverage in the SH. Apparently, the Earth system has mechanisms that adjust cloud patterns in a way that limits hemispheric differences in the energy balance. Because of this, there is very small interannual variability in the planetary albedo, whose magnitude is only a few per cent of that of seasonal variations (Stevens and Schwartz, 2012; Voigt et al., 2013; Stephens et al., 2015).

Models that are part of CMIP5 do not reproduce this observed hemispheric symmetry of the reflected flux (Fig. 1.21a). The main reason is that they do not reproduce the hemispheric difference in radiation reflected by

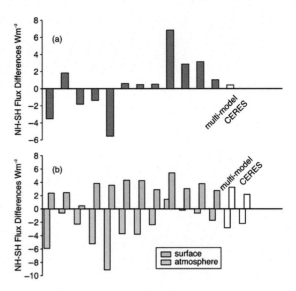

FIGURE 1.21: (a) The NH-SH difference in the TOA reflective flux from thirteen years of CERES data, from individual CMIP5 models and their multi-model mean. (b) The SH-NH difference in atmospheric and surface contributions to the TOA flux are shown in blue and orange, respectively. Figure from Stephens et al. (2015) after fig. 6 in Voigt et al. (2013). Copyright © 2015 John Wiley & Sons, Inc

the atmosphere (Fig. 1.21b). This implies that models have different cloud patterns and fail to reproduce the mechanisms that control clouds (Chapter 7). The amount, patterns and timing of precipitation also deviate from observations, and lead to errors in the distributions of latent heating. Combined with differences in diabatic heating from cloud–radiation interactions, biases in the large-scale circulation will arise. Errors in surface precipitation also affect the coupling between the land and the atmosphere, and have consequences for regional circulations and the hydrological cycle. Hence, it might not come as a surprise that errors in clouds affect the sensitivity of Earth's climate as a whole, and different climate models predict a different increase in global mean temperature (Chapter 13).

Underlying these uncertainties is the inability of global models to simulate the wide range of scales and processes that clouds entail (Section 1.2). Limited by computing power, the processes that shape clouds and their effects must be represented in models by empirical relationships, or **parameterisations**. Principally, all the processes that take place on scales smaller than the model grid remain unresolved and need such a parameterisation (Chapter 6). The first parameterisations were included in numerical simulations of the life cycle of a thunderstorm, which required a simple model or relationship to connect the formation of precipitation to the vertical air motions and water vapour distributions inside clouds. This led to the first parameterisation of the effects of convective clouds (Manabe et al., 1965), as well as of cloud microphysics (Kessler, 1995). Since then, parameterisations of many processes have been developed, and climate models have grown increasingly complex. Many levels of details have been added, for instance, the chemical composition of aerosols or the different pathways in which cloud droplets grow to precipitation-sized drops.

The effect of a single parameterisation can be difficult to understand, and each parameterisation introduces an uncertainty. Even more difficult is to understand the effect of two parameterisations when they are coupled in a model. When developing a parameterisation, this is often not carefully thought through, and also difficult to test. This makes errors introduced by a single parameterisation hard to trace, and is why the behaviour of climate models has become increasingly difficult to understand.

Another issue is that parameterised quantities in models do not necessarily represent the same quantity in observations, or are not even measurable. An example of such a parameter is the entrainment rate in parameterisations of cumulus convection. In models that do not have a fine enough grid spacing to resolve convection, an entrainment rate needs to be prescribed to model how the mass transported by convection changes with height. Bulk entrainment rates are often derived from the output of high-resolution simulations (LES or SRMs) of a field of convective clouds. But it is unclear whether a single bulk entrainment rate suffices to represent all convective cloud fields. For instance, when clouds aggregate or organise themselves on scales smaller than the model's grid spacing, the net vertical transport of mass might be very different from when clouds do not aggregate.

The spatial organisation of clouds is another example. Research on cloud dynamics (Section 1.2.2.2) has long recognised the importance of cloud organisation, and has shown a clear connection between clouds, their organisation and the momentum of the flow, such as in fast travelling squall lines. Organised cloud systems may have a different vertical transport of heat, moisture and momentum, and they can also have different latent heating and radiative heating rates. In turn, these can drive different (local) circulations. Hence, organisation may be crucial for climate, and there is an emerging desire to include such processes in climate models. But how and which processes shape the organisation of clouds is not fully understood. For instance, it is not understood whether clouds self-organise or whether their organisation is triggered by changes in the large-scale environment, or a combination of both. Organisation is something that is easy to spot by eye on visible satellite imagery, when the images span a large enough area, such as in some of the photos in Fig. 1.1. But it is unclear how to quantify organisation and on what scales, and therefore it remains unclear how to include organisation in models.

Indeed, one of the biggest challenges for climate models is coupling the cloud thermodynamic, microphysical and radiative processes to horizontal and vertical motions that accompany cloud systems and large-scale atmospheric circulations (Stevens and Bony, 2013). In other words: one must take into account all the different views of clouds at the same time in order to study their effect of climate. As a consequence of the limitations in representing clouds in climate models, their response to changes in atmospheric composition (e.g., carbon dioxide) varies significantly, even for bulk measures such as global mean temperature. Climate feedbacks that involve clouds have been identified as one of the major sources for the uncertainties in predicting the future of our climate (Bony et al., 2006; Dufresne and Bony, 2008).

Global modelling of clouds will of course greatly benefit from the ever-increasing computing power, which reduces the dependency on parameterisations and makes it possible to resolve both clouds and the large-scale circulation at the same time. But even 'high' resolution simulations only crudely represent processes such as turbulence and microphysics. Therefore, the only way forward is to continually improve our conceptual understanding of clouds and their effect and to constrain models with observed behaviour of clouds. This requires a number of efforts, which include, among others, reducing the gap between the modelling and observation community, by combining modelling and measurement efforts under the umbrella of international programmes, and designing field campaigns, as well as ground-based and space-borne observing systems to address model uncertainties and test new ideas.

1.4 Book Outline

This book is meant to guide and inspire a future generation of scientists to ask and answer important questions about how clouds matter to the Earths' climate, which – as the first consequences of global warming are manifesting themselves – has never been more important. The book will do so by reviewing the present understanding of clouds and their interaction with the climate system in four different parts.

Chapter 2 outlines how clouds are a turbulent dispersion of condensate in a multi-phase flows, which requires an understanding of the relevant laws of fluid dynamics and thermodynamics. Chapter 3 describes how microphysical processes inside clouds shape their appearance and will discuss warm clouds, cold clouds and rain formation. Chapter 4 describes how clouds interact with electromagnetic radiation and modulate energy flows through the Earth system. This theoretical basis in Chapters 2–4 aims to set the foundation upon which other parts of the book build. These include a discussion of the different tools to study clouds, including models and observations (Chapters 5–7), the clouds that occur in different climate regimes (Chapters 8–10) and the interaction between clouds with other processes in the Earth system (Chapters 11–13).

Chapter 5 discusses the use of simpler conceptual models with reduced complexity to study cloud–climate interactions. A variety of models are introduced that cover different cloud types. Chapter 6 deals with the concept of parameterisation of processes in large-scale models and the uncertainties associated with them. How to evaluate models by making use of observations is discussed in Chapter 7.

Tropical and subtropical clouds are covered in Chapter 8, which focuses on the interaction between clouds and atmospheric dynamics, and how these drive much of the variability in climate in the tropics. Unlike the tropics where precipitation and heating processes play a major role, clouds in mid latitude and polar regions are more affected by the rotation of the Earth, and are described in Chapters 9 and 10. How clouds interact with aerosols and with the land surface follows in Chapters 11 and 12. Finally, Chapter 13 addresses the behaviour of clouds under climate change and introduces methods to decipher this behaviour in climate models, along with a discussion on precipitation patterns, and how these might change as the climate warms.

Further Reading

A comprehensive overview of the climate system and climate variability, as well as the atmospheric circulations present on Earth is in the textbook *Global Physical Climatology* by Dennis Hartmann (2016). A shorter read, with limited equations, which provides great insight into the atmospheric processes that determine our climate, including clouds, is the textbook *Atmosphere, Clouds and Climate* by David Randall (2012).

An overview of our current understanding of the impact of aerosols and clouds on climate, and the modelling uncertainties that still exist, is in Chapter 7 ('Clouds and Aerosols') in *Climate Change 2013: The Physical Science Basis*, written by Olivier Boucher and co-authors (2013). The review chapter 'Trends in Observed Cloudiness and Earth's Radiation Budget' by Joel Norris and Anthony Slingo (2009) is part of a book collection about clouds in the perturbed climate system. This particular chapter raises the important issue that our field has to push for future observation systems that have the stability and longevity to measure long-term variations in cloudiness and the radiation budget with improved precision and accuracy before current observing systems will be taken down.

Excellent books written from the perspective of cloud and storm dynamics which explain how different clouds and cloud systems form and interact with the large-scale flow include *Cloud Dynamics* by Robert Houze (1993) and *Storm and Cloud Dynamics* by William Cotton and co-authors (2010). The dynamics of non-precipitating and precipitating convective clouds are described in *Atmospheric Convection*, by Kerry Emanuel (1994). As the title reveals, this textbook also provides a deep background in convection.

Some lighter reading is provided in a number of monographs. The book *Radar and Atmospheric Science: A Collection of Essays in Honor of David Atlas* (Wakimoto and Srivastava, 2003) contains a collection of monographs about radar meteorology in honour of David Atlas, who started his career in the military during the Second World War, and who spurred the use of radar for atmospheric science and weather prediction throughout his career. Another monograph, *Cloud Systems, Hurricanes, and the Tropical Rainfall Measuring Mission (TRMM)* by Tao and Adler (2013) provides a good review on convective clouds and their role in tropical circulations, as a tribute to Joanne Malkus.

Exercises

(1) Describe which of the cloud types in Table 1.1 occurs most frequently and which of the cloud types has the largest cloud amount-when-present in each of the following four regions:

(i) The mid latitude storm tracks, including all ocean between 60°S and 70°S during December/January/February (DJF)

(ii) The subtropical oceans, including all ocean between 20°S and 30°S during DJF

(iii) The tropical rain bands over land, including all land between 0° and 10°N during June/July/August (JJA)

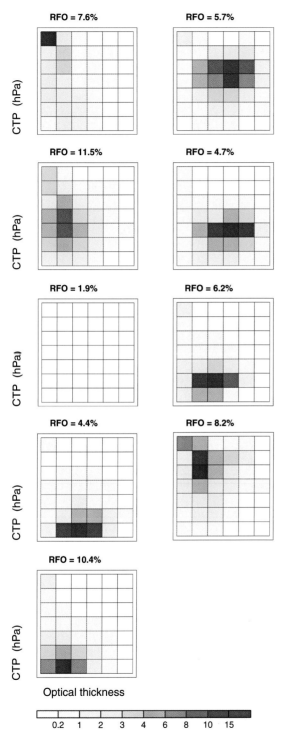

FIGURE 1.22: The CTP-τ histograms of nine of twelve cloud states in frequency of occurrence. The origin of these histograms is described in Section 1.2.1.3. As part of Exercise 4, these nine histograms need to be matched with their corresponding global occurrence maps in Fig. 1.23. RFO denotes relative frequency of occurrence.

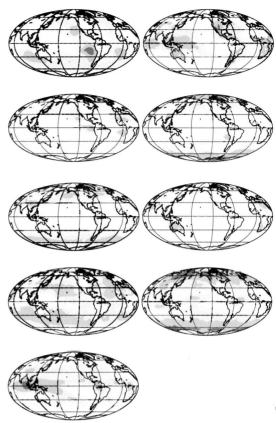

FIGURE 1.23: The geographical distribution (in frequency of occurrence) of nine cloud states, which need to be matched with their corresponding CTP-τ histograms (Fig. 1.22) as part of Exercise 4.

(iv) The tropical rain bands over ocean, including all ocean between 0° and 10°N during JJA

Use the zonally averaged data from the gridded land and ocean maps (5° latitude-longitude resolution) that can be downloaded from the website: www.atmos.washington.edu/CloudMap/. Explain the differences between the four regions using your knowledge of the general circulation of the atmosphere.

(2) Layers of low-, mid- and high-level cloud can occur simultaneously and overlap each other. Their degree of overlap will determine the total cloud cover as projected onto the surface. Different methods for calculating cloud overlap exist which relate the cloud amount at a every given height level to the cloud cover. Such methods are derived and tested using ground-based remote sensing, and used for instance in parameterisations of cloudiness in models. Cloud cover (cc) is calculated from cloud amount (ca) at two levels

k and l as follows when assuming either *minimum* overlap, *maximum* overlap or *random* overlap:

$$cc_{min} = \min(1, ca_k + ca_l) \tag{1.3}$$

$$cc_{max} = \max(ca_k, ca_l) \tag{1.4}$$

$$cc_{rand} = ca_k + ca_l - ca_k ca_l \tag{1.5}$$

Assume three layers of cloud: low-, mid- and high-level cloud. Use the zonally averaged amount of low, mid and high cloud from the surface climatology used in Exercise 1 as an estimate for ca_l, ca_m and ca_h. For each of the four regions defined in Question 1 calculate cc_{min}, cc_{max} and cc_{rand}. Compare these cloud covers with the zonally averaged total cloud cover given by the surface climatology. Is the atmosphere closer to minimum, maximum or random overlap? Why?

(3) Explain why the real cloud overlap in nature is likely to be larger or smaller than that calculated from the surface-based climatology.

(4) The CTP-τ histograms of nine of the twelve cloud states that emerge from a clustering algorithm applied to daily-average histograms are shown in Fig. 1.22 (the histograms of three cloud states that correspond to deep cumulus (Cb), cumulus (Cu) and stratocumulus (Sc) are shown in Section 1.2.1.3, Fig. 1.9). Match these histograms with their corresponding maps in Fig. 1.23. Describe the cloud types and their location in the low, mid or upper troposphere. Note that some cloud states represent different stages of clouds in Howard's cloud classification, which include, for instance, transition regions of one cloud type into another.

Part I Cloud Fundamentals

2 Clouds as Fluids

Bjorn Stevens and A. Pier Siebesma

From a fluid-dynamical point of view, clouds are a dilute dispersion of condensate in a multiphase and multicomponent turbulent flow. Their description thus adopts the language of thermodynamics and fluid dynamics. The multiphase (ice–liquid–vapor) and multicomponent (dry air and water) essence of the flow complicates the thermodynamic and fluid-dynamical description of fluids, which is usually presented for single component, single phase flows, as for liquid water, or simply dry air. In this chapter, an overview of the main equation systems and concepts used to describe clouds and cloudy flows is presented. It is assumed that the reader has a basic background in atmospheric thermodynamics and fluid mechanics, and this text attempts to build on this background to introduce the special elements related to both topics as applied to potentially cloudy systems and as required by subsequent chapters. For readers relatively new to particular topics, Further Reading is suggested at the end to help fill in what is otherwise a dense presentation of the material.

2.1 Thermodynamics

2.1.1 A Multicomponent Multiphase System

The atmosphere, or air, as we experience it, is a multicomponent gas in which a great variety (if not great amount) of fine-scale particulate matter is suspended. The gas phase constituents include several major gases (nitrogen, oxygen, argon) which through the current era have existed in a relatively fixed proportion to one another. To a large degree these determine the thermodynamic properties of 'dry air', that is an ideal mixture composed of 78.08 % N_2, 20.95 % O_2 and 0.934 % Ar, by volume. Additionally, the atmosphere contains variable vapours such as carbon dioxide and water, along with a host of seemingly minor gases (e.g., neon, helium, nitrous oxide, ozone, methane, sulphur compounds, organics), some of

which can be important for determining the radiative properties of the atmosphere and for air quality. Of the variable constituents, water is the most striking as it ranges from abundances that vary over many many orders of magnitude, from nearly zero in the coldest regions of the upper troposphere, to as much as 4 % by volume over very warm bodies of water. Because of its proclivity to change phase and the manner in which these phases change affect the local temperature on the one hand, and foster diverse interactions with radiant energy on the other, water indelibly marks motions in the lower atmosphere on all time and spatial scales. It is hard to think properly about atmospheric motions, let alone clouds, without considering how water is coupled to them. In this sense, the simplest, accurate description of the dynamic atmosphere comprises at least two-components, dry air and water, with one component (water) admitting multiple phases.

The basic thermodynamic properties of the atmosphere thus depend on its component parts. Typically these are defined in terms of their mass m such that for an equilibrium system, four constituents of the moist atmosphere can be defined, dry air, vapour, liquid water and solid (ice) water, denoted by the Roman subscripts d, v, l and i, respectively. The total mass of the system is thus given by sum of the constituent masses such that

$$\mathcal{M} = \mathcal{M}_d + \mathcal{M}_v + \mathcal{M}_l + \mathcal{M}_i. \tag{2.1}$$

The specific mass, rather than say the mixing ratio or a molar concentration, is used to describe the amount of matter (Table 2.1). The normalised or specific mass of a component x is denoted by $q_x = \mathcal{M}_x/\mathcal{M}$. We distinguish between equilibrium condensed phases associated with clouds, which evolve with the thermodynamic state in a more or less reversible way, and larger hydrometeors that do not. Larger hydrometeors, like rain drops and most forms of ice, develop through irreversible microphysical processes such as the collision and coalescence of water droplets or by rapid vapour deposition in conditions of

TABLE 2.1: Mass variables

$\mathcal{M}$	Mass
ρ	Density
q_{x}	Specific mass of some constituent x
p	Thermodynamic pressure ($p_{\mathrm{d}} + p_{\mathrm{v}}$)
p_{x}	Partial pressure of some constituent x

very high supersaturation. These are more difficult to approximate as an equilibrium phase. Because they are larger, non-equilibrium phases of water in the atmosphere are also more dilute and short-lived. Their presence, which we largely ignore, requires the introduction of a more expansive view of the thermodynamic constituents within a moist atmosphere – for instance, by accounting for the mass (and perhaps temperature and velocity) of rain and most forms of ice. Non-equilibrium phases of water are discussed in terms of the microphysical processes in Chapter 3.

Molar descriptions are useful for describing chemical reactions. The use of mass mixing ratios, where the constituent masses are described in terms of their mass fraction relative to dry air, $\mathcal{M}_x/\mathcal{M}_{\mathrm{d}}$, is often adopted by descriptions which introduce water as an additional constituent. The mixing ratio approach is also helpful when considering the possibility that different components of a mass element have their own velocity. In the case of the mixing ratio, the normalising mass is invariant in so long as the basic flow describes the motion of the dry air. Because the focus in this chapter is on a thermodynamic description for an equilibrium system, we adopt the specific humidity description.

Thermodynamic systems in local equilibrium are thus identified with mass elements, sometimes referred to as fluid- or air-parcels. Formally, the concept gains validity for mass elements small enough that the volume they occupy encompasses a scale much smaller than the scale over which thermodynamic properties vary, but much larger than the mean free path. Diffusion rapidly homogenises the atmosphere on scales smaller than the Kolmogorov length scale, $l_{\mathrm{K}} = (\nu^3/\varepsilon)^{1/4}$, where ν is the viscosity of the atmosphere and ε is the turbulent dissipation rate. In vigorous cumulus clouds, the dissipation rate may approach $0.05\ \mathrm{m}^2\ \mathrm{s}^{-3}$ which, given a kinematic viscosity $\nu = 1.5 \times 10^{-5}\ \mathrm{m}^2\ \mathrm{s}^{-1}$, implies that variations in thermal properties are not present on scales less than $l_{\mathrm{K}} = 0.5\ \mathrm{mm}$. This is several thousand times larger than the mean free path of an air molecule, making the concept of an air parcel a useful one.

In developing the present thermodynamic description of a cloudy atmosphere, we make several assumptions. The atmosphere is assumed to be comprised of dry air and water. Both the dry air and water in its vapour form are assumed to be ideal gases (non-interacting point particles) in an ideal mixture. The condensate phase is taken to be an ideal liquid, so that it is incompressible, and it's volume fraction is considered to be negligibly small, so that the total volume $\mathcal{V}$ of the system is seen by both the gas and vapour constituents – that is, $v_{\mathrm{v}} = \mathcal{V}/\mathcal{M}_{\mathrm{v}}$ denotes the specific volume of the vapour. The specific heats of the different constituents are considered to be small, surface effects – for instance, at the condensate/vapour interface are neglected, as are electromagnetic interactions.

The assumption that the condensate volume fraction is vanishingly small arises from the diluteness of clouds. Typically, the specific condensate mass is less than $1\ \mathrm{g\ kg}^{-1}$, and, given the approximately thousand-fold increase in density in the condensate versus vapour phases (for typical atmospheric pressures), this implies that the volume fraction of condensate in the atmosphere is of the order of 10^{-6}. The diluteness of condensate can challenge the concept of an equilibrium thermodynamic system, as on the Kolmogorov scale the condensate is not continuously distributed. To get around this difficulty, one often imagines an air parcel as being on the scale of around $1\ \mathrm{m}^3$. Strictly speaking, volumes of air this large cannot be thought of in terms of a single temperature, but the error of this approximation is typically much less than those associated with other approximations invoked in the description of such systems.

2.1.2 A Notational Challenge

A particular challenge of describing moist atmospheric systems is finding a suitable notation. Many symbols are overloaded. As an example, the symbol v is often used to denote the specific volume in thermodynamic systems, or the second component of the velocity vector in fluid-dynamical systems. The Roman form, v of the same letter is used to denote the form of a subscript 'vapour', and sometimes 'virtual'. The symbol s may be used to denote static energy or entropy or in subscript form a saturated state, or process, or simply a surface quantity.

This chapter is not the proper place to attempt to systematically overhaul an overloaded notation, but some slight deviations are introduced. For instance, density temperature terminology is favoured over the more old-fashioned 'virtual' temperature, and script fonts

TABLE 2.2: Subscript notation

d	Dry air
v	Water vapour
l	Liquid water
i	Solid water (ice)
c	Condensate
t	Total water
s	Saturated state, or process
e	Equivalent (all condensate) reference state
ℓ	Liquid-free (all vapour) reference state
p	Isobaric process
0	Some specified reference state

are introduced for many of the quantities of classical thermodynamics, for example, $\mathcal{M}$ and $\mathcal{V}$ for mass and volume, respectively. Following the usual convention, lower case denotes a mass specific, or intensive, quantity, so that v denotes the specific volume $\mathcal{V}/\mathcal{M}$. An exception to this rule is in the case of the specific mass, which following conventions in the atmospheric sciences is denoted by q, rather than m. δQ, however, denotes heating. Because no special symbol for the specific heating is introduced, no conflict arises with the use of q to denote specific mass.

For entropy, the text defers to the thermodynamic tradition and adopts the symbol $\mathcal{S}$, or s for the specific form. Clausius is said to have adopted the symbol $\mathcal{S}$ for entropy in honour of Sadi Carnot, a tradition this chapter upholds. Likewise the classical notation is also used to denote enthalpy by $\mathcal{H}$, or h for the specific form. In some later chapters, s and h may, for traditional reasons, be used to represent the dry and moist static energy, respectively, which in this chapter are denoted by η_d and η_e. But in these later chapters there should be no risk of confusion with the entropy and enthalpy. The Gibbs potential is given by $\mathcal{G}$, whose specific form g is not to be confused with g the gravitational acceleration. Following these conventions, a subtle distinction arises between the script font of particular symbol and the italic font, for example, v denotes specific volume, rather than v. The text also reserves the *italic* font for variables, that is, measures of something, such as t for time. Abbreviations, even if only one letter long, are written in the Roman font. Hence, q_t is the total water specific humidity, and the subscript t is written in the Roman font because in this usage it abbreviates total water. In an attempt to minimise subsequent confusion, the common subscript abbreviations used in this chapter, and throughout the book, are presented in Table 2.2.

2.1.3 Equation of State

Taking an air parcel to be comprised of an ideal mixture of ideal gases, perhaps in the presence of condensate, the equation of state is that for an ideal gas of variable composition, such that

$$p = p_d + p_v = \left(\frac{\mathcal{M}_d R_d}{\mathcal{V}} + \frac{\mathcal{M}_v R_v}{\mathcal{V}} \right) T, \tag{2.2}$$

where R_d and R_v are the specific gas constants of 'dry air' and water vapour, respectively. Using the subscript c and subscript t to denote the total amount of condensate, and total amount of water (irrespective of phase), respectively (Table 2.2), the specific mass of total water can be expressed as

$$q_t = q_v + q_l + q_i = q_v + q_c. \tag{2.3}$$

Defining the density of the gaseous/vapour mixture as $\rho = \mathcal{M}\mathcal{V}^{-1}$ allows one to formulate the equation of state as

$$p = \rho R T \tag{2.4}$$

where the specific gas constant depends on the amount and distribution of water,

$$R = R_d + q_v R_v - q_t R_d. \tag{2.5}$$

The ratio of the gas constants often arises, albeit in different forms, which makes it helpful to define two constants that depend only on this ratio:

$$\epsilon_1 = \frac{R_d}{R_v} \approx 0.622 \quad \text{and} \quad \epsilon_2 = \left(\frac{1}{\epsilon_1} - 1 \right) \approx 0.608. \tag{2.6}$$

Many derivations are aided by simply accepting that the gas constants and specific heats depend on the composition of the fluid. The density temperature T_ρ

$$p = \rho R_d T_\rho \quad \text{where} \quad T_\rho = T \left(1 + \epsilon_2 q_v - q_c \right) \tag{2.7}$$

makes the sensitivity to composition explicit. In the absence of water, the density temperature is the air temperature, otherwise it can be interpreted as the temperature of a dry air parcel having the same density and pressure as the given air parcel.

In the older literature, the density temperature is often called the virtual temperature, although some authors use both terms, distinguishing between them based on whether or not condensate effects are included. Here the term density temperature is preferred as it makes physical reference to the specific quantity that this particular temperature is meant to help describe.

Given the pressure, the density temperature determines the density and thus is important to the concept of the buoyancy, or effective acceleration, of a fluid parcel in the presence of a gravitational field. From a scale analysis of the vertical momentum equation, the buoyancy of a fluid parcel can be measured by the extent to which its density differs from a background or reference density. For instance, assume that locally the density is given in terms of a deviation from such a reference state density ρ_0 such that

$$\rho = \rho_0 + \rho' \tag{2.8}$$

where the $'$ denotes a deviation. In this case, the buoyancy b has units of acceleration and can be defined as

$$b \equiv -g\frac{\rho'}{\rho_0} \approx g\left(\frac{T'}{T_0} + \frac{R'}{R_0}\right) = g\frac{T'_\rho}{T_{\rho,0}}. \tag{2.9}$$

The approximation of the buoyancy in terms of the density temperature follows from the assumption that the relative change in pressure is small compared to the relative change in density, that is, $p'/p_0 \ll T'/T_0$, and thus Eq. (2.7) identifies the density temperature as the dynamically relevant variable for describing fluid motions.

2.1.4 The First Law and Its Consequences

For an atmospheric system it proves useful to use temperature and pressure to describe the state of the system. The choice of pressure rather than volume, as is more customary in the description of laboratory systems, arises because the pressure is fixed externally by the weight of the surroundings.[1]

With pressure as a thermodynamic coordinate, the First Law becomes

$$d\mathcal{H} = \delta Q + \mathcal{V}dp \tag{2.10}$$

where δQ denotes an infinitesimal amount of heating of the system, and $\mathcal{H}$ is the enthalpy, or heat function. Because $\mathcal{H}$ is an extensive variable, its specific value h can be written as a linear combination of the mass weighted contribution of the constituent enthalpies:

$$h = q_d h_d + q_v h_v + q_l h_l + q_i h_i. \tag{2.11}$$

The laws of thermodynamics only constrain *changes* in enthalpy, which means that the actual value of enthalpy

---

[1] As a result, the isobaric specific heat of some substance x is written as c_{p_x}. Dropping the subscript p simplifies the notation, but may cause confusion in a few places. For example, the isobaric specific heat for vapour c_{p_v}, can be mistaken for c_v, which is commonly used to denote an isometric specific heat, that is, one at constant volume. But, as indicated earlier, isometric processes are rarely relevant to the atmosphere, and are therefore not introduced. Keeping this in mind should help minimise confusion. Table 2.3 provides values of the specific heats for dry air and the different phases of water.

TABLE 2.3: Common thermodynamic constants, where c_{p_x} denotes the isobaric specific heat of either dry air, vapour, liquid or ice. Isobaric specific heat capacities for water phases are given at the triple point. Reference values of entropy and phase-change enthalpies are given at $T = 273.15$ K and at the standard pressure $p_\theta = 1{,}000$ hPa.

Quantity	Value	Unit
R_d	0.2870	kJ kg^{-1} K^{-1}
R_v	0.4615	kJ kg^{-1} K^{-1}
c_{p_d}	1.005	kJ kg^{-1} K^{-1}
c_{p_v}	1.865	kJ kg^{-1} K^{-1}
c_l	4.219	kJ kg^{-1} K^{-1}
c_i	2.097	kJ kg^{-1} K^{-1}
$s_{d,ref}$	6.783	kJ kg^{-1} K^{-1}
$s_{v,ref}$	10.321	kJ kg^{-1} K^{-1}
$l_{v,ref}$	2500.9	kJ kg^{-1}
$l_{f,ref}$	333.4	kJ kg^{-1}

can only be given relative to a reference value, and can thus differ depending on what one adopts for this reference value. In Eq. (2.11), the reference state enthalpies are hidden in the specific enthalpies of the various constituents.

Different phases of matter, for instance, liquid water versus water vapour, differ in their specific enthalpy, so that phase changes imply a change in the enthalpy of the system, which may, for instance, be used to do work. These are often referred to as latent heats, that is, an amount of 'heat' (rather enthalpy) that is realised only through a change of phase, but is otherwise latent. For water, with one crystalline (ice) phase these phase-change enthalpies are, in specific form, denoted as

$$\ell_v = h_v - h_l \tag{2.12}$$

$$\ell_l = h_l - h_i. \tag{2.13}$$

The first, the specific enthalpy of vaporisation, is the enthalpy required to vaporise a unit mass of liquid, often it is called the latent heat of condensation. The second, the specific enthalpy of fusion, is the enthalpy released by the freezing of a unit mass of liquid, and is often called the latent heat of fusion. Both are positive and their sum is the specific enthalpy of sublimation ℓ_s. The naming convention is not entirely consistent, but follows historical usage

Usually, it is assumed that the specific enthalpies depend only on T. Actually, this is a property of ideal fluids and perfect gases, the latter directly from Joule's classic free expansion experiment, so this assumption is equivalent to approximating the atmosphere as a perfect gas and condensate as an ideal fluid, which for practical purposes is a very good approximation. In this case, for some constituent x,

$$\mathrm{d}h_\mathrm{x} = c_{p_\mathrm{x}}\mathrm{d}T, \qquad (2.14)$$

where the constant c_{p_x} is called the isobaric specific heat for the constituent x in its gaseous phase. Thus it follows that

$$\mathrm{d}\ell_\mathrm{v} = (c_{p_\mathrm{v}} - c_\mathrm{l})\mathrm{d}T, \qquad (2.15)$$

which is known as Kirchoff's relation and also holds for the other phase-change enthalpies. For the various phases of water and for dry air it is also safe to assume that c_{p_x} is constant, at least within the homosphere (below 100 km). In this case,

$$h_\mathrm{x}(T) = h_\mathrm{x}(T_0) + c_{p_\mathrm{x}}(T - T_0) \qquad (2.16)$$

and,

$$\ell_\mathrm{v}(T) = \ell_\mathrm{v}(T_0) + (c_{p_\mathrm{v}} - c_\mathrm{l})(T - T_0). \qquad (2.17)$$

Because c_{p_v} is less than c_l, this implies that the enthalpy of vaporisation decreases with temperature, and would (if the specific heats really were temperature invariant) vanish at a critical temperature.

From the previous discussion, it may have been inferred that in so far as we speak of the enthalpy of a system, or some other quantity that might depend on the system enthalpy, we must specify what we have adopted for the reference state. In a system that does not admit phase changes, there is little need to take care as to what one assumes for the reference state enthalpies, and this is more of a formal requirement. If phase changes are permitted, differences in the reference enthalpies accompany changes in phase of the matter, and demand more care.

2.1.5   Enthalpies

In the literature one often encounters closed form expressions for the enthalpy, or related variables like static energies. It is often not clear that these expressions depend on an assumed reference state. Historically such expressions have been derived for liquid–gas systems. In the following, these expressions are derived in a manner that makes the assumed reference state explicit, and in so doing clarifies the relationships among the different enthalpies. To align the definitions with the historical development of the subject it is also assumed that there is only one condensed phase, that corresponding to liquid. A more general treatment, to account for a solid phase, is presented in Section 2.1.11.1.

For a closed system, $\delta q_\mathrm{t} = \delta q_\mathrm{d} = 0$. These constraints reduce the degrees of freedom in Eq. (2.11) and are accounted for by rewriting Eq. (2.11) in terms of q_t. To do so requires substituting $1 - q_\mathrm{t}$ for q_d and eliminating either q_v, or q_l in favour of q_t. Eliminating q_l and making use of the definition of the phase-change enthalpies allows Eq. (2.11) to be recast as:

$$h = (1 - q_\mathrm{t})h_\mathrm{d} + q_\mathrm{t}h_\mathrm{l} + q_\mathrm{v}\ell_\mathrm{v}. \qquad (2.18)$$

For some (not-necessarily infinitesimal) perturbation about a reference state, the enthalpy change can thus be written as

$$\Delta h = (1 - q_\mathrm{t})\Delta h_\mathrm{d} + q_\mathrm{t}\Delta h_\mathrm{l} + \Delta(q_\mathrm{v}\ell_\mathrm{v}). \qquad (2.19)$$

Assuming a reference state temperature $T_{0,\mathrm{e}}$ such that all the condensate is in the form of liquid (which can only be approximately true), yields the following expression for the enthalpy of a reference state,

$$h_{0,\mathrm{e}} = (1 - q_\mathrm{t})h_\mathrm{d}|_{0,\mathrm{e}} + q_\mathrm{t}h_\mathrm{l}|_{0,\mathrm{e}}. \qquad (2.20)$$

By fixing the (arbitrary) values of the constituent specific enthalpies at the reference temperature to be

$$h_\mathrm{d}(T_{0,\mathrm{e}}) = c_{p_\mathrm{d}}T_{0,\mathrm{e}} \quad \text{and} \quad h_\mathrm{l}(T_{0,\mathrm{e}}) = c_\mathrm{l}T_{0,\mathrm{e}}, \qquad (2.21)$$

the enthalpy $h = \Delta h + h_{0,\mathrm{e}}$ can be written as

$$h_\mathrm{e} = c_{p_\ell}T + q_\mathrm{v}\ell_\mathrm{v}. \qquad (2.22)$$

Here the subscript e has been added to the enthalpy as a reminder of the reference state with respect to which it has been defined, and

$$c_{p_\ell} = c_{p_\mathrm{d}} + q_\mathrm{t}(c_\mathrm{l} - c_{p_\mathrm{d}}) \qquad (2.23)$$

denotes the specific heat of the system in this reference state.

Alternatively, eliminating q_v in Eq. (2.11) and adopting a reference state temperature $T_{0,\ell}$ wherein all the water is in the vapour phase with the constituent specific enthalpies are fixed so that

$$h_{0,\ell} = (1 - q_\mathrm{t})h_\mathrm{d}|_{0,\ell} + q_\mathrm{t}h_\mathrm{v}|_{0,\ell} = c_{p_\ell}T_{0,\ell} \qquad (2.24)$$

where

$$c_{p_\ell} = c_{p_\mathrm{d}} + q_\mathrm{t}(c_{p_\mathrm{v}} - c_{p_\mathrm{d}}), \qquad (2.25)$$

yields another form for the enthalpy. It is called the liquid-water enthalpy and is denoted by subscript ℓ:

$$h_\ell = c_{p_\ell}T - q_\mathrm{l}\ell_\mathrm{v}. \qquad (2.26)$$

The subscript e or ℓ serves as a reminder of which reference state has been adopted. The former, which we call the 'equivalent' reference state, is somewhat more common, both because it has a somewhat longer history and because it has some advantageous properties – although thermodynamically, particularly if the ice phase is considered, it makes less physical sense. The two enthalpies, h_e and h_ℓ, differ from one another by a constant, reflecting their different reference states.

2.1.6 Entropy

The Second Law postulates the existence of an entropy state function S, defined by the property that in equilibrium the state of the system is that which maximises the entropy function. Such a function has the property that $\delta Q \leq T\,dS$, with the equality sign holding for reversible transformations. For this case, the First Law can be written in the form

$$d\mathcal{H} = T\,dS + \mathcal{V}\,dp, \tag{2.27}$$

thereby identifying the entropy and the pressure as the independent variables in the enthalpy formulation.

As an extensive state function, the entropy, like the enthalpy, can be decomposed into its constituent parts:

$$s = q_d s_d + q_v s_v + q_l s_l + q_i s_i. \tag{2.28}$$

Unlike the enthalpy, the absolute entropy is not arbitrary to within a constant value. The Third Law specifies that the entropy must go to zero as T goes to zero. Hence the reference entropies cannot be arbitrarily specified. This has consequences for the description of irreversible processes.

For an ideal gas, such as dry air, the specific form of Eq. (2.27) can be integrated to yield an expression of s written in terms of a reference entropy, so that, for instance,

$$s_d = s_{d,0} + c_{p_d} \ln(T/T_0) - R_d \ln(p_d/p_0), \tag{2.29}$$

where $s_{d,0}$ is the reference entropy of dry air at the temperature T_0 and pressure p_0. As in the derivation of an expression for the enthalpy, it is assumed that the specific heats are constant between T and T_0. An analogous expression can be derived for s_v. For the condensed phases, the condensate is assumed to be ideal so that changes in pressure do not contribute to the entropy. Reference values for the entropy of dry air and water vapour at standard pressure and temperature are given in Table 2.3. Reference entropies of the condensed phases can be derived from the reference values of the phase-change enthalpies and the reference value for water vapour.

A general expression for the composite entropy can thus be derived with respect to the chosen reference state. Here again the basic ideas are developed for a system that does not allow a solid (ice) phase. Relative to a system in an 'equivalent' reference state, wherein all the water mass

is in the condensed phase and for which the pressure is the standard pressure, $p_\theta = 1{,}000\,\text{hPa}$,

$$s = s_{e,0} + c_{p_\ell} \ln(T/T_0) - R_e \ln(p_d/p_\theta) + q_v(s_v - s_l), \tag{2.30}$$

with $s_{e,0} = s_{d,0} + q_t(s_{l,0} - s_{d,0})$, and c_{p_ℓ} defined as before. The only gas-phase constituent in the reference state is dry air, hence the gas constant $R_e = (1 - q_t)R_d$. The value of $s_{e,0}$ is thus determined by the amount of water in the system and the reference state temperature and pressure, denoted by T_0 and p_θ, respectively.

There are two ways to look at Eq. (2.30). Given a completely specified reference state, it provides an expression for the entropy. This was the sense in which it was derived. Alternatively, one can use this equation to ask what would the reference state temperature need to be, for the system in the reference state configuration (as specified through the pressure, amount and distribution of water mass) to have the same entropy as in the given state. In this case $s_{e,0}$ is set equal to s and Eq. (2.30) becomes an equation for T_0 conditioned on the choice of reference state. For the choice of the equivalent reference state, this alternative application of Eq. (2.30) leads to the interpretation of T_0 as the temperature the system would attain if all of its water was reversibly condensed, and then separated mechanically from the gas but maintained in thermal equilibrium with the dry air as the system was brought reversibly to the reference states pressure – a process that is easier to imagine than to realise.

In the absence of ice processes, an expression corresponding to Eq. (2.30), but for the liquid-free reference state, follows analogously as

$$s_\ell = s_{\ell,0} + c_{p_\ell} \ln(T/T_0) - q_d R_d \ln(p_d/p_{d,\theta})$$
$$- q_t R_v \ln(p_v/p_{t,\theta}) - q_l(s_v - s_l), \tag{2.31}$$

with $s_{\ell,0} = s_{d,0} + q_t(s_{v,0} - s_{d,0})$. The pressures, $p_{d,\theta}$ and $p_{t,\theta}$, assume a reference state that is sub saturated but with the same mass of total water and dry air as in the actual state, so that $p_\theta = p_{d,\theta} + p_{t,\theta}$. Physically, the reference state temperature is that which the system would attain if reversibly brought to the reference (liquid-free) state pressure (by convention p_θ), which requires that the vapour pressure be less than the saturation vapour pressure at this temperature, so that any condensate that may initially be in the system transforms to vapour.

2.1.7 The Clausius–Clapeyron Equation

For a closed isobaric and isothermal system, it follows from Eq. (2.27) that for a reversible process

$$0 = d\left(\mathcal{H} - TS\right), \tag{2.32}$$

which introduces the Gibbs free energy, or Gibbs potential, as $G = \mathcal{H} - TS$, that is, the energy available to do work in an isothermal and isobaric system. From this definition it follows that the difference in the Gibbs energy of two constituents is related to the differences in their enthalpies and entropies, so, for example,

$$g_v - g_l = h_v - h_l - T(s_v - s_l). \tag{2.33}$$

From the postulates of thermodynamics, whereby in equilibrium $\mathcal{H}$ and T adopt values that maximise S, it follows that the Gibbs free energy of a system in equilibrium is a minimum. The condition that, for a closed isothermal and isobaric system, the Gibbs free energy is a minimum determines the partitioning between two phases of matter, say liquid water and water vapour, in equilibrium, whereby for this equilibrium partitioning the vapour state can be said to be saturated, a condition that in the ensuing presentation is denoted by subscript s. The minimisation of the free energy determines the phase partitioning because it requires that the specific Gibbs energy of each phase must be equal, that is, $g_v = g_l$, otherwise a redistribution of the mass between the phases could lower the total Gibbs energy. This property can be used to derive the temperature dependance of the phase partitioning as follows.

If the temperature of the system changes, this implies a change in the Gibbs energy, such that

$$dg = dh - sdT - Tds = vdp_v - sdT, \tag{2.34}$$

likewise

$$dg_v = v_v dp_v - s_v dT \quad \text{and} \quad dg_l = v_l dp_v - s_l dT. \tag{2.35}$$

But, because the maintenance of equilibrium requires that $dg_v = dg_l$, it follows that for such a transformation the vapour pressure changes with temperature as,

$$dp_v = \frac{s_v - s_l}{v_v - v_l} dT. \tag{2.36}$$

This is the Clapeyron equation describing how vapour pressure changes with temperature.

Clapeyron's equation can be cast in a simpler form by substituting for v_v from the ideal gas law in the denominator of the fraction on its right-hand side, and additionally noting that $v_l \ll -v_v$ so that $v_v - v_l \approx v_v$, and by rewriting the entropy differences in terms of the vaporisation enthalpy. This latter step is accomplished by realising that as far as the numerator on the right-hand side of Eq. (2.36) is concerned, for a saturated system $g_v = g_l$, which from Eq. (2.34) implies $s_v - s_l = \ell_v/T$. On the basis of these insights, Clausius showed that Clapeyron's equation can be written in the form

$$d(\ln p_v) \approx \frac{\ell_v}{R_v T} d(\ln T), \tag{2.37}$$

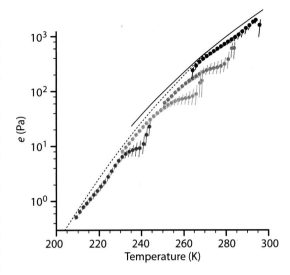

FIGURE 2.1: Saturation vapour pressure over liquid (solid line) and ice (dashed line). Filled circles and lines show vapour pressure in the atmosphere, arranged according to temperature for different pressure levels (900 hPa, black; 700 hPa, blue; 500 hPa, orange; 300 hPa, red). At $T = 0\,°C$, the saturation vapour pressure is 610.15 Pa. At $T = -30\,°C$, the saturation vapour pressure over liquid water is 50.8 Pa compared to 38.0 Pa over ice at the same temperature. Saturation with respect to liquid for $T < 0\,°C$ is relevant because super-cooled water is often present in the atmosphere, with homogeneous nucleation of ice particles first occurring at about $T = -38\,°C$.

which has come to be known as the Clausius–Clapeyron equation. Fig. 2.1 shows that the Clausius–Clapeyron equation very effectively delimits the distribution of water throughout the atmosphere. The atmosphere sits atop a reservoir of water, which endeavours to bring the air above it into saturation. If the amount of moisture exceeds the saturation value it condenses, and condensate is effectively removed by precipitation from the system. Hence the saturation specific humidity limits the amount of water in the atmosphere, as seen in Fig. 2.1. Because Eq. (2.37) so strongly controls the distribution of water in the atmosphere, if one had to single out a particular equation as being the most important for the functioning of Earth's climate, it would be this equation.

2.1.8 Potential Temperatures

In the atmospheric sciences, there is a tradition of using temperature variables to measure the system's entropy. These are usually called potential temperatures as they measure the temperature the system would have to have in some specified reference state for the entropy of this state to be identical to that of the given state. It thus follows that these temperatures are invariant (for a closed system) under an isentropic process, but their properties and absolute values depend on the specification of the reference state.

For the equivalent reference state, equating θ_e with the value of T_0 chosen so that $s_{e,0} = s$ in Eq. (2.30) implies that

$$c_{p_\ell} \ln \theta_e = c_{p_\ell} \ln T - R_e \ln(p_d/p_\theta) + q_v(s_v - s_l), \quad (2.38)$$

with $p_\theta = 1{,}000$ hPa denoting standard pressure.

Eq. (2.38) can be recast in a more familiar form by expressing the pressure of the dry air in terms of the total pressure and the specific humidity, and by expressing the vapour–liquid entropy difference in terms of the latent heat. For the former,

$$p_d = p \left(\frac{R_e}{R} \right). \quad (2.39)$$

For the latter, the entropy difference at the end of Eq. 2.38 is expressed relative to the vapour entropy in saturation, so that

$$s_v - s_l = s_v - s_s + s_s - s_l$$
$$= s_v - s_s + (\ell_v/T). \quad (2.40)$$

Here the last expression arises because the condition of phase equilibrium is the equality of the specific Gibbs enthalpies of the phases, so that $s_v - s_l = (h_v - h_l)/T = \ell_v/T$. From Eq. (2.29), the difference between the vapour and saturation vapour entropy is measured by the difference in the partial pressures

$$s_v - s_s = -R_v \ln \left(\frac{p_v}{p_s} \right) \quad (2.41)$$

where p_v/p_s defines the relative humidity. So that with Eqs (2.39)–(2.41), Eq. (2.38) can recast as

$$\theta_e = T \left(\frac{p_\theta}{p} \right)^{\frac{R_e}{c_{p_\ell}}} \Omega_e \exp \left(\frac{q_v \ell_v}{c_{p_\ell} T} \right). \quad (2.42)$$

Here the term

$$\Omega_e = \left(\frac{R}{R_e} \right)^{\frac{R_e}{c_{p_\ell}}} \left(\frac{p_v}{p_s} \right)^{\frac{-q_v R_v}{c_{p_\ell}}} \approx 1 \quad (2.43)$$

has been introduced as a separate factor because, by virtue of the smallness of q_t ($\ll 1$), it depends only very weakly on the thermodynamic state. Eq. (2.42) is a complicated expression presented in all its fullness, but it is rarely used in this form for practical applications. For many purposes far simpler expressions capture much of the essential physics. These, simpler expressions and the assumptions they imply, are discussed in Section 2.1.11.2.

Choosing instead T_0 so that the liquid-free reference state has the same entropy as the given state introduces the liquid-water potential temperature as

$$\theta_\ell = T \left(\frac{p_\theta}{p} \right)^{\frac{R_\ell}{c_{p_\ell}}} \Omega_\ell \exp \left(-\frac{q_l \ell_v}{c_{p_\ell} T} \right), \quad (2.44)$$

where

$$\Omega_\ell = \left(\frac{R}{R_\ell} \right)^{\frac{R_\ell}{c_{p_\ell}}} \left(\frac{q_t}{q_v} \right)^{\left(\frac{q_t R_v}{c_{p_\ell}} \right)} \quad (2.45)$$

with $R_\ell = R_d(1 + \epsilon_2 q_t)$, c_{p_ℓ} is given by Eq. (2.26) and once more p_θ is the standard pressure. Here too, simpler expressions that provide a reasonably good approximation to θ_ℓ are discussed in Section 2.1.11.2.

In the absence of condensate Eq. (2.45) simplifies to a quantity resembling the familiar dry potential temperature

$$\theta = T \left(\frac{p_\theta}{p} \right)^{R/c_p}, \quad (2.46)$$

where R and c_p depend on composition of the fluid, and hence q_t. For the particular case of dry air, we denote θ by θ_d as a reminder that $c_p = c_{p_d}$ and $R = R_d$. The non-dimensional pressure describing the proportionality between temperature and potential temperate arises frequently and is known as the Exner function,

$$\Pi \equiv \left(\frac{p}{p_\theta} \right)^{R/c_p}. \quad (2.47)$$

Unlike θ, both θ_e and θ_ℓ are conserved under isentropic transformations of moist air in a way that accounts for the isentropic phase changes between vapour and liquid. For this reason, they are often adopted as thermodynamic variables. They differ from one another in that moisture contents, particularly in the lower troposphere, cause θ_e to be substantially larger than θ, while θ_ℓ is typically only slightly less, if at all, than θ by virtue of the typically small amounts of condensate suspended in the air. These differences are most readily evident by recognising that many of the terms in Eqs (2.42) and (2.44) only contribute small corrections to much simpler forms of the equations, so that

$$\theta < \theta_e \approx \theta \exp \left(\frac{q_v \ell_v}{c_{p_d} T} \right) \quad (2.48)$$

and

$$\theta \geq \theta_\ell \approx \theta \exp \left(\frac{-q_l \ell_v}{c_{p_d} T} \right). \quad (2.49)$$

Further discussion of the advantage of one or the other choice of potential temperature is provided at the end of Section 2.1.11.3.

2.1.9 Static Energies

In an atmosphere absent of horizontal pressure gradients, and with the vertical distribution of pressure in hydrostatic balance, the change in pressure following parcel displacements follows the geopotential ϕ as $v\mathrm{d}p = -\mathrm{d}\phi$. In this case, the adiabatic (literally *no heating*) form of the First Law, Eq. (2.10), becomes

$$0 = \mathrm{d}\left(\hat{h} + \phi\right). \tag{2.50}$$

This defines the static energy $\eta = \hat{h} + \phi$ as an adiabatic invariant of parcel displacements in such an atmosphere. The name arises because η measures the total energy a parcel has were it static – the kinetic energy is not accounted for. In an analogy to the potential temperatures, it could just as well be called the potential enthalpy, that is, the enthalpy a parcel would have were it adiabatically brought to the surface given some specification of the reference enthalpy. The restriction that η is only conserved for an atmosphere in which pressure varies hydrostatically in the vertical, and not at all horizontally might seem restrictive, but the vertical distribution of pressure is (especially on larger scales) well approximated by hydrostatic balance and at least in the tropics horizontal pressure gradients are very small. So that in the more general setting η is very nearly conserved and is often treated as if it were conserved.

Neglecting for a moment corrections implied by the failure of these assumptions, or the effects of an ice phase, the form of the static energy depends on which reference state the enthalpy is referred to. Adopting the equivalent reference state, in which case $\hat{h} = \hat{h}_{\mathrm{e}}$ leads to the following definition of the static energy,

$$\eta_{\mathrm{e}} = c_{p_\ell}T + \ell_{\mathrm{v}}q_{\mathrm{v}} + \phi. \tag{2.51}$$

For the liquid-free reference state, $\hat{h} = \hat{h}_\ell$ and

$$\eta_\ell = c_{p_\ell}T - \ell_{\mathrm{v}}q_{\mathrm{l}} + \phi. \tag{2.52}$$

The static energy defined in terms of the equivalent reference state η_{e} is called the moist static energy; η_ℓ is called the liquid-water static energy.

Both η_{e} and η_ℓ are exactly conserved for adiabatic transforms of the moist system if pressure only varies vertically, and, then, hydrostatically. In the absence of water both become identical to the dry static energy,

$$\eta_{\mathrm{d}} = c_{p_{\mathrm{d}}}T + \phi. \tag{2.53}$$

The names, moist static energy versus equivalent potential temperature or the liquid-water static energy and liquid-water potential temperature, emerged historically and not

in relation to one another. Because the equivalent reference state contains all the moisture in the condensed form, a more informative terminology would be to refer to the condensation and evaporation potential temperatures (Eqs (2.42) and (2.44)) and the condensation and evaporation potential enthalpies (Eqs (2.51) and (2.52)), respectively.

2.1.10 Heat Engines and Maximum Entropy Production

2.1.10.1 Heat Engines

Many thermodynamic processes can be interpreted in terms of a heat engine, which converts heat into work. The atmosphere is one of them. The amount of work that can be done by a heat engine is limited to that which can be done by a reversible heat engine, that is, the Carnot cycle. Hence, the idea of heat engines arises in many places as a way to bound the work that can be done by a system.

The Carnot cycle, shown schematically in a fashion applicable to the atmospheric circulation (Fig. 2.2), is an ideal process composed of four stages. In the first stage, the system extracts an amount of energy Q_1 from a reservoir at some temperature T_1, increasing the entropy by the amount $\mathrm{d}S = Q_1/T_1$. In the second stage, the system performs an amount of work $W_{12} > 0$ on its environment without a change in entropy, until it reaches the temperature of a second reservoir at some temperature T_2, where $T_2 < T_1$. In the third stage, the system loses an amount Q_2 of energy to its environment at a temperature T_2, decreasing its entropy by the amount Q_2/T_2. Finally, in the fourth stage, an amount W_{21} of work is done on the system to isentropically return it to its initial state at temperature T_1. For a reversible process, the entropy

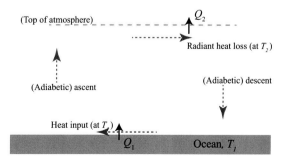

FIGURE 2.2: The atmosphere as a heat engine, where air is heated by an amount Q_1 over the ocean at some effective temperature T_1 and loses energy (Q_2) radiatively at some cooler temperature T_2. It is assumed to ascend and descend adiabatically.

gained in the first stage equals what is loss in the third stage so that $Q_2/T_2 = Q_1/T_1$. More generally, if entropy is produced by irreversible processes, then

$$\frac{Q_1}{T_1} = \frac{Q_2}{T_2} - \Delta S_{\mathrm{irr}}, \qquad (2.54)$$

where $\Delta S_{\mathrm{irr}} > 0$ accounts for the entropy by irreversible processes.

The net work done over the course of one cycle is given by $W = W_{1,2} - W_{2,1}$. By conservation of energy,

$$W = Q_1 - Q_2 = Q_1 \left(1 - \frac{T_2}{T_1}\right) - T_2 \Delta S_{\mathrm{irr}}. \qquad (2.55)$$

The term in the parentheses defines the efficiency of the Carnot cycle, it defines how much of the heat extracted from the warmer reservoir can be converted to work in an irreversible process. From the Second Law, the entropy production by irreversible processes is positive, so that irreversible processes further limit how much of the energy extracted from the warm reservoir can be used to do work. The atmosphere does not do work on the Earth or space, but one can imagine the system doing work by converting heating into the kinetic energy of the circulation (as schematically shown in Fig. 2.2), which is then removed from the system (either by dissipation or through conversion to another form of energy, for instance, wind power) to maintain a steady state.

More generally one can apply the question of how much work can be done to the question of the magnitude of the irreversible entropy production by a variety of processes. If no work is done, or any work that is done simply balances dissipation, and the system is stationary, $Q_2 = Q_1 = \Delta Q$, and from Eq. (2.54),

$$\Delta S_{\mathrm{irr}} = \Delta Q \left(\frac{1}{T_2} - \frac{1}{T_1}\right). \qquad (2.56)$$

An equation of this form can thus be used constrain how much entropy production the system can support. For instance, when applied to the tropical circulation, one can think of energy being added to the system through surface fluxes at $T_1 = T_{\mathrm{sfc}}$ and energy being lost to space through radiation, and an effective radiative temperature of T_{rad}. Taking ΔQ to be about $240\,\mathrm{W\,m^{-2}}$, $T_{\mathrm{rad}} = 255\,\mathrm{K}$ and $T_{\mathrm{sfc}} = 300\,\mathrm{K}$ determines the irreversible entropy production to be $0.14\,\mathrm{W\,m^{-2}\,K^{-1}}$. The challenge then becomes to associate ΔS_{irr} with specific processes in the hope that Eq. (2.56) can be used to bound the strength of these processes.

As an example, assume that the dissipation of kinetic energy, to create power for wind turbines, was the main source of irreversible entropy production. Then the Carnot cycle could be used to bound the strength of the circulation

driven by adding energy to the fluid at some temperature T_1 and extracting it at a temperature T_2. All that would be needed would be a relationship between the rate of dissipation and the strength of the circulation.

This discussion has taken for granted that Q is given. In principle, the amount of heating depends also on the circulation, which extracts energy more or less efficiently from the heat reservoir, that is, stronger winds increase the flux of moisture and internal energy from the surface. Such system's can arrive at stationary solutions which differ in terms of the amount of power they produce. Often it is found that nature tends to prefer the solutions that maximise power, an idea closely related to the idea of maximum entropy production.

2.1.10.2 Entropy Production by Irreversible Processes

The relationship between the potential temperatures and entropy becomes evident upon substitution of θ_e back into Eq. (2.30), whereby

$$\theta_e = T_0 \exp\left(\frac{s - s_{e,0}}{c_{p\ell}}\right), \qquad (2.57)$$

wherein T_0 now adopts some fixed value which also fixes $s_{e,0}$ independently of the value of s. The differential of Eq. (2.57) is

$$c_{p\ell}\frac{\mathrm{d}\theta_e}{\theta_e} = c_{p\ell}\frac{\mathrm{d}T_0}{T_0} + \mathrm{d}s - \mathrm{d}s_{e,0} - \left(\frac{s - s_{e,0}}{c_{p\ell}}\right)\mathrm{d}c_{p\ell}, \quad (2.58)$$

which could have just as easily been formed by taking the differential of Eq. (2.30).

For a closed system, q_t is constant, hence composition-dependent reference properties ($s_{e,0}, c_{p\ell}$) are (along with T_0) also constant, so that

$$\mathrm{d}s = c_{p\ell}\mathrm{d}(\ln \theta_e) = c_{p\ell}\mathrm{d}(\ln \theta_\ell). \qquad (2.59)$$

However, if the relative composition of the system changes, through a change in q_t, then those reference properties, which depend on the composition of the system, also change and these equalities no longer hold. Instead,

$$\mathrm{d}s = c_{p\ell}\mathrm{d}(\ln \theta_e) \\ + \left[(c_1 - c_{p_d})\ln(\theta_e/T_0) + (s_{1,0} - s_{d,0})\right]\mathrm{d}q_t. \qquad (2.60)$$

From this equation, the importance of mixing of water substance arises both directly through the term $\mathrm{d}q_t$, and indirectly though its contributions to changes in $\ln\theta_e$.

Even in a closed system, in which no work is done at the boundaries, the approach to equilibrium, for instance by mixing air masses at different temperatures, or with different amounts of constituent matter because it happens in a state of disequilibrium, irreversibly increases the entropy. In a system with condensate, or experiencing phase changes, evaporation in subsaturated air or condensation in supersaturated air is also a non-equilibrium processes, and associated with an irreversible production of entropy. Also, dissipative processes, for instance, the dissipation of kinetic energy also is a source of entropy, and enthalpy. As it turns out, apart from radiative processes, most of the entropy production in the atmosphere is associated with the mixing of moisture, and that fraction which is associated with the dissipation of kinetic energy comes from a surprising source. Hydrometeors falling through the atmosphere reach a terminal velocity when the drag they experience balances their gravitational acceleration. The drag represents work being done on the fluid by the hydrometeors, work which is dissipated in the wake of the droplet and which is substantially larger than the dissipation of kinetic energy from large-scale fluid motions.

2.1.11 Further Thoughts on Thermodynamic Variables

2.1.11.1 Incorporating the Ice Phase

In the previous discussion, the ice phase was neglected because the static energies and the potential temperatures are usually defined, for historical reasons, to only incorporate the liquid phase. Neglecting the ice phase also makes it easier to understand the basic concepts underlying the construction of conserved moist thermodynamic variables such as θ_e or η_ℓ.

Once the basic idea of the moist entropies and enthalpies is clear, however, it is possible to generalise their definition to incorporate the ice phase. Doing so results in the following generalisation of Eqs (2.22) and (2.26) so that the static energies become

$$\eta_e = c_{p_\ell} T + \ell_v q_v - \ell_f q_i + \phi, \tag{2.61}$$

$$\eta_\ell = c_{p_\ell} T - \ell_v q_l - \ell_s q_i + \phi. \tag{2.62}$$

The presence of ice also encourages the definition of an ice-only reference state, wherein $h_{d,0} = h_{i,0} = 0$, so that for a system in thermal equilibrium,

$$\eta_i = c_i T + \ell_s q_v + \ell_f q_l + \phi. \tag{2.63}$$

with $c_i = c_{p_d} + q_t(c_i - c_{p_d})$. This form of the moist static energy is thermodynamically more sensible than Eq. (2.61), as a reference state in which all the water exists as solid (ice) condensate is asymptotically more accessible than the 'equivalent' state wherein all water is in the liquid form. The reference state corresponding to Eq. (2.62) is

even easier to work with as it is accessible absolutely, not just asymptotically, that is, there are (atmosphere-like) temperatures at which all the condensate will exist as vapour, where as all of the vapour is extinguished in favour of ice only as the temperature goes to zero.

Introducing expressions for the potential temperatures that account for the ice phase is somewhat more involved. Analogous to the generalisation of the enthalpy implicit in Eq. (2.61), the expression for the entropy becomes

$$s = s_{e,0} + c_{p_\ell} \ln(T/T_0) - R_e \ln(p_d/p_\theta)$$
$$+ q_v(s_v - s_l) - q_i(s_l - s_i). \tag{2.64}$$

The entropy difference between the condensed phases is calculated by assuming each is an ideal condensate, so that its entropy is independent of pressure. Integrating the entropy for each phase from its value at the triple-point temperature T_* and noting that at the triple point $s_l - s_i = \ell_f/T_*$, it follows that

$$s_l - s_i = \frac{\ell_f}{T_*} + (c_l - c_i) \ln\left(\frac{T}{T_*}\right). \tag{2.65}$$

Given this expression for the entropy difference between the condensed phases, the expression for the generalisation of the equivalent potential temperature to account for the ice phase thus introduces an additional multiplicative factor in the definition of θ_e.

$$\left(\frac{T}{T_*}\right)^{-q_i \frac{(c_l - c_i)}{c_{p_\ell}}} \exp\left(-\frac{q_i \ell_f}{c_{p_\ell} T_*}\right). \tag{2.66}$$

The generalization of the the the equivalent potential temperature to account for ice, thus becomes

$$\theta_e = T \left(\frac{p_\theta}{p}\right)^{\frac{R_e}{c_{p_\ell}}} \left(\frac{T}{T_*}\right)^{-q_i \frac{(c_l - c_i)}{c_{p_\ell}}} \Omega_e \exp\left(\frac{q_v \ell_v}{c_{p_\ell} T} - \frac{q_i \ell_f}{c_{p_\ell} T_*}\right). \tag{2.67}$$

It reduces to the traditional definition of θ_e, as expected, in the absence of ice.

2.1.11.2 More Approximate Descriptions

The application of the laws of thermodynamics to any real system necessarily involves approximations. The atmosphere is no exception. Even as developed up until this point, a great number of approximations have been made. Not only have ideal mixtures of ideal constituents been assumed, but it is assumed that the condensate phase to occupy zero volume. Some assumptions have not even been mentioned, for instance, the neglect of surface effects or of the effects of impurities in the condensate phase; electric dipole and magnetic moments of the matter under consideration have also been neglected.

Of the assumptions that have been made, many are tenable only in a certain range of temperatures. For example, if the specific heats were actually constant irrespective of temperature, then Kirchoff's relation and the requirement that the vaporisation enthalpy be zero at absolute zero implies that $\ell_{v,0} = (c_l - c_{p_v})T_0$, which from Table 2.3 is clearly not the case. Variations in the specific heats with temperature are crucial to explain the value the enthalpy of vaporisation attains at ambient temperatures. For a perfect gas, the specific heats are related to the degrees of freedom of a molecule over which thermal energy is equally distributed. However, not all of the degrees of freedom of molecules in the atmosphere are accessible at all temperatures, and even for simple molecules degrees of freedom are frozen out at lower temperatures, which explains why the specific heat for the diatomic constituents of dry air such as N_2 and O_2 are not influenced by vibrational degrees of freedom, which only become important at much higher temperatures.

Notwithstanding the departures from exactness entailed by all of this approximation, for many applications, the bookkeeping entailed for a fully two component description of the atmosphere as a mixture of water and dry air may unnecessarily complicate the physical description and obscure physical relationships. For the study of clouds and cloud processes, latent heating is essential, but for many questions the influence of moisture on the specific heat (and even the gas constant) may not be important. With the assumption that $c_l = c_{p_v} = c_{p_d}$,

$$\eta_e \approx \eta_d + \ell_{v,0} q_v, \tag{2.68}$$

$$\eta_\ell \approx \eta_d - \ell_{v,0} q_\ell. \tag{2.69}$$

Here the vaporisation enthalpy has been replaced by its value at the triple point consistent with the fact that in assuming $c_{p_v} = c_l$, Kirchoff's relation implies that ℓ_v is independent of temperature. For many purposes, as shall be evident through the remainder of this book (and even the later sections of this chapter). Eqs (2.68) and (2.69) are preferred as they simplify the analysis and more clearly illustrate physical ideas that don't depend on differences in the gas constants and specific heats.

Similarly, simple expressions for the potential temperatures require two additional assumptions: (i) difference between the gas constants of water vapour and dry air can be neglected; (ii) moisture variations are only important in so far as they multiply the phase-change entropy. Adopting these additional assumptions leads to the the following approximations to the potential temperatures

$$\theta \approx \theta_d, \tag{2.70}$$

$$\theta_e \approx \theta_d \exp\left(\frac{q_v \ell_v}{c_{p_d} T}\right), \tag{2.71}$$

$$\theta_\ell \approx \theta_d \exp\left(\frac{-q_l \ell_v}{c_{p_d} T}\right), \tag{2.72}$$

and in so doing helps highlight the close relationship between the potential temperatures and the static energies.

In addition, exact expressions for the saturation vapour pressure over liquid or solid (ice) surfaces do not exist. This has given rise to a rich literature of approximations (as reviewed by Murphy and Koop, 2005). A formula that for many purposes strikes a good balance between accuracy and simplicity, was suggested by O Tetens in 1930, and reformulated later by Bolton (1980) as

$$p_s = 6.112 \exp\left[\frac{17.67(T - 273.15)}{T - 29.65}\right]. \tag{2.73}$$

For $238.15\,\text{K} < T < 308.15\,\text{K}$, Eq. (2.73) differs by less than 0.5 % from the more accurate, albeit much more complex, formulations reviewed by Murphy and Koop. It is similarly accurate to what one would derive by assuming that ℓ_v varied from its triple-point value linearly in T following Kirchoff's relation. Similar, expressions for ice and their relation to more complex expressions are explored further in the Exercises at the end of this chapter.

Although approximations such as those already outlined are often adopted, care is warranted. For instance, neglecting differences in the gas constants implies that moisture fluctuations no longer contribute to density fluctuations, that is, that $T_\rho = T$. In the tropical boundary layer, moisture fluctuations are much larger than temperature fluctuations and contribute to roughly half the variability in T_ρ. Deriving a consistent thermodynamic framework requires an asymptotic approach, for instance, by expanding the equations about a small parameter equal to the difference between the specific heats and gas constants so as to better appreciate at which order they contribute to one or the other expression. Such an analysis would then identify what level of thermodynamic description is necessary to describe a system for a particular application, but has yet to be developed.

2.1.11.3 Choice of Thermodynamic Variables

The question often arises as to which thermodynamic variables are most appropriate to adopt for the study of a given process. There is a tradition of calling these thermodynamic coordinates as they define the space within which thermodynamic processes are studied, something that is especially evident in the diagrammatic methods reviewed in the next section. Enthalpy-based variables are well suited to the treatment of mixing processes because they are so naturally linked to the extensive variables, and, except for the dissipation of kinetic energy, there are no sources of enthalpy through irreversible processes in the isobaric system.

It would appear that solving for the evolution of the potential temperature θ_e rather than the entropy s_e would make no difference. However, mixing processes are an irreversible source of entropy production which must be explicitly accounted for if the entropy equation is being solved. These are implicitly included when the potential temperature equation is solved. Consider the mixing between two air masses, so that the mixed state can be denoted by an overbar, identifying intensive properties as the (appropriate) average of the intensive properties of the two systems. The difference between averaging entropies, compared to potential temperatures, becomes readily apparent when it is considered that $\ln \overline{\theta} \geq \overline{\ln \theta}$. The homogenisation of θ by a system that solves for θ increases the entropy (which is proportional to $\ln \theta$ of the system as one would expect for an isobaric process which conserves enthalpy. In this sense, θ behaves like an enthalpy variable.

Deciding whether the condensed (equivalent), η_e or η_i or the evaporated (condensate-free) η_ℓ, reference state representation is more or less favourable depends on what is being described. Traditionally, systems in which liquid water is explicitly accounted for favour η_ℓ because it reduces to the dry static energy in the absence of condensate. Systems in which water vapour is the other thermodynamic coordinate tend to favour η_e representations, in part because η_e is, given the approximations in Eq. (2.68), roughly constant even for open systems in which precipitation converges or diverges from a parcel of air. In the more exact representations of η_e, precipitation affects the value of the total water, and hence c_{p_ℓ}, and thus acts as an enthalpy source. But these effects are usually small and as a result θ_e or η_e are widely used to describe large-scale systems, particularly in so far as they involve precipitation as, for instance, discussed in Chapter 8, and θ_ℓ or η_ℓ find favour in studies of non-precipitating boundary layer clouds, as discussed in Chapter 5.

2.2 Thermodynamic Diagrams

Thermodynamic diagrams are used to represent the state of a system, as well as thermodynamic processes. An example of such a process might be how temperature changes as air rises in the absence of heating, that is, *adiabatically* with or without condensational processes. Such diagrams can say much about the state of the atmosphere and how it might have come into being as they can quickly convey a wide range of quantitative information to the trained eye. This information can be helpful in determining the subsequent evolution of the atmosphere and is routinely used by weather forecasters to evaluate the likelihood of different events, ranging from fog formation to the energy available for deep convective overturning.

Many of these diagrams predate the widespread use of digital computers. Even though today it is possible to compute many quantities directly, thermodynamic diagrams are still very useful for encapsulating the state of the atmosphere and remain widely used. The most common thermodynamic diagram is the skew-T diagram, and its cousin the Tephigram. Other diagrams include the Clapeyron, the Emagram or Neuhoff, and the Stüve diagram. Less familiar, but also useful is the Paluch diagram, and moist-static energy diagrams.

One difficulty that all diagrams share is that they are two-dimensional, and the most compact description of the state of the atmosphere encompasses three dimensions, for instance, $\{T, p, q_t\}$. Two ways have been devised for getting around this problem. One is to recast moisture as a temperature variable, as is done in constructing the dew-point temperature. The other is to link moisture to temperature and pressure by a saturation assumption. Examples of both approaches are evident throughout this book.

2.2.1 Skew-T and Related Diagrams

The skew-T diagram adopts temperature T and $\ln p$ as its thermodynamic coordinates. The logarithm of pressure is chosen for the vertical coordinate rather than the pressure itself because in an isothermal atmosphere height varies with $\ln p$, and hence for a realistic temperature profile the ordinate is roughly proportional to height. Isotherms are skewed at an angle of about 45° from the vertical, so T varies along vertical lines with $\ln p$. The exact angle of skewness is chosen so that the isentropes of a dry atmosphere and isotherms are orthogonal at $1,000$ hPa and 0 °C, as illustrated in Fig. 2.3. Skewing the isotherms in this fashion thus better differentiates isotherms from isentropes.

Closely related to the skew-T, and somewhat predating it, is the tephigram, literally the $T \varphi$ gram, where φ was originally used to denote potential temperature. Hence, in the tephigram the coordinate system is orthogonal, with T measured by the abscissa and $\ln \theta_d$ by the ordinate. Usually tephigrams are right-rotated so that the ordinate becomes roughly proportional to $\ln p$ and hence height.

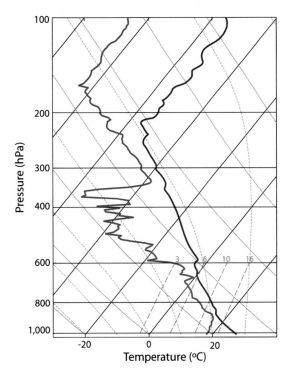

FIGURE 2.3: Skew-T ln p diagram: saturated pseudo-adiabats (grey, short dashed) are almost perpendicular to the abscissa near the surface and curve to become parallel to dry adiabats (grey solid) as p decreases. Equisaturation lines are shown by long dashes, in grey. Plotted is the 3 July 2009 sounding from Cabauw in the Netherlands, prior to a period of severe weather. The blue line denotes the locus of dew-point temperatures measured at different pressures as the sounding rose through the atmosphere, the red line the temperature.

This follows because along a dry adiabat the First Law dictates that

$$\ln \theta - \ln T = -\frac{R_{\mathrm{d}}}{c_p} \ln p + \text{const.} \qquad (2.74)$$

Hence, associating the y-axis with $\ln \theta$ and the x-axis with T, isobars can be shown to satisfy an equation of the form $y = \ln x + C$. At temperatures of practical interest (which, measured in Kelvin, are large) the curvature in the isobars (which vary with $\ln T$) in such a space is small, so that if the diagram is appropriately rotated, isobars become approximately horizontal. Hence, tephigrams and skew-T diagrams are very similar, the chief difference being that isobars are slightly curved on the former and adiabats are slightly curved on the latter.

Both the skew-T and the tephigram are derived from the emagram, which uses an orthogonal T-ln p coordinate system. The emagram is thought to be the first thermo-dynamic diagram to be routinely used to visualise and quantify the state of the atmosphere. It was proposed in the late nineteenth century by Heinrich Hertz, who is mostly known for his contributions to understanding electromagnetic radiation.

Isobars on a tephigram and the isentropes of a dry atmosphere on a skew-T diagram are examples of fundamental lines. Fundamental lines are isolines whose shape is decided by the thermodynamic coordinates. Other examples of fundamental lines include equisaturation curves and isotherms. In a dry atmosphere, adiabats and isentrope are used interchangeably to describe a line of constant θ_{d}. Through a coordinate transform, the fundamental lines of one thermodynamic diagram can serve as the thermodynamic coordinates of another. The relative orientation of the fundamental lines on the skew-T diagram help visualise how different processes are related to one another, and are evident upon a closer inspection of Fig. 2.3. Isobars are horizontal lines, which decrease logarithmically upwards so that the relation between height and distance along the ordinate is nearly linear. The isotherms are straight lines, slanted at roughly 45° towards the upper right, and are plotted in 20 K bands. Dry adiabats (lines of constant θ_{d}) are thin grey lines with negative slope and slight convexity, which intersect the isotherms nearly perpendicularly. Pseudo-adiabats, which we define later, are thin, dotted curves that are approximately vertical at high pressure and temperatures, but curve to the left and become nearly parallel to dry adiabats at cold temperatures. Equisaturation lines, describing how the dew-point temperature (defined later) varies with $\ln p$ and T given constant specific humidity, are marked by short, dashed lines, and inclined to the left of the isotherms.

To visualise how moisture is distributed in the atmo-sphere, on a skew-T diagram one plots the dew point of the air, as shown by the blue line in Fig. 2.3. The dew-point temperature is the temperature air would have if it were cooled to the point of saturation, and as such it depends only on the moisture content of the air and the ambient pressure. If the dew-point temperature equals the temperature, this implies that the air is saturated, $p_{\mathrm{v}} = p_{\mathrm{s}}$. Thus, the dew-point depression, measured as the difference between the temperature and the dew-point temperature, measures the relative humidity, so that in Fig. 2.3 the air is relatively dry between 500 hPa and 350 hPa, and nearly saturated at around 900 hPa and again near 680 hPa. For an adiabatic process, the dew-point temperature follows an equisaturation curve. Hence, the close alignment between the dew-point temperature and lines of equisaturation along with the alignment of tem-perature with lines of constant potential temperature is

TABLE 2.4: Frequently used temperatures

	Name	Equation
T	Temperature	–
θ	Potential T	(2.46)
T_ρ	Density T	(2.7)
θ_ρ	Density θ	(2.46) with T_ρ
θ_e	Equivalent θ	(2.42); (2.71) as approx.
θ_s	Saturation θ_e	θ_e with $q_v = q_s$, $q_l = 0$
θ_ℓ	Liquid water θ	(2.44); (2.72) as approx.
$\theta_{\ell,\rho}$	Density θ_ℓ	(2.44) with T_ρ

a signature of a layer that is well mixed by turbulence, as, for instance, is the case near the surface (lower $150\,\mathrm{hPa}$) in Fig. 2.3.

Another useful way to visualise the state of the atmosphere is in terms of θ_e and its value if the atmosphere were saturated at the same temperature, plotted versus height, or temperature, in the atmosphere. Using temperature as a vertical coordinate is particularly helpful for climate change studies because, for reasons discussed in Chapter 13, the top of the troposphere more closely maintains a constant temperature than it does a constant height. The value that θ_e would adopt were its specific humidity set equal to the saturation specific humidity, without changing the temperature or pressure, is called the saturation equivalent potential temperature. It is denoted by θ_s and defined as

$$\theta_s = T\left(\frac{p_\theta}{p}\right)^{\frac{R_s}{c_s}} \Omega_s \exp\left(\frac{q_s \ell_v}{c_s T}\right),\qquad(2.75)$$

with

$$\Omega_s = \left[1 + \frac{q_s R_v}{R_s}\right]^{\frac{R_s}{c_s}},\qquad(2.76)$$

where $c_s = c_{p_d} + q_s(c_l - c_{p_d})$ and $R_s = (1 - q_s)R_d$. The saturation equivalent potential temperature adds yet another temperature to the bewildering array of temperatures adopted by meteorologists (Table 2.4). Because q_s depends only on T and p, lines of constant θ_s are fundamental lines on a skew-T diagram, as well as the other thermodynamic diagrams already discussed, and define the pseudo-adiabat. The word pseudo arises because formally the process corresponding to constant θ_s is similar to a reversible adiabat, but the removal of condensate upon condensation, as implied by the use of q_s instead of q_t, implies a loss of condensate enthalpy by the system, hence it is not truly adiabatic. The difference between θ_s and θ_e measures the subsaturation.

2.2.2 Isentropes, Adiabats and Lapse Rates

For the moist system, irrespective of whether the moisture condenses, the state vector is three-dimensional. In addition to the thermodynamic variables required to describe the dry system, an additional variable is required to specify the composition of the system, q_t for instance. An assumption that constrains the third degree of freedom is thus required to render the state of the system as a point in a two-dimensional space. This third degree of freedom also leads to a distinction between isentropes and adiabats. In a dry system, where the composition of the system is fixed, an adiabatic process is an isentropic process. In a moist system, where the composition of the system weights the different contributions of water versus dry air to the system entropy, isentropes and adiabats are not the same. In a dry atmosphere, the loss of mass does not change its specific entropy, the loss of condensate from a moist system does.

In the dry system, isentropes (or adiabats) are described by lines of constant potential temperature θ_d. In the moist system, isentropes are described by lines of constant θ_e. The rate at which the temperature falls off (or lapses), as a parcel is lifted to lower pressure, or greater altitude, is called the lapse rate. In a dry atmosphere, the decrease of temperature with height along a line of constant θ_d is called the dry adiabatic lapse rate Γ_d. From the enthalpy form of the First Law, for an adiabatic process,

$$0 = \mathrm{d}\hbar_d - v\mathrm{d}p.\qquad(2.77)$$

For hydrostatic changes in pressure, and given the definition of the dry enthalpy, this implies that $c_{p_d}\mathrm{d}T + \mathrm{g}\mathrm{d}z = 0$ which yields the dry adiabatic lapse rate as

$$\Gamma_d \equiv -\frac{\mathrm{d}T}{\mathrm{d}z} = \frac{\mathrm{g}}{c_{p_d}}\qquad(2.78)$$

In a moist, but *unsaturated*, atmosphere, an analogous process yields a slightly modified lapse rate of g/c_{p_ℓ}. But because c_{p_ℓ} depends on q_t, the lapse rate depends on the composition of the system, and its derivation assumed that c_{p_ℓ} is constant, that is, the system is closed. It thus is best described as the unsaturated moist isentropic lapse rate.

The *saturated* moist isentropic lapse rate is derived similarly, but starting from the form of the First Law valid for a saturated system,

$$0 = \mathrm{d}\hbar_e - v\mathrm{d}p \quad\text{where}\quad \hbar_e = c_{p_\ell}T + \ell_v q_s,\qquad(2.79)$$

where the expression for $\hbar_e$ is taken from Eq. (2.22). One could have equivalently started from Eq. (2.38), and assumed that θ_e is constant, but doing so makes the derivation of the lapse rate more cumbersome. For an isentropic process c_{p_ℓ} is constant, and the main complication arises from the term involving the saturation specific humidity q_s as

$$q_s = \frac{p_s}{p - p_s} \epsilon_1 (1 - q_t) \tag{2.80}$$

Irrespective of the term in parentheses, which is only a minor correction, q_s depends on both p and (through p_s) T. To simplify the notation, let's define the partial derivatives

$$\beta_p \equiv -\frac{\partial \ln q_s}{\partial \ln p} = \frac{R}{q_d R_d} \approx 1, \tag{2.81}$$

$$\beta_T \equiv \frac{\partial \ln q_s}{\partial \ln T} = \frac{\ell_v}{R_v T} \beta_p \approx \frac{5,400\,\text{K}}{T}. \tag{2.82}$$

Adapting this notation to Eq. (2.79) allows us to write the isentropic form of the First Law as

$$\left[c_p + \ell_v \left(q_s \frac{\beta_T}{T} \right) \right] dT - \left(1 + q_s \ell_v \frac{\beta_p}{RT} \right) v dp = 0, \tag{2.83}$$

where $c_p \equiv c_{p_d} q_d + c_{p_v} q_s + c_l q_l$ is the specific heat for the composite system, which we have assumed to be saturated. Substituting with the hydrostatic relation for $v dp$, the lapse rate (or decrease of temperature with height) for a saturated isentropic process follows as

$$\Gamma_s \equiv -\frac{dT}{dz}\bigg|_{\theta_e} = \gamma \Gamma_d, \tag{2.84}$$

with

$$\gamma \equiv \frac{c_{p_d}}{c_p} \left[\frac{1 + q_s \beta_T \left(\frac{R_v}{R} \right)}{1 + q_s \beta_T \left(\frac{\ell_v}{c_p T} \right)} \right]. \tag{2.85}$$

In deriving Eq. (2.85), β_T is expressed in terms of β_p using Eq. (2.82), and fusion enthalpy has been neglected. The saturated moist isentropic lapse rate (or saturated isentrope) thus depends on q_t as well as T and p.

The dimensionless lapse rate γ measures the relative role of internal energy compared to the vaporisation enthalpy in doing the work required to lift a parcel. It decreases with temperature from a value very near unity at the colder temperatures of the upper troposphere, where little water is available to condense, to a value near 0.4 at 300 K (Fig. 2.4), with an inflection point near 280 K.

In discussing lapse rates, terminology can be confusing. Γ_d is called the dry adiabatic lapse rate, that is, the lapse rate of dry air following an adiabatic process. In contrast, Γ_s is the saturated isentropic lapse rate: 'isentropic' because its derivation additionally assumes that mass was conserved, so that q_t is constant, which is more stringent then simply assuming that no heat is added; and 'saturated' because it was assumed that $q_t \geq q_s$ so that $q_v = q_s$.

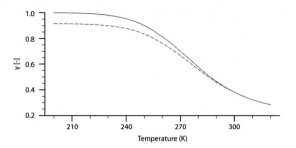

FIGURE 2.4: Non-dimensional lapse rate γ following Eq. 2.85 for a fixed atmospheric pressure of 1,000 hPa. The dashed curve shows the pseudo-adiabat (for which it is assumed that condensed water is removed) and the solid curve the saturated isentropic (or reversible moist adiabat). The pseudo-adiabat asymptotes to a value less than unity because of the specific heat of condensed water, whose contribution depends on the temperature of the air at initial saturation, here taken as 300 K.

To contrast with the dry adiabat it is tempting and common to call Γ_s the moist adiabat, or reversible moist adiabat. The former is common,[2] but can be confusing because the moist system need not be saturated, and an adiabatic process need not be isentropic. The latter is more precise, but uses two words ('reversible' and 'adiabatic') when one ('isentrope') would work just fine. Given that systems are in general moist, it would be both simpler and more sensible to talk about saturated and unsaturated isentropes.

For large parcel displacements, an increasing fraction of the initial vapour will be converted to condensate, and the assumption that this condensate remains in the parcel becomes increasingly untenable. A better approximation would be to adjust Γ_s, at every temperature and pressure, to that which it would have if the system were just saturated, that is, $q_t = q_s$. This is equivalent to assuming that condensate is immediately lost as precipitation, hence its enthalpy content can no longer contribute to the expansional work done by the rising parcel, which leads to a greater change in temperature with decreasing pressure, that is, a larger lapse rate. Such a process is irreversible and non-adiabatic and so the resultant lapse rate, denoted $\Gamma_{\bar{s}}$, is often called the pseudo-adiabat, see Fig. 2.4. A straightforward calculation of the difference between pseudo-adiabats and saturated isentropes shows that they

---

[2] There is a certain tradition of imprecision in the usage of the phrase 'moist adiabat' as it is mostly intended to draw distinction to a dry adiabat. In this sense, its usage (also elsewhere in this book) is agnostic to fine distinctions arising from whether the process is strictly reversible, the fineness of the thermodynamic approximation or how precisely, if at all, ice-phase processes are accounted for.

differ most appreciably in the upper troposphere. This is to be expected because T decreases with height, hence q_s decreases and q_l (whose effects are neglected in the pseudo-adiabat) increases correspondingly. Overall, the differences in T tend to be less than 0.5 K below 400 hPa, increasing to as much as 3 K–5 K between 100 hPa and 200 hPa.

One advantage of the pseudo-adiabat is that it depends only on T and p and is representable as a fundamental line on a standard atmospheric thermodynamic diagram. Additionally, it more naturally accommodates an ice phase. Whereas accounting for freezing (which was not done earlier) would lead to an isothermal layer in the saturated isentrope at the triple-point temperature, the pseudo-adiabat is only discontinuous in its derivative at this point. Nonetheless, the saturated isentrope (accounting for ice formation, and thus computed numerically rather than with Eq. (2.85)) defines a thermodynamic limit – following any adiabatic process, an air parcel rising to a given altitude must be colder than its saturated isentropic value. In the lower troposphere, deviations from the saturated isentrope are most likely to be associated with mixing processes, simply by virtue that the absolute humidity differences between ascending parcels and their environment are largest there. In the upper troposphere, precipitation and the retardation of ice formation contribute the most to differences between rising parcels and their saturated isentropic value. An interesting footnote is that the temperature dependence of γ, which is important for climate change, is greatest at temperatures around 280 K, which are characteristic of the present-day lower tropical troposphere.

2.2.3 Soundings

Soundings in the atmosphere are measurements of its state as a function of altitude. Generally they are made with sensors lofted by a balloon, or dropped with a parachute. In either case the sensor package drifts with the mean wind as it rises (or falls) and uses global positioning systems to measure the wind vector, and *in situ* sensors to measure temperature and relative humidity along its trajectory. And other quantities in some special cases, for example, ozone. An example of a standard meteorological sounding is shown in Fig. 2.3, taken from a summer day in the Netherlands, foreshadowing severe weather. Plotted are two lines, the red line demarcates the temperature and the blue line demarcates the dew-point temperature. Sometimes wind-vectors are plotted alongside the sounding, but not in the present case.

A variety of thermodynamic processes can be inferred from a sounding. For instance, well-mixed layers (or dry-adiabatic processes) will have the dew point follow the equisaturation line, and temperature will, in the absence of condensation, follow the dry adiabat. This is more or less the case near the surface (pressures greater than 900 hPa) in the sounding plotted in Fig. 2.3. It suggests that adiabatic lifting of surface air would lead to condensation (an equal temperature and dew point) at a pressure near 900 hPa. This level, where air lifted isentropically condenses, is called the lifting condensation level, or LCL. Further ascent would follow the saturation isentrope, here approximated by the pseudo-adiabat which falls off somewhat less rapidly with height than the temperature in the sounding. This implies that air parcels condensing at the LCL and rising along the saturated adiabat will be warmer than their environment and, modulo the contribution of condensate to density, more buoyant. Only at about 250 hPa does the environmental temperature begin to increase again, in association with the tropopause, limiting the buoyancy of saturated ascent from the LCL. Above this level the atmosphere is more or less isothermal, and one is in the stratosphere. Some of the other interesting features in this sounding will be discussed in subsequent sections.

2.3 Convective Instability

Archimedes is credited with having developed the concept of buoyancy. According to the Archimedean Principle, an object will rise if its mass is less than that of the fluid it displaces. If it is denser, and hence has a greater mass than the fluid it displaces, it will sink. The rising or sinking is a result of an imbalance from between the pressure and the gravitational force which either allow the denser object to do work on its environment as it sinks, or the environment to do work on the object as it rises. This imbalance between the hydro-static pressure and the gravitational force is measured by the buoyancy, as introduced in Eq. (2.9).

In this section, the buoyancy is used to explore the convective stability of different configurations of a static fluid. First, the stability of a layer to infinitesimal displacements within that layer is considered. Next, the stability of two layers whose properties change discontinuously at an interface, for instance, a cloud boundary, is explored. Finally the response of the stability to finite displacements is considered. One of the things that makes moist fluids fascinating is the variety of convective instabilities that they support.

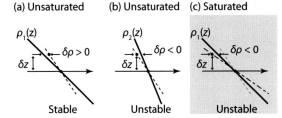

(a) Unsaturated (b) Unsaturated (c) Saturated

FIGURE 2.5: Schematic showing how buoyancy changes for an infinitesimal isentropic displacement δ of a fluid parcel relative to the environmental density profile $\rho(z)$ within the layer and denoted by the solid line. Panels (a) and (c) share the same density stratification, $\rho_1(z)$, but the fluid in (c) is saturated. Panels (a) and (b) both show an unsaturated case, but $\rho_2(z)$ in panel (b) decreases less strongly in height than $\rho_1(z)$ in panel (a). Lines also denote unsaturated (dashed) and saturated (dashed and dotted) isentropic density change with height.

2.3.1 Infinitesimal Displacements within a Layer

For a homogeneous fluid, an infinitesimal displacement δ of a fluid parcel will be unstable if the buoyancy arising from a displacement is in the same sense of the displacement. For instance, if an upward displacement is accompanied by a relative increase in buoyancy, the displacement will be be amplified. The buoyancy due to an infinitesimal displacement is proportional to the difference between the density change that would arise from an isentropic displacement of a fluid parcel (denoted by subscript s to denote the idea of a reversible adiabatic, or isentropic, displacement) and the environmental density gradient, such that

$$b = -g \frac{\rho(z+\delta z)\big|_s - \rho(z+\delta z)}{\rho(z)} \qquad (2.86)$$
$$= -g \left(\partial_z \ln \rho \big|_s - \partial_z \ln \rho \right) \delta z.$$

If the environmental density decreases with height less than the isentropic change in the density of a displaced fluid parcel, then the fluid parcel will, after a small upward displacement, find itself less dense than the environmental density at that same level and begin to accelerate upward. This situation is shown in panels (b) and (c) in Fig. 2.5 If the environmental density decreases with height more than it does for an isentropic change, than the fluid parcel will find itself more dense than the environment and accelerate downwards, as depicted in panel (a). In this situation, the environmental density profile is said to be stable and in the absence of dissipation the displaced fluid parcel will oscillate about its equilibrium level with the frequency N, where

$$N^2 = g \left(\partial_z \ln \rho \big|_s - \partial_z \ln \rho \right). \qquad (2.87)$$

N is called the Brunt–Väisälä frequency. In the case that $N^2 < 0$, the fluid is said to be convectively unstable. Because the density change following an isentropic displacement of a fluid parcel is a thermodynamic property of the fluid, N is given by the environmental stratification and the state of the fluid, in particular whether or not it is saturated. Hence, and as shown in Fig. 2.5, a given density profile which is stable if the fluid is unsaturated may be unstable if the fluid is saturated. A density gradient that is stable in the case that the fluid is saturated but unstable in the case the fluid is unsaturated is called conditionally unstable. To make clear whether or not the density gradients in a fluid are being evaluated in comparison to saturated versus unsaturated isentropic parcel displacements, N_s is used to define the former and N_d the latter.

If it is assumed that the pressure felt by the disturbed parcel adjusts instantaneously to the pressure at the new height, $z + \delta z$, then pressure differences between the displaced parcel and its environment vanish. In this case it is sufficient to consider

$$\frac{\rho'}{\rho} = -\frac{T'_\rho}{T_\rho} = -\frac{T'}{T} - \frac{R_v}{R}q'_v + \frac{R_d}{R}q'_t. \qquad (2.88)$$

In the absence of saturation, $q'_v = q'_t$ and q_t is conserved for the parcel displacement. Environmental gradients of q_t do, however, contribute to the density difference between the displaced parcel and the environment, such that

$$N^2 = g \left[\frac{1}{T} \left(\partial_z T - \partial_z T\big|_s \right) + \left(\frac{R_v - R_d}{R} \right) \partial_z q_t \right]. \qquad (2.89)$$

The difference between the isentropic lapse rate and the environmental lapse rate can be expressed in terms of the entropy, or potential temperature, gradient, such that the Brunt–Väisälä frequency for the unsaturated fluid becomes

$$N^2 = c_p \Gamma \frac{d \ln \theta}{dz} + g \left(\frac{R_v - R_d}{R} \right) \frac{dq_t}{dz}. \qquad (2.90)$$

In deriving Eq. (2.90), $c_p \Gamma$, where Γ denotes the adiabatic lapse rate, substitutes for g in the first term on the right-hand side. Writing N^2 in terms of Γ anticipates the derivation of its expression for a saturated layer. In the completely dry case, the expression simplifies further because the second term vanishes, Γ reduces to Γ_d and c_p reduces to c_{p_d}.

For a saturated layer, changes in q_v vary as q_s. Changes in q_s can be related to temperature fluctuations through the Clausius–Clapeyron equation, as described by Eq. (2.82). Pressure fluctuations make no contribution because the displaced parcel is assumed to adjust to the pressure in the environment, so that terms depending on the difference between its pressure and that of the environment vanish. Accounting for the temperature-related q_s variations as well as gradients in q_t thus results in the following expression for the Brunt–Väisälä frequency in a saturated layer:

$$N_s^2 = g\left[\frac{1}{T}\left(1 + q_s\frac{R_v}{R}\beta_T\right)(\partial_z T - \partial_z T|_s) - \frac{R_d}{R}\partial_z q_t\right].$$
(2.91)

As in the dry case, the difference between the actual lapse rate and the saturated isentropic lapse rate is, following Eq. (2.38), proportional to the vertical gradient of the value of θ_e in the saturated layer. Hence the Brunt–Väisälä frequency in a saturated layer, can be expressed analogously to the expression for unsaturated layers, such that

$$N_s^2 = c_p\Gamma_s\frac{d\ln\theta_e}{dz} - g\frac{R_d}{R}\frac{dq_t}{dz}.$$
(2.92)

As is expected, N_s^2 reduces to N_d if q_t vanishes. Because $d\theta_e/dz < d\theta/dz$, the saturated Brunt–Väisälä frequency is less than that of the dry fluid, and perturbations that would be stable for an unsaturated fluid ($N^2 > 0$) may be unstable ($N_s^2 < 0$) for a saturated fluid.

The difference between N^2 and N_s^2 define three stability regimes of the atmosphere. A region of absolute stability and instability, irrespective of the saturation state of the atmosphere, and between them a region where the atmosphere is said to be conditionally unstable. A conditionally unstable layer is one which would be unstable if it were saturated but stable if it is unsaturated. These different stability regimes, demarcated roughly by the sign of Γ_d and $\Gamma_{\tilde{s}}$, are illustrated schematically in Fig. 2.6, as they would appear in relationship to one another in a skew-T diagram. This demarcation is only rough because N^2 and N_s^2 also depend on moisture, and thus are not fundamental

lines on a skew-T. Because Γ_s decreases in magnitude with increasing temperature, as roughly illustrated by the tendency of pseudo-adiabats in the skew-T diagram (Fig. 2.4) to become more perpendicular to dry adiabats at warmer temperatures, the domain of conditional instability increases with temperature. It is easier to destabilise a warm atmosphere.

2.3.2 Stability across Interfaces

In the atmosphere very sharp gradients are frequent, and arise from different air masses coming together to form a contact discontinuity. Perhaps the best example of this is a cloud. In such situations the stability of the interface can be explored by comparing the change of density across the interface, or the effect of mixing on the fluid properties on one or the other side of the interface. Both situations are depicted schematically in Fig. 2.7 for the most interesting case of when one fluid is saturated and the other is not.

For the case of an overlying unsaturated fluid separated from an underlying saturated fluid by an interface, as depicted in Fig. 2.7, the stability of the interface depends on whether or not the upper fluid (layer 2) is more or less dense than the lower layer. If the upper fluid is denser, that is, $\Delta\rho > 0$ the interface will be unstable. This condition can be expressed in terms of the thermodynamic variables $\{h, q_t\}$. Assume that the differences between the fluid states is small so that,

$$\frac{\Delta\rho}{\rho} \approx -\left(\frac{\Delta T}{T} + \frac{\Delta R}{R}\right),$$
(2.93)

where $R = R_d + q_v R_v - q_t R_d$, hence its change reflects changes in fluid composition. For the saturated fluid, $q_v = q_s$ and for the unsaturated fluid, $q_v = q_t$. The change in the temperature and gas constants across the layers can be expressed in terms of changes in q_t and h_e given

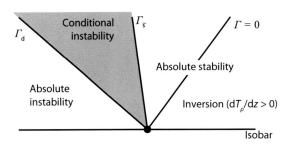

FIGURE 2.6: Different regions of atmospheric stability, as delineated by the fundamental lines demarcating different lapse rates on a skew-T diagram.

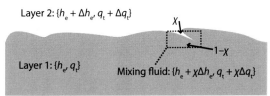

FIGURE 2.7: The interface between an unsaturated fluid overlying a saturated fluid and their thermodynamic state. Because the boundary between the fluids is assumed to be infinitesimally thin, there is no difference in the pressure between the two layers.

$$\Delta h_e \approx (\Delta c_{p_\ell})T + c_{p_\ell}\Delta T + (\Delta \ell_v)q_v + \ell_v \Delta q_t + \ell_v q_c \tag{2.94}$$

and

$$\Delta R \approx R_v(\Delta q_t + q_c) - R_d \Delta q_t. \tag{2.95}$$

In deriving these expressions it is assumed that the state of the upper-layer fluid can be expressed in terms of a Taylor series expansion about the state of the lower-layer fluid. Such an approximation is only roughly correct as for many situations Δq_t is of the same order as q_t, nonetheless it gives some insight into the stability of the layer.

Rearranging Eqs (2.94–2.95) results in an expression for the density difference between the two layers as

$$\frac{\Delta \rho}{\rho} = \left[\frac{\ell_v}{c_{p_e}T} - \frac{R_v}{R}\right](\Delta q_t + q_c)$$
$$+ \left[\frac{c_1 - c_{p_d}}{c_{p_e}} + \frac{R_d}{R}\right]\Delta q_t - \frac{\Delta h_e}{c_{p_e}T}. \tag{2.96}$$

For $\delta\rho > 0$, the stratification of the two layers is unstable, hence the instability criterion can, assuming $\Delta q_t < 0$ and $q_c \ll |\Delta q_t|$, be written as

$$\frac{\Delta h_e}{\ell_v \Delta q_t} > \kappa_1, \tag{2.97}$$

where

$$\kappa_1 \approx 1 + \frac{T}{\ell_v}\left[(c_1 - c_{p_d}) - \left(\frac{R_v - R_d}{R}\right)c_{p_e}\right]. \tag{2.98}$$

For conditions typical of the subtropics, $\kappa_1 \approx 1.3$. If one adopts the common approximation of neglecting differences between the specific heats, $\kappa_1 \approx 1$. In either case it means that for the situation depicted in Fig. 2.7, the moist enthalpy, or equivalently the moist static energy, must decrease across the interface; decreasing more strongly the greater the change in q_t. Eq. (2.98) can be thought of as a form of conditional instability, whereby an increase in the moist static energy across the layer is sufficient for the interface to be stable in the absence of mixing.

The case in which the two layers mix introduces additional possibilities. Isobaric mixing of fluid elements from the two layers as depicted in Fig. 2.7 can, through non-linearities in the equation of state, lead to a mixture whose density is not bounded by the temperatures of the constituent air masses. Consider how small, isobaric, changes in the density, defined in Eq. (2.93), depend on changes in the thermodynamic coordinates $\{h_e, q_t\}$. We first consider the case where the mixing fraction χ of the unsaturated fluid is small, so that the mixture remains saturated. This case differs from the static situation in that now $q_v = q_s$

which changes with T. Denoting the small change by δ, so for instance $\chi\Delta h_e = \delta h_e$, this implies that

$$\delta R = R_v \delta q_s - R_d \delta q_t = R_v \frac{q_s}{T}\beta_T \delta T - R_d \delta q_t. \tag{2.99}$$

Similarly, an expression involving q_s arises in expanding the expression for δh_e such that

$$\frac{\delta\rho}{\rho} = \left[\frac{c_1 - c_{p_d}}{c_p + \frac{\beta_T \ell_v q_s}{T}}\left(1 + \frac{R_v}{R}\beta_T q_s\right) + \frac{R_d}{R}\right]\delta q_t - \frac{\delta h_e}{c_{p_d}T}\gamma. \tag{2.100}$$

For the usual case in which saturated air mixes with drier air so that $\delta q_t < 0$, then this implies that density fluctuations will be negative if

$$\frac{\Delta h_e}{\ell_v \Delta q_t} > \kappa_2, \tag{2.101}$$

where

$$\kappa_2 = \frac{c_{p_d}T}{\gamma\ell_v}\left[\frac{c_1 - c_{p_d}}{c_p + \frac{\beta_T \ell_v q_s}{T}}\left(1 + \frac{R_v}{R}\beta_T q_s\right) + \frac{R_d}{R}\right] \tag{2.102}$$

is the buoyancy reversal parameter. Because we have assumed that Δq_t is negative, this implies that Δh must be sufficiently negative for mixtures to be more dense than their constituent components. For conditions typical of subtropical stratocumulus, $\kappa_2 \approx 0.56$.[3] Because $\kappa_2 < \kappa_1$, this implies the existence of stable interfaces, which as a result of mixing processes, for instance by diffusion, become unstable.

The situation whereby

$$\kappa_1 > \frac{\Delta h_e}{\ell_v \Delta q_t} > \kappa_2 \tag{2.103}$$

defines the case of buoyancy reversal, expressed here in the thermodynamic coordinates $\{h_e, q_t\}$, but because for an isobaric process $\Delta\eta = \Delta h$, the buoyancy reversal criterion is equivalently formulated in terms of moist static energy. It says that for η'_e sufficiently negative, the accompanying density fluctuation will be less than zero, assuming the perturbed fluid parcel remains saturated.

Physically this condition just expresses the fact that the evaporation of condensate from the saturated layer, when it mixes with the drier layer, causes cooling in the mixture that can more than offset the warming from the increase in dry enthalpy that accompanies the mixing. This leads to the curious phenomenon of a mixing induced instability. This type of instability is thought to be important for stratocumulus decks in the very dry subtropics, and likely inhibits them from becoming too deep (e.g., as discussed in Chapter 5).

---

[3] If, as is common, the thermodynamics is developed without considering differences in specific heats, then in such a system $\kappa_2 = 0.23$.

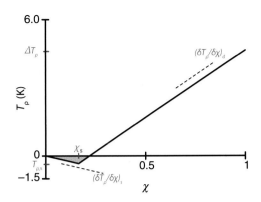

FIGURE 2.8: Mixing diagram showing how the buoyancy of a saturated layer changes upon mixing with a fraction χ of unsaturated air for two air masses whose buoyancy differs by $\Delta b = 0.154$. The saturation mixing fraction χ_s is the mixture with the minimum buoyancy, $b_s = -0.011$. The different rate of change of buoyancy with mixing fraction is illustrated by the dashed lines for saturated and unsaturated processes, respectively.

The situation for buoyancy reversal based on the state of a typical subtropical stratocumulus cloud is illustrated with the help of Fig. 2.8. Here one sees that for small mixing fractions, the density temperature decreases, reaching its most negative value when the mixing fraction takes on a critical value associated with that mixture which just evaporates all the condensate contributed by the saturated component of the mixture. This is called the saturation mixing fraction χ_s. For $\chi > \chi_s$, the mixed parcel warms within increasing χ. The figure shows that for typical stratocumulus layers there is not sufficient liquid water to support strongly negatively buoyant mixtures, that is, χ_s is small.

Thus, in the atmosphere all conditionally unstable interfaces are also subject to buoyancy reversal. As saturated air rises through the environment, the cloud–clear-air interface is necessarily destabilised by mixing processes at the cloud front. As the cloudy air rises, clear air mixes at the cloud edge in ways that are very effective of consuming the available potential energy of the rising plume. This makes clouds very effective mixing entities, something that is hidden from view by their visual appearance.

2.3.3 Subcritical Instabilities and Convective Available Potential Energy

A subcritical instability is one in which the basic state is only unstable to perturbations greater than a certain amplitude. There is a tradition of considering the stability of finite displacements of fluid parcels as a way of measuring the stability of the atmosphere. Here the thinking

is that disturbances, for instance a spreading gust front, can bring layers of the atmosphere to a state, and into an environment, where they become unstable.

The susceptibility of the atmosphere to finite amplitude parcel displacements is often measured by the convective available potential energy, or CAPE. CAPE measures the amount of work the atmosphere is capable of doing on a parcel lifted to its level of free convection (LFC). Denoting the CAPE by the symbol $\mathcal{A}$:

$$\mathcal{A} = \int_{z_\mathrm{f}}^{z_\mathrm{n}} b \, \mathrm{d}z, \tag{2.104}$$

where b is the buoyancy, as defined in Eq. (2.9), and the limits of integration are z_f, the LFC, and z_n, the level of neutral buoyancy (LNB). This is the level at which the parcel lifted following some specified process, for instance, along a saturated isentrope or pseudo-adiabat, ceases to be buoyant relative to the environment. Most of the time z_n is near the tropopause.

Substituting Eq. (2.9) into Eq. (2.104) and using the hydrostatic equation to replace the integration in height by an integration in pressure yields

$$\mathcal{A} = \int_{p_\mathrm{n}}^{p_\mathrm{f}} R_\mathrm{d} T'_\rho \, \mathrm{d}(\ln p). \tag{2.105}$$

Thus, on the skew-T diagram (e.g., Fig. 2.3), $\mathcal{A}$ is roughly equal to the area between the environmental temperature and the dashed line. 'Roughly' because the potential acceleration depends also on the available moisture, so as to account for differences between T and T_ρ.

CAPE as defined by Eq. (2.104) depends sensitively on the properties of the parcel being lifted and the manner in which it is lifted. Small differences in the initial state of a parcel can lead to large differences in $\mathcal{A}$. For instance, a slight drying of the boundary layer in Fig. 2.3 will raise the LCL and decrease the temperature of the saturated adiabat, and raise the LFC, thereby reducing $\mathcal{A}$.

Because $\mathcal{A}$ describes the work the atmosphere can do on a parcel, or alternatively the potential energy available to a convecting parcel, it can be related to the maximum kinetic energy. That is, it bounds the amount of kinetic energy a parcel could have, thus defining a velocity scale:

$$w_\mathrm{max} = \sqrt{2\mathcal{A}}, \tag{2.106}$$

which is another measure of the intensity of convection.

2.3.3.1 Other CAPE-like Measures

CAPE is the most common measure of the atmosphere to support overturning through the instability of saturated ascent. Whether or not CAPE is present in the atmosphere

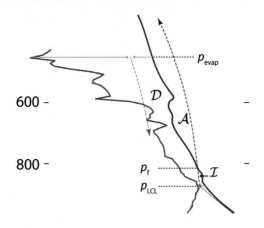

FIGURE 2.9: Illustration of different types of CAPE using the sounding of Fig. 2.3. The CAPE is illustrated by the area between the red dashed line showing the saturated adiabatic lapse rate for a parcel rising from the LCL and the solid red line denoting the environmental temperature. The convective inhibition is defined by the negative area between p_f an the ρ_{LCL}. The blue dashed line shows the temperature of air formed by saturating very dry air near 500 hPa, and maintaining it in saturation as it descends, and the area between the blue dashed line and the environmental temperature demarcates downdraft CAPE.

depends on the properties of the parcel being lifted and the process whereby it is lifted. Even then, many other factors come into play in deciding on the capacity of the atmosphere to overturn. The amount of work that must be invested to access the CAPE of a sounding is highly variable. In some cases one does not have to do a large amount of work on a parcel before the atmosphere starts returning the favour. To measure the work that must be invested to realise CAPE, another parameter, the convective inhibition (or CIN), is introduced. It is the analogue to CAPE but measures the amount of work that must be done to lift a parcel from some reference level p_* to its LFC:

$$\mathcal{I} = -\int_{p_*}^{p_f} R_d T_\rho' \, d(\ln p). \tag{2.107}$$

Thus, the ability of the atmosphere to do work on an air parcel depends not on CAPE alone but also on other factors such as CIN. In Fig. 2.9, CIN corresponds to the area between the environmental temperature and the saturated isentrope rising from the LCL to the LFC.

Another CAPE-like measure of the atmosphere is called downdraft CAPE, or $\mathcal{D}$. It and the other forms of CAPE are illustrated in Fig. 2.9. $\mathcal{D}$ measures the stability of saturated downward displacements of air first brought to its wet-bulb temperature by evaporation of falling rain. The wet-bulb temperature (sometimes denoted T_w) is bounded by the dew-point temperature

and the actual temperature. It is the temperature that one gets by isobarically bringing air to saturation through evaporation. Because water is evaporated into the air, the air both cools and moistens, increasing its mixing ratio while decreasing its temperature. Hence,

$$\mathcal{D}(p_i) = \int_{p_{\text{evap}}}^{p_{\text{sfc}}} R_d T_\rho' \, d(\ln p), \tag{2.108}$$

where here $T_\rho(p_{\text{evap}}) = T_w$ and T_ρ' is the difference between the density temperature of a parcel brought to saturation by evaporating water into it and the environmental value of T_ρ. Physically, $\mathcal{D}$ measures the stability of air to evaporation of rain. In environments with large values of $\mathcal{D}$, vigorous downdrafts can be formed by evaporating water (from precipitation) into a dry ambient environment. The analogy to buoyancy reversal, whereby negatively buoyant parcels can be created through isobaric mixing, should be apparent.

2.3.3.2   Caveats on CAPE

As alluded to earlier, the actual CAPE of a given atmospheric sounding is not a number without ambiguity. This ambiguity stems mostly from the varied definitions associated with it. Earlier, CAPE has been defined as the positive area on the thermodynamic diagram. Others define it as the *net* positive area associated with a parcel lifted dry adiabatically from the surface to the LNB. Others compute the CAPE associated with a parcel characterised by the *mean* properties of the lower 10 hPa–50 hPa of the atmosphere. Yet others adjust the surface properties of the sounding to reflect what they anticipate conditions will be like at some particular point in time. Because it measures a conditional process, different definitions arise naturally, and these can be both quantitatively and qualitatively different, as how much energy is available to a parcel raised to its level of free condition depends very much on the parcel being lifted.

The thermodynamic processes which govern the evolution of the state of the parcel above the LCL also play a large role in determining CAPE. For instance, parcels in which ice forms will have different values of CAPE than parcels in which ice is not allowed to form. Likewise, rising parcels usually mix with their environment; in fact, the more unstable they are, the more they are likely to mix. Thus, CAPE can be seen as a function of the state of a parcel, the environment and the specified type of thermodynamic process for the rising parcel. These ambiguities do not diminish the value of a measure like CAPE, but they do indicate that if it is to be used quantitatively, the particular use of the concept must be made precise.

2.3.4 Slice Method

The ability of CAPE to characterise the potential energy available to convection is based on a number of idealisations. These include: (i) that the parcels being lifted follow the specified thermodynamic process; (ii) that the parcels being lifted are characteristic of the air from which the convection actually develops; (iii) that the response of the environment can be neglected. More refined measures of convective instability attempt to address one or more of these limitations. Most notable among these is the slice method introduced by Jacob Bjerknes, one of the pioneers in the development of meteorology as a branch of physics, in the 1930s. Unlike in parcel theory, where infinitesimal parcels are assumed to move through a quiescent environment, Bjerknes's slice method considers the finite size of convective motions and hence the compensating downward motions they induce. By doing so, it identifies an effective stability, which depends on the area fraction occupied by the ascending motion.

The starting point for this method is to assume a convecting atmosphere at some reference height z_0 above cloud base. It is envisioned that in some sufficiently large area, the fraction of the convecting area is a robust quantity which can be denoted by a. Integrating the vertical velocity over the convecting region allows one to define a mean convective velocity w_c. Mass continuity across the reference level allows one to express the mean subsiding velocity of the environment w in terms of (a, w_c):

$$aw_c + (1 - a)w = 0. \tag{2.109}$$

In convective modelling, the quantity aw_c is related to the convective mass flux $\mathcal{M}_c$ by the density, that is,

$$\mathcal{M}_c = \rho a w_c. \tag{2.110}$$

In what follows, the density is assumed to be constant, and hence dropped from the discussion. This is consistent with the treatment of shallow convection, wherein density differences are small, and Eq. (2.109) is a statement of mass conservation.

Assuming that the air rising in the convecting region is rising along a saturated isentrope, while the subsiding air follows a dry adiabat, the temperature difference between the convecting region and the environment at the reference level can be expressed as follows:

$$T_c(z_0) - T(z_0) \approx T_{01} - \Gamma_s \Delta z_\uparrow - (T_{02} + \Gamma_d \Delta z_\downarrow), \tag{2.111}$$

where both $\Delta z_\uparrow$ and $\Delta z_\downarrow$ are defined as positive

$$T_{01} = T(z_0) + \Delta z_\uparrow \Gamma, \tag{2.112}$$

$$T_{02} = T(z_0) - \Delta z_\downarrow \Gamma, \tag{2.113}$$

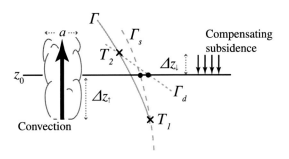

FIGURE 2.10: Situation modelled by the slice method. The convecting air temperature cools following the saturated adiabat Γ_s as it rises a distance $\Delta z_\uparrow$. The compensating subsidence covers a greater area and descends a distance $\Delta z_\downarrow$ and warms following the dry adiabat Γ_d. The convection is energetically inhibited because the response of the environment stabilises the atmosphere.

and $\Gamma = -dT/dz$ is the environmental lapse rate. Note that this implies that the model is most appropriate if $\Delta z_\uparrow$ is associated with the distance above the cloud base.

These relations (illustrated schematically in Fig. 2.10) describe the temperature in the convecting region as the temperature that the environmental air at a distance $\Delta z_\uparrow$ below z_0 would have it if were lifted to z_0 along a saturated isentrope (denoted Γ_s) while the temperature of the environment at z_0 is that which the environmental air a distance $\Delta z_\downarrow$ above the reference level would have were it brought dry adiabatically to z_0. The distances $\Delta z_\uparrow$ and $\Delta z_\downarrow$ are given by $w_c \Delta t$ and $-w \Delta t$, respectively, with the sign convention chosen to ensure that a positively measured downward displacement leads to a negative, or downward, velocity.

Substituting from previous yields an expression for the temperature difference in terms of a, w_c and Γ,

$$\frac{T_c - T}{\Delta t} = w_c(\Gamma - \Gamma_s) - w(\Gamma - \Gamma_d) \tag{2.114}$$

$$= w_c \left[\Gamma - \Gamma_s + \frac{a}{1 - a}(\Gamma - \Gamma_d) \right]. \tag{2.115}$$

Thus, the criterion for convective instability $T_c - T > 0$ is equivalent to requiring that

$$\Gamma > \Gamma_s + a(\Gamma_d - \Gamma_s) > \Gamma_s. \tag{2.116}$$

This requirement is more severe than that given by the parcel method, or for the case of infinitesimal displacements in the absence of a humidity gradient. The physical difference being that the compensating downward motion stabilises the environment. However, in the limit when a goes to zero, the original criterion, $\Gamma > \Gamma_s$, which one derives from parcel theory, is recovered.

A further implication of this result is that moist convection is most unstable if its fractional area is smallest.

Physically this is not surprising, as for a given w_c, vanishing a implies that the compensating environmental motion is minimised and hence the environment is stabilised the least. This type of analysis might help explain the spacing of convective systems, or why nature prefers to concentrate convection over rather small areas. Mathematically this is apparent by considering the neutral limit of Eq. (2.115):

$$\Gamma - \Gamma_s(1 - a) - a\Gamma_d = 0. \tag{2.117}$$

Hence, for a given unstable lapse rate, only for

$$a \le \frac{\Gamma - \Gamma_s}{\Gamma_d - \Gamma_s} \tag{2.118}$$

will the atmosphere be convectively unstable.

2.4 Fluid Dynamics

For studying conservation laws and force balances in flows, the concept of the substantial derivative, denoted

$$D_t := \partial_t + v \cdot \nabla, \tag{2.119}$$

needs to be mastered. It describes how a quantity changes following its motion, here denoted by v, with ∇ as the gradient operator. In an inertial coordinate system in a gravitational field, $g = (0, 0, -g)$, the laws governing the conservation of mass, momentum and energy of the two component fluid can (with some assumptions to be discussed later) be expressed as

$$D_t\rho = -\rho\nabla \cdot v, \tag{2.120}$$

$$D_t v = -\frac{1}{\rho}\left[\nabla p + \mu\nabla^2 v\right] + g, \tag{2.121}$$

$$D_t s_\ell = \frac{1}{\rho T}\left[\mu(\nabla v) : (\nabla v) - \nabla \cdot F_\hbar\right], \tag{2.122}$$

$$D_t\rho_t = -\rho_t\nabla \cdot v. \tag{2.123}$$

Eq. (2.120) describes conservation of total mass, where the density, $\rho = \rho_d + \rho_t$, depends on the constituent densities for dry air and total water. The second is the Cauchy equation, describing Newton's law for a fluid parcel, with velocity $v = (u, v, w)$. Surface forces acting on the parcel are written in terms of the pressure gradient (for isotropic forces). Deviatoric surface forces are assumed to behave in a diffusive fashion described by the dynamic viscosity μ. The only body force is that arising from the gravitational acceleration g. The Second Law includes the divergence of an enthalpy flux $F_\hbar$, associated with radiative and conductive enthalpy transfer, and heating through the dissipation of kinetic energy, which is positive definite and proportional to the dynamic viscosity. Water

mass conservation is described in terms of the total water density ρ_t by Eq. (2.123).

Eqs (2.120)–(2.123) can be derived by assuming that the two components of the fluid constitute an ideal mixture, and the velocity is the mass averaged velocity of the two components of the fluid, and the ideal gas law provides an additional constraint to close the system. The system holds in the case of a multicomponent fluid in the limit where the condensed phase must be assumed to be comprised of sufficiently small particles so that it can be described as a continuous field, which diffuses and flows, and has the same temperature, as the other constituents of the fluid. These assumptions become difficult to maintain for the development of a precipitate phase, wherein large particles are, by definition, falling through the flow, and their thermal inertia can lead to large temperature differences to that of the mean flow. An approximate way of treating such situations will be discussed at the end of the next section. An introduction to a literature that provides a more exact description, by individually tracking the motion of the different fluid components and exchanges of momentum and energy between them, is provided in the section on Further Reading.

2.4.1 Soundproof Equations

The previous set of equations are rarely used as a basis for the investigation of fluid processes in meteorology or oceanography. The main reason is that despite the many simplifications in their derivation, they still include a phenomenon not thought to be essential for many processes to which the equations may be applied – sound waves. Compared to most processes of interest, sound waves (acoustic modes) are very fast, and mostly serve to adjust the pressure field so as to maintain some form of incompressibility. For most applications of meteorological interest, assuming the flow satisfies an incompressibility condition following its evolution (rather than trying to simulate the adjustment process) makes the equations easier to handle. Such equation sets are usually arrived at by neglecting the contribution of pressure fluctuations on density in the continuity equation, such that it can be written in terms of a soundproof density ρ_a,

$$D_t\rho_a = -\rho_a\nabla \cdot v, \tag{2.124}$$

with different forms of incompressibility arising based on how ρ_a is approximated, for example, Table 2.5.

The most restrictive of these arises from the Boussinesq approximation, but it is also the most widely used and the most straightforward for outlining the main form

TABLE 2.5: Soundproof equation sets

Assumption	Name
$\rho_a = \rho_0$	Boussinesq
$\rho_a = \rho_0(z)$	Anelastic
$\rho_a = \dfrac{\rho_0(z)\theta_0(z)}{\theta(x,y,z,t)}$	Pseudo-incompressible

of the scaling arguments to arrive at a soundproof equation set. Therefore, the derivation for the Boussinesq case is sketched out here.

The Boussinesq approximation involves solving for the velocity v, density ρ, potential temperature θ_ℓ, pressure p and total-water specific humidity q_t which are assumed to vary about a dry and resting reference state $\{\rho_0, \theta_0, p_0(z)\}$. Only the reference state pressure varies, and then only vertically.

In the limit of small perturbations, which for the moment we denote by a prime,

$$\rho'(p, T_\rho) \approx \left(\frac{\partial \rho}{\partial p}\right)_0 p' + \left(\frac{\partial \rho}{\partial T_\rho}\right)_0 T'_\rho \qquad (2.125)$$

so that

$$\frac{\rho'}{\rho_0} = \frac{p'}{p_0} - \frac{T'_\rho}{T_{\rho,0}}. \qquad (2.126)$$

The principle assumptions of the Boussinesq system are that density perturbations ρ' are, relative to the mean state, small, such that

$$\varepsilon \equiv \frac{\rho'}{\rho_0} \ll 1, \qquad (2.127)$$

and that relative pressure perturbations are in most cases of interest much smaller than relative temperature perturbations. With these two assumptions,

$$\frac{\rho'}{\rho_0} \approx -\frac{T'_\rho}{T_{\rho,0}}. \qquad (2.128)$$

Eq. (2.128) is a good approximation because the pressure is an integral quantity, in that through the hydrostatic balance it depends on the integral of the temperature field through the atmosphere. So, local temperature perturbations are expected to be much larger than pressure perturbations. Alternatively, this approximation follows because $p'/p_0 \sim M^2$ where the Mach number M is much less than unity, where temperature fluctuations are influenced by entropy fluctuations, which are independent (but generally larger) than the Mach number.

Consider the case when the flow is governed by a single velocity-scale U and length-scale H (which imply a timescale H/U) determined by the inertial motions. In this case, the continuity Eq. (2.120) can be non-dimensionalised by the time, velocity and distance scales (with non-dimensional quantities denoted by tilde),

$$(\tilde{t}, \tilde{v}, \tilde{x}) \quad \text{where} \quad t = H\tilde{t}/U, \quad v = \tilde{v}U, \quad x = H\tilde{x},$$

such that

$$\tilde{D}_t(\varepsilon) + (1 + \varepsilon)\tilde{\nabla} \cdot \tilde{v} = 0, \qquad (2.129)$$

where $\tilde{D}_t$ and $\tilde{\nabla}$ denote the non-dimensional form of the convective (substantial) derivative and gradient operator, for example, Eq. (2.119). With Eq. (2.127), the leading order balance is simply non-divergence, which in terms of dimensional variables becomes:

$$\nabla \cdot v = 0. \qquad (2.130)$$

To analyse the balances in the momentum equation, Eq. (2.121) can be multiplied by the density, here expanded as $\rho_0 + \rho'$, and a similar procedure followed, such that:

$$(1 + \varepsilon)\tilde{D}_t v = -\tilde{\nabla}(p_0 + p') + (1 + \varepsilon)g + \frac{1}{\mathcal{R}_e}\tilde{\nabla}^2 v, \quad (2.131)$$

where $\tilde{p}' = (H/\rho_0 U^2)p'$, and $\tilde{g} = (0, 0, -gH/U^2)$. $\mathcal{R}_e \equiv UH/\nu$ is the Reynolds number, with the kinematic viscosity $\nu = \mu/\rho$. $\mathcal{R}_e$ is typically very large, 10^8, for geophysical flows. The leading order balance in this equation is $dp_0/dz = -\rho_0 g$, that is, hydrostatic balance, which is here expressed in dimensional form, so that

$$D_t v = -\frac{1}{\rho_0}\nabla p' - \left(\frac{T'_\rho}{T_{\rho,0}}\right)g + \nu\nabla^2 v. \qquad (2.132)$$

A simple application of the scale analysis outlined earlier would suggest that the last term, which involves the viscosity, should be negligible. This is not the case, as turbulence leads to the production of velocity fluctuations across very small scales, much smaller than used to scale the pressure gradients, and these ensure that the viscous term contributes at leading order. The form of Eq. (2.132) makes clear the manner in which fluctuations in density temperature T_ρ drive density fluctuations. The term $(-T'_\rho/T_{\rho,0})g$ has units of acceleration, it is a reduced gravity and is called the buoyancy term and denoted as $b = (0, 0, b)$, whereby $b = gT'_\rho/T_{\rho,0}$. The last term in Eq. (2.132) is the diffusion, which acts on small-scale velocity fluctuations.

The thermodynamic equation, Eq. (2.122), when written in terms of θ_ℓ, becomes at leading order,

$$D_t\theta_\ell = \frac{\theta_0}{c_{p_\ell}T_0(z)}\left[\nu(\nabla v) : (\nabla v) - \nabla \cdot F_h\right]. \qquad (2.133)$$

Given the continuity equation, the expression for water mass conservation can be written in terms of the specific humidity, as

$$D_t q_t = 0. \tag{2.134}$$

Taken together, Eqs (2.130), (2.132), (2.133) and (2.134) define the moist Boussinesq equations. They form a closed set given an equation of state. The equation of state is given in the form of the buoyancy function $b(T, p, q_t)$. Both T and p are diagnostic functions. T follows from the definition of θ_l, given $p' + p_0$, and p' is given by taking the divergence of the momentum equation, in which case the incompressibility of the flow, Eq. (2.130), results in an equation for $\nabla^2 p$ which can be inverted for p'.

Allowing for non-equilibrium condensate phase introduces additional complexity and usually involves a greater degree of approximation. For instance, these phases will introduce entropy source terms, and in general will not follow the flow. Increasingly large hydrometeors, like hail, will have temperatures that increasingly depart from that of the ambient flow. The hydrometeors also exert a force on the flow, but this is accounted for through the condensate loading term that appears in the equation for the density temperature, for example, Eq. (2.7). In the subsequent discussion, we also drop the primes, taking it for granted that the thermodynamic quantities denote departures from the reference state. So doing facilitates an exploration of how these departures in turn depart from an expected value as is customarily done in the analysis of turbulence.

2.4.2   Quasi-static (Primitive) Equations

For large-scale atmospheric flows, the quasi-static equations describe the dynamics to a good degree of approximation. In someways these are the most original dynamic equations for the study of atmospheric motions, having been derived by Vilhelm Bjerknes, Jacob Bjerknes's father and in someways the father of modern meteorology, as the basis for numerical weather prediction at the turn of the last century. When written with pressure as a vertical coordinate, these equations take on a particularly convenient form, and formally resemble the equations used to study small-scale flows. The starting point for deriving the equations is again Eqs (2.120)–(2.123). Considerations of large-scale flows in a non-inertial reference frame (e.g., the rotating Earth), cannot, almost by definition, neglect the Coriolis acceleration, $f \times v$, and are frequently written in a coordinate system appropriate to the sphere (e.g., spherical coordinates) or some segment of a sphere

(i.e., by mapping the cartesian coordinate system conformally onto a Mercator projection). For completeness, in the discussion that follows the apparent acceleration from the Coriolis force is represented through the inclusion of the vertical component of the inertial frequency $f = \{0, 0, f\}$, but a Cartesian geometry is retained for simplicity. Consistent with the scale of the analysis, viscous transport terms are neglected.

The crucial assumption of the quasi-static equations is that the circulations are shallow, equivalently that the vertical-length scale is comparable to the depth of the troposphere and is much smaller than the horizontal-length scale. This implies that the vertical velocities first appear at a higher order. To follow the implications of this assumption, let us represent the velocity vector as a two-dimensional (horizontal) vector on a geopotential surface v_ϕ and a component w perpendicular to this surface,

$$v = (v_\phi(x, y, z, t), w(x, y, z, t)),$$

where $w \ll v_\phi$. For this scaling, the leading order momentum equations (Eq. (2.121)) become

$$D_t v_\phi = -f \times v - \frac{1}{\rho} \nabla_\phi p, \tag{2.135}$$

$$0 = -\frac{1}{\rho} \frac{\partial p}{\partial z} - g, \tag{2.136}$$

where ∇_ϕ denotes the two-dimensional (x, y) gradient operator along surfaces of constant geopotential. These equations can be combined with the thermodynamic equation (Eq. (2.122)) and continuity equations (Eqs (2.120) and (2.123)) to form a closed set; although, because w has been eliminated (Eq. (2.136)), further manipulation is required to derive an equation for w.

Rather than following this path, it proves convenient to rewrite the equations with pressure as a vertical coordinate, as this greatly simplifies the equations at the expense of the boundary conditions which become more challenging – the pressure velocity ω, unlike the kinematic velocity w, does not generally vanish at the surface. The hydrostatic constraint, (Eq. (2.136)), forms the basis for such a transformation, as a hydrostatic atmosphere pressure effectively measures the mass of the atmosphere above a given point in space, so it can be thought of as a mass coordinate. Upon integration, and by virtue of the fact that ρ is positive definite,

$$p(x, y, z, t) = p(x, y, 0, t) - g \int_0^z \rho(x, y, \zeta, t) \, d\zeta$$

is a strictly decreasing function of z, with ζ substituting for z in the integral to avoid confusion.

In carrying out the transformation, it proves useful to note that for some generic field ϑ,

$$\vartheta = A(x, y, z, t) = A(x, y, z(x, y, p, t), t) \qquad (2.137)$$

$$= B(x, y, p, t), \qquad (2.138)$$

where A and B will generally have different forms. Writing terms in this manner helps illustrate that

$$\left(\frac{\partial \vartheta}{\partial x}\right)_z = \frac{\partial A}{\partial x}$$

will not in general equal

$$\left(\frac{\partial \vartheta}{\partial x}\right)_p = \frac{\partial B}{\partial x}$$

as the terms on the left-hand side imply a change in the variable along a constant height surface, and the terms on the right-hand side imply a change along an isobaric surface. Rather,

$$\left(\frac{\partial \vartheta}{\partial x}\right)_z = \left(\frac{\partial \vartheta}{\partial x}\right)_p + \left(\frac{\partial z}{\partial x}\right)_p \frac{\partial \vartheta}{\partial z} \qquad (2.139)$$

$$= \left(\frac{\partial \vartheta}{\partial x}\right)_p + \frac{1}{g}\left(\frac{\partial \phi}{\partial x}\right)_p \frac{\partial \vartheta}{\partial z}, \qquad (2.140)$$

where the final expression is written in terms of the geopotential ϕ. Similar forms follow for the partial derivatives with respect to the other coordinates, y and t. The subscript z and p are used to denote in which coordinate system the differentiation is taking place. To discriminate between the vertical velocity in the z-coordinate system, $w \equiv D_t z$, the pressure velocity is defined by the symbol ω such that $\omega \equiv D_t p$.

Applying this coordinate transformation greatly simplifies the continuity equation, making it divergence free,

$$\nabla \cdot \boldsymbol{v} = \nabla_p \cdot \boldsymbol{v}_p + \partial_p \omega = 0. \qquad (2.141)$$

This makes physical sense, because working in terms of pressure, a mass coordinate, mass conservation implies that the mass field must be divergence-free. Applying the transformation to the pressure gradient's terms in the horizontal momentum equations transforms these to linear functions of the geopotential, such that

$$D_t \boldsymbol{v}_p = -\boldsymbol{f} \times \boldsymbol{v} - \nabla_p \phi, \qquad (2.142)$$

where, in pressure coordinates, the differential operator D_t takes the form

$$D_t = \partial_t + \boldsymbol{v}_p \cdot \nabla_p.$$

Eq. (9.1) shows that adopting pressure as the vertical coordinate eliminates density from the momentum equations.

For large-scale systems it is common to work with temperature as the thermodynamic coordinate, this is convenient in pressure coordinates where the pressure term in the enthalpy form of the thermodynamic equation (Eq. (2.10)) can simply be replaced by the vertical velocity, such that

$$D_t T = -\frac{1}{c_{pe}}\left[\omega \partial_p \phi - \nabla \cdot \boldsymbol{F}_h - Q_{\text{cnd}}\right], \qquad (2.143)$$

where a term arising from changes in ℓ_v following the flow has been neglected by virtue of the fact that $q_v \ll 1$. The net condensation heating Q_{cnd} on the right-hand side of Eq. (2.143) thus is the only term that remains to represent the effect of phase changes on the enthalpy, and it is proportional to changes in the water vapour specific humidity,

$$D_t q_v = -\frac{Q_{\text{cnd}}}{\ell_v}. \qquad (2.144)$$

Eqs (2.141)–(2.144) are the quasi-static equations, although the modern literature often refers to them as the primitive equations in pressure coordinates, one could just as well call them the Bjerknes equations. Both the Boussinesq and the quasi-static equations in pressure coordinates share a common structure. In both cases, the approximations or coordinate transforms render the momentum sources linear in thermodynamic quantities, and in both cases the flow is purely solenoidal, that is, non-divergent. As a practical matter, this non-divergence is, however, handled very differently in the different systems. In the quasi-static equations, ω becomes a diagnostic quantity, except at the surface. For both systems the equations are non-linear as a consequence of the quadratic form introduced through the advective derivative. This non-linearity has the effect that upon averaging the equations, for instance, to eliminate smaller scales, additional terms, involving correlations between fluctuating quantities, are introduced into the governing equations. This is discussed further later on.

For moist flows, the primitive equations may be supplemented by explicit equations for the condensate mass, of the form $D_t q_c = C$, similar to what was proposed for the moist Boussinesq equations. Likewise, by casting Eqs (2.143) and (2.144) in enthalpy form, neglecting variations in c_{pe} and ℓ_v, and summing, a moist form of the conservation law emerges,

$$D_t \eta_e = \nabla \cdot \boldsymbol{F}_h + \left(\partial_t + \boldsymbol{v}_p \cdot \nabla_p\right)\phi. \qquad (2.145)$$

The equation illustrates that in addition to the divergence of the enthalpy flux $\boldsymbol{F}_h$, the time rate of change of the geopotential following the isobaric component of the flow contributes to a change of the moist static energy, as anticipated in the derivation of the static energy. Often the latter term will be neglected, in which case the equivalent moist static energy η_e becomes an adiabatic invariant of the flow, much like θ_ℓ for the moist Boussinesq system.

2.5 Reynolds Averaged Equations

Reynolds averaging is a particular type of averaging that facilitates the isolation of large-scale degrees of freedom. For a variety of reasons it makes sense to restrict oneself to some subset of, usually larger, scales when studying fluid motion. Computationally, this is motivated by a desire to limit the degrees of freedom that must be solved. Theoretically, some assumptions are more readily justified (i.e., that pressure is everywhere hydrostatic) when small scales are neglected. And on simple practical grounds, the smallest scales of the flow, on the order of the Kolmogorov scale, are simply not relevant for many questions. Even in turbulent flows, the fluctuations associated with a particular realisation of a flow are often not of interest, rather the mean properties.

Let the Reynolds average of a quantity be denoted by an overbar. The overbar is defined through a filtering, or averaging, procedure with the following properties:

$$\overline{\overline{\varphi}} = \overline{\varphi} \quad \text{and} \quad \overline{\overline{\varphi}\vartheta} = \overline{\varphi}\,\overline{\vartheta}, \tag{2.146}$$

where φ is some field, and ϑ is a second field, which could also be identical to φ. The filter is further assumed to commute with other linear operators, such as differentiation in the time or space domain. Defining the deviation from the Reynolds averaged by the prime, so that,

$$\varphi = \overline{\varphi} + \varphi' \tag{2.147}$$

implies that

$$\overline{\varphi'} = 0 \quad \text{and} \quad \overline{\overline{\varphi}\varphi'} = 0. \tag{2.148}$$

Even with these properties, an application of the Reynolds average to a non-linear quantity, such as the convective derivative, introduces additional terms. For the case of the substantial, or advective, derivative, which, given the assumption of incompressibility, Eq. (2.130) can be written in flux form, such that

$$\overline{\mathrm{D}_t u} = \partial_t \overline{u} + \nabla \cdot (\overline{v}\,\overline{u}) + \nabla \cdot \left(\overline{v'u'}\right). \tag{2.149}$$

The last term on the right-hand side had no counterpart in the original equations, and is the divergence of a quantity called the Reynolds stress, or in the case that u is a scalar, the Reynolds flux. For small-scale flows the Reynolds fluxes act in all directions; in large-scale flows it is often assumed that the averaging is over much larger areas in the horizontal than in the vertical. In this case it can be assumed that the flow is sufficiently homogeneous that the

horizontal fluxes can be neglected relative to the resolved quantities, that is,

$$\nabla_p \cdot \left(\overline{v_p}\,\overline{u}\right) \gg \nabla_p \cdot \left(\overline{v'_p u'}\right),$$

and only the vertical transport by fluctuating quantities is retained.

2.5.1 Types of Filtering

Reynolds averaging is a conceptual procedure, where the filtering of the equations is thought of in terms of an average over an ensemble of realisations of a flow. For most applications of interest, a particular flow realisation is solved, for instance, by simulation, and filtering is employed to separate small from large scales. In the case that such filtering is applied in spectral space, with a wavenumber cut-off filter that truncates all scales smaller than a specified scale, the Reynolds averaging rules can also be shown to hold. However, many flows are solved in the physical domain, and in conditions (i.e., over limited areas) where spectral filtering is not possible. Here the filter that is used to separate large from small scales is usually defined implicitly, through the numerical methods employed to solve the equation. These methods distort scales near the grid scale and truncate all the scales smaller than the grid scale. But filters based on a local stencil in physical space (like a running average) are not spectrally sharp, and thus influence a range of scales at, and larger than, the grid scale. For such filters the Reynolds averaging rules usually cannot hold. The additional terms that arise from these imperfections of the filters are normally neglected or assumed small compared to the errors in the parameterisation of the terms that are retained.

2.5.2 Turbulence Kinetic Energy Equation

Clouds are usually turbulent, and an application of Reynolds averaging can be used to illustrate the basic mechanisms which control the production and dissipation of turbulence. The resultant equation, called the turbulence kinetic energy (TKE) equation plays an important role in the parameterisation of clouds and boundary-layer processes. Because TKE is conceptualised as being concentrated in eddies whose scale is much smaller than a scale height, a starting point for its derivation is the Boussinesq equations. An equation for the Reynolds averaged velocity can be derived by writing the velocity equation, (Eq. (2.132)), in terms of fluctuating and averaged quantities, averaging and then applying the Reynolds averaging rules. So doing results in an equation of the form

$$\overline{D_t}\overline{\boldsymbol{v}} + \overline{(\boldsymbol{v}'\cdot\nabla)\,\boldsymbol{v}'} = -\frac{1}{\rho_0}\nabla\overline{p} + \overline{\boldsymbol{b}} + \nabla\cdot\overline{\boldsymbol{\tau}}, \qquad (2.150)$$

where here the viscous term in Eq. (2.121) has been written in terms of the stress tensor, which is defined as twice the dynamic viscosity times the rate-of-strain tensor,

$$\boldsymbol{\tau} = \nu\left[(\nabla\boldsymbol{v}) + (\nabla\boldsymbol{v})^T\right]. \qquad (2.151)$$

The additional term on the left-hand side of Eq. (2.150) arises because the substantial derivative for the Reynolds averaged flow is only defined with respect to the averaged flow, that is,

$$\overline{D_t} \equiv (\partial_t + \overline{\boldsymbol{v}}\cdot\nabla). \qquad (2.152)$$

Subtracting Eq. (2.150) from Eq. (2.132) results in an equation for the velocity fluctuations of the form

$$\overline{D_t}\boldsymbol{v}' - \overline{(\boldsymbol{v}'\cdot\nabla)\,\boldsymbol{v}'} = -\frac{1}{\rho_0}\nabla p' + \boldsymbol{b}' + \nabla\cdot\boldsymbol{\tau}' - \boldsymbol{r}', \qquad (2.153)$$

where

$$\boldsymbol{r}' = (\boldsymbol{v}'\cdot\nabla)\,\boldsymbol{v}' + (\boldsymbol{v}'\cdot\nabla)\,\overline{\boldsymbol{v}}. \qquad (2.154)$$

Taking the inner product between Eq. (2.153) and $\boldsymbol{v}'$ and averaging results in a scalar equation for the small-scale kinetic energy (or TKE),

$$\overline{e} = \tfrac{1}{2}\overline{\boldsymbol{v}'\cdot\boldsymbol{v}'} \qquad (2.155)$$

such that

$$\overline{D_t}\overline{e} = -\frac{1}{\rho_0}\nabla\cdot\overline{(\boldsymbol{v}'p')} + \overline{\boldsymbol{v}'\cdot(\nabla\cdot\boldsymbol{\tau}')} + \overline{w'b'} - \overline{\boldsymbol{v}'\cdot\boldsymbol{r}'}. \qquad (2.156)$$

From basic relations in vector calculus and the rules of Reynolds averaging, the second and fourth terms on the right-hand side can be readily simplified, yielding the common form of the equation for TKE, wherein the evolution of $\overline{e}$ following the mean flow is given as the balance between the turbulent transport, production and dissipation of turbulence energy,

$$\overline{D_t}\overline{e} = \nabla\cdot\boldsymbol{\mathcal{T}} + \boldsymbol{\mathcal{P}} - \varepsilon. \qquad (2.157)$$

The individual terms in this balance are given respectively as

$$\boldsymbol{\mathcal{T}} = \left[\overline{\boldsymbol{v}'e'} + \frac{1}{\rho_0}\overline{\boldsymbol{v}'p'} + \overline{\boldsymbol{v}'\cdot\boldsymbol{\tau}'}\right] \qquad (2.158)$$

$$\boldsymbol{\mathcal{P}} = \overline{w'b'} - \overline{\boldsymbol{v}'u'}\cdot\partial_x\overline{\boldsymbol{v}} - \overline{\boldsymbol{v}'v'}\cdot\partial_y\overline{\boldsymbol{v}} - \overline{\boldsymbol{v}'w'}\cdot\partial_z\overline{\boldsymbol{v}} \qquad (2.159)$$

$$\varepsilon = \nu\overline{(\nabla\boldsymbol{v}')\cdot(\nabla\boldsymbol{v}')}. \qquad (2.160)$$

An important limit of these equations is when the averaging is anisotropic, such that the horizontal scale is much larger than the vertical scale of the averaging, as might be the case when considering boundary layers

which are homogeneous along the boundary. This is also the case in large-scale models where grid cells may be thousands of times larger in the horizontal direction than in the vertical direction. If one further assumes that the energy is contained in scales much smaller than the largest (horizontal) averaging scale, then it is often justified to assume homogeneity in these directions, so that terms involving horizontal derivatives may be neglected. In this case, the transport and production terms simplify to the following:

$$\nabla\cdot\boldsymbol{\mathcal{T}} = \partial_z\left[\overline{w'e'} + \frac{1}{\rho_0}\overline{w'p'}\right] \qquad (2.161)$$

$$\boldsymbol{\mathcal{P}} = \overline{w'b'} - \overline{u'w'}\partial_z\overline{u} - \overline{v'w'}\partial_z\overline{v} \qquad (2.162)$$

where viscous transport has also been neglected.

2.5.3   Buoyancy Production in Multiphase Flows

The buoyancy term can be related to perturbations in thermodynamic variables, as buoyancy fluctuations are carried by both moisture and temperature fluctuations. For instance, given θ_ℓ and q_t as thermodynamic coordinates, the buoyancy flux can be written in terms of contributions from each, such that

$$\overline{w'b'} = \frac{g}{T_{\rho,0}}\left(a_{\theta_\ell}\big|_0\,\overline{w'\theta'_\ell} + a_q\big|_0\,\overline{w'q'_t}\right), \qquad (2.163)$$

where for a thermodynamic coordinate ϑ,

$$a_\vartheta \equiv \frac{\partial T_\rho(\vartheta,\ldots)}{\partial\vartheta}. \qquad (2.164)$$

This partial derivative provides a more formal basis for defining the Brunt–Väisälä frequency as $N^2 = (g/z)a_\phi$.

A definition of a first-order phase transition is that the partial derivatives describing the material properties of the system are discontinuous at the phase boundary. The familiar example being the compressibility of liquid being manifestly different than that of vapour. Hence, the partial derivatives relating perturbations in the density temperature to those in the state variables will take on different values according to whether the flow is saturated or not. For the case of an unsaturated flow,

$$a_{\theta_\ell} \approx \frac{\theta}{T} \quad\text{and}\quad a_{q_t} \approx T\epsilon_2. \qquad (2.165)$$

These equations imply that for unsaturated flow, for instance, for the surface fluxes in the convective boundary layer, sensible heat fluxes are about fifteen times more efficient, per unit of surface cooling, at accelerating the flow, compared to moisture fluxes. However over great expanses of the tropical oceans latent heat fluxes are more than fifteen-times larger, meaning that latent heat fluxes contribute as much, or more, to the surface buoyancy flux as do sensible heat fluxes.

In saturated flow, $q_v = q_s(T, p)$ and $q_c = q_t - q_s$. Hence, in saturated air fluctuations in the specific humidity terms that appear in the expression for T_ρ will result from fluctuations in θ_ℓ, likewise, fluctuations of q_t will result in fluctuations in T, so that

$$a_{\theta_\ell} \approx \left(\frac{T}{\theta}\right)\gamma \quad \text{and} \quad a_{q_t} = T\left[\left(\frac{\ell_v}{c_p T}\right)\gamma - 1\right], \quad (2.166)$$

where γ, the ratio of the saturated to dry adiabatic lapse rate, was defined in Eq. (2.85). Given typical values, this implies that, unlike for unsaturated flows, moisture fluctuations are very efficient in generating buoyancy fluctuations in saturated flows, but entropy fluctuations are not.

At temperatures characteristic of subtropical stratocumulus, the saturated value of a_{q_t} exceeds the unsaturated value by about a factor of five. Following γ, this ratio will decrease with increasing temperatures, but not as strongly as humidity fluctuations are expected to increase. Hence the relative importance of humidity, as compared to θ_1, fluctuations is expected to increase with warming. The relative effectiveness of moisture fluctuations, in saturated versus unsaturated situations, makes physical sense, because an increase in q_t is, everything else being the same, associated with condensation, which warms the parcel, thereby increasing its buoyancy. Similar reasons explain why a_{θ_ℓ} can be as much as a factor of two smaller in saturated, relative to unsaturated conditions. Overall moisture fluxes are an important source of boundary-layer turbulence in tropical regions, and an essential source in saturated flows, that is, clouds.

2.5.4 Effective Heat and Moisture Sources, and the Gross Moist Stability

In this section we present some concepts that have been developed to understand the gross, or bulk, properties of large-scale flows. The starting point for doing so are the quasi-static equations, written in a form that neglects the isobaric transport of geopotential. This is generally a good approximation in the tropics, where horizontal pressure gradients are weak. In this case the quasi-static thermodynamic equations for a large-scale moist flow can be written for the dry static energy η_d and the specific humidity q_v as

$$\overline{D_t\eta_d} = -\partial_p \overline{\omega'\eta_d'} + Q_{cnd} + Q_{rad} = Q_1, \quad (2.167)$$

$$\overline{D_t\overline{q}_v} = -\partial_p \overline{\omega'q_v'} - \frac{Q_{cnd}}{\ell_v} = -\frac{Q_2}{\ell_v}. \quad (2.168)$$

Here the overbar denotes a Reynolds average, and horizontal homogeneity is assumed. The source terms

in the dry static energy budget are associated with condensational Q_{cnd} and radiative heating Q_{rad}, respectively. This formulation of the thermodynamic equations separates the large-scale adiabatic processes from the small-scale and diabatic processes, which are then grouped together in single terms, the apparent enthalpy source Q_1 and the apparent moisture sink Q_2. As could be anticipated from the equation for the moist static energy (neglecting changes in the geopotential and vaporisation enthalpy following along the isobaric flow), the sum of Eq. (2.167) and the enthalpy form of Eq. (2.168) shows that

$$\overline{D_t\overline{\eta}_e} = Q_{rad} - \partial_p \overline{\omega'\eta_e'} = Q_1 - Q_2. \quad (2.169)$$

This equation provides a powerful constraint for many diagnostic studies, but also for modelling (see, for instance, Chapter 5). For instance, by integrating over the troposphere, from some p_{top} to p_{sfc} it demonstrates that on large scales, where the net import or export of moist static energy by the advective flow can be neglected, changes in the moist static energy are driven by surface moist static energy fluxes (which are dominated by evaporation, equivalently precipitation) and the net transfer of radiant energy out of the column, that is,

$$\langle D_t\eta_e \rangle = \langle Q_{rad} \rangle + \left.\overline{\omega'\eta_e'}\right|_{sfc}, \quad (2.170)$$

where the angle brackets denote a vertical integral, so that in pressure coordinates,

$$\langle \vartheta \rangle = \frac{1}{\Delta p} \int_{p_{top}}^{p_{sfc}} \overline{\vartheta}(\ldots, p)\, dp, \quad (2.171)$$

where $\Delta p = p_{sfc} - p_{top}$ denotes the pressure depth of the integral and Reynolds averaging is assumed implicit in the definition of the vertical average.

Expanding the first term on the left-hand side of Eq. (2.170) shows that the

$$\partial_t\langle\eta_e\rangle + \langle v_p \cdot \nabla_p\eta_e\rangle + \langle\omega\partial_p\eta_e\rangle$$
$$= \langle Q_{rad} \rangle + (\Delta p)^{-1}\left.\overline{\omega'\eta_e'}\right|_{sfc}. \quad (2.172)$$

The vertical pressure velocity can be decomposed using Galerkin methods, a typical example of which is the projection of the continuous vertical structure of a variable onto orthogonal basis functions. Such a transformation, in terms of ω is equivalent to writing

$$\omega = \sum_{i=1}^{\infty} \hat{\omega}_i(x, y, t)\Omega_i(p), \quad (2.173)$$

where the basis functions are given by the Ω_i. For the simple case in which the variance of the vertical motion is well represented by a single vertical basis function Ω_1,

$$\omega = \hat{\omega}_1(x, y, t)\Omega_1(p) + \omega^*. \quad (2.174)$$

In this case, the vertically averaged moist static energy equation becomes

$$\partial_t\langle\eta_e\rangle + \langle v_p \cdot \nabla_p\eta_e\rangle + \langle\omega^*\partial_p\eta_e\rangle + \langle\omega\rangle G(\eta_e)$$
$$= (\Delta p)^{-1}\,\overline{\omega'\eta_e'}\Big|_{\text{sfc}} + \langle Q_{\text{rad}}\rangle, \quad (2.175)$$

which introduces the gross vertical averaging operator G. It is defined such that for an arbitrary variable ϑ,

$$G(\vartheta) \equiv \langle\Omega_1\partial_p\vartheta\rangle. \quad (2.176)$$

In Eq. (2.172), the term $G(\eta_e)$ has come to be called the gross moist stability. If, in Eq. (2.175), to the extent that the first three terms on the left-hand side are small and can be neglected relative to the other terms, an equation for the amplitude of the gravest mode of vertical motion can be derived in terms of the energy input into the column and its gross moist stability, such that

$$\hat{\omega}_1 \approx \frac{1}{G(\eta_e)}\left[(\Delta p)^{-1}\,\overline{\omega'\eta_e'}\Big|_{\text{sfc}} + \langle Q_{\text{rad}}\rangle\right]. \quad (2.177)$$

This equation helps illustrate the physical meaning of $G(\eta_e)$ as it defines the proportionality between the moist enthalpy fluxes that power the flow and the strength of the mean vertical circulation. It is the moist analogue to the balance in clear skies between the dry static stability, the local radiative heating and the vertical velocity at a point.

A similar analysis may be applied to the moisture equation. In this case, advective fluxes are small compared to evaporative fluxes (as might be expected on sufficiently large scales), a simple balance arises such that

$$P - E_{\text{sfc}} - \langle\omega\rangle G(q_v) \approx 0, \quad (2.178)$$

where E_{sfc} denotes the surface evaporation, $P = \langle Q_{\text{cnd}}\rangle$ is the net precipitation and $G(q_v)$ defines a gross moisture lapse rate. Some authors prefer to directly relate precipitation to energy fluxes, and in so doing they define a different form of the gross moist stability as the ratio

$$\frac{Q_1}{Q_2} - 1. \quad (2.179)$$

This is sometimes called the normalised gross moist stability, and its apparent similarity to the efficiency of heat engines is a notational accident. Defining the normalised gross moist stability in this fashion avoids having to make assumptions about the structure of the circulation when discussing the relationship between fluxes of energy and precipitation. These and related ideas are discussed further in Chapter 5, they also prove to be important in Chapter 13, where they are applied to understanding patterns of precipitation change in the tropics.

2.6 Turbulence

2.6.1 Phenomenology of Turbulence

A hallmark of turbulent flows is that they distribute energy over a broad range of scales. This can be thought of in terms of a Fourier representation of their energy spectrum, as energy is densely distributed across scales through the non-linearities of inertial interactions. In a flow where only one scale is forced, for instance, through an instability, the resulting fluid motions excites other scales of motions through non-linear interactions and secondary instabilities. This results in a rich range, or spectrum, of length scales becoming evident, for instance, as revealed through a Fourier decomposition of the motion which quantifies how much variability is carried by a particular scale of motion. These interactions have a purpose, as they transport inviscid invariants of the flow (e.g., energy) from scales where energy is excited to scales where molecular processes operate efficiently and can thus dissipate the energy. This type of transport across scales is called a cascade, and it is illustrated schematically in Fig. 2.11. According to this view, energy is injected into the system at a macroscopic scale l_0 that drives the largest eddies, inviscidly transferred to smaller eddies in the inertial subrange, and finally dissipated into heat at the Kolmogorov scale l_K.

The description of fully developed turbulence as a cascade of interacting eddies dates back to the work of Lewis Fry Richardson in the early part of the twentieth century, and the self-similar character of the eddies have been confirmed in numerous experiments. Yet, as of today, there is still no deductive theory that derives the observed self-similar scaling laws from the Navier–Stokes equations. Fortunately, much of the basic behaviour of fully

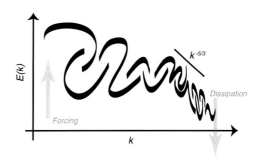

FIGURE 2.11: Cartoon of energy cascade whereby energy 'cascades' inviscidly from large scales to small scales where it is then dissipated. Here the scale of the motion is denoted by its wave number $k = 2\pi/l$ where l is the wavelength.

developed turbulence characterised by a high Reynolds number can be understood on a simple phenomenological basis.

For a turbulent flow in a stationary state, the TKE equation can be approximated as a balance between the rate of energy injection E_I at the macroscale and energy dissipation rate ε at the microscale

$$\partial_t e \approx E_I - \varepsilon \approx 0. \qquad (2.180)$$

Consider now an eddy of size l with a relative velocity v_l be defined as a velocity difference over a distance l and that is characterised by an eddy turnover time through

$$t_l \sim \frac{l}{v_l}, \qquad (2.181)$$

where the symbol $\sim$ means 'equal up to an order of magnitude'. The main hypothesis, stated in the celebrated 1941 paper by Andrej N. Kolmogorov, is that the rate of energy transfer cascading from the largest eddies down to the smallest eddies is constant and approximated by the mean dissipation rate ε. The transfer of energy in the inertial range $l_K < l < l_0$ can be intuitively interpreted as the flux of turbulent kinetic energy $e \sim v_l^2$ from an eddy of size l cascading down to the smaller eddies.

Taking the eddy turnover time t_l as the typical timescale during which the eddy breaks up into smaller eddies (Fig. 2.12), the transfer of energy can be written as

$$\frac{e_l}{t_l} \sim \frac{v^3}{l} \sim \varepsilon \sim \text{cnst}, \qquad (2.182)$$

so that

$$v_l \sim \varepsilon^{1/3} l^{1/3} \propto l^{1/3}, \qquad (2.183)$$

which is shorthand for one of the main results of Kolmogorov's 1941 theory, K41. A number of key properties directly follow from K41. First, K41 shows that the velocity field in turbulence is self-similar and described with an exponent $h = 1/3$. This implies that the gradient of the velocity field would be singular everywhere for

small l if there would not be a Kolmogorov dissipation scale where diffusion becomes dominant. Diffusive effects can be neglected only if the eddy turnover time t_l is much smaller than the diffusion timescale $t_{l,\text{diff}} \sim l^2/\nu$, but, because the diffusion time scale goes to zero faster than the eddy turnover time, there will always be a scale at which diffusion becomes dominant. Equating the diffusion and the eddy turnover time gives as an estimate for this Kolmogorov dissipation scale

$$l_K \sim \left(\frac{\nu^3}{\varepsilon}\right)^{1/4}. \qquad (2.184)$$

The range of scales between the outer scale at which the energy is injected and the Kolmogorov scale is dictated by the integral-scale Reynolds number $\mathcal{R}_e = UL/\nu$ where U denotes the relative velocity scale related to the largest eddies. By estimating the energy dissipation rate as the energy transfer rate at the outer scale, that is, $\varepsilon \sim U^3/L$ and using Eq. (2.184), the ratio between L and l_K scales with the Reynolds number as

$$\frac{L}{l_K} \sim \mathcal{R}_e^{3/4}. \qquad (2.185)$$

A typical convective atmospheric boundary layer of $L \sim 1$ km depth and a typical relative velocity scale at this outer scale of $1\,\text{m s}^{-1}$ is characterised by a $\mathcal{R}_e \sim 10^8$, which justifies the value presented in the scale analysis of Section 2.4.1. This value of $\mathcal{R}_e$ implies a Kolmogorov scale of the order of 1 mm. This simple example shows the enormous range of spatial scales, six orders of magnitude, over which the K41 scaling behaviour of the velocity field extends. The scaling behaviour K41 can be easily reformulated into the famous 5/3 energy spectrum by taking the Fourier transform of the turbulent kinetic energy $e_l \sim v_l^2$

$$e(k) \sim \varepsilon^{2/3} k^{-5/3} = \alpha_K \varepsilon^{2/3} k^{-5/3}, \qquad (2.186)$$

where the wave number k can be associated to the inverse eddy size, that is, $k \sim 1/l$. The energy spectrum demonstrates that the largest eddies are the most energetic. Universality requires that the proportionality constant α_K be independent of the particular flow. This prediction is well supported experimentally, with $\alpha_K = 1.6$ being called the Kolmogorov constant. At present there exists no accepted theory which explains the value of Kolmogorov's constant. The lack of such a theory is the quintessential problem of turbulence.

As will be demonstrated in Chapter 5, the self-similarity of atmospheric turbulence allows the possibility of realistically numerical modelling at relative coarse resolutions of around 100 m. Such large eddy simulations (LESs) resolve only the largest, most energetic eddies,

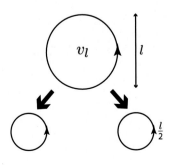

FIGURE 2.12: Schematics of the breakdown of a turbulent eddy of size l into smaller eddies.

whereas the smaller unresolved eddies are parameterised in terms of the resolved large eddies through the use of K41.

2.7   Models of Clouds and Circulation

For the most part, the equations required to describe clouds are known. This means that a great deal of insight into the role of clouds in the climate system can be attained by exploring the behaviour of these equations under different circumstances. This approach, whereby equation systems describing cloud systems are used to construct virtual laboratories from which insights are derived, is the dominant research methodology at the present time.

Two challenges however limit the insights that can be derived from these approaches, and continue to necessitate a strong connection with observations. One is that not all the equations are known. In particular, on microscopic scales many of the interactions among particles, and between the gas and particle phases especially with respect to ice processes, are still poorly understood. This issue is even more manifest when it comes to describing the interaction of the atmosphere with other components of the climate system, for instance, the biosphere. The other challenge is that even if the equations were all known, they would encompass far too many degrees of freedom to ever contemplate solving, even given computers with computing capacities millions of times more powerful than what is presently possible. These challenges mean that the simulation systems used as virtual fluid-dynamical laboratories of cloudy atmospheres are necessarily approximate.

The approximations made to render a fluid-dynamical description of a cloudy atmosphere computationally tractable are threefold: (i) some truncation of scale; (ii) an aggregated description (parameterisation) of the behaviour of particulate matter and condensate; (iii) a simplification in the treatment of radiative transfer. In addition, to close the equations inevitably involves developing approximate models that describe interactions with other components of the climate system, for instance, the surface, as well as the behaviour of degrees of freedom that, because of the scale truncation, are not explicitly represented. The different types of cloud-dynamical models are distinguished mostly by the first approximation, namely at what scale they truncate the fluid-dynamical equations. Fig. 2.13 illustrates the range of scales within the atmosphere, processes associated with different scales and the names associated with models that truncate their equations at a given scale. Because the range

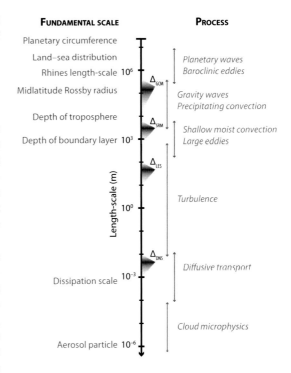

FIGURE 2.13: Physical length-scales in the atmosphere, and processes associated with them. Different classes of models are here characterised by the size Δ of their grid mesh, which is related to the smallest scales that they can represent.

of processes included by a particular truncation differs depending on the scale of the truncation, these different types of models differ, as discussed later, qualitatively from one another.

2.7.1   Direct Numerical Simulation

Simulations based on fluid-dynamical descriptions that resolve eddies down to the Kolmogorov scale are called direct numerical simulations, or DNSs. DNS is the most fundamental description of fluid-dynamical transport, but because computational restrictions limit the largest space and time scales that can be represented, it is also the most special purpose. Advances in computing power has, however, begun to make it possible to use DNS over large enough domains that it has become applicable to problems of interest to atmospheric scientists, up to and including boundary-layer scales. Because it is applied to problems whose largest scales are still relatively small, DNS is often based on the Boussinesq approximation, which on the scales usually represented can be asymptotically justified. But for some DNS the full compressible equations, or even two-fluid equations, may be applied. In DNS, a main hypothesis is that of Reynolds number similarity, which states that once a sufficiently large Reynolds number is achieved, then important statistical properties – like vertical profiles of mean and variances of temperature and specific humidities – cease to depend on the Reynolds number. Even in the case that Reynolds number similarity is a good assumption, DNS remains limited by many of the same limitations of models that more approximately represent the fluid-dynamical equations, namely simplifications in the coupling to realistic boundaries, or necessarily approximate descriptions of microphysical processes and radiative transfer. Despite these limitations which DNS shares with other types of cloud circulation models, DNS is emerging as not just an exciting method for studying classic boundary layer flows but also for studying cloud mixing processes and the interactions among turbulence, cloud-microphysical processes and radiant energy transfer.

2.7.2 General Circulation Models

General circulation models, or GCMs, are usually based on the primitive equations, and applied to the global atmosphere in a way designed to explicitly represent the main mechanism, baroclinic instability, of meridional heat transport in the mid-latitude troposphere (see Chapter 10). Because baroclinic eddies have a scale of about 1,000 km, they can be well resolved by models using a grid whose mesh-size is on the order of a few hundreds of kilometres or finer (Fig. 2.13). GCMs are a cornerstone of numerical weather prediction, and also serve as the dynamical core of the atmosphere in global climate models. The identical initialism (GCM) for global climate models and GCMs – upon which they are often based – is a source of ambiguity, here it is restricted to mean the latter. In GCMs, the vertical heat transport, associated with a cascade of processes ranging from near-surface turbulent eddies, to boundary-layer circulations, to deep convective clouds, and even cloud systems are not at all resolved, and must be approximated through parameterisations. The approximate, and generally inadequate, representation of vertical heat and moisture transport and of cloud processes in GCMs also influences their ability to properly represent radiant energy transfer through the atmosphere, something which is vital for an adequate representation of the climate system. Over the years, a great deal of effort, as reviewed in Chapter 6, has been expended in better understanding how the process of vertical heat and momentum transport by eddies (and waves) is related to the mean state of the larger-scale flow that GCMs explicitly represent. Their chief limitation

relative to smaller-scale models is the uncertainty of these processes, and their importance for the large-scale flow. This is especially true in the tropics, where moist cumulus convection and radiant energy transfer interact strongly with the circulation. The main advantage of the GCMs is that they provide a closed fluid-dynamics description of the transient dynamics of the large-scale flow, whose statistical properties define Earth's climate.

2.7.3 Storm-Resolving Models

Storm-resolving models, SRMs, typically forgo a global description so as to allow for sufficiently fine scales to resolve the vertical overturning of the troposphere. This then permits them to crudely, but explicitly, represent convective storms that reach through the depth of the troposphere – even when such storms adopt commensurately fine horizontal scales, as for the case of precipitating deep convection in the tropics. This implies that an SRM must have a spatial resolution commensurate with the depth of the tropical troposphere, and hence grid-mesh spacings in the order of a few kilometres. The requirement of such a fine grid mesh has, until only recently, limited the domain of SRMs to be in the order of hundreds to a few thousands of kilometres. SRMs are often based on the anelastic equations, but increasingly, as they are being applied globally, solve the fully compressible equations.

The great advantage of SRMs is that their explicit representation of at least the gravest modes of convective heat transport in the troposphere allows for a more consistent representation of circulation features that often accompany such storms, from cold pools and gust fronts, to stratiform cloud shields. It also more naturally allows these circulation systems to be coupled with parameterisations of cloud microphysical processes and radiant energy transfer. Unlike a GCM, whereby most parameterisations must be coupled with one another, in SRMs, at least for the deepest clouds, parameterisations of cloud microphysical processes and radiation can interact with circulations that the model explicitly, albeit crudely, resolves. In the literature, SRMs adopt a cacophony of names: for instance, cloud resolving models (CRMs) cloud system resolving models or convection permitting models (CPMs). Some authors distinguish between CRMs and CPMs by virtue of how they parameterise yet finer scale processes. Because the distinguishing factor of SRMs is their ability to resolve storms, meaning primarily the circulation (wind) systems and geometric features of tropical convection, but not necessarily the finest scales of associated clouds or convection, the SRM nomenclature is adopted in this book.

By definition, an SRM does not include a parameterisation of deep convection. However, vertical heat transport arises on many other scales, for instance, from the large eddies within the atmospheric boundary layer, or from shallow convective clouds which might reach a depth of a few kilometres, and must be parameterised in an SRM, a process that shares many of the complexities and pitfalls of the parameterisation of deep convection, albeit in a way that is hopefully less strongly coupled to the large-scale flow. Historically, when applied to domains of limited area, SRM simulations also needed to parameterise (or specify) the large-scale flow, for instance, by relaxing it back to some presumed state, or by nudging it to lateral boundary conditions that are taken from another source. SRMs are, however, increasingly being applied to study large-scale problems in geophysical fluid dynamics, and in so doing provide a better foundation for understanding the ways in which cloud processes affect these dynamics.

2.7.4 Large Eddy Simulation

LES differs from SRM simulations in that it endeavours to resolve also the large eddies responsible for the heat and momentum transport within the atmospheric boundary layer, which for convective situations may be between 500 m and a few kilometres. This implies a spatial resolution of about 500 m, and hence a grid mesh size of about 100 m or less, which is still much larger than the Kolmogorov scale. As long as the large eddies are well resolved, LES has the advantage that the smaller-scale motions which are not resolved can often be assumed to be representable in a universal way based on an understanding of the phenomenology of homogeneous and isotropic turbulence, as discussed in Section 2.6.1. Because LES resolves the turbulent circulations on the cloud scale, it is thought to provide an adequate description of mixing processes associated with cumulus clouds, and to provide a strong foundation for linking radiative transfer to explicitly resolved clouds, and microphysical processes to explicitly resolved circulations. However, even when run at resolutions of a few metres, some mixing processes remain poorly represented within LES, especially those in regions of strong stratification, as one finds atop stratocumulus-topped boundary layers (Chapter 5) and in nocturnal or stately stratified boundary layers. Such a fine spatial resolution also precludes a consideration of many meso- and large-scale processes, as historically LES domains have been on the order of kilometres. This means that LES usually neglects, or at least distorts, the way in which processes on the small scale influence the mesoscale

in ways that may feed back and influence the small scales. Like SRMs, LESs are also using equation sets that allow them to simulate flows over a deeper atmosphere and are being applied to ever larger domains, thereby providing insight into how cloud and boundary-layer processes induce and influence circulations on mesoscales (10–100 km).

2.8 Outlook

Although it might seem that the foundations of a fluid dynamical description of clouds is more or less worked out, there are several interesting and active areas of open research. In terms of the thermodynamics, basic questions relating to a consistent but simplified representation of non-equilibrium components of multiphase flows remain to be clarified. This involves not only thermal effects, for instance, associated with multiphase particles (melting snow and graupel), but also dynamic effects when particles decouple from the flow. Only recently has research begun to attempt a full accounting of the entropy production in association with non-equilibrium phases and much work remains to be done, for instance, in association with the entropy production by cloud microphysical processes.

Better understanding how to measure the energy available to convection, even in idealised frameworks, is an area where more research would be beneficial. Is there an optimal definition of convective available potential energy, the state to which a convecting atmosphere attempts to relax or the timescale on which they relax? In this context many basic problems related to the stability of convecting atmospheres, and the implied asymmetries between upward (saturated) and downward (unsaturated) motion remains open. Here the advancing power of DNSs to explore instabilities in moist atmospheric flows and the nature of turbulence in multiphase atmospheric flow is very promising. At the same time, more research is needed on conceptual models capable of better elucidating constraints on moist systems arising from the Second Law.

An exciting open area of research, called forth by increasing computing power, is how to create soundproof equation systems that are energetically consistent on very large domains, and which consistently represent multiphase thermodynamics. Most asymptotic approaches to these problems remain focused on dry atmospheres and fail to consistently represent moist and microphysical processes. On the larger scales, the moist static energy framework is promising and a better understanding of how the energetics of the flow couple to the circulation using this framework is an exciting and popular area of research, but here the role of shallow circulations, which such a framework endeavours to neglect, emerges as a key question.

Advances in computing power are beginning to blur the lines between traditional modelling frameworks. Recent years have witnessed the advent of hybrid models that embed LESs or SRMs in GCMs, through an approach referred to as super-parameterisation (Chapter 6). Even more recently, very highly parallelised computations run on tens and even hundreds of thousands of processors have begun to make it possible to perform LESs on grid scales of a hundred metres, across domains of more than a thousand kilometres, for periods of weeks, or to run SRMs over the entire globe, thereby superseding GCMs, for periods of days to weeks. Along with DNSs, the ability of these very computationally intensive approaches to encompass the interaction of a wide range of scales make them very exciting tools that are beginning to shed new light on how clouds and circulations interact.

Further Reading

Section 2.1 Thermodynamics

The subject is treated in many introductory meteorological texts, but in most cases poorly. The best introductory guide to classical thermodynamics is given by the short series of lectures by Enrico Fermi (1956), which is published by Dover Publications. This book provides a readable and elegant presentation of basic thermodynamic concepts from a classical perspective. Although the treatment is general, many of the particular concepts of atmospheric thermodynamics, for instance, the derivation of the potential temperature or the Clausius–Clapeyron equation, are readily extracted from this book. The definitive modern treatment of classical thermodynamics, with extensions to statistical mechanics, is provided in Herbert B. Callen's beautiful book entitled *Thermodynamics and an Introduction to Thermostatics* (Callen, 1985), in particular for its more elegant and modern presentation of the laws and their relationship to the existence of quantities like entropy or enthalpy. For a discussion of particular concepts related to atmospheric thermodynamics, and in particular the special role played by moisture, students and practitioners, will be well served by the classic (but out of print) book by Iribarne and Godson (1981) or the introductory chapters of *Atmospheric Convection* by Kerry Emanuel (1994). The latter has particularly clear discussions of the moist potential temperatures.

Pascal Marquet's (2011, 2017) papers on moist entropy and entropy potential temperatures are closely related to the topics introduced in this chapter, the development, application and subsequent discussion of his ideas are recommended for readers interested in a deeper understanding of these topics. Papers by Pauluis and Held (2002a, 2002b) on the entropy budget of radiative convective equilibrium are an excellent and definitive introduction to an entropy view on the atmospheric circulation. The paper by Kleidon and Renner (2013) provides a very clear overview of the use of entropy and maximum power concepts to understand the hydrological cycle.

Section 2.2 Thermodynamic Diagrams

Visualising the state of the atmosphere, and the development of thermodynamic processes on thermodynamic diagrams can be a useful way to develop understanding. In addition to Iribarne and Godson (already mentioned), a paper by Böing et al. (2014) provides insight into the use of thermodynamic diagrams for the study of mixing processes.

Section 2.3 Convective Instability

This section is presented in a rather untraditional way, starting from the small-scale mixing process associated with what has come to be known as buoyancy reversal. This basic concept has a long history in the literature, but a beautiful and influential presentation of the ideas can be found in a 1980 paper by David A. Randall (1980). The title alludes to the fact that buoyancy reversal is related to traditional ideas of convective instability, the treatment and different measures of which are well presented in the aforementioned book by Emanuel. Buoyancy reversal has always been thought to be an important concept for stratocumulus clouds, and recent work on the topic by Juan Pedro Mellado (2010, 2017) provides the most insightful contributions on the topic from the point of view of fundamental fluid dynamics, as this work has helped clarify long-standing uncertainty related to the role of buoyancy reversal in cloud mixing.

Section 2.4 Fluid Dynamics

Fluid dynamics is a vast topic. A very good basic text on basic fluid mechanics is that by Kundu and Cohen (2002) for which the early editions are perfectly adequate. For a complete treatment of multiphase flows, including discussion of the many intricacies associated with the treatment of particles, the article by Peter Bannon (2002) is a good reference. In deriving the Boussinesq or other soundproof equation systems, asymptotic techniques are employed. For a guide to asymptotic approaches in fluid mechanics, through which many of the equation sets used to describe flows in the atmosphere and ocean can be derived, the reader is referred to a review article by Rupert Klein (2010). Although work on these topics might seem academic, the question of sound-proof equation sets that relax the requirements on the basic state are coming increasingly into focus as numerical simulation is capable of solving flows with an increasing number of degrees of freedom, which allows for a treatment of small scales over very large domains wherein the background state might be expect to vary greatly.

For an introduction to Reynolds averaging (or filtering approaches in general) and turbulence, most books on turbulence will provide a good introduction. Perhaps the best book on the topic with an atmospheric perspective in mind (for a consideration of Reynolds averaging, and higher order equations) is the book by Wyngaard (2010). Stephen Pope's book (*Turbulent Flows*, 2000) has become a standard text on turbulence, but the beauty of the subject is perhaps better captured by the Uriel Frisch's book (*Turbulence, the Legacy of A. N. Kolmogorov*, 1996), which influenced the presentation of the turbulence cascade in Section 2.6.1. G. I. Barenblatt's book (*Scaling, Self Similarity, and Intermediate Asymptotics*, 1996) is an excellent introduction to similarity theory. None of these texts adequately cover moist, or multiphase, turbulent flows as appropriate for clouds.

The discussion of effective heat and moisture sources originates in the pioneering analysis of Michio Yanai, beginning with his 1973 paper (Yanai et al., 1973). The idea of gross moist stability was introduced by Neelin and Held (1987). A useful and more general review is provided by David Raymond et al. (2009).

Exercises

(1) Show that choosing a reference state enthalpy for both h_e and h_ℓ determines the reference state vaporisation enthalpy. What is $\ell_{v,0}$ if one assumes that $h_l(T_{0e}) = c_l T_{0,e}$ and $h_v(T_{0,\ell}) = c_{p_v} T_{0,\ell}$?

(2) Show that if in equilibrium h and T adopt values that maximise the entropy function s, then this implies that the Gibbs potential g is minimised.

(3) Assuming that the specific heats are constant, use the values in Table 2.3 to show that the reference entropy for liquid water at $0\,°C$ is $3.518\,\text{kJ}\,\text{kg}^{-1}\,\text{K}^{-1}$ and that for solid ice at the same temperature is $2.298\,\text{kJ}\,\text{kg}^{-1}\,\text{K}^{-1}$.

(4) If the energy is removed by internal, turbulent dissipation ($\mathcal{E}$) then this represents an additional source

TABLE 2.6: Constants for Tetens's formula, as given in Eq. (2.189), for saturation vapour pressure over liquid water and ice.

State	a	b (K)
Water	17.2693882	35.86
Ice	21.8745584	7.66

of heating that must be accounted for, that is, the First Law becomes

$$W = \mathcal{E} + Q_1 - Q_2 \tag{2.187}$$

and the Second Law

$$\frac{Q_1}{T_1} = \frac{Q_2}{T_2} - \frac{\mathcal{E}}{T_\mathcal{E}} - \Delta S_{\text{irr}}. \tag{2.188}$$

Show how in steady state if the dissipation happens near the lower boundary, at temperature $T_\mathcal{E} = T_1$, this implies a system that does more work than a Carnot engine. Why doesn't this contradict the Second Law?

(5) In deriving the Clausius–Clapeyron equation, it is assumed that $\ell_v/T = s_v - s_l$. Demonstrate that this is true in the case of a saturated system.

(6) Tetens's formula originally formulated the saturation vapour pressure as

$$p_s = p_s(T_*) \exp\left[\frac{a(T - T_*)}{T - b}\right], \tag{2.189}$$

where $T_* = 273.16\,\text{K}$ is the triple-point temperature, so that $p_s(T_*) = 610.78\,\text{Pa}$. The fitting parameters a and b are given in Table 2.6 for saturation over liquid water and ice, respectively. Compare this form to that given in Eq. (2.73).

(7) Derive an expression for the saturation pressure over liquid by assuming: (i) ℓ_v is constant and (ii) by allowing ℓ_v to vary with T following Kirchoff's relation. Compare these to Tetens's formula as given in Eq. (2.73) and the reference fit by Goff and Gratch as summarized by Murphy and Koop (2005).

(8) Derive Ω_ℓ, in Eq. (2.45), which defines the liquid-water potential temperature.

(9) Fill in the steps of the derivation of the ice-liquid potential temperature as given by Eq. (2.67).

(10) Starting from Eq. (2.38), derive Eq. (2.83), stating all assumptions.

(11) By deriving an equation for the equisaturation line on a skew-T diagram, calculate the change in the dew-point depression with height for adiabatic ascent. Use this to derive an expression for the LCL of a parcel with a given dew-point depression.

TABLE 2.7: Symbols and key variables used in this chapter

Symbol	Description	Units (SI)
$v, \mathcal{V}$	Specific volume, volume	$\mathrm{m^3\,kg^{-1}}$, $\mathrm{m^3}$
$s, \mathcal{S}$	Specific entropy, entropy (Eq. (2.28))	$\mathrm{J\,kg^{-1}\,K^{-1}}$, $\mathrm{J\,K^{-1}}$
$h, \mathcal{H}$	Specific enthalpy (Eq. (2.11))	$\mathrm{J\,kg^{-1}}$, J
$g, \mathcal{G}$	Specific Gibbs free energy, Gibbs free energy ($\mathcal{G} = \mathcal{H} - T\mathcal{S}$)	$\mathrm{J\,kg^{-1}}$, J
δQ	Heating	J
a_ϑ	$\partial_\vartheta T_\rho$ (Eq. (2.164))	–
b	buoyancy (Eq. (2.85))	$\mathrm{m/s^2}$
c_{p_e}	Equivalent state c_p (Eq. (2.22))	$\mathrm{J\,kg^{-1}}$
c_{p_ℓ}	Liquid-free state c_p (Eq. (2.26))	$\mathrm{J\,kg^{-1}}$
$\bar{e}$	TKE (turbulence kinetic energy)	$\mathrm{m^2\,s^{-2}}$
η_d	Dry static energy	$\mathrm{J\,kg^{-1}}$
η_e	Moist static energy (Eq. (2.51))	$\mathrm{J\,kg^{-1}}$
η_ℓ	Liquid-water static energy (Eq. (2.52))	$\mathrm{J\,kg^{-1}}$
η_s	Saturation η_e	$\mathrm{J\,kg^{-1}}$
κ_1	Interfacial stability criterion	–
κ_2	Buoyancy reversal parameter (Eq. (2.102))	–
k	Wavenumber, $2\pi/$wavelength	–, or $\mathrm{rad\,m^{-1}}$
l	Size of an eddy	m
E	Turbulence energy spectral density	$\mathrm{m\,s^{-2}}$
N^2	Brunt–Väisällä frequency (Eq. (2.89))	$\mathrm{s^{-1}}$
N_s^2	N^2 in saturated air (Eq. (2.91))	$\mathrm{s^{-1}}$
Q_1	Apparent enthalpy source	$\mathrm{J\,kg^{-1}\,s^{-1}}$
Q_2	Apparent moisture sink	$\mathrm{kg\,kg^{-1}\,s^{-1}}$
Q_cnd	Net condensational heating	$\mathrm{J\,kg^{-1}\,s^{-1}}$
Q_rad	Radiative heating	$\mathrm{J\,kg^{-1}\,s^{-1}}$
R_e	Equivalent-state gas constant	$\mathrm{J\,kg^{-1}\,K^{-1}}$
R_ℓ	Liquid-state gas constant	$\mathrm{J\,kg^{-1}\,K^{-1}}$
Γ_s	Saturated adiabatic lapse rate	$\mathrm{K\,m^{-1}}$
Γ_d	Dry adiabatic lapse rate	$\mathrm{K\,m^{-1}}$
α_K	Kolmogorov constant	–
β_ϑ	$(\vartheta/q_\mathrm{s})\,\partial_\vartheta q_\mathrm{s}$, e.g., Eqs (2.81) and (2.82)	–
ϵ_1	$R_\mathrm{d}/R_\mathrm{v} \approx 0.622$	–
ϵ_2	$R_\mathrm{v}/R_\mathrm{d} - 1 \approx 0.608$	–
ε	Dissipation rate of TKE	$\mathrm{m^2\,s^{-3}}$
γ	$\Gamma_\mathrm{s}/\Gamma_\mathrm{d}$ (Eq. (2.85))	–
χ	Mixing fraction	–
μ	Dynamic viscosity	$\mathrm{N\,s\,m^{-2}}$
p_θ	Reference pressure, 10×10^5 Pa	Pa
ϕ	Geopotential, gz	$\mathrm{J\,kg^{-1}}$
l_K	Kolmogorov microscale	m
ν	Kinematic viscosity	$\mathrm{m^2\,s^{-1}}$
ϑ	Generic variable	$[\vartheta]$
a	Convective area fraction (Fig. 2.10)	–
$\mathcal{M}$	Convective mass flux (Eq. (2.110))	$\mathrm{kg\,m^{-2}\,s^{-1}}$
$\mathcal{A}$	CAPE (Eq. (2.105))	$\mathrm{J\,kg^{-1}}$
$\mathcal{I}$	CIN (Eq. (2.107))	$\mathrm{J\,kg^{-1}}$

TABLE 2.7: (Continued)

Symbol	Description	Units (SI)
$\mathcal{D}$	Downdraft CAPE (Eq. (2.108))	$\mathrm{J\,kg^{-1}}$
$\mathcal{G}(\vartheta)$	Gross lapse rate operator	$\mathrm{Pa^{-1}}\;[\vartheta]$
$\mathcal{P}$	Mean production of TKE	$\mathrm{J\,kg^{-1}\,s^{-1}}$
$\mathcal{T}$	Mean flux of TKE	$\mathrm{J\,m\,kg^{-1}\,s^{-1}}$
F_h	Diabatic enthalpy flux	$\mathrm{J\,m\,s^{-1}\,kg^{-1}}$
τ	Stress tensor	$\mathrm{m^2\,s^{-2}}$
v	Velocity vector (u, v, w)	$\mathrm{m\,s^{-1}}$
$\mathcal{R}_e$	Reynolds number UH/v	–

(12) Assume that surface fluxes are well represented by the bulk aerodynamic formula,

$$\overline{w'\vartheta'} = C\|v_{10\mathrm{m}}\|\,(\vartheta_{10\mathrm{m}} - \vartheta_{\mathrm{sfc}}),$$

where ϑ is a generic scalar quantity, like temperature or moisture. Estimate the height of the LCL given the surface latent and sensible heat fluxes, assuming a drag coefficient C of 0.001. State any additional assumptions that you may have to make to arrive at a solution. Does the expression depend on wind speed or sea-surface temperature?

(13) Assuming that in the tropics the temperature profile follows the saturated isentrope corresponding to saturation over the warmest surface waters, if $\delta\eta_e$ measures the difference between the surface moist static energy and that in the free troposphere, derive an expression for $\delta\eta_e$ as a function of the deviation from the warmest sea-surface temperatures. Does this expression depend on the temperature of the warmest waters, or only on the temperature difference?

(14) Derive an expression for the rate at which q_1 increases with height for adiabatic ascent.

(15) Assuming an atmospheric thermal structure wherein $\eta_{e,s} = 335\,\mathrm{kJ\,kg^{-1}}$ assuming $c_{P_e} = c_{P_d}$, at what altitude is q_s less than q_{CO_2} assuming a pre-industrial concentration of carbon dioxide of 280 ppm.

(16) Derive an energy principle for the Boussinesq system.

(17) For a more general fluid the deviatoric forces can be expressed more generally in terms of a stress tensor τ so that the viscosity term in Eq. (2.121) is instead written in terms of the divergence of this stress tensor, cf Eq. (2.150). Using the definition of the incompressible stress tensor, demonstrate that the dissipation term (Eq. 2.160), is non-negative. For these manipulations it proves useful to use tensor notation, such that

$$\tau = \mu\left[\frac{\partial u_i}{\partial x_j} + \frac{\partial u_j}{\partial x_i}\right].$$

(18) In Eq. (2.145), show that the term involving the gradient of the geopotential can be related to the change in the kinetic energy of the flow. How can one interpret the energetic effect of the time derivative of the geopotential?

(19) By adopting typical pressure gradients in the tropics, or the extra tropics, calculate to what extent it is reasonable to neglect geopotential gradients along pressure surfaces when calculating the evolution of η_e in Eq. (2.145).

(20) Using the reanalysis, calculate the empirical orthogonal functions of the monthly mean vertical velocity field in the tropical atmosphere over the ocean. How much variance does the leading mode ω_1 explain? What is the wind profile associated with this vertical mode? How do the EOFs vary if one compares EOF calculated over climatological subsidence versus convective regions?

3 Clouds as Particles

Hanna Pawlowska and Ben Shipway

Chapter 2 demonstrated the importance of dynamics and thermodynamics in the development of clouds. While these processes are fundamental to cloud formation and evolution, they are only the beginning of the story. It is the cloud microstructure and its interaction with the radiation that influences the shapes and appearance of clouds; for example, the distinct characteristic cauliflower edges of a cumulus congestus cloud are associated with liquid droplets that quickly evaporate when mixed with the subsaturated environment, while the blurred streaks of a cirrus cloud are a consequence of longer-lived ice particles.

Further than just an aesthetic influence on clouds, the details of the microphysical properties of the cloud affect its future development and its interaction with the environment. Interactions with radiation can lead to variations in absorption, emission and scattering of heat and light and, as a result, have far reaching consequences for global energy budgets. Thermal interactions through redistribution of heat and moisture can modify both the local behaviour of clouds and have cumulative effects on large-scale atmospheric circulations. Chemical interactions result in removal, generation and transformation of aerosols and gases. Interaction with the electrosphere leads to the redistribution of electric charge and the dramatic discharge in the form of lightning strikes. Each interaction is highly dependent on the shape, size, number and material phase of the collective particles within the cloud.

This chapter considers some of the fundamentals of the microphysical processes that determine the properties of cloud particles. Section 3.1 summarises some of the language that is used to describe and characterise the myriad collections and combinations of particles that are to be found across all types of clouds. Section 3.2 takes a specific look at warm clouds which make up the vast majority of precipitating clouds to be found in the Earth's atmosphere and in Section 3.3 the effect that the ice phase has on cloud microphysics is discussed. The symbols used are summarised at the end of this chapter (Table 3.6).

3.1 Description of Cloud Microphysics

3.1.1 Particle Types and Distribution

Water is the only common compound that can be found naturally occurring in its vapour, liquid and solid states. Cloud elements exemplify this property, and in their various forms can be composed of liquid alone (warm, or liquid phase), solid (ice phase) or a mixture of the two (mixed phase – a terminology that takes for granted that water vapour is a prerequisite for all cloud formation). In addition to this, the dipolar nature of water molecules means that, as a solid, they form crystalline structures. Although water is known to exist in fifteen different stable crystalline arrangements, at the temperatures and pressures common to the formation of ice in clouds, only hexagonal ice exists in a stable form.[1] However, this hexagonal structure leads to ice which can grow into a variety of shapes, or habits (see Figure 3.1). Additional collisions between particles further extends the myriad of possibilities of particle shape and size that together make up a cloud.

The vast diversity of particle morphology makes it often useful to characterise individual particles by some simple metric, such as a maximum length scale (e.g., D) or a parameter describing the asymmetry of non-spherical particles (e.g., ϕ), while collections of particles can then be described using the statistics of these parameters.

---

[1] Although recent evidence suggests that metastable *cubic* and *stacking-disordered* ice may be influential in the early stages of ice formation.

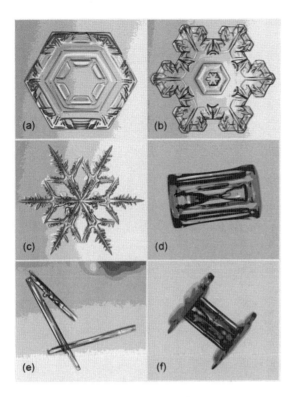

FIGURE 3.1: Various habits of ice particles: (a) a simple plate crystal, (b) a plate with complex symmetry, (c) a stellar dendrite, (d) a hexagonal columnar crystal, (e) needles, (f) a capped column. Printed with permission from snowcrystals.com

the number of particles per unit bin size, $n_i = N_i/\Delta r_i$, in which case n_i is expressed in units $cm^{-3}\,\mu m^{-1}$. For many purposes, the particle number density function is described by a continuous distribution expressed by an analytical function. We will use a notation $n(r)$, where $n(r)dr$ is the number of particles in the infinitesimal size interval $(r, r + dr)$.

If, as in most cases, the distribution cannot be described by an analytical function, plotting $n(r)$ versus r provides a useful way to show how particles are distributed with size.

Liquid particles are traditionally divided with respect to their sizes into three groups: the smallest particles with sizes in the range up to about $30\,\mu m$ (radius) are termed *cloud droplets* or simply *droplets*. Larger particles, up to about $300\,\mu m$ are called *drizzle drops*. All particles that are larger than that are termed *raindrops*. This division is based not only on sizes but also reflects the processes that are important in their formation and growth. Cloud droplets grow mostly by condensation of water vapour. This mechanism is inefficient for the generation of larger particles that are brought into being by collision and coalescence processes. Often drizzle drops will evaporate in the subsaturated sub-cloud layer before reaching the ground.

While liquid particles are predominantly spherical (large rain drops will tend to become more oblate as they fall), ice particles come in a wide variety of shapes or *habits* (see Figs. 3.1 and 3.2). While the majority of ice crystals observed have irregular shapes, often the growth of regular crystalline structures is observed. At temperatures warmer than about $-22\,°C$, the characterisation of habit variation is reasonably well constrained by temperature and supersaturation. The picture is less clear at temperatures colder than $-22\,°C$, although at these temperatures irregular crystals and polycrystals tend to dominate.

The regular crystal forms can be broadly characterised into primary habits, plates or columns, and secondary habits, needles, dendrites or rosettes. The primary structures that form are generally a function of temperature, while in air, which is supersaturated with respect to water, secondary features will form. Roughly speaking, plate-like crystals will form in the ranges $-1\,°C$ to $-3\,°C$ and $-10\,°C$ to $-22\,°C$, while columnar structures will grow at the intermediate temperatures $-3\,°C$ to $-10\,°C$. If the air near the crystal is supersaturated with respect to water, the columns will grow to form long needles at temperatures approaching $-6\,°C$, sectored plates form between $-9\,°C$ to $-12\,°C$ and $-16\,°C$ to $-20\,°C$ and dendritic forms can grow at temperatures near $-15\,°C$. Following a

For the purposes of measurement, theory and modelling, it is convenient to consider the particle size distribution (PSD), sometimes referred to as the particle spectrum. The PSD summarises the number of particles of a given characteristic size within a given volume of cloud. When considering spherical droplets, the PSD is usually described by a discrete series of pairs of numbers (N, r) where N is a number of particles having radius r in a unit volume. N is usually expressed in cm^{-3} and r in μm. Diameter can also be used to characterise particle size. To retain consistency with other textbooks, we use predominantly radius to describe liquid particles, while later in the chapter (and in other chapters in this book), diameter will be more convenient. For non-spherical particles, the dependent variable may be a characteristic length such as the maximum diameter.

In this discrete representation, size classes of a given width (Δr_i) are defined and N_i represents a number of particles that have a radius in a given interval (or 'size-bin'), occupying the range (r_i, $r_i + \Delta r_i$), where i is the bin number. Often it is more practical to use a continuous function, the particle number density (n_i), that is,

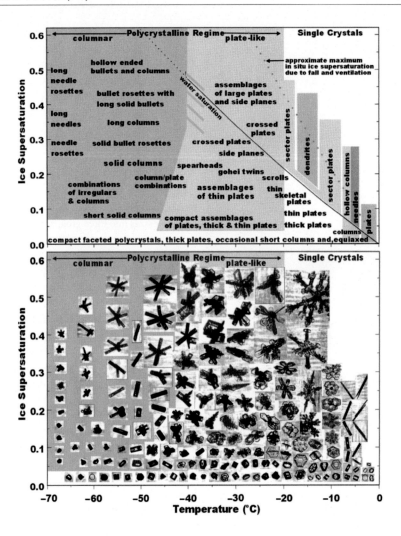

FIGURE 3.2: Habit diagram of Bailey and Hallett (2009) drawn from laboratory observations. The different regimes vary significantly with temperature, with supersaturation playing a further role in developing additional features.

comprehensive and careful examination of laboratory and *in situ* observations, Bailey and Hallett (2009) produced a diagram describing the different growth regimes which is reproduced here in Fig. 3.2.

Although ice crystals and snowflakes can be well categorised into these dominant growth regimes, it should be re-emphasised that the symmetry of crystal growth is easily disturbed through collisions that lead to the aggregation and break up of particles and subsequently more irregular growth patterns.

3.1.2   Cloud Microphysical Parameters

The PSD encapsulates a complete description of the number of each size of particle in a given spectrum.

However, in order to characterise the microphysical properties of a cloud volume, it is often only necessary to have information about the statistical *moments* of the density function $N(r) = n(r)/N_{\text{total}}$, where N_{total} denotes the total number, that is,

$$N_{\text{total}} = \sum_i N_i = \int_0^\infty n(r)\, \mathrm{d}r. \tag{3.1}$$

The moments are defined by the weighted integrals over the full spectrum as follows:

$$\langle r^j \rangle = \frac{1}{N_{\text{total}}} \sum_i N_i r_i^j$$

$$= \int_0^\infty r^j N(r)\, \mathrm{d}r, \quad \text{where} \quad j = 1, 2, \dots \tag{3.2}$$

TABLE 3.1: Moments of PSD

Moment order	Parameter	Application
1	Mean radius $\bar{r} = \langle r \rangle$	–
2	Mean surface radius $r_{\mathrm{srf}} = \langle r^2 \rangle^{1/2}$	Short-wave extinction $\beta_{\mathrm{ext}} = Q_{\mathrm{ext}} \pi N r_{\mathrm{srf}}^2$
3	Mean volume radius $r_{\mathrm{vol}} = \langle r^3 \rangle^{1/3}$	Liquid water content $\mathrm{LWC} = \frac{4}{3} \pi \rho_{\mathrm{l}} N r_{\mathrm{vol}}^3$ specific mass $q_{\mathrm{l}} = \frac{\mathrm{LWC}}{\rho} = \frac{4}{3} \pi \frac{\rho_{\mathrm{l}}}{\rho} N r_{\mathrm{vol}}^3$
6	No name	Radar reflectivity

TABLE 3.2: Typical sizes and concentrations of cloud particles

Particle	Radius range	Concentration (m^{-3})
Cloud droplets	1–30 μm	10^6–10^9
Drizzle drops	30–500 μm	10^5–10^6
Raindrops	1 mm	10^2–10^3
Snow crystals	0.01–5 mm	10^3–10^8
Snow aggregates	1–20 mm	10^2–10^4
Graupel	1–5 mm	10^1–10^3
Hailstones	5–20 mm	10^1–10^3

We will see how different statistical moments of the PSD are used to describe cloud microphysical properties.

Usually, the statistical moments are calculated separately for each identified microphysical species (i.e., droplets, rain, hail, etc.) In Table 3.1 we list the most commonly used moments of a PSD.

The mean surface radius, mean volume radius and 6th moments of the PSD are widely used because they refer directly to measurable microphysical parameters. Extinction (i.e., absorption and scattering of light) of a given volume of cloud depends on a total cross-section of particles present in that volume; see Section 4.1.1.1. It can be expressed by a product of the number concentration and the mean surface radius ($\beta_{\mathrm{ext}} \propto N r_{\mathrm{srf}}^2$). Liquid water content refers to the total mass of water contained within all the cloud particles present in a given volume of air; it depends therefore on cloud droplet number concentration and mean volume radius. Last but not least, the 6th moment is used to describe reflectivity as measured by radar. It is worth noting that very often the cloud microphysical properties are referred to for a given mass of moist air instead of a volume of air; the total mass of water contained within all cloud particles present in a given mass of moist air is referred to as the specific mass of liquid water content, $q_{\mathrm{l}} = \mathrm{LWC}/\rho$, where ρ is the density of the (moist) air.

Measurements of these microphysical parameters can then provide indirect information about the PSD. Typical values of particle sizes and concentrations are given in Table 3.2.

The radiative interactions with clouds are determined from these moments in different ways for long-wave and the short-wave radiation (see Chapter 4 for the discussion of extinction efficiency). Interaction with radiation is

characterised by the cloud optical thickness (τ_{c}), which represents the total extinction of a cloud column (see Eq. (4.5)). The extinction efficiency Q_{ext} depends on the size parameter $x = 2\pi r / \lambda$, where r is the particle size and λ is the wavelength (λ), however, it varies little for visible light and typical particle sizes so can be reasonably approximated as a constant, $Q_{\mathrm{ext}} \approx 2$ (see discussion in Section 4.1.1.1). To determine the cloud optical depth, it is useful to introduce a new parameter, r_{eff}, the effective radius. The effective radius is defined as

$$r_{\mathrm{eff}} = \frac{\langle r^3 \rangle}{\langle r^2 \rangle} \tag{3.3}$$

and is in fact proportional to the ratio of the liquid water content to the extinction ($r_{\mathrm{eff}} \propto \frac{q_{\mathrm{l}}}{\sigma_{\mathrm{ext}}}$; see Table 3.1). Using r_{eff} one can write the expression for the cloud optical thickness as follows:

$$\tau_{\mathrm{c}} = \frac{3}{4} \frac{Q_{\mathrm{ext}}}{\rho_{\mathrm{l}}} \int_{z_{\mathrm{cb}}}^{z_{\mathrm{ct}}} \rho \frac{q_{\mathrm{l}}}{r_{\mathrm{eff}}} \, \mathrm{d}z, \tag{3.4}$$

where z_{cb} and z_{ct} are cloud-base and cloud-top height, respectively. The effective radius is not directly measurable and must be deduced from extinction measurements or else calculated from the PSD. Without such measurements, the expression for cloud optical thickness is cumbersome. A simple and commonly used approach to overcome that problem is to assume that the effective radius takes a constant value that differs between continental and maritime clouds. These values correspond to the value of a mean radius or mean volume radius close to the top of the cloud. Cloud radiative properties are very sensitive to the microphysics in the upper part of a cloud. Defining

$$W = \int_{z_{\mathrm{cb}}}^{z_{\mathrm{ct}}} \rho q_{\mathrm{l}} \, \mathrm{d}z \tag{3.5}$$

as the total liquid water content within the column (a parameter often called the liquid water path, LWP), and assuming $Q_{\text{ext}} = 2$ and r_{eff} is a constant, Eq. (3.4) takes a simpler form:

$$\tau_c = \frac{3}{2\rho_1} \frac{W}{r_{\text{eff}}}. \tag{3.6}$$

In the long-wave regime, the extinction efficiency cannot be approximated by a constant value; it is rather proportional to the size parameter. It can be shown that a cloud's optical depth depends only on the LWP.

The mean radius, mean surface radius and mean volume radius are equal only for a mono-disperse distribution of particles (i.e., all particles having the same size). In poly-disperse spectra (i.e., distributions with particles of different sizes), the spectral width results in a bias between each of these parameters. This bias is often accounted for by introducing a correction factor k with $k = (r_{\text{vol}}/r_{\text{eff}})^3 = (r_{\text{srf}}/r_{\text{vol}})^6$ that varies from 0.67 ± 0.07 in continental air masses to 0.80 ± 0.07 in maritime air. Introducing this coefficient k allows us to rewrite Eq. (3.4) for the optical thickness of the cloud in the form:

$$\tau_c = Q_{\text{ext}} \left(\frac{3\sqrt{\pi}}{4\rho_1} \right)^{2/3} \int_{z_{\text{cb}}}^{z_{\text{ct}}} (kN)^{1/3} (\rho q_1)^{2/3} \, dz. \tag{3.7}$$

3.1.3   Observation and Measurements

Clouds are a common phenomenon in nature. Observing them is a vital part in many applications, especially weather forecasting. Responding to motions in the atmosphere, they become a visible manifestation of moist physical processes. Although microphysical processes determine the structure of a cloud, obtaining a satisfactory picture of cloud microphysics is arguably one of the most difficult tasks. There are multiple reasons.

Direct observation can only be achieved by *in situ* measurements. It can be done by airborne research – onboard instrumented aircraft or instrumented platforms introduced into a cloud (e.g., suspended below a helicopter or lofted on a balloon or kite), or from an observational tower high enough to reach into clouds. Each of these techniques has its advantages and disadvantages. Measurements performed by instruments installed on board an aircraft give a picture of cloud properties along an aircraft trajectory. It is possible to sample different parts of a cloud, characterise the extended cloud field and observe temporal evolution. Aircraft also allow *in situ* observations to be made in very remote regions of the troposphere.

The amount of data collected during a research flight is huge, however getting a good picture of the cloud deduced from these measurements can be challenging.

Microphysics, being a small-scale phenomenon, requires characterisation at length scales in the order of centimetres. Timescales required are also short, and it is a challenge for instruments used for airborne research to perform measurements fast enough to give representative answers. Observations from instrumented platforms or towers are perhaps easier because these instruments don't have to operate at such a high frequency. However, with static instruments one can only get insight into a limited part of a cloud: it is a Eulerian observation.

Indirect cloud observations directed at retrieving cloud microphysical properties include lidar, radar and microwave measurements performed either from the ground, from satellites or onboard instrumented aircraft. They measure different radiative properties and need supplementary a priori assumptions to retrieve their microphysical properties. Their big advantage over airborne measurements is that they are able to provide a very long series of observations.

3.2   Warm Clouds

Clouds are a special type of aerosol that come into being when air becomes supersaturated with respect to water. Cloud droplets grow by diffusion and condensation of water vapour onto the droplet surface. Small droplets move with the air flow in the cloud, but once droplets become large enough to move with respect to the air, they start to collide with one another and sometimes merge (coalesce). Large drops fall below the cloud base and evaporate in the subsaturated environment below the cloud base. The chain of processes that lead to the formation of droplets, drizzle and rain will be described in the next sections.

3.2.1   Gibbs Free Energy in a Transient System

Vapour pressure is the key to determine the creation and subsequent growth of cloud droplets. In the previous chapter, the Gibbs free energy of a system was introduced and used to derive the Clausius–Clapeyron equation which describes how vapour pressure varies with temperature. This equation, however, is only applicable to the preferred equilibrium state of a system. In a transient system where mass is transferred from one phase to another, for example, vapour to liquid or liquid to ice, work must be done to form clusters of molecules. This additional work creates an energy barrier which must be overcome before the adjustment to the new equilibrium can take place.

To examine such a change we must add a term to the change in Gibbs energy such that

$$\mathrm{d}g = v\mathrm{d}p - s\mathrm{d}T + \mu_j\mathrm{d}n_j. \tag{3.8}$$

In this latter term, μ_j is used to denote the chemical potential, which represents the change in g with the change in the quantity, n_j, of a particular phase of the system, given a constant temperature and pressure. Here j might denote vapour, liquid or ice phases (v, l, i, respectively). Considering a change in the system under the condition that $\mathrm{d}p = 0$ and $\mathrm{d}T = 0$, that is, isobaric and isothermal, the change in dg due to a number n of molecules bonding onto a single cluster of molecules could be written as

$$\Delta g = (\mu_l - \mu_v)n + F, \tag{3.9}$$

where the chemical potential per molecule in the vapour and condensed phases are given to be μ_v and μ_l, respectively, and F represents the surface free energy of the condensed cluster. The term F will be proportional to the exposed surface area of the cluster, and so for the early stages of particle growth can be considered to be roughly isotropic or spherical. Thus,

$$\Delta g = (\mu_l - \mu_v)n + \lambda n^{\frac{2}{3}}, \tag{3.10}$$

with some constant of proportionality,[2] λ. Given Eq. (3.10), and noting that at saturation $\mu_l = \mu_v$ and for supersaturated conditions, $\mu_l > \mu_v$, we can plot the curve of Δg as a function of the number of molecules forming an initial cluster (Fig. 3.3). It is clear from this figure that for saturated conditions, Δg continues to increase as the cluster size increases, thus under these conditions any new clusters that form will be unstable and rapidly break apart. For supersaturated conditions, the curve for Δg exhibits a maximum for clustering of a critical size, beyond this a decrease of free energy occurs as the cluster gains mass and so can grow uninhibited. We note that the higher the supersaturation, the lower the critical cluster size and the easier it is for droplets to form.

In supersaturated conditions typically seen in the atmosphere, the critical cluster size is usually much larger than the size of clusters which form through random fluctuations in the vapour phase. For homogeneous conditions many hundreds of per cent supersaturation are required to form liquid water droplets spontaneously. Thus, liquid water droplets in clouds require some catalyst to overcome the apparent energy barrier.

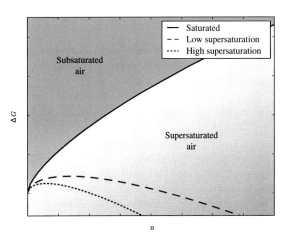

FIGURE 3.3: Gibbs free energy change from the clustering of n molecules. Clusters larger than a critical size (maximum of each curve) are able to grow unimpeded. (Adapted from *The Physics of Clouds* by Mason (1971))

3.2.2 Activation

The catalysts which facilitate droplet formation in the atmosphere are primarily submicron particles called cloud condensation nuclei (CCN). The process that leads to the creation of cloud droplets is called activation. Activation is analogous to the nucleation of water droplets in that the free-energy barrier must be overcome. However, the liquid phase already exists (via deliquescence), so activation is not a phase-nucleation phenomenon, rather it represents a change from stable to unstable growth in response to increasing ambient humidity. The concept of activation is crucial to our understanding of how aerosol particles act as CCN and establish the initial microstructure of clouds. Surface tension and the presence of soluble material in the droplet, respectively increase and decrease the saturation vapour pressure over the droplet. The saturation vapour pressure in equilibrium with pure water, and in the absence of surface tension associated with the liquid, is denoted by e_l. It is a function of the temperature T only, and described by the Clausius–Clapeyron equation – Eq. (10.1). More generally, the equilibrium saturation vapour pressure over an aqueous solution droplet, with radius r, is denoted by e. The ratio of the equilibrium vapour pressure over the solution droplet to that of pure liquid water, $S_e = e/e_l$, is described by the Köhler equation:

$$S_e = a_w \exp\left(\frac{2\sigma_l}{r\rho_l R_v T}\right). \tag{3.11}$$

Here σ_l is the surface tension coefficient for liquid water, R_v is the gas constant for water vapour, a_w is the water activity that depends on a mole fraction from the solute

---

[2] For the assumption that the cluster is spherical with radius r, $F = 4\pi r^2 \sigma$ with σ the specific surface energy; for meteorological conditions, $\sigma = 0.072\,\mathrm{J\,m^{-2}}$.

(aerosol) particle. The modification of the saturation vapour pressure by the water activity depends on the ratio of the number of moles of dissociated solute compared to the number of moles of pure liquid water, such that

$$a_{\mathrm{w}} = \exp\left(-\nu_{\mathrm{a}}\frac{n_{\mathrm{a}}}{n_{\mathrm{l}}}\Phi_{\mathrm{a}}\right), \qquad (3.12)$$

where n_{a} and n_{l} are the number of moles of solute (aerosol, subscript a) and water (liquid, subscript l), respectively, ν_a is the dissociativity of the solute aerosol (in the case of NaCl, $\nu = 2$). The factor Φ_{a} is the 'practical osmotic coefficient' – a fitting factor, usually less than one – that reflects the departure of the ideal description known as Raoult's law. The ratio $n_{\mathrm{a}}/n_{\mathrm{l}}$ is usually expressed either by the mass of the solute matter (m_{a}) or by an equivalent dry radius r_{a} so that $m_{\mathrm{a}} = (4/3)\pi r_{\mathrm{a}}^3 \rho_{\mathrm{a}}$. The mass of the water is given by $m_{\mathrm{l}} = (4/3)\pi(r^3 - r_{\mathrm{a}}^3)\rho_{\mathrm{l}}$. Assuming that $n_{\mathrm{a}} \ll n_{\mathrm{l}}$ (which is the usual case) and therefore $r_{\mathrm{a}} \ll r$, we see that

$$\frac{n_{\mathrm{a}}}{n_{\mathrm{l}}} = \frac{r_{\mathrm{a}}^3 \rho_{\mathrm{a}} M_{\mathrm{l}}}{r^3 \rho_{\mathrm{l}} M_{\mathrm{a}}}, \qquad (3.13)$$

where M_{a} and M_{l} are molecular mass of solute and water, respectively.

Combining Eqs (3.11)–(3.13), the Köhler equation can further be expressed in a form:

$$S_e = \exp\left(\frac{A(T)}{r} - \frac{B_{\mathrm{a}}}{r^3}\right), \qquad (3.14)$$

where $A = 2\sigma_{\mathrm{l}}/(\rho_{\mathrm{l}}R_{\mathrm{v}}T)$ is a weak function of temperature and the solute effect of the condensation nucleus is expressed by B_{a}, which depends on the size of the aerosol as measured by its dry radius r_{a}, and its chemical composition, such that $B_{\mathrm{a}} = r_{\mathrm{a}}^3 \nu(\rho_{\mathrm{a}}/\rho_{\mathrm{l}})(M_{\mathrm{l}}/M_{\mathrm{a}})\Phi_{\mathrm{a}}$. The first term in Eq. (3.14) is called the curvature (or Kelvin) term since it describes the surface tension effect over a curved surface; the second term is called the solute term and it describes the effect of the decrease of saturation equilibrium due to the presence of soluble matter.

Assuming that both terms of the exponent are much smaller than one, the Köhler equation can be presented in a simpler form:

$$S_e(r, B_{\mathrm{a}}, T) \approx 1 + \frac{A(T)}{r} - \frac{B_{\mathrm{a}}}{r^3}, \qquad (3.15)$$

where $S_e - 1$ denotes the *equilibrium supersaturation* over the solution droplet. The Köhler equation is a basis for understanding the activation of cloud droplets which grow by diffusion in a supersaturated environment (Fig. 3.4). For smaller droplets, the solute term proportional to $1/r^3$ in Eq. (3.15) dominates (dotted line in Fig. 3.4),

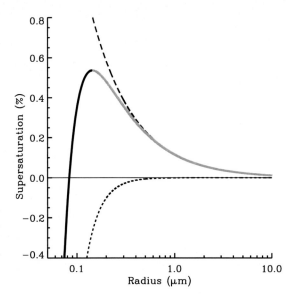

FIGURE 3.4: Köhler curve. Equilibrium supersaturation for a solution droplet formed on a sodium chloride (NaCl) condensation nucleus of dry radius 0.02 μm. Dashed line represents the curvature (Kelvin) term and dotted line represents solute effect of the Köhler equation. The black, solid part of the line indicates supersaturation for the stable equilibrium, the grey, solid part of the line corresponds to the unstable equilibrium.

while for larger drops the curvature term proportional to $1/r$ becomes more important (dashed line in Fig. 3.4). The resultant effect is shown by the solid black and grey curve. It is straightforward to show from Eq. (3.14) that for r going to zero, the supersaturation becomes zero which implies that in very dry atmospheres, small solution drops can exist in equilibrium. For very small solution droplets, the equilibrium supersaturation is below zero.

The solid black line in Fig. 3.4 corresponds to the stable equilibrium for solute droplets. If in given environmental conditions a droplet that is initially in equilibrium with the environment becomes a little bigger, then the actual environmental saturation becomes smaller than equilibrium and the droplet evaporates until the equilibrium is reached, and vice versa. If the relative humidity is increased by a small amount, the droplet will grow until it reaches equilibrium. This process of increasing the relative humidity and allowing the droplet to grow to its equilibrium size will continue until the environment becomes saturated and even beyond by few tenths of a per cent until the critical supersaturation is reached that corresponds to the peak in the Köhler curve – in the example given in Fig. 3.4, a supersaturation of 0.53 % corresponds to a critical radius of 0.14 μm.

The critical values of radius and saturation are given by:

$$r_{\text{crit}} = \sqrt{3B_a/A} \qquad (3.16a)$$

$$S_{\text{crit}} = 1 + \sqrt{4A^3/27B_a}. \qquad (3.16b)$$

Up to this point droplets will grow only if the relative humidity is increased. If the relative humidity slightly increases its critical value, enabling the droplet to grow beyond the critical size, the equilibrium saturation decreases and the equilibrium is unstable (grey part of the solid curve in Fig. 3.4). Any change in the environmental saturation causes the droplet to grow or evaporate but to deviate from the initial conditions.

When smaller than r_{crit} and in stable equilibrium a droplet is called a haze particle. Haze particles reside interstitially between the cloud drops. Droplets that grow beyond the critical conditions are called activated. Smaller CCN need higher supersaturation to be activated. Also, CCN having bigger molecular mass need higher supersaturation. Typical CCN have dry sizes of tenths of a micron and activate at supersaturations on the order of 0.1%. Describing a whole diversity of atmospheric aerosols that can serve as CCN by attributing them their sizes, chemical composition, dissociativity and 'practical osmotic coefficient' can be unnecessarily complicated. Therefore, the water activity a_w (Eq. (3.12)) is often expressed by only one parameter κ and effective dry radius of the solute aerosol, r_a (or condensation nuclei) in a form:

$$a_w = \frac{r^3 - r_a^3}{r^3 - r_a^3(1 - \kappa)}. \qquad (3.17)$$

This is called the κ-Köhler approach.[3] Atmospheric particle matter is typically characterised by $\kappa > 0.2$, with lower values sometimes observed for particular locations and periods. Lower values of κ indicate less hygroscopic, or less CCN-active behaviour. $\kappa = 0$ is for pure water. The number and sizes of activated droplets depend generally on two factors: (i) the physicochemical properties of the solute (number, size, chemistry) and (ii) the dynamical generation of supersaturation. The physiochemical properties of the solute depend on the type of aerosol (natural or anthropogenic), whereas supersaturation appears as a result of adiabatic cooling when the moist air is lifted upward or is due to the radiative cooling of extended layers of the atmosphere. For a given thermodynamic condition, stronger vertical motion results in higher supersaturation, allowing smaller droplets to be activated. In convective clouds, activation processes occur in a layer up

to 20–30 m above the cloud base. Activated droplets grow quickly by vapour diffusion, reducing supersaturation and inhibiting activation of any new CCN. However, anywhere in a cloud where the supersaturation again increases, new CCN can be activated. This will be discussed in Section 3.2.6.2.

A useful way to describe the cloud-forming propensity of an aerosol population is by its *activity spectrum* which represents the number of particles per unit volume (concentration) that are activated to become cloud droplets, expressed as a function of supersaturation. The activated nuclei (CCN) are the subset of the total aerosol population upon which water condenses in a warm cloud. The aerosol size is often described by a lognormal distribution or a combination of two or three different lognormal distributions

$$n(r_a) = \frac{N_a}{\sqrt{2\pi} r_a \ln \sigma_g} \exp\left(-\frac{\ln^2(r_a/r_{a,g})}{2\ln^2 \sigma_g}\right), \qquad (3.18)$$

where $n(r_a)$ is the number density of aerosol, N_a is total dry aerosol concentration, σ_g is the geometrical standard deviation and $r_{a,g}$ the median radius of the distribution. For a given supersaturation, only aerosol particles with radii bigger than the critical radius for that supersaturation (see Eqs (3.16a) and (3.16b)) are activated. Therefore, the concentration of activated particles ($N_{\text{CCN}}(S)$) for a given supersaturation is given by an integral of Eq. (3.18):

$$
\begin{aligned}
N_{\text{CCN}}(S) &= \int_{r_{\text{crit}}}^{\infty} n(r_a)\, dr_a \\
&= \frac{N_a}{2}\left[1 + \text{erf}\left(\frac{\ln(r_{a,g}/r_{\text{crit}})}{\sqrt{2}\ln \sigma_g}\right)\right],
\end{aligned} \qquad (3.19)
$$

where $\text{erf}(x)$ denotes the Gauss error function of x. Examples of activation spectra calculated using Eq. (3.19) are shown in Fig. 3.5.

3.2.3 Condensational Growth

Activated droplets grow by vapour diffusion. For a single drop in a uniform field of water vapour this process is described by the following equation:

$$\frac{dm}{dt} = 4\pi r D_v \left(\rho_v - \rho_{v,e}\right), \qquad (3.20)$$

where m is a mass of a droplet, D_v is diffusivity of water in air, ρ_v is the density of water vapour in the far droplet environment and $\rho_{v,e}$ is the density of water vapour in equilibrium at the droplet surface. Its value can be inferred using the ideal gas law and the Köhler equation (Eq. (3.15)). The mass of a drop of radius r is $m = 4/3\pi\rho_l r^3$

---

[3] For details, see Petters and Kreidenweis (2007).

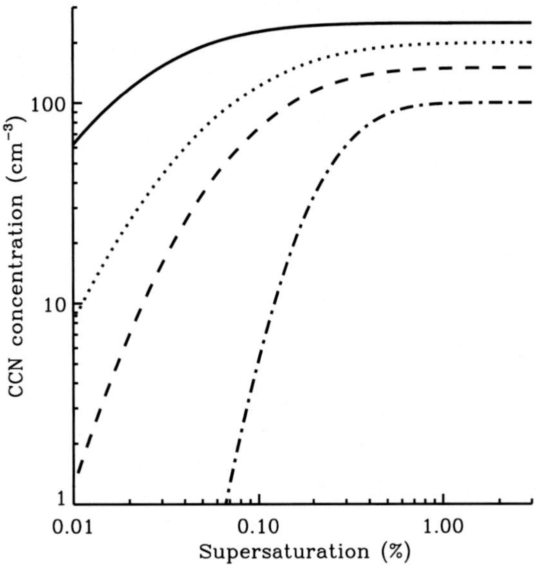

FIGURE 3.5: Activation spectra calculated using Eq. (3.19). They are calculated using $\kappa = 0.6$. The values of the median radii r_{s0} are 0.18 μm, 0.08 μm, 0.065 μm and 0.035 μm, shown by solid, dotted, dashed and dotted-dashed lines, respectively. The respective dispersion σ_g are 2.15, 2.15, 1.9 and 1.47. For convenience of comparison, CCN(S) is calculated with the dry aerosol concentration, N_a of 250, 200, 150 and 100 cm^{-3} from the upper to the lower curves.

from which it follows that $dm = 4\pi\rho_1 r^2\, dr$, therefore, Eq. (3.20) reads:

$$\begin{aligned}
\frac{dr}{dt} &= \frac{1}{r}\frac{D_v e_1}{R_v T \rho_1}\frac{(p_v - e)}{e_1} \\
&\approx \frac{1}{r}\frac{D_v e_1}{R_v T \rho_1}\left(S - 1 - \frac{A}{r} + \frac{B_a}{r^3}\right),
\end{aligned} \tag{3.21}$$

where $S = p_v/e_1$ is the saturation vapour pressure ratio relative to an equilibrium over pure water, and the approximate form comes from the application of Eq. (3.15) to e/e_1. If we note that for large enough drops the curvature and solute corrections to the saturation vapour pressure are negligible and define $F_D(T) = \frac{R_v T \rho_1}{D_v e_1}$, then the equation takes a very simple form:

$$\frac{dr}{dt} = \frac{1}{r}\frac{S-1}{F_D}. \tag{3.22}$$

The growth of a drop radius is inversely proportional to its radius, indicating that while large drops accumulate more mass than small drops, their radius changes slower, and small drops grow faster.

In deriving Eq. (3.21) we assumed that the temperature was constant everywhere. Condensation of vapour onto the droplet surface leads to an increase of the temperature due to the condensation enthalpy. The diffusion of heat

away from the droplet is described by an equation similar to Eq. (3.20):

$$\frac{dQ}{dt} = 4\pi r K (T_r - T), \tag{3.23}$$

where T is ambient temperature, T_r is the temperature at the drop's surface and K is the coefficient of thermal conductivity of air. For the steady-state growth process, the heat transport must be balanced by the heat release due to the condensation at the drop surface. Eqs (3.20) and (3.23) were first derived by James Clerk Maxwell in 1875; and so the theory of the steady-state growth of spherical water drops is called Maxwell theory. Eqs (3.20) and (3.23) have to be solved numerically. However, there exists a useful analytical approximation for calculating the rate of condensational growth of a drop. It results in the following equation:

$$r\frac{dr}{dt} = \frac{(S - 1) - \frac{A}{r} + \frac{B_a}{r^3}}{[F_D + F_k]} \tag{3.24}$$

with

$$F_D(T) = \frac{R_v T \rho_1}{D_v e_1} \quad\text{and}\quad F_k(T) = \frac{\ell_v \rho_1}{kT}\left(\frac{\ell_v}{R_v T} - 1\right). \tag{3.25}$$

The term $F_D(T)$ is associated with vapour diffusion; $F_k(T)$ is the term associated with heat conduction and ℓ_v the specific enthalpy of vaporisation (Table 2.3). The diffusion factor is of approximately the same magnitude as the conductivity factor and is $O(1 \times 10^{10}\,\text{s}\,\text{m}^{-2})$.

When drops are large they fall with an appreciable velocity (see Section 3.2.4). The assumption that the vapour field surrounding each drop is spherically symmetrical becomes invalid. The water vapour field around the drop is distorted and the heat and mass transfer increase, analogous to the way the wind enhances evaporation from the land/sea surface. These effects are incorporated into the theory of diffusional growth by multiplying the diffusion and conductivity coefficient by an appropriate ventilation coefficient (f_v) that is described empirically as a function of Reynolds number $(\mathcal{R}_e = 2r u_\infty(r)/\nu$, where u_∞ describes the terminal drop velocity and ν is kinematic viscosity). The ventilation factors need not be the same for vapour (diffusivity) and heat (conductivity) transfer, however, they are often assumed to be so as a convenient approximation so that $D_v'/D_v = K'/K = f_v > 1$:

$$f_v = \begin{cases} 1.00 + 0.09\,\mathcal{R}_e & 0 \le \mathcal{R}_e \le 2.5 \\ 0.78 + 0.28\,\mathcal{R}_e^{1/2} & \mathcal{R}_e \ge 2.5. \end{cases} \tag{3.26}$$

D_v' and K' are effective diffusivity and effective conductivity of air that replace D_v and K in Eqs (3.20), (3.23) and (3.25). Evaluating f_v for typical conditions shows that the ventilation factor is negligible for droplets smaller than 10 μm in radius, $f_v \approx 1.06$ for $r = 20$ μm and increases to $f_v = 1.25$ for $r = 40$ μm. The growth by condensation of drops this large is usually negligible compared with the mass transfer through collision coalescence (see Fig. 3.7). Hence, the ventilation effect is negligible for condensational growth, but becomes very significant in the evaporation of raindrops.

For droplets which are large enough to neglect the curvature and solute terms in Eq. (3.24) (radii bigger than 5 μm; see Fig. 3.4) the growth rate is primarily determined by the degree of supersaturation $(S − 1)$ and the droplet size. For a given supersaturation the rate of growth of a droplet surface is constant. In fact, Eq. (3.20) can be written in a form: $dr^2/dt \propto (S − 1)$ and implies that the radius of smaller droplets increases faster than the radii of bigger drops. This leads to the narrowing of a droplet size distribution if diffusional growth is the only active process.

Observations performed *in situ* in convective clouds don't provide common evidence of the existence of very narrow spectra. Early measurements suffered from a lack of instrumental precision, but the most recent observations still see little evidence. A cloud parcel in higher levels in a cloud may be a mixture of parcels that although grew solely by water vapour diffusion might have different history, that is, their growth occurred in different supersaturation conditions, therefore droplet spectra might be larger than predicted by the adiabatic growth theory (this issue will be discussed in Section 3.2.6.2). Nevertheless, because of the amount of time it requires (given the low supersaturation within clouds) condensational growth alone is not able to account for the larger ($r \approx 20$ μm) drops thought necessary to initiate the collision-coalescence process. The precise mechanisms that lead to the creation of drops large enough to fall below the cloud base and form precipitation remain uncertain given the relatively short lifetimes of some clouds.

3.2.4 Terminal Fall Speeds of Drops and Droplets

To describe many cloud microphysical processes, an accurate knowledge of the terminal fall velocity of cloud droplets and raindrops is needed. This information is also necessary when attempting to infer particle sizes, and thus rain rates, from Doppler radar data. For large raindrops it is also important to know how their shape is modified by pressure forces on their surface as they move through the air.

TABLE 3.3: Analytic formulae for droplet fall velocity in three different Reynolds ($\mathcal{R}_e$) number regimes. Droplet radius is in cm and terminal velocity is in cm s^{-1}.

$r < 30$ μm	$u_\infty = 1.19 \times 10^6 r^2$
40 μm $< r < 600$ μm	$u_\infty = 8.00 \times 10^3 r$
600 μm $< r < 2$ mm	$u_\infty = 2.01 \times 10^3 r^{1/2}$

A complete basis for the calculation of terminal velocity can be obtained from the Navier–Stokes equations of motion for the air flowing past the drops and the motion of the water inside the drop. The terminal velocity of a particle is reached when the drag forces exerted on a drop by the surrounding air become sufficiently large to balance the gravitational force. With such a balance, the terminal velocity u_∞ can be expressed as

$$u_\infty = \frac{2}{9} \frac{r^2 g \rho_1}{(C_D \mathcal{R}_e/24) \rho v}, \tag{3.27}$$

where $\mathcal{R}_e$ is Reynolds number, C_D is the drag coefficient, ρ_1 is water density, ρ is air density and v is kinematic viscosity of air.

Generally speaking, since Reynolds number is itself a function of the fall speed of the particle, Eq. (3.27) represents an implicit equation for u_∞ and needs to be solved iteratively once information about C_D is known. However, at very low Reynolds numbers ($\mathcal{R}_e \ll 1$), in the Stokes regime, the drag is proportional to the speed (rather than square of speed) and so $C_D \mathcal{R}_e/24 = 1$. In this situation the terminal velocity can be derived exactly and it takes the form

$$u_\infty = \frac{2}{9} \frac{r^2 g \rho_1}{v \rho}. \tag{3.28}$$

Some effort has been made to extend the exact solution for the regime with Reynolds number close to unity. However, most descriptions are based on the fitting of functional forms to experimental laboratory data. Several functions can be found in the literature, all proposing different formulae for different regimes of Reynolds numbers that are conveniently translated into drop sizes. For small cloud droplets, typically smaller than 20 μm in diameter, the Stokes solution already presented provides a good approximation. For intermediate drop sizes from a few tens of microns to about a millimetre, the fall velocity is proportional to r. For very large drops, where the movement can induce an internal circulation within a drop, a much weaker dependence on the drop's radius is expected (proportional to $\sqrt{r}$). Analytical formulae for terminal fall speed are given in Table 3.3.

TABLE 3.4: Empirical terminal velocity parameters for the expression in Eq. (3.29). Here l is mean free path of air molecules, with a reference value of $l_0 = 6.62 \times 10^{-6}$ cm and η is dynamic viscosity with reference value of $\eta_0 = 1.818 \times 10^{-5}$ kg m^{-1} s^{-1}. Reference pressure and density are $p_0 = 1{,}013.25$ hPa, and $\rho_0 = 1.204$ kg m^{-3}. All quantities taken from table 2 of Beard (1977, 1978)

$$1\,\mu m \leq r \leq 20\,\mu m$$

$j_{\max} = 4$

$c_{1,4} = \{10.5035, 1.08750, -0.133245, -0.00659969\}$

$f = (\eta_0/\eta)[1 + 1.255\,l/r(\text{cm})]/[1 + 1.255\,l_0/r(\text{cm})]$

$l = l_0(\eta/\eta_0)(p_0\,\rho_0/p\,\rho)^{\frac{1}{2}}$

$$20\,\mu m \leq r \leq 3\,mm$$

$j_{\max} = 8$

$c_{1,4} = \{6.5639, -1.0391, -1.4001, -0.82736\}$

$c_{5,8} = \{-0.34277, -0.083072,$
$\qquad -0.010583, -0.00054208\}$

$f = 1.104\epsilon_s + [1.058\epsilon_c - 1.104\epsilon_s] \cdot$
$\qquad [6.2146 + \ln r(\text{cm})/5.01] + 1$

$\epsilon_s = (\eta_0/\eta) - 1$

$\epsilon_c = (\rho_0/\rho)^{\frac{1}{2}} - 1$

$\eta \approx 1.832 \cdot 10^{-5}\{1 + 0.00266[T(\text{K}) - 296]\}\,[\text{kg m}^{-1}\,\text{s}^{-1}]$

$\rho \approx 0.348\,p(\text{hPa})/T(\text{K})\,[\text{kg m}^{-3}]$

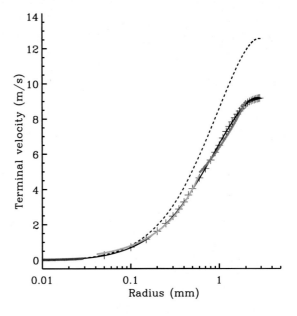

FIGURE 3.6: Terminal velocity. Thin black lines show velocities calculated using Eq. (3.29) with coefficient from Table 3.4 and with adjustment coefficients as formulated in Table 3.4. The black continuous line is for temperature 20 °C and pressure 1,000 hPa; the black dotted line (bigger velocities) is for −10 °C and 500 hPa. Thick grey lines show velocities calculated using analytic formulae given in Table 3.3. The plus signs overlying the black continuous line show terminal fall speed from observational data given by Gunn and Kinzer (1949)

To accommodate solutions to Eq. (3.27) for higher Reynolds numbers (i.e., larger droplets), series solutions can be developed to fit laboratory data for the drag coefficient. A good fit to the laboratory data is described by the following expression:

$$u_\infty(r,T,p) = f(T,p) \cdot \exp\left[\sum_{j=0}^{j_{\max}} c_j(\ln(2r))^j\right], \qquad (3.29)$$

where here u_∞ is the terminal drop velocity in cm s^{-1} and r is the drop radius in cm. The c_j as well as the correction factor f for ambient pressure and temperature depends on the size of the drop and are given in Table 3.4. Fig. 3.6 presents terminal velocity as given by Eq. (3.29) in atmospheric conditions $p = 1{,}000$ hPa and $T = 20$ °C, and aloft, that is, for $p = 500$ hPa and $T = -10$ °C. Velocities based on the analytic formulae as given in Table 3.3 are also shown.

As discussed in Chapter 6, for a population of particles, mixing-ratio-weighted terminal velocity, number-concentration-weighted terminal velocity and reflectivity-weighted terminal velocity can be defined by integrating the u_∞ over the distribution and weighting by the respective moment. This is required, for instance, to describe the effect of droplet sedimentation on a particular moment of a distribution.

3.2.5   Collision and Coalescence

In a population of droplets of differing sizes, the differences between the relative fall velocities can lead to collisions. Large droplets fall faster than small ones; potentially sweeping up those lying in their path. Due to the motion of the air around the larger drop, the fraction of droplets that will collide with a falling drop is reduced as small droplets are swept aside in the airstream around the drop. The ratio of the number of droplets that will collide to the total number of droplets that lie in the path of the falling drop is called the *collision efficiency* and depends upon both the size of the collecting and the collected drops.

Assuming a collision does occur between two drops, this does not necessarily mean they will stick to one another, that is, coalescence is not guaranteed. When a pair of drops collide they may bounce apart or they may collide and remain united. For larger colliding drops other scenarios are equally possible, as a temporary unification followed by a separation or break up into several smaller drops. The ratio of the number of coalescences to the number of collisions is called the *coalescence efficiency*.

The product of the *collision efficiency* and the *coalescence efficiency* defines the *collection efficiency*. The term *collision coalescence* is used to refer to the process of two drops coming into contact and merging into a single larger one. Here it is assumed that collisions between droplets and/or drops occur due to the differences in their terminal velocities (i.e., gravitational collision coalescence). Turbulent collisions will be discussed in Section 3.2.6. Typically, many collision-coalescence events are needed to produce a single drizzle drop. For instance, 125 cloud droplets of 10 μm diameter are needed to form one 50 μm drizzle drop.

The concept of gravitational collection is rather simple. As a large drop of radius R falls with velocity U_∞, it will collect smaller droplets of radius r and velocity u_∞. The volume of a cylinder that is swept out by the larger drop and overlaps with the smaller drop in a unit time is given by $\pi (R + r)^2 \|U_\infty - u_\infty\|$. If N_r is a concentration of droplets of radius r, and $E_c(R, r)$ is the collection efficiency that reflects the possibility that the collection volume is modified by the flow, then the rate of mass increase of the larger drop is:

$$\frac{dM}{dt} = N_r \left(\frac{4}{3}\pi r^3 \rho_l \right) K(R, r), \tag{3.30}$$

where $K(R, r)$ is a gravitational collision kernel defined as

$$K(R, r) = \pi (R + r)^2 \|U_\infty - u_\infty\| E_c(R, r). \tag{3.31}$$

If collected drops are much smaller than the collector drop, then by setting $u_\infty = 0$ and $R + r \approx R$ the change of larger drop size reads:

$$\frac{dR}{dt} = \rho \frac{\overline{E_c}\rho}{4\rho_l} q_l U_\infty, \tag{3.32}$$

where $\overline{E_c}$ is an average collection efficiency. Since the collection efficiency increases with the radius of collecting drop R and terminal velocity increases with radius, the rate of growth proceeds more and more rapidly as drop size increases. Fig. 3.7 schematically compares the condensational (diffusional) growth rate with the collision-coalescence growth rate as a function of radius. The condensational growth rate decreases with radius (following Eq. (3.24)) while collision coalescence growth rate increases with radius. Fig. 3.7 shows a 'gap' between approximately 10–30 μm where drop growth rates by both mechanisms are very slow. For precipitation to form as

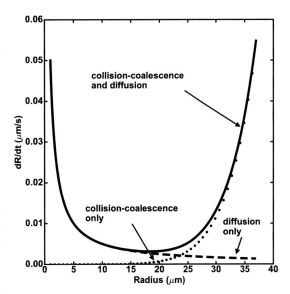

FIGURE 3.7: Drop growth rate by condensation and collision coalescence. The dashed line represents growth by diffusion only, and the dotted line represents growth by collision coalescence only, while the solid curve represents the combined growth rate. Condensational growth rate decreases with increasing radius, while collision-coalescence growth rate increases with increasing radius. (Adopted from fig.8.6, Curry and Webster (1999))

rapidly as is observed, drops must somehow reach a size of 25 μm (radius) more rapidly. Different mechanisms that accelerate that process will be discussed in Section 3.2.6.

More generally, the physics of the interactions between drops and droplets can be described by a kernel K, defined as the probability that, in a unit interval of time and in a unit volume of air, a drop with radius R will collide and merge with a drop of radius r (Eq. (3.31)). This kernel considers the volume swept out by two colliding drops without hydrodynamic interactions and includes additional effects (e.g., hydrodynamic interactions) through the collection efficiency $E_c(R, r)$. $E_c(R, r)$ is a product of collision (E_{coll}) and coalescence (E_{coal}) efficiencies:

$$E_c(R, r) = E_{coll}(R, r) E_{coal}(R, r), \tag{3.33}$$

with the former including hydrodynamic interactions and the latter considering the fact that not all collisions result in drop (or droplet) merging.

Typically it is assumed that the coalescence efficiency is close to unity. In fact, its value is uncertain to within a factor of two, and only qualitative information about the latter is known. It is thus often assumed that the collision efficiency subsumes the effect that might cause the coalescence efficiency to depart from unity. Indeed, present research suggests that the largest gap in our understanding of the collection efficiency is not due to the poor understanding of coalescence efficiency, but rather in how turbulent motions modify the collision efficiency.

Because initial growth by collision coalescence involves droplets grown by condensation (i.e., particles of relatively small sizes), the collection efficiency for such droplets is of particular interest. With these small droplets, Stokes flow approximations apply and the flow around the droplets is laminar. Both theoretical and experimental studies stress that the most efficient collection occurs when the size ratio between collected and collector drop is in the range 0.4–0.8. Two effects play a role in this behaviour: (i) small droplets have little inertia and therefore are easily deflected by the flow around the collector drop (i.e., low collision efficiency) and (ii) relative velocity of drops with similar sizes reduces the probability of collisions. There is, however, some evidence that the applicable range of the Stokes flow model may be less than is often assumed. Traditionally, the model has been assumed valid for $r \leq 30\,\mu$m. But while it is true that for single spheres of such radii the Stokes drag closely approximates the actual drag, for two spheres the effect of inertial acceleration in the fluid can cause a significant differential rate of fall even for considerably smaller sizes. The like-sized drops of sizes bigger than $20\,\mu$m are therefore super-efficient at collection, which is associated with wake-capture effects, wherein the interaction of the droplet wake introduces a collision that would not have otherwise been expected.

A useful way to introduce the concept of the collection process is to separate a population of droplets in two groups: larger drops acting as collectors and smaller droplets being collected. Any two drops may interact with each other and collide with some probability, but we are interested in how in the population of cloud droplets, the number of droplets of a given mass x evolves. The coalescence of a drop of mass $x - y$ with a drop of mass y will create a drop of mass x. If we associate $x - y$ with the mass of the larger drop, then y can vary between 0 and $x/2$. Evolution of the population of drops having mass x can be described by the Smoluchowski coalescence equation, named after Marian Smoluchowski who was the first to derive it:

$$\frac{\partial}{\partial t} n(x, t) = \int_0^{x/2} n(x - y, t) n(y, t) K(x - y, y) \, \mathrm{d}y$$
$$- n(x, t) \int_0^\infty n(y, t) K(x, y) \, \mathrm{d}y. \quad (3.34)$$

The first term on the right-hand side is the gain term that describes all collision-coalescence events per unit time that result in the formation of a drop of mass x. The second term, referred to as the loss term, describes all collision-coalescence events per unit time between drops with mass x and all other drops. Drop break up (either spontaneous or due to collisions) is not considered in Eq. (3.34). The kernel $K(x, y)$ describes the rate at which drops with mass x coalesce with drops of mass y.

Eq. (3.34) is often called the stochastic collection equation, although it is a purely deterministic equation and does not contain any stochastic elements. The reason for this is because Eq. (3.34) can be interpreted as the mean field representation of a stochastic process, analogous to the way diffusion is the mean field representation of Brownian motion; the latter being a stochastic process whose net effect can be described deterministically. For including collision and coalescence of droplets in cloud models an analytic solution to the collection equation – Eq. (3.34) – is desirable. A collection kernel for the range of droplet and drop sizes relevant to clouds was approximated in polynomial form by Alexis B. Long, and is often used in cloud modelling. It is given by

$$K(x, y) = \begin{cases} k_{\mathrm{c}}(x^2 + y^2), & \max(x, y) \leq x^* \\ k_{\mathrm{r}}(x + y), & \text{otherwise.} \end{cases} \quad (3.35)$$

where $x^* = 2.6 \times 10^{-9}$ g corresponds to a drop radius of $40\,\mu$m separating cloud droplets from rain droplets. For K in units of $\mathrm{cm}^3\,\mathrm{s}^{-1}$, the constants are $k_{\mathrm{c}} = 9.44 \times 10^9\,\mathrm{cm}^3\,\mathrm{g}^{-2}\,\mathrm{s}^{-1}$ and $k_{\mathrm{r}} = 5.78 \times 10^3\,\mathrm{cm}^3\,\mathrm{g}^{-1}\,\mathrm{s}^{-1}$, as proposed by Long (1974). The Long kernel does not vanish for mono-disperse droplet size distribution (i.e., when $x = y$). Instead, it emphasises the role of larger drops. Eq. (3.35) provides a simple and concise, and therefore convenient, description of the collection process (for the case of no turbulence).

3.2.6 Effects of Turbulence on Droplet Growth

As was presented in the previous section it is difficult to explain the rapid growth of cloud droplets in the size range 15–$30\,\mu$m in radius because neither the diffusional nor the gravitational collision-coalescence mechanism is effective (see Fig. 3.7). Several mechanisms have been proposed to explain this issue, including the entrainment of dry environmental air into the cloud, turbulent fluctuations of the water vapour supersaturation and collision coalescence enhanced by turbulence.

3.2.6.1 Effect of Turbulence on Differential Growth
Small-scale turbulence produces perturbations in velocity, temperature and moisture fields that influence droplet growth rate. However, the effects of turbulence on the

droplet growth are small, if not insignificant. It can be understood through the rapid rearrangement of droplet positions. Cloud droplets that grow faster at one instant (because the local concentration is small and the supersaturation large) subsequently move to a region where the local concentration is high and the supersaturation is low. Such changes occur rapidly, typically in less that a few seconds, and the effects average out over time relevant for droplet growth in a cloud (several minutes or longer). The key point is that diffusional growth is reversible, in contrast to irreversible growth by collision coalescence, as discussed later on.

3.2.6.2 Entrainment and Mixing Effect

Entrainment of subsaturated environmental air and a subsequent mixing leads to the observed reduction of the liquid water content of a cloud. Mixing dilutes the concentration of droplets and thus reduces the competition for available water vapour and enhances the growth rate of some favoured droplets. Thermodynamically, the mixing of subsaturated air requires net evaporation. The key issue is whether the dilution results in the reduction of only the droplet sizes (*homogeneous mixing*), only the droplet concentration (*extremely inhomogeneous mixing*) or both the concentration and size (*inhomogeneous mixing*). The way the mixing proceeds depends on the relative magnitude of the timescales for turbulent mixing (or turbulent homogenisation) (τ_{mix}) and the timescale for droplet evaporation (τ_{evap}). The timescale for mixing may be defined as a turnover time of a turbulent eddy, that is, $\tau_{mix} \equiv l u(l)$, where $u(l)$ is characteristic velocity of an eddy of a scale l. In clouds τ_{mix} takes values between a few seconds and a few dozens of seconds. The droplet evaporation timescale can be estimated from Eqs (3.24) and (3.25) as $\tau_{evap} = Ar^2/(1 - \text{RH})$, where RH is the relative humidity of the dry environmental air and $A = (F_D + F_k)/2$ is of the order of 10^{10}s/m^2. For a droplet with radius $r = 10\,\mu\text{m}$ in the air with RH = 5 % the droplet evaporation timescale equals 1 s; for RH = 70 % $\tau_{evap} \approx 3\,\text{s}$, and for RH = 99 % $\tau_{evap} \approx 100\,\text{s}$. Homogeneous mixing takes place when the timescale for turbulent mixing is much shorter than the droplet evaporation timescale. In such cases, all droplets are exposed to the same conditions during the process of evaporation, and their sizes decrease according to Eq. (3.24). The droplet concentration remains unchanged, unless the smallest droplets evaporate completely. Droplet concentration is however smaller than in the previously unmixed cloud parcel because of the dilution due to the entrainment of droplet-free environmental air. In the opposite limit, when the timescale for turbulent mixing is much

longer than the droplet evaporation timescale, extremely inhomogeneous mixing is thought to take place. Before homogenisation of the whole mixture, restoration of the saturated conditions is achieved in a portion of the fluid by evaporation of all droplets present in that volume. In the whole mixture, the concentration is reduced, but the droplet sizes remain unchanged. In fact, droplets present in originally unmixed portion of cloud are diluted by mixing with the saturated air.

If in the case of extremely inhomogeneous mixing the subsequent homogenisation within the undiluted air is not rapid enough, further upward motion will result in the supersaturation production and in the activation of additional cloud droplets, thus broadening the droplet size distribution overall. Otherwise condensation occurs on the depleted remaining droplets, thus reducing the competition for the available vapour, and favouring the growth of larger droplets.

The evaporation timescale τ_{evap} is smaller in polluted clouds where for a given liquid water content the droplet concentration is higher and droplet sizes smaller than in clean clouds. Therefore, in polluted clouds – with all other conditions remaining the same – the mixing scenarios are closer to homogeneous than extremely inhomogeneous.

Cloud entrainment (cumulus entrainment in particular) discussed so far is associated with large-scale (i.e., not much smaller than the cloud itself) motions that bring subsaturated cloud-free environmental air into the cloud. It also involves a wide range of scales characterising the turbulent transport of cloudy air upward within a cumulus cloud. As a result, droplets in the vicinity of a single point follow different trajectories through a turbulent cloud. The key point is that both large and small eddies play equally important roles; large eddies, with the supersaturation field close to the saturation value, affect the growth of cloud droplets differently in different conditions (i.e., droplet concentration and size, vertical velocity); small-scale eddies, incapable of affecting the droplet size distribution as described earlier, are needed to allow droplets to move from one large eddy to another. This process is referred to as large-eddy hopping, in analogy to island hopping (i.e., crossing an ocean by a series of shorter journeys between islands). A concept of large-eddy hopping is presented in Fig. 3.8. Large-eddy hopping results in different histories of droplets arriving at a given point in a cloud and affects the observed size distribution.

In summary, small-scale turbulence alone cannot lead to a significant broadening of the cloud droplet size distribution, but is the key to large-eddy hopping because it allows individual droplets to move from one large eddy to

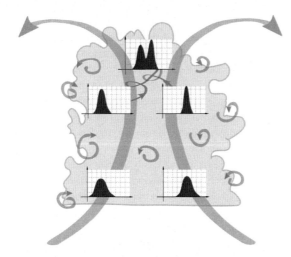

FIGURE 3.8: Formation of cloud droplet spectra involving the large-eddy hopping mechanism. Large arrows show the cloud-scale flow associated with cloud development. Turbulent eddies fill the volume of the cloud and provide different growth histories for droplets arriving at a given location higher in the cloud. Note that some droplets may evaporate at cloud edges and then be reactivated when entrained back into the cloud. Mixing droplet spectra that follow different flow trajectories lead to wide (e.g., multimodal) spectra in the upper part of the cloud. Figure thanks to Katarzyna Nurowska

another. Different activation and growth histories lead to a significant spectral broadening.

3.2.6.3 *Effect of Turbulence on Collision and Coalescence*

Over the past fifteen years, an increasing number of studies have been reported in both the engineering and atmospheric literature concerning the collision rate of inertial particles in a turbulent flow. The turbulent transport effect concerns the effect of local shear and air acceleration on the relative fluctuating motion of droplets. Inertial particles tend to accumulate in local regions of low vorticity and high strain rate owing to an inertial bias or the centrifugal effect. Sedimenting particles bias their trajectories towards regions of downdraft fluid motion around vortices and could settle significantly faster that the terminal velocity because of this preferential sweeping mechanism.

For droplets with a radius falling in the size gap (15–20 μm), the collision efficiency has a sensitive dependence on the radius and any alteration of the far-field (separation larger than approximately ten radii) and near-field (separation less than one radius) relative motion. Turbulence alters the distributions of the angle of approach and relative velocity for colliding droplets; thus quantitatively we expect turbulence to modify the collision efficiency.

Simulations using direct numerical simulation for droplets with sizes falling in the size gap found that turbulence can significantly enhance the collision efficiency. The enhancement depends nonlinearly on the size ratio of colliding droplets and grows with increasing dissipation rate. The net enhancement (compared to the pure gravitational collision efficiency) ranges from factors of 1 to 5. It becomes larger when the colliding droplets are rather different in size or are nearly equal in size. When droplets are nearly equal in size, the gravitational kernel is small owing to small collision efficiency. When droplets are very different in size, that gravitational kernel may also be small owing to small collision efficiency. Therefore, air turbulence plays an important role in enhancing the gravitational collision kernel when the collision efficiency is small.

Recent studies using the turbulent collection kernel showed that the turbulence effects could reduce the rain initiation time by 25–40 % for conditions typical of clouds in a range of environments. The effect is more pronounced near the cloud top where dissipation rates are higher.

3.3 Cold Clouds

When the air within a cloud reaches temperatures below 0 °C, there exists the potential for ice particles to form. The consequences of such a change in phase are numerous;[4] (i) the additional latent heat release within the cloud will elevate its buoyancy and invigorate the turbulent motions, (ii) since equilibrium vapour pressure over water is greater than that over ice, cloud ice particles can continue to grow in conditions unfavourable to the growth of liquid droplets, leading to further microphysical changes to the thermodynamic balance of the cloud and enhancing the efficiency of precipitation generation, (iii) the radiative characteristics of ice and liquid drops are different such that, for equivalent size and concentration of particles, liquid water will more efficiently scatter shortwave radiation and emit long-wave radiation, (iv) unlike liquid drops which tend to be roughly spherical in shape, ice particles grow in numerous different shapes (or habits) and, as a result, different particles can have very different physical characteristics and (v) due to their more stable structure, ice particles can grow to be orders of magnitude larger than liquid drops, leading to more massive, faster falling and potentially damaging hydrometeors (e.g., hail).

---

[4] We also note that the ice phase plays an important part in the electrification of clouds. We shan't discuss this here, but refer the reader to texts recommended for further reading.

Much of the theory behind ice formation and evolution shares a common basis with that of liquid droplets. However, cold cloud processes are even more enigmatic than their warm phase counterparts and, despite considerable theoretical, experimental and observational effort over the past century or more, there remains much to be understood about when and where ice particles will form.

3.3.1 Homogenous Nucleation

Section 3.2.1 discussed the energy barrier associated with the clustering of molecules which proves to be prohibitive to the spontaneous homogeneous nucleation of liquid water droplets at supersaturations that are to be found in Earth's atmosphere. A similar barrier exists and inhibits the spontaneous crystallisation of vapour molecules to form ice particles. Indeed, Ostwald's rule of stages[5] suggests that in order to transform to the ice phase, molecules in the vapour phase will first form supercooled liquid water (i.e., liquid water at temperatures below 0 °C). Given such a pathway, the energy barrier associated with the homogeneous formation of ice must exceed that for liquid water and so homogeneous ice formation from the vapour phase is even less likely than liquid formation.

However, within pre-existing supercooled liquid droplets, the process of ice embryo formation becomes slightly different to that from the vapour phase and is more readily achieved. Unlike clustering of molecules within the vapour phase, liquid to ice clustering requires only the reorientation of the molecules to form the hydrogen-bonded crystal structure of the ice. Since ice is the preferred phase at temperatures below 0 °C, and since the crystallisation process becomes ever more efficient as the supercooling increases, ice embryos will form and initiate the freezing of the droplet in cold enough conditions. Experiments suggest that pure water droplets of diameter less than 5 μm will freeze spontaneously at temperatures below −38 °C, while larger droplets will freeze at slightly warmer temperatures due to the increased probability of random cluster formation. Thus, clouds that reach these temperatures will, with a high probability, freeze any liquid droplets that have not previously frozen.

In very cold conditions where the air is subsaturated with respect to water, unactivated solution droplets may freeze homogeneously (bypassing the intermediate step of

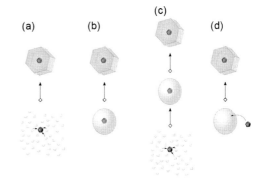

FIGURE 3.9: Schematic description of the different mechanisms of heterogeneous nucleation. (a) Deposition freezing – vapour molecules deposit directly onto the ice nucleus (IN), (b) immersion freezing – supercooled water containing an IN freezes, (c) condensation freezing – at water saturation, an intermediate liquid droplet forms on the IN which then freezes, (d) contact freezing – an IN impinges on the surface of a liquid drop, initiating the freezing process.

liquid droplet formation). However, due to the presence of the dissolved ions which depress the freezing temperature, this freezing process will occur at temperatures lower than that for pure water. Thus, as is common in cirrus clouds, ice particles can form homogeneously at humidities below water saturation, but at temperatures colder than −38 °C.

3.3.2 Modes of Heterogeneous Nucleation

Just as CCN can provide a site on which liquid water droplets can form and transcend the energy barrier of critical cluster formation, heterogeneous ice nucleation can be initiated by the presence of ice nuclei (IN). Unlike droplet formation for which only one mechanism can occur — that is, the direct flux of vapour onto the CCN – nucleation of ice by IN can be initiated in a variety of ways, which are illustrated in Fig. 3.9 and described here.

3.3.2.1 Mechanisms of Heterogeneous Freezing

In situations where the environmental temperature is very cold and the air is supersaturated with respect to ice, certain aerosol particles can provide a suitable site onto which molecules in the vapour phase can directly bond. Such a mechanism, referred to as deposition freezing, is only likely to take place in conditions which are subsaturated with respect to water. At water saturation, an intermediate step by which a liquid droplet first forms and then subsequently freezes is more likely. This mode of freezing is referred to as condensation freezing.

In the case of supercooled water droplets, two additional mechanisms by which IN are able to initiate

---

[5] Ostwald's rule of stages represent a rule of thumb rather than a law based on fundamental thermodynamic considerations, but has been shown to be robust at temperatures warmer than −100 °C.

freezing are commonly observed. The first involves the IN which are already immersed within the liquid droplet, either through the initial droplet formation having been mixed with the CCN, or through subsequent impaction scavenging[6] of the IN by the droplet. In this mode of freezing, referred to as immersion freezing, water molecules within the droplet collect on active sites on the IN. If the IN is large enough and provides favourable sites of sufficient size to generate an embryo that exceeds the critical size to overcome the energy barrier, then the droplet will freeze.

In a similar manner, IN which impact upon the surface of a supercooled liquid droplet can serve to act as a nucleus for ice formation. In this case, the mode of freezing is referred to as contact freezing. Experimental studies suggest that the contact freezing process can take place at warmer temperatures than that typically observed for immersion freezing. In part this may be because salts contained within the droplet may spoil the favourable nucleating sites of the IN (while simultaneously depressing the freezing temperature of the droplet). In addition, temperatures at the surface of a drop may be colder than in the interior. Such mechanisms are speculative and a comprehensive understanding of this process remains to be determined.

Noting this increased likelihood of freezing when an IN meets the surface of the droplet, an additional mechanism for freezing of supercooled droplets has been proposed. In this mode, known as evaporation or inside-out contact freezing, an IN immersed within a droplet migrates to the inner interface of the surface of the droplet. This is most likely to happen as a droplet begins to shrink, that is, in conditions subsaturated with respect to water as the droplet evaporates and loses mass. However, observations of such freezing mechanisms in nature are currently lacking.

3.3.3 Ice Nuclei

Numerous different aerosol particles of varying chemical composition and having various different sources of origin can be found in the atmosphere. Despite many years of experimental and observational research, it is still not clear what makes an efficient IN and which materials most commonly provide the nucleus for the glaciation of clouds. This said, if an aerosol particle possesses certain properties, it is more likely to be efficient as an IN:

(In)solubility: If the aerosol particle is soluble, then it will likely disintegrate as it absorbs water. The fragmented substrate then provides a less rigid surface for ice germ formation, while dissolved salts will depress the effective freezing temperature. Thus, typically, good IN tend to be highly water-insoluble.

Size: In order to overcome the energy barrier to nucleation, the IN must have size greater than or comparable to the critical ice germ size. *In situ* observations and laboratory experiments generally support this assertion. It should also be noted that there is often a correlation between size of a particle and its chemistry, thus it is possible that particles with suitable chemistry for germ formation tend to be those that are largest. There are known exceptions to this rule; for example, some organic materials with very small diameters (0.01 μm) have been shown to have good nucleating ability at relatively warm temperatures (−8 °C).

Chemistry: In order for an ice germ to form on the surface of a particle, it must have hydrogen bonds available to form the crystal lattice structure of the ice. An IN with molecules having similar bond strength, polarity and rotational symmetry to that of ice at its surface will more readily initiate the freezing process.

Crystallography: In addition to the chemical properties of the IN surface, the geometrical arrangement of molecules can play an important role in germ formation, especially if the molecular arrangement closely mirrors that of the ice. A good example of this is silver iodide which has a hexagonal structure and has been shown to nucleate ice at temperatures as warm as −4 °C.

Water adsorption: Many solids have the property of being able to adsorb[7] water molecules onto their surface. This is the result of physical and chemical mechanisms, but is facilitated at lower temperatures (and higher humidities). IN will typically have distinct localised sites on the surface which are able to adsorb water. The adsorption of a layer of water molecules can then initiate the ice nucleation process. The number and temperature dependence of the active sites on a particle will characterise its ability to nucleate ice.

---

[6] Impaction scavenging refers to the collision of aerosol particles with liquid or ice particles through Brownian diffusion, gravitational inertial capture, turbulent inertial capture or thermophoresis and diffusiophoresis.

[7] Adsorption is a surface process as opposed to absorption whereby the water permeates through the volume of a particle.

While this list gives a reasonable overview of the properties that make ice nucleation more likely, these are neither necessary nor sufficient to classify a given particle as being a good or bad IN.

It is also worth mentioning the role of biogenic particles in ice nucleation. Little is known about the global prevalence of biological aerosol although sources can be found over vegetative land masses and oceanic surfaces. Experiments conducted in the 1970s suggested that ice nucleation was much more efficient in droplets containing soil samples when the organic content was higher. It was subsequently demonstrated that bacteria were the primary agent in the nucleation process, enabling ice formation at temperatures as high as −4 °C – much warmer than temperatures at which most minerals can act efficiently.

Part of the reason that bacteria are efficient at nucleating ice arises from protein structures which they can form on their outer membrane. The repetitive structure of these proteins provides a suitable site onto which the water can crystallise. It is not clear why the bacteria produce these structures, but at sub-zero temperatures, there may be some benefit to the organism through the latent heat which is released. Other organic particles, such as pollens and fungi, have been shown to have good ice nucleating properties at relatively warm temperatures, but even less is known about these.

3.3.3.1 Stochastic versus Singular

The nucleation process is inherently a random process, dependent on the self-organisation of water molecules to form an ice embryo. However, there remains some debate as to whether the way in which IN catalyse the freezing process is predominantly a stochastic or a deterministic process.

In a stochastic description, each IN facilitates the pathway, but doesn't alter the underlying random process of ice embryo formation. As a result, similar droplets containing similar IN will have an equal chance of freezing at a given temperature. From this point of view, one would expect a population of droplets exposed to a fixed temperature to freeze over a period of time, with more droplets freezing as time goes by. Such time-dependent freezing has been observed in laboratory experiments.

By contrast, the singular description of freezing behaviour considers each IN to have a deterministic temperature at which an embryo can form and the drop instantly freeze. Using this description of the process, there would be no time-dependent behaviour for droplets maintained at a constant temperature and further cooling

would be required in order for more droplets to freeze. Contradictory experiments have further been shown to demonstrate this alternative viewpoint.

It is likely that the specific processes are a combination of both singular and stochastic behaviour and a number of reconciliatory hypotheses have been proposed. These generally make use of the idea that each IN, rather than being characterised by a single freezing temperature, possesses a number of distinct active sites on its surface. While each site might behave in a deterministic manner, the overall behaviour of the particle and moreover a collection of many particles would appear to be stochastic.

3.3.4 Diffusional Growth

Once ice particles are formed, they can then grow by vapour diffusion in a supersaturated environment. This growth takes place in a similar way to the condensation process that controls the growth of liquid droplets, and, similar to Eq. (3.20), can be written as

$$\frac{dm}{dt} = 4\pi C D_v (\rho_{v,\infty} - \rho_{v,c}). \tag{3.36}$$

We note that Eq. (3.36) differs from Eq. (3.20) in the pre-factor C, which (by analogy with electrostatics) is termed the capacitance, and for spherical ice particles $C = r$, that is, we recover the full form for Eq. (3.20). However, since ice particles are generally not spherical, the capacitance can be represented by a range of expressions which depend on the habit of the ice particles in question. Table 3.5 lists some of the representations of capacitance that can be found in the literature. For simple shapes, for example, spheres or spheroids, theoretical expressions for the capacitance can be taken from the literature on electrostatic capacitance and these apply equally well to the Brownian diffusion problem. For more complicated geometries, such as those found in ice crystals and aggregates, finding a closed expression is more challenging.

Westbrook et al. (2008) used a numerical random-walk algorithm to simulate the paths of individual water molecules around ice crystals of different shapes. A summary of some of their findings is shown in Table 3.5. It is interesting to note that their calculations for ice aggregates (see Section 3.3.6) are relatively insensitive to the shapes of the constituent components of the aggregated snowflake (i.e., variation of $C/D_{\max}$ is between 0.25 and 0.28). This is in agreement with the observational study, and is valuable information when attempting to represent the bulk statistics of the countless number of aggregated crystals found in nature.

TABLE 3.5: Representations of capacitance C for different ice particle geometries. Here D denotes the dominant dimension, which for round objects is the diameter, hexagonal column the cross-section and for rosettes and aggregates the maximum dimension. For the prolate spheroid, a and b denote the major and minor axes, and ϕ denotes the aspect ratio.

Shape	Capacitance	Reference
Sphere	$\frac{1}{2}D$	McDonald (1963), theoretical
Thin disc	$\frac{D}{\pi}$	McDonald (1963), theoretical
Prolate spheroid	$\sqrt{a^2-b^2}/\ln\left[(a+\sqrt{a^2-b^2})/b\right]$	McDonald (1963), theoretical
Circular cylinder	$0.3185D(1+0.868\phi^{0.76})$	Smythe (1962), theoretical
Hexagonal columns	$0.29D(1+0.95\phi^{0.75})$	Westbrook et al. (2008), numerical
Six-point bullet rosette	$0.4\phi^{0.25}D$	Westbrook et al. (2008), numerical
Aggregates	$0.26D$	Field et al. (2008), observations

As with condensational growth and evaporation in the liquid phase, particles which are moving relative to their environment experience an enhancement in the rate of transfer of mass onto the particle (or away from the particle in subsaturated conditions). The resulting increase in effective diffusivity is represented by the ventilation coefficient and takes the same form of that described by Eq. (3.26).

3.3.4.1 Wegener–Bergeron–Findeisen Process

The coexistence of water in all three phases brings about a rich variety of clouds and cloud behaviour. A particular mechanism which results from this coexistence is the Wegener–Bergeron–Findeisen (WBF) process (named after three early twentieth-century scientists who observed and developed the theory to describe the phenomenon).

Due to the simple fact that at a given temperature the equilibrium vapour pressure over water is greater than that over ice, there will be tendency for the ice phase to grow at the expense of the liquid phase in regions of mixed phase clouds where ice and liquid particles coexist. Consider the case that the vapour content is such that the environment is supersaturated with respect to water, then both liquid and ice will grow, removing vapour until liquid saturation is reached. At this point, because the air is still saturated with respect to ice, the ice will continue to grow and deplete the vapour. This subsequent reduction in vapour leaves the environment subsaturated with respect to water and the liquid drops will start to evaporate. This latter process will continue with the ice particles removing vapour and the liquid particles evaporating until no liquid remains. A dramatic illustration of the WBF process are hole-punch clouds (Fig. 3.10)

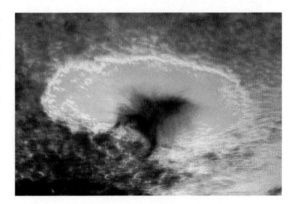

FIGURE 3.10: Demonstration of Wegener–Bergeron–Findeisen (WBF) mechanism. A 'hole-punch' cloud is generated when ice is formed (often on the surface of an aircraft penetrating the cloud) in a layer of supercooled liquid water. The WBF mechanism rapidly evaporates the liquid cloud and deposits vapour on the ice particles (which here can be seen as fall streaks).

3.3.5 Terminal Fall Speeds of Ice Particles

In order to make progress with interpreting Eq. (3.27), the drag associated with a falling particle of characteristic radius r can be equated with that of a projected volume or 'cushion' of air which surrounds the particle and experiences only pressure-induced drag and no friction (Fig. 3.11). Boundary-layer theory suggests that such a cushion, having projected radius $r + \delta$, is determined by the relationship

$$\frac{\delta}{r} = \frac{\delta_0}{\sqrt{\mathcal{R}_e}}, \tag{3.37}$$

with δ_0 a constant. Given δ and taking the frictionless drag coefficient of the ensemble of particle plus boundary layer

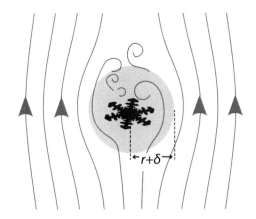

FIGURE 3.11: Near to a body, viscous drag may dominate, but a purely pressure-induced drag can be considered from a projected 'cushion' of air surrounding the particle (represented here by the grey, shaded region) and having radius $r + \delta$.

as the constant C_0, we find that the coefficient of drag for the original particle is now

$$C_D = C_0 \left(1 + \frac{\delta_0}{\sqrt{\mathcal{R}_e}} \right)^2 . \qquad (3.38)$$

The balance equation between gravitational acceleration and drag can be rearranged to define the Davies or Best number X as

$$X \equiv C_D \mathcal{R}_e^2 = \frac{2 \rho_L v g D^2}{A \rho \nu^2} . \qquad (3.39)$$

The important aspect of the non-dimensional number, X, is that it is dependent only on the shape and size of the particle and not on its fall speed. Since Eq. (3.38) now provides information about the dependence of C_D on the Reynolds number, this X–$\mathcal{R}_e$ relationship can be solved to find

$$\mathcal{R}_e \equiv \frac{D u_\infty}{\nu} = \left(\frac{\delta_0^2}{4} \right) \left[\left(1 + \sqrt{\frac{16X}{\delta_0^4 C_0}} \right)^{1/2} - 1 \right]^2 . \qquad (3.40)$$

For spherical particles, C_0 and δ_0 are well approximated as 0.29 and 9.06 respectively. These values are also consistent in the Stokes' limit ($24 C_D / \mathcal{R}_e \sim 1$) for small values of the Reynolds number. For irregular ice particles, however, observational studies suggest the X–$\mathcal{R}_e$ relationship Eq. (3.40) departs from measured values particularly for larger particles (larger X). Such a departure can be argued as being due to the limitations of the boundary-layer theory as air can pass through more complicated ice aggregate structures and effectively reduce δ, while non-spherical shapes will also alter the pressure-induced drag term coefficient C_0. To circumvent these issues, Mitchell

and Heymsfield (2005) have suggested that Eq. (3.40) be extended with an additional term aX^b, which can also be used as an additional correction for turbulence.

3.3.6 Collection Processes, Riming and Ice Multiplication

Differences in fall speeds between ice particles can lead to collisions and gravitational collection in the same way as for liquid drops. As with the liquid phase, we might describe collisions involving ice particles using ideas relating to the area of sweep-out of a collecting particle, the efficiency of collisions and the efficiency of collection once a collision occurs. Unlike the liquid phase, however, collisions between particles of the ice phase or of mixed phases can involve a number of additional physical mechanisms. Quantification of collision-coalescence processes are thus poorly understood.

3.3.6.1 Collisions between Ice Particles

The efficiency with which two ice particles stick to one another on collision is a strong function of temperature. At warmer temperatures, approaching 0 °C, the surface of an ice particle may develop a thin quasi-liquid film on its surface. This liquid surface is more amenable to sticking with particles with which it comes into contact, but will be less 'sticky' as temperature decreases. As a result, aggregation of ice particles due to this surface adhesion generally tends to happen at temperatures warmer than around −10 °C.

Once the primary bond is formed between particles, a sintering process can act to reinforce the 'neck' between the two particles. In this process, the sintering is thought to primarily involve particles in the vapour phase depositing around the contact point. These may either originate from the environment in high humidity conditions or else be evaporated from the neighbouring sections of the ice particles before re-depositing.

Observations suggest that snow flake dimensions tend to reach a maximum at temperatures near −1 °C. Such a maximum is consistent with the idea that sticking efficiency increases with warmer temperature. However, a second maximum has been observed at temperatures from −12 °C to −17 °C. This second maximum coincides with the preferred regime for ice structures with dendritic features to form. Such dendritic forms are prone to interlocking with one another on collision, thus leading to higher collection efficiencies.

A third mechanism for enhancing collection efficiencies relates to the local electric fields surrounding ice particles. Electric fields may serve to align plate-like crystals and although the induced attraction between particles is weak, it is thought to be strong enough to extend contact times long enough for sintering processes to act and reinforce the contact areas. Generally, a high electric field, such as that found in very active thunderstorms, is required for such a mechanism to occur.

3.3.6.2   Collisions between Ice Particles and Liquid Drops (Riming)

In mixed-phase clouds, where supercooled liquid and ice particles coexist, gravitational collection will result in collisions between particles of different phases. The freezing of the supercooled liquid on contact with the ice particle is known as riming. Two collection mechanisms may arise in such a scenario; if the ice particle is large and the liquid droplet is small, then the ice particle acts as the collector as the droplets follow the flow around the crystal; if the liquid droplet is large compared to the ice particles, then the capture mechanism is reversed. As with the hydrodynamics of liquid–liquid interactions, the efficiency of collection is highly dependent upon the sizes of both the collector and the collected particles, with additional complications arising due to the different drag behaviour of the different ice habits.

3.3.6.3   Ice Multiplication

In situ and remote observations of clouds frequently reveal regions of a cloud having concentrations of ice particles which are orders of magnitude more numerous than the IN upon which they might form. This apparent paradox is explained by processes in which ice particles, after their initial embryos are formed on IN, may break up into many smaller fragments. These fragments may then themselves grow through diffusional or collection processes.

The direct fragmentation of relatively light and slow falling dendrites can occur when impacted by more dense and faster-falling graupel particles. Fragments of dendrites are frequently observed by *in situ* measurements, although often these may be the spurious result of impaction of the snow particles on the observing equipment.[8]

A more efficient and likely dominant ice multiplication process is that due to rime-splintering, sometimes known as the Hallet–Mossop process, when relatively large cloud droplets (roughly >25 μm) freeze onto an impinging ice particle, the pressure forces generated due to the thermal expansion can cause the frozen or partially frozen droplet

to shatter. Demonstrations show that, at a temperature of −5 °C, up to 350 ice splinters are formed for every milligram of rimed liquid. The efficacy of splinter production falls off to zero at warmer temperatures of about −2.5 °C and colder temperatures of about −7.5 °C. It is speculated that at warmer temperatures the freezing process is too slow, allowing the liquid particle to flow over the ice surface, while at colder temperatures the freezing process is too rapid.

Given that ice itself acts as the best nucleus for further glaciation of clouds, an efficient ice multiplication process will to a certain extent lessen the impact that interstitial IN may have on the ice particle numbers and subsequent cloud evolution. Because of this, it is important to conceptually separate the role that aerosol (i.e., IN and CCN) play in their interaction with, and controls on, warm and cold clouds.

3.3.7   Graupel and Hail

As pristine ice particles become more heavily rimed, they lose their original angular shapes, becoming covered in layers of frozen droplets. Such a heavily rimed ice particle is known as graupel. Typically, graupel particles will be more rounded than lightly rimed or pristine ice, taking on a spherical or conical appearance.

In vigorous convective updraughts with high liquid-water contents, the rate of riming can be such that graupel particles will form into hailstones. Such conditions are most often observed in mid-latitude convective clouds over continental land masses, where the altitude of the melting level is lower than in tropical regions. The lifetime of a hailstone within a convective cloud can be considerable as it is carried up by the updraughts before falling back through the cloud under gravity. We note that the fall speed of a large hailstone may be $O(10 \text{ ms}^{-1})$, similar to the typical strength of updraught found in convective clouds. During its journey through a cloud, the hailstone will experience a range of thermodynamic conditions; changes in both temperature and the ice and liquid environments through which it passes. This can lead to changes in the structure of the particle as it grows (see Fig. 3.12).

As a hailstone travels through very cold regions containing ice and snow, collisions will lead to a build up of relatively low density ice on its surface. Collisions with supercooled liquid water droplets in a very cold environment will also lead to instantaneous freezing and the capture of small pockets of air within the frozen hailstone. These small pockets will efficiently reflect light and so

---

[8]   Recent advances in the design of probe tips has reduced the occurrence of such 'shattering' events, leading to more reliable imaging.

FIGURE 3.12: A large hailstone with alternate layers of opaque and transparent ice.

appear as opaque white regions in Fig. 3.12. This mode of growth of the hailstone is known as 'dry' growth. In contrast, if conditions are warmer or supercooled liquid water contents are higher, the collection and subsequent enthalpy release through freezing of a sufficient mass of droplets can raise the temperature of the surface of the hailstone above $0\,°C$. If this happens, the liquid surface of the hailstone may flow to fill any pockets of air. This mode of growth is referred to as 'wet' growth and leads to the transparent (darker) regions, as can be seen in Fig. 3.12.

3.4   Interplay of Microphysics with Macrophysics

It is important to understand how microphysics can influence cloud evolution and impact upon its environment, but equally important to understand to what extent this influence dominates or is dominated by macrophysical effects (i.e., variations on scales much larger than the microscale). When considering individual clouds, there is often a simple cause and effect that can be observed and which relates to one or several of the microphysical processes discussed. However, in clouds or cloud systems which evolve over a long period or have closely coupled interactions between the microphysics, dynamics and radiation, the precise pathway between microphysical processes and the macrophysical properties can be harder to determine.

The processes discussed in Section 3.2.6.2 provide one such example of the interaction between local microphysical processes and the turbulent eddies of the cloud. As another example of a simple impact, we might consider

the influence of aerosol on precipitation formation within a single shallow cumulus or a short-lived stratocumulus cloud which exists at warm temperatures (above $0\,°C$). Chapter 11 provides a more comprehensive discussion of the impact of aerosol on clouds, but for this example we would expect an increase in the number of soluble aerosol particles to result in an increase in the number of activated cloud droplets as air rises through the cloud base. Since droplets are more numerous, for a given source of moisture the mean size of the droplets resulting from condensational growth will be smaller, this effects both the timescale of the mixing and the coalescence processes which could in turn influence the cloud macrostructure.

However, one might also use the same scenario to demonstrate that the simple pathway is not always the most likely. If, in this example, the driving dynamics provides a moisture source which is large or a cloud which is very deep (but for simplicity still below the $0\,°C$ level), then it may have had enough time to generate a few raindrop-sized particles. In this case, with sufficient cloud water, the accretion mechanism can rapidly take over from the condensation mechanism in growing the cloud droplets to produce precipitation (cf., Fig. 3.7). By becoming more dominant, the accretion reduces the impact that the increased aerosol concentration has on the rate and efficiency of precipitation production. The expected impact due to the simple pathway is thus much weaker than expected.

Another example of the complex pathways that can exist between microphysical behaviour and resulting macrophysical outcome can be seen in mixed phase stratocumulus clouds which are frequently seen in polar regions. These clouds have very important climate impacts as a result of their influence on the amount of radiation reaching the surface. They are typically long-lived, sometimes persisting for several days and of particular note is their propensity to maintain a supercooled liquid water layer near the top of the cloud. This layer, due to its liquid phase, provides more long-wave cooling than the ice cloud and so generates more turbulent energy which helps to prevent the cloud from dissipating. However, if one were to look at the microphysical processes in isolation, the expectation would be that the WBF process would rapidly remove the liquid droplets in favour of the ice phase. Instead the complicated interaction with both the dynamics of the cloud and the radiation, as well as other microphysical processes such as particle sedimentation prevent the WBF process from dominating and allow the supercooled layer to persist. The resultant mixed-phase cloud arises from a combination of numerous thermodynamic, microphysical, turbulent and

TABLE 3.6: Symbols and key variables used in this chapter

Symbol	Description	Units (MKS)
Particle properties		
r	Particle radius	m
r_a	Equivalent dry radius	m
r_{vol}	Mean volume radius	m
r_{srf}	Mean surface radius	m
r_{eff}	Effective radius	m
$n(r)$	Density of size distribution	$m^{-3}m^{-1}$
Thermodynamic properties		
p	Pressure	Pa
T	Temperature	K
ρ	Density of the (moist) air	$kg\,m^{-3}$
v	Specific volume	$m^3\,kg^{-1}$
ρ_l	Water density	$kg\,m^{-3}$
e	Saturation vapour pressure	Pa
e_l	Saturation vapour pressure in equilibrium with pure water	Pa
S_e	Supersaturation	–
ℓ_v	Specific enthalpy of vaporisation	$J\,kg^{-1}$
$\mathcal{R}_e$	Reynolds number	–
F	Surface free-energy of the condensed cluster	$J\,kg^{-1}$
$\mathcal{g}$	Gibb's energy	$J\,kg^{-1}$
μ_j	Chemical potential of phase (indexed by j)	$J\,kg^{-1}$
s	Entropy	$J\,kg^{-1}$
Cloud properties		
N	Particle concentration	cm^{-3}
LWC	Liquid-water content	$kg\,m^{-3}$
q_l	Specific mass of the liquid water	$kg\,kg^{-1}$
W	LWP	$kg\,m^{-2}$
σ_{ext}	Short-wave extinction	m^{-1}
Q_{ext}	Extinction efficiency	–
τ_c	Cloud optical thickness	–
Particle formation and growth		
r_{crit}	Critical value of radius	m
S_{crit}	Critical mass of saturation	–
σ_l	Surface tension coefficient for liquid water	$J\,m^2$
D_v	Diffusivity of water in air	$m^2\,s^{-1}$
k	Coefficient of thermal conductivity of air	$W\,m^{-1}\,K^{-1}$
v_a	Dissociativity of the solute aerosol	–
Φ_a	'Practical osmotic coefficient'	–
a_w	Water activity	–
κ	Parameter in the κ-Köhler approach	–
f_v	Ventilation coefficient	–
C_D	Dragg coefficient	–
$E_c(R,r)$	Collection efficiency	–
$K(R,r)$	Collection kernel	$m^3\,s^{-1}$
C	Capacitance	m

radiative processes, with a dependence on the background winds, moisture, temperature and aerosol properties.

These examples demonstrate the need for a holistic view when attempting to understand the role of microphysics on clouds in the climate system.

Further Reading

For a comprehensive and more in-depth reference to cloud microphysics, the text by Pruppacher and Klett, *Microphysics of Clouds and Precipitation*, published by Springer Netherlands (2010) is highly recommended. The book provides an exhaustive and detailed presentation of all subjects related to cloud microphysics including aerosol, water drops, snow and crystal formation and growth processes. The book also extends to cloud chemistry and electricity. It is incontestably *the* reference on cloud and precipitation microphysics.

A textbook *Physics and Chemistry of Clouds* by Dennis Lamb and Johannes Verlinde (2011) provides a quantitative yet approachable path to learning the inner workings within clouds. It is a very useful reference text for advanced students, researchers and professionals.

For those who seek a more introductory overview of cloud microphysics, the book by Rogers and Yau, *A Short Course in Cloud Physics* (1996) should be very useful.

An Introduction to Clouds: From the Microscale to Climate by Ulrike Lohmann, Felix Lüönd and Fabian Mahrt (2016) is a comprehensive and up-to-date introductory textook on clouds. It provides a fundamental understanding of clouds, ranging from cloud microphysics to the large-scale impacts of clouds on climate.

A review of aerosol-cloud issues published in *Nature* in a paper entitled 'Untangling Aerosol Effects on Clouds and Precipitation in a Buffered System' by Bjorn Stevens and Graham Feingold (2009) is also worth reading.

Section 3.1 Observation and Measurements
An extensive review of airborne measurement techniques is given in the European Facility for Airborne Research (www.eufar.net)-initiated text book, *Airborne Measurements for Environmental Research: Methods and Instruments* by Wendisch and Brenguier (2013).

Section 3.2 Warm Clouds
Cloud activation is a very complex process, however, in many introductory meteorological textbooks, it is treated in very basic way. A very good discussion of aerosol activation is given by Khvorostyanov and Curry (2006). The κ-Köhler method for representing the activation process is described by Petters and Kreidenweis (2007).

A comprehensive description of terminal fall speed velocity is given by Beard (1977).

A review paper by Grabowski and Wang (2013) discusses many still poorly resolved cloud formation processes and demonstrates a need to fully understand the condensation-coalescence bottleneck in warm rain formation.

Section 3.3 Cold Clouds
For the myriad forms of ice crystals and snowflakes, Bailey and Hallett (2009) provide a comprehensive summary of the dominant growth regimes. Many photographs of these particles can also now be found through an image search on the Internet.

For a discussion on the singular versus stochastic growth mechanisms, the reader is directed to various papers in the literature by Durant and Shaw (2005), Möhler et al. (2006), and Connolly et al. (2009).

Section 3.4 Interplay of Microphysics with Macrophysics
Other chapters in this book will provide information on many other aspects that influence and are influenced by microphysics when considering clouds in the climate system. However, a short review of Arctic mixed-phase clouds that highlights the complexity of these interactions is given by Morrison et al. (2012).

Exercises

(1) *Cloud radiative properties:* Show that in the long-wave regime the cloud optical thickness depends mostly on the LWP.

(2) *Cloud radiative properties:* Derive Eq. (3.7) based on extinction and optical thickness definitions (see Section 3.1.2 and Chapter 4).

(3) *Droplet activation:* Plot the Köhler curve (Eqs (3.11) and (3.17)) for different values of r_a and κ. For sufficiently big droplets (let's call them R), the equilibrium saturation converges to 100 %. It means that the curvature and solute terms in the Köhler equation become unimportant. For given κ values, plot a relation between dry aerosol radii and R.

(4) *Droplet activation:* In the κ-Köhler approach (Eqs (3.11) and (3.17)) the aerosol soluble matter is represented by two parameters: r_a – the dry radius of aerosol and κ. The critical supersaturation, that is, the supersaturation that has to be attained for a particle to become a cloud droplet as a function of r_a and κ. Calculate and plot a relationship of the critical supersaturation as a function of dry radius r_a for $\kappa = 0.0$, 0.1, 1.0. See fig. 1 in Petters and Kreidenweis (2007).

(5) *Condensational growth:* For a given value of temperature, solve Eq. (3.24) and plot temporal evolution of droplets radii for different initial radii and different values of supersaturation (assumed constant during the process). Observe how the width of initial size distribution evolves with time and for different values of supersaturation.

(6) *Terminal fall-speed relationships:* The buoyancy corrected gravitational acceleration of a liquid drop is given by

$$mg - F_b = (\rho_L - \rho)vg, \qquad (3.41)$$

where v is droplet volume ($v = \frac{4}{3}\pi r^3$), and the drag force experienced by a falling particle is

$$F_D = \frac{1}{2}C_D\rho V^2 A, \qquad (3.42)$$

where $A = \pi r^2$ is the drop's cross-section and V is the drop velocity. Derive the relationship in Eq. (3.27). (HINT: assume $\rho_L \gg \rho$.)

(7) Starting from Eq. (3.20), show that for constant vapour density in the near and far field and constant uniform temperature, the rate of change of the radius of a drop can be written as

$$r\frac{dr}{dt} = \frac{D_v}{\rho_L R_v T}e_l(T)(S-1). \qquad (3.43)$$

Solve this to derive an expression for the radius as a function of time.

Now consider a 10 μm supercooled droplet which falls or is mixed from its saturated environment into a subsaturated environment with relative humidity of 70 %. The temperature of the drop's environment is 270 K. Given a fixed value for the diffusivity of water as $2.2 \times 10^{-5}\,\mathrm{m^2s^{-1}}$ and approximating $e_l(270\,\mathrm{K}) \approx 500\,\mathrm{Pa}$, calculate the time it takes for the droplet to completely evaporate in its new environment.

(8) Now consider a thin hexagonal ice particle, which we approximate by a thin circular disc with an aspect ratio leading to the mass–radius relationship $M = 152r^3$. The mass of the particle is the same as the drop in the previous example and it experiences an environment with the same temperature and vapour content. Using the capacitance for a thin disc in Table 3.5, calculate how long it takes the ice particle to sublimate. In this case, at 270 K, the saturation vapour pressure over hexagonal ice is roughly 3 % smaller than that over liquid.

(9) Following on from the previous two examples, find the relative timescales for complete evaporation/sublimation of a supercooled droplet and a thin plate of ice at different temperatures. (You can do this using empirical formulae for the vapour pressures, or else approximate given that the vapour pressure over ice is less than that over liquid water by approximately 14 %, 25 %, 37 % and 50 % at temperatures of 260 K, 250 K, 240 K and 230 K, respectively.)

4 Clouds as Light

Robert Pincus and Hélène Chepfer

Clouds interact strongly with light: the electromagnetic radiation emitted by the sun and by Earth and its atmosphere. Clouds are easy to see because they are by far the most opaque and variable component of the atmosphere over large swaths of the electromagnetic spectrum. This means clouds can modulate energetic flows, changing local heating rates and affecting the fluxes of energy at the surface and the top of the atmosphere. The same strong interaction forms the basis for remote sensing, the process of inferring information about the state of the clouds from their effect on the radiation field. Knowledge about how clouds affect the radiation budget, and hence the climate, on large scales is available primarily from estimates of cloud properties obtained through remote sensing.

This chapter assumes general background knowledge about radiation including basic terminology; the sources of solar (short-wave) and terrestrial (long-wave) radiation, their spectral distribution, and the modes of interaction with the atmosphere; and the basic links between the planet's top-of-the-atmosphere radiation budget and the system's temperature. Discussion focuses on cloud–radiation interactions but these occur in the context of rich, spectrally complicated interactions of radiation with atmospheric gases.

4.1 Microphysics, Material Properties, and the Effect of Cloud Elements on Radiation

As discussed in Section 3.1.2, clouds are fundamentally a collection of particles suspended in the atmosphere. It is this microphysical state of a cloud – the distribution of particle sizes, shapes and thermodynamic phase – that determines the strength and nature of the small-scale interaction with radiation. The parameters controlling this interaction are collectively called the cloud single-scattering or optical properties; they describe the

location-dependent medium through which radiation propagates. As described in Section 4.2, it is the combination of these properties with cloud macrophysical state – primarily extent and location within the atmosphere – that determines the cloud radiative properties, and hence clouds' impact on the radiation field.

4.1.1 Principles of Interaction

4.1.1.1 Fundamental Interaction: Extinction

The strength of the interaction between individual cloud particles and radiation is determined by two quantities: the geometric cross-section C_{geo} of the particles and the ratio of this size to the wavelength λ of radiation in question. The relationship is described by the size parameter $x = \pi D / \lambda$, where D is a (linear) size characterising the particle cross-section (i.e., $C_{geo} = \pi D^2 / 4$ for spheres).

Each particle interacting with radiation diminishes or *extinguishes* some of the radiation incident on it according to the extinction cross-section C_{ext}. Fig. 4.1a shows the extinction efficiency $Q_{ext} = C_{ext} / C_{geo}$ as a function of size parameter x using the properties of water at wavelength $\lambda = 0.65\,\mu$m. Three regimes are evident:

- The *geometric optics* regime, $x \gg 1$, in which $Q_{ext} \approx 2$ regardless of x.
- The *Mie regime*, $x = O(1)$, in which Q_{ext} varies about 2 with changes in x.
- The *Rayleigh regime*, in which the particle is small relative to the wavelength ($x \ll 1$) and extinction efficiency $Q_{ext} \propto x^4$.

Interactions between particles and radiation are efficient, in that Q_{ext} is near its asymptotic value, in all but the Rayleigh regime. The asymptotic value of Q_{ext} is 2, rather than 1, because diffraction around the edges of all but the smallest particles has roughly as large an influence on the radiation as do bulk interactions with the particle itself.

99

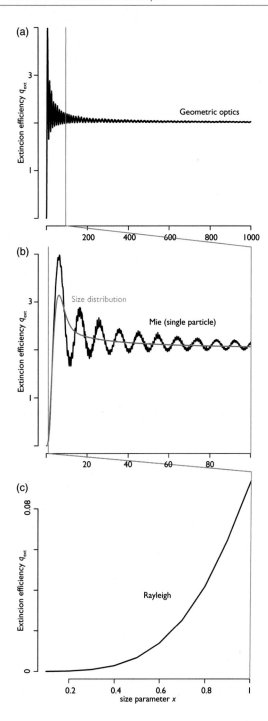

FIGURE 4.1: Extinction efficiency for spheres as a function of size parameter x over a range of scales of x (a–c). Calculations use an index of refraction ($n = 1.33 + 10^{-7}i$) similar to that of pure water at visible wavelengths (0.4–0.7 μm). The blue line in (b) shows extinction integrated over a drop size distribution where x is computed using the effective radius re; all other results are for a single particle size.

Typical sizes for cloud particles are $O(10\,\mu m)$ for liquid drops and $O(100\,\mu m)$ for ice crystals (see Table 3.2), while the wavelengths associated with the bulk of solar and thermal infrared radiation are 0.4–0.8 μm and 8–12 μm, respectively, for the most energetic spectral regions. Size parameters for clouds in the atmosphere are therefore large ($x \gg 1$) and interactions efficient (Q_{ext} near 2) for both solar and infrared radiation.

Extinction efficiency can be computed for individual particles using methods described in Section 4.1.3. The total strength of interaction for each particle is determined by the extinction cross-section $C_{ext}(D) = Q_{ext}(D)C_{geo}(D)$. To compute the strength of the overall interaction between radiation and a cloud with a distribution of particle sizes $n(D)$ (Section 3.1.1, where the same quantity is formulated in terms of radius, r rather than diameter, D), the contributions of all particles must be accounted for to produce the bulk extinction coefficient

$$\beta_{ext} = \int_0^\infty C_{ext}(D)\, n(D)\, dD = N\langle C_{ext}(D)\rangle, \qquad (4.1)$$

where C_{ext} and β_{ext} have units of squared length and inverse length, respectively. N is the particle concentration and brackets denote an integral over particle sizes, that is, $\langle\cdot\rangle = 1/N \int_0^\infty \cdot\, n(D)\, dD$ (see Eq. 4.1).

The effective radius r_{eff} is often used to describe the particle size distribution when computing single-scattering properties. For spherical cloud droplets $r_{eff} = \langle r^3\rangle/\langle r^2\rangle$ (as in Eq. (3.3)); more generally, effective radius is the value of $r = D/2$ that reproduces the ratio of volume to surface area for the distribution. Effective radius and its equivalents for non-spherical particles are useful because single-scattering properties including β_{ext} depend far more strongly on r_{eff} than on the details of $n(D)$.

4.1.1.2 The Fate of Extinguished Radiation: Absorption and Scattering

Radiation interacting with (extinguished by) a cloud may be either scattered or absorbed. The degree to which non-diffracted radiation meets one fate or the other depends in part on the material properties of the cloud drops: absorption increases with the imaginary part of the complex index of refraction. The optical properties of clouds are dominated by the index of refraction of water. As can be seen in Fig. 4.2, the complex part is small at visible wavelengths (e-folding lengths for absorption are of $O(10\,m)$) but relatively large at thermal infrared wavelengths, where absorption is $\sim 10^7$-times larger. Thus, scattering dominates the transfer of solar radiation in cloudy skies while thermal infrared radiation is dominated by absorption and

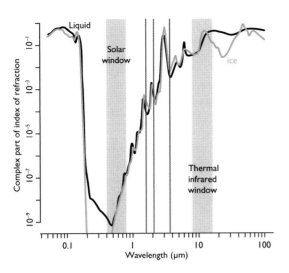

FIGURE 4.2: Complex part of the index of refraction of pure water in liquid and ice phases as a function of wavelength. Absorption is very low in the solar atmospheric window where most sunlight arrives (roughly 0.4–0.8 μm, left grey band), which is why clouds appear white to our eyes, but water and hence clouds absorb quite strongly in the thermal infrared (right grey band). The three purple lines indicate wavelengths used in the remote sensing of droplet size described in Section 4.4.2.3.

emission. In practice, nearly all cloud particles nucleate on some form of aerosol, but the approximation that pure water determines cloud optical properties is quite good.

Extinction is the sum of the scattering and absorption, that is, $\beta_{ext} = \beta_{sca} + \beta_{abs}$, for a collection of cloud particles. The relative importance of scattering is expressed through the single-scattering albedo

$$\omega_0 = \beta_{sca}/(\beta_{sca} + \beta_{abs}) = \beta_{sca}/\beta_{ext}. \qquad (4.2)$$

Scattering is conservative ($\omega_0 \approx 1$, meaning all light is reflected rather than absorbed) for clouds over most of the visible portion of the spectrum, which is why clouds appear white to our eyes. The diffracted radiation that amounts to roughly half the extinction efficiency for each particle is always scattered, so that $\omega_0 = 0.5$ even at wavelengths at which absorption is strong.

4.1.1.3   The Anisotropy of Scattering

Scattering redistributes radiation preferentially in specific directions determined by the size, shape and complex index of refraction of the particle. (Shape is most relevant for ice crystals, and includes both the macroscopic habit and the microscale surface roughness.) For a round or randomly oriented particle, so almost but not quite always in Earth's atmosphere, this redistribution can be expressed purely as a function of the scattering angle Θ between

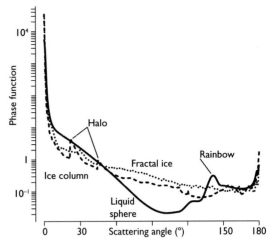

FIGURE 4.3: Scattering phase functions for a distribution of cloud droplets (solid line), regular hexagonal ice columns (dashed) and irregular fractal ice. All phase functions have a strong, narrow forward scattering peak due to diffraction.

FIGURE 4.4: Rainbows below precipitating shallow clouds over the subtropical Atlantic ocean. Photo: Bjorn Stevens

the incident and scattered radiation, where $\Theta = 0$ indicates direct forward scattering. The unit-less probability distribution $P(\Theta)$ is called the scattering phase function and is normalised so that $\int_{1}^{-1} P(\mu) \, d\mu = 4\pi$, where $\mu \equiv \cos \Theta$. Examples are shown in Fig. 4.3. Phase functions for individual size parameters exhibit finely detailed structure as a function of angle, varying in magnitude by five or more orders of magnitude. Phase functions become smoother when aggregated over size or shapes but robust phenomena emerge nonetheless, including a strong forward-scattering peak for liquid clouds and two maxima at the root of lovely phenomena: one, in round drops, at $\Theta = 138°$, that is responsible for rainbows (see Fig. 4.4), and another, present in hexagonal ice crystals at $\Theta = 22°$, that leads to halos.

On rare occasions, the redistribution of radiation depends explicitly on the absolute direction. One example occurs when hexagonal crystals in cirrus clouds become horizontally oriented by high-speed laminar winds in the upper troposphere. In this circumstance, the crystals act as tiny mirrors and preferentially reflect solar radiation at specular angles (i.e., at an angle of reflection equal to the angle of incidence relative to the plate orientation).

For many calculations, especially the calculations relevant for energy transport described in Section 4.2.1, the complete phase function is more detail than is needed and $P(\Theta)$ may be summarised by its first moment:

$$g = \frac{\int_{-1}^{1} \mu P(\mu) \, d\mu}{\int_{-1}^{1} P(\mu) \, d\mu}. \tag{4.3}$$

The asymmetry parameter g is the fraction of light scattered into the forward ($\mu > 0$) hemisphere, relative to the direction of the incident radiation. Typical values are about 0.85 in liquid-water clouds and 0.75 for large, randomly oriented, non-spherical (aggregate) ice crystals, values much closer to fully forward scattering ($g = 1$) than to isotropic scattering ($g = 0$).

4.1.1.4 Cloud Particles in Context

The interaction of light and clouds is determined by the scattering and extinction cross-sections of each particle; this slightly complicates the equations needed to determine the bulk optical properties of a mixture of particles (different drop sizes, say, or a mixture of ice and liquid clouds). Using i to denote individual components, the relations are

$$\beta_{ext} = \sum_i \beta_{ext,i}$$

$$\omega_0 = \sum_i \omega_{0,i} \beta_{ext,i} / \beta_{ext} \tag{4.4}$$

$$P(\Theta) = \sum_i P(\Theta)_i \omega_{0,i} \beta_{ext,i} / \omega_0.$$

Eq. (4.4) makes it clear why clouds, when they are present, dominate the flow of radiation through the atmosphere relative to condensed materials. Aerosol particles are one or more orders of magnitude smaller than cloud particles, leading to both smaller cross-sections and, in the infrared, to reduced extinction efficiencies. Aerosols are frequently far more absorbing than clouds, however, and can affect the vertical distribution of heating by absorbing sunlight. Precipitation particles are substantially larger than cloud particles (Table 3.2) but the concentrations are so much smaller than the impact on both solar and infrared radiation can normally be neglected.

4.1.2 Polarisation: Oriented Light

Radiation is an electromagnetic wave and polarisation the variable that describes the preferred orientation, if any, of this wave as a function of time. Much of the radiation in the atmosphere has no preferred orientation: radiation emitted by the sun, atmosphere and land surface is unpolarised. Scattering by cloud particles and gases can modify the polarisation state: Rayleigh scattering by gases, for example, increases the degree of linear polarisation (Polarised sunglasses reduce the glare from the sky by selectively filtering this light). The state of polarised light is described using the Stokes vector (I, Q, U, V) in which the elements refer to radiance (see Section 4.2.1) and the degrees of horizontal, vertical and circular polarisation, respectively. The polarised analogue of the phase function $P(\Theta)$ is the 4×4 Müller matrix $\mathbf{M}(\Theta)$. The first element of $\mathbf{M}(\Theta)$ describes the redistribution of radiance, that is, it is precisely $P(\Theta)$. The polarisation state of radiation does not impact the energy transported by radiation but can be exploited to infer cloud properties via remote sensing (see Section 4.4.3).

4.1.3 Tools for Computing Optical Properties

The quantities β_{ext}, ω_0, and $P(\Theta)$ or $\mathbf{M}(\Theta)$ are collectively called cloud single-scattering or optical properties. These properties are intrinsic, meaning that they are defined for an ensemble of cloud particles and do not depend on the cloud's extent or location in the atmosphere. The properties are governed by the microphysical state of the cloud, including the distribution of cloud particle size and shape, and by the wavelength of radiation through the size parameter and the complex index of refraction.

Determining cloud optical properties given these parameters is at least conceptually straightforward, although numerical methods vary in complexity according to the degree of asymmetry of the particles and the size parameter. Several methods solve the Maxwell equations governing the electromagnetic field inside and outside the particle and apply boundary conditions at the interface. The Lorenz–Mie theory applies to spheres of any size parameter and variants can be used for coated or concentric spheres; the electric fields are expanded as a set of spherical Bessel functions. The equations are relatively simple. Asymmetric ice particles are more challenging. Rotationally symmetric particles can be treated with the T-matrix method which generalises the Mie approach by using two sets of coupled equations. The properties of very large particles of arbitrary shape, such as large ice crystals at visible wavelengths, can be computed with geometric optics methods that use bulk Fresnel relations to trace all possible paths through a particle. Smaller size parameters require time-consuming approaches including the discrete dipole approximation and finite difference time domain methods, both of which discretise the domain in space and solve sets of equations describing the time-dependent electric field.

All methods for computing optical properties share the assumption that cloud particles are separated by distances much larger than the particle size (see Chapter 3). This implies that interaction between radiation and individual particles is independent ('incoherent') so that the optical properties can be computed for each particle class independently and summed.

4.2   From Location, Extent and Local Properties to the Radiation Field

Intrinsic optical properties – extinction coefficient β_{ext}, single scattering albedo ω_0 and scattering phase function $P(\Theta)$ or its first moment g – determine the interaction of radiation with clouds on the small scale. The large-scale radiation field depends how these properties are distributed in space. The (extrinsic) integral measures of these properties, described later, are collectively called cloud radiative properties. It is these properties and the distribution of temperature that determine the impact of clouds on the radiation field.

4.2.1   Definitions

Clouds, more than any other component of the atmosphere, exhibit great variability in all spatial directions. To keep the problem tractable, it is usual to simplify the problem to consider radiation traveling through a slab medium that varies only in the vertical dimension. (The impact of this one-dimensional assumption is revisited in Section 4.2.5.) Because the radiation field depends on (wavelength-dependent) optical properties, we define the vertical coordinate τ in terms of the unit-less optical depth:

$$\tau = \int_0^z \beta_{ext}(z)\, dz. \qquad (4.5)$$

It is usual in atmospheric problems to define $\tau = 0$ to be the top of the atmosphere. The symbol τ is sometimes also used to denote the optical depth (integrated extinction) of some portion of the atmosphere, so it's common to see τ used to refer to 'high cloud optical depth', for example.

Where relevant, the rest of the chapter uses τ_c to indicate cloud optical depth.

For unpolarised radiative transfer, the fundamental unit of radiation is radiance $I(\tau, u, \phi)$: the rate per unit area of electromagnetic energy flowing in a particular direction in a particular location, with direction described by the cosine of the polar angle u ($u > 0$ for downward transport) and azimuthal angle ϕ and location specified by the vertical coordinate τ. Radiance is defined across some spectral interval so that the units are $\mathrm{W\,m^{-2}\,sr^{-1}\,\mu m^{-1}}$ (the units for spectral interval may vary). Radiance is the first element of the Stokes vector mentioned in Section 4.1.2.

(In many treatments the cosine of the polar angle is denoted μ; u distinguishes the polar angle describing the direction in which radiation is propagating from the scattering angle appearing in Section 4.1.1.3.)

The total rate of energy transported across a given level by radiation per unit area is called the irradiance F and can be divided into upward and downward components, each describing the energy in the hemisphere flowing normal to a horizontal (constant τ) plane:

$$
\begin{aligned}
F^{\uparrow}(\tau) &= \int_0^{2\pi}\!\!\int_0^{-1} I(\tau, u, \phi) u\, du\, d\phi \\
F^{\downarrow}(\tau) &= \int_0^{2\pi}\!\!\int_0^{1} I(\tau, u, \phi) u\, du\, d\phi.
\end{aligned}
\qquad (4.6)
$$

Irradiance has units of $\mathrm{W\,m^{-2}\,\mu m^{-1}}$ but is often integrated over spectral intervals and expressed in units of $\mathrm{W\,m^{-2}}$. The terms intensity and flux are synonyms for radiance and irradiance, respectively.

Radiation incident on a layer can be absorbed or scattered into other directions; this impact is described with bulk quantities called absorptance $\mathcal{A}$, reflectance $\mathcal{R}$ and transmittance $\mathcal{T}$. Energy conservation requires that $\mathcal{A} + \mathcal{R} + \mathcal{T} = 1$. $\mathcal{R}$ and $\mathcal{T}$ describe the ratio of radiation returning from or passing through a layer to the incident radiation, normally in the absence of internal sources of radiation. Because scattering can change the direction of radiation, $\mathcal{R}$ and $\mathcal{T}$ are fundamentally functions of direction (e.g., $\mathcal{R}(u, \phi, u', \phi')$), but may be integrated over polar and/or zenith angles and applied to irradiance. The two-stream approximation, described in more detail in Section 4.2.4, provides a useful if rough approximation for hemispherically averaged quantities in the absence of absorption:

$$
\begin{aligned}
\mathcal{R} &\approx (1 - g)\tau / (2 + (1 - g)\tau) \\
\mathcal{T} &\approx 2 / (2 + (1 - g)\tau).
\end{aligned}
\qquad (4.7)
$$

The plane albedo α describes the hemispherically integrated reflectance of a system. Each of these quantities is spectrally dependent through the dependence on τ.

The ultimate source of all radiation, whether arising from the Sun or Earth, is thermal emission. The spectral distribution of emission from a perfectly black body depends on temperature T and is given by the Planck function

$$I_B(\lambda, T) = \frac{2hc^2}{\lambda^5} \frac{1}{\exp(hc/\lambda k_B T) - 1}, \qquad (4.8)$$

where h is Planck's constant k_B the Boltzmann constant and c the speed of light. Setting the derivative of Eq. (4.8) with respect to wavelength to 0 produces Wien's law ($T\lambda_{max} = C \approx 2.9 \times 10^{-3}$ m), which shows that warmer objects emit at shorter wavelengths.

Emissivity ϵ_λ is a spectrally dependent bulk measure that describes the efficiency with which a medium emits radiation. Perfect black bodies are defined by $\epsilon_\lambda = 1$; grey bodies are those with spectrally uniform emissivity less than 1. Clouds are roughly grey through most of the thermal infrared spectrum. Energy conservation requires that absorptance and emissivity be equal ($\mathcal{A} = \epsilon$) at every wavelength, a relationship known as Kirchoff's Law. The emissivity of an atmospheric layer is determined by the absorption optical depth $\tau_{abs} = (1 - \omega_0)\tau$ as

$$\epsilon = 1 - e^{-\tau_{abs}}. \qquad (4.9)$$

4.2.2 The Radiative Transfer Equation

How does the radiation field change in space? Consider a beam of radiation traveling through a differential unit of atmosphere $d\tau$ at polar angle $\arccos u$. As described in Section 4.1.1, the beam will be depleted by scattering and absorption and augmented by scattering and emission (the complement of absorption). The change in radiance can be expressed in the differential form of the radiative transfer equation (RTE)

$$u \frac{dI(\tau, u, \phi)}{d\tau} = -I(\tau) + S(\tau). \qquad (4.10)$$

The internal source term $S(\tau)$ includes isotropic thermal emission from the atmosphere itself (gases, clouds, aerosols) following Eq. (4.8) as well as the scattering of radiation into direction (u, ϕ) from other directions, that is,

$$S(\tau) = (1 - \omega_0(\tau))I_B(T(\tau))$$
$$+ \frac{\omega_0(\tau)}{4\pi} \int_0^{2\pi} \int_{-1}^1 P(\tau, \Theta)I(\tau, u', \phi') \, du' d\phi', \qquad (4.11)$$

where the scattering angle Θ depends on the directions of the outgoing (u, ϕ) and incoming beams (u', ϕ') according to standard trigonometric relations.

The solution to Eq. (4.10) shows that the radiance at any given location τ is the integral of the contributions from all locations the beam has passed through

$$I(\tau, u, \phi) = I(\tau_0)e^{-(\tau-\tau_0)/u}$$
$$+ \int_{\tau_0}^{\tau} e^{-(\tau-\tau')/u} S(\tau') \, d\tau' \qquad (4.12)$$

The first term on the right-hand side of Eq. (4.12) is the radiation incident at the boundary τ_0 (i.e., Earth or ocean surface for upwelling radiation and the top of the atmosphere for downwelling radiation) and attenuated along the path between the boundary and level τ. The second term is the total (emission plus scattering) source at τ' which is then attenuated before arriving at level τ. The arguments to the exponentials are always negative because u and $\tau - \tau'$ are always of the same sign. This equation is fully implicit and therefore difficult to solve because the radiance field $I(\tau, u, \phi)$ appears on both sides of the equation.

4.2.3 Simplifications: Two Disjoint Regimes

Radiation in the Earth's atmosphere comes from thermal emission by the Sun ($T \sim 5{,}800$ K) and from the Earth ($T \sim 288$ K) and its atmosphere. Because the temperatures of these two sources are so different, there is essentially no spectral region in which radiation from both sources is equally important (the maximum overlap is near $\lambda \approx 4\,\mu$m). This makes it possible to refer separately to the longwave, or infrared, or thermal radiation emitted by the Earth, and the short-wave, or solar, radiation emitted by the Sun.

Treating the two regimes independently allows for helpful simplifications. One is that the boundary conditions are very different: solar radiation arrives at the top of the atmosphere as a collimated beam as opposed to the essentially isotropic emission of the surface in the infrared. Equally important is that, to a good approximation, emission can be neglected when considering solar radiation, and scattering can be neglected in calculations of infrared radiation as long as the distinction between extinction and absorption is respected.

One consequence is that the impact of clouds on the shortwave radiation field varies over a much wider range of optical depths than in the long-wave. As is apparent from Eq. (4.9), $\epsilon \to 1$ for $\tau_{abs} \gtrsim 3$, so that even moderately optically thick clouds are essentially opaque to infrared radiation and further increases in τ do not change the radiation field. (This is why emissivity is often a more useful measure than optical depth for problems in the infrared.) Eq. (4.7), on the other hand, shows that $\mathcal{T}$ and $\mathcal{R}$ continue to change with τ for values as large as 100, a value realized by only a very small proportion of clouds on Earth.

Clouds co-exist with aerosols in a gaseous atmosphere with rich spectral structure but they are the optically thickest component of the atmosphere for large portions of the solar and thermal infrared spectrum, especially in the 'atmospheric windows' (0.4–0.8 μm in the solar and 8–12 μm in the infrared). This makes it useful to examine the behaviour of radiative transfer in atmospheres with clouds alone, neglecting the contributions of aerosols and gases.

In the thermal infrared, the complex part of the index of refraction of water is high and the single-scattering albedo of water quite low (see Fig. 4.2). In this circumstance, scattering may be neglected and the RTE can be simplified by ignoring the radiation scattered almost directly forward by diffraction, and writing the RTE neglecting scattering using the absorption optical depth

$$I_{\mathrm{IR}}(\tau_{\mathrm{abs}}, u, \phi) = I_{\mathrm{IR}}(\tau_0) e^{-(\tau_{\mathrm{abs}} - \tau_0)/u}$$
$$+ \int_{\tau_0}^{\tau} e^{-(\tau_{\mathrm{abs}} - \tau'_{\mathrm{abs}})/u} I_B(T(\tau')) \, d\tau'_{\mathrm{abs}}. \tag{4.13}$$

Thermal infrared radiance (and irradiance) is governed by the temperature T and absorption optical depth τ_{abs}. Boundary conditions are straightfoward: there is almost no downwelling infrared radiation at the top of the atmosphere ($I_{\mathrm{IR}}(\tau = 0, u > 0, \phi) = 0$), while upwelling radiation at the surface is the product of the Planck function (Eq. (4.8)) at the surface temperature and the spectrally dependent bulk surface emissivity.

In the solar window, clouds and gases are almost entirely non-absorbing ($\omega_0 \approx 1$) so that the RTE can be approximated as

$$I_{\mathrm{solar}}(\tau, u, \phi)$$
$$= I_{\mathrm{solar}}(\tau_0) e^{-(\tau - \tau_0)/u} + \frac{1}{4\pi} \int_{\tau_0}^{\tau} e^{-(\tau - \tau')/u}$$
$$\times \left(\int_0^{2\pi} \int_1^1 P(\tau', \Theta) I_{\mathrm{solar}}(\tau', u', \phi') \, du' \, d\phi' \right) d\tau'. \tag{4.14}$$

From this equation, it can be seen that the visible radiance (and irradiance) is governed by the scattering phase function $P(\Theta)$ and optical depth τ and is independent of cloud temperature. The boundary condition for solar radiation at the top of the atmosphere is a collimated beam traveling in a single direction $u_0 > 0, \phi_0$, that is, $I_{\mathrm{vis}}(0, u, \phi) = S_0 \delta(u_0) \delta(\phi_0)$, where $S_0 \approx 1361 \, \mathrm{W} \, \mathrm{m}^{-2}$ is the solar constant and δ the Kroenecker delta function.

It is sometimes useful to think of visible radiation as the sum of a direct component, containing radiation which has not been scattered, and a diffuse component. The RTE for the direct component has no internal source (although the direct beam is a source for the diffuse component). After applying the boundary condition, the solution to Eq. (4.12) is

$$I_{\mathrm{vis}}^{\mathrm{dir}}(\tau, u, \phi) = I_{\mathrm{vis}}(0, u, \phi) e^{-\tau/u_0}. \tag{4.15}$$

This relationship is referred to as the Beer–Lambert law.

4.2.4 Tools for Solving the RTE

The complete RTE (Eq. (4.12)) describes the distribution of radiance at a single wavelength as a function of angle through the depth of the atmosphere. Numerical techniques exist to solve this implicit integral equation and even, given enough computational effort, the far more complicated three-dimensional version of the equation. Still, many important problems admit substantially simpler approximations. Section 4.2.3 described one set of simplifications, namely the neglect of scattering for longwave calculations and of internal sources for short-wave calculations. Eq. (4.13), in fact, fully describes the radiance field in the absence of scattering if the optical properties of the atmosphere and the temperature are known as a function of height.

The solution to the RTE for scattering atmospheres, Eq. (4.14), is more complicated because the radiance field appears on both sides of the equation. The widely used discrete ordinates method approaches the problem by expanding the azimuthal (ϕ) dependence of the radiance field into uncoupled Fourier modes and solving the now-simplified scattering integral for each mode at discrete polar angles (u). An alternative is the adding method. The angular distribution of radiation reflected from $\mathcal{R} = \mathcal{R}(u, \phi, u', \phi')$, and transmitted through $\mathcal{T}$, a very optically thin layer can be computed assuming that radiation scatters at most one time within the layer. The combined properties of two such layers can be determined by computing the infinite series describing all possible interactions between them. Schematically

$$\mathcal{R}_{1+2} = \mathcal{R}_1 + \mathcal{T}_1 \mathcal{R}_2 \mathcal{T}_1 + \mathcal{T}_1 \mathcal{R}_1 \mathcal{R}_2 \mathcal{R}_1 \mathcal{T}_1 + \cdots$$
$$= \mathcal{R}_1 + \mathcal{T}_1^2 \mathcal{R}_2 / (1 - \mathcal{R}_1 \mathcal{R}_2), \tag{4.16}$$

where the series (in practice, a series of matrix multiplications at discrete values of u, ϕ) converges because all elements of $\mathcal{R}$ and $\mathcal{T}$ are always less than one. The properties of arbitrarily many layers can be computed by repeating the process.

Angular detail may also be unnecessary, as when radiation calculations are performed to compute energy transfer. In this case, it is normally more effective to develop approximate equations describing irradiance rather than computing the radiance field and integrating the result using Eq. (4.6). The approximate equations are broadly known as 'two-stream' methods because the radiance field is assumed to have no azimuthal structure and to depend very simply on the polar angle (e.g., isotropic in each hemisphere, or varying linearly with u, which is known as the Eddington approximation). Analytic equations describe the reflectance and transmittance of layers given arbitrary values of τ_c, ω_0 and g; layers of different properties can be combined using the scalar equivalent of the summation performed in Eq. (4.16). Well-posed two-stream methods reproduce the hemispheric average of angularly detailed calculations to within a few per cent.

The accuracy of two-stream treatments for clouds can be greatly increased by accounting for the very strong peak in the scattering phase function for cloud particles caused by diffraction. Because the peak is restricted to within a few degrees of direct forward scattering (see Fig. 4.3), a useful approximation is to treat the phase function as the sum of a δ-function peak of magnitude f in the forward direction and some smoother phase function $P^*(\Theta)$, that is,

$$P_\delta(\Theta) \approx 2f\delta(0) + (1-f)P^*(\Theta).$$

A new set of 'δ-scaled' optical properties can be derived by requiring the asymmetry parameters of $P(\Theta)$ and $P_\delta(\Theta)$ to be identical. This yields

$$\tau^* = (1 - \omega_0 f)\tau$$
$$\omega_0^* = (1-f)\omega_0/(1-\omega_0 f) \qquad (4.17)$$
$$g^* = (g-f)/(1-f),$$

where reduced optical thickness and decreased forward scattering combine to produce the same reflection and transmission. (The equation describing radiation in a medium that only scatters and emits, Eq. (4.13), is in fact the RTE expressed in terms of optical properties delta-scaled using $f = 1$, $\omega_0 = 1/2$.) In these equations, f is a free parameter although it's common to assume $f = g^2$. The δ-scaling in Eq. (4.17) is usually applied to irradiance calculations but related approaches are available for radiance computations.

4.2.5 The Limits of Accuracy

Computational methods introduce errors through discretisation and simplifying approximations. In most circumstances, however, the accuracy of radiative transfer calculations is limited in practice by the amount of computational effort that can be expended and by limited knowledge about the state of the atmosphere. In some cases, this limited accuracy is not related to clouds at all. For example, spectrally integrated broadband irradiance calculations are required to compute the total heating rates by radiation in the atmosphere and at the surface. Such calculations are needed in weather and climate models. The main difficulty with broadband integration is the enormous variability in the opacity of gases with wavelength. Reference 'line-by-line' models with very high spectral resolution can reproduce radiation measurements to within tenths of a per cent when the temperature and humidity distribution is well characterized. Such models are very computationally expensive, however, and approximate parameterisations for treating this complicated spectral dependence (often using so-called k-distributions) are normally developed. These greatly reduce computation while modestly increasing error. Different problems require different trade-offs between computational cost and accuracy.

For other problems, such as estimating the effect of clouds on Earth's radiation field (Section 4.5.2), accuracy is bounded primarily by limited knowledge of the state of the atmosphere, especially the time-varying three-dimensional distribution of cloud physical properties.

In many cloudy conditions, the limit to accuracy is more fundamental, with roots in the one-dimensional assumption that underlies the RTE – Eq. (4.12). The physical properties of clouds relevant for radiation (e.g., condensed water path, drop size) vary enormously across a wide range of scales. This poses two distinct difficulties. The first relates strictly to the assumption of homogeneity. The solution to the RTE is non-linear in many parameters so that, in clouds with spatially varying optical properties, the radiance or irradiance field computed using spatially averaged properties is not equal to the average of the field computed using variable properties. The second difficulty arises because radiation does not, in fact, restrict its travels to the vertical dimension. Objects including clouds cast shadows, reflect from their sides and otherwise cause net horizontal transfers of radiation that are not captured by the one-dimensional equation. The short-wave radiance field, especially, may be influenced by the optical properties of regions a kilometre or more away. These 'three-dimensional radiative transfer' effects are largest where gradients in optical properties are strongest, for example, in deep convective clouds with large vertical extent and complicated shapes, and in shallow cumulus clouds that have small horizontal extent and so a large ratio of cloud edge to cloud interior volume. Computational methods – Monte Carlo integration or a few explicit methods – are available to solve the three-dimensional analogue of the RTE. These are very computationally expensive but show that three-dimensional radiative transfer can change estimates of domain-averaged radiation fields by an order 10% (local effects, for example, in shadows, can of course be much higher).

4.3 Clouds' Impact on the Radiation Field

Radiation is a form of energy; a system the emits more radiation than it absorbs must cool unless energy is added to it by other means. Within the atmosphere, for example, changes with height in $F^{net} = F^{\downarrow} - F^{\uparrow}$ imply a gain or loss of radiant energy, which is related to the local heating rate through

$$\frac{\partial T}{\partial t} = -\frac{1}{\rho c_p}\frac{\partial F^{net}}{\partial z}, \tag{4.18}$$

where ρ is the density and c_p the isobaric specific heat capacity of the (moist) atmosphere at location z.

4.3.1 Cloud Radiative Effect

Clouds can have an enormous impact on the distribution of net irradiance at all layers in the atmosphere because they are so optically thick. This motivates a desire to summarise the 'cloud radiative effect' (CRE) as the difference in between all-sky and clear-sky irradiance

$$CRE = F^{net}_{all-sky} - F^{net}_{cloud-free-sky}, \tag{4.19}$$

so that temperature change and CRE have the same sign.

The concept of the CRE can be applied to long- or short-wave irradiance or to their sum. It is most commonly computed at the top of the atmosphere to assess the impact of clouds on Earth's overall energy budget, but the concept applies equally well at the surface; the difference between top of the atmosphere and surface values describes cloud radiative heating within the atmosphere. Given estimates of the three-dimensional distributions of clouds (e.g., from active remote sensing instruments, Section 4.4.3), it is possible to resolve the vertical profile of the CRE, that is, the impact of clouds on local heating rates through the depth of the atmosphere.

Because the CRE is determined relative to clear skies, the impact also depends on the state of the non-cloudy atmosphere and surface. One implication is that clouds

with the same physical properties can have very different radiative effects depending on the underlying surface. Optically thick clouds have a large short-wave impact at the top of the atmosphere when they are over the ocean, for example, but little impact over bright surfaces such as snow, ice and desert. This is illustrated in more detail in the next section.

In general, clouds are brighter and colder than Earth's surface so their effect at the top of the atmosphere is to cool the system by reducing absorbed short-wave radiation and to warm the planet by reducing the outgoing long-wave radiation. These effects are opposed at the surface, where decreased solar radiation in the presence of clouds acts to cool but increased downwelling radiation acts to warm. The net impact depends on many factors, including the cloud albedo (determined primarily by optical thickness), the temperature contrast with the surface (determined in part by vertical position in the atmosphere), and the properties of the atmosphere and surface.

Observations of the cloud radiative impact, described in Section 4.5 show that the globally averaged total (short-wave + long-wave) CRE is negative, that is, that clouds act to cool Earth.

4.3.2 Limiting Behaviour: Optically Thick and Thin Clouds

Useful intuition about the CRE can be developed by looking at the limits of very thick and very thin clouds while ignoring any impacts of the non-cloudy atmosphere, focusing on the window regions (Section 4.2.3) in which clouds dominate the radiation field.

In the long-wave, where scattering plays a small role, the upwelling radiance above a cloud depends on the emissivities (ϵ_c, ϵ_s) and temperatures (T_c, T_s) of the cloud and the surface:

$$I^{\uparrow}_{IR} \approx (1 - \epsilon_c)\epsilon_s I_B(T_s) + \epsilon_c I_B(T_c), \tag{4.20}$$

where $\epsilon_s I_B(T_s)$ represents emission by the surface at temperature T_s; this radiation is then modified by emission at cloud temperature T_c. Surface temperatures are almost always higher than cloud temperatures, so $I_B(T_s) > I_B(T_c)$.

Optically thin clouds for which $\tau_c \ll 1$, have, by definition, a small impact on the radiation field that can be treated as a linear perturbation to the radiation at the boundary. Since $\epsilon_c \approx \tau_c$ for small τ_c, Eq. (4.20) becomes

$$I^{\uparrow}_{IR} \approx (1 - \tau_c)\epsilon_s I_B(T_s) + \tau_c I_B(T_c). \tag{4.21}$$

As a result, the radiative effect of thin clouds depends on $T_s - T_c$, ϵ_s and $\epsilon_c \approx \tau_c$.

Short-wave radiation above thin clouds can be approximated by considering only single scattering events. Approximating $P(\Theta) \approx 4\pi g \delta(0) + (1 - g)$ provides an expression for the upwelling short-wave radiation at the top of the atmosphere traveling in direction μ_s

$$
\begin{aligned}
I_{\text{vis}}^{\uparrow} &\approx (1 - g)\tau_c I_0^{\downarrow} + I_{\text{vis,s}}^{\uparrow} \\
&\approx (1 - g)\tau_c I_0^{\downarrow} + e^{-\tau_c/\mu_0}\rho_s e^{-\tau_c/\mu_s} I_0^{\downarrow} \\
&\approx I_0^{\downarrow}((1 - g)\tau_c + e^{-\tau_c/(\mu_0+\mu_s)}\rho_s),
\end{aligned}
\tag{4.22}
$$

where $I_0^{\downarrow}$ is the solar incident light at the top of the atmosphere at solar zenith angle μ_0 and ρ_s is the surface reflectance. One consequence is that the presence of thin clouds over highly reflective (large ρ_s) surfaces like ice or desert does not strongly influence the radiation leaving the atmosphere while their impact over dark surfaces ($\rho_s \to 0$) like oceans can be much more noticeable.

Above and below very optically thick clouds ($\tau_c \gg 1$, $\epsilon_c \approx 1$) the long-wave radiance is close to the Planck emission at the local cloud temperature:

$$
\begin{aligned}
I_{\text{IR}}^{\uparrow} &\approx I_B(T_{\text{cloud-top}}) \\
I_{\text{IR}}^{\downarrow} &\approx I_B(T_{\text{cloud-base}}).
\end{aligned}
\tag{4.23}
$$

Thus, in optically thick clouds, the emitted radiation is solely a function of cloud temperature, which higher clouds emitting less (and hence having a smaller value of brightness temperature T_b) than lower clouds. This can be exploited to estimate the temperature at cloud top as described in Section 4.4.2.1. It also implies that the properties of the surface have no influence on the radiation above the cloud.

In the short-wave, large values of optical thickness imply large amounts of multiple scattering, which means that upward radiance increases roughly logarithmically with optical depth (Eq. (4.7)). As with the infrared, upward short-wave radiance is independent of the properties of the underlying surface. Beneath thick clouds, the direct beam is entirely extinguished and even the downwelling radiation contains no signature of the angular location of the sun: radiance is isotropic in ϕ and smooth in u.

4.4   Remote Sensing: Observing Clouds with Radiation

As the previous sections have illustrated, the impact of clouds on the radiation field in the atmosphere depends on their microphysical properties, through the influence of particle size and shape on single-scattering properties, and macrophysical properties, through vertical location, which influences temperature, and extent, which influences optical thickness. CREs on larger scales therefore depend on the distribution of these properties in time and space. The most complete estimates of these distributions available on large temporal and spatial scales are derived using remotely sensed observations from satellites.

4.4.1   Principles for Remote Sensing Cloud Properties

Remote sensing is the process of interpreting radiation measurements, localised in direction and spectral interval, to infer information about the state of the system affecting the radiation. For the remote sensing of clouds information and/or assumptions about the non-cloud atmosphere and surface are combined with modelling to isolate the cloud's impact on the observed radiance; this influence can then be interpreted in terms of the cloud's physical properties. One implication is that individual observations must be classified as cloudy or not-cloudy based on the information available. This precursor step of 'cloud detection' is tightly coupled to the range of properties observable with a given system (see also the discussion in Section 4.5.1.1).

Remote sensing of cloud properties exploits the change in the radiation field caused by the presence of clouds relative to the clear sky, that is, the CRE. The process is made robust by seeking measurements that are affected strongly by clouds and weakly, or not at all, by other components of the atmosphere. This is why most cloud properties retrieved from satellites use observations of radiances in the atmospheric windows (see Section 4.2.3). Extracting quantitive information requires substantial amounts of modelling, including constructing physical models of the atmosphere and the clouds themselves and then using these to predict the impact of those clouds on the radiation measurements, using the techniques described in previous sections. The difference between the expected clear-sky signal and observed cloudy-sky signal can then be interpreted in terms of the parameters of the cloud model.

Remote sensing measurements are fundamentally of radiance, localized in space and direction. For radiation fields with strong anisotropy, this implies the models used to predict the measurement require explicit angular resolution, which rules out simple methods sufficient for computing irradiance (e.g., two-stream methods, as described in Section 4.2.4) in favour of more complete and computationally costly methods like discrete ordinates.

TABLE 4.1: Selected satellite observations of clouds (and precipitation)

Name	Spectral detail	Primary cloud quantities	Epoch
Passive systems			
ISCCP	2 (vis, thermal infrared (IR))	τ_c, p_{ct}	1983–
PATMOS-X	3 (vis, near-IR, thermal IR)	τ_c, p_{ct}	1983–
MODIS	36 (vis to thermal IR)	τ_c, p_{ct}, r_{eff}	1999–
CERES	broadband (solar, IR)	long wave, short wave irradiance at the top of the atmosphere	2000–
SSM/I	4 (microwave, 3 polarisations)	Liquid-water path	1987–
AIRS	hyperspectral (thermal IR)	$\overline{\epsilon_c}, p_{ct}, r_{eff}$ for ice clouds	2000–
Active systems			
CloudSat	94 GHz	Radar reflectivity Water content (derived)	2006–
CALIPSO	0.532 μm and 1.024 μm	Lidar backscatter, polarisation	2006–
Active systems for precipitation			
Tropical Rainfall Measuring Mission (TRMM)	14 GHz	Radar reflectivity, precipitation	1997–2015
Global Precipitation Measurement	14 GHz and 35 GHz	Radar reflectivity, precipitation	2014–
Prospective systems			
EarthCare	7 (vis, near-IR, thermal IR)	τ_c, p_{ct}, r_{eff}	2021ğ–
	2 (broadband vis, IR)	Radiative fluxes	
	94 GHz, Doppler	Radar reflectivity, vertical motion Water, rain, snow, content (derived)	
	0.355 μm, Doppler	Extinction, vertical motion	

In this table τ nominally refers to optical thickness at visible wavelengths (normally 0.67 μm to reduce the impact of Rayleigh scattering by gases). A data set may be derived from a specific instrument (Atmospheric Infrared Sounder (AIRS), Moderate-Resolution Imaging Spectroradiometer (MODIS), Clouds and Earth's Radiant Energy System (CERES), Special Sensor Microwave/Imager (SSM/I)) aboard one or more platforms, or from a suite of instruments aboard a single platform (CloudSat, Cloud-Aerosol Lidar and Infrared Pathfinder Satellite Observation (CALIPSO)), or from a time-varying set of observations from different instruments and platforms (International Satellite Cloud Climatology Project (ISCCP), Pathfinder Atmospheres-X (PATMOS-X)). ISCCP and PATMOS-X apply similar but independently produced algorithms to a small number of measurements, but ISCCP prefers observations from geostationary satellites where they are available in order to observe the diurnal cycle, while PATMOS-X relies solely on sun-synchronous polar orbiters so that the distribution of viewing and illumination angles is more spatially uniform. $\overline{\epsilon}$ denotes area-averaged emissivity which may be expressed in other ways (effective emissivity, effective cloud cover). AIRS is one of several instruments providing high-spectral resolution measurements in the infrared. Cloud detection and phase assignment are precursors to any physical retrieval algorithm. EarthCare, planned for launch in 2019, combine four cloud- and radiation-oriented instruments on a single platform.

Remote sensing observations are interpreted using the one-dimensional, plane-parallel homogeneous RTE. Both inhomogeneity and net horizontal transport of radiation are known to be present in clouds at the scales relevant to satellite observations, however, and using this equation where it is not strictly valid (Section 4.2.5) introduces errors that can be difficult to assess. One concrete implication is described in Section 4.5.1.1.

This section describes some of the methods used to derive information about clouds from remote sensing measurements in a generic way, emphasizing the common principles underlying remote sensing and cloud radiative impacts. Each of the methods described later, however, is used to produce one or more of the observational data sets described in Table 4.1.

Methods for remotely sensing clouds can be categorized as *passive* methods, for which the source of radiation is either the sun, Earth or the atmosphere, and *active* methods that provide their own source of radiation.

4.4.2 Passive Measurements

As described in Section 4.3, clouds have a strong impact on both solar and infrared radiation leaving the atmosphere, so it's natural to exploit these impacts to infer information about cloud properties. Indeed, visible and infrared observations were the earliest obtained from satellites, starting with the Nimbus weather satellites launched in the 1960s.

4.4.2.1 Cloud-Top Pressure from Thermal Emission

At wavelengths between roughly 8 μm and 14 μm, radiance is approximately given by Eq. (4.13) and an observed radiance can be inverted to determine the brightness temperature T_b, the temperature a perfect blackbody would need to have in order to emit the observed radiance. That is, given a measurement of monochromatic infrared radiance I, the brightness temperature can be inferred from the emission, Eq. (4.8).

If cloud emissivity can be determined (or is assumed), T_b can be converted to a physical temperature; cloud pressure can then be determined from cloud temperature if the profile of temperature as a function of pressure is known. The temperature or pressure retrieved in this way is usually described as being the value at cloud top and denoted p_{ct}. In fact, the value represents a weighted average through the first three or so optical depths, as measured from the top of the cloud, which can be quite different than the actual cloud top in tenuous, optically thin or multi-layered clouds. It is, however, the pressure consistent with the cloud impact on the top-of-the-atmosphere outgoing long-wave radiation.

4.4.2.2 Cloud Optical Thickness from Reflected Solar Radiation

As Eq. (4.14) demonstrates, the amount of solar radiation leaving the top of the atmosphere depends on τ, $P(\Theta)$ and ω_0. Formally one would then require three independent observations to estimate these three parameters. The process can be simplified by choosing a wavelength at which liquid and ice are essentially non-absorbing so that ω_0 is known to be near 1. It's then common to assume a microphysical model for all clouds (i.e., a drop size distribution for liquid clouds, or a possibly temperature-dependent distribution of sizes, habits and surface roughness for ice clouds), use these to estimate $P(\Theta)$ a priori, and use a single measurement of reflected solar radiance to infer τ_c.

Reliance on a microphysical model means that determination of the thermodynamic phase (liquid or ice) is a precursor to the inference of optical thickness. Methods for phase determination range from simple thresholds based on brightness temperature in the infrared window to methods exploiting the spectrally dependent difference in the complex index of refraction of liquid and ice, depending on what information is available from a given observing system. If the phase can't be clearly determined, estimates of optical thickness are themselves uncertain, but given accurate phase determination optical thickness results for liquid clouds are fairly robust because $P(\Theta)$ is not particularly sensitive to the details of realistic size distributions. Ice particles are substantially more variable, which leads to greater uncertainty.

4.4.2.3 Cloud Particle Size from Reflected Solar Radiation

This approach to estimating optical thickness can be extended to estimate cloud particle size by adding an additional measurement at a wavelength in which absorption by condensed water is significant and absorption by the clear atmosphere negligible ($\lambda > 1$ μm, roughly and with some exceptions). This pairing of absorbing and non-absorbing wavelengths exploits the fact that reflection arises at interfaces and absorption in the interior of media, so that measurements at non-absorbing wavelengths are sensitive to total area $\sim \langle D^2 \rangle$, the vertical integral of which is proportional to τ, while measurements at absorbing wavelengths sensitive to total volume $\sim \langle D^3 \rangle$. Given one measurement of each type, it is therefore possible to simultaneously estimate τ and ω_0 or, using a microphysical model to provide links among these parameters and g, to estimate τ and r_{eff}. Particle sizes estimated this way are weighted towards the top of the cloud, especially in optically thick clouds. Simultaneous estimates of τ and r_{eff} can be combined to estimate liquid or ice water paths through $\tau \approx \frac{3}{2\rho_1} \frac{LWP}{r_{eff}}$ (as in Eq. (3.4)), but this indirect inference does not represent additional information and is subject to bias from systematic variations of particle size with height. Observations of absorption have historically been made at ~ 3.7 μm, although these must account for thermal emission by the cloud itself. More recent sensors make use of observations at 2.1 μm and 1.6 μm, although lower single-scattering albedos at these wavelengths mean that interpretation can be more strongly compromised when the clouds being observed are far from the plane-parallel homogeneous model.

4.4.2.4 Liquid Water Path from Microwave Emission and Absorption

A range of remote sensing techniques use measurements in spectral regions far from the peaks of solar or terrestrial emission to estimate cloud properties. A useful example is the estimation of a liquid-water path from microwave measurements. In the microwave, at frequencies from roughly 20 GHz to 200 GHz (wavelengths of 0.15–1.5 cm), emission and absorption is due primarily to oxygen, water vapour and liquid water. Liquid water is quite absorptive while ice scatters but does not absorb much (which is why microwave ovens are so effective at cooking food but so poor at melting ice cold enough to have no liquid layer.) Microwaves are much larger than cloud droplets. In the Rayleigh regime, absorption efficiency Q_{abs} increases linearly with size parameter so the absorption coefficient C_{abs} increases as the third moment of the drop size distribution – that is, proportional to the mass concentration. This sensitivity allows direct estimates of a column-integrated liquid-water path from passive microwave observations.

The emissivity of land surfaces at these frequencies is quite variable so that retrievals are done only over the ocean. The ocean surface is partly reflective in the microwave, while the atmosphere is semi-transparent so that the top-of-the-atmosphere radiance includes contributions emitted directly by the atmosphere, emitted by the surface and attenuated, and radiation emitted by the atmosphere, reflected from the surface, and attenuated through the atmosphere. It's common to assume that atmospheric emission occurs at nearly the same temperature (T_s) as surface emission, which is reasonable given that water vapour and liquid clouds are near the surface. Expressing radiance as a linear function of brightness temperature (a good approximation in the microwave) we have

$$T_b = T_s(1 - \mathcal{T}) + \epsilon T_s \mathcal{T} + T_s(1 - \mathcal{T})(1 - \epsilon)\mathcal{T}$$
$$= T_s(1 - \mathcal{T}^2(1 - \epsilon)), \tag{4.24}$$

where ϵ is the emissivity of the sea surface and $\mathcal{T}$ the transmissivity of the atmosphere along the viewing path.

In the microwave, ϵ depends on wind speed, through the distribution of wave sizes, but also on the polarisation state, so that reflection and emission from the ocean surface is polarised while the atmospheric absorption and emission depend only on the wavelength through the absorption coefficient. As a result, two or more pairs of polarised top-of-the-atmosphere brightness temperature measurements, made at frequencies with different vapour and liquid absorption coefficients, can be inverted to estimate column-integrated water vapour and liquid-water paths. Such estimates date to the early 1990s and provide one of the longest time series available from satellites. Microwave radiance measurements at the top of the atmosphere fundamentally depend on all liquid water in the column including both clouds and rain. Separate estimates of cloud- and rain-water paths from such measurements are possible only by applying further assumptions.

4.4.3 Active Measurements: Lidar and Radar

Passive remote sensing methods exploit radiation emitted naturally by the sun or the earth/atmosphere system. Active sensors, in contrast, use their own source to send out very short pulses of radiation and measure the timing and intensity of the returning 'echo' to high accuracy. The characteristics of the return provide information about the cloud properties that can be located precisely as a distance from the sensor by measuring the time-of-flight. This allows active instruments to provide detailed information on the vertical distribution of clouds in the atmosphere.

4.4.3.1 Commonalities

Active remote sensing measurements rely on scattering, which is important in the visible and microwave portions of the spectrum used by lidar and radar instruments, respectively. Lidar and radar instruments are designed with very narrow fields of view so that returns are dominated by radiation that is scattered once and returns directly in the direction from which it came. This radiation is extinguished along the path in both directions, so that the power p at the detector returned from distance d is

$$p(d) = \frac{K}{d^2}\alpha_{bck}(d)\exp\left(-2\eta\int_0^d \beta_{ext}(d')\,dd'\right)$$
$$= \frac{K}{d^2}\alpha_{bck}(d)\exp(-2\eta\tau), \tag{4.25}$$

which is essentially the RTE in a scattering media in a single direction $P(\Theta = \pi)$. Here K describes instrument characteristics, including the initial pulse strength and receiver gain, and $\alpha_{bck}(d) = \beta_{sca}(d) \times P(\Theta = \pi, d)/4\pi$ is the backscatter coefficient. Despite the narrow field of view, radar and lidar signals do contain some radiation that has been scattered multiple times. Multiple scattering implies longer path lengths than single scattering, so the signal takes longer to arrive at the detector and is interpreted as coming from larger ranges. The multiple-scattering factor η decreases from unity (for single scattering) to lower values when the number of scattering increases. When the optical depth is large ($\eta\tau \gtrsim 3$), the signal becomes too small to measure and the beam is said to be fully attenuated.

Extinction and scattering provide information about particle numbers and sizes while polarisation can be used to infer information about particle shape. Many lidars and radars exploit this by emitting linearly polarised radiation and interpreting the signal using a version of Eq. (4.25) in which $P(\Theta)$ is replaced by other elements of the Müller matrix (Section 4.1.2). Light reflected from spherical particles retains its polarisation after a single scattering event, so linear polarisation in the returned signal implies that the target is made of spherical (i.e., liquid) particles, while reduced polarisation ratios imply non-spherical particles, including ice and, for lidars, aerosols. This phase identification is substantially more certain than those made on the basis of spectral signatures in passive measurements.

4.4.3.2 Differences

Although they rely on the same general principles, the different wavelengths at which lidars and radars operate have profound implications. Lidars for atmospheric research normally use lasers at ultraviolet and visible wavelengths so that clouds, aerosols and Rayleigh scattering from gases all contribute to the signal. Visible wavelengths are substantially smaller than cloud particles so that the backscatter efficiency $Q_{bck} \approx 1$ (Section 4.1.1.1) and the returned signal is dominated by the integrated projected area of the cloud particles (β_{sca}). Extinction is relatively high and water clouds, especially, are frequently so optically thick that lidar signals are completely attenuated before reaching the cloud base.

Radar relies on the same principles as lidar but uses microwave radiation with wavelengths of a few millimetres or longer. This puts cloud droplets in the Rayleigh regime in which $Q_{bck} \propto x^4$, so that α_{bck} is sensitive to the sixth moment of the drop size distribution. This means that the largest cloud particles dominate the signal and the smallest are essentially invisible. Radar wavelengths can therefore be chosen to be sensitive to different populations of cloud or precipitation particles. Popular wavelengths for cloud radars are roughly 3.2 mm and 8.6 mm (frequencies of 94 GHz and 35 GHz, referred to as W and Ka bands, respectively). These radars attenuate very rapidly in the presence of strong precipitation. Radars used to probe precipitation employ centimetre-scale wavelengths (often 2–3.2 cm, Ku-band radars with frequencies around 13 GHz) to which the clouds themselves are essentially transparent.

The passive remote sensing observations described in Section 4.4.2 are vertically integrated measures of cloud properties where, for most quantities, the integral is weighted towards values at the cloud top. One consequence is that any high clouds in a scene 'mask' lower clouds, making it difficult to determine the true distribution of properties for all clouds in the atmosphere. Active instruments do not suffer from this limitation as long as the clouds are not thick enough to entirely attenuate the signal (i.e., essentially always for cloud radars). Existing active instruments, on the other hand, are restricted to a single near-nadir viewing direction, so that samples sizes are much smaller than from passive instruments.

4.4.4 Sampling and Orbits

Satellite observations are useful in observing the large-scale distribution of cloud properties because platforms in orbit can observe large geographic areas in relatively short periods and so can provide essentially the only global view. Different orbits allow for different sampling strategies, which may be useful in trying to capture, for example, the strong diurnal cycle exhibited by clouds in some regimes. Three orbits are commonly used by Earth-observing satellites:

Sun-synchronous orbits are low-altitude ($\sim$700 km) high-inclination orbits that precess by roughly $1\,^{\circ}\,\mathrm{d}^{-1}$ so that the local equator-crossing time is fixed throughout the year. Sun-synchronous are sometimes referred to as 'polar' but, in fact, most such orbits pass poleward of about 80° rather than directly over the pole. Depending on the width of the instrument swath, the entire globe may be sampled within a few days.

Geostationary orbits are high-altitude ($\sim$36,000 km) orbits with a 24-hour period. Satellites placed in this orbit above the equator appear stationary with respect to Earth's surface. A single satellite can observe half the globe continuously, making it possible to completely sample the diurnal cycle. Viewing angles for sensors in geostationary orbit are fixed function of geographic location; where the viewing angle is large (e.g., poleward of about 60°, or equally far longitudinally from the sub-satellite point), the view is distorted.

Tropical orbits are low-altitude orbits with low inclination angles. This allows the low latitudes to be viewed at different times of day and the diurnal cycle sampled over the course of several weeks or months by a single satellite, making it especially useful for observing, for example, tropical precipitation.

The orbit from which observations are made, along with the sampling characteristics of the sensor, determine the distribution of viewing and illumination geometries at which a given location will be sampled. The cloud properties inferred from measurements depend on the these angles. Some of this sensitivity is the result of simple geometry: broken clouds appear to cover more area when viewed from the side, and oblique views provide a longer path length through the atmosphere, resulting in a higher contrast between clouds and the clear atmosphere. This implies that the spatial distribution of cloud properties reported from a given platform depends in part on the choice of orbit.

The horizontal scale and sampling density of satellite observations varies significantly, from scales of a few tens of metres for active remote sensing instruments to tens of kilometres for passive microwave or broadband radiation measurements. Observations at these scales often include a variety of clouds, whose horizontal scales range from hundreds of meters in shallow cumulus to hundreds of kilometers in extended stratiform systems.

4.5 The Global Distribution of Clouds and Their Radiative Effects

As described in Section 4.2, clouds' impact on the top-of-the-atmosphere radiation budget depends on their spectrally dependent optical properties: the long-wave brightness temperature, the long-wave emissivity, which depends on optical depth, and the short-wave albedo, which also depends on asymmetry parameter and single-scattering albedo. Section 4.4 describes remote sensing methods for determining these parameters. It is possible to estimate the distribution of the CRE by using these properties, as obtained from satellites, in radiative transfer calculations.

Satellite observations of clouds were first made in the 1960s and have steadily increased in number and diversity so that, as this book is written, every remote sensing method described in Section 4.4 is being routinely applied. Table 4.1 surveys some widely used sets of observations. These observations form the basis for our understanding of how clouds are distributed around the planet and what their effects are on the energy budget.

4.5.1 Observations of the Global Distribution of Cloudiness

4.5.1.1 On the Concept of Cloud Fraction

The concept of cloud fraction was originally developed as a way to roughly quantify observations of clouds made by human observers (see Section 1.1.1). From a remote sensing perspective, cloud fraction represents the proportion of individual observations whose observed characteristics differ significantly from the signal expected from a cloud-free sky. But the detectability of clouds depends on the

physical characteristics of the observing system: cold, optically thin cirrus, for example, are much easier to detect with thermal emission or active lidar measurements than in measurements of solar reflection, while optically thick boundary-layer clouds are essentially invisible to centimetre-wavelength radars despite an enormous impact on reflected short-wave radiation. Cloud detection is further affected by the angles at which the clouds are viewed and illuminated by the sun, as well as the relative sizes of the sensor spatial resolution and the clouds. More human considerations also play a role: the decision about where to draw the line between clouds and clear skies is necessarily somewhat arbitrary and will reflect the application at hand, particularly the decision as to whether the detection should favour identifying all measurements affected by clouds (a 'clear-sky conservative' approach) or only those measurements that are entirely cloudy ('cloud-conservative'). Small and/or thin, and therefore hard-to-detect, clouds are common in the atmosphere, so relatively small changes in detection thresholds, instrument sensitivity, and so on can have significant impacts on this integral measure.

This implies that cloud fraction is not an inherent property of the atmosphere but rather one defined relative to some observing or modelling system. Indeed, the view of cloudiness one gets is strongly coloured by the instruments used to detect those clouds. Fig. 4.5 illustrates this sensitivity by showing estimates of high ($p_{ct} \lessapprox 440\,\mathrm{hPa}$), mid-level, and low ($p_{ct} \gtrapprox 680\,\mathrm{hPa}$) cloudiness as determined by two observing systems, one exploiting single infrared and visible-wavelength radiances and one using high spectral-resolution infrared observations. Understanding the full distribution of cloudiness requires synthesising these disparate views of the world's clouds. For example, estimates from this particular visible/thermal infrared technique assign much more of the cloudiness to the middle of the atmosphere than is inferred from the infrared observations. The latter, which are consistent with lidar observations, are sensitive to thin clouds, while calculations of the CRE using properties from bi-spectral methods are consistent with more direct observations (see Section 4.8). This implies that many high clouds in the atmosphere are optically thin and frequently overlie brighter, lower clouds.

4.5.1.2 The Distribution of Clouds and the General Circulation

At a more qualitative level, the signature of the planetary-scale circulation can been seen in the distribution of clouds (see Fig. 4.6). Clouds are frequent in regions of deep convection, including the Inter Tropical Convergence Zone (ITCZ), South Pacific Convergence Zone and over the

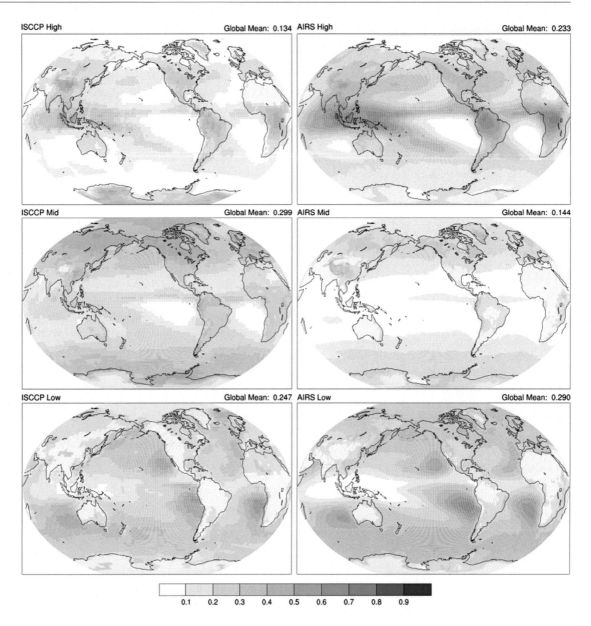

FIGURE 4.5: Estimates of the total frequency of clouds in three broad layers of the atmosphere with boundaries at 440 hPa and 680 hPa, as inferred from two remote sensing systems. The specific observations are the ISCCP and AIRS listed in Table 4.1.

Maritime Continent. Clouds are rare over deserts like the Sahara and Australia, but frequent over the mid-latitude storm tracks (which, in the southern hemisphere, essentially circles the globe). Even the marine stratocumulus regions at the eastern edges of subtropical oceans, the result of shallow boundary layers trapped by large-scale subsidence over cold, equator-ward currents, are visible on the planetary scale.

The distribution of clouds through the depth of the atmosphere (Fig. 4.7) also shows the signature of the general circulation. The uppermost level in which clouds are frequent tracks the variation in tropopause depth with latitude. The ITCZ appears as a narrow band of cloudiness; here the relatively low cloud fractions reflect clouds which cover a relatively small area but, when present, extend through much or all of the troposphere. The ITCZ is, in the global mean, slightly north of the geographic equator. Clouds rarely extend above 2.5 km in the subtropical subsidence regions (e.g., the Trades), while cloudiness is ubiquitous in the mid latitudes. Most of the clouds are

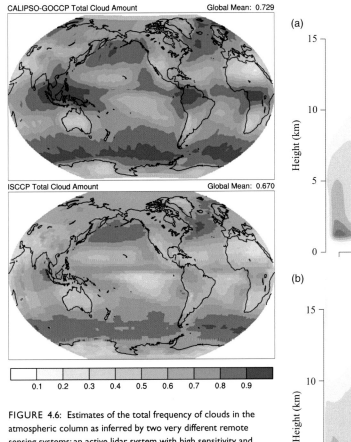

CALIPSO-GOCCP Total Cloud Amount   Global Mean: 0.729

ISCCP Total Cloud Amount   Global Mean: 0.670

0.1   0.2   0.3   0.4   0.5   0.6   0.7   0.8   0.9

FIGURE 4.6: Estimates of the total frequency of clouds in the atmospheric column as inferred by two very different remote sensing systems: an active lidar system with high sensitivity and limited spatial sampling, and a system using only passive measurements in the visible and thermal infrared windows (Calipso and ISCCP, respectively; see Table 4.1). The signature of the general circulation is evident in both estimates despite the large difference in total cloud amount, which derives mostly from the lidar's sensitivity to tenuous clouds.

visible to millimetre-wavelength radars, as plotted in the top panel; adding higher-sensitivity lidars in the bottom panel reveals the presence of thin, diffuse ice clouds high in the atmosphere.

4.5.2 Observations of the Top-of-the-Atmosphere CRE

The CRE can be computed given the three-dimensional distribution of clouds as obtained from remote sensing measurements, as just described. A more direct approach is to observe the irradiance difference between clear and cloudy skies by making broadband radiance observations in place of the narrowband measurements used to estimate cloud properties. These radiance observations must still be mapped to energetically relevant irradiances through

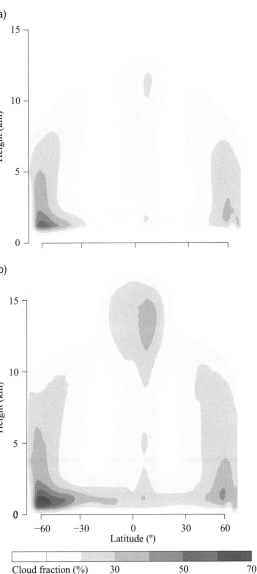

Cloud fraction (%)   30   50   70

FIGURE 4.7: The zonal-mean distribution of cloud fraction through the depth of the atmosphere, as observed by (a) radar and (b) by a combination of radar and the more sensitive lidar. This distribution, too, reflects the general circulation, and affects the flow of energy within and at the boundaries of the atmosphere.

angular distribution functions (empirical models might be chosen for each scene based, say, on complementary observations) and still further modelling (perhaps based on spectrally limited but temporally resolved observations) is needed to account for temporal variations. The CRE can then be determined by subtracting the average irradiance from the average of the cloud-free irradiance observations.

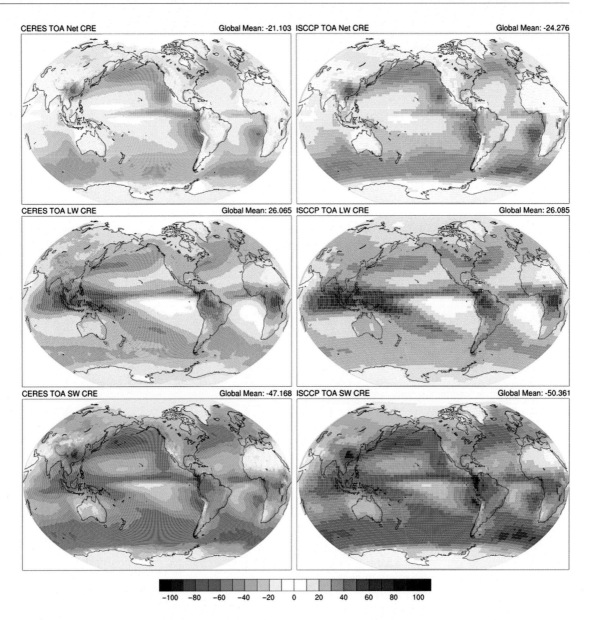

FIGURE 4.8: Top-of-the-atmosphere net CRE and its long- and short-wave components as determined from broadband observations (left column) and using broadband calculations based on narrow-band based on narrow-band estimates of cloud properties. Deep clouds in the Tropics and storm tracks have a strong effect on the long-wave budget, while bright clouds in these locations and the subtropical oceans have a strong effect on the short-wave budget. The net effect is to cool the planet by roughly 20 W m^{-2}.

This is the approach taken by the CERES[1] (Clouds and Earth's Radiant Energy System) project (although the idea is general). In practice, the CERES instrument is normally flown with other narrowband instruments

used to make high spatial resolution estimates of cloud properties (e.g., τ, p_{ct}) inside each of the relatively large CERES footprints (roughly 20 km at nadir for satellites in sun-synchronous orbits); the joint distribution of these properties is used along with ancillary information to classify each scene among a finite number of types. CERES instruments are able to change their orientation with respect to the satellite platform so that viewing geometry can be sampled over time. Empirical angular

---

[1] The name CERES refers both to an instrument, four of which have been placed in orbit between 1997 and 2013, and also to the mission, including the algorithms used to interpret the measurements. CERES is the successor to the ERBE (Earth Radiation Budget Experiment) the was begun in 1984.

distribution models for each scene type, including clear skies, are built from this distribution of broadband radiance observations. Observations from geostationary satellites are used to interpolate irradiance values between the relatively sparse observations.

The annual-mean CRE at the top of the atmosphere, as determined from twelve years of broadband observations, is shown in the left column of Fig. 4.8. The longwave CRE (annual, global mean $26\,\mathrm{W\,m^{-2}}$) is smaller than the short-wave effect ($-47\,\mathrm{W\,m^{-2}}$) so that clouds reduce the radiation absorbed by the planet by about $-21\,\mathrm{W\,m^{-2}}$ in the global annual mean. The substantial spatial structure in the distribution of net CRE and its components reflects the distribution of clouds themselves. Deep clouds in the Tropics, for example, have a small net effect on top-of-the-atmosphere CRE but large contributions to both the long- and short-wave portions of the radiation budget, indicating that they are, on average, cold and bright relative to the surface, as one expects from clouds produced by convection that frequently fill the troposphere. Regions of persistent, extensive low clouds, such as the stratocumulus decks in the eastern subtropical ocean basins, show a small long-wave CRE, since the clouds are so close to the surface temperature, but a large impact on short-wave radiation. Bright clouds with a strong short-wave cloud effect are also prevalent in the mid-latitude oceans, where the shallower tropopause means that long-wave effect cannot compensate completely, so that the net effect is small but negative.

As an aside, note that inferring CRE using the difference between observed clear- and all-sky irradiance differs from the approach used in models, which perform two radiation calculations, one of which excludes the impact of clouds on surface and top-of-the-atmosphere irradiances. Because clouds are associated with high relative humidity, clear-sky results from models are computed in systematically moister environments than can be observed. Modelling studies have suggested that this impacts estimates of clear-sky irradiance and CRE by 1–$2\,\mathrm{W\,m^2}$.

The right-hand column of Fig. 4.8 shows the CRE computed based on retrievals of cloud properties (the particular data are obtained from the ISCCP project listed in Table 4.1). The clear-sky conditions used in these retrievals (profiles of temperature and surface values of albedo, emissivity and temperature) are used in broadband radiative transfer calculations with and without clouds; those clouds are characterized using cloud top pressure and optical thickness determined from observations in the infrared and solar windows (Sections 4.4.2.1 and 4.4.2.2). The main difference with respect to CERES is an increase in the estimate of the short-wave radiative effect of about $3\,\mathrm{W\,m^{-2}}$ in the global annual mean that propagates to the estimate of net CRE. In general, though, estimates of cloud properties from the top of the atmosphere are consistent with estimates of clouds' impact on the radiation budget in the same location, indicating that the processes underlying the interaction are well understood.

4.6 Closing Energy Budgets

Clouds' impact on the top-of-the-atmosphere radiation is the most obvious influence on climate because it affects the equilibrium temperature of the planet. But the presence of clouds also affects the energy budgets within the atmosphere and at the surface, as described in this chapter and assessed in Fig. 1.7. At the surface, clouds reduce the amount of incoming sunlight and increase downwelling long-wave radiation. Their role within the atmosphere is more complicated. Clouds act to reduce the amount of solar radiation available for absorption by water vapour, for example, but multiple scattering also increases the overall path length travelled by radiation within the atmosphere. Thus, computing the fluxes at the surface and within the atmosphere requires more detailed knowledge about the vertical distribution of cloud (and atmospheric) radiative properties than do estimates at the top-of-the-atmosphere.

4.6.1 Clouds and Energy Flows through the depth of the Atmosphere

This need was one of the motivations for developing space-borne active remote sensors, and it has yielded much better observational estimates of clouds' impact on the surface and atmospheric energy budgets (Fig. 4.9). Clouds warm the surface substantially at high latitudes where low temperatures and small water vapour concentrations make the atmosphere optically thin in the infrared. In regions where deep convection is common, however, clouds cool the surface because the reduction of solar radiation has a larger impact than increasing the emissivity of the moist, warm atmosphere. Subtropical marine stratocumulus, which have a large impact at the top of the atmosphere, have almost no effect at the surface because the reduction of incoming solar energy is nearly balanced with the increase in downwelling long-wave radiation. The estimated surface CRE is only slightly larger than the effect at the top of the atmosphere, so that the overall effect of clouds on the atmosphere is small ($\sim 2\,\mathrm{W\,m^{-2}}$) although regional effects can be as large as $50\,\mathrm{W\,m^{-2}}$.

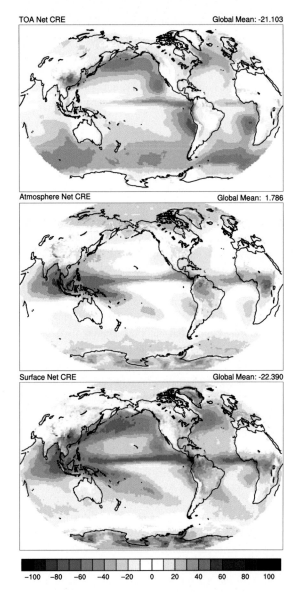

FIGURE 4.9: Net CRE at the surface and within the atmosphere. The underlying measurements are of broadband radiance at the top of the atmosphere (from CERES, see Table 4.1). Additional observations, including those from space-borne lidar and radar, and empirical models are used to map radiance to irradiance, to interpolate the broadband observations through the diurnal cycle, and to estimate the vertical distributions of clouds in the atmosphere based on passive measurements, so that the estimates are substantially less direct than the estimates at the top of the atmosphere (Fig. 4.8).

4.6.2 Synthesizing Complementary Measurements

The remote sensing techniques described in Sections 4.4.2 and 4.4.3 are distinguished by factors including

the source of the observed radiation and by the relative importance of emission and scattering at the wavelengths being observed. Each individual measurement, therefore, provides a unique but limited view. This leads naturally to the idea of combining measurements with different characteristics to obtain a richer view of clouds in the Earth system. The use by CERES of temporally resolved narrowband observations to estimate diurnal variations in the radiation budget is an example of this kind of synthesis. Another is shown in Fig. 4.7, where synthesising high-sensitivity lidar and high-coverage radar observations provides for a much more complete view of the vertical distribution of clouds than could be obtained from either kind of sensor alone.

The value of making complementary observations has been highlighted by the A-Train, a satellite constellation in a sun-synchronous orbit making daytime observations in early afternoon. The constellation includes platforms with a range of passive sensors (including the MODIS, CERES, and AIRS instruments described in Table 4.1 as well as platforms carrying the first operational space-borne lidar (CALIPSO) and radar (CloudSat)). The active measurements in particular have provided new insights into relationships among clouds, radiation and precipitation – the most crucial components of the energy budget through the depth of the atmosphere. The EarthCare platform (Table 4.1), set to launch in 2022 seeks to develop a deeper and more accurate understanding of the role of clouds in modulating these energy flows.

4.6.3 Frontiers

Attempting to close the energy budget within the atmosphere and at the surface highlights the frontiers of current observational capabilities. The balance between the sum of latent and sensible heating and net radiative cooling of the atmosphere is a strong physical constraint, but attempts to reconstruct this balance from observations have not yet been entirely successful. This is partly because each component of the budget is assessed with different remote sensing techniques: latent heating can be inferred from the amount of precipitation reaching the ground, as estimated by cloud and precipitation radars, while estimates of the net atmospheric cooling by radiation rely on top-of-the-atmosphere irradiance measurements, profiles from active remote sensors and radiative transfer modelling.

It is perhaps unsurprising that the biases and conditional errors in these independent estimates leave substantial gaps in the budget. Thus, although theories are well established for the interaction of radiation with clouds (Sections 4.1 and 4.2) and subsequent impact on the irradiance distribution within the atmosphere (Section 4.3), the limited information available from remote sensing (Section 4.4) and subsequent lack of detail about the four-dimensional distribution of cloud properties (Section 4.5) limit the ability to fully characterise the flow of energy through the climate system, even as clouds play a central role in modulating those flows.

Further Reading

There are a great many textbooks on the fundamentals of radiative transfer. Grant Petty, *A First Course on Atmospheric Radiation* (Petty, 2006) is a self-published textbook that emphasises physical understanding with less mathematical detail than most texts. Craig Bohren and Eugene Clothiaux, *Fundamentals of Atmospheric Radiation* (Bohren and Clothiaux, 2006) is a personal favourite, quirky and insightful.

Sections 4.1 Microphysics, Material Properties, and the Effect of Cloud Elements on Radiation

H. C. van de Hulst, *Light Scattering by Small Particles* (van de Hulst, 1980) is the fundamental text for single scattering, reprinted from a 1957 monograph by Dover in 1980. Michael Mishchenko, Joop Hovenier and Larry Travis, *Light Scattering by Nonspherical Particles: Theory, Measurements, and Applications* (Mishchenko et al., 2000) is the analogue to van de Hulst for the far-more-complicated problem of non-spherical particles.

Sections 4.2 From location, Extent, and Local Properties to the Radiation Field

Much earlier in his career, Craig Bohren wrote one of the most lucid descriptions of the physics multiple scattering (Bohren, 1987), which gives a quick path to understanding the interaction of clouds and solar radiation. Jacqueline Lenoble, *Atmospheric Radiative Transfer* (Lenoble, 1993) is a classic text that gives the detailed formulations of the RTEs in various useful situations.

Sections 4.4 Remote Sensing: Observing Clouds with Radiation

Graeme Stephens, *Remote Sensing of the Lower Atmosphere* (Stephens, 1994) remains a reasonably current and quite useful introduction to the physical principles of remote sensing. As might be expected from the originator of CloudSat, the treatment of radar remote sensing is particularly good. Claus Weitkamp

is the editor of *Lidar: Range-Resolved Optical Remote Sensing of the Atmosphere* (Weitkamp, 2005). The book covers a broad range of topics including the remote sensing or aerosols and clouds. Michel Capderou, *Handbook of Satellite Orbits* (Capderou, 2014), provides an exceedingly complete description of orbital mechanics. The review article by Claudia Stubenrauch and colleagues (Stubenrauch et al., 2013) assesses the distribution of cloudiness provided by a wide range of remote sensing observations and highlights the degree to which the view depends on the viewer.

Sections 4.5 The Global Distribution of Clouds and Their Radiative Effects

Norman Loeb and Bruce Wielicki, *Earth's Radiation Budget* (Loeb and Wielicki, 2015) is an instructive, up-to-date CRE contribution to *Encyclopaedia of Atmospheric Sciences*. As discussed in depth in Chapter 13, changes to the CRE with warming are at the root of cloud feedbacks. A review article by Paulo Ceppi and colleagues (Ceppi et al., 2017) provides a focused view of a some of the well and less-well understood aspects of cloud feedbacks.

Sections 4.6 Closing Energy Budgets

Tristan L'Ecuyer led an effort (L'Ecuyer et al., 2015) to close the Earth's energy budget by considering the uncertainty in observations of each term. That the problem remains so hard highlights the importance of a diverse and complete observing system.

Exercises

(1) Use typical particle size and concentration information as gleaned from Chapter 3 to assess whether the assumption that cloud particles are separated by distances much larger than the particle size (Section 4.1.3) is reasonable.

(2) What's a reasonable cloud depth for a water cloud to completely attenuate a lidar signal? What about ice?

(3) In what wavelength ranges are liquid cloud particles in the Rayleigh, Mie and geometric optics regimes? What about ice clouds?

(4) What if all the cloud particles where smaller than $1 \, \mu m$? What if water absorption was not minimum in the visible domain?

(5) Derive the lidar equation (Eq. (4.25)) from the RTE for a scattering medium (Eq. (4.14)).

For the next set of problems, consider three clouds with varying micro- and macro-physical properties.

TABLE 4.2: Symbols and key variables used in this chapter

Symbol	Description	Units (MKS)
D	Particle diameter	μm
$n(D)$	Size distribution	m^{-3}
λ	Wavelength	μm
x	Size parameter	–
Θ	Scattering angle	°
μ	Cosine of the scattering angle Θ	–
u	Cosine of the zenith angle θ	–
ϕ	Azimuth angle	°
T	Temperature	K
p	Pressure	hPa
p_{ct}	Cloud-top pressure	hPa
Cloud optical properties		
Q_{ext}	Extinction efficiency (sca for scattering, abs for absorption)	–
C_{ext}	Extinction cross-section	$μm^{-2}$
β_{ext}	Extinction coefficient	m^{-1}
w_0	Single scattering albedo	–
$P(\Theta)$	Scattering phase function	–
g	Asymmetry factor	–
$M(\Theta)$	Müller matrix	–
Radiative variables		
I	Radiance (intensity)	$W\,m^{-2}\,sr^{-1}\,μm^{-1}$
I, Q, U, V	Stokes vector	$W\,m^{-2}\,sr^{-1}\,μm^{-1}$
$I_B(\lambda, T)$	Blackbody radiance emission	$W\,m^{-2}\,sr^{-1}\,μm^{-1}$
T_b	Brightness temperature	K
F	Irradiance (flux)	$W\,m^{-2}\,μm^{-1}$ or $W\,m^{-2}$
F_{net}	Net irradiance (net flux)	$W\,m^{-2}$
τ	Optical depth	–
τ_c	Cloud optical depth	–
ϵ	Emissivity	–
$\mathcal{R}$	Reflectance of the atmosphere	–
$\mathcal{T}$	Transmissivity of the atmosphere	–
β	Backscatter coefficient	$m^{-1}\,sr^{-1}$
α	Albedo	–
CRE	Cloud radiative effect	$W\,m^{-2}$

Stratus particle concentration $N = 400\,cm^{-3}$, effective particle size $r_{eff} = 5\,μm$, vertical extent $\Delta z = 600\,m$, temperature $T = 6\,°C$.

Cirrus $N = 10\,cm^{-3}$, $r_{eff} = 25\,μm$, $\Delta z = 200\,m$, $T = -40\,°C$.

Cumulonimbus $N = 70\,cm^{-3}$, $r_{eff} = 8\,μm$ for $T > 0\,°C$ and $50\,μm$ for $T \leq 0\,°C$, $\Delta z = 7000\,m$, $T_{base} = 6\,°C$.

For all clouds, assume a constant lapse rate $\Gamma = -6.5\,K\,km^{-1}$ and a surface temperature of $T_{sfc} = 28\,°C$.

The cumulonimbus cloud is deep enough to rain, which will affect the profile of liquid and ice water content.

(6) Compute the infrared and visible cloud radiative parameters for each cloud, assuming that all particles are round:

  (a) Compute the liquid or ice water path and the visible optical depth.

  (b) Estimate the infrared cloud optical depth and cloud emissivity

(7) Compute the resulting radiation fields:
  (a) Compute the amount of direct solar radiation that will reach the ground after passing through the cloud.
  (b) Compute the infrared irradiance emitted by these clouds, and compare it with the irradiance emitted by the surface.
(8) Imagine now that the ice spheres for the cirrus cloud are replaced with hexagonal columns of length $L = 10\,\mu m$ and width $D = 5\,\mu m$.
  (a) Express the radius r of a sphere of equivalent volume as a function of L and R.
  (b) Recompute the optical properties and resulting radiation field for the cirrus cloud composed of hexagonal crystals, assuming that the amount of condensed water remains constant.
(9) Compute the short wave, long wave and total cloud-radiative effect of each cloud at the top of the atmosphere and at the surface. Do status and cirrus clouds warm or cool the Earth system/the surface when they are located above a warm ocean? What over an ice sheet (some assumptions will be necessary)? What if the cirrus cloud is located above the stratus cloud?

Part II Clouds and Modelling

5 Conceptualising Clouds

Stephan de Roode and Roel Neggers

This chapter is dedicated to the formulation and interpretation of conceptual and bulk models for describing (parts of) Earth's climate system. Of particular interest are theoretical and conceptual models of cloudy and convective layers, that are useful for investigating cloud-coupled circulation systems within the atmosphere. In mathematics, bulk models often have the shape of a *weak solution* of a (set of) equations, which is the solution of an integral over a certain parameter range of interest. Understanding such weak solutions of differential equations can provide insight, because the complexity is reduced and the transparency of the system of equation is perhaps enhanced. At the same time, key interactions between individual components in the system can be maintained, often through terms defined at the integral boundaries. In this chapter we look at techniques for deriving weak solutions of the governing equations of the atmosphere, focusing on thermodynamic quantities. The use of such weak solutions has had a long tradition in boundary-layer meteorology, but also in tropical meteorology, then considering the troposphere. These two regimes are the main focus of this chapter.

In Section 5.1, the typical behaviour of various thermodynamic state variables in the tropics will briefly be reviewed. This behaviour inspires the application of certain simplifications in the associated budget equations. In Section 5.2, these budget equations will be formulated, forming the foundation of some well-known and often-used bulk models. The subsequent Sections 5.3 and 5.4 then describe and interpret the application of bulk modelling to (i) the free troposphere and (ii) the planetary boundary layer (PBL).

5.1 State Variables for Bulk Models

Temperature T and water vapour specific humidity q_v are both thermodynamic state variables, in that they describe the state of the atmosphere at any given moment. Water can occur in various phases, such as water vapour, clouds and precipitation. This section is dedicated to exploring what the behaviour of thermodynamic state variables in nature might imply for the configuration of conceptual bulk models. Ideally, the formulation of such models should somehow be tailored to this behaviour; it is instructive to spend some thought on this beforehand. This section can thus be considered a necessary preparation for the next sections that are dedicated to the formulation and interpretation of bulk models. In addition, we explore what the use might be of formulating in terms of conserved variables. These variables are combinations of temperature and humidity, formulated so that they are conserved for adiabatic displacements of fluid parcels. This aspect makes them potentially interesting for bulk studies. The thermodynamic state variables as used in this chapter have already been introduced in Chapter 2. Use is made of simplified expressions for some conserved variables.

5.1.1 Weak Temperature Gradients

Fig. 5.1a shows an instantaneous snapshot of the global temperature field on the 500 hPa isobar on 21 March 2014 at 12:00 UTC, as obtained from the European Centre for Medium-Range Weather Forecasts (ECMWF) analysis at T1279 horizontal resolution. The tropics can be defined as the region (roughly between 30°S and 30°N) where in the free troposphere, horizontal temperature gradients are weak. This region is confined to these latitudes because there the Coriolis force is small, and winds cannot balance pressure gradients arising from thermal differences. The existence of the weak temperature gradient (WTG) can be understood by means of a scale-analysis of horizontal and vertical components of the momentum equation for different regimes of rotation, making use of the hydrostatic **125**

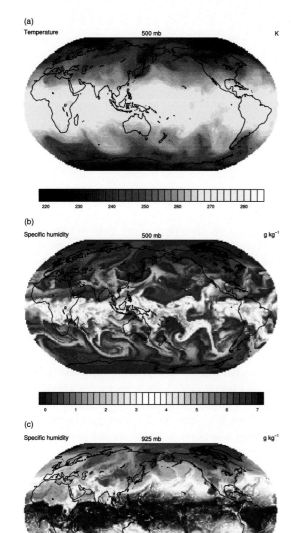

FIGURE 5.1: Instantaneous global temperature and water vapour fields on 21 March 2014 at 12:00 UTC, as obtained from the ECMWF T1279 analysis: (a) temperature [K] at 500 hPa, (b) water vapour specific humidity [g kg^{-1}] at 500 hPa and (c) at 925 hPa. Note that a different colour scale is used in the two humidity plots, for visualisation.

approach. We start by writing the vertical component of momentum equation in hydrostatic balance,

$$\frac{\partial p}{\partial z} = -\rho g. \tag{5.1}$$

A simple scale-analysis is then performed, using l_v as the depth-scale of the troposphere. The horizontal difference

in pressure between two locations at height l_v can then be expressed as a function of the horizontal difference in density,

$$\delta p \sim g l_v \delta \rho. \tag{5.2}$$

The next step is then to scale the horizontal pressure fluctuation using the horizontal component of the momentum equation,

$$\underbrace{\frac{D\vec{U}}{Dt}}_{\frac{U^2}{l_h}} + \underbrace{f\hat{k} \times \vec{U}}_{fU} = \underbrace{-\frac{1}{\rho}\nabla p}_{\frac{\delta p}{\rho l_h}}, \tag{5.3}$$

where $f = 2\Omega \sin \phi$ is the Coriolis parameter, dependent on the Earth's angular velocity Ω and the latitude ϕ. For scaling this equation, we choose l_h as the typical horizontal length-scale and U as the horizontal velocity scale. As a result, the inertial term is scaled as U^2/l_h, the Coriolis force as fU and the pressure gradient force as $\delta p/\rho l_h$.

At this point, it is instructive to consider the balance between forces for two typical regimes of rotation. To this purpose, two relevant dimensionless numbers are introduced. The Rossby number is defined as the ratio of the inertial to the Coriolis term,

$$\mathcal{R}_o = \frac{U}{f l_h}, \tag{5.4}$$

while the Froude number can be defined as the ratio of the vertical inertial term to the gravitational acceleration,

$$\mathcal{F}_r = \frac{U^2}{g\, l_v}. \tag{5.5}$$

With $U = 10$ m s^{-1}, $g = 10$ m s^{-2} and $l_v = 10$ km, the Froude number is of the order 10^{-3}. In the geostrophic regime in the mid latitudes (subscript g) the Coriolis parameter f is large, and $\mathcal{R}_o \ll 1$. As a consequence, the Coriolis force approximately balances the pressure gradient force in Eq. (5.3). Combining this with Eq. (5.2) gives

$$\left(\frac{\delta\rho}{\rho}\right)_g \sim \frac{fUl_h}{g\, l_v} \sim \frac{\mathcal{F}_r}{\mathcal{R}_o}. \tag{5.6}$$

In the tropical regime (subscript tr), however, f is much smaller, and $\mathcal{R}_o \gtrsim 1$. So the Coriolis term is relatively small in Eq. (5.3), and the pressure gradient term is more or less balanced by the inertial term. Again combining with Eq. (5.2) gives

$$\left(\frac{\delta\rho}{\rho}\right)_{tr} \sim \frac{U^2}{g\, l_v} \sim \mathcal{F}_r. \tag{5.7}$$

Comparing Eqs (5.6) and (5.7) shows that in the tropics the horizontal density fluctuations have to be much smaller compared to the mid latitudes. And, as is clear from Eq. (2.128), density perturbations are directly related to temperature perturbations. This finally explains why horizontal temperature gradients in the tropics are much smaller compared to the mid latitudes. In a rotating coordinate system, the apparent forces can balance the pressure gradient force. But without rotation the apparent forces disappear, so pressure (and with it temperature) gradients cannot be balanced. This creates motions that quickly act to eliminate the gradients. A consequence of this result is that in the tropics large horizontal pressure gradients cannot exist. In practice, the temperature field stays constant because gravity waves, triggered by convection, are efficient in quickly spreading out temperature anomalies caused by convection.

The existence of WTG has some consequences for the interaction between tropical convection and the larger-scale flow. In fact, as will be discussed later, the WTG condition puts constraints on the large-scale circulation in the tropical troposphere that creates opportunities for creating bulk models. Note that the WTG approach does not hold in the boundary layer, where friction makes circulations less efficient in rapidly adjusting the fluid so as to balance the pressure gradients.

5.1.2 Atmospheric Rivers

Fig. 5.1b shows an instantaneous snapshot of the mid-tropospheric water vapour field on the 500 hPa isobar. It immediately becomes clear that the WTG approach does not apply to the water vapour field. The instantaneous map shows very strong local gradients, due to the existence of filament-like structures extending from a narrow band of high values in the tropics far out into the mid latitudes. This shape quite resembles the spatial structures one can observe in numerical or laboratory experiments, in which a passive scalar is advected around between a source at the equator and a sink at poles. The associated filaments of water vapour contribute significantly to the horizontal transport between the tropics and the poles. For these reasons, these structures are sometimes referred to as 'atmospheric rivers'; a famous example is the 'pineapple express', a popular name for atmospheric rivers starting near the Hawaiian archipelago and reaching the Californian coast.

It is instructive to compare the mid-tropospheric water vapour field to that closer to the surface. Fig. 5.1c shows the water vapour field on the 925 hPa isobar, which is close to the typical height of the atmospheric mixed layer over tropical oceans. Apart from showing that water vapour values at low levels are much higher than in the free troposphere, what also becomes clear is that the field is much more constant. The horizontal gradients are much weaker, in that sense more resembling the temperature field.

What explains the different behaviour of the water vapour field in the mid troposphere and near the surface? As will be discussed in Section 5.4, the existence of a PBL plays an important role in this. The quasi-equilibrium (QE) state of the PBL is closely tied to the surface properties through the surface fluxes of enthalpy and water; as a result, the constancy of surface fields such as the sea-surface temperature (SST) is imprinted on the water vapour field in the PBL. In contrast, transport of water vapour into the free troposphere above is much more sporadic and incidental. This only happens in a significant way in isolated areas, small in area coverage, where tropospheric-deep convection is active. There the deep convection acts to transport the high PBL values into the free troposphere. These occurrences of intense vertical mixing, as well as significant precipitation generation, can then be interpreted as the main sources and sinks of free-tropospheric water vapour, respectively. The main area where deep convection occurs almost continuously is the Inter Tropical Convergence Zone (ITCZ). In between the ITCZ and the mid latitudes, so long as sharp vertical gradients do not arrive, water vapour is often advected around like a passive scalar.

The observed decoupling in water vapour between the PBL and the free troposphere could be of use in the configuration of bulk models. It can be argued that any representation of this behaviour requires a system with at least two independent layers; any interaction between them can then occur at their interface (PBL top). For water vapour, one could then use a bulk mixed-layer model (MLM) such as will described in detail in the last part of this chapter.

5.1.3 Vertical Structure

Other interesting aspects of the tropical thermodynamic state can be established from radiosonde profiles. Fig. 5.2 shows the temperature and water vapour soundings measured at Lihue, Hawaii, on 26 March 2014 at 12 UTC. What catches the eye in Fig. 5.2a is that the temperature structure in the troposphere, situated between approximately 2 and 10 km high, more or less follows a reversible moist adiabat. Apparently, the reversible moist adiabat captures the tropospheric thermodynamic state to a reasonable degree. This is not a coincidence, but in fact is

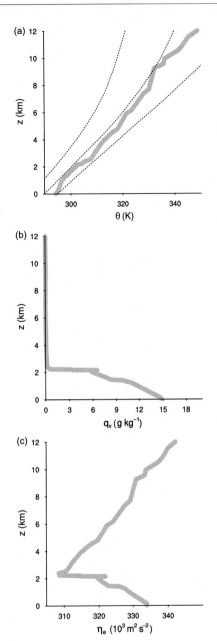

FIGURE 5.2: Atmospheric soundings at Lihue, Hawaii, obtained on 26 March 2014 at 12:00 UTC. (a) Potential temperature θ, including several reversible moist adiabats (i.e., saturated isentropes, shown as dotted lines) for reference. (b) Water vapour specific humidity q_v. (c) Moist static energy η_e. Radiosonde data courtesy of the University of Wyoming

a robust feature throughout the tropics and subtropics. This is again related to the WTG effect; gravity waves are efficient in quickly spreading the typical vertical structure associated with deep convection (i.e., the reversible moist adiabat) horizontally, throughout the tropics.

The second important aspect visible in Fig. 5.2 concerns the vertical distribution of water vapour. The highest values can be found in the lowest kilometre or so, while the air above 2 km is significantly drier. These two layers are separated by a thin layer of decreasing water vapour but increasing temperature. Because temperature increasing with height is counter to its normal pattern, this is sometimes referred to as temperature inversion, or simply an inversion, and is frequent in the subtropics where the ocean surface is cold compared to the convecting regions of the tropics, which then set the thermal structure of the free troposphere. This sounding thus supports the conceptual picture emerging from Fig. 5.1 that the atmospheric boundary layer (ABL) and free troposphere behave like distinctly different air masses as far as temperature and humidity are concerned.

Opportunities exist for bulk models to make use of these observed typical vertical structures of temperature and humidity in the tropics. For example, one could consider assuming certain fixed vertical structures for the thermodynamic state variables. For the free tropospheric temperature, an obvious choice would be the reversible moist adiabat; in contrast, for the potential temperature and water vapour in the PBL, the prime candidate would be a well-mixed (constant) structure. Thus, applying fixed vertical structures in the model effectively reduces the number of free levels in the vertical dimension to one; on the other hand, information on the vertical structure that is required for example by a radiative transfer scheme is still maintained.

5.1.4 Moist Static Energy

A variable that is a combination of the thermodynamic state variables q_v and T is the moist static energy η_e. In Chapter 2, this variable has been defined exactly; in this chapter we make use of the approximation Eq. (2.68), for convenience repeated here;

$$\eta_e \approx c_p T + \phi + \ell_v q_v. \tag{5.8}$$

The moist static energy is the sum of the dry static energy $\eta_d = c_p T + \phi$ and the latent heat release associated with condensation of all present water vapour. This makes η_e approximately conserved for phase changes of atmospheric humidity, such as formation and evaporation of clouds and precipitation. This can be understood by considering that a reduction of the humidity term $\ell_v q_v$ due to the formation of clouds is exactly countered by an increase of the temperature term $c_p T$ due to the release of latent heat associated with condensation.

Interpretation in terms of moist static energy can facilitate the investigation of feedback mechanisms between diabatic heat sources such as convection and the larger-scale circulation, as will be illustrated in the next sections. Another useful aspect of moist static energy as a variable is illustrated by Fig. 5.2c, showing a vertical profile of η_e that can be considered typical for the subtropical marine trade-wind areas. Most profiles of moist static energy feature a maximum in the boundary layer, followed by a minimum in the very dry air atop the shallow cloud layer (e.g., just above 2 km in the present sounding), topped by a gradual increase with height in the free troposphere and stratosphere. The lower maximum is associated with the humidity maximum in the PBL. Above the PBL the humidity is much smaller, and there the gradual increase of η_e with height mainly reflects the behaviour of the dry static energy η_d. In this context, it is instructive to imagine the properties of a convective plume that rises from the mixed layer. Because the moist static energy is approximately conserved for phase changes, plume properties can only change by mixing with the environment. Undiluted plumes therefore show up in η_e diagrams as straight vertical lines; the more intense the mixing, the quicker the plume properties will approach the mean profile. This feature makes conserved variables such as moist static energy useful for studies and subsequent modelling of mixing processes in the atmosphere. Examples are the mixing between the PBL and the free troposphere, and the mixing between convective clouds and their environment.

5.2 Bulk Models

Bulk-layer models, also known as slab models, are systems of equations that describe the height-averaged properties of a layer of finite depth that is bounded by interfaces or surfaces. The associated equations can be either prognostic (time-dependent) or diagnostic (time-constant). The concept of bulk modelling is schematically illustrated in Fig. 5.3. Somehow, the layers of interest behave as one entity; a good example is the well-mixed layer in the PBL that more or less has a constant potential temperature and water vapour specific humidity throughout its depth. Bulk models are good at describing layers where most variations can be captured by a single degree of freedom. The bulk model approach can still be applicable when the internal structure is not constant, as long as it is known; a good example of such layers is the free troposphere in the tropics for temperature, the profile of which is typically close to the moist adiabat (see Fig. 5.2a).

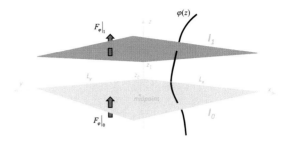

FIGURE 5.3: Schematic illustration of a horizontal bulk model. Variables x, y and z are the horizontal and vertical spatial coordinates. The lower interface l_0 (light grey) and upper interface l_1 (dark grey) of the bulk model are situated at heights z_0 and z_1, respectively. The vertical profile of an arbitrary variable $\varphi(z)$ is indicated in black, while the kinematic vertical flux F_φ at the interface is indicated by the black arrows.

The common behaviour throughout the layer motivates reducing the complexity in a model formulation, by formulating equations in terms of vertically averaged properties instead of properties at multiple layer-internal levels. Even if the vertical variations in the system are important and worth describing, they can be complex, and a less ambitious theory may be satisfied with simply describing the evolution of mean quantities. This procedure, involving vertical integration, eliminates one dimension from the system. Because of the bulk behaviour of the layer, such models might still capture the key behaviour of the layer as a whole, as well as its interaction with other components of the climate system. A common purpose of bulk models is to reduce model complexity to a bare minimum; or to simplify the problem so that it becomes more tractable, accepting that not everything will, in the end, be described. On the one hand, the model should be complex enough to capture the main behaviour, while on the other, hand the model should be simple enough to get rid of unimportant second-order phenomena. In practice, a balance has to be struck between model complexity and model transparency. How the balance is struck depends on the purpose of the study, on the nature of the problem of interest, and on the computational facilities that are available to the researcher.

In this section the basic equations of bulk models will be formulated. In the following sections, the concept will then be applied to:

- The free troposphere, situated between the top of the PBL and the tropopause (Section 5.3).
- The PBL, including associated clouds, situated between the Earth's surface and the inversion capping the PBL (Section 5.4);

5.2.1 Vertical Integration

The purpose is to formulate a bulk model that is generally applicable, and can be applied to any geophysical fluid dynamics problem of interest. The general configuration is schematically illustrated in Fig. 5.3. Suppose a horizontal layer in a three-dimensional spatial domain is situated between two heights z_0 and z_1. The sizes (L_x, L_y) then represent a horizontal domain around a midpoint (x_i, y_i). The vertical profile of an arbitrary variable φ is also shown. The kinematic flux F_φ of this variable at the two interfaces are indicated as vertically pointing black arrows. By convention, fluxes are positive upwards.

5.2.1.1 Averages and Cumulatives

Vertical integrals are the foundation of any bulk model. They are a key tool for summarising the mean behaviour of a layer as a single entity, and for thus reducing the complexity of a model describing this. More precisely stated, we consider vertical integrals of an arbitrary variable φ over the height range between z_0 and z_1 that is covered by the layer of interest. In case of a vertical *average*, the integral is normalised by the layer depth,

$$\langle \varphi \rangle = \frac{1}{\Delta z} \int_{z_0}^{z_1} \varphi(z)\, dz, \qquad (5.9)$$

where $\Delta z \equiv z_1 - z_0$ is the depth of the layer in height coordinates. Alternatively, one could use pressure p as the vertical coordinate,

$$\langle \varphi \rangle = \frac{1}{\Delta p} \int_{p_1}^{p_0} \varphi(p)\, dp. \qquad (5.10)$$

From this point onwards, the brackets indicate a vertical average. The integration variable of choice is assumed to depend on the context.

In case of vertical *cumulatives*, the integral is not normalised. Some particular cumulatives are frequently used in meteorology. Good examples are the 'integrated water paths', such as the liquid-water path (LWP) W, as already defined in Eq. (3.5), and the total column water vapour path (also known as precipitable water (PW) or the water-vapour path),

$$\text{PW} = \int_{z_0}^{z_1} \rho_v(z) q_v(z)\, dz, \qquad (5.11)$$

A 'derived cumulative' is the total cloud cover, expressing which fraction of the horizontal domain is covered by any cloud above,

$$\text{TCC} = \frac{1}{L_x L_y} \int_x \int_y \mathcal{I}\,[\,\text{W}(x, y)\,]\, dx\, dy, \qquad (5.12)$$

where $\mathcal{I}$ is an indicator function similar to the Heaviside function,

$$\mathcal{I}(x) = \begin{cases} 0 & \text{if } x \leq 0 \\ 1 & \text{if } x > 0. \end{cases} \qquad (5.13)$$

For this reason the TCC is sometimes also referred to as the 'projected cloud cover'.

Note that vertical averaging in mass (i.e., pressure) space was already introduced in Eq. (2.171). In parts of this chapter, integrals are calculated in physical (i.e., height) space, rather than mass space. When integrating in physical space it is necessary to include the density ρ_v in the definition of some of the aforementioned cumulatives.

5.2.1.2 Leibniz's Rule of Integration

Prognostic budget equations typically include partial derivatives in both z and t. In case these terms are vertically integrated, Leibniz's rule of differentiation under an integral becomes relevant. This rule states that the partial derivative inside an integral can be transferred outside the integral in case both the variable and its derivative are continuous,

$$\frac{d}{dt} \left(\int_{z_0(t)}^{z_1(t)} \varphi(z, t)\, dz \right) = \int_{z_0(t)}^{z_1(t)} \frac{\partial}{\partial t} \varphi(z, t)\, dz \\ + \varphi(z_1, t) \frac{dz_1}{dt} - \varphi(z_0, t) \frac{dz_0}{dt}. \qquad (5.14)$$

The last two terms on the right-hand side are non-zero in case the boundaries of integration are not time-constant. In bulk models for the atmosphere, this is the case when the heights of its interfaces change with time. A good example is the deepening of the PBL as a result of top-entrainment of free tropospheric air. For now, a detailed discussion of this situation is postponed to Section 5.4; for simplicity, in the remainder of this section the layer interfaces are assumed to be constant in both time and horizontal space. As a result the last two terms on the right-hand side vanish, and only the first term remains. This significantly simplifies the equation, but still maintains sufficient complexity to allow explanation of the main aspects of bulk models in general.

5.2.2 The Bulk Budget

As discussed in Section 2.4.2, the quasi-static (primitive) equations for an arbitrary conserved variable φ can be written as

$$\frac{D\varphi}{Dt} = S_\varphi^{\text{net}}, \qquad (5.15)$$

where the left-hand side is the total derivative and S_φ represents the net source of φ due to all other processes. The total derivative can be expanded, and a distinction can be made between flux-driven sources and non-flux-driven sources. Adopting Eq. (2.149) and writing in tensor notation then gives

$$\frac{\partial \varphi}{\partial t} + u_i \frac{\partial \varphi}{\partial x_i} = -\frac{\partial F_{\varphi,i}}{\partial x_i} + S_\varphi, \qquad (5.16)$$

where F_φ is the net kinematic flux of variable φ, which is defined as the dynamic flux of a variable (i.e., its flow rate through a unit area) divided by air density ρ. The advection and flux terms are written using tensor notation, so that each in principle represents three terms, one for each spatial direction. For the atmosphere flux, $F_{\varphi,i}$ represents transporting processes such as radiation (R), convection (C) and precipitation (P),

$$F_\varphi = F_\varphi^R + F_\varphi^C + F_\varphi^P, \qquad (5.17)$$

where for the purpose of formulating bulk models only the vertical flux components are considered, leaving out subscript i to simplify notation. Term S_φ then represents all remaining sources and sinks such as phase changes of water, which, depending on the thermodynamic state variable, may or may not be conserved.

Height-averaging of Eq. (5.16) then yields the prognostic equation for the bulk average $\langle \varphi \rangle$,

$$\frac{\partial \langle \varphi \rangle}{\partial t} + \left\langle u_i \frac{\partial \varphi}{\partial x_i} \right\rangle = -\frac{F_\varphi|_1 - F_\varphi|_0}{\Delta z} + \langle S_\varphi \rangle, \qquad (5.18)$$

where $F_\varphi|_0$ and $F_\varphi|_1$ stand for the kinematic fluxes at the lower and upper interfaces z_0 and z_1, respectively. Note that on the left-hand side, the Leibniz's rule was applied to bring the time-derivative outside the height-integral. In the vertical averaging of the flux term, the horizontal components were assumed to average out to zero.

Eq. (5.18) is the foundation of most bulk models applied in climate science. The individual terms can be interpreted as follows. The second term on the left-hand side represents the vertically averaged tendency due to advection *within* the layer. This term can be significant whenever a significant gradient of φ exists across the layer. The best example is arguably large-scale subsidence in combination with a substantial lapse rate of φ. However, for simplicity the internal advection term is often set to zero, by assuming layer-internal homogeneity.

The first term on the right-hand side of Eq. (5.18) is often referred to as the *flux divergence* term. When interpreting this term, it is important to consider that it acts on the *difference* between the two interface fluxes. This means that both upward and downward fluxes can produce a positive tendency, as long as the transport into the layer exceeds the transport out of it. This is illustrated in

(a) *Upward fluxes*

$\partial_t \langle \varphi \rangle > 0$

(b) *Downward fluxes*

$\partial_t \langle \varphi \rangle > 0$

(c) *Local interaction*
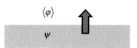
$\langle \varphi \rangle$
ψ

(d) *Remote interaction*
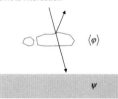
$\langle \varphi \rangle$
ψ

FIGURE 5.4: Schematic illustration of four situations of a bulk layer and its interface fluxes, including (a) flux convergence featuring upward fluxes, (b) flux convergence featuring downward fluxes, (c) local interaction between adjacent layers through the interface flux and (d) remote interaction between separated layers through a vertically transporting agent.

Figs. 5.4a and 5.4b. The flux-divergence term is commonly used to link the height-dependence of vertical transport (e.g., convection, precipitation and radiation) to the time-development of layer-average properties.

5.2.3 Equilibration

With the bulk budget for a bulk layer established, the next step is to gain more understanding of its behaviour. One wonders which terms dominate, and how all terms might over time balance out. This process is often referred to as *equilibration*.

The right-hand side of the budget S_φ^{net} always represents the net source of each variable, including processes involving vertical fluxes and other remaining processes,

$$S_\varphi^{net} = -\frac{\partial F_\varphi}{\partial z} + S_\varphi. \qquad (5.19)$$

Vertically integrating Eq. (5.19) then gives

$$\langle S_\varphi^{net} \rangle = -\frac{F_\varphi|_1 - F_\varphi|_0}{\Delta z} + \langle S_\varphi \rangle. \qquad (5.20)$$

Equilibrium is defined as the state in which all processes are in balance, corresponding to a system that has reached a steady state. As a result,

$$\frac{\mathrm{D}\langle\varphi\rangle}{\mathrm{D}t} = \langle S_\varphi^{\mathrm{net}}\rangle = 0. \tag{5.21}$$

When only flux-related processes are considered, the vertical integration over the depth of an arbitrary layer then yields

$$F_\varphi|_0 = F_\varphi|_1. \tag{5.22}$$

In this case, the net incoming flux at the bottom is exactly balanced by the net outgoing flux at the top; in other words, what goes into the layer must also come out of it. When the system adjusts and finds a new equilibrium, the net fluxes at the boundaries are in balance again by definition, but one of the thermodynamic state variables might have changed.

While Eqs (5.21)–(5.22) describe an equilibrium state, what is not specified is how the system actually approaches and reaches this state. This depends on the nature of the participating processes, and their interaction. Different processes in a system can act on very different timescales. In case one is slow and the others act fast and in the opposite direction, the fast processes always 'beat' the slow and continuously acting process. A state then establishes in which the fast processes 'respond' to the slow process by quickly counteracting and removing the tendency created by the slow process. This behaviour has analogies with the interaction between a driver and a responder. In the cartoon illustrated in Fig. 5.5, the processes that make the grass grow play the role of

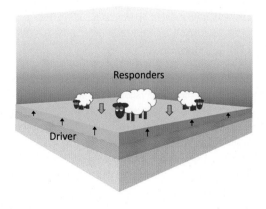

FIGURE 5.5: Schematic illustration of the concept of QE, as applied to the length of grass in a meadow with grazing sheep. The driving process is the combined effect of sun, rain and soil, which makes the grass grow (black arrows). The responding process is the flock of sheep, continuously reducing the length of the grass (blue arrows).

driver, by slowly but continuously producing fuel for the sheep (the responders). When an area is grazed bare, the sheep move on, allowing the grass to recover. The eventual effective height of the grass can be referred to as a 'quasi-equilibrium' state; at any given time the responding processes are in approximate balance with the driving process, yet a small net tendency might remain that causes a trend on much longer timescales. In this situation, the individual tendencies associated with the driving and responding processes are both much larger than the net tendency.

In the atmosphere, radiation can act on pretty long timescales, on the order of thirty days or more. The turnover timescales associated with dry and moist convection, and the associated precipitation, are typically much shorter, in the order of minutes and hours. In the fight between slow radiative processes and convection the latter usually wins; in that case, the slow radiation can thus be interpreted as the driving process to which convection responds. Convection acts to quickly overturn the gradual but persistent buildup of instability by the slow radiation.

In case some processes act on similar timescales, it is not clear a priori how, or if at all, a stable equilibrium will be reached. For example, radiation does not always act slowly; new research shows that radiation can act on the timescale of the circulation, by which the two become more tightly coupled. Also, clouds are coupled to both the slow driving process (radiation) and the fast responding process (convection), involving complex dependencies and feedbacks. Through their significant impact on both long- and short-wave radiation, clouds can make the radiation react on the same short timescales as are typical of the processes behind cloud formation. These feedbacks complicate the interpretation and simulation of the cloud-climate system, and are one of the reasons why the prediction of future climate is such a challenging topic in the atmospheric sciences. This topic is explored further in Chapter 13.

5.2.4 External Interactions

Bulk layers can interact with other components in the Earth's climate system. These components then somehow 'feel' the presence of the bulk layer. In a climate system model of reduced complexity, these interactions have to be parameterised. Such interactions can take place locally, between adjacent layers, or remotely, when two layers are separated in space.

The simplest way of describing the interaction between two layers is by means of a single-variable function, acting on bulk properties of a layer. Suppose ψ is a characteristic of a second layer at a certain height z that is affected by the presence of our bulk layer. Then the dependency takes the form

$$\psi \sim \mathcal{F}\left(\langle\varphi\rangle\right), \qquad (5.23)$$

where $\mathcal{F}$ is the functional form containing the behaviour of the dependency.

Two types of interactions can be distinguished, as schematically illustrated by Figs. 5.4c and 5.4d. *Local interactions* can take place between adjacent layers, through the interface flux. What leaves one layer must go into the next. The flux can be an exchange between the layers or with the bounding surface, modelled with an exchange velocity, as for instance in the bulk aerodynamic parameterisation of surface fluxes, depending on the difference between the properties of the two layers,

$$F_\varphi|_0 = -C_\varphi \; |\vec{v}|_s \; [\langle\varphi\rangle - \varphi_0], \qquad (5.24)$$

where $|\vec{v}|_s$ is the amplitude of the wind near the interface, φ_0 the value at the layer interface and C_φ an exchange velocity which scales with the mean wind at the interface, which can be considered typical for this process. For example, the bulk flux parameterisation is a key component of the bulk MLM model for the PBL, and will be described in more detail in Section 5.4.

Remote interactions between two bulk layers represent impacts of one layer on another through a transporting agent that is capable of carrying properties across the intermediate space that separates the two layers. In the atmosphere, such vertical transporters can include radiation, convection and precipitation. For example, consider a subtropical boundary layer featuring a warm homogeneous cloud layer. Eq. (3.6) gives a simple relation between the optical thickness on the one hand and the ratio of the adiabatic LWP W (a cumulative) to the effective radius of the cloud droplets r_e on the other. The more optically thick the clouds are, the less radiation penetrates to the layers situated below the cloud deck. Other examples include the dependence of the downward short-wave irradiance at the surface on the total cloud cover, and the dependence of surface precipitation on cloud amount in layers aloft. The radiative equilibrium (RE) model is built on this type of interaction (see Section 5.3.1).

A special example in which both local and remote interactions play a role is the surface energy budget,

$$R^{\text{net}} + H + E + G = 0, \qquad (5.25)$$

where R^{net} represents the net radiative flux at the surface, H and E are the surface sensible and latent heat fluxes, and G the ground heat flux (all fluxes positive upwards). Local interactions include the surface latent and sensible heat fluxes and the ground heat flux, while remote interactions include the net radiative energy flux. The surface energy budget plays a key role in interactions between the surface and the atmosphere.

5.2.5 Preserving Internal Structures

In some cases, the internal structure of a bulk layer might matter. A good example is internal vertical advection, as already mentioned in the discussion of the associated term in Eq. (5.18). Another example is radiative transfer. For example, the terrestrial long-wave radiation emitted at each level depends on its local temperature. The vertical transfer of solar radiation through a bulk layer depends on the internal cloud overlap, gas and water vapour (see Chapter 4 for more details). To capture this behaviour, it is necessary to retain more information about the vertical structure in the system of equations.

One way in this can be achieved in a bulk model is through *separation of variables*. This technique makes use of the fact that some vertical structures are typical for a certain layer and variable of interest. Good examples are the vertical structure of temperature in the tropical free troposphere and the atmospheric clear convective boundary layer (CBL), and the triangular shape typical of condensed water in warm stratocumulus clouds (e.g., see Fig. 5.25). Separation of variables is a common technique used to solve partial differential equations and is the basis for their numerical integration in models. When their vertical degrees of freedom can be limited, their behaviour is often easier to understand. So a bulk model really differs from a general circulation model (GCM), in that only one vertical degree of freedom is allowed.

Separation of variables comes down to a decomposition of variables depending on (x, y, z, t) into (x, y, t)-dependent variables on the one hand and (z)-dependent basis-functions on the other,

$$\varphi(x, y, z, t) = \varphi_{\text{r}}(z) + \sum_{i=1}^{N} A_i^\varphi(z) \; \varphi_i(x, y, t). \qquad (5.26)$$

Here subscript r indicates a reference profile. Variable A_i^φ represents a set of N basis functions that are independent of time, each representing a typical vertical structure of φ. Variable φ_i is a weight associated with basis function i, dependent only on (x, y, t). The computational benefit of separation of variables becomes clear when Eq. (5.26) is vertically integrated,

$$\langle\varphi\rangle(x, y, t) = \langle\varphi_{\text{r}}\rangle + \sum_{i=1}^{N} \langle A_i^\varphi\rangle \; \varphi_i(x, y, t). \qquad (5.27)$$

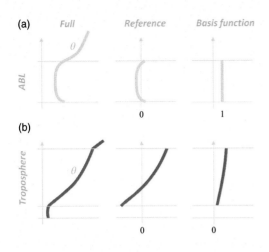

FIGURE 5.6: Schematic illustration of separation of variables. (a) shows the application to the atmospheric mixed layer, while (b) shows the free troposphere.

The weight φ_i now appears outside the vertical integral, so that the three-dimensional spatial (x, y, z) domain has been reduced to a limited number N of two-dimensional (x, y) fields. Note however that the impact of vertical structure is still maintained, through the constants $\langle A_i^{\varphi} \rangle$.

The use of separation of variables in bulk modelling for studying the climate system is schematically illustrated in Fig. 5.6. For example, the typical concave structure of the potential temperature θ in the PBL could be represented in the reference profile. Using a constant basis function with height then accounts for this vertical structure. This can in turn influence parameterised exchanges between the bulk layer and its surroundings, such as the surface flux described by Eq. (5.24).

Bulk models for the free troposphere also benefit from this approach. For example, the typical moist-adiabatic vertical temperature structure could be represented in the reference term. The basis function would then be designed a bit differently, now increasing with height; this is required in order to capture the changing slope of the moist adiabat with temperature (also visible in Fig. 5.2), an effect that is of crucial importance when considering climate change. A consequence of the separation of variables is that associated terms also become separated; in the case of the free troposphere, its internal vertical advection term in Eq. (5.18) can then be written in terms of products of (x, y, t) fields and scalars representing height-averaged (p) fields. A final example of the use of reference vertical structures in the free troposphere is in parameterisations of convective adjustment, as will be discussed later in this chapter.

5.3 Bulk Models for the Free Troposphere

The typical behaviour of the thermodynamic state variables in the troposphere discussed in Section 5.1 motivates making simplifications in key components of the primitive equations that govern the thermodynamic state of and circulation in the free troposphere. Basically, this technique comes down to removing dependencies in variables that are observed to be very weak anyway. The idea of such bulk models is to only attempt to describe specific aspects, such as the mean state. This does not imply that the other aspects are deemed unimportant, but that if the aim is to only understand the evolution of the mean state, we might not have to consider them.

This section attempts to give the reader an overview of both historic and recently formulated bulk models for the troposphere that can be used to investigate cloud-climate feedbacks. Models that will be discussed include globally averaged models, but also tropical circulation models of reduced complexity. The idea is not to present a complete description of models here; instead, the focus lies on the key assumptions that lie at the basis of the simplified formulations, and thus define the associated models.

5.3.1 Pure RE

This subsection and the next explore equilibria in the temperature budget for simplified representations of the *globally averaged* Earth-atmosphere system. This means that variables do not depend on latitude or longitude, and the advection term in the prognostic budget equation Eq. (5.18) is zero. Because radiation (as already explored in Chapter 4) is generally considered to be the driver in the energy budget, it makes sense to start with a purely radiatively driven system. Once this system is better understood, the next step is then to add convection, in a simplified way. Simplified systems based on an energy budget in equilibrium are known as energy balance models (EBM). These models are often formulated in terms of boundary fluxes, because in equilibrium a balance has to exist between the fluxes entering and leaving the system. The shift in balance between individual budget terms associated with two different equilibria can give insight into the nature of this shift. For example, with an EBM one could investigate how and why the atmosphere would warm in a future climate (i.e., a new equilibrium) as a result of a doubling of carbon dioxide in the atmosphere.

The first equilibrium models considered the equilibrium temperature T of the Earth-atmosphere system with radiation being the only acting process (for simplicity

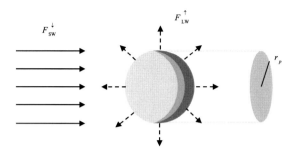

FIGURE 5.7: Schematic illustration of the radiative fluxes in global equilibrium models. Short- (SW) and long-wave (SW) irradiance are indicated by the solid and dashed arrows, respectively. The shadow area is indicated as a grey disc, with r_p the planetary radius. Figure adapted from Hartmann (2016)

convection is ignored). This means $F_T^{\text{net}} = F_T^R$, which in turn is expanded into four more terms, all irradiances,

$$F_T^R = F_{\text{LW}}^\uparrow + F_{\text{LW}}^\downarrow + F_{\text{SW}}^\uparrow + F_{\text{SW}}^\downarrow, \tag{5.28}$$

where the subscripts LW and SW refer to the long- (or infrared, terrestrial) and short-wave (solar) irradiance. The arrows indicate the direction of the fluxes (upward or downward). The long- and short-wave irradiances act together to drive the system towards a certain equilibrium. Fig. 5.7 summarises the general configuration of irradiances on the global scale. The short-wave radiative energy that enters the Earth's atmosphere every second can be written as the solar flux density integrated over the area of the planet exposed to that flux, sometimes referred to as the 'shadow area',

$$\rho c_{p_d} \int_A F_{\text{SW}}^\downarrow \, dA = S_0 \pi r_p^2, \tag{5.29}$$

where A is the surface of the Earth, $S_0 = 1{,}361$ W m^{-2} is the solar constant and r_p is the radius of the planet. For simplicity, it is assumed that no solar radiation is absorbed by the atmosphere. In addition, a fraction $\alpha_p \approx 0.3$ of this flux, also known as the 'planetary albedo', is assumed to be reflected by the Earth's surface and the atmosphere. The long-wave irradiance is dependent on the temperature of the emitting body. In contrast to the short-wave irradiance, the long-wave irradiance is emitted over the surface of the whole planet, following the Stefan–Boltzmann law,

$$\rho c_p \int_A F_{\text{LW}}^\uparrow \, dA = \sigma T^4 4\pi r_p^2, \tag{5.30}$$

where $\sigma = 5.67 \cdot 10^{-8}$ W m^{-2} K^{-4}. What temperature T in Eq. (5.30) represents depends on the configuration of the Earth-atmosphere system. For example, the efficiency of the absorption of long-wave radiation by the atmosphere co-determines this value; an absorption efficiency less

(a)

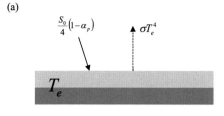

(b)

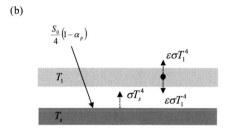

(c)

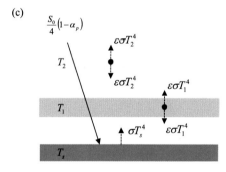

FIGURE 5.8: Schematic illustration of three discretizations of the globally averaged Earth-atmosphere system: (a) whole earth, (b) single-layer atmosphere and (c) two-layer atmosphere. The surface is plotted in dark grey, the atmospheric layers in lighter grey. The incoming short- and emitted long-wave fluxes are indicated by the solid and dashed arrows, respectively.

than 1 lets long-wave radiation emitted by the Earth escape to space. To obtain insight into this question, the radiative flux balance is solved for various configurations, ranging from a zero-dimensional system (the whole Earth) to a multi-layer atmosphere consisting of many free layers. This is illustrated in Fig. 5.8. For each configuration, the equilibrium radiative fluxes will be estimated, yielding RE temperatures for all components of the system (in this case layers).

5.3.1.1   Whole Earth

The simplest possible system consists of only one layer, representing the planet as a whole, including both planet and atmosphere (see Fig. 5.8a). Assuming equilibrium implies a balance between the net incoming solar flux $F_{SW}^{\downarrow}$ and the outgoing long-wave flux $F_{LW}^{\uparrow}$. Equating Eq. (5.29) to Eq. (5.30) and accounting for the planetary albedo α_p then gives

$$\frac{S_0}{4}\left(1 - \alpha_p\right) = \sigma T_e^4, \tag{5.31}$$

where T_e is also known as the 'effective emission temperature'. Rearranging gives

$$T_e = \left[\frac{S_0}{4\sigma}\left(1 - \alpha_p\right)\right]^{\frac{1}{4}}. \tag{5.32}$$

Using the reference values of S_0 and α_p as stated earlier gives $T_e = 255$ K ($-18\,°C$). Comparison to the observed global mean surface temperature of 288 K ($15\,°C$) shows that the effective emission temperature resulting from the whole-Earth model is much lower. Apparently, in reality the atmosphere makes it difficult for the surface to radiate energy as effectively, because some of it is absorbed by the atmosphere and radiated back down. To properly represent this capability in the model, it is required to represent the atmosphere separately, with at least one layer.

5.3.1.2 Single-Layer Atmosphere

The degrees of freedom are now increased by moving from a zero-dimensional to a one-dimensional configuration. In some EBMs, the extra dimension is introduced in the form of a latitudinal dependence; here we choose to introduce height-dependence. In that sense we are still considering a globally averaged framework. Such bulk models have a limited number of interacting components, such as an atmosphere, Earth, ocean or ice shelf.

Including more independent layers in the atmosphere might yield better, more realistic estimates of surface temperature T_s. The idea is that this somehow introduces a 'greenhouse effect' into the system. Having additional layers in the system that are transparent for short-wave radiation but emit long-wave radiation differently depending on temperature and emissivity ensures that effectively energy is radiated back downwards by the atmosphere. This then results in a warmer equilibrium state at lower levels. The simplest option is to insert a single bulk atmosphere with temperature T_1. The ground surface is assumed to emit long-wave radiation like a perfect blackbody, so that its emissivity $\epsilon = 1$. The atmosphere is allowed to have an emissivity less than 1. This yields two equations with two unknowns (T_s and T_1). For the surface, the radiative balance becomes (with sources on the left-hand side and sinks on the right-hand side)

$$\frac{S_0}{4}\left(1 - \alpha_p\right) + \epsilon\sigma T_1^4 = \sigma T_s^4, \tag{5.33}$$

where it is assumed that all downward long-wave flux is absorbed at the ground surface. For the bulk atmosphere (layer 1) this balance becomes

$$a\sigma T_s^4 = 2\epsilon\sigma T_1^4, \tag{5.34}$$

where a is the efficiency coefficient of the absorption of long-wave radiation by the atmospheric layer. The factor 2 on the right-hand side results from the fact that the atmosphere emits in two directions, upwards and downwards (see Fig. 5.8b). The solution for the two unknowns T_s and T_1 can be written in terms of the effective emission temperature T_e,

$$T_s = \left[\frac{2}{2 - a}\right]^{\frac{1}{4}} T_e, \tag{5.35}$$

$$T_1 = \left[\frac{1}{\epsilon}\frac{a}{2 - a}\right]^{\frac{1}{4}} T_e. \tag{5.36}$$

Note that a special case of solution Eq. (5.36) could be obtained by assuming that the atmosphere also emits like a blackbody, $\epsilon = 1$, by which ϵ would disappear as a variable in the solution. Instead, we apply the assumption of thermal equilibrium (Kirchhoff's law) which states that absorptivity equals emissivity ($a = \epsilon$), which gives

$$T_s = \left[\frac{2}{2 - \epsilon}\right]^{\frac{1}{4}} T_e, \tag{5.37}$$

$$T_1 = \left[\frac{1}{2 - \epsilon}\right]^{\frac{1}{4}} T_e. \tag{5.38}$$

Fig. 5.9a shows these solutions for various values of ϵ. What is clear is that the single-layer bulk atmosphere model can yield better estimates of the surface temperature ($T_s = 288$ K for $\epsilon = 0.77$). For $\epsilon = a = 1$ all upward long-wave flux is absorbed by the bulk atmosphere, so that $T_1 = T_e$. In contrast, for $\epsilon = a = 0$ all of the long-wave radiation emitted by the Earth escapes to space, so that $T_s = T_e$. In general, for a less efficiently absorbing atmosphere (i.e., smaller a), both temperatures decrease; note that their difference (a crude estimate of the vertical lapse rate of temperature and water) remains more or less constant in this model.

5.3.1.3 Two-Layer Atmosphere

The single-layer atmosphere model provides the crudest possible estimate of the atmospheric lapse rate; the profile consist of two points only. To improve the vertical structure of the atmospheric temperature profile, a second atmospheric layer is now included in the model, with the highest layer having the highest index (see Fig. 5.8c). This two-layer configuration could for example be applied to describe the boundary layer and the free troposphere. The system now consists of three equations with three

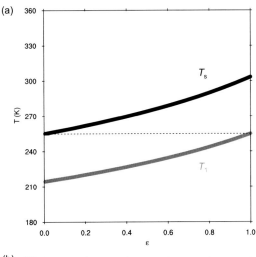

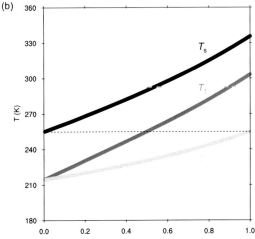

FIGURE 5.9: RE solutions of (a) the single-layer model and (b) the two-layer model, for $T_e = 255$ K and for a range of values of the emissivity ϵ. Surface temperature T_s is plotted in dark grey the atmospheric temperature T_1 in medium grey and T_2 in light grey. The Earth's effective emission temperature T_e is indicated by the dashed line, for reference.

unknowns (T_s, T_1 and T_2). For simplicity, all atmospheric layers are assumed to have the same absorptivity a and emissivity ϵ. The budget equation of each layer now contains two terms representing the absorption of long-wave radiation emitted by the two other layers, which introduces dependence on their temperature. For the surface this gives

$$\frac{S_0}{4}\left(1 - \alpha_p\right) + \epsilon\sigma T_1^4 + (1-a)\,\epsilon\sigma T_2^4 = \sigma T_s^4. \tag{5.39}$$

The third term on the left-hand side contains $(1-a)$, expressing that this is the part of the downward long-wave flux emitted by layer 2 that has not been absorbed

by layer 1. The first atmospheric layer receives long-wave flux from above and below,

$$a\,\sigma T_s^4 + a\,\epsilon\sigma T_2^4 = 2\,\epsilon\sigma T_1^4. \tag{5.40}$$

Finally, the top atmospheric layer receives long-wave flux from the surface and the lower layer,

$$(1-a)a\,\sigma T_s^4 + a\,\epsilon\sigma T_1^4 = 2\,\epsilon\sigma T_2^4. \tag{5.41}$$

This system can still be solved analytically, which is left as an exercise for the reader.

Fig. 5.9b shows the solution of Eqs (5.39)–(5.41) for a range of emissivities. Compared to the one-layer atmosphere configuration, the vertical profile is now better resolved, consisting of three points. The temperature of the lower layer is always 'sandwiched' between the surface and top-layer value. This is to be expected in this configuration, because the input of energy into the system occurs through the lower boundary; solar radiation is absorbed by the surface, which is then distributed upward by long-wave radiative transfer. A second observation is that the vertical structure of temperature in the atmospheric column changes significantly over the range of emissivities, with the gradient being larger aloft or near the surface depending on the value of the emissivity.

In many ways, the temperature profile obtained with this model is not yet realistic. First, note that in this two-layer atmosphere model, $T_s = 288$ K for $\epsilon = a = 0.48$, which is a lower value than that obtained with the single-layer atmosphere model. However, this solution depends heavily on the choice of assuming the same absorptivity in all atmospheric layers. In reality, many gases, especially water vapour and when present water condensate, determine the net efficiency of the long-wave absorption in a layer. Second, the temperature still decreases monotonically with height in this setup, while in nature the temperature is observed to actually increase with height above the tropopause. One reason is that absorption of short-wave radiation by ozone directly heats the stratosphere, a process that is not accounted for in this setup. But the relative coldness of the troposphere compared to the stratosphere is also caused by the efficient cooling of the troposphere due to the presence of water vapour, which the stratosphere lacks. It is clear that yet more degrees of freedom need to be added to the model to reproduce these observed features, which will be addressed next.

5.3.1.4 Full Profile Calculations
The final step is to drastically increase the number of layers in the model, so that it can in principle better reproduce important features in the vertical temperature structure. In that sense, the model becomes a simple version of a what is

called a single-column model (SCM). These models can be columnar grid versions of GCMs in which only sub-grid parameterisations are active. In other words, there is no exchange (advection) of air between columns. The only (vertical) transfer of momentum, enthalpy and water is done by parameterised physical processes such as turbulence, convection, precipitation and radiation. Note that of all these processes radiation need not be parameterised, using for example spectroscopy and ray tracing methods. Radiative parameterisations that are used for efficiency reasons can thus be related to the exact calculations.

Early RE models based on full-profile calculations made use of iterative procedures to let the system of layers equilibrate from some initial state towards a steady-state solution. For each layer, including the surface, the temperature evolves in such a way so that eventually the incoming and outgoing radiative fluxes exactly compensate. These simulations use prescribed profiles of radiative constituents, each active in the long- and short-wave channels in its own unique way. These constituents can include water vapour, carbon dioxide, ozone, aerosols and clouds. Their impact on radiation has already been discussed in Chapter 4, but will be briefly summarised here. Water vapour is the most important greenhouse gas, by efficiently absorbing and emitting long-wave radiation. Carbon dioxide is a relatively minor greenhouse gas (compared to water vapour), but it is the most abundant of the long-lived greenhouse gases, and thus has an important controlling effect on climate. Ozone mainly resides in the stratosphere, and interacts with ultra-violet radiation to act as a local heat source. The impact of aerosol and clouds is more complex, and will not be considered for now.

Fig. 5.10a shows the temperature profiles of various RE simulations with an SCM of a present-day GCM. In the RE simulations, the radiation scheme is the only active component of the sub-grid physics; all the other components (e.g., turbulence, convection and clouds) are switched off. The profiles shown represent the equilibrium state after two years of simulation. The control simulation reflects a standard cloud-free atmosphere that includes present-day profiles of important greenhouse gases such as water vapour, carbon-dioxide and ozone. The simulation in which ozone is removed illustrates that this gas alone is responsible for the typical increase of temperature with height above the tropopause.

Although the full-profile RE calculations produce much improved vertical temperature profiles compared to the single- and double-layer models, there are still problems with the pure RE model. For example, compared to radiosonde observations, the RE state is

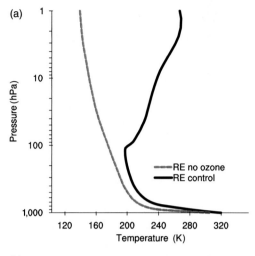

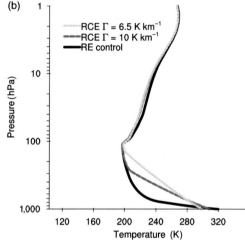

FIGURE 5.10: RE and radiative-convective equilibrium (RCE) temperature profiles as obtained with the European Community Earth System SCM. The RE control experiment (solid black) reflects the default present-day atmospheric composition as used in the integrated forecasting system Cycle 31r1. In (a), an additional RE experiment without ozone is shown, while in (b), two RCE experiments are included, each using a different tropospheric lapse rate Γ. All profiles reflect the model state after one year of time-integration, when equilibrium is reached.

pretty reasonable in the stratosphere, but is far too cold in the middle to lower troposphere. The fact that the water vapour scale height is about 2 km suggests that water vapour plays an important role in driving this cold bias. The strongly overestimated temperature lapse rate in the lower troposphere has implications for the atmospheric stability. Comparison with the dry adiabatic temperature profile (see Fig. 5.10) reveals that the tropospheric temperature profile as produced by the RE model is absolutely unstable. In this situation,

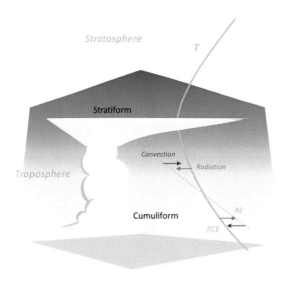

FIGURE 5.11: Schematic illustration of a system in RCE. Featuring are the basic two shapes of deep convective clouds, with cumuliform clouds accommodating the transporting updrafts, and stratiform, anvil-shaped clouds typically situated near the tropopause. The RCE temperature profile is shown in solid orange, with the RE lapse rate is indicated by the dashed orange line. The effects of radiation and convection on the tropospheric temperature profile are indicated by the blue and red arrows, respectively.

convective overturning can be expected, the impact of which should somehow be included in the model.

5.3.2 Radiative-Convective Equilibrium

The RE model is now expanded with a simple representation of convection that acts to quickly remove a convective instability that develops in the tropospheric part of profile. The equilibrium states produced by this model are referred to as radiative-convective equilibria (RCEs). Fig. 5.11 gives a schematic illustration of a system in RCE. While the radiative transfer continuously drives the tropospheric profile towards an unstable state, convection arises and gravity waves adjust the non-convecting portion of the atmosphere. More indirect impacts of convection on the heat budget can also be identified. The first is the greenhouse effect of any redistributed humidity. The second indirect effect is the formation of clouds associated with convection. In the deep tropics, stratiform anvil clouds can form in the upper part of the troposphere, just below the tropopause. These tropical clouds, produced by convective towers, can be optically thick and cover significant areas. The associated high albedo can be efficient in reflecting short-wave radiation, while the high cloud condensate values can act to 'trap' energy through the long-wave part

of the electromagnetic spectrum. Note that a distinction needs to be made between long- and short-wave opacity. In the long-wave clouds become optically thick (i.e., $\epsilon = 1$) for very small amounts of condensate, whereas their albedo in the short-wave α increases more gradually as a function of condensate.

The first step in understanding RCE is to ignore clouds, and to only consider the vertical redistribution of enthalpy. In the context of the budget, Eq. (5.16) applied to dry static energy η_{d}, this process involves formulating a model that produces the vertical profile of convective flux $F_{\eta_{\mathrm{d}}}^{\mathrm{C}}(z)$. Perhaps the simplest class of convective models, often referred to as convective adjustment models, makes use of the concept of relaxation towards a known convective reference profile of dry static energy, $\eta_{\mathrm{d}}^{\mathrm{C}}$,

$$S_{\eta_{\mathrm{d}}}^{\mathrm{C}} = -\frac{\partial}{\partial z} F_{\eta_{\mathrm{d}}}^{\mathrm{C}} = \frac{\eta_{\mathrm{d}}^{\mathrm{C}} - \eta_{\mathrm{d}}}{t_{\mathrm{adj}}}, \tag{5.42}$$

where $S_{\eta_{\mathrm{d}}}^{\mathrm{C}}$ is the diabatic heating tendency due to convection and t_{adj} is the timescale typical of the convective adjustment. The rate at which the temperature profile is adjusted towards the reference profile equates to the vertical divergence of convective flux of static energy. Eq. (5.42) describes a Newtonian relaxation process, featuring a typical timescale t_{adj}. In case t_{adj} is infinitely small, the adjustment is instantaneous, which in effect is equivalent to setting the profile to the reference value. In case the adjustment timescale is finite, the model state is allowed to depart from the reference state. The relaxation process then always acts to nudge the model state towards the reference state. In this setup, the eventual steady-state profile thus describes a QE state, as already briefly discussed in Section 5.2.3.

Making the choice for the adjustment model comes with the obligation to define a reference profile $\eta_{\mathrm{d}}^{\mathrm{C}}$. An adiabatic structure is a good choice, as it represents a neutral state for convection, when all instability is exactly removed. It can be expected that convection ceases automatically when the potential energy for overturning has been depleted. In that sense, (pseudo-)adiabats can be argued to be a reasonable choice as a reference profile towards which convection pushes the atmosphere. In case convection does not involve condensation, a dry (pseudo-)adiabatic structure can be used. This would be applicable to the well-mixed clear CBL. In case convective clouds form, the moist (pseudo-)adiabat would be an appropriate choice. A more in-depth discussion on adiabats can be found in Chapter 2.

In early RCE models, the convective adjustment is assumed to act so fast that it allows simply setting the temperature lapse rate of the tropospheric part of the profile to the reference value. Convective adjustment is assumed to only act in the troposphere; in effect, this comes down to applying RCE below the tropopause, and RE above. Fig. 5.10b shows the temperature profile resulting from RCE experiments with the SCM. Shown are various options for convective overturning; from no convection (i.e., pure RE), via cloud-free convection (using the dry pseudo-adiabatic temperature lapse rate of $9.8 \, \mathrm{K \, km^{-1}}$) to cloudy convection (using the observed temperature lapse rate of $6.5 \, \mathrm{K \, km^{-1}}$). Using the dry adiabatic lapse rate improves the tropospheric lapse rate but it is still too large. Using the observed lapse rate yields a reasonable tropopause height, and represents the most realistic representation of the atmospheric temperature profile so far.

One wonders what controls the observed tropospheric lapse rates of temperature and humidity. At low latitudes (i.e., the tropics) the tropospheric temperature is typically observed to be reasonably close to the moist pseudo-adiabat (see Fig. 5.2). This suggests that moist convective adjustment is the dominant mechanism that reacts to the radiative driving of the atmosphere; this process would automatically cease when the lapse rate approaches the moist adiabat. In contrast, at high latitudes baroclinic waves compete with smaller-scale convection in the vertical redistribution of enthalpy. In these mixed regimes, it is not so clear what reference profiles would be appropriate for the adjustment model. In general, the same applies to humidity, where one may use a fixed relative humidity (RH). A motivation for this choice is that the humidity field is to some degree controlled by changes in RH; for example, supersaturation produces condensate which removes water, while subsaturation near the surface produces evaporation which adds water vapour. If the circulation does not change dramatically, one thus expects the atmosphere to adjust to a thermodynamic state where the distribution of humidity with temperature is more or less unchanged.

Alternatives exist for the simple adjustment model in the representation of convective transport $F_s^C(z)$ in RCE studies. For example, one could let the sub-grid convection scheme of a GCM represent the convective overturning term in the enthalpy budget. These convection schemes often have the adjustment model at their foundation, but typically have grown more complex over the years. An attractive, more recent option is to actually resolve the convective flux using a storm-resolving model (SRM) or a large eddy simulation (LES) model. These models apply such high resolutions that they resolve convective dynamics and clouds on a domain of limited

dimensions, from which the kinematic convective flux $F_s^C(z)$ can be sampled. It has become increasingly popular to run three-dimensional models of RCE which can also resolve the large-scale circulations associated with the convective adjustment.

5.3.3 Box Models

The RE and RCE models as discussed in Sections 5.3.1 and 5.3.2 are usually formulated to describe a horizontally averaged system. In other words, no latitudinal or longitudinal dependence exists in the associated budget equations, and only the vertical dimension is maintained to some degree. *Box models* are bulk models of a slightly different nature, in which different compartments of the climate system are represented as separate entities that may or may not interact. A strict definition of a box model is that it should have zero (spatial) dimensions. However, a more loose definition can also be adopted, by allowing for some spatial reference of where each entity is located.

One example of a box model is the two-layer representation of the Earth-atmosphere system as discussed in Section 5.3.1.2, with each layer acting as a heat reservoir that can interact with the other. To this two-box system one could add more compartments. These can simply have the shape of additional layers, but can also represent spheres outside the system of interest, such as oceans or sea ice. Another possibility is that boxes represent separate areas, for example, one tropical and one extra-tropical. The interaction between boxes depends on their definition; for example, in a system where boxes represent separate areas, their interaction would represent exchange by advection, while in an atmosphere-ocean system, they would represent surface fluxes.

Of particular interest to this book are box models that are designed to investigate cloud-climate feedbacks in the tropics. This sub-class of models is schematically illustrated in Fig. 5.12, and is designed to describe the Walker–Hadley cell that spans the subtropical and tropical marine areas on the globe. This large-scale circulation roughly consists of a narrow ascending branch over the tropical warm pool featuring deep convective clouds, and a broad descending branch over the tropical eastern part of the oceanic basin where the SST is lower. Throughout the whole domain, a PBL exists that accommodates the low-level trade-wind flow, featuring stratocumulus clouds in the upstream areas where SSTs are lower and shallow fair-weather cumulus clouds in the downstream areas. Tropical circulations are described in detail in Chapter 8.

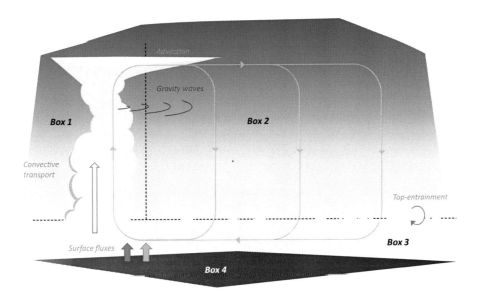

FIGURE 5.12: Box model interpretation of the Walker–Hadley cell in the tropics. Four main boxes are distinguished: Box 1, the warm pool ascent region; Box 2, the tropospheric subsidence region; Box 3, the PBL; and Box 4, the oceanic mixed layer. Possible exchange processes between boxes are indicated in blue, as discussed in the text. Figure inspired by Betts and Ridgway (1988, 1989)

The box models designed for the Walker–Hadley cell are conceptually built on the global RCE models, the key difference being that the boxes do not represent a global mean but an area of limited horizontal extent. Single-box models have been formulated that describe only the subsiding branch; the narrow deep convective area is not modelled explicitly. Simplified representations of the low-level stratocumulus and cumulus clouds can be added. This further refines the first-order representation of the atmospheric water cycle as introduced by RCE, now including evaporation and precipitation, and allows investigation of the impact of the radiative coupling of boundary-layer clouds on the equilibrium state.

It can be argued that a two-component approach better fits the main tropospheric structure of the Walker–Hadley cell, with one box representing the narrow, ascending deep convective area and the other the wide subsiding region. The two-box approach would represent a step up in model complexity, and thus enables more feedbacks to occur in the system. The two-box setup has been used to address the big question: what regulates the temperature in the tropics, and which processes are responsible for counteracting the warming associated with tropical precipitation? This role is sometimes referred to as the 'thermostat' function. Some studies have suggested that the cooling effect of optically thick clouds is responsible, while others rather point at the clear-sky long-wave cooling in the wide

descending branch as the dominating process (the so-called furnace-radiator fin model).

5.3.4   Bulk Circulation Models

An intermediate between GCMs and box models are models with only a few layers. These models generally reduce the vertical degrees of freedom by using bulk approximations to the vertical structure of the atmosphere. In other words, the dependence on the horizontal coordinates x and y in the primitive equations is retained, but the number of vertical degrees of freedom is reduced to only a few. Maintaining horizontal spatial dependence implies that such models can be used to describe horizontal circulations (as explored in more detail in Part III of this book). On the other hand, the use of a few essential vertical modes makes sure that a fair amount of understanding can still be obtained. For this reason, these models are sometimes referred to as 'intermediate complexity circulation models', but will here be referred to as 'bulk circulation models'. The defining features of two such models will be discussed in this section, including WTG models and QE Tropical Circulation Models (QTCM).

5.3.4.1   WTG Models
The use of the WTG approach (described earlier in Section 5.1.1) for bulk modelling of the tropics can be illustrated

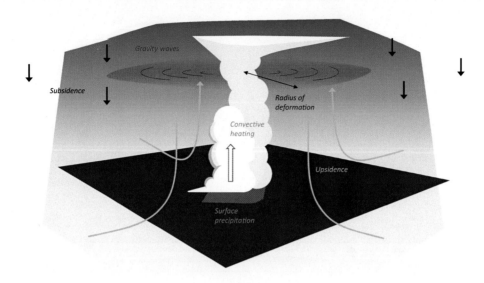

FIGURE 5.13: Schematic illustration of the effect of gravity waves triggered by precipitating deep convection in the tropics, and the response of the larger-scale circulation, in the WTG framework. The scene can be interpreted as an isolated warm pool with rising motion (or 'upsidence') induced by convective heating within an area defined by the radius of deformation, surrounded by a subsidence field induced by radiative cooling.

by considering the budget equation Eq. (5.15) for dry static energy η_d, written in expanded form and now using pressure as the vertical coordinate,

$$\frac{\partial \eta_d}{\partial t} + \vec{U}_h \cdot \nabla_h \eta_d + \omega \frac{\partial \eta_d}{\partial p} = S_{\eta_d}^{\text{net}}, \tag{5.43}$$

where ω is the vertical pressure velocity. In the tropics, the vertical advection term dominates over the local tendency and the horizontal advection on the left-hand side, reflecting the validity of the WTG assumption. Thus, ignoring the first two terms on the left-hand side leaves a dominant balance between diabatic heating terms and the vertical advection term,

$$\omega \frac{\partial \eta_d}{\partial p} = S_{\eta_d}^{\text{C}} + S_{\eta_d}^{\text{R}} + S_{\eta_d}^{\text{L}}, \tag{5.44}$$

where the diabatic heating processes include convective transport (C), radiative transport (R) and latent heating due to phase changes of water (L), respectively. Eq. (8.6) is sometimes referred to as the 'WTG equation'. In the tropics, in areas of significant convection, the turbulent-convective heating plus the latent heating typically dominate over the radiative heating term so that $S_{\eta_d}^{\text{C}} + S_{\eta_d}^{\text{L}} \gg S_{\eta_d}^{\text{R}}$. Also note that averaging of this heating over the troposphere yields

$$\langle S_{\eta_d}^{\text{C}} + S_{\eta_d}^{\text{L}} \rangle = \frac{g}{\Delta p} (P + H), \tag{5.45}$$

where P and H are the surface precipitation and sensible heat fluxes, respectively, and Δp the pressure depth of the troposphere (see Section 5.2.1 for a discussion on vertical integration). The ratio $g/\Delta p$ is introduced using the hydrostatic approach, which allows writing the surface energy fluxes in their familiar dynamic form, expressed in W m^{-2}. This gives for the WTG (Eq. (8.6))

$$\left\langle \omega \frac{\partial \eta_d}{\partial p} \right\rangle = \frac{g}{\Delta p} (P + H). \tag{5.46}$$

When finally assuming that the vertical gradient of η_d is more or less constant, a close link is suggested between the large-scale vertical motion ω on the one hand and precipitation P on the other.

The link between the larger-scale vertical motion ω and the precipitation as suggested by the WTG equation can be interpreted in two ways. In the classic interpretation, atmospheric convection responds to the conditions (i.e., instability) created by the larger-scale circulation. The opposite view, which is commonly thought to apply in the tropics, is that the diabatic heat sources play the role of the controlling process, and larger-scale circulation actually responds to this. This perspective is illustrated in Fig. 5.13. Suppose a diabatic heat source such as precipitating deep convection causes a local temperature anomaly. Because in the weakly rotating regime strong horizontal temperature gradients are not allowed to exist, this temperature perturbation is quickly smoothed out over a large area by gravity waves triggered by the convection. To balance the energy budget, the larger-scale flow has to respond, by creating ascent which in turn cools the column. Through mass-continuity, here written in pressure coordinates,

$$\nabla_h \cdot \vec{U}_h + \frac{\partial \omega}{\partial p} = 0, \qquad (5.47)$$

mid-tropospheric ascent is associated with low-level convergence and upper-level divergence.

Through the WTG constraint (Eq. (5.46)), the convective system thus manages to affect the larger-scale flow in which it is embedded. An important implication is that precipitation and vertical motion become very closely tied to one another through the moisture budget, hence prescribing vertical motion in effect determines the precipitation. This makes it hard to use LES or SRMs to study deep convection using the traditional framework. The WTG theory was developed as a way to get around this problem and thereby provide a more stringent test of how models respond to large-scale forcings, that is, radiation or surface fluxes.

The WTG equation also creates opportunities for bulk modelling. One example is WTG constrained single-column modelling. In the classic SCM approach, thermodynamic state variables like temperature and humidity are treated prognostically, while the large-scale subsidence is prescribed. This is problematic in case the parameterised processes such as convection are able to affect the large-scale subsidence on short timescales. By using the WTG equation, the large-scale subsidence is instead treated as a diagnostic variable, as a function of the parameterised processes that depend on the prognostic state variables such as s and q. Such feedbacks are not captured by classic SCM studies. Taking this even one step further, one could apply WTG to a whole grid of adjacent SCMs that cover the tropics, which would yield a simple circulation model.

The two views on the coupling between convection and circulation differ in the direction of the response, a situation resembling the classic chicken and egg problem. What both perspectives share, however, is the assumption that the coupling between deep convection and circulation is tight. This tightness, combined with the uncertainty about the direction in which this response mechanism works, is one of the reasons why convective parameterisation is generally considered such a difficult and complex problem. Convective parameterisation is discussed in much more detail in Chapter 6.

It should be noted that analogies exist between the WTG approach for the tropics and the quasi-geostrophic (QG) approach for the mid latitudes. By applying the WTG approach, the effect of gravity waves

is represented implicitly. Similar to the QG model, these waves are assumed to act infinitely fast, and are not explicitly represented in the framework. The main difference is that the WTG model assumes a balance in the enthalpy equation, for weakly rotating conditions, while QG assumes a balance in the momentum equation, for strongly rotating conditions.

5.3.4.2 QE Tropical Circulation Models: Separation of Variables

An alternative simplification that can be made in the primitive equations concerns the vertical structure of the state variables, as already discussed in Sections 2.5.4 and 5.2.5. The fact that in the tropics the tropospheric temperature profile is always close to the moist adiabat motivates separating state variables depending on (x, y, p, t) into (x, y, t)-dependent variables on the one hand and assumed (p)-dependent basis-functions on the other. For temperature and humidity, this could be written as

$$T = T_r(p) + A_1^T(p) \, T_1(x, y, t), \qquad (5.48)$$

$$q = q_r(p) + A_1^q(p) \, q_1(x, y, t). \qquad (5.49)$$

Here subscript r refers to some reference state, while the numeric index indicates the mode of the vertical basis function. Multiple modes could in principle be included, for simplicity only one is included here. In case of temperature, T_r is assumed to follow the moist adiabatic profile, and is independent of time and location. Function $A_1^T(p)$ then carries the shape of the moist adiabat perturbation per temperature perturbation $T_1(x, y, t)$, as illustrated in Fig. 5.6b. For humidity, the reference structure could be assumed to depend on the saturation value.

Similarly, for horizontal velocity $\vec{v}$, one could write

$$\vec{v} = \sum_{i=0}^{1} V_i(p) \, \vec{v}_i(x, y, t). \qquad (5.50)$$

This way a barotropic $(i = 0)$ and a baroclinic $(i = 1)$ velocity mode can be maintained, which together determine for a large part the observed vertical structure of the horizontal wind. The vertical structure of the barotropic mode would be constant,

$$V_0(p) = 1. \qquad (5.51)$$

The vertical structure of the baroclinic mode $V_1(p)$ depends on the vertical structure of the temperature perturbation $a_1(p)$. This can be explained as follows. The existence of a horizontal temperature gradient in a layer between two isobars (the definition of baroclinicity) implies a horizontal gradient in geopotential, through the hypsometric equation. This corresponds to a pressure

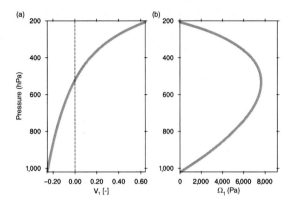

FIGURE 5.14: Vertical structures of (a) horizontal velocity (V_1, unit-less) and (b) vertical velocity (Ω_1, in Pa) associated with baroclinic motions in the tropics.

gradient, which is a fundamental force in the momentum equation. The smallness of the Coriolis parameter f in the tropics means that pressure gradients cannot exist for long, as already discussed in Section 5.1.1, and a circulation develops to balance the momentum equation. This mechanism explains the strong link between temperature perturbations and baroclinic velocity. Without deriving its functional form, the baroclinic velocity structure $V_1(p)$ is shown in Fig. 5.14a.

The use of vertical structures for the baroclinic horizontal velocity component implies a vertical structure for the vertical velocity component. This can be understood by integrating the mass continuity equation in pressure coordinates Eq. (5.47),

$$\int_p^{p_s} \frac{\partial \omega}{\partial p} \, dp' = \omega(p_s) - \omega(p) = -\int_p^{p_s} \nabla_h \cdot \vec{v} \, dp', \quad (5.52)$$

where p' is the integration variable and the vertical velocity at the surface $\omega(p_s)$ can for simplicity be assumed to be zero. The pressure velocity $\omega(x, y, p, t)$ is then expressed in terms of the horizontal divergence of the baroclinic velocity mode $\vec{v_1}$ and an associated vertical velocity basis function $\Omega_1(p)$,

$$\omega = \int_p^{p_s} \nabla_h \cdot \vec{v} \, dp' = -\Omega_1(p) \, \nabla_h \cdot \vec{v_1}, \quad (5.53)$$

where

$$\Omega_1(p) = -\int_p^{p_s} V_1(p') \, dp'. \quad (5.54)$$

Function $\Omega_1(p)$ carries the typical vertical structure of the tropospheric pressure velocity profile, which peaks near the middle of the troposphere (See Fig. 5.14b). In case of mid-tropospheric large-scale ascent, this is associated with low-level convergence and upper-level divergence;

the sign of V_1 is chosen such that $\nabla \cdot \vec{v_1}$ is positive for this situation. This makes $\Omega_1(p)$ positive definite. Unlike in Section 2.5.4, where one starts by assuming a vertical velocity structure, here it is connected to the assumed structure of the horizontal motion.

The maintenance of (x, y, t)-dependent variables allows the simplified system of equations to be applicable as a tropical circulation model. Such bulk circulation models for the tropics are often referred to as a QTCM, for the following reasons. Moist adiabats are at the basis of the model, which makes this setup most applicable to the tropics, and perhaps not so much to the extra-tropics. QE is then introduced by letting convective adjustment Eq. (5.42) act on small temperature departures from the moist adiabat. QTCMs have shown significant skill in reproducing tropical climate dynamics and precipitation patterns, and have been actively used in investigating climate change. An attractive aspect of the reduced degrees of freedom is that the computational load of a global simulation is similarly reduced, which greatly facilitates scientific research compared to full-complexity climate models. The theory behind QTCMs should be interpreted as a hypothesis as to how well the tropical circulation can be understood in terms of one vertical (deep) mode. But sometimes more than one mode are required. For example, recent research has suggested that many tropical phenomena of interest (e.g., convective aggregation, the Madden–Julian oscillation and the ITCZ) depend on the interaction between shallow and deep circulations.

The adoption of predefined vertical structures in the primitive equations can help in gaining insight into tropical dynamics. For example, a useful concept is the *gross moist stability* $\mathcal{G}$. As already discussed in Section 2.5.4, it can be derived by substituting Eq. (5.53) into the moist static energy budget equation and integrating vertically,

$$\langle \partial_t \eta_e \rangle + \langle \vec{v} \cdot \nabla \eta_e \rangle + \mathcal{G}^1 \nabla \cdot \vec{v_1} = \langle S_{\eta_e}^{\text{net}} \rangle. \quad (5.55)$$

The gross moist stability $\mathcal{G}^1$ originates from the vertical advection term. It carries information on the vertical structure of the vertical velocity field associated with the divergence of the baroclinic mode of the horizontal wind,

$$\mathcal{G}^1 \equiv \langle \Omega_1 (-\partial_p \eta_e) \rangle. \quad (5.56)$$

The superscript 1 indicates that only the baroclinic mode of the wind is considered. Note that in Eq. (5.55) $\mathcal{G}^1$ is multiplied by an expression of divergence that by convention is positive for low-level convergence, as follows from the definition of function $V_1(p)$, as shown in Fig. 5.14a. This explains the appearance of the minus sign in Eq. (5.56), a slight difference with definition Eq. (2.176).

The gross moist stability contains information on the vertical gradient of the moist static energy, and is thus an expression of stability. It can be interpreted as representing the work that the large-scale circulation has to perform in order to overcome the existing moist static stability in an atmospheric column. The gross moist stability thus determines how strong a circulation a given input of energy can support. In that sense it represents the stiffness of the system, and can thus be interpreted as a stability. In principle, any vertically integrated budget equation contains something like a gross stability; in case of η_d this is called the *gross static stability* $G^1_{\eta_d}$, while for q this is called the *gross moisture stratification* or alternatively the *gross moisture lapse rate* G^1_q. Substituting Eq. (5.8) into Eq. (5.56) gives

$$G^1 = G^1_{\eta_d} - G^1_q, \tag{5.57}$$

with

$$G^1_{\eta_d} \equiv \langle \Omega_1 (-\partial_p \eta_d) \rangle, \tag{5.58}$$
$$G^1_q \equiv \langle \Omega_1 (L_v \partial_p q_v) \rangle. \tag{5.59}$$

Note that the minus sign in Eq. (5.59) has been left out by convention. By considering the typical buildup of the vertical profiles in Fig. 5.2, it can be understood that all gross stability terms should be positive. This is even true for the total G^1, because the negative contribution by G^1_q (due to the humidity accumulation in the PBL) is always overcompensated by the positive contribution by $G^1_{\eta_d}$ (due to the increase of η_d with height throughout the troposphere). This is illustrated in Fig. 5.15, based on one month of global fields as obtained from weather analyses generated by a numerical prediction model. Note that the $G^1_{\eta_d}$ field is pretty flat, as it must be in the presence of WTGs, so that the local minimum in G^1 in the central tropics is mainly due to the maximum in G^1_q. The latter reflects the accumulation of low-level humidity in those areas.

In the tropics, the gross moist stability gives insight into the constraints placed by the moist static energy budget on the larger-scale circulation, and with it the precipitation field. For example, comparing Fig. 5.15c to Fig. 5.15d shows that the local minima in G^1 match the regions of strong precipitation to a high degree. Why is this the case? As a thought experiment one could further simplify the budget equation Eq. (5.55) by neglecting the horizontal advection and by assuming equilibrium, which yields

$$\nabla \cdot \vec{v_1} = \frac{\langle S^{\mathrm{net}}_{\eta_e} \rangle}{G^1}. \tag{5.60}$$

The left-hand side represents the low-level convergence of the large-scale circulation, while the right-hand side is the ratio of the net moist static energy forcing to the gross moist stability. It is instructive to interpret the left-hand side as responding to the right-hand side. With G^1 being positive definite, the sign of the circulation response then depends on the sign of the net forcing, while its amplitude depends on the magnitude of both the forcing and G^1. For the moment, assuming the forcing constant and positive (a good assumption for the deep tropics), this means that a maximum in convergence is purely due to a minimum in G^1. In general, one could conclude that given a positive forcing, strong convergence is required in regions where the gross moist stability is small, and vice versa. Interpreting the precipitation field in the tropics as a proxy for large-scale low-level convergence finally explains its resemblance with minima in G^1. Such heating-induced low-level convergence is also illustrated schematically in Fig. 5.13. To summarise, the circulation always acts to balance the moist static energy budget for the bulk troposphere, with its response being constrained by the net forcing and the existing gross moist stability.

It is instructive to compare Eq. (5.55) to the vertically integrated WTG Eq. (5.46). Due to the assumption of a fixed vertical structure for $\vec{v_1}$, the gross moist stability G^1 has been separated from the horizontal divergence term in Eq. (5.55). Assuming the vertical gradient to be constant, as was done in the interpretation of the WTG equation, is then no longer required. In addition, (i) through the use of moist static energy, the impact of water vapour is automatically accounted for and (ii) the horizontal advection term is still maintained, which could be important in bulk studies where lateral inflow of moist static energy is important.

5.4 A Bulk Model for Vertically Well-Mixed Boundary Layers

Fig. 5.16 shows a distinct layer of haze, whose top appears very sharp and very homogeneous in the horizontal directions. The photographer took advantage of rising thermals that buoyed his glider up to the top of the haze layer. The figure also shows typical observations of the vertical thermodynamic structure of a clear convective atmosphere during daytime. The temperature profile follows a dry adiabat lapse rate up to the thermal inversion layer in which the temperature strongly increases. This thermal inversion layer can have depths of just several tens of metres, and sharply marks the relatively warm and dry free tropospheric air from the air in the layer below. The potential temperature and the specific humidity in the boundary layer are fairly well mixed which is due to the stirring and homogenising effect of turbulence. For this reason,

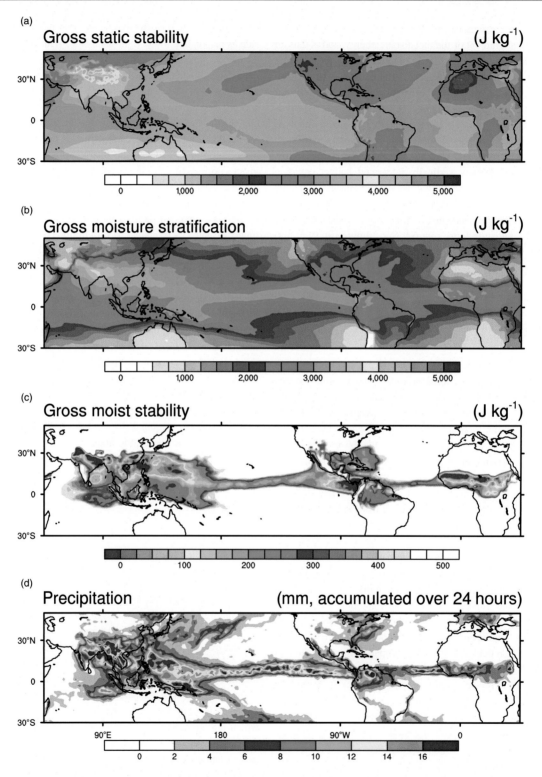

FIGURE 5.15: Monthly means maps of (a) the gross static stability M_{η_d}, (b) the gross moisture stratification M_q, (c) the gross moist stability M and (d) the surface precipitation, for July 2014. The plotted data was derived from a combination of ECMWF analyses and short-range (3-hour) forecasts. Note that in for the gross moist stability the colour scale is reversed to highlight the minima in M.

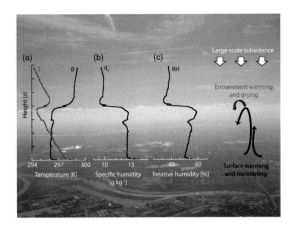

FIGURE 5.16: Tethered balloon observations of (a) the temperature T (red line) and the potential temperature θ, (b) the specific humidity q, and (c) the RH. The data were collected on 23 August 2001, between 10:12 and 10:35 UTC, at Cabauw in the Netherlands. The data have been superposed on a representative photograph of a CBL above the city of Rotterdam as observed from a glider by Adriaan Schuitmaker on another date.

the CBL is often called the mixed layer. It is frequently observed over land during daytime as a result of a heating of the ground surface by solar radiation which triggers convective plumes. Note that the RH is not constant in the mixed layer but increases with height. From the photograph it can be seen that high RH values at the top of the mixed layer can give rise to a distinct layer of haze. This is due to aerosol deliquescence, that is, the process of aerosols absorbing moisture from the atmosphere until they dissolve in the absorbed water. This gives rise to a more efficient scattering of light particularly at RH values above 80% (see Section 3.2.2 and Section 11.3). Another indicator of a high RH is the presence of a few scattered shallow clouds.

The thermal inversion layer acts like some kind of lid in the sense that it prohibits thermals to rise much further, which explains why air pollutants that are emitted near the ground surface tend to accumulate in the mixed layer especially on clear days, as subsidence and warm air aloft inhibit vertical mixing through the troposphere. In the remainder of the text, the terms inversion or inversion layer are loosely used to refer to the thermal inversion layer. If rising plumes penetrate the inversion layer, their negative buoyancies at this height will strongly damp their upward vertical velocities. In fact, they will ultimately sink, thereby mixing some warm and dry air from just above the inversion layer into the boundary layer, a process which is referred to as entrainment. The thermodynamic evolution of the boundary layer can actu-

ally be straightforwardly calculated from the fluxes at the ground surface and the top of the mixed layer. Such a bulk approach, which is called an MLM, considers the vertically integrated budget equations for heat and moisture, thereby effectively treating the CBL as a single slab (Tennekes, 1973).

The MLM not only allows the prediction of possible formation of clouds at the top of the mixed layer, but it will be explained that it also can be applied to study the temporal evolution of the cloud-topped boundary layer. The applicability of the MLM can even be pushed further to investigate how large-scale conditions such as the surface temperature or the large-scale divergence of the wind control the stratocumulus amount. Last, it will be discussed that if we relax the assumption of well-mixedness, the concept of a single-slab atmospheric layer is also a useful approach for shallow-cumulus cloud layers.

5.4.1   Evolution of the Boundary-Layer Depth

The boundary-layer depth h depends on the large-scale vertical velocity $\overline{w}$ and the entrainment velocity w_{e},

$$\frac{\mathrm{d}h}{\mathrm{d}t} = w_{\mathrm{e}} + \overline{w}_h. \tag{5.61}$$

This tendency equation is actually another way of expressing conservation of mass, where $w_e > 0$ represents a measure of the volume of air that is mixed per unit horizontal area per unit time into the boundary layer by turbulent eddies. The large-scale subsidence $\overline{w}$ is driven by a large-scale divergence of the horizontal winds (D),

$$\overline{w}(z) = -\int_0^z D(z)\mathrm{d}z = -\int_0^z \left(\frac{\partial \overline{u}}{\partial x} + \frac{\partial \overline{v}}{\partial y}\right)\mathrm{d}z. \tag{5.62}$$

This means that if there is a divergence in the horizontal air flow (i.e., a net outflow, which is typically present under conditions of a high pressure system), conservation of mass dictates that the removal of mass out of the boundary layer must be accompanied by a compensating large-scale subsiding velocity, $\overline{w} < 0$, which acts to contract and lower the boundary-layer top. As an analogy, one may compare this process with opening the drain in a bathtub such that the removal of water causes its surface level to decrease or a bucket with holes in its sides.

5.4.2   The Vertically Integrated Budget Equation for Conserved Variables

The bulk model equations can be derived from the Reynolds averaged budget equation for any arbitrary

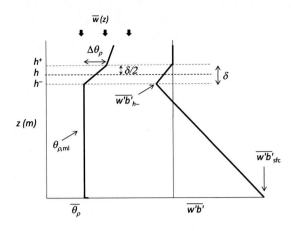

FIGURE 5.17: Schematic of the vertical mean profiles of θ_ρ and the buoyancy flux $\overline{w'b'}$ in the CBL. Adapted from VanZanten et al. (1999)

conserved variable φ such as q_t and θ_ℓ. This is derived in Section 2.5 and is repeated here for convenience,

$$\frac{\partial \overline{\varphi}}{\partial t} = -\overline{u}_h \frac{\partial \overline{\varphi}}{\partial x_h} - \overline{w} \frac{\partial \overline{\varphi}}{\partial z} - \frac{\partial \overline{w'\varphi'}}{\partial z} - \frac{\partial F_\varphi^{\mathrm{src}}}{\partial z}, \qquad (5.63)$$

where F_φ^{src} represents the horizontal slab mean value of the vertical flux of, for example, drizzle or radiation. The roman subscript h indicates the two horizontal dimensions and should not be confused with the boundary-layer depth (italic h).

Fig. 5.17 shows a schematic of the CBL structure for θ_ρ, where the subscript ml is used to denote the mixed-layer mean value of any quantity. An MLM assumes that conserved variables are vertically well mixed in the boundary layer, and that the depth of the inversion layer δ is infinitesimally thin.

As a first step towards the MLM equations, we will integrate Eq. (5.63) from the surface (sfc) to the top of the boundary layer h^- to give

$$\int_0^{h^-} \frac{\partial \overline{\varphi}}{\partial t} \mathrm{d}z = -\overline{w'\varphi'}_{h^-} + \overline{w'\varphi'}_{\mathrm{sfc}} - F_\varphi^{\mathrm{src}}|_{h^-} + F_\varphi^{\mathrm{src}}|_{\mathrm{sfc}}. \qquad (5.64)$$

Note that because of our assumption of a constant value of $\overline{\varphi}$ in the mixed layer, the subsidence term has vanished. We have ignored the mean horizontal advection term by assuming horizontally homogeneous conditions. In MLMs, the surface flux is usually computed from a simple bulk formula, like Eq. (5.24). At this point, the flux at the top of the boundary layer $\overline{w'\varphi'}_{h^-}$ needs to be specified.

It is possible to express $\overline{w'\varphi'}_{h^-}$ as a function of the entrainment rate. Here we will take a formal approach for its derivation, but in the exercises one can find some suggestions for a simplified derivation. We will again

integrate the budget Eq. (5.63), but now from the base (h^-) to the top (h^+) of the inversion layer,

$$\int_{h^-}^{h^+} \frac{\partial \overline{\varphi}}{\partial t} \mathrm{d}z = -\int_{h^-}^{h^+} \overline{w} \frac{\partial \overline{\varphi}}{\partial z} \mathrm{d}z$$
$$+ \overline{w'\varphi'}_{h^-} - F_\varphi^{\mathrm{src}}|_{h^+} + F_\varphi^{\mathrm{src}}|_{h^-}. \qquad (5.65)$$

Above the thermal inversion layer there is no turbulence such that $\overline{w'\varphi'}_{h^+} = 0$.

Because the inversion-layer height is time dependent, we will need to invoke Leibniz's rule of integration according to Eq. (5.14),

$$\int_{h^-}^{h^+} \frac{\partial \overline{\varphi}(z,t)}{\partial t} \mathrm{d}z = \frac{\mathrm{d}}{\mathrm{d}t} \left[\int_{h^-}^{h^+} \overline{\varphi}(z,t) \mathrm{d}z \right] - \Delta \varphi \frac{\mathrm{d}h}{\mathrm{d}t}, \qquad (5.66)$$

where we neglected the time rate of change in the thickness of the inversion layer relative to the time rate of change in its height. This is mathematically equivalent to assuming a more stringent condition of a fixed inversion layer depth $\delta = h^+(t) - h^-(t)$. We define the operator $\Delta_{z_1}^{z_2} \varphi \equiv \varphi_{z_2} - \varphi_{z_1}$ to denote the difference of φ between two arbitrary heights z_1 and z_2. However, for the special case of the inversion jump, we will simply use Δ,

$$\Delta \varphi \equiv \overline{\varphi}_{h^+} - \overline{\varphi}_{h^-}. \qquad (5.67)$$

If $\overline{\varphi}$ is assumed to vary linearly with height in the inversion layer, $\overline{\varphi} = \varphi_{\mathrm{ml}} + \Delta \varphi (z - h^-)/\delta$, then we can directly evaluate the following integral,

$$\frac{\mathrm{d}}{\mathrm{d}t} \left[\int_{h^-}^{h^+} \overline{\varphi}(z,t) \mathrm{d}z \right] = \delta \frac{\partial(\varphi_{\mathrm{ml}} + \Delta\varphi/2)}{\partial t}, \qquad (5.68)$$

and likewise the integral including the large-scale subsidence,

$$\int_{h^-}^{h^+} \overline{w} \frac{\partial \overline{\varphi}}{\partial z} \mathrm{d}z = \overline{w}_h \Delta \varphi, \qquad (5.69)$$

where we used $\overline{w}_{h^-} \approx \overline{w}_h$. With aid of Eqs (5.61), (5.66), (5.68) and (5.69) the budget Eq. (5.65) becomes,

$$\delta \frac{\partial(\varphi_{\mathrm{ml}} + \Delta\varphi/2)}{\partial t} = \overline{w'\varphi'}_{h^-} + w_e \Delta\varphi - \Delta F_\varphi^{\mathrm{src}}. \qquad (5.70)$$

Eq. (5.70) can be strongly simplified by taking the limit $\delta \to 0$ to yield

$$\overline{w'\varphi'}_{h^-} = -w_e \Delta\varphi. \qquad (5.71)$$

This expression is often referred to as the 'flux-jump relation' and states that the turbulent flux at the top of the boundary layer is proportional to the entrainment rate w_e and the inversion jump of φ.

These considerations make it possible to now express the MLM budget Eq. (5.64) as

$$\frac{\mathrm{d}\varphi_{\mathrm{ml}}}{\mathrm{d}t} = \frac{w_e \Delta\varphi + C_d U(\varphi_{\mathrm{sfc}} - \varphi_{\mathrm{ml}}) - \Delta_{\mathrm{sfc}}^{h^+} F_\varphi^{\mathrm{src}}}{h}. \qquad (5.72)$$

An essential ingredient in the derivation of this budget equation is the inversion layer depth $\delta \rightarrow 0$, which is sometimes referred to as the zeroth-order jump model. We further notice that the tendency of φ_{ml} is inversely proportional to the boundary layer depth h.

Eqs (5.61) and (5.72) comprise the MLM equations. The change in the vertically integrated value of φ is governed by the difference of the turbulent fluxes at the surface and the top of the boundary layer, and the presence of diabatic source terms.

If the large-scale mean wind structure is known, the entrainment rate needs to be specified to close the MLM equations. Because for the CBL the scaling behaviour of entrainment has been very well established from observations and numerical models like DNS and LES, we will therefore first explore the applicability of the MLM to this case. Next, it we will be shown that the MLM is a very powerful tool to understand the evolution and the dynamics of the stratocumulus-topped boundary layer (STBL). Despite the fact that shallow-cumulus cloud layers are not vertically well mixed, it will be shown that, like the previous sections on bulk models for the troposphere, the bulk boundary-layer model equations can also be applied to understand its steady-state solutions, not unlike what was done for the troposphere as a whole in Section 5.3.

5.4.3 The Dry CBL

Since the early sixties, many studies have been devoted to understand the temporal evolution of the mixed layer. Early MLMs were applied to the ocean mixed layer in which the buoyancy depends on the salinity and temperature. The dynamical structure of the buoyancy-driven mixed layer in the ocean is identical to the clear CBL, such that the MLM is applicable to both systems.

Fig. 5.18 shows a typical example of the mixed-layer height h evolution over land as observed at the meteorological measurement station at Cabauw in the Netherlands. In the early morning, h rapidly deepens with time, after which it reaches an approximate steady state during the afternoon. During the night, a two-layer structure is seen. The lower layer is associated with the nocturnal stable boundary layer which is cooled from the surface. Just below 2 km, the observations indicate the residual mixed layer, which is a result of convective mixing and entrainment during the preceding daytime period.

To illustrate the effect of convection on the boundary-layer evolution and the entrainment flux, Fig. 5.19 presents LES results of the horizontal slab mean profiles for θ_ρ and $\overline{w'b'}$ at intervals of 2 hours. The case was

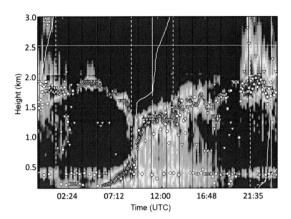

FIGURE 5.18: CBL evolution as observed from a lidar at Cabauw in the Netherlands, 5 May 2008. Local time is two hours later than UTC. The red colours can be interpreted as the top of the mixed layer. Plot kindly provided by Henk Klein Baltink (KNMI (Royal Netherlands Meteorological Institute))

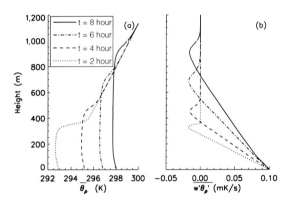

FIGURE 5.19: LES results of a CBL driven by a constant surface buoyancy flux. The lines show the results at t = 2, 4, 6 and 8 hrs for (a) θ_ρ and (b) $\overline{w'\theta'_\rho}$.

driven by a constant surface buoyancy flux and the large-scale subsidence was set to zero. The LES results show that at the top of the CBL, $\overline{w'b'}$ is a constant fraction of its surface value,

$$\overline{w'b'}_h = -A\overline{w'b'}_{sfc}. \tag{5.73}$$

The minimum buoyancy flux appears to be 0.2 times the surface flux value. However, if we ask what value of A would be required for a zeroth-order model, which assumes an infinitely thin thermal inversion layer depth, to evolve similarly to the LES then a much larger value is required. In our LES of the dry CBL, it would be close to 0.4. However, a commonly used value is $A = 0.2$. LES studies show that in the presence of wind shear across the inversion layer, the entrainment rate will be enhanced compared to the shear-free case (Pino et al., 2003). Substitution in the flux-jump relation (Eq. (5.71)) gives

$$w_e = A \frac{\overline{w'b'}_{\mathrm{sfc}}}{\frac{g}{\theta_0} \Delta\theta_\rho}, \tag{5.74}$$

which allows the expression of the budget equation for θ_ρ and the boundary-layer depth evolution as, respectively,

$$\frac{\mathrm{d}\theta_{\rho,\mathrm{ml}}}{\mathrm{d}t} = \frac{(1+A)\overline{w'\theta'}_{\rho\mathrm{sfc}}}{h}, \tag{5.75}$$

$$\frac{\mathrm{d}h}{\mathrm{d}t} = A \frac{\overline{w'b'}_{\mathrm{sfc}}}{\frac{g}{\theta_0} \Delta\theta_\rho} + \overline{w}_h. \tag{5.76}$$

For negligibly small vertical variations in the large-scale divergence $\overline{w}(z) = -Dz$, and subsequently the equilibrium boundary-layer depth h_{eq} can be obtained from

$$h_{\mathrm{eq}} = \frac{A}{D} \frac{\overline{w'b'}_{\mathrm{sfc}}}{\frac{g}{\theta_0} \Delta\theta_\rho}. \tag{5.77}$$

The equilibrium mixed-layer height is thus controlled by the large-scale divergence D, the surface buoyancy flux and the thermal inversion stability as measured by the jump $\Delta\theta_\rho$. Note that although h can be in a stationary state, the turbulent fluxes of heat at the surface and at the top maintain a warming tendency of the mixed layer.

Fig. 5.19 actually shows that in the absence of large-scale subsidence and under the influence of a positive surface buoyancy flux, the boundary-layer height keeps growing. An approximate solution for the boundary-layer growth can be obtained with aid of the time derivative of the inversion jump,

$$\frac{\mathrm{d}\Delta\theta_\rho}{\mathrm{d}t} = \frac{\mathrm{d}\overline{\theta}_{\rho h^+}}{\mathrm{d}t} - \frac{\mathrm{d}\theta_{\rho,\mathrm{ml}}}{\mathrm{d}t}. \tag{5.78}$$

The value of $\overline{\varphi}_{h^+}$ depends on the height of the boundary layer, so its change depends on the change in height of the boundary layer, and hence entrainment,

$$\frac{\mathrm{d}\overline{\varphi}_{h^+}}{\mathrm{d}t} = w_e \frac{\mathrm{d}\overline{\varphi}}{\mathrm{d}z} = w_e \gamma_\varphi, \tag{5.79}$$

where γ_φ denotes the vertical gradient of $\overline{\varphi}$ in the free troposphere, which we will assume to be constant in time. Because the subsidence advects the height of the inversion layer, but does not modify the magnitude of the inversion jump, Eq. (5.79) also applies to cases with large-scale subsidence. Substitution of this expression in Eq. (5.78), and using Eqs (5.61) and (5.75), gives

$$h \frac{\mathrm{d}\Delta\theta_\rho}{\mathrm{d}t} = \gamma_{\theta_\rho} h \frac{\mathrm{d}h}{\mathrm{d}t} - (1+A)\overline{w'\theta'}_{\rho\mathrm{sfc}}. \tag{5.80}$$

According to Eq. (5.78), the jump $\Delta\theta_\rho$ increases due to entrainment (since the free troposphere is stably stratified,

$\gamma_{\theta_\rho} > 0$), but this increase is to some extent counteracted by the warming of the mixed layer due to the turbulent fluxes of heat at the surface and the top of the mixed layer. If we assume that temporal variations in $\Delta\theta_\rho$ are negligibly small, it directly follows that for a constant surface flux, the boundary layer depth must grow with the square root of time,

$$h(t) = \sqrt{h_0^2 + \frac{2\overline{w'\theta'}_{\rho\mathrm{sfc}}}{\gamma_{\theta_\rho}}(1+A)t}, \tag{5.81}$$

with h_0 the mixed-layer depth at $t = 0$. A strong growth of h is promoted in case of large values of the surface buoyancy flux or a weak static stability of the free troposphere, that is, small γ_{θ_ρ}. The assumption that the inversion jump $\Delta\theta_\rho$ is constant in time is a rather strongly idealised boundary condition, and the errors associated with this approximation typically amounts to errors in the boundary-layer depth of about 20%.

The MLM can be applied to assess whether clouds will develop at the top of a CBL. Let us consider a reference case which has the following initial conditions: a mixed-layer height $h = 300$ m, a constant potential temperature and specific humidity in the mixed layer, respectively, $\theta_{\mathrm{ml}} = 288$ K and $q_{v,\mathrm{ml}} = 9$ g kg^{-1}, with inversion jumps $\Delta\theta = 5$ K, $\Delta q_v = -5$ g kg^{-1}. Furthermore, we set $\gamma_\theta = 6$ K km^{-1} and we assume a constant specific humidity in the free troposphere $q_{v,\mathrm{ft}} = 4$ g kg^{-1}. We prescribe constant surface fluxes, SHF = LHF = 100 W m^{-2}, and a surface pressure of 1,018 hPa. We note that fluctuations of θ_ρ can be well approximated in terms of θ' and q_v'. For example, from a linearisation we can express $\overline{w'\theta'_\rho}$ in terms of the given surface fluxes,

$$\overline{w'\theta'_\rho} \approx \overline{w'\theta'} + \epsilon_2 \overline{\theta} \, \overline{w'q_v'}, \tag{5.82}$$

with $\overline{w'\theta'}_{\mathrm{sfc}} = \Pi \, \mathrm{SHF}/(\overline{\rho}c_p)$, $\overline{w'q_v'}_{\mathrm{sfc}} = \mathrm{LHF}/(\overline{\rho}\ell_v)$ and Π the Exner function which translates the potential temperature to temperature (see Eqs (2.46) and (2.47)).

The MLM results for the reference case are shown in Fig. 5.20, in addition to three sensitivity experiments. Because the entrainment parameterisation holds for a clear atmosphere, the MLM simulation is stopped in case clouds are detected from the RH at the top of the mixed layer (RH$_{\mathrm{top}}$). For each example shown, the mixed-layer height grows with time, and is largest for the reference case. The presence of large-scale subsidence slightly diminishes the growth of h. In case of a very small sensible heat flux of 10 Wm^{-2}, temporal changes in the mixed-layer height and its potential temperature are very small. The doubling of $\Delta\theta_\rho$ has a distinct effect on h as it reduces the entrainment rate by a factor of 2. However, according

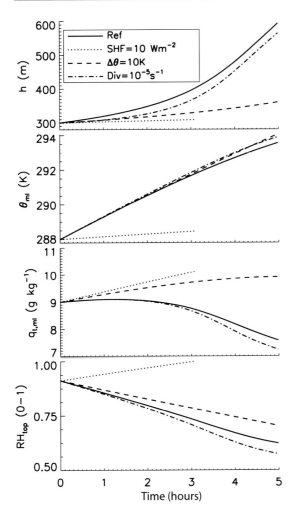

FIGURE 5.20: MLM results of the mixed-layer height h, the mixed-layer values of the potential temperature θ_{ml} and the specific humidity $q_{v,ml}$ and the RH at the top of the mixed layer $RH_{top} = 1$ as a function of time. The conditions for the reference case and the three sensitivity experiments are explained in the text. The simulation with the reduced SHF was stopped at the moment that clouds appeared at the top of the mixed layer as diagnosed from $RH_{top} = 1$.

large value of $\Delta \theta_\rho$, its warming trend is sufficient to avoid the saturation of air.

Before we move on to applying the MLM to clouds, we will first rewrite the CBL entrainment parameterisation. For buoyancy-driven turbulence, it is convenient to define a convective velocity scale w_*,

$$w_*^3 = 2.5 \int_0^h \overline{w'b'} \, \mathrm{d}z. \tag{5.83}$$

This allows the expression of the entrainment relation for the CBL Eq. (5.73) in the following non-dimensional form,

$$\frac{w_e}{w_*} = \frac{A}{\mathrm{Ri}_*}, \tag{5.84}$$

with the convective Richardson number Ri_* defined by,

$$\mathrm{Ri}_* \equiv \frac{gh}{\theta_0} \frac{\Delta \theta_\rho}{w_*^2}. \tag{5.85}$$

Eq. (5.84) can be rewritten as,

$$w_e = A \frac{w_*^3}{\frac{gh}{\theta_0} \Delta \theta_\rho}. \tag{5.86}$$

Because w_* is the vertical integral of the buoyancy flux, which itself depends on the entrainment flux of buoyancy, Eqs 5.84 and (5.86) are implicit relations for the entrainment rate. Both relations are actually equivalent to Eq. (5.74), where the latter allows the calculation of the entrainment velocity directly. The non-dimensional formula for the entrainment velocity has appeared to be convenient for use in STBLs. In particular, the effects of processes that take place in clouds, for example, radiation and phase changes of water, are partly accounted for in the convective velocity scale. However, the additional physics also require a modification of the entrainment efficiency factor, as we will see in the following discussion.

5.4.4 The STBL

Extended fields of STBLs are frequently observed off the coasts of California, Peru and Angola. The conditions in these areas are characterised by relatively low SSTs and a large-scale subsidence which warming tendency exceeds the radiative cooling in the free troposphere. This configuration favours the formation of a strong temperature inversion that traps the moisture that is evaporated from the surface. Fig. 5.21 shows a remarkable example of a stratocumulus cloud deck which boundaries appear to coincide with California's coastal region. As part of the First International Satellite Cloud Climatology Project Regional Experiment (FIRE) aircraft observations were

to Eq. (5.71), the entrainment flux of heat remains almost similar to the reference case value, so the stronger stability of the thermal inversion layer has hardly any effect on the heating rate of the mixed layer. Interestingly, the mixed-layer specific humidity can either be smaller or larger at the end of the simulation compared to the initial condition. The cases which have a relatively small entrainment rate tend to moisten. The question whether clouds develop can be answered from the time series of RH_{top} which shows that only the case with the reduced SHF becomes cloud-topped. Although $q_{v,ml}$ also increases for the case with a

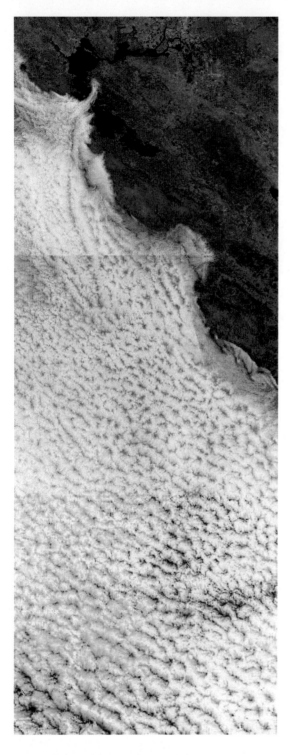

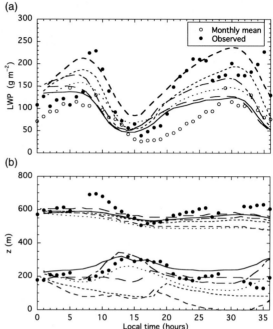

FIGURE 5.22: (a) shows the modelled LWP (lines) and (b) presents the cloud-base and cloud-top heights from six LES models as a function of time for 14 and 15 July 1987 (from 0 to 48 hours local time). The solid dots are the hourly mean observed values and the open circles are the hourly monthly mean values. Adapted from Duynkerke et al. (2004) reproduced with permission from Wiley

collected in this cloud deck, and these findings will be used to illustrate the presented theory.

The persistence and distinct diurnal cycle of stratocumulus clouds is nicely illustrated from the monthly mean LWP as observed from San Nicholas Island shown in Fig. 5.22. During the night, the LWP tends to increase due to a strong long-wave radiative cooling at the top of the cloud layer. The long-wave radiative cooling and latent heating effects in the cloud layer are the main processes that drive turbulence and support vertically well mixed STBLs. During daytime absorption of solar radiation will tend to heat the cloud layer which may promote a thinning of the cloud layer. The heating of the cloud layer may also be stronger than that of the sub-cloud layer giving rise to a weakly stable layer between the sub-cloud and cloud layers. This in turn acts to reduce the vertical upward transport of moisture causing a distinct vertical gradient of the humidity.

Observations collected during the FIRE measurement campaign were modelled with six LES models (Duynkerke et al., 2004). The SST, large-scale subsidence velocity and the horizontal advective tendencies were prescribed, whereas radiative transfer and turbulence were computed by the models. The modelled diurnal cycle of the LWP is fairly well captured by the LES models, with an increasing LWP during the night and vice versa during daytime. However, the variation among the models in the LWP is quite considerable. After 5 hours simulation time, the LWP difference is more than 50 $g\,m^{-2}$. An inspection of the cloud boundaries shows much smaller variations in the modelled cloud-top heights than in the cloud-base height. The maximum differences of the modelled entrainment rates compared to the mean value of all models are of the order $1\,mm\,s^{-1}$, which is equivalent to a tendency of the cloud-top height of only $3.6\,m\,h^{-1}$. One may ask if such rather small deviations in the entrainment rate explain the spread cloud LWP? To answer this question we will apply the MLM. In addition, we will apply the MLM to explore how the stratocumulus equilibrium state solutions are controlled by the large-scale forcing conditions, which is of interest to explore how future climate change may affect the presence of subtropical stratocumulus. We use these questions to illustrate the application of the MLM, and its ability to advance intuition and understanding.

5.4.4.1 Stratocumulus Evolution

We will use the conserved variables θ_ℓ and q_t for the stratocumulus MLM, which budget equations are presented here for convenience,

$$h\frac{d\theta_{\ell,\mathrm{ml}}}{dt} = w_e\Delta\theta_\ell + \overline{w'\theta'_\ell}_{\mathrm{sfc}} - \Delta^{h^+}_{\mathrm{sfc}}F^{\mathrm{src}}_{\theta_\ell}, \qquad (5.87)$$

$$h\frac{dq_{t,\mathrm{ml}}}{dt} = w_e\Delta q_t + \overline{w'q'_t}_{\mathrm{sfc}} - \Delta^{h^+}_{\mathrm{sfc}}F^{\mathrm{src}}_{q_t}. \qquad (5.88)$$

Recall that the liquid-water potential temperature θ_ℓ was derived in Eq. (2.44), here we use the approximate form given in Eq. (2.72). Here the source term may represent the vertical flux of radiation or drizzle. Any other entropy or enthalpy variables could be used in place of θ_ℓ.

To illustrate the large impact of the entrainment rate on the stratocumulus evolution, Fig. 5.23 shows MLM results for three different entrainment rates of 5, 6 and $7\,mm\,s^{-1}$, respectively, which are representative for the range of the LES entrainment results during the first night. The mean LES values of the surface turbulent fluxes were taken as input, and horizontal advection of heat and moisture as well as long-wave radiative fluxes were included in the source terms. The MLM results are in good agreement with the LES results, and nicely show that the height of the inversion layer is only slightly affected by the small entrainment differences, as opposed to the cloud-base height which shows much larger variations.

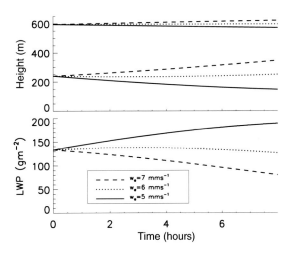

FIGURE 5.23: The cloud-base and cloud-top height, and the cloud LWP during the first night of the EUROCS (European Cloud Systems) FIRE stratocumulus case for three different entrainment velocities as obtained from the MLM.

Because LWP $\propto (h - z_b)^2$, the LWP is quite sensitive to the entrainment rate, particularly through its effect on the cloud-base height.

The most critical element of single slab bulk models is the parameterisation of the entrainment rate. The possible use of the CBL entrainment parameterisation Eq. (5.86) for stratocumulus clouds has been tested from aircraft observations. The entrainment rate is typically assessed with the aid of the flux-jump relation (Eq. (5.71)). Fig. 5.24 compares the entrainment rate in stratocumulus with the parameterisation (Eq. (5.86)) for the CBL. It is immediately clear that the entrainment relation that was obtained for the CBL gives significantly smaller values than the experimental results obtained in stratocumulus. The much larger values of the entrainment in stratocumulus can be mainly explained by the strong long-wave radiative cooling at the cloud top. In addition, phase changes of water in the cloud layer provide local sources of turbulence.

The effect of evaporative cooling at the cloud top may be incorporated in the entrainment parameterisation (Eq. (5.86)) through a modified efficiency factor A_{NT} (Nicholls and Turton, 1986),

$$A_{\mathrm{NT}} = A\left[1 + a_2\left(1 - \frac{\Delta_m}{\Delta\theta_\rho}\right)\right], \qquad (5.89)$$

where the subscript NT refers to Nicholls and Turton who proposed this form, with

$$\Delta_m = 2\int_0^1 (\theta_\rho(\chi) - \overline{\theta_\rho})d\chi \qquad (5.90)$$

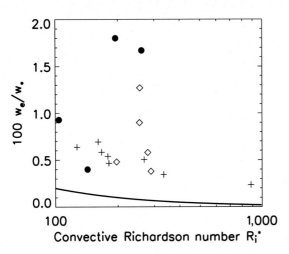

FIGURE 5.24: The entrainment parameterisation for a clear CBL (solid line) and results obtained from aircraft in stratocumulus clouds. The filled circles, open diamonds and plus symbols indicate results from the ASTEX (the Atlantic Stratocumulus Transition Experiment), North Sea and the Second Dynamics and Chemistry of Marine Stratocumulus field study stratocumulus measurement campaigns (de Roode and Duynkerke, 1997; Faloona et al., 2005; Nicholls and Turton, 1986)

representing the mean excess of θ_ρ with respect to its cloud-top mean value for mixtures of air from just above the thermal inversion layer and cloud top following the mixing diagram shown in Fig. 2.8, with the mixing fraction $\chi = \frac{m_{ft}}{m_{ft}+m_{ct}}$ a measure of the ratio of air originating from the cloud top 'ct' and the free troposphere 'ft'. All possible mixing fractions are assumed to have an equal probability to occur. In the absence of cloud liquid water $\Delta_m = \Delta\theta_\rho$, such that the parameterisation yields the correct entrainment rate for a clear CBL. By contrast, in the case that cloudy air is mixed with relatively warm and dry inversion layer air, evaporative cooling will enhance the entrainment efficiency A_{NT} because $\Delta_m < \Delta\theta_\rho$ and Δ_m will decrease for drier inversion layers. Furthermore, cloud droplet sedimentation removes liquid water near the cloud top which tends to diminish the effect of evaporative cooling. If this is taken into account, it is found that $a_2 \approx 15$ (Bretherton et al., 2007). The effects of radiative cooling and condensation are included in the convective velocity scale w_*. In practice, this implies that to compute the entrainment rate one needs as input the surface fluxes and the inversion jumps of heat and moisture, the cloud base and cloud top heights, and the flux profiles of the radiation and drizzle. Various studies have shown that the Nicholls and Turton parameterisation can fairly well reproduce entrainment rates from observations and LES studies.

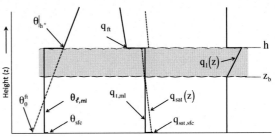

FIGURE 5.25: Schematic representation of the STBL. The MLM framework assumes that conserved thermodynamic quantities are vertically well mixed between the surface and the base of the thermal inversion layer. The surface fluxes of heat and moisture are proportional to the difference between their surface and their mixed-layer values. In the free troposphere, the potential temperature increases with height. It may be expressed in terms of an extrapolated surface value θ_0^{ft}. For simplicity, a constant value for the free tropospheric specific humidity is used in this example. The well-mixed character of the boundary layer enforces an adiabatic cloud liquid-water specific humidity lapse rate. Adapted from de Roode et al. (2014) reproduced with permission from Wiley

5.4.4.2 Stratocumulus Equilibrium State Solutions

Equilibrium state solutions offer insight into how large-scale forcing conditions control the stratocumulus cloud amount and they are of interest to assess how the stratocumulus amount may respond to a global warming scenario. In this highly idealised approach, a diurnally averaged radiative forcing needs to be applied, whereas in the free troposphere the tendencies due to subsidence, large-scale advection and radiation must balance.

Fig. 5.25 depicts a schematic setup for the vertical profiles of θ_ℓ, q_t and q_l in a MLM. The free tropospheric potential temperature can be expressed as

$$\theta_{ft} = \theta_0^{ft} + \gamma_\theta z, \text{ for } z \geq h^+, \quad (5.91)$$

with γ_θ the vertical gradient of the potential temperature above the inversion layer. The quantity θ_0^{ft} represents the extrapolated free tropospheric potential temperature at the surface.

Fig. 5.26a presents steady-state solutions of the boundary-layer depth for a suite of experiments with different values for the SST and the large-scale divergence, the latter being constant with height. Only non-precipitating stratocumulus cases were considered. The results presented in Fig. 5.26a were obtained with the entrainment rate calculated with the Nicholls and Turton parameterisation. The results show that as expected small values of D allow for deep boundary layers. The overall deepening of h with increasing SST can be explained by a weaker stability of the thermal inversion layer, since in all experiments the same free tropospheric state was

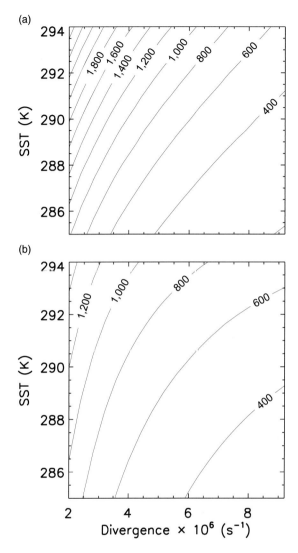

FIGURE 5.26: The boundary-layer height h as a function of the SST and the large-scale divergence of the horizontal wind, and constant values $q_{t,ft} = 4\,\mathrm{g\,kg^{-1}}$, $\theta_0^{ft} = 292\mathrm{K}$, $U = 7\,\mathrm{m\,s^{-1}}$ and $\Delta F_{rad} = 0.035\,\mathrm{m\,K\,s^{-1}}$ which corresponds to a difference in the net radiative flux of $40\,\mathrm{W\,m^{-2}}$ across the boundary layer. (a) shows the results as obtained with the Nicholls and Turton entrainment parameterisation, and the results in (b) were obtained with a highly simplified parameterisation for the entrainment rate Eq. (5.92) with $\eta = 0.8$.

used. The basic structure of the results can be rather well captured if we use a strongly simplified parameterisation for the entrainment rate

$$w_e = \eta \frac{\Delta F_{rad}}{\Delta \theta_\ell} \qquad (5.92)$$

with η an entrainment efficiency factor for stratocumulus. This expression is analogous to the CBL entrainment closure but now with the forcing comprised by the change

in the net radiative flux across the cloud layer divided by ρc_p such that ΔF_{rad} has units $\mathrm{m\,K\,s^{-1}}$. One could describe Eq. (5.92) as a linearisation of a more complex entrainment law to emphasise the important role of thermal stratification and radiative forcing. As an illustration, we will use an efficiency factor $\eta = 0.8$, which is much higher than A. This choice is based on the findings in Fig. 5.24 and on the fact that radiative cooling and entrainment warming are often found to dominate the heat budget, which only has a very small contribution from the surface heat flux. Note that Eq. (5.92) only depends on the longwave radiative cooling and unlike the Nicholls and Turton parameterisation does not include buoyancy effects in the cloud layer due to phase changes of water.

In the last two decades, observations collected during large field campaigns such as the Second Dynamics and Chemistry of Marine Stratocumulus field study (DYCOMS II) and DNS and LES modelling experiments based on these cases have strongly increased our understanding of the entrainment process. The validity of the LES results is typically established from a detailed comparison with observations, similarly to the examples shown in Figs. 5.22 and 5.28. The DYCOMS II Flight 1 stratocumulus model intercomparison case (Stevens et al., 2005) has appeared to be a particular challenging case for LES models. This case satisfies the buoyancy reversal criterion (see Section 2.3.2) such that entrainment of relatively warm and dry air and subsequent mixing with cloud-top air gives rise to negatively buoyant parcels. As this will generate turbulence, which in turn drives entrainment, this positive feedback has been speculated to lead to a rapid break-up of the stratocumulus cloud deck. However, the observations showed a persistent stratocumulus cloud deck. The LES modelling results of the DYCOMS II case showed a large sensitivity to the details of the sub-grid model, in particular with regards to how they represent mixing across the cloud top. It was found that if the spurious sub-grid mixing at cloud top is reduced the observations could be reproduced with some fidelity.

Fig. 5.27 presents the equilibrium state results for some other boundary-layer properties, some of which can be straightforwardly interpreted with aid of the analytical steady-state solution of the MLM,

$$\varphi_{ml} = \varphi_{sfc} + \frac{w_e(\varphi_{h^+} - \varphi_{sfc}) - \Delta_{sfc}^{h^+} F_\varphi^{src}}{w_e + C_d U}. \qquad (5.93)$$

For $\theta_{\ell,ml}$ the entrainment rate can be eliminated with aid of the entrainment toy model to yield,

$$\theta_{\ell,ml} = \theta_{sfc} + \frac{(1 - \eta)\Delta F_{rad}}{C_d U}. \qquad (5.94)$$

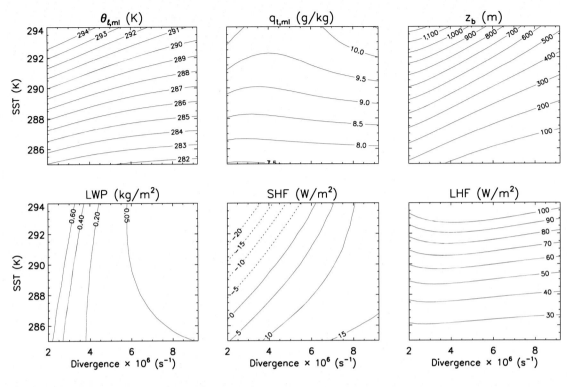

FIGURE 5.27: Equilibrium state solutions of the STBL as obtained with the Nicholls and Turton entrainment parameterisation and the same large-scale forcing conditions as in Fig. 5.26. The panels show the mixed layer values of θ_ℓ and q_t, the cloud-base height, the LWP, and the sensible and latent heat fluxes.

For the case of $\eta = 1$, which means that entrainment warming is balancing radiative cooling, $\theta_{\ell,\mathrm{ml}} = \theta_{\mathrm{sfc}}$. The equilibrium state solution for the surface heat flux is equal to

$$C_\mathrm{d} U (\theta_{\mathrm{sfc}} - \theta_{\ell,\mathrm{ml}}) = (1 - \eta) \Delta F_{\mathrm{rad}}. \qquad (5.95)$$

Because the solutions assume an equilibrium state, more entrainment of warm inversion layer air implies that the input of heat from the surface has to be reduced subsequently. In case the warming due to entrainment exceeds the radiative cooling, the surface heat flux will even become negative. This situation is found in the regime of small large-scale divergence and large SST. By contrast, more entrainment of dry air at the cloud top needs to be compensated by an enlarged surface evaporation. This notion follows from the solution for $q_{t,\mathrm{ml}}$ which reads,

$$q_{t,\mathrm{ml}} = q_{\mathrm{sat,sfc}} + \frac{q_{t,\mathrm{ft}} - q_{\mathrm{sat,sfc}}}{1 + \frac{C_\mathrm{d} U}{Dh}}, \qquad (5.96)$$

with which we can obtain an analytic expression for the surface evaporation,

$$C_\mathrm{d} U (q_{\mathrm{sat,sfc}} - q_{t,\mathrm{ml}}) = \frac{C_\mathrm{d} U (q_{\mathrm{sat,sfc}} - q_{t,\mathrm{ft}})}{1 + \frac{C_\mathrm{d} U}{Dh}}. \qquad (5.97)$$

For deep boundary layers, in particular for $C_\mathrm{d} U / Dh \to 0$, the surface evaporation is maximised, $C_\mathrm{d} U (q_{\mathrm{sat,sfc}} - q_{t,\mathrm{ml}}) = C_\mathrm{d} U (q_{\mathrm{sat,sfc}} - q_{t,\mathrm{ft}})$. It is also interesting to note that for a relatively dry boundary layer, surface evaporation will be larger, and the cloud base will be relatively high.

Although we are looking at equilibrium state solutions, the general boundary-layer properties we have found with the MLM are in a pretty good agreement with the observed structure in the main stratocumulus regions as mentioned before. Typically, the SST tends to increase equatorwards, with the boundary layer typically becoming deeper, the surface evaporation increasing, and the surface sensible heat flux remaining rather small.

Motivated by climate modelling results that show an overall decrease in the number of low clouds under a future global warming scenario (see Chapter 13), LES was applied to study equilibrium states of the STBLs. To mimic a perturbed climate, the large-scale forcing conditions were perturbed by a simultaneous warming of the sea surface and free troposphere. The findings suggest that the stratocumulus cloud layer will slightly thin. Because this will increase the solar radiative flux reaching the ground surface, the change in the cloud layer thickness thus imposes a positive feedback to the enhanced greenhouse effect. Climate modelling results also suggest a weakening of the strength of the Hadley cell circulation in a future climate. If the subsidence is reduced in the perturbed climate simulations, LES results predict that the stratocumulus cloud layer will thicken, a tendency which is consistent with the MLM results shown in Fig. 5.26.

Similar experiments with SCMs have indicated that the spread among different models is very large, which is due to the fact that parameterisations in large-scale models are often forced to compensate large-scale errors associated with a failure to resolve even the mean structure, which most parameterisations assume is known. In addition, the vertical resolution applied in climate models is too coarse to capture the inversion layer and its sharp jumps in temperature in humidity. Different SCMs give opposite cloud feedbacks for a global warming scenario. For example, models with a strong positive entrainment response to a warming scenario tend to yield thinner clouds and vice versa. Such a cloud thinning is accompanied by a stronger increase in the surface evaporation than in models that give a cloud thickening, a finding which is consistent with the MLM experiments.

We conclude that idealised studies with bulk models like the MLM can help to improve our understanding of the dependency of the stratocumulus cloud system to the prevailing large-scale forcing conditions. Despite the fact that we are dealing with vertically integrated properties, and that the relevant processes like vertical turbulent transport and long-wave radiative cooling occur at scales smaller than the boundary-layer depth, we will next demonstrate that the MLM framework can also be used to diagnose the vertical fluxes of heat, moisture and buoyancy. This is a very useful approach for interpreting results from more complex models and measurements.

5.4.4.3 Turbulent Flux Profiles in Stratocumulus

The temporal evolution of the ABL is strongly controlled by entrainment. Following Eq. (5.86), to obtain the entrainment velocity, it is necessary to calculate the buoyancy flux profile. Before we will apply the MLM framework to explain the vertical profile of the buoyancy flux in stratocumulus in detail, we will first inspect Fig. 5.28 which shows the buoyancy flux and the vertical velocity variance profiles as obtained from LES experiments of the STBL. Compared to the clear CBL in which the buoyancy flux depends linearly on height, the buoyancy flux profile in the STBL looks much more complex. The figure shows that the buoyancy flux tends to

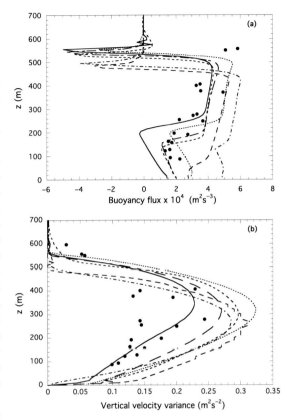

FIGURE 5.28: Aircraft observations and six LES model results of (a) $\overline{w'b'}$ and (b) $\overline{w'w'}$ for the EUROCS FIRE stratocumulus case during night-time. Adapted from Duynkerke et al. (2004) reproduced with permission from Wiley

increase strongly near cloud-base height, which is caused by the latent heat release in saturated updrafts. At the top of the cloud we find a negative buoyancy flux which is due to the entrainment of warm and dry air from just above the inversion layer into the boundary layer. Just below this minimum we find another large jump in the buoyancy flux which can be explained by the long-wave radiative cooling that drives the formation of negatively buoyant downdrafts. It is also found that models with small values for the buoyancy flux are accompanied with large negative values at the cloud top. Actually, models with the largest entrainment fluxes have the smallest vertical velocity variance values in the boundary layer. This finding, and the presence of the jumps in the buoyancy flux profile, can both be elegantly explained with the MLM. To this end, we will first introduce the concept of quasi-steady-state fluxes. This refers to a state in which the tendencies of the mean values of scalars are constant with height, which implies that the shape of their vertical profiles is invariant.

5.4.4.4 Quasi-Steady-State Flux Profiles

The MLM assumption that conserved variables are constant with height in the boundary layer puts a strong, though useful, constraint on the turbulent flux profiles. It is most convenient to illustrate the effect of diabatic source terms on flux profiles of conserved variables, so we will first discuss the height dependency of the fluxes for q_t and θ_ℓ. Once their flux profiles are known, the buoyancy flux can be straightforwardly derived. This exercise will be invaluable to the understanding of cloud dynamics.

If $\overline{\varphi}$ is constant with height in the mixed layer at any time, we can write

$$\frac{\partial}{\partial z}\frac{\partial \overline{\varphi}}{\partial t} = \frac{\partial}{\partial t}\frac{\partial \overline{\varphi}}{\partial z} = -\frac{\partial^2(\overline{w'\varphi'} + F_\varphi^{\mathrm{src}})}{\partial z^2} = 0, \quad (5.98)$$

where in the second term we changed the order of differentiation. Integration gives

$$\frac{\partial(\overline{w'\varphi'} + F_\varphi^{\mathrm{src}})}{\partial z} = c_{\varphi,1} = -\frac{\partial \overline{\varphi}}{\partial t}, \quad (5.99)$$

from which it follows that the tendency of $\overline{\varphi}$ is constant with height. Integrating once more we get

$$\overline{w'\varphi'} + F_\varphi^{\mathrm{src}} = c_{\varphi,1}z + c_{\varphi,2}. \quad (5.100)$$

In the absence of the source term, the values of the two integration constants $c_{\varphi,1}$ and $c_{\varphi,2}$ can be straightforwardly obtained from the surface and top boundary conditions,

$$\overline{w'\varphi'}(z) = \overline{w'\varphi'}_{\mathrm{sfc}}\left(1 - \frac{z}{h}\right) - w_e \Delta\varphi \frac{z}{h}, \quad (5.101)$$

where we invoked the flux-jump condition for the flux at the boundary-layer top. Note that this linear flux profile has actually been used in the discussion of the CBL.

Let us now include the effect of long-wave radiative cooling at the cloud top which is a sink term in the budget equation for θ_ℓ. Fig. 5.29 shows that the long-wave radiative flux changes strongly near the top of the stratocumulus cloud top. Let us express this flux jump by ΔF_{rad} across a layer with depth δ_{rad},

$$F_{\mathrm{rad}}(z) = 0 \qquad\qquad 0 < z < h - \delta_{\mathrm{rad}},$$

$$F_{\mathrm{rad}}(z) = \Delta F_{\mathrm{rad}}\left(\frac{z - h + \delta_{\mathrm{rad}}}{\delta_{\mathrm{rad}}}\right) \quad h - \delta_{\mathrm{rad}} \le z < h. \quad (5.102)$$

At the surface $z = 0$, it follows from Eqs (5.100) and (5.102) that the integration constant $c_{\theta_\ell,2} = \overline{w'\theta'}_{\ell\mathrm{sfc}}$. Likewise, from Eq. (5.100) and the flux-jump relation we obtain for $z = h$,

$$c_{\theta_\ell,1} = \frac{-w_e\Delta\theta_\ell + \Delta F_{\mathrm{rad}} - \overline{w'\theta'}_{\ell\mathrm{sfc}}}{h}. \quad (5.103)$$

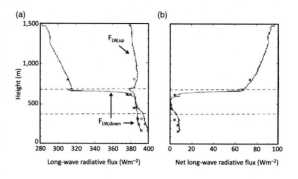

FIGURE 5.29: Aircraft observations of the upward and downward long-wave radiative fluxes collected during (a) flight A209 of the ASTEX experiment and (b) the net long-wave radiative flux (right). Adapted from Duynkerke et al. (1995), reproduced with permission from the AMS

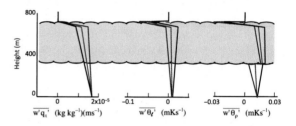

FIGURE 5.30: Schematic representation of $\overline{w'q_t'}, \overline{w'\theta_\ell'}$ and $\overline{w'\theta_\rho'}$. The three different line styles, black, red and blue, indicate three cases with different entrainment velocities, small, reference and large, respectively.

We can now express $\overline{w'\theta_\ell'}$ as a function of height,

$$\overline{w'\theta_\ell'}(z) = \overline{w'\theta_\ell'}_{\mathrm{sfc}}\left(1 - \frac{z}{h}\right)$$
$$- (w_e\Delta\theta_\ell - \Delta F_{\mathrm{rad}})\frac{z}{h} - F_{\mathrm{rad}}(z). \quad (5.104)$$

Recalling that $F_{\mathrm{rad}}(z) = 0$ up to $h - \delta_{\mathrm{rad}}$, we will now assume that the radiative flux divergence takes place across an infinitely thin layer, that is $\delta_{\mathrm{rad}} \to 0$, which allows us to write

$$\overline{w'\theta_\ell'}(z) = \overline{w'\theta_\ell'}_{\mathrm{sfc}}\left(1 - \frac{z}{h}\right) - (w_e\Delta\theta_\ell - \Delta F_{\mathrm{rad}})\frac{z}{h}. \quad (5.105)$$

Fig. 5.30 schematically depicts vertical fluxes for a nocturnal non-drizzling stratocumulus case that is in a quasi-steady state. In the absence of drizzle, $\overline{w'q_t'}$ must vary linearly with height in order to maintain a vertically well-mixed layer, with the vertical gradient of the flux controlled by its values at the surface and the top of the boundary layer. The figure also shows that following the flux-jump relation, a larger entrainment rate causes larger fluxes of q_t and θ_ℓ at the top, and vice versa. If the entrainment flux of q_t is larger than the surface flux, the boundary layer tends to become drier with time. The flux of θ_ℓ exhibits a strong jump at the top of the boundary layer (see Eq. (5.104)), where its decrease with height indicates a strong warming tendency. This is a direct consequence of the long-wave cloud-top radiative cooling which must be partly compensated by a warming due to turbulent transport of heat. In other words, turbulence acts to redistribute the radiative cooled air throughout the entire boundary-layer depth, thereby ensuring tendencies of θ_ℓ that do not depend on height.

5.4.4.5 Buoyancy Flux Profile

Fig. 5.30 also shows $\overline{w'\theta'_\rho}$, which at first glance may seem to be completely disconnected from the fluxes of θ_ℓ and q_t. However, by realising that following Eq. (2.7) we can express θ_ρ as

$$\theta_\rho = \theta(1 + \epsilon_2 q_v - q_l) \qquad (5.106)$$

and with the use of Eqs (2.163) and (2.164) we can express $\overline{w'b'}$ as

$$\begin{aligned} \overline{w'b'} &= A_d\overline{w'\theta'_\ell} + B_d\overline{w'q'_t}, \\ \overline{w'b'} &= A_m\overline{w'\theta'_\ell} + B_m\overline{w'q'_t}, \end{aligned} \qquad (5.107)$$

with the dry thermodynamic coefficients A_d and B_d for unsaturated air and likewise moist coefficients A_m and B_m to be used in the cloud layer.

The $\overline{w'b'}$ profile shown in Fig. 5.30 is a direct result of Eq. (5.107) with appropriate values for the thermodynamic coefficients. In the sub-cloud layer, the contribution of the moisture flux to the buoyancy flux is usually rather small. By contrast, phase changes of water and the associated latent heating effects are incorporated mainly in the $\overline{w'q'_t}$ contribution to $\overline{w'b'}$, and act to promote positive buoyancy fluxes in the cloud layer. The buoyancy flux due to entrainment at the top of the cloud layer is negative, meaning that relatively warm inversion layer air is mixed downwards by entraining eddies. Long-wave radiative cooling in the upper part of the cloud layer promotes the formation of cold downdrafts.

The examples of the buoyancy flux profiles shown in the figure illustrate their dependency on the entrainment rate. It appears that a larger entrainment velocity tends to reduce the buoyancy flux values in the boundary layer. The LES results show that smaller buoyancy fluxes are accompanied by reduced values of the vertical velocity variance.

If the buoyancy flux at the top of the sub-cloud layer becomes negative, surface-driven convective thermals may not be able to reach the cloud layer. This may reduce the transport of heat and moisture from the ground surface to the cloud layer, a situation which is called 'decoupling'. If this is the case, the cloud layer will tend to become

warmer (in terms of θ_ℓ or s_ℓ) and drier compared to the sub-cloud layer. The difference in the thermodynamic states between the sub-cloud and cloud layers is found to be strongly correlated with the boundary-layer depth (Park et al., 2004), and consequently this condition prohibits to treat the boundary layer as a single mixed layer. With the aid of the entrainment parameterisation, Eq. (5.84), and the buoyancy flux according to Eq. (5.107), it is possible to derive an analytical expression that states under which conditions the buoyancy flux in the sub-cloud layer becomes negative. In general, decoupling is supported for a large value for the surface evaporation, a small radiative cooling of the cloud top, a relatively high cloud-base height and a large value for the entrainment efficiency factor A_{NT} (Bretherton and Wyant, 1997). This can be qualitatively understood from Fig. 5.30 which shows that a large value for w_e acts to decrease the buoyancy flux in the sub-cloud layer. Surface evaporation plays a key role since it controls the rate with which latent heat is released in the cloud layer, thereby providing the fuel for the production of turbulence which in turn drives the entrainment. Second, the negative value of $\overline{w'\theta'_\ell}$ at the cloud top due to entrainment is counteracted by the process of long-wave radiative cooling. If the radiative flux divergence near the cloud top is relatively small compared to the entrainment flux of θ_ℓ, this causes a layer with negative values for $\overline{w'\theta'_\ell}$ from just below the base of the radiatively cooled layer. If this layer with negative $\overline{w'\theta'_\ell}$ extends down to the sub-cloud layer, and if its absolute value is larger than the positive moisture flux contribution to the buoyancy flux, then according to Eq. (5.107), a negative buoyancy flux near the top of the sub-cloud layer can be expected. Note that this condition is more easily satisfied if the sub-cloud layer is relatively deep. Last, the entrainment efficiency may be enhanced if the free troposphere is relatively dry, as this will promote a stronger evaporative cooling when this air is entrained into the top of the cloud layer.

5.4.5 Bulk Models for Shallow Cumulus Clouds

So far we have considered well-mixed layers. Because shallow cumulus clouds develop in a conditionally unstable cloud layer, which has a temperature gradient between the dry and moist adiabatic lapse rate, the MLM framework is not applicable to this system. Fig. 5.31 shows the mean state for a boundary layer with shallow cumulus clouds as diagnosed from LES results of the Barbados Oceanographic and Meteorological Experiment (BOMEX) case. In the sub-cloud layer, θ_ℓ and q_t are

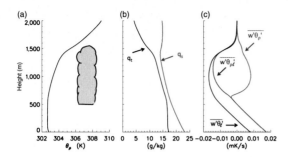

FIGURE 5.31: Vertical profiles of the hourly mean values of the (a) density temperature, (b) the total and (c) saturated specific humidities, and vertical fluxes of θ_ℓ, θ_ρ and $\theta_{\rho\ell}$ for a cumulus-topped boundary layer as obtained from a LES of the Barbados Oceanographic and Meteorological Experiment (BOMEX) case (Siebesma and Cuijpers, 1995)

constant with height, but in the cloud layer they do clearly exhibit vertical gradients. The boundary layer is topped by a thermal inversion layer.

The dynamical structure of the sub-cloud layer is very similar to that of the CBL, and it is sometimes called the surface-based mixed layer. For example, the buoyancy flux at its top is slightly negative, and the vertical profile of the vertical velocity variance looks similar to the CBL. At the top of the sub-cloud layer cumulus clouds develop in negatively buoyant rising thermals that are driven from a warm and moist surface. If the latent heat release is sufficient, and if their vertical velocity is high enough, they can reach the level of free convection, which is defined as the height at which they become positively buoyant. Only the strongest thermals will develop into cumulus clouds that can rise up to the inversion layer, which is manifested by their low cloud fraction typically being about 0.1–0.2, and cloud core areas are even a small fraction of that, typically around 0.02 or 0.03. Despite their small cloud fraction, cumulus clouds play an important role in transporting relatively cold and moist sub-cloud layer air into the free troposphere.

5.4.5.1   *Steady-State Solutions*

Like we did for stratocumulus, we will now explore the role of some large-scale forcing conditions on the bulk properties of shallow-cumulus cloud layers that are in an equilibrium state. However, we will now need to consider a system which consists of a vertically well-mixed sub-cloud layer, and a conditionally unstable cloud layer that is capped by a thermal inversion layer. An essential difference with the MLM is that for shallow cumulus we will not apply an entrainment closure at the top of the cloud layer, but instead we will use the fact that the buoyancy flux at the top of the sub-cloud layer is a constant fraction of its

surface value. The values for the cloud-top entrainment flux will now be obtained from the assumption of quasi-steady-state fluxes.

Let us integrate the budget equation from the surface to the height just above the inversion layer that caps the cloud layer (h^+) and at which the turbulent flux vanishes (Schalkwijk et al., 2013),

$$\int_{z=0}^{h^+} \frac{\partial \overline{\varphi}}{\partial t} dz = \int_{z=0}^{h^+} \overline{\varphi} \frac{\partial \overline{w}}{\partial z} dz$$
$$- \varphi_{h^+} \overline{w}_{h^+} + \overline{w'\varphi'}_{\mathrm{sfc}} - \Delta_{\mathrm{sfc}}^{h^+} F_\varphi^{\mathrm{src}}, \quad (5.108)$$

where we used the chain rule for integration and $\overline{w}_{\mathrm{sfc}} = 0$. It will appear convenient to use the fact that in the mixed layer $\overline{\varphi}(z) = \varphi_{\mathrm{ml}}$, which allows us to write

$$\int_{z=0}^{h^+} \overline{w} \frac{\partial \varphi_{\mathrm{ml}}}{\partial z} dz = -\int_{z=0}^{h^+} \varphi_{\mathrm{ml}} \frac{\partial \overline{w}}{\partial z} dz + \varphi_{\mathrm{ml}} \overline{w}_{h^+} = 0. \quad (5.109)$$

If we add the zero-value expression in the middle of this equation to the steady-state budget Eq. (5.108), we get,

$$0 = \overline{w'\varphi'}_{\mathrm{sfc}} - \overline{w}_{h^+}(\varphi_{h^+} - \varphi_{\mathrm{ml}})$$
$$+ \int_{z_{\mathrm{LCL}}}^{h^+} [\overline{\varphi}(z) - \varphi_{\mathrm{ml}}] \frac{\partial \overline{w}}{\partial z} dz - \Delta_{\mathrm{sfc}}^{h^+} F_\varphi^{\mathrm{src}}. \quad (5.110)$$

The first and last terms in Eq. (5.110) are also present in the MLM Eq. (5.72) and represent the input of heat or moisture by their respective turbulent surface fluxes and diabatic forcings such as radiation and drizzle. The second term differs from the flux-jump relation (Eq. (5.71)) in the sense that the entrainment rate is replaced by the absolute value of the subsidence. Another difference with the MLM is that the inversion jump is replaced by the difference between the value just above the inversion layer φ_{h^+} and the sub-cloud layer value φ_{ml}. This is due to the fact that the cumulus cloud layer is not vertically well mixed. In the case it would be vertically well mixed, the second term could actually be expressed by the flux-jump relation. Likewise, the third term is due to subsidence acting on a mean vertical gradient. Note that the lower boundary of integration is shifted from the surface to the lifting condensation level (LCL) as between those heights $\overline{\varphi} = \varphi_{\mathrm{ml}}$. The mean vertical gradients of θ_ℓ and q_{t} in the cloud layer are not known. One way to circumvent this problem is to capture the effect of subsidence in one single free parameter β,

$$\beta = 1 - \frac{1}{\overline{w}(h)} \int_{z_{\mathrm{LCL}}}^{h^+} \frac{\overline{\varphi}(z) - \varphi_{\mathrm{ml}}}{\varphi_{h^+} - \varphi_{\mathrm{ml}}} \frac{\partial \overline{w}}{\partial z} dz, \quad (5.111)$$

with which we can express the budget equation as

$$\beta \overline{w}(h)(\varphi_{h^+} - \varphi_{\mathrm{ml}}) = \overline{w'\varphi'}_{\mathrm{sfc}} - \Delta_{\mathrm{sfc}}^{h^+} F_\varphi^{\mathrm{src}}. \quad (5.112)$$

For a cumulus cloud layer that is warmer and drier than the sub-cloud layer, $\beta < 1$. This means that the input of free tropospheric air at the top of the boundary layer by subsidence will be reduced. This makes sense since a part of the subsidence warming and drying is actually taking place in the cloud layer, and it is precisely this effect which is incorporated in the β term. If cloud-layer gradients are ignored by setting $\beta = 1$, the steady-state budget equation becomes identical to the MLM. Most importantly, from Eq. (5.112) solutions can obtained that compare reasonable well with LES results, in particular in terms of the sub-cloud and cloud-layer depths, and sub-cloud values of θ_ℓ and q_{t}.

The external diabatic forcing needs to be prescribed. We will consider non-precipitating cumuli and to ensure a steady-state free troposphere, we will implicitly apply a height-dependent cooling rate which balances the warming due to subsidence. Moreover, in the boundary layer, radiative cooling is a necessary sink of heat that is needed to balance the heating by the turbulent flux. Last, we assume that in the lower part of the atmosphere the large-scale divergence is constant with height.

For $\beta = 1$ the steady-state solution can be obtained with the use of the prescribed forcings and the bulk surface flux parameterisation (Eq. (5.24)),

$$\theta_{\ell\mathrm{ml}} = \theta_{\ell\mathrm{sfc}} + \frac{\theta_{\ell 0}^{\mathrm{ft}} - \theta_{\ell\mathrm{sfc}} - \frac{1}{2}h\gamma_{\theta_\ell}}{1 + \frac{C_d U}{Dh}}, \qquad (5.113)$$

where we assumed a bulk radiative cooling that is balanced by subsidence warming according to $\Delta_{\mathrm{sfc}}^{h^+} F_{\theta_\ell}^{\mathrm{src}} = \frac{1}{2}Dh^2\gamma_{\theta_\ell}$. We note that this solution is analogous to the one found for the stratocumulus mixed layer (Eq. (5.93)). However, we have not yet specified an entrainment relation at the cloud top. Therefore, at this point we cannot determine the boundary-layer depth h. Fortunately, it turns out to be possible to solve the shallow-cumulus budget equations for the special case of a steady state. In particular, we will make use of the fact that aircraft observations and LES results strongly suggest that the cumulus sub-cloud layer is dynamically very similar to the CBL, for example as shown from the vertical flux of θ_ρ shown in Fig. 5.31. In the bulk model, we will apply the entrainment closure at the top of the sub-cloud layer where the buoyancy flux has its minimum value and which height approximately coincides with the LCL,

$$\overline{w'b'}_{\min} = -A\overline{w'b'}_{\mathrm{sfc}}. \qquad (5.114)$$

The similarity between the cumulus sub-cloud layer and the CBL is pinned down through the value of the entrainment efficiency. The LCL can be diagnosed from the solutions of $\theta_{\ell,\mathrm{ml}}$ and $q_{\mathrm{t,ml}}$ and is a function of h, which is unknown. We will therefore invoke the steady-state heat

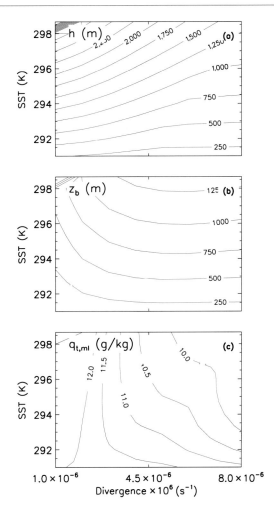

FIGURE 5.32: Equilibrium state solutions of the (a) boundary-layer depth, (b) cloud-base height and (c) the total water specific humidity in the sub-cloud layer for a cumulus-topped boundary layer for the following large-scale conditions, $q_{\mathrm{t,ft}} = 0$ g kg^{-1}, $\theta_0^{\mathrm{ft}} = 290$ K, $U = 10$ ms^{-1} and a mean vertical potential temperature gradient of 6 K km^{-1} in the free troposphere.

budget for the sub-cloud layer, where a balance between the radiative cooling and warming due to turbulence puts the following constraint on the LCL,

$$(1 + A)\overline{w'\theta'_{\rho\ell}}_{\mathrm{sfc}} = \frac{1}{2}D\gamma_\theta z_{\mathrm{LCL}}^2, \qquad (5.115)$$

where in the absence of cloud liquid water, $\theta_\rho = \theta_\ell(1 + \epsilon_2 q_{\mathrm{v}})$. With the aid of Eq. (5.113) we can compute $\theta_{\ell,\mathrm{ml}}(h)$ and $q_{\mathrm{t,ml}}(h)$ and diagnose $z_{\mathrm{LCL}}(h)$. The combination that satisfies the sub-cloud layer heat budget, Eq. (5.115), then provides the steady-state solution, if any.

Fig. 5.32 shows analytical solutions of the bulk cumulus model in a phase space with dimensions large-scale divergence and the SST. We find that the cloud-top height is predominantly controlled by the SST. The vertical stability decreases for a higher SST, allowing the clouds to grow deeper. If the SST becomes smaller than 290.6 K, no cumulus solutions are found. Higher cloud-base values are found for increasing values of the SST. The LWP has very small values for relatively large values of the large-scale divergence.

5.4.5.2 Mixing-Line Models

In mixing-line models, the vertical structure of the cumulus-topped boundary layer is expressed as

$$\overline{\varphi}(z') = \varphi_{ml}[1 - f(z')] + f(z')\overline{\varphi}_{h^+}, \tag{5.116}$$

with the normalised vertical coordinate $z' = (z - z_{LCL})/(h - z_{LCL})$. The vertically integrated budget equation includes the following two parameters,

$$\alpha = \int_0^1 f(z') dz', \tag{5.117}$$

and

$$\beta' = \int_{z_{LCL}}^h \frac{\partial f}{\partial z} \frac{\overline{w}(z)}{\overline{w}|_h} dz. \tag{5.118}$$

For a perfectly vertically mixed layer, $\alpha = 0$ and $\beta' = 1$. If the boundary of integration is replaced from z_{LCL} to h, the mixing-line model yields the same bulk equations as we applied for the dry CBL and stratocumulus. For a shallow cumulus case, these parameters can be prescribed or they may be diagnosed from LES results. In the latter case, the mixing-line model well captures the sensitivity of the boundary-layer structure, except for very high free tropospheric moisture contents.

5.4.6 Stratocumulus to Cumulus Transitions

Fig. 5.33 depicts a cumulus cloud that penetrates the stratocumulus cloud layer above. This situation is frequently observed in the STBL and may occur if absorption of solar radiation in the cloud layer causes it to become warmer (in terms of θ_ℓ or s_ℓ) compared to the sub-cloud layer. However, observations indicate that in particular deep STBLs, including nocturnal ones, do rather generally exhibit a two-layer structure, as depicted in Fig. 5.33, with the sub-cloud and cloud layers each being vertically well mixed. The difference between the sub-cloud and cloud-layer thermodynamic properties have been found to increase with increasing depth of the STBL (Wood and Bretherton, 2004) that favours the decoupling of the cloud layer from the sub-cloud layer (Bretherton and Wyant, 1997). The stable stratification that is present between the sub-cloud and cloud layer hinders surface-driven thermals to reach

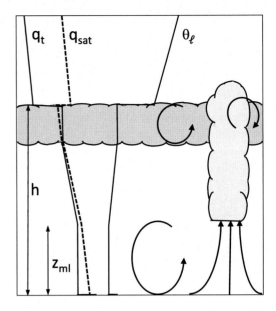

FIGURE 5.33: Schematic of a cumulus cloud penetrating a stratocumulus cloud layer. The cumulus cloud has its base near the top of a vertically well mixed sub-cloud layer.

the stratocumulus, except for some that become saturated at the top of the sub-cloud layer. The latent heat release allows them to rise further as shallow cumulus clouds, thereby feeding the stratocumulus cloud deck with moist sub-cloud layer air. The key question for this situation is whether the stratocumulus cloud layer will thicken, thin or even break up?

Because of its distinct two-layer structure, we cannot apply the MLM to the cumulus under stratocumulus regime, while the cumulus bulk model is not suitable either. Because large-scale models typically calculate the turbulent transport in shallow cumulus and stratocumulus clouds with different parameterisation schemes, they have difficulties in faithfully representing the mixed regime with both cloud types present. LES models however do not suffer from this problem as they resolve eddies larger than the grid size. LES runs of the stratocumulus to cumulus transition are often performed in a Lagrangian sense, which means that an air mass is followed while it is being advected from the subtropics to the tropics by the trade winds over increasing SSTs.

The break-up of the stratocumulus deck appears to be strongly controlled by the humidity and temperature just above the inversion layer. The quantity κ takes both into account and is defined as

$$\kappa \equiv 1 + \frac{c_p}{\ell_v} \frac{\Delta\theta_\ell}{\Delta q_t}. \tag{5.119}$$

It is found that for sufficiently large κ, stratocumulus will break up, which will be the case for a weak thermal stability of the inversion layer in terms of a small $\Delta\theta_\ell$, or a large negative humidity jump across the inversion layer. Although there is no well-defined threshold value for κ that predicts stratocumulus break-up, it can be understood that a large value will promote thinning of the cloud layer. First of all, a weak stability of the inversion layer in terms of a small temperature jump supports large entrainment rates. According to the flux-jump relation Eq. (5.71), for a given entrainment rate, the turbulent flux of dry air from just above the inversion layer into the cloud layer will be larger for a larger value of Δq_t. A large value for κ can thus be interpreted as a condition that promotes a strong influx of dry air from just above the inversion layer into the cloud layer. During the stratocumulus transition, the humidity jump will tend to increase because of the increasing saturation specific humidity at the sea surface.

The evolution of the cloud layer can be well described by the MLM provided that it is applied to the stratocumulus cloud layer only. The lower boundary conditions need to be specified at the cloud-base height instead of at the ground surface. The tendency of the liquid water can be diagnosed from the tendencies of heat and moisture, which in turn can be used to calculate the change of the cloud-layer thickness, or equivalently the LWP.

Another application of the MLM to the stratocumulus transition concerns the thermodynamic evolution of the sub-cloud layer and the surface heat fluxes (de Roode et al., 2016). As an illustration we will present an idealised case following an air mass during its advection by the mean horizontal wind for which the surface temperature increases linearly with time at a rate α_T,

$$\mathrm{SST}(t) = \mathrm{SST}_0 + \alpha_T t, \tag{5.120}$$

with SST_0 the initial value. We aim to explain why observations show that the surface flux of θ_ρ remains rather small while the LHF tends to increase during the stratocumulus transition. If we approximate the surface value of θ_ρ to be controlled only by the temperature we can write

$$\theta_{\rho,\mathrm{sfc}}(t) = \theta_0 + \alpha_\theta t, \tag{5.121}$$

with θ_0 its initial value and α_θ its time rate of change. Furthermore, we neglect the small gradual growth of the sub-cloud layer depth z_{ml} by using a constant value, we neglect any tendencies caused by the radiative fluxes and we focus on non-precipitating clouds. If shallow cumulus clouds are present under a stratocumulus cloud deck, both observations and LES results indicate that the sub-cloud-layer structure is dynamically very similar to the CBL, in the sense that the buoyancy flux at the top of the sub-cloud

layer is a fixed fraction of the surface value. Using the bulk surface flux formula, Eq. (5.24), we can express the budget equation for $\theta_{\rho,\mathrm{ml}}$ in the sub-cloud layer as,

$$\frac{\partial\theta_{\rho,\mathrm{ml}}}{\partial t} = \frac{\theta_0 + \alpha_\theta t - \theta_{\rho,\mathrm{ml}}}{\tau_{\theta_\rho}}, \tag{5.122}$$

with $\tau_{\theta_\rho} \equiv \frac{z_{\mathrm{ml}}}{(1+A)C_d U_{\mathrm{ml}}}$ and U_{ml} the horizontal wind speed in the sub-cloud layer. Its solution reads

$$\theta_{\rho,\mathrm{ml}}(t) = \theta_{\rho,\mathrm{sfc}}(t) - \alpha_\theta \tau_{\theta_\rho} + C_1 \exp^{-t/\tau_{\theta_\rho}}, \tag{5.123}$$

with C_1 a proportionality factor that follows from the initial conditions,

$$C_1 = \theta_{\rho,\mathrm{ml},0} - \theta_0 + \alpha_\theta \tau_{\theta_\rho}. \tag{5.124}$$

The last term in Eq. (5.123) can be interpreted as a memory term which tends towards zero for increasing time. With the aid of the surface flux bulk formula, we find for $t \to \infty$

$$\overline{w'\theta'_\rho}_{\mathrm{sfc}} = \frac{\alpha_\theta z_{\mathrm{ml}}}{1+A}. \tag{5.125}$$

Interestingly, this solution is constant with time. For $\alpha_\theta = 2\,\mathrm{K\,day}^{-1}$ and $z_{\mathrm{ml}} = 500\,\mathrm{m}$ we obtain $\overline{w'\theta'_\rho}_{\mathrm{sfc}} = 0.01\,\mathrm{m\,K\,s}^{-1}$ which is a representative value for the stratocumulus transition regime.

For the sub-cloud layer humidity budget we need to take into account that the surface saturation humidity is dictated by Clausius–Clapeyron, which we will use in approximated form (Stevens, 2006),

$$q_{\mathrm{sat,sfc}}(\mathrm{SST}) = q_0 \exp\left[\frac{\ell_v}{R_v \mathrm{SST}_0^2}(\mathrm{SST} - \mathrm{SST}_0)\right], \tag{5.126}$$

with SST_0 and q_0 the initial values, and with the use of Eq. (5.120) we can express $q_{\mathrm{sat,sfc}}$ as a function that increases exponentially with time according to,

$$q_{\mathrm{sat,sfc}}(t) = q_0 e^{t/\tau_{\mathrm{CC}}}, \tag{5.127}$$

with the timescale $\tau_{\mathrm{CC}} = \frac{R_v \mathrm{SST}_0^2}{\ell_v \alpha_T}$. The budget equation for $q_{\mathrm{t,ml}}$ can be written as

$$\frac{\partial q_{\mathrm{t,ml}}}{\partial t} = -\frac{q_{\mathrm{t,ml}}}{\tau_q} + \frac{q_0}{\tau_q} e^{t/\tau_{\mathrm{CC}}}, \tag{5.128}$$

with the timescale $\tau_q \equiv \frac{z_{\mathrm{ml}}}{(1-r_{q_t})C_d U_{\mathrm{ml}}}$. Here for compact notation r_{q_t} indicates the vertical turbulent flux of q_t at the top of the sub-cloud layer normalised by its surface value and gives a measure of the fraction of the surface humidity flux that is transported out of the sub-cloud layer by the cumuli. LES results of the stratocumulus transition show that r_{q_t} is affected by the diurnal cycle of solar radiation. During daytime, the solar heating of the cloud layer causes a weakening of the cumulus moisture transport to the stratocumulus which results in an increase of sub-cloud layer moisture, and vice versa during night-time. Here we will analyse the behaviour of the system as a function of r_{q_t}. If we apply a constant value of r_{q_t} we obtain the following solution,

$$q_{\mathrm{t,ml}}(t) = \frac{q_0}{1 + \frac{\tau_q}{\tau_{\mathrm{CC}}}} \exp^{t/\tau_{\mathrm{CC}}} + C_2 \exp^{-t/\tau_q}, \qquad (5.129)$$

with

$$C_2 = q_{\mathrm{t,ml,0}} - \frac{q_0}{1 + \frac{\tau_q}{\tau_{\mathrm{CC}}}}. \qquad (5.130)$$

The last term on the right-hand side of Eq. (5.129) tends towards zero for increasing time. Let us consider the first term on the right-hand side of Eq. (5.129). For the case in which the cumuli do not transport any humidity out of the sub-cloud layer, $r_{q_t} = 0$, we have a maximum moistening rate of the sub-cloud layer. However, since the moistening rate of the sub-cloud layer is less than at the surface, $q_{\mathrm{t,ml}}(t) < q_{\mathrm{sat,sfc}}(t)$, we conclude from the bulk surface flux formula, Eq. (5.24), that the LHF increases with time. If $r_{q_t} = 1$, the flux divergence of the humidity is zero, $q_{\mathrm{t,ml}}$ will be constant in time, and the LHF will increase faster than for $r_{q_t} = 0$. The key point here is that the stronger the ventilation of moisture from the sub-cloud layer to the stratocumulus, the faster the LHF will tend to increase.

5.5 Outlook and Future Directions

In this chapter, bulk modelling techniques have been applied to parts of the atmosphere that have a significant vertical extent, such as the turbulent CBL and the troposphere. The degrees of freedom in the vertical dimension have been purposely reduced, maintaining only a few essential modes to still preserve some degree of realism. This can enhance the model transparency, making these systems easier to understand. The researcher can thus focus better on what controls the evolution of the layer mean thermodynamic state, without being side-tracked by second-order phenomena. This is essential for gaining insight into basic system-internal feedback mechanisms, and can elucidate the role played by individual components such as enthalpy, water and clouds, as well as their interaction with the circulation. The bulk budget equations have been derived through the vertical integration across the layer of interest, maintaining information on its vertical structure whenever that is thought important. Through Leibniz's rule of integration, the interface terms can be accounted for that arise due to the potentially changing position of the layer boundaries with time.

The conceptual bulk modelling of the atmosphere has had a long and successful history. However, it is far from redundant. Increasingly, bulk techniques are applied to make progress in areas of the atmospheric sciences where this is currently most required. This in particular concerns water and clouds in the climate system, their impact on radiation and their coupling to the circulation (as discussed further in Parts III and IV of this book). Climate model predictions still suffer from large uncertainties, which are at least partially related to cloud parameterisation. At best, such parameterisations reflect our current understanding of cloud processes, which has been built up over decades (as discussed in the next chapter). Evidently, the way to make significant progress is to first gain more insight into cloud processes, and then to improve parameterisations accordingly. In that sense, bulk model interpretations have proven to be a powerful tool to learn about clouds from observational data sets and fine-scale model simulations.

For example, bulk modelling has been useful for investigating the process of entrainment of air into low-level stratocumulus clouds. What controls this process, and which laws govern it, are questions that are still not fully answered. As discussed in Section 5.4, an advantage of bulk mixed-layer modelling is that top-entrainment naturally appears as an interface term in the budget equations, through the Leibniz's rule of integration. While subsidence also contributes to the bulk mass budget, various recent field-campaigns in the subtropical marine subsidence areas have successfully provided reliable estimates of large-scale subsidence, using new measurement strategies. As a result, the entrainment can be diagnosed as a residual from the observed net deepening of the mixed layer.

Recent research has also shed more light on bulk entrainment into Arctic air masses in transformation. The air overlying the boundary layer is often observed to be more humid, associated with intrusions of warm air masses into the Arctic area. This means that entrainment tends to maintain boundary-layer stratocumulus instead of dissipating them, an intriguing reversal of roles compared to the subtropical subsidence areas. With Arctic amplification at full steam, this link between entrainment and Arctic clouds is a topic that is intensively researched at the moment. Arctic clouds are discussed in-depth in Chapter 10.

Conceptual modelling is also actively used to deal with the sheer variety of cloud shapes and sizes that can co-exist within a single cloud field. This is illustrated by Fig. 5.34, showing a photo of a typical cloud scene in the

FIGURE 5.34: A heterogeneous cloud scene over the subtropical Atlantic, as observed on 11 December 2013 at 19:45 UTC from the German research airplane HALO (high altitude and long range research aircraft) during the NARVAL field-campaign (next-generation aircraft remote sensing for validation studies). Multiple cloud types co-exist in close proximity, featuring fair-weather cumulus, stratocumulus, cumulus congestus, deep cumulus and cirrus sheets associated with cumulus anvils. Figure courtesy of Bjorn Stevens

subtropical Atlantic that features fair-weather cumulus, stratocumulus and deeper cumulus in close proximity. Modelling transitions between those regimes, or modes, is an active research field, for example between stratocumulus to cumulus and shallow cumulus to congestus. Questions that are asked include, what controls stratiform cloudiness atop cumulus? What is the role of precipitation in the formation of cold pools, and what is the role of shallow clouds in the process of convective aggregation? Bulk interpretations of such modes as observed or diagnosed in high-resolution simulations is helping to understand and represent the impact of this cloud heterogeneity in circulation models.

The interaction between clouds and circulation has become a topic that is intensively researched in recent years. Conceptual bulk modelling can provide constraints that can help pick apart these interactions, and isolate the role of small-scale (cloud) processes. For example, the role of shallow circulations within the boundary layer in the organisation and aggregation of convective clouds has been successfully modelled in a bulk mixed-layer

setup. Also, secondary circulation impacts on momentum transport in general can be understood and modelled conceptually. In the tropics, the RCE and WTG constraints are now frequently applied in local fine-scale simulations and SCMs to represent (to some extent) the interaction between small-scale processes and the larger-scale circulation.

Further Reading

Section 5.1 State Variables for Bulk Models

While in this chapter the thermodynamic state variables are only described in a general way, focusing on behaviour that is useful for bulk modelling, for a much more fundamental description we refer to Chapter 2. A recent review about thermodynamic state variables in climate science was written by Stevens and Bony (2013), titled 'Water in the atmosphere', in which the interplay between water, air circulation and temperature in Earth's changing climate are discussed.

The identification of the WTG in the tropics and its explanation are often considered to be cornerstones of tropical meteorology. It is exciting to revisit the literature that describes this early thinking, such as the book *Tropical Meteorology* by Riehl (1954), and the paper 'A note on large-scale motions in the tropics' by Charney (1963). It is instructive to realise that even in the current golden age of supercomputing, such analytic concepts of atmospheric behaviour are still very relevant.

The decoupling of humidity between the low-level boundary layer and the free troposphere in the tropics, and the resulting emergence of atmospheric rivers, is addressed in the landmark paper by Pierrehumbert and Yang (1993). They describe how isolated vertical mixing of humidity can explain the formation of elongated filaments of humidity on isentropic surfaces that resemble the atmospheric rivers observed in nature.

Moist static energy is by now widely used in scientific studies of clouds in the atmosphere. One recent example relevant for the context of this book on clouds and climate is the study by Brient and Bony (2013). Use is made of the bulk moist static energy budget of the ABL to interpret the sign of the low-cloud feedback as predicted by a climate model under global warming.

Section 5.3 Bulk Models for the Free Troposphere

Many conceptual bulk models of clouds in the troposphere have been formulated, dedicated to a large variety of topics. The idea of this chapter has been to only discuss the basic theoretical concepts at their foundation. Only a few recent studies are highlighted below, which may serve as an access point for the reader to the extensive literature dedicated to this topic.

In the article 'Atmospheric moist convection' by Stevens (2005), various forms of moist convection in the (lower) troposphere are reviewed through a consideration of a number of typical regimes. Theoretical concepts are considered that help interpret observed convective phenomena and their interaction with the larger-scale flow.

The concept of box models as discussed in this chapter has been around for a few decades. Betts and Ridgway (1989) presented a key paper in which they develop a box model perspective of the Hadley circulation in the tropics that includes an ABL, a framework that still defines much of our thinking today. Pierrehumbert (1995) makes use of an idealised radiative-dynamic two-column model of the tropical Walker circulation to investigate the balance between the radiative cooling in the subtropical dry subsidence areas and the convective warming in the tropical ascent areas. Note that the use of box models is much

wider than the atmospheric sciences; in particular, in chemistry such approaches are very common.

The idea of gross moist stability has been pioneered by Neelin and Held (1987), while the associated QTCMs have been introduced by Neelin and Zeng (2000). The impact of the boundary-layer–troposphere coupling on the tropical circulation, in particular the width and strength of the ITCZ, was investigated by Neggers et al. (2007) by implementing a bulk MLM in a QTCM.

The concepts of WTG and RCE as discussed in this chapter are frequently used in recent research on clouds and climate change. For example, Zhang et al. (2013) make use of the WTG approach to construct cases of subtropical marine low-level clouds in present and future climate, to be used or LES or single-column modelling. Bellon and Sobel (2010) combine WTG and a QTCM SCM to investigate equilibration processes in the Hadley circulation. A similar large-scale constraint introduced relatively recently is the radiative advective equilibrium, which has been applied by Cronin and Jansen (2016) to Arctic situations and by Dal Gesso and Neggers (2017) to single-column modelling on climate timescales during the subtropical dry season at Barbados.

A string of recent papers on the spatial aggregation of convective cloud populations reflects the current interest in this topic. Many make use of an RCE setting. For example, Bretherton et al. (2005) and Muller and Held (2012) investigate deep convective self-aggregation with a cloud-resolving model in RCE, focusing on the mechanisms responsible for its onset and maintenance. An excellent recent review of research on convective aggregation in numerical simulations was written by Wing et al. (2017).

Section 5.4 Bulk Model for Vertically Well-Mixed Boundary Layers

In 'Bulk boundary-later concepts for simplified models of tropical dynamics', Bjorn Stevens (2006) presents a more elaborate discussion on bulk models of cloud-topped boundary layers. The text book the *Atmospheric Boundary Layer* by Jordi Vilà (2015) and colleagues includes MLM software and it offers a nice discussion on the chemical composition of the ABL. The prediction of cloud formation over land with aid of a bulk model is explained in the study 'Relative humidity as an indicator for cloud formation over heterogeneous land surfaces' by Chiel van Heerwaarden and Vilà (2008). The question as to what extent drizzle and aerosol concentrations control the stratocumulus cloud amount is investigated in Rob Wood's paper 'Cancellation of aerosol indirect effects in marine stratocumulus through cloud thinning' (2007). His paper 'Stratocumulus clouds' (2012) gives a thorough overview on the current knowledge of this cloud type. A critical component of the MLM is the entrainment closure. The article 'Cloud-top entrainment in stratocumulus clouds' by Juan Pedro Mellado (2017) reviews some recent advances on stratocumulus cloud-top entrainment and discuss some remaining challenges. A nice practical application of a MLM using large-scale forcing conditions extracted from climate models to study the stratocumulus climate feedback is presented in 'CMIP3 subtropical stratocumulus cloud feedback interpreted through a mixed-layer model' by Caldwell et al. (2013). The evolution from a well-mixed boundary layer to one that is decoupled is explained by Chris Bretherton and Matthew Wyant Bretherton (1997). By inspection of the buoyancy flux in the sub-cloud layer, they present a criterion that involves the ratio between the surface latent heat flux and the radiative cooling at the top of the cloud layer. The idea is that a strong evaporation drives a strong entrainment of warm air due to the release of latent heat in the cloud layer. Decoupling will be favoured if the radiative cooling is not strong enough to compensate for the entrainment warming effect. In 'On the growth of layers of nonprecipitating cumulus convection' Stevens (2007) concludes that the cumulus-topped boundary layer grows linearly in time. To explain these findings, he makes use of the similarity of the cumulus sub-cloud layer to the dry CBL, which puts a strong constraint on the vertical profile of the buoyancy flux.

Exercises

(1) In Section 5.1.1, the WTG in the tropics was explained using a scale analysis of the momentum equation. A WTG can be expressed in terms of a horizontal temperature perturbation δT. Note that in Chapter 2 temperature perturbations were already related to density perturbations, through Eq. (2.7). With this equation as a starting point, use Poisson's equation (or the first law of thermodynamics in adiabatic form) and two additional assumptions to show that in certain conditions horizontal perturbations in density are directly related to horizontal perturbations in potential temperature,

$$\frac{\delta \rho}{\rho} \sim -\frac{\delta \theta}{\theta}. \tag{5.131}$$

(2) In this exercise we consider the equilibrium state of the tropical atmosphere as a result of radiative transfer only. Use a planetary albedo of 0.3.

  (a) First consider the radiative balance of the Earth without atmosphere, as schematically illustrated

in Figs. 5.7 and 5.8a. Using the previously laid out equations and parameters, derive an expression for the effective emission temperature T_e by equating the incoming solar short-wave flux to the outgoing infrared long-wave flux. Calculate its value and compare it to the observed global mean surface temperature $T_s = 288$K. What does this say about this simple model? What is still missing?

  (b) Next we introduce an extra layer in the model, representing an atmosphere with temperature T_A (see Fig. 5.8b). Assume that the atmosphere is transparent for short-wave radiation and opaque for long-wave radiation, meaning that it absorbs all long-wave radiation. Also, assume that all layers emit radiation like a blackbody. By considering the equilibrium flux budget at the top of the atmosphere (TOA), show that

$$T_A = T_e. \tag{5.132}$$

Similarly, use the equilibrium flux budget for the surface layer to show that

$$T_S = 2^{1/4} T_e = 303 \text{ K}. \tag{5.133}$$

Compare this value to the one from (a), what can we say about the role of the atmosphere in the Earth's energy budget?

  (c) Finally, we introduce a third layer into the system (see Fig. 5.8c). By considering the TOA flux budget, show again that the temperature of the top layer (2) is

$$T_2 = T_e. \tag{5.134}$$

Use the flux budgets for the middle layer (1) and surface (s) to show that

$$T_1 = 2^{1/4} T_e, \tag{5.135}$$

$$T_S = 3^{1/4} T_e. \tag{5.136}$$

What is the vertical structure of temperature in this system? Is it realistic?

(3) By applying Kirchhoff's law, which states that in thermal equilibrium the long-wave absorbtivity a should equal the emissivity ϵ, eliminate a in Eqs (5.39)–(5.41). In addition, allow each atmospheric layer to have its own emissivity (ϵ_1, ϵ_2). Then solve this system analytically.

(4) Consider the surface energy balance.

  (a) Name and discuss the various terms in the surface energy balance.

(b) Make a sketch of the diurnal cycle (i.e., daily time development over 24 hours) of each of these terms on a clear day over land.

(c) Give the definition of the Bowen ratio.

(d) Describe all four components of the net radiative flux at the surface

(e) Discuss how boundary layer clouds might affect these components.

(5) Let us consider an idealised spherical planet with no atmosphere, warmed by a distant sun with a solar irradiance S = 1,360 Wm^{-2}. The surface of this planet is uniform and in RE with the incoming solar radiation. The surface of the planet is black except for some areas that are covered by perfectly white daisies, which are distributed randomly over the surface of the planet so that the planet looks uniformly grey when viewed from space. Hence, the albedo of such a planet is equal to the areal coverage of daisies.

(a) Estimate how the average surface temperature T_G of this planet depends on the areal coverage of daisies, knowing that the surface emissivity is assumed to be equal to unity. Hint: use the Stefan–Boltzmann law and consider that the planet is in a steady state and there is no atmosphere.

(b) Make a figure to illustrate the dependency found in (a). How do the results change for a stronger sun (e.g., S = 1,460 Wm^{-2})?

(c) Let us consider a second case in which the planet is not in a steady state. The average surface temperature is T_G = 30C and the area covered by daisies is 50% of the total surface of the planet. The radiative energy is all emitted as latent and sensible flux (i.e., the soil flux is zero). Knowing that the Bowen ratio is β = 0.7, what are the values of the sensible and latent heat fluxes?

(d) Right above the daisies, a horizontal wind is blowing with U = 3 m s^{-1}. Calculate the temperature and humidity at that height, which corresponds to p = 1,000 hPa. Assume that the surface is just saturated with water vapour and use a bulk coefficient for heat and moisture both equal to 2.5 × 10^{-3}. Do you expect the temperature and humidity of air to be higher or lower for a larger surface of the planet covered by daisies? Why?

(6) The flux-jump relation Eq. (5.71) is an essential closure of the MLM. Consider Fig. 5.35, showing vertical profiles of the concentration of some species C in the boundary layer and free troposphere, indicated with subscripts ml and ft, respectively, at times t (solid line) and $t + dt$ (dashed line). Assume that there is no surface flux and that changes in the boundary layer

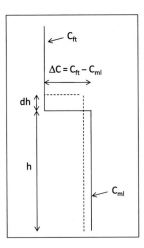

FIGURE 5.35: Concentration of a species C in the well-mixed boundary layer and lower free troposphere.

are only due to entrainment, so there is no subsidence. Show that the flux-jump relation directly follows from a vertical integration of C from $z = 0$ to $z = h + dh$ at times t and $t + dt$. Apply conservation of C and express the change in C_{ml} in terms of $dh/dt = w_e$.

(7) The original definition for the convective velocity scale was proposed for a CBL, $w_*^3 = \overline{w'b'}_{sfc}h$, which differs from Eq. (5.83). The buoyancy flux in the CBL can be expressed as a linear function of height. Show that these expressions are equivalent. Which entrainment to surface flux ratio for the buoyancy flux A is implicitly assumed?

(8) Parameterisations for stratocumulus entrainment have been inspired by studies on the CBL. Consider a CBL in the free convection limit, which means that there is no generation of TKE by wind shear. Apply a vertical integration of the steady-state TKE equation from the surface to just above the boundary layer.

(a) Show that the vertical integral of the viscous dissipation term is maximum if there is no entrainment.

(b) For which flux ratio value A does the viscous dissipation vanish?

(9) Consider a STBL with $\theta_{\ell,ml}$ = 289.0 K and $q_{t,ml}$ = 9.0 g kg^{-1}. The surface pressure is 1,017.8 hPa.

(a) Diagnose the cloud-base height (possible refer to Chapter 2 for instruction).

(b) Just above the inversion, θ = 297.5 K and q_t = 1.5 g kg^{-1}. The long-wave radiative flux divergence is 70 Wm^{-2}, the heat and moisture fluxes are 0.01 mKs^{-1} and 3·10^{-5} kg kg^{-1}ms^{-1}, respectively. Compute the entrainment rate that gives a negative buoyancy flux just below the cloud base.

TABLE 5.1: Symbols and key variables used in this chapter

Symbol	Description
δp	Synoptic pressure difference scale (horizontal)
$\delta \rho$	Synoptic density difference scale (horizontal)
$\delta \theta$	Synoptic potential temperature difference scale (horizontal)
U	Synoptic horizontal wind scale
L	Synoptic length scale (horizontal)
l_v	Tropospheric depth scale
Fr	Froude number $\mathrm{Fr} = U^2/(gD)$
Ro	Rossby number $\mathrm{Ro} = U/(fL)$
φ	Arbitrary variable
z_0	Base height of bulk layer
z_1	Top height of bulk layer
Δz	Bulk layer depth $\Delta z = z_1 - z_0$
p_s	Surface pressure
p_t	Tropopause pressure
Δp	Tropospheric pressure depth $\Delta p = p_s - p_t$
LWP	Liquid-water path
TCVW	Total column water vapour
TCC	Total cloud cover
L_x, L_y	Horizontal domain sizes
$I(x)$	Indicator function
S_φ	Source of φ
F_φ	Kinematic flux of φ
C_φ	Exchange velocity of φ at the surface
R^{net}	Dynamic surface net radiative flux
H	Dynamic surface sensible heat flux
E	Dynamic surface latent heat flux
G	Dynamic surface ground heat flux
P	Dynamic surface precipitation flux
N	Number of modes in the separation of variables
$\varphi_r(z)$	Reference vertical profile of φ (height coordinates)

Symbol	Description
$A_i^\phi(z)$	i-th basis function of φ (height coordinates)
$F_{\mathrm{LW}}^\uparrow$	Kinematic long-wave radiative flux (upward)
$F_{\mathrm{LW}}^\downarrow$	Kinematic long-wave radiative flux (downward)
$F_{\mathrm{SW}}^\uparrow$	Kinematic short-wave radiative flux (upward)
$F_{\mathrm{SW}}^\downarrow$	Kinematic short-wave radiative flux (downward)
S_0	Solar constant
r_p	Earth's radius
α_p	Planetary albedo
T_e	Effective emission temperature
T_s	Surface temperature
ϵ	Long-wave emissivity
a	Efficiency of long-wave absorption
t_{adj}	Typical timescale of convective adjustment
$V_0(p)$	Barotropic basis function of $\vec{v}$ (pressure coordinates)
$V_1(p)$	Baroclinic basis function of $\vec{v}$ (pressure coordinates)
$\Omega_1(p)$	Baroclinic basis function of ω (pressure coordinates)
$\mathcal{G}^1$	Gross moist stability (baroclinic)
$\mathcal{G}^1_{\eta_d}$	Gross static stability (baroclinic)
$\mathcal{G}^1_q$	Gross moisture stratification (baroclinic)
A	Entrainment efficiency of the CBL
A_{NT}	Entrainment efficiency of the STBL
D	Large-scale divergence of the horizontal wind
h	Boundary-layer depth
w_e	Entrainment velocity
γ_φ	Vertical gradient of φ
w_*	Convective velocity scale
Ri_*	Convective Richardson number

Summary of Symbols and Other Conventions

Table (5.1) provides a reference of additional symbols used throughout the chapter. Different types of averaging are also invoked. Angle brackets, $\langle \vartheta \rangle$ are preferred for vertically averaged, or bulk quantities (in this case the vertical average of ϑ) and the simple over-bar is used to denote Reynolds, or Ensemble, averaging. In this chapter, the vertical averaging is over the depth of the troposphere, in other chapters it proves convenient to average over the depth of the boundary layer.

6 Parameterising Clouds

A. Pier Siebesma and Axel Seifert

Cloud processes act on a wide range of spatial scales, ranging from the synoptic scale cloud systems of around 1,000 km all the way to the cloud droplet growth processes that occur on scales that can be as small as 1 μm. Global numerical weather and climate models nowadays operate at numerical resolutions in the range 10–50 km. This implies that the majority of the cloud processes are not resolved by the numerical grid of these large-scale models. Such unresolved processes, usually referred to as sub-grid processes, include turbulent transport of heat, moisture and momentum, condensation and evaporation processes in clouds, the formation of precipitation and the interaction of clouds with radiation. All these processes are crucial for the global hydrological and energy cycle and need to be included adequately in large-scale models. These sub-grid processes are usually approximated by parameterisations in terms of the numerically resolved large-scale variables. Parameterisations can be viewed as more process-specific versions of the conceptual models discussed in Chapter 5. Many of the systematic errors in climate and weather models can be attributed to uncertainties in the parameterised cloud-related processes. Likewise, as will be shown in Chapter 13, the spread in climate model sensitivity also depends critically on the choice of parameter values in cloud and convection parameterisations. Therefore, any major progress in reducing systematic model errors and uncertainty in climate model sensitivity can only be expected through improving the description of sub-grid processes in general and cloud-related processes especially.

This chapter starts with an introduction to the concept of parameterisation and its impact on a global scale followed by a discussion on the desired level of complexity. The bulk of this chapter describes current parameterisation concepts of turbulence, dry and moist convection, cloud processes and their underlying physics. The final section is dedicated to current and future challenges of parameterisations of cloud related processes.

6.1 The Concept of Parameterisation

A complete numerical description of the atmospheric dynamics would require global direct numerical simulations (DNSs) with a millimetre grid-mesh in order to resolve turbulent eddies down to the Kolmogorov scale. However, even if computational power would increase at the same pace as it did over the last decades, global simulations at this resolution will not be possible within the coming 100 years. This implies that in the foreseeable future, large-scale models will be operating with a grid-mesh that will at best only resolve part of the many cloud-related processes.

Parameterisation can be viewed as an attempt to account for those processes that are not resolved by the grid-mesh of large-scale models, by means of a simplified statistical description, usually expressed in terms of variables and processes that are numerically resolved. For a cumulus convection parameterisation, for instance, this implies a statistical representation of the vertical transport of heat, moisture and momentum including the thermodynamics associated with all the condensation processes of a cumulus cloud ensemble that populates a grid box of a large-scale model. Parameterisations of the planetary boundary layer aim to describe the turbulent transport by all the unresolved turbulent eddies in a statistical manner while the task of a cloud parameterisation is to estimate the cloud fraction and the mass and phase of cloud condensate in a grid box. So, in all these examples, parameterisations do not predict the behaviour of individual components of the phenomena but rather the overall effect of an ensemble of many of these components in a statistical sense.

The analogy with the kinetic theory of gases is useful here. The Maxwell–Boltzmann distribution of the microscopic particle speeds in an ideal gas gives a highly accurate statistical parameterisation in terms of the macroscopic average kinetic energy or, equivalently,

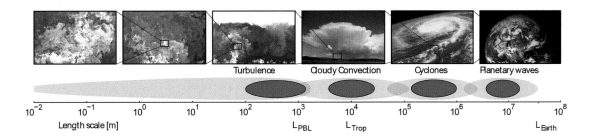

FIGURE 6.1: Illustration of the atmospheric scales of motion. Common atmospheric phenomena are categorised according to their typical length scale (blue shades roughly indicate where the phenomena's energy is most often concentrated). Important length scales are L_{PBL} and L_{trop}, the typical height of the planetary boundary layer and that of the troposphere, respectively.
Figure from Schalkwijk et al. (2015), Copyright © 2015 American Meteorological Society (AMS)

temperature. There are two important prerequisites for this famous and first-ever statistical law in physics to hold. First, the number of molecules has to be large enough so that a statistical treatment can be applied and second, the gas has to be in thermal equilibrium. Similar conditions apply for parameterisations. For instance, cumulus convection parameterisations rely on the assumption that the number of convective clouds in a grid box of a large-scale model is large enough to justify a statistical approach. Furthermore, a well-defined statistical parameterisation is only possible if the rate of change of the large-scale variables is slow enough to give the cloud ensemble sufficient time to adjust to the large-scale state. Only in the case of such a quasi-equilibrium is it reasonable to expect a one-to-one correspondence between the resolved variables and the parameterised response of the sub-grid processes. For parameterisations in current climate models with a resolution of 50 km, both these requirements are fulfilled for boundary-layer turbulence and shallow cumulus convection that operate on the kilometre scale and can adjust fast enough to be in quasi-equilibrium. For deep convective clouds that can extend over tens of kilometres, these requirements are already debatable for current climate model resolutions and will do so increasingly with further refined resolutions.

On the other hand, the resolution of a large-scale model needs to be fine enough to resolve the synoptic disturbances that drive the cloud processes that need to be parameterised. Ideally, this requires a scale separation between the unresolved small-scale three-dimensional processes and the resolved quasi-two-dimensional large-scale dynamics that drive these processes. Such a scale separation would offer an optimal model resolution fine enough to resolve the large-scale dynamics and coarse enough to statistically parameterise the remaining small-scale processes. Unfortunately, there is no such

thing as a scale separation or a spectral gap within the wide spectrum of atmospheric motions. Instead, as schematically illustrated in Fig. 6.1, there is a continuum of dynamical processes from the largest synoptic scales all the way down to the Kolmogorov scale at the millimetre scale where atmospheric turbulent kinetic energy is dissipated into heat.

Nevertheless, there are a few typical length scales that are worth mentioning and are relevant for all parameterisations. As indicated in Fig. 6.1, at horizontal resolutions finer than the depth of the troposphere ($L_{\mathrm{trop}} \simeq 10\,\mathrm{km}$) it becomes possible to resolve convective overturning by deep cumulus clouds. Not coincidentally, this is roughly the resolution where models start to replace their hydrostatic constraint by a prognostic equation of the vertical velocity. As mentioned in Chapter 2, such non-hydrostatic models operating at resolutions in the range of 1–5 km are referred to as storm-resolving models (SRMs). At these resolutions, parameterisations for the deep convective cumulus towers are usually switched off, but parameterisations for shallow cumulus convection and three-dimensional turbulence are still active. Only at resolutions of around 100 m, substantially finer than the depth of the planetary boundary layer ($L_{\mathrm{PBL}} \approx 1\,\mathrm{km}$), the largest three-dimensional turbulent eddies are resolved. At these resolutions, the unresolved smaller turbulent eddies can be parameterised in terms of the larger resolved eddies in a reliable way, using the well-known scaling properties of three-dimensional turbulence. Such large-eddy simulations (LESs) require only additional parameterisations of cloud microphysical processes.

Finally, a few words on the vertical resolution used in large-scale models. Cloud and convection parameterisations estimate the vertical transport of heat, moisture and momentum and resulting vertical profiles of cloud-related variables such as cloud fraction, condensed water and precipitation. A sufficiently high vertical resolution is an important prerequisite for an accurate representation of the vertical distribution of these variables. Near the surface in the planetary boundary layer where the turbulent processes are the most vigorous and subjected to rapid changes, a vertical resolution of 10–100 m is required. At higher altitudes, the vertical resolution can be decreased to a few hundred metres near the tropopause, and to even coarser resolutions in the stratosphere. A minimum number of around 100 vertical model levels is therefore desirable, a requirement which is only incidentally met by large-scale models.

6.2 Overview of the Parameterised Processes

The basic dynamical framework of large-scale models is based on the conservation laws of momentum, mass, energy and water as approximated by the primitive equations that were derived in Section 2.4.2. These equations are valid for small parcels of air but can not be applied directly to the grid-scale-averaged properties of air associated with grid volumes in a large-scale model. To derive appropriate averaged equations for an arbitrary variable φ, a horizontal averaging filter operator in physical space is adopted

$$\overline{\varphi}(x,y) = \frac{1}{\Delta x \Delta y} \int_{x-\frac{1}{2}\Delta x}^{x+\frac{1}{2}\Delta x} \int_{y-\frac{1}{2}\Delta y}^{y+\frac{1}{2}\Delta y} \varphi(x',y')\mathrm{d}x'\mathrm{d}y', \quad (6.1)$$

where Δx and Δy are the averaging intervals in the horizontal directions associated with the grid size of the model. Applying this operator to the conservation equations of dry static energy s_d, water vapour q_v and condensed water q_c and decomposing each variable into a mean and fluctuating part, as outlined in Section 2.5 gives,[1]

$$\overline{D_t}\overline{s}_\mathrm{d} = -\partial_p\overline{\omega's'_\mathrm{d}} + \ell_\mathrm{v}(c-e) + Q_\mathrm{rad} \quad (6.2\mathrm{a})$$

$$\overline{D_t}\overline{q}_\mathrm{v} = -\partial_p\overline{\omega'q'_\mathrm{v}} - (c-e) \quad (6.2\mathrm{b})$$

$$\overline{D_t}\overline{q}_\mathrm{c} = -\partial_p\overline{\omega'q'_\mathrm{c}} + (c-e) - G, \quad (6.2\mathrm{c})$$

where the terms on the right-hand side represent the unresolved sub-grid processes that require a parameterised approach: the vertical divergence of eddy fluxes, the condensation and evaporation rates $(c-e)$, the autoconversion rate from condensed water to rain G and the temperature tendency due to radiation Q_rad. The horizontal eddy flux terms are ignored in Eq. (6.2) because the horizontal

resolution is much coarser than the vertical resolution in large-scale models.[2] As a result, the sub-grid terms on the right-hand side are restricted to a grid column of the model while the horizontal interaction is achieved only through the resolved advective terms. While this is a reasonable approximation for large-scale models down to horizontal resolutions of 10 km, this assumption gradually breaks down at finer resolutions. At the turbulent eddy resolving resolutions used in LES both the horizontal and vertical sub-grid eddy flux terms become equally important.

The set of equations in Eq. (6.2) represents a still common situation in large-scale models without separate prognostic variables for the liquid and the ice condensed phase. In that case, ℓ_v represents the effective vaporisation enthalpy for an ice–water mixture expressed as an empirical function of temperature. Models with more complex cloud microphysics parameterisations do have separate prognostic equations for the liquid and the ice mass while large-scale models with simpler microphysics do not even have a prognostic equation for the condensed phase at all. In the latter case, a diagnostic balance between the three parameterised processes on the right-hand side of Eq. (6.2c) is assumed.

Spatial horizontal resolutions of $O(100\,\mathrm{km})$ used in large-scale models allow for numerical time steps that can be as large as $O(10^4\,\mathrm{s})$. Since this timescale is larger than the time it takes for rain, snow and hail to reach the surface, hydrometeors are often treated diagnostically in climate models. Therefore, in the absence of evaporation and melting processes, the surface precipitation flux P_s can be directly related to the autoconversion rate G as an integral over the depth of atmosphere of the latter

$$P_s = \int_0^{p_s} G\frac{dp}{g} \equiv \langle G \rangle. \quad (6.3)$$

In many of the upcoming discussions, it will be more convenient to work in height coordinates rather than in pressure coordinates. The turbulent flux divergence terms in Eq. (6.2) can be easily transformed back into height coordinates by using $\omega \simeq -\rho g w$ and assuming hydrostatic balance, so that

$$\partial_p\overline{\omega'\varphi'} \simeq \rho^{-1}\partial_z\rho\overline{w'\varphi'} \simeq \partial_z\overline{w'\varphi'}, \quad (6.4)$$

where the Boussinesq approximation has been applied in the last step, which is only valid for layers much shallower than the scale height of the atmosphere of around 8 km.

---

[1] In Chapter 2, η is used to as a symbol for static energy, rather than s which is used in Chapter 2 to denote entropy. Throughout this chapter we will use again the more traditional notation s for the static energies.

[2] In reality, horizontal diffusion is applied between the adjacent grid boxes in large-scale models, but more for numerical than for physical reasons as to prevent accumulation of energy at the grid scale which might occur through energy cascading downward from the large-scales.

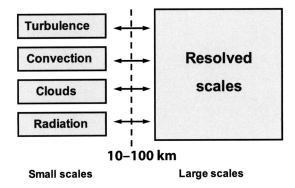

FIGURE 6.2: Depiction of the interaction between resolved and parameterised unresolved cloud-related processes (convection, turbulence, clouds and radiation) in traditional climate models.

The usefulness of moist conserved variables introduced in Chapter 2 is easily illustrated by reducing the set of equations in Eq. (6.2) to two budget equations for the moist static energy[3] $s_e = s_d + \ell_v q_v$ and the total water specific humidity $q_t \equiv q_v + q_c$

$$\overline{D_t \bar{s}_e} = -\partial_p \overline{\omega' s'_e} + Q_{rad} \tag{6.5a}$$

$$\overline{D_t \bar{q}_t} = -\partial_p \overline{\omega' q'_t} - G. \tag{6.5b}$$

The only remaining sub-grid terms in Eq. (6.5) are the turbulent flux divergence terms, a radiative heating term Q_{rad} and the autoconversion term G.

In a Eulerian form, the budget equations for $\varphi \in \{s_d, s_e, q_t, q_v, q_c\}$ can be written in a more schematic form as

$$\partial_t \overline{\varphi} = (\partial_t \overline{\varphi})_{res} + (\partial_t \overline{\varphi})_{sub\text{-}grid}, \tag{6.6}$$

where the resolved part is represented by the large-scale advection terms

$$(\partial_t \overline{\varphi})_{res} = -\bar{u}\partial_x \overline{\varphi} - \bar{v}\partial_y \overline{\varphi} - \overline{\omega}\partial_p \overline{\varphi} \tag{6.7}$$

and the unresolved sub-grid part is given by the right-hand side of Eqs (6.2) and (6.5).

Traditionally, the sub-grid processes in large-scale models processes are treated by four distinct different parameterisation schemes that predominantly interact with the mean resolved variables as indicated schematically in Fig. 6.2. These parameterisations are (i) a boundary-layer turbulence parameterisation, (ii) a cumulus convection parameterisation, (iii) a cloud

[3] Alternatively, the liquid-water static energy $s_\ell \equiv s_d - \ell_v q_c$ could also have been used here. Note that static energies are related to their corresponding potential temperatures approximately via $\theta_x \approx s_x/c_p$ with $x \in \{d, e, \ell\}$. See Chapter 2 for further details.

parameterisation and (iv) a radiative transport parameterisation. Heating tendencies of each of these parameterisations of a typical climate model EC-Earth (Hazeleger et al., 2010) are displayed in Fig. 6.3 as a function of latitude and height, averaged over the period 1988–2008 and using prescribed sea-surface temperatures (SSTs). Although the precise contributions and partitioning of these parameterised tendencies vary from model to model, this example does illustrate the typical contributions and spatial distributions of these four parameterisation types, as outlined later.

Boundary-layer turbulence schemes describe exchange of heat, moisture and momentum between the surface and the atmospheric surface layer and the further turbulent transport of these quantities deeper into the planetary boundary layer. Early boundary-layer schemes only included turbulent transport in the clear boundary layer but more recent boundary-layer schemes also incorporate well-mixed stratocumulus-topped boundary layers. Fig. 6.3d shows a warming in the lowest kilometres of the atmosphere, especially in the low latitudes. This is the result of the turbulent mixing of the surface sensible heat flux into the planetary boundary layer. Similarly, through vertical transport of moisture from surface evaporation, the net annual mean effect of the boundary-layer turbulence is a moistening of the boundary layer, most strongly over the oceans at low latitudes.

Cumulus convection schemes parameterise the vertical transport of heat, moisture and momentum by cumulus clouds out of the boundary layer deeper into the atmosphere. Through the latent heat release of condensation processes, the air in cumulus towers becomes buoyant with respect to the environment, causing strong vertical transport of these quantities that can extend all the way to the tropopause. Cumulus convection is the main mechanism for vertical transport in the atmosphere out of the planetary boundary layer. It is also a main source of precipitation. The annual averaged heating profile of the convection scheme displayed in Fig. 6.3a shows that the strongest heating occurs in the tropics throughout the whole troposphere with maximum values of up to $7\,\mathrm{K\,day^{-1}}$. In the subtropics, both the vertical extent and the intensity of the convective heating is limited by the subsiding branch of the Hadley circulation. The secondary peak of the heating profile over the mid latitudes is due to cumulus convection in the storm tracks. The cooling tendency of the convection scheme in the boundary layer is predominantly the result of the ventilation of heat out of the boundary layer into the free atmosphere.

An unrealistic property of most convection schemes is that they do not prognose or diagnose the associated

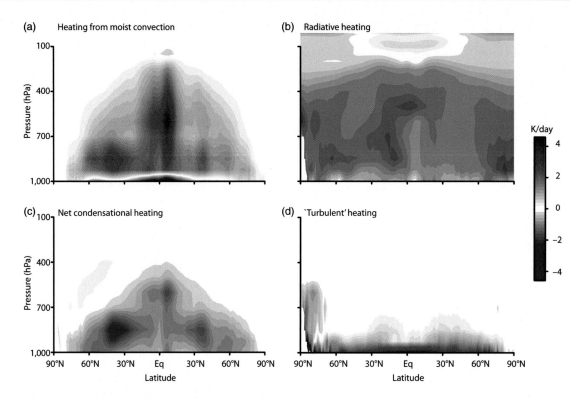

FIGURE 6.3: Multi-annual mean of the heating tendencies due to the various physical parameterisations in a typical climate model (EC-Earth [Hazeleger et al., 2010]) as a function of latitude and height. The heating tendencies due to: (a) cumulus convection, (b) radiative processes, (c) net condensation remaining cloud processes and (d) boundary-layer turbulence processes.
Figure courtesy: Carlo Lacagnina

cloud fraction and cloud condensate. Instead, this is usually determined by a *cloud scheme* that has the task of estimating all the relevant cloud properties such as cloud fraction, cloud water/ice and cloud droplet number density for all cloud types, convective and non-convective. Cloud schemes also diagnose the precipitation amounts for the non-convective clouds, sometimes referred to as large-scale precipitation. The net heating effect of the cloud scheme in Fig. 6.3c is relatively small and overall negative. This is because most of the condensational heating is already taken into account by the convection scheme. Nevertheless, the role of the cloud scheme is crucial as it provides the cloud properties that are needed in the *radiation parameterisation* to determine the cloud radiative effects. The radiative heating as calculated by the radiation scheme is shown in Fig. 6.3b and displays a rather uniform cooling of 1–2 K d^{-1} throughout the whole troposphere. For more details on the radiative transfer calculations, that form the heart of any radiation scheme, the reader is referred to Chapter 4.

The division of the cloud-related processes over the boundary-layer, convection and cloud parameterisation schemes might appear to be odd for the neophyte entering

the field: the turbulent flux divergence terms are shared between the boundary-layer scheme and the convection scheme while condensation and precipitation processes occur in all three parameterisations. The reasons for this particular division are merely historical. The first-generation large-scale models in the 1960s, equipped with only a few vertical model levels, used transfer relations between the surface and the lowest model level, moist adiabatic adjustment and a prescribed cloud climatology. As increasing computer resources allowed for more vertical model levels and a higher horizontal resolution, the main efforts over the last three decades of the last century have been invested in maturing these simple parameterisation into more complex and, without any doubt, more realistic statistical representations of the cloud-related processes. The transfer relations have evolved into boundary-layer schemes, the moist adjustment into cumulus convection schemes and the prescribed cloud climatologies into more flexible cloud schemes. Only during the last ten years more consideration has been given to the interactions between these parameterisations and to more unified formulations.

6.3 The Design of Parameterisation

A well-designed parameterisation should provide an accurate description of the influence of the sub-grid processes in terms of the resolved state in a computationally efficient manner. In practice, this is done by designing physically plausible equations that describe these sub-grid processes in terms of the resolved processes. These equations usually contain parameters that are uncertain and need to be constrained by observations and/or by theoretical considerations. As the number of these parameters N_p increases, it also becomes increasingly more difficult to constrain the parameterisation. Therefore, the number of uncertain parameters N_p should be kept as low as possible in a well-designed parameterisation.

As an example, consider the parameterisation of vertical transport of heat and moisture in the cloud-free boundary layer. Recall from Eq. (6.2) that, in the absence of radiative and condensational processes, the truncated equations for $\varphi \in \{\theta, q_v\}$ read as

$$\overline{D_t \varphi} = -\partial_z \overline{w' \varphi'}. \tag{6.8}$$

The most basic parameterisation is to assume that the turbulence flux is proportional and down the vertical gradient of $\overline{\varphi}$

$$\overline{w' \varphi'} = -K \partial_z \overline{\varphi} \tag{6.9}$$

by introducing an eddy-diffusivity (ED) K. The justification of this name can be understood by substituting the parameterisation Eq. (6.9) in Eq. (6.8), demonstrating that an ED approach approximates the turbulent transport term as a diffusion process where the intensity of mixing is determined by K. This ED parameterisation is completed by finding an appropriate expression for K.

The simplest choice would be to approximate K by a constant K_0 that could be determined by simultaneously measuring turbulent fluxes and the corresponding vertical gradient of the $\overline{\varphi}$ under a wide range of different atmospheric conditions. This is an example of a simple, computationally efficient parameterisation that is easy to constrain by observations since $N_p = 1$. Unfortunately it is not very accurate: K is close to zero under stable atmospheric conditions in the absence of turbulence but can be large under unstable atmospheric conditions in a convective boundary layer. Therefore, the use of a *constant* ED K_0 is too restrictive and hence inaccurate. Usually, turbulent mixing is therefore parameterised using an ED formulation in which K depends on atmospheric stability and shear. This introduces extra parameters so that N_p will increase, but are still reasonably easy to constrain with observations.

Despite the fact that ED parameterisations are among the most popular parameterisations for boundary-layer turbulence in large-scale models, they are also rather phenomenological. The underlying implicit assumption that turbulent transport can always be approximated as a local diffusion process is questionable, as will be demonstrated in the next section.

A more systematic parameterisation approach is to derive prognostic budget equations for the turbulent flux and derive parameterisations from these second-order moments. Prognostic equations for higher-order moments and their correlations can be obtained using the same procedure outlined in Section 2.5.2 for deriving the turbulence kinetic energy (TKE) equation. For the turbulent fluxes of $\varphi \in \{\theta, q_v\}$ this gives

$$\overline{D_t w' \varphi'} = \underbrace{-\overline{w'^2} \partial_z \overline{\varphi}}_{G} + \underbrace{\overline{b' \varphi'}}_{B} - \underbrace{\frac{1}{\rho_0} \overline{\phi' \partial_z p'}}_{P} - \underbrace{\partial_z \overline{w'^2 \varphi'}}_{T},$$

$$\tag{6.10}$$

with $b' = \beta \theta'_\rho$ where $\beta = g/T_{\rho,0} \approx 0.033$ denotes the buoyancy parameter. The terms on the right-hand side represent the mean gradient production (G), the buoyancy production (B), the pressure covariance (P) and the turbulent transport (T) of the turbulent flux. Contributions due to vertical wind shear and dissipation have been neglected.

Note that in deriving budget equations of second-order statistics, third-order terms like the transport term are introduced. In general, prognostic equations for the nth moment introduce new unknown terms of the $(n + 1)$th order. This is the famous closure problem mentioned in Chapter 2 and already visible in the prognostic Eq. (6.8) of $\overline{\varphi}$ that contains an unknown second-order term. A parameterisation where nth order terms are used to approximate higher order terms is referred to as an nth order closure. In this parlance, the ED parameterisation (Eq. (6.9)) constitutes a first-order closure, while the third-order transport term in Eq. (6.10) requires a second-order closure which is usually done through a down-gradient formulation

$$\overline{w'^2 \varphi'} \approx -K_{w\phi} \partial_z \overline{w' \varphi'}. \tag{6.11}$$

In addition, Eq. (6.10) also introduces new cross-correlation terms such as the pressure-covariance term that requires additional parameterisations.

Higher-order closures use more prognostic equations. In case of a first-order closure, there are three first-order equations that prognose the mean state of $\overline{\theta}$, $\overline{q}_v$ and $\overline{w}$. In the case of a second-order closure, there are six additional prognostic equations for the second-order moments, that is, $\overline{w'^2}$, $\overline{\theta'^2}$, $\overline{q_v'^2}$, $\overline{w'\theta'}$, $\overline{w'q_v'}$ and $\overline{q_v'\theta'}$. The number of degrees of freedom N_f, here simply defined as the number of prognostic equations, increases with the order of the closure. A larger number of degrees of freedom N_f allows for a more complete description of the sub-grid processes but also increases the number of uncertain parameters N_p requiring additional observational constraints.

Increasing N_f does not always imply an increase of N_p. As a thought experiment, one could consider using a DNS in each grid box of a large-scale model and deduce the turbulence fluxes from these simulations. In this case, $N_p = 0$, as there are no uncertain parameters left that require observational constraints, which makes it more a benchmark rather than a parameterisation. This brute force attempt requires a DNS that prognoses complete three-dimensional fields numerically from which the turbulent flux for the large-scale model can be diagnosed. The number of degrees of freedom N_f, proportional to the number of grid boxes of such DNS scales as

$$N_f \approx \left(\frac{L}{l_K}\right)^3 \sim \mathcal{R}_e^{9/4}, \qquad (6.12)$$

where L denotes the grid size of the large-scale model, l_K the Kolmogorov scale, and where Eq. (2.185) has been used in the last step. For $L = 10\,\text{km}$ and $l_K \approx 1\,\text{mm}$, this implies $N_f \sim 10^{24}$ degrees of freedom, clearly an unfeasable numerical task in any foreseeable future. The aim of parameterising turbulent transport is of course to find a computational more efficient description with lesser degrees of freedom.

One way to decrease N_f, without deteriorating the quality of the DNS benchmark too much, is to degrade the resolution to $10\,\text{m}$, thereby changing the DNS into an LES framework, still accurate enough to resolve the largest turbulent eddies while the smaller sub-grid eddies can be parameterised realistically with an ED parameterisation. Further reduction of N_f can be obtained by using two-dimensional LES and/or reducing the resolution further to the kilometre scale. This approach, known as superparameterisation is applied in large-scale models to parameterise deep cumulus convection by using a two-dimensional SRM.[4]

A distinguishing property of these type of parameterisations is that they provide information on the spatio-temporal coherence of the sub-grid turbulent processes, in contrast with more traditional statistical parameterisations in which all the spatial sub-grid information is lost.

[4] The use of an SRM as a superparameterisation was pioneered by Grabowski and Smolarkiewicz (1999). An overview of the method and its applications can be found in Randall et al. (2016).

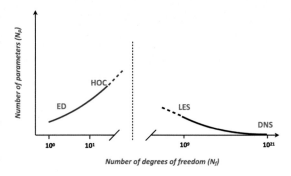

FIGURE 6.4: Options for parameterisation design of the planetary boundary layer in terms of the number of degrees of freedom N_f and the number of tunable parameters N_p that need to be constrained by observations. ED are among the simplest parameterisations. Higher-order closures (HOCs) have a larger number of degrees of freedom but also require more parameters that need to be constrained. On the other side of the spectrum, DNS can be considered as a hypothetical benchmark for parameterisations while coarse-grained versions such as LES and SRM can be used as superparameterisations that take into account the spatio-temporal coherence.

The discussed options for parameterisation design are schematically summarised in Fig. 6.4 in terms of the number of degrees of freedom N_f and the number of parameters N_p required by the parameterisation. The former is a measure of the computational cost of the parameterisation while the latter can be considered as a measure of how complicated the parameterisation is. Simple parameterisations using low order closures are well constrained (N_p is small), computationally inexpensive (N_f is small), but the often oversimplified relationships between the resolved state and the sub-grid process may lead to poor parameterisations. Parameterisations with a larger number of degrees of freedom N_f, deliver in principal more complete and adequate parameterisations but are also more difficult to constrain. A superparameterisation approach provides the most complete representation of the sub-grid processes and is relatively well constrained, but is computationally expensive. One recurrent theme in parameterisation development is finding the appropriate level of complexity in order to design a parameterisation that is accurate, well constrained and still computationally feasible.

6.4 Parameterisation of Boundary-Layer Turbulence

The planetary (or atmospheric) boundary layer can be loosely defined as the part of the atmosphere that directly

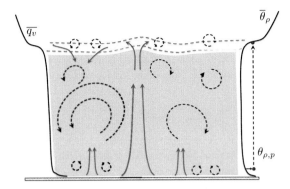

FIGURE 6.5: Schematic picture of the clear convective boundary layer topped by an inversion.

responds to surface-induced forcings on timescales of an hour or less. These forcings are predominantly buoyancy and shear production of turbulence. Buoyancy production originates from the heating of the surface by the sun resulting in buoyant, rising thermals that transport heat and moisture from the surface into the atmosphere. Shear production, whose strength is proportional to the vertical gradient of the horizontal wind in the boundary layer is a direct result of the frictional drag of the Earth's surface. Both production terms mix air in the boundary layer, thereby producing turbulence and forming an effective transport mechanism for heat, moisture and momentum.

During sunny, fair-weather conditions, buoyancy will be the dominant turbulence production mechanism. In this case, the boundary layer is unstable and is said to be in a state of free convection. Due to the strong turbulent mixing, such a convective boundary layer is characterised by well-mixed profiles of potential temperature and specific humidity, schematically sketched in Fig. 6.5. During night-time, when the surface becomes colder than the overlying atmosphere, the boundary layer becomes stable and turbulent production is strongly suppressed. In the presence of strong winds under overcast conditions or during night, shear is the dominant turbulence production mechanism. In those cases, the boundary layer is neutral and in a state of forced convection.

The state of the planetary boundary layer is strongly modulated by the diurnal cycle due to the insolation of the sun, especially over land. During daytime, the convective boundary layer fights its way into the stable free troposphere and grows in depth towards heights of one to several kilometres, depending on the strength of the solar insolation and the atmospheric stability. Due to the growth of the convective boundary layer into the stable troposphere, a thin strongly stable layer or inversion layer is formed at the interface between the turbulent

boundary layer and the quiescent free troposphere. Across this interface, through the punching of the turbulent air from the boundary layer into the free troposphere, the warm and dry tropospheric air is entrained into the boundary layer. This top-entrainment process therefore also contributes to the growth of the convective boundary layer. Parameterised descriptions of top entrainment have been described in detail in Chapter 5.

If the boundary layer grows deep enough, the moist rising thermals can become supersaturated at the top of the boundary layer and clouds will form, usually shallow cumulus clouds, as seen in Fig. 6.9. As a result, a conditional unstable cloud layer is created by these cumulus clouds overlying an unsaturated well-mixed subcloud layer and topped by an inversion. Over sea, as illustrated in Fig. 5.21, the subcloud layer is around 500–700 m deep and often topped by solid stratocumulus deck, a relatively thin cloud layer, several hundred metres thick that forms below a strong inversion. In contrast with a cumulus-topped boundary layer, this stratocumulus cloud layer is still well mixed in terms of moist conserved thermodynamic variables (see Fig. 5.25). If the stratocumulus-topped boundary layer grows too deep, it becomes unstable and breaks up into a cumulus-topped boundary layer, the most abundant cloud type over the subtropical oceans.

The cloud-controlling conditions under which one may expect a clear, cumulus-topped or stratocumulus-topped boundary layer and transitions between these different boundary layer regimes have been extensively discussed in Chapter 5 using conceptual simplified bulk models. This section will demonstrate how these cloud-controlling conditions can be used in designing general-purpose parameterisations for the use in well-mixed clear and stratocumulus-topped boundary layers. Most large-scale models parameterise boundary-layer turbulence using an ED approach, given by Eq. (6.9). Numerous different ED formulations have been proposed. This is in part due to the fact that diffusion will always remain an oversimplification for real turbulent flows. Therefore, no single ED formulation can adequately parameterise the transport associated with the wide range of manifestations of atmospheric turbulence.

In general, the ED can be written, mainly on dimensional grounds, as a product of a turbulent velocity scale w_t and a turbulent length scale l_t,

$$K = c_k \, w_t \, l_t, \tag{6.13}$$

where c_k is a dimensionless constant of the order of unity. The parameterisation challenge is to find useful estimates for these scales. This is not trivial since these scales are not well defined and only their product can be constrained by flux-gradient observations. A qualitative definition for l_t that can be used as guidance is the dominant turbulent eddy size at a given height z, while w_t can be viewed as the orbital velocity of this eddy. The main motivation for these definitions are that the coherent eddy sizes determine the distance over which a parcel can transport heat, moisture and momentum before losing its properties by mixing. For the convective boundary layer, as seen in Fig. 6.5, this implies small length scales near the surface and the inversion where the eddy size is limited to its distance to the surface and the inversion height while large length scales are present in the middle of the convective boundary layer where the transport is dominated by the large turbulent eddy structures.

Parameterisation of the cumulus-topped boundary layer is usually part of the cumulus convection scheme that also represents the deeper penetrative cumulus convection. Therefore, the parameterisation of the cumulus-topped boundary layer will be postponed until Section 6.5.

6.4.1 The Surface Layer

The surface layer is the region of the boundary layer close to the surface, loosely defined as the lowest 10% of the boundary layer. It is the place where the surface fluxes are determined and where the ED approach has been most successful through the use of Monin–Obukhov (MO) similarity theory. It is based on the assumptions that turbulent fluxes are approximately constant in the surface layer and that the size of the turbulent eddies are limited by the height z over the surface. The results of MO similarity theory are used to derive parameterised expressions for the surface fluxes in terms of mean profiles.

To apply similarity theory, characteristic values of velocity, temperature and humidity in the surface layer are needed. A relevant velocity scale in the surface layer can be defined as the square root of the magnitude of the near-surface vertical momentum flux[5]

$$u_* \equiv \left[\overline{u'w'}^2_{\text{srf}} + \overline{v'w'}^2_{\text{srf}} \right]^{1/4}, \qquad (6.14)$$

which is usually referred to as the friction velocity. For notational simplicity, it will be assumed that the mean horizontal wind in the surface layer is aligned positively along the x-axis, that is, $\bar{v}_h = (\bar{u}, 0)$, in which case the friction velocity simplifies to

$$u_* = \left[-\overline{u'w'}_{\text{srf}} \right]^{1/2}, \qquad (6.15)$$

[5] Strictly speaking, momentum fluxes include by definition the density, that is, $\rho \overline{u'w'}$. We will however also refer to momentum fluxes in the absence of the density term.

where the minus sign is included, because the surface momentum flux is always negative for a positive horizontal velocity.

Characteristic temperature and humidity scales can be defined similarly in terms of the surface sensible and latent heat flux

$$\theta_* \equiv -\overline{w'\theta'}_{\text{srf}}/u_*, \qquad q_{v*} \equiv -\overline{w'q'_v}_{\text{srf}}/u_*. \qquad (6.16)$$

The simplest form of turbulent transfer in the surface layer can be obtained under neutral conditions. In that case, the buoyant forces are negligible and the only driving force of turbulence is the shear production described by u_* and z. Applying the ED approach (Eq. (6.9)) under the assumption of a constant momentum flux in the surface layer gives

$$K \partial_z \bar{u} = u_*^2 \quad \text{with} \quad K = \kappa u_* z, \qquad (6.17)$$

where the ED is described by Eq. (6.13), with the turbulent eddy size equal to the distance to the surface, that is, $l_t = z$, and the turbulent velocity scale given by the friction velocity. Eq. (6.17) can easily be rewritten in the more familiar flux-gradient form

$$\partial_z \bar{u} = \frac{1}{\kappa} \frac{u_*}{z}, \qquad (6.18)$$

where κ is the von Karman constant for which a measured value of approximately 0.4 has been found. Integration of Eq. (6.18) gives the famous logarithmic wind profile

$$\bar{u} = \frac{u_*}{\kappa} \ln \left(\frac{z}{z_0} \right). \qquad (6.19)$$

The integration constant z_0 is by definition the height at which the velocity vanishes and is usually referred to as the roughness length. This length scale is a measure of the roughness of the surface and ranges from values of 2×10^{-4} m over calm seas to values of 2×10^{-2} m for short grass, 10^{-1} m for crop fields and can be larger than 1 m for towns and forests.

Using the same procedure, similar flux-gradient expressions can be obtained for scalar fields $\varphi \in \{\theta, q_v\}$

$$\bar{\varphi} - \bar{\varphi}_{\text{srf}} = \frac{\varphi_*}{\kappa} \ln \left(\frac{z}{z_\varphi} \right), \qquad (6.20)$$

indicating that potential temperature and humidity have similar logarithmic profiles under neutral conditions but can have a *different* roughness length z_φ. Typical values of the roughness length for heat and moisture over land are only 10–30% of the values for the roughness length z_0 of momentum.

Eqs 6.19 and 6.20 can be rearranged in a more useful form for surface flux parameterisations of momentum heat and moisture

$$\overline{w'\varphi'}_{\mathrm{srf}} = -C_\varphi \overline{u} \left(\overline{\varphi}(z) - \overline{\varphi}_{\mathrm{srf}} \right) \tag{6.21a}$$

$$\overline{u'w'}_{\mathrm{srf}} = -C_\mathrm{m} \overline{u}^2, \tag{6.21b}$$

where C_m, the drag coefficient and C_φ, the bulk transfer coefficient of $\varphi \in \{\theta, q_\mathrm{v}\}$ can be expressed under neutral conditions as

$$C_{\mathrm{m,n}}(z) = \frac{\kappa^2}{\left[\ln (z/z_0) \right]^2} \tag{6.22a}$$

$$C_{\varphi,\mathrm{n}}(z) = \frac{\kappa^2}{\left[\ln (z/z_0) \ln \left(z/z_\varphi \right) \right]}. \tag{6.22b}$$

Eq. (6.21) provides a parameterisation for the surface fluxes, but has been derived assuming neutral conditions ($d\overline{\theta}_\rho/dz \approx 0$) so that no dependency on the stability is taken into account in Eq. (6.22).

In the case of (un)stable conditions, the surface buoyancy flux

$$\overline{w'b'}_{\mathrm{srf}} \equiv \beta \overline{w'\theta'}_{\rho,\mathrm{srf}} \equiv -\beta u_* \theta_{\rho*} \tag{6.23}$$

will influence the flux-gradient relation (Eq. (6.18)). MO similarity conjectures that the dimensionless gradients in the surface layer can be described by universal gradient functions Φ that are a function of a dimensionless scaling parameter ξ that only depends on the friction velocity u_*, the height z and the buoyancy flux through $\theta_{\rho*}$. Such a scaling parameter ξ can be constructed on dimensional grounds as

$$\xi \equiv \frac{\kappa z \beta \theta_{\rho*}}{u_*^2} \equiv \frac{z}{L}, \tag{6.24}$$

which defines the Obukhov length scale L. The formulation of the dimensionless gradients (Eq. (6.18)) can be generalised to

$$\frac{\kappa z}{u_*} \partial_z \overline{u} = \Phi_\mathrm{m} (\xi) \qquad \frac{\kappa z}{\varphi_*} \partial_z \overline{\varphi} = \Phi_\mathrm{h} (\xi), \tag{6.25}$$

where different universal gradient functions for heat and momentum have been introduced. The Obukhov length can be interpreted as the height above the surface where the buoyancy production starts to dominate over the shear production. Note that L is positive for stable conditions, negative for convective conditions and infinite for neutral cases. Therefore, ξ can be viewed as a measure of the stability at height z in the surface layer. For neutral conditions, ($\xi \simeq 0$) stability corrections can be neglected, implying $\Phi(0) = 1$, as to make Eq. (6.25) consistent with Eq. (6.18).

The gradient functions have been well-determined empirically during numerous field experiments over the last sixty years. For unstable conditions ($\xi < 0$), a formulation that fits observations and also obeys the

expected asymptotic behaviour in the free convective limit ($\xi \to -\infty$) is

$$\Phi_\mathrm{h} = \mathrm{Pr_n} \left(1 + \gamma_\mathrm{h} |\xi|^{\frac{2}{3}} \right)^{-\frac{1}{2}} \quad \Phi_\mathrm{m} = \left(1 + \gamma_\mathrm{m} |\xi|^{\frac{2}{3}} \right)^{-\frac{1}{2}}, \tag{6.26}$$

where $\mathrm{Pr_n} \simeq 0.95$ is the Prandtl number for neutral conditions and $\gamma_\mathrm{h} \simeq 7.9$ and $\gamma_\mathrm{m} \simeq 3.6$ (Wilson, 2001).

Rewriting Eq. (6.25) in the ED form of Eq. (6.17) leads to

$$K_\mathrm{h} = \kappa z u_* \Phi_\mathrm{h}^{-1} \qquad K_\mathrm{m} = \kappa z u_* \Phi_\mathrm{m}^{-1} \tag{6.27}$$

and illustrates the effect of the stability corrections on the eddy diffusivities of heat and momentum: unstable conditions ($\Phi < 1$) lead to larger EDs. It also illustrates that under unstable conditions the ED of heat is larger than for momentum as can be expressed by the turbulent Prandtl number $\mathrm{Pr_t} \equiv K_\mathrm{m}/K_\mathrm{h}$.

The effect of the stability corrections on the profiles can be seen by integrating Eq. (6.25)

$$\overline{u} = \frac{u_*}{\kappa} \left(\ln(z/z_0) - \Psi_\mathrm{m}(\xi) \right) \tag{6.28a}$$

$$\overline{\varphi} - \overline{\varphi}_{\mathrm{srf}} = \frac{\varphi_*}{\kappa} \left(\ln(z/z_0) - \Psi_\mathrm{h}(\xi) \right), \tag{6.28b}$$

where stability profile functions Ψ are introduced that are related to the gradient functions through $\Phi \equiv 1 - \xi(\partial\Psi/\partial\xi)$ so that $\Psi > 0$ for unstable conditions and $\Psi < 0$ for stable conditions.

The drag and transfer coefficients can now be generalised to non-neutral conditions as

$$C_\mathrm{m}(z) = \frac{\kappa^2}{\left[\ln (z/z_0) - \Psi_\mathrm{m}(\xi) \right]^2} \tag{6.29a}$$

$$C_\varphi(z) = \frac{\kappa^2}{\left[\ln (z/z_0) - \Psi_\mathrm{m}(\xi) \right] \left[\ln \left(z/z_\varphi \right) - \Psi_\mathrm{h}(\xi) \right]}. \tag{6.29b}$$

For strong convective conditions, these coefficients can become up to two- to three-times larger compared to neutral conditions. The reduction of the drag and transfer coefficients under stable conditions can be even much stronger but will not further discussed here.[6] In general, the surface flux parameterisation, Eq. (6.21), with the coefficients given by Eq. (6.29) works well in an average sense. Individual measurements of these coefficients however do give rise to deviations from the mean up to 50 %.

---

[6] Hogstrom (1996) provides a review of the basic characteristics of the surface layer for both stable and unstable conditions.

6.4.2 The Clear Boundary Layer

The assumptions that lead to the ED formulation (Eq. (6.27)) do not hold anymore outside the surface layer. Therefore, in the outer layer, that is, the part of the boundary layer outside the surface layer, other methodologies are required to parameterise the ED. In all methods discussed in this section, the TKE plays a central role in estimating the turbulent velocity scale.

6.4.2.1 TKE Closure

Many large-scale models carry a prognostic equation for the TKE $\frac{1}{2}\bar{e} \equiv \overline{(u'^2 + v'^2 + w'^2)}$ and use its square root as an estimate of the turbulent velocity scale w_t in Eq. (6.13) so that

$$K_h = c_k \sqrt{\bar{e}}\, l_t, \qquad K_m = \text{Pr}_t\, K_h. \qquad (6.30)$$

For this purpose, a parameterised version of the TKE Eq. (2.155) is needed

$$\overline{D_t \bar{e}} = \underbrace{K_m \left[(\partial_z \bar{u})^2 + (\partial_z \bar{v})^2 \right]}_{S}$$
$$+ \underbrace{\overline{w'b'}}_{B} + \underbrace{\partial_z (K_m \partial_z \bar{e})}_{T} - \underbrace{\epsilon_e}_{D}, \qquad (6.31)$$

where the transport term T from Eq. (2.161) has been parameterised as down-gradient diffusion of $\bar{e}$ and the shear production terms in Eq. (2.162) as down-gradient diffusion of momentum. For the clear boundary layer, the buoyancy flux can be parameterised as down-gradient diffusion of the density potential temperature

$$\overline{w'b'} = -\beta K_h\, \partial_z \bar{\theta}_\rho, \qquad (6.32)$$

where $\beta = g/T_{\rho,0}$ is the buoyancy parameter already introduced in the context of Eq. (6.10). The dissipation term (D) is usually parameterised as Newtonian relaxation of the TKE

$$\varepsilon_e = c_\varepsilon \frac{\bar{e}}{\tau}, \qquad (6.33)$$

with c_ε representing a dimensionless coefficient of order unity. The relaxation time τ can be related to the turbulence length scale as $\tau \simeq l_t / \bar{e}^{\frac{1}{2}}$, by using the Kolmogorov assumption that the transfer of TKE in the inertial subrange, is equal, up to a constant, to the energy dissipation as expressed in Eq. (2.182).

The TKE approach (Eq. (6.30)) needs to be completed with an estimate for the turbulence length scale l_t. A frequently used length scale is

$$\frac{1}{l_t} = \frac{1}{\kappa z} + \frac{1}{\lambda_0} + \frac{1}{c_e\, e^{1/2}/N}, \qquad (6.34)$$

with $c_e \simeq 0.1$, where N, the Brunt–Väisälä frequency, is used as a measure of stability, and where $\lambda_0 \simeq 300\,\text{m}$ sets the scale for the largest turbulent eddy in the boundary layer. The motivation of this length-scale formulation is that it interpolates in convective conditions ($N \simeq 0$) between the length scale in the surface layer (Eq. (6.27)) and a prescribed maximum length scale λ_0 in the boundary layer, while under stable conditions the length scale outside the surface layer is dominated by the third term in Eq. (6.34). Consistent with the surface layer (Eq. (6.26)), a turbulent Prandtl number Pr_t is used to yield a larger ED for heat and humidity than for momentum.

Eqs (6.30)–(6.34) forms a closed set from which the ED can be solved and is usually referred to as a 1.5-order, because the TKE is the only second-order moment that is prognosed. TKE-based ED parameterisations give reliable values for w_t that are well constrained by the energetics of the TKE Eq. (6.31), but they all suffer from uncertainties originating from arbitrary choices of the turbulent length-scale formulations.

6.4.2.2 First-Order Closure

A simpler alternative for parameterising the ED can be obtained by assuming a diagnostic balance in Eq. (6.31) between the production terms B and S and the destruction term D, thereby neglecting storage and transport terms. Using the parameterisation Eq. (6.33) for D, gives a closed expression for $\sqrt{\bar{e}}$ that can be substituted in Eq. (6.30) so that

$$K_m = l_t^2 |\partial_z \bar{v}_h|\, F(\text{Ri}_f), \qquad F = c_F\, (1 - \text{Ri}_f)^{\frac{1}{2}}, \qquad (6.35)$$

where $\bar{v}_h$ is the horizontal velocity vector and where a flux Richardson number $\text{Ri}_f \equiv -B/S$ has been introduced as a measure of stability. All the proportionality factors have been absorbed in c_F. The ED for heat is related through the turbulent Prandtl number via $K_m = \text{Pr}_t K_h$.

The result (Eq. (6.35)) is consistent with the earliest mixing length theories proposed by Prandtl in 1925. This first-order closure works reasonable well for shear-driven neutral to weakly stable boundary layers ($\text{Ri}_f > -1$), when a local balance of TKE is an acceptable approximation

6.4.2.3 K-profile Method

Under unstable conditions, the transport term in the TKE equation cannot be neglected, and consequently Eq. (6.35) is not applicable anymore. A simple alternative for the convective boundary layer is to prescribe a profile of the ED K (Troen and Mahrt, 1986). It is based on observations that K is small near the surface and the inversion height z_i and peaks in the middle of the convective boundary layer. An expression that fulfils these conditions and matches with surface layer similarity is

$$K_{h,m} = \kappa z u_* \Phi_{h,m}^{-1} \left(1 - \frac{z}{z_i}\right)^p, \quad (6.36)$$

where the exponent p is in the range $1.5 \leq p \leq 2.5$. In case of a convective boundary layer with unstable conditions ($\xi \ll 0$), the gradient functions Φ scale as $\|\xi\|^{-1/3}$. This allows to rewrite Eq. (6.36) in terms of the convective velocity scale $w_* \equiv (z_i \overline{w'b'}_{\text{srf}})^{1/3}$, introduced in Eq. (5.83)

$$K_{h,m} = c_{h,m} w_* z_i \left(\frac{z}{z_i}\right)^{\frac{4}{3}} \left(1 - \frac{z}{z_i}\right)^p. \quad (6.37)$$

All proportionality constants have been absorbed in the constant $c_{h,m}$. Eq. (6.37) shows that the maximum value of K in the middle of the convective boundary layer is determined by the product of the inversion height z_i and the convective velocity scale w_*.

The K-profile strongly depends on the inversion height z_i which is usually determined using the parcel concept, introduced in Chapter 2 to mimic the behaviour of a rising plume. The parcel is initialised at the lowest model level with the average temperature and humidity of the atmosphere at that level plus a temperature excess $\Delta\theta$, as indicated in Fig. 6.5. This temperature excess, which is proportional to the surface sensible heat flux, represents the sub-grid variability of the temperature and takes into account that the plume originates from near surface air that is warmer than its average value. The parcel is lifted along an unsaturated isentrope until it has reached its zero buoyancy level which defines the inversion height, as illustrated in Fig. 6.5. Because the density potential temperature is a conserved variable for unsaturated isentropic processes, the inversion height z_i can be determined from the implicit equation

$$\overline{\theta}_\rho(z_i) = \overline{\theta}_\rho(z_1) + \Delta\theta, \quad (6.38)$$

where $\overline{\theta}_\rho(z_1)$ is the average density temperature at the lowest model level somewhere in the surface layer. More sophisticated methods may also include dilution due to mixing of the parcel with the environment during the ascent.

6.4.3 The Stratocumulus-Topped Boundary Layer

If the boundary layer is sufficiently humid and topped by a strong inversion, it can develop into a stratocumulus-topped boundary layer. The characteristics of such a stratocumulus-topped boundary have been extensively described in Chapter 5. It is often well-mixed in terms of the moist-conserved variables q_t and θ_ℓ and capped

by a strong inversion through a strong increase of θ_ℓ and decrease of q_t. The upper part of this boundary layer is saturated, causing a solid cloud deck with a liquid-water content that is increasing with height, as dictated by the saturated isentrope.

A direct application of any of the ED formulations described in the previous section fails since the thermodynamics of the condensational effects are not taken into account. The use of a K-profile method with an unsaturated isentropic parcel ascent, described by Eq. (6.38), diagnoses an inversion height z_i around the cloud base rather than at the cloud top, as seen in Fig. 6.6a. Within the TKE approach, application of Eq. (6.32) will diagnose a negative buoyancy production in the cloud layer, leading to destruction of turbulent kinetic energy. In both cases, the transport of heat and moisture in the stratocumulus layer is inhibited and limited to the dry subcloud layer. As a result, the stratocumulus will be cut off from any moisture supply from the surface and will dry out. The moisture released from the surface will be distributed over a shallower boundary layer that will become too moist. Therefore, generalised formulations are required, leading to enhanced EDs in the stratocumulus layer as a result of the condensational heating effects.

The K-profile method can be generalised by allowing condensation of the isentropic parcel ascent. Above its lifting condensation level (LCL), the parcel will follow a saturated isentrope until it has found its level of neutral buoyancy (LNB) as an estimate for the inversion height z_i seen in Fig. 6.6b. As a result, cloud-top height is diagnosed as z_i and using this estimate in the prescribed K-profile described by Eq. (6.36) will enhance mixing in the stratocumulus layer.

The TKE approach can be generalised by extending the parameterised buoyancy production term (Eq. (6.32)) to saturated conditions. This can be done by writ-

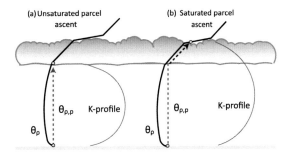

FIGURE 6.6: Determination of the inversion height in a stratocumulus-topped boundary layer by (a) a dry parcel ascent and (b) a moist parcel ascent that takes condensational effects into account.

ing the buoyancy flux as a linear combination of the fluxes of the moist conserved variables θ_ℓ and q_t, as in Eq. (5.107), and applying a down-gradient ED approach to these fluxes

$$\overline{w'b'}_u = -K_h\beta\left(A_u\partial_z\overline{\theta}_\ell + B_u\overline{\theta}\partial_z\overline{q}_t\right) \tag{6.39a}$$

$$\overline{w'b'}_s = -K_h\beta\left(A_s\partial_z\overline{\theta}_\ell + B_s\overline{\theta}\partial_z\overline{q}_t\right), \tag{6.39b}$$

where the subscript u denotes unsaturated conditions and subscript s saturated conditions. The thermodynamical coefficients have been derived in Section 2.5.3 and can be expressed as

$$A_u = 1 + \epsilon_2\overline{q}_t \approx 1, \qquad B_u = \epsilon_2 \approx 0.608 \tag{6.40a}$$

$$A_s = \gamma, \qquad B_s = \frac{\ell_v}{c_p\overline{T}}\gamma - 1, \tag{6.40b}$$

where γ is the non-dimensional lapse rate, given in Eq. (2.85) and displayed in Fig. 2.4, and can be interpreted as the ratio between the saturated and the unsaturated lapse rates. Note that the buoyancy parameterisation for unsaturated air given by Eq. (6.39a) is equivalent to Eq. 6.32. For warm temperatures ($\gamma \approx 0.4$), B_s can be up to four-times larger than B_u, causing a strong increase of the buoyancy flux around the cloud base, as seen in Fig. 5.30. For cold temperatures ($\gamma \approx 1$), this factor can increase to a factor of more than ten.

In the case of a partially cloudy layer, when only a fraction a_c is cloudy, it is tempting to approximate the total buoyancy flux by a linear combination of the saturated and the unsaturated contributions, weighted by the cloud fraction

$$\overline{w'b'} = a_c\overline{w'b'}_s + (1 - a_c)\overline{w'b'}_u. \tag{6.41}$$

It can be shown that Eq. (6.41) is exact, in the case when the joint probability distribution of θ_l, q_t and w is a multivariate Gaussian distribution (Sommeria and Deardorff, 1977). This is an acceptable approximation for stratocumulus, where the cloud fraction a_c is close to unity, but breaks down in the case of cumulus clouds, as will be demonstrated in Section 6.5.

Once the ED have been determined using Eq. (6.30), it can be used to parameterise the turbulent flux divergence. For the clear boundary, the ED approach can be applied to the fluxes of q_v and s_d (or equivalently θ) in Eq. (6.2) which will tend to make well-mixed profiles of these fields in the convective boundary layer.

Unfortunately, applying an ED approach to the fluxes of s_d, q_v and q_l in Eq. (6.2) in the case of stratocumulus will lead to erroneous down-gradient fluxes with even the wrong sign, as can be seen in Fig. 6.7. Turbulent liquid-water fluxes in stratocumulus layers are positive

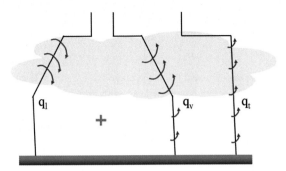

FIGURE 6.7: Schematic illustration of how application of an ED approach on non-conserved variables can lead to wrong estimates of the turbulent transport.

and upward, while an ED approach applied on the gradient of the liquid-water profile will diagnose a negative downward flux. Therefore, the ED approach needs to be applied to the fluxes of moist conserved variables, such as q_t and s_e (or equivalently θ_l) in Eq. (6.5). The consequence is that only the tendencies of the moist conserved variables can be obtained. Since most large-scale models usually use non-conserved prognostic variables, a diagnostic separation of q_t and θ_l into q_v, q_l and θ is required. This separation is usually achieved by the cloud scheme and the used methods will be discussed in Section 6.6.

6.4.4 Top Entrainment

One of the most important boundary-layer processes that requires parameterisation is the entrainment of dry and warm air from the free troposphere across the inversion into the boundary layer. This top-entrainment process influences the warming, drying and growth of the boundary layer. As demonstrated in Chapter 5, top entrainment depends on the energetics of the boundary layer such as the surface buoyancy flux and the radiative cooling, but also on the inversion strength defined by the change of temperature and humidity across the inversion. The underlying physics of top entrainment is not well captured by simple ED parameterisations such as a K-profile method, where the entrainment flux depends on the rather arbitrary choice of the assumed shape of the mixing profile, the vertical resolution and the determination of the inversion height z_i by the test parcel. In case of the more physically based turbulent kinetic energy parameterisation, the top entrainment still depends critically on the choice of the constant c_e in Eq. (6.34), which determines the length scale under stable conditions.

A more constraining alternative is to explicitly prescribe the ED at the top of the boundary layer in such a way that it obeys the entrainment rules discussed in Chapter 5. Several different top-entrainment formulations for the dry and stratocumulus-topped boundary layer have been discussed in Chapter 5 in terms of the entrainment velocity w_e and will not be repeated here. The entrainment velocity is related to the turbulent flux across the inversion of a variable $\varphi \in \{\theta_l, q_t\}$ by

$$\overline{w'\varphi'} = -w_e \Delta\varphi, \qquad (6.42)$$

where $\Delta\varphi$ denotes the change in value of φ across the inversion. Substituting Eq. (6.42) in the down-gradient flux parameterisation Eq. (6.9) shows that an ED

$$K_e = w_e \Delta z \qquad (6.43)$$

describes an entrainment flux across the inversion consistent with the the entrainment velocity w_e that has been derived from a top-entrainment parameterisation.

6.4.5   Local versus Non-local Transport

Although the convective clear and stratocumulus-topped boundary layer can be assumed to be well mixed to first order, a closer inspection of the thermal structure shows a more detailed structure. Fig. 6.8 shows the typical mean vertical profile of the potential temperature θ of a dry convective boundary layer from an LES. To emphasise the vertical structure, the minimum temperature has been subtracted. The y-axis has been non-dimensionalised by rescaling z with the inversion height z_i. The surface layer is strongly super-adiabatic due to the upward sensible heat fluxes at the surface. The degree of the instability is weakening until the middle of the boundary layer which can be considered to be well mixed. The upper half of the convective boundary layer is slightly stable and finally becomes strongly stable at the inversion.

The corresponding turbulent heat flux profile, normalised by its surface heat flux Q_0, is shown in the same Fig. 6.8. It decreases linearly with height, so it can be well described by

$$\overline{w'\theta'} = Q_0 \left(\frac{z_i - z}{z_i} \right) + E \frac{z}{z_i}, \qquad (6.44)$$

where E denotes the entrainment flux at z_i. The linear decreasing heat flux warms the boundary layer uniformly, as can be understood from Eq. (6.8), so that the shape of

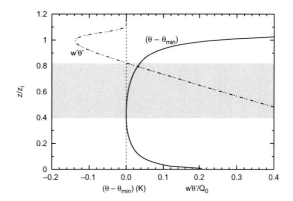

FIGURE 6.8: Vertical profile of the potential temperature θ minus its minimum value in the convective boundary layer from an LES. The dashed-dotted line indicates the corresponding profile of the vertical turbulent heat flux $\overline{w'\theta'}$ normalised by its surface value Q_0. The grey shaded area indicates where the flux $\overline{w'\theta'}$ is counter to the gradient of θ.

the θ-profile is not changing in time. In such a case, the system is by definition in a quasi-steady state. In the upper half of the CBL, indicated by the shaded zone in Fig. 6.8, the potential temperature profile is stable, while the heat flux is still positive. This is the counter-gradient zone where the flux moves against the gradient of the profile, a process that cannot be represented by a local ED approach.

The counter-gradient transport is due to the strongly non-local character of the buoyancy force which drives the heat flux. A test parcel, initiated in the surface layer is unstable under positive temperature perturbations and will start to ascent. Due to the positive buoyancy force, proportional to the density temperature difference between the parcel and its environment, the parcel will accelerate in the vertical direction. When the parcel reaches the upper part of the convective boundary, the buoyancy force is still positive, so that the parcel can continue to rise, even though the local gradient of the potential temperature profile is against the direction of the flux, as illustrated in Fig. 6.5. The parcel continues to rise *against* the local gradient to its LNB in the inversion or even slightly beyond if it has accumulated enough kinetic energy to overshoot this level.

A pragmatic way to parameterise a non-local positive heat flux also in the upper part of the convective boundary layer is to add a counter-gradient term to the local ED

$$\overline{w'\theta'} = -K\partial_z\overline{\theta} + \overline{w'\theta'}_{\mathrm{nl}} \,;\; \overline{w'\theta'}_{\mathrm{nl}} = K\gamma_{\mathrm{cg}}, \qquad (6.45)$$

where the counter-gradient parameter γ_{cg} quantifies the strength of the non-local transport. This way a net positive upward turbulent transport can be parameterised, even if the local down-gradient eddy diffusion is negative.

The non-local formulation of Eq. (6.45) can be related to the prognostic flux Eq. (6.10) by parameterising the pressure-covariance term P in Eq. (6.10) as a relaxation process

$$P \simeq \frac{\overline{w'\theta'}}{\tau_\theta}, \qquad (6.46)$$

where τ_θ is the return to the isotropy timescale for the turbulent flux. The standard local ED formulation (Eq. (6.9)) follows from assuming a local equilibrium between the gradient production term and the parameterised pressure-covariance term (Eq. (6.46)) with an ED

$$K = \sigma_w^2 \tau_\theta, \qquad (6.47)$$

consistent with Eq. (6.13), assuming that the length scale l_t is related to the isotropy timescale via $\sigma_w l_t \simeq \tau$. The nonlocal term in Eq. (6.45) can be obtained from the prognostic flux Eq. (6.10) by including the buoyancy term to the equilibrium condition, which leads to an identification of the counter-gradient parameter as

$$\gamma_{cg} = \beta \overline{\theta'_\rho \theta'} \sigma_w^{-2}. \qquad (6.48)$$

Analytical quasi-steady solutions of Eq. (6.8) where the flux is parameterised by Eq. (6.45) are explored in Exercise 2. Without the counter-gradient term (i.e., $\gamma_{cg} = 0$), the quasi-steady solutions are absolutely unstable in almost the whole boundary layer. In that case, only the limit of infinitely large EDs provide solutions that are well mixed. Quasi-steady solutions that include a non-local counter-gradient parameter, on the other hand, show an unstable lower part and a stable upper part of the convective layer, in better agreement with observations and LES results. Non-local transport plays a crucial role in cumulus convection which is one of the reasons why a local ED approach is of very limited value in case of cumulus parameterisations.

6.4.6 LESs of the Atmospheric Turbulence

So far, ED parameterisations have been discussed for models that use a resolution much coarser than the depth of the boundary layer. In the case of LES, the resolution is of the order of 100 m so that the largest eddies are numerically resolved. For such resolutions under convective conditions, the turbulent length scale is determined by the grid-mesh size

$$l_t = (\Delta x \Delta y \Delta z)^{\frac{1}{3}}. \qquad (6.49)$$

For stable situations, the turbulent length scale can be substantially smaller than the used grid-mesh size. In that

case, a length scale similar to the one appearing in the last term of the right-hand side of Eq. (6.34) is used.

Application of the length-scale formulation (Eq. (6.49)) in the TKE closure (Eq. (6.30)) or in the first-order closure (Eq. (6.35)) and applied to all three spatial directions are widely used subgrid formulations in LESs for atmospheric applications. Note that the choice of the length-scale parameterisation (Eq. (6.49)) provides a scale-aware ED formulation. The heuristic scaling arguments in Section 2.6 of turbulence in the inertial subrange indicates that the TKE scales as $\bar{e} \sim l_t^{2/3}$. This implies that the ED (Eq. (6.30)) with a length-scale formulation (Eq. (6.49)) scales as

$$K \sim l_t^{4/3}, \qquad (6.50)$$

demonstrating that the ED decreases with resolution in accordance with the famous 4/3 Richardson law.

6.5 Parameterisation of Cumulus Convection

Radiative long-wave cooling tends to push the atmosphere slowly towards a state of conditional instability, thereby making it amenable for cumulus convection. The interplay between large-scale subsiding motions and boundary-layer turbulence, however, can create stable boundary-layer inversions that can delay or even inhibit the onset of cumulus convection. This inhibition allows for the accumulation of conditional available potential energy (CAPE), a concept introduced in Section 2.3.3, that builds up capacity for moist convection until it is initiated by some trigger mechanism. In the case of surface-driven cumulus convection, triggering can be initiated by thermals that have absorbed enough sensible and latent heat from the underlying surface to reach the LCL, break through the inhibition layer where the updraft is negatively buoyant and rise to the level of free convection (LFC), as seen Fig. 6.9. From that point onwards, the cumulus convection can further grow until it reaches its LNB. Cloud-top height is usually close to this LNB, or marginally higher due to overshooting as a result of their accumulated kinetic energy. Alternatively, moist convection can also be triggered dynamically through large-scale convergence, a common mechanism in the case of mid-latitude frontal systems.

As explained in Chapter 2, cumulus convection organises itself through strong narrow updrafts, driven by the latent heat release of condensation, that cover only a small horizontal fraction a_u of the atmosphere while the environment is characterised by weak downward motions,

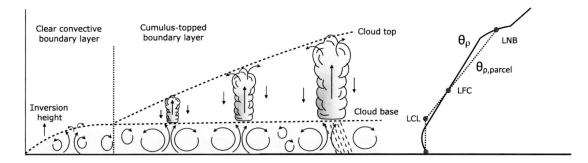

FIGURE 6.9: Left: Schematic illustration of the development of the dry convective boundary layer into a cumulus-topped boundary layer. Right: An idealised profile of the density temperature of the cumulus-topped boundary layer (full line), along with a profile of an adiabatic parcel lifted from near the surface (dotted line). The height were the adiabatic parcel becomes supersaturated is identified as the LCL, the first level beyond the LCL where the parcel becomes positively buoyant is indicated as the LFC. The level where the parcel becomes again negatively buoyant near the inversion of the cloud-topped boundary layer is marked as the LNB.

usually referred to as compensating subsidence. If the cumulus towers become deeper than typically 3 km, the liquid-water loading is large enough to initiate precipitation. The precipitation intensity increases with the cloud depth and eventually will be strong enough to generate saturated downdrafts. These downdrafts transport relatively cold air downwards back into the subcloud layer. If these downdrafts reach the bottom of the subcloud layer, they spread over the surface in the form of a density current that laterally propagates away from the rainfall. These density currents often appear in circular patterns, which are known as cold pools. Whereas the outflow boundaries of the cold pools are often associated with high relative humidity (RH), the downdrafts cause the inflow of cold and dry air in the interior of the cold pools. As a result, triggering of cumulus convection is suppressed within the cold pools but enhanced at the outflow boundaries, resulting in a fascinating mesoscale organisation mechanism of cumulus clouds.

A cumulus convection parameterisation has the daunting task of representing the effect of all these processes in terms of the resolved variables on the scale of numerical grid. To the extent it is possible, this is usually accomplished by breaking down the problem in a number of tasks. First, a trigger mechanism needs to be formulated; does cumulus convection occur in a grid box? Second, a model is required that represents the transport of a cumulus cloud ensemble that populates the grid box. Lastly, the effect of the cloud ensemble on the mean state, in terms of the tendencies of the resolved variables needs to be quantified. This last step is usually referred to as the 'principal closure'.

Early convection parameterisations used in the first-generation climate models were convective adjustment schemes. In its simplest form, moist convection is triggered when the air becomes supersaturated and the temperature lapse rate becomes unstable to saturated perturbations. In that case it is assumed that cumulus convection acts to adjust the temperature profile back to its saturated isentrope, while the supersaturated moisture is removed as precipitation. Although such adjustment schemes are simple, computationally inexpensive, thermodynamically consistent and preventing the atmosphere to become gravitationally unstable there are also drawbacks. They dry and cool the lower troposphere too much and produce precipitation that is too intense. Conceptually, they are also less attractive since they only represent the overall effects of moist convection, which makes it hard to add specific detailed processes such as up- and down-drafts, mixing processes and cloud microphysics. Nevertheless, they are still used in simple climate models of intermediate complexity.

In 1969 Akio Arakawa[7] introduced a new way to conceptualise cumulus convection (Arakawa 1969). His approach explicitly parameterises the condensation and transport terms for heating and moistening appearing at the right-hand side of Eq. (6.2). It is essentially a mass flux concept based on an entraining plume model. It describes how cumulus clouds can modify the environment while condensation only takes place inside clouds and provides a closure based on a quasi-equilibrium assumption. In

[7] Akio Arakawa (1927–), a Japanese scientist at the University of California, Los Angeles (UCLA), has played a leading role in the development of atmospheric GCMs from the beginning. His main contributions are the formulation of finite difference schemes that respect the physical principles of energy and enstrophy conservation and the development of the theory for the interaction of a cumulus cloud ensemble with the large-scale environment. Both developments have influenced the design of practically all numerical weather and climate models.

1974, this scheme was further developed in a seminal paper by Arakawa and his student Wayne Schubert. Many of the cumulus parameterisations today are still based on the principles and assumptions formulated in what will be referred to as the Arakawa–Schubert scheme (Arakawa and Schubert, 1974). The following subsections will discuss underlying assumptions, approximations and developments of mass flux parameterisations for cumulus convection.

6.5.1   Mass Flux Concept

The general idea of mass flux models is to simplify the turbulence flux in terms of one or a few simple up- and downdrafts that can approximate the convective transport of a cumulus cloud ensemble. The simplest way is to decompose a general circulation model (GCM) grid box at each vertical model level into two parts: a convective updraft part with the fractional area a_u consisting of all saturated convective updrafts and a complementary environmental fraction with a fractional area $1 - a_u$. Updraft averages of the variable $\varphi \in \{\theta_\ell, \theta_\rho, q_t, q_v, \ldots\}$ are defined as the spatial averaged values over the convective subdomain and are related to their environmental counterpart via

$$\overline{\varphi} = a_u \varphi_u + (1 - a_u)\, \varphi_e, \tag{6.51}$$

where the subscripts u and e indicate respectively the updraft and the environmental averages.

Within the mass flux concept, it is assumed that the difference between the updraft and the environmental averages is much larger than the typical fluctuations within the updrafts and the environment. This supports the use of a top-hat approximation, that neglects all the fluctuations within the updrafts and the environment, as seen in Fig. 6.10. The top-hat approximation allows the simplification of the turbulent flux as

$$\overline{\rho w' \varphi'} \approx \rho \frac{a_u}{1 - a_u} \left(w_u - \overline{w}\right)\left(\varphi_u - \overline{\varphi}\right). \tag{6.52}$$

Cumulus updrafts tend to organise in narrow updrafts, as explained in Chapter 2, with updraft velocities much larger than the grid box averaged vertical velocity $\overline{w}$. Therefore, Eq. (6.52) can be further simplified to

$$\overline{\rho w' \varphi'} \approx \mathcal{M}_u \left(\varphi_u - \overline{\varphi}\right), \quad \mathcal{M}_u \equiv \rho a_u w_u, \tag{6.53}$$

where a bulk updraft mass flux $\mathcal{M}_u$ has been defined, identical to the one introduced in Eq. (2.110). Eq. (6.53) constitutes the most basic bulk mass flux approximation.

The validity of the mass flux approximation has been well confirmed by observations and LES results for non-precipitating cumulus convection. This is illustrated in Fig. 6.11 which shows the turbulent flux of q_t from an

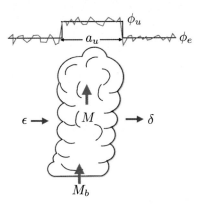

FIGURE 6.10: Top: Schematics of a top-hat approximation, where the variable φ is approximated by its average value φ_u in the cloudy updraft and by its average value φ_e in the environment. Bottom: Schematics of the entraining/detraining plume model derived in Section 6.5.2.

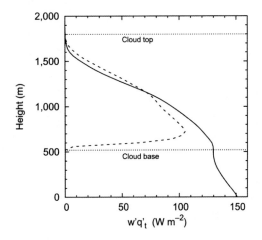

FIGURE 6.11: Turbulent flux of q_t from an LES of marine shallow cumulus convection as observed during BOMEX. The solid line represents the simulated turbulent flux while the dashed line represents the parameterised flux as in right-hand side of Eq. (6.53) with all the terms directly diagnosed from the same simulation.

LES of a marine subtropical cumulus-topped boundary layer observed during the Barbados Oceanographic and Meteorological Experiment (BOMEX) (Siebesma et al., 2003). In the same figure, the mass flux approximation of Eq. (6.53) is displayed based on the mass flux $\mathcal{M}_u$, the sampled cloudy updraft field $q_{t,u}$ and the domain averaged $\overline{q}_t$, diagnosed from the same LES. This example illustrates that Eq. (6.53) provides an excellent parameterisation for the turbulent transport in the cloud layer of shallow cumulus convection, provided that adequate estimates of the mass flux and the cloudy updrafts can be made.

What makes the mass flux concept work so well in the case of cumulus convection and why is it not used

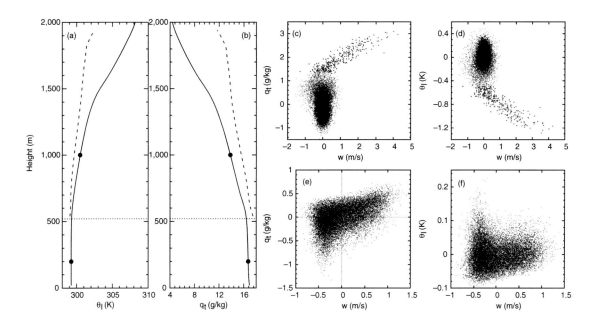

FIGURE 6.12: Profiles and scatter plots based on an LES of shallow cumulus convection. (a) and (b) show the domain-averaged profiles of θ_l and q_t in solid lines while the dashed lines represent the averaged values of these variables in the cloudy updraft. (c) and (d) show scatter plots of the vertical velocity w versus q_t and θ_l, respectively, at a height of 1,000 m in the middle of the cloud layer. Domain-averaged values of w, q_t and θ_ℓ, have been subtracted. The red dots represent saturated (cloudy) grid points that are positively buoyant while the blue points represent cloudy grid points that are negatively buoyant. The small black dots represent unsaturated grid points. (e) and (f) show the similar scatter plots as in (c) and (d) in the middle of the subcloud layer at a height of 200 m. The strong bimodal character of the scatter plots in the cloud layer illustrates why the mass flux approximation (Eq. (6.53)) works so well in the cloud layer.

as well in the subcloud layer and the clear convective boundary layer? To answers these questions, it is useful to have a closer look at the turbulent structure of the cumulus-topped boundary layer. Fig. 6.12a and b display the domain-averaged profiles of q_t and θ_l from an LES, based on the same BOMEX case on a horizontal domain, representative for a grid box of a GCM. The profiles show the classic structure of a marine shallow cumulus-topped boundary layer: a well-mixed subcloud layer in the lowest 500 m, followed by a cloud layer, unstable for saturated perturbations but stable for unsaturated perturbations and topped by an inversion layer, characterised by a strong increase of θ_ℓ and a strong decrease of q_t.

The convective subcloud layer has a similar structure as the clear convective boundary layer and is characterised by upward transport of relative moist and positive buoyant air originating from the surface layer and downward transport of dryer and negative buoyant air. This is illustrated in Fig. 6.12e and f with scatter diagrams of the vertical velocity w versus q_t and θ_ℓ in the middle of the subcloud layer at 200 m obtained from the LES. Each point in the diagram represents the value of temperature, humidity and vertical velocity in a grid box of 50×50 m of the

simulation. Domain average values of w, q_t and θ_ℓ have been subtracted so that the turbulent flux at this height can be simply obtained by summing up the product of the x and y values of all the dots in the scatter plot. The scatter diagrams indeed demonstrate positive correlations of the vertical velocity w with q_t and, to a lesser extent, with θ_l indicating upward transport of warm and moist air and downward transport of cooler and dryer air. The shape of the joint probability distribution function (PDF) also suggests that there is a continuum of up- and down-drafts in the subcloud layer, ranging from weak to strong.

These characteristics are fundamentally different in the cumulus cloud layer. The strongest thermals in the subcloud layer reach the LCL and are fuelled by condensational heating which allow them to further accelerate along a saturated isentrope in an environment that is stable for unsaturated updrafts. As a result, the cumulus layer is characterised by narrow strong cloudy saturated updrafts venting moist air out of the subcloud layer into an otherwise quiescent free atmosphere. This behaviour is clearly visible in Fig. 6.12c and d where scatter diagrams in the middle of the cloud layer at 1,000 m are shown. The coloured dots indicate grid points from the simulation that are saturated, contain condensed water and can be considered to be cloudy. The colour indicates whether these cloudy points are positively buoyant (red) or negatively buoyant (blue) with respect to the environment. The scatter plot demonstrates that most of the saturated upward transport is represented by a small fraction of positive buoyant moist cloud elements (red dots) while the environment appears to be dominated by random fluctuations with a small net subsiding motion. The resulting joint PDF is bimodal and strongly skewed: one peak corresponding to the buoyant saturated updrafts of the cloud ensemble and another peak corresponding to slowly subsiding environmental air. It is due to the conditional instability and the resulting bimodal structure of the PDF that the mass flux approximation works so well for cumulus convection. The absence of this clear seperation between up- and down-drafts in the subcloud layer makes this part of the atmosphere less suitable for this approach.

Convection parameterisations that use Eq. (6.53) as a starting point are referred to as *bulk* mass flux models. Note that w_u and a_u only matter through $\mathcal{M}_u$ as far as the turbulent transport is concerned so that individual parameterisations of these variables are not required, provided that an acceptable parameterisation of $\mathcal{M}_u$ can be formulated. The approximations $a_u \ll 1$ and $|w| \ll w_u$ can only be made if the size of the grid cell is much larger than the individual sizes of the cloudy updrafts. For resolutions finer than 10 km, additional parameterisations of a_u and w_u are required.

In the case of deep convection, the bulk mass flux approximation is often extended with a bulk downdraft term associated with precipitation

$$\rho\overline{w'\varphi'} \approx \mathcal{M}_u\left(\varphi_u - \overline{\varphi}\right) + \mathcal{M}_d\left(\varphi_d - \overline{\varphi}\right), \tag{6.54}$$

where the average values of the mass flux and the variable φ in the cloudy downdrafts are denoted by $\mathcal{M}_d$ and φ_d. Including downdrafts in convection schemes allows for the possibility to parameterise downward transport of moist cool air back into the boundary layer.

More complex decompositions are possible and also used. Spectral mass flux models subdivide the cloud ensemble further into sub-ensembles of clouds that have a similar cloud top height, labelled by the subscript i

$$\rho\overline{w'\varphi'} \approx \sum_i \mathcal{M}_{u,i}\left(\varphi_{u,i} - \overline{\varphi}\right). \tag{6.55}$$

This more general formulation has the advantage of utilising different updraft plumes for deeper and shallower clouds at the same time, but also requires a larger degree of complexity that needs to be constrained by parametric assumptions. One of the first published mass flux schemes

of Arakawa and Schubert in 1974 was in fact a spectral mass flux scheme.

In the upcoming subsections, mainly mass flux parameterisations based on the simple bulk decomposition of Eq. (6.53) will be considered as they adequately introduce the main concepts.

6.5.2   Cloud Updraft Equations

To design a bulk mass flux parameterisation, additional updraft equations for φ_u and $\mathcal{M}_u$ are required. A convenient starting point is the conservation law for φ in the flux form within the anelastic approximation

$$\partial_t \varphi + \nabla_h \cdot \left(\boldsymbol{v}_h \varphi\right) + \frac{1}{\rho}\partial_z\left(\rho\, w\varphi\right) = F_\varphi, \tag{6.56}$$

where all the source and sink terms are collected in F_ϕ and where the subscript h is a shorthand notation for the horizontal velocity components, that is, $\nabla_h = (\partial_x, \partial_y)$ and $\boldsymbol{v}_h \equiv (u, v)$. We assume a grid box with a horizontal area A in which cloudy updrafts at height z and time t occupy an area $A_u(z,t)$. To derive updraft equations for φ, Eq. (6.56) is integrated over the area $A_u(z,t)$, which gives, after applying Leibniz's integral rule and the Gauss divergence theorem

$$\partial_t\left(a_u\varphi_u\right) + \frac{1}{A}\oint_{\text{bnd}} u_b\,\varphi\,dl + \frac{1}{\rho}\partial_z\left(a_u\rho\,\overline{w\varphi}^u\right) = a_u F_{\varphi,u} \tag{6.57a}$$

with $u_b = \hat{\boldsymbol{n}}\cdot(\boldsymbol{v} - \boldsymbol{v}_b),$ \tag{6.57b}

where $a_u = A_u/A$ is the fractional updraft area, $\hat{\boldsymbol{n}}$ is an outward-pointing unit vector perpendicular to the boundary between the updrafts and the environment, $\boldsymbol{v}$ is the 3D velocity vector at the boundary between the updrafts and the environment and $\boldsymbol{v}_b$ the 3D velocity vector of the moving boundary. The second term on the left-hand side describes the exchange of φ across a boundary that is changing with time in size and form. Note that it is the velocity u_b, *relative* to the velocity $\boldsymbol{v}_b$ of the boundary itself, that determines the exchange term. This way it is guaranteed that there is no exchange across the boundary diagnosed when the cloud updrafts are simply being advected by the wind. For the special case, $\varphi = 1$ and $F_u = 0$, Eq. (6.57a) reduces to the continuity equation of the updraft ensemble

$$\partial_t a_u + \frac{1}{A}\oint_{\text{bnd}} u_b\,dl + \frac{1}{\rho}\partial_z\rho\,a_u w_u = 0 \tag{6.58}$$

and describes how the updraft fraction a_u depends on the net lateral mass inflow across the boundary and the vertical mass flux divergence.

Eq (6.58) forms a good starting point to introduce the mixing between the cloud updrafts and their environment. This mixing process is usually described by entrainment and detrainment. Entrainment rates E are associated with the inflow rate of mass into a region (the cloudy updraft ensemble in the present case) while detrainment rates D describe the complementary outflow of air from the cloud updrafts into the environment. It is therefore natural to define these rates respectively as

$$E_\mathrm{d} \equiv -\frac{\rho}{A} \oint_{\mathrm{bnd},u_\mathrm{b}<0} u_\mathrm{b}\, dl \qquad (6.59\mathrm{a})$$

$$D_\mathrm{d} \equiv +\frac{\rho}{A} \oint_{\mathrm{bnd},u_\mathrm{b}>0} u_\mathrm{b}\, dl, \qquad (6.59\mathrm{b})$$

so that entrainment is defined as that part of the line integral where there is inflow of mass into the updraft region and detrainment as the complementary part where there is mass outflow into the environment. The subscript d has been attached to indicate that the exchange rates result from a *direct* and exact evaluation of the boundary, as indicated by Eq. (6.59). As a result of this definition, the lateral growth of a cloudy updraft area by condensation will be diagnosed as entrainment while shrinking by evaporation will be diagnosed as detrainment.

Although it is relatively easy to determine $E_\mathrm{d} - D_\mathrm{d}$ as a residual from Eq. (6.58), it is by no means trivial to determine the entrainment and detrainment terms separately, neither in field or laboratory experiments, nor in cloud resolving numerical simulations.[8] In addition, even if E_d and D_d are diagnosed correctly, they can not be used directly in Eq. (6.57a) because the entrainment and detrainment terms are convoluted with φ. A pragmatic solution is to use a top-hat approximation in the same spirit as for the mass flux approximation (Eq. (6.53)). In doing so, bulk entrainment and detrainment rates E_u and D_u can be defined that transport the *average* environmental properties into the cloud updrafts and vice versa. The exchange term in Eq. (6.57a) can be written in terms of these effective rates as

$$\frac{\rho}{A} \oint_{\mathrm{bnd}} u_\mathrm{b}\, \varphi\, dl = D_\mathrm{u}\varphi_\mathrm{u} - E_\mathrm{u}\varphi_e. \qquad (6.60)$$

These bulk entrainment and detrainment rates, defined by Eq. (6.60), are the relevant rates for parameterisations that use the bulk mass flux approximation (Eq. (6.53)).

Substituting Eq. (6.60) in Eq. (6.57a), applying the mass flux approximation to the flux divergence term and assuming a steady state gives

---

[8] Numerical techniques to derive direct entrainment and detrainment rates can be found in Romps (2010) and Dawe and Austin (2011).

$$\partial_z \left(\mathcal{M}_\mathrm{u}\varphi_u\right) = E_\mathrm{u}\overline{\varphi} - D_\mathrm{u}\phi_u + \rho a_\mathrm{u}F_{\varphi,\mathrm{u}} \qquad (6.61\mathrm{a})$$

$$\partial_z \mathcal{M}_\mathrm{u} = E_\mathrm{u} - D_\mathrm{u}. \qquad (6.61\mathrm{b})$$

More specifically, using the source and sink terms of the non-conserved variables $\varphi \in \{s_\mathrm{d}, q_\mathrm{v}, q_\mathrm{c}\}$ given by Eq. (6.2) leads to

$$\partial_z \left(\mathcal{M}_\mathrm{u}\, s_{\mathrm{d},\mathrm{u}}\right) = E_\mathrm{u}\bar{s}_\mathrm{d} - D_\mathrm{u}s_{\mathrm{d},\mathrm{u}} + \ell_\mathrm{v}\rho c \qquad (6.62\mathrm{a})$$

$$\partial_z \left(\mathcal{M}_\mathrm{u}\, q_{\mathrm{v},\mathrm{u}}\right) = E_\mathrm{u}\bar{q}_\mathrm{v} - D_\mathrm{u}q_{\mathrm{v},\mathrm{u}} - \rho c \qquad (6.62\mathrm{b})$$

$$\partial_z \left(\mathcal{M}_\mathrm{u}\, q_{\mathrm{c},\mathrm{u}}\right) = E_\mathrm{u}\bar{q}_\mathrm{c} - D_\mathrm{u}q_{\mathrm{c},\mathrm{u}} + \rho c - \rho G. \qquad (6.62\mathrm{c})$$

If there is no prognostic equation for condensed water q_c, the detrained condensed water is assumed to evaporate instantaneously, that is, $\rho e = D_\mathrm{u}q_{\mathrm{c},\mathrm{u}}$ outside the cloud updrafts. Eqs (6.61b)–(6.62c) form the basis for most bulk mass flux convection schemes that are used nowadays in large-scale models. Generalisation to spectral schemes and extensions with downdrafts are straightforward.

The updraft Eqs (6.62a)–(6.62c) can be rewritten more conveniently in an entraining plume form by defining fractional entrainment ε and detrainment rates δ as

$$E_\mathrm{u} \equiv \varepsilon \mathcal{M}_\mathrm{u} \qquad D_\mathrm{u} \equiv \delta \mathcal{M}_\mathrm{u}. \qquad (6.63)$$

Eliminating D_u by using the continuity Eq. (6.61b), the entraining plume equations of the moist conserved variables θ_l and q_t take the simple form

$$\partial_z\theta_{\ell,\mathrm{u}} = -\varepsilon \left(\theta_{\ell,\mathrm{u}} - \overline{\theta}_\ell\right) + \ell_\mathrm{v}\rho G/\mathcal{M}_\mathrm{u} \qquad (6.64\mathrm{a})$$

$$\partial_z q_{\mathrm{t},\mathrm{u}} = -\varepsilon \left(q_{\mathrm{t},\mathrm{u}} - \overline{q}_\mathrm{t}\right) - \rho G/\mathcal{M}_\mathrm{u} \qquad (6.64\mathrm{b})$$

$$\partial_z \ln \mathcal{M}_\mathrm{u} = \varepsilon - \delta. \qquad (6.64\mathrm{c})$$

Eqs (6.64a) and (6.64b) indicate that, without precipitation, the moist conserved variables within the updrafts are only affected by entrainment processes. The physical interpretation is simple and straightforward. In the absence of entrainment, that is, $\varepsilon = 0$, the updraft model illustrates that $\theta_{\ell,\mathrm{u}}$ and $q_{\mathrm{t},\mathrm{u}}$ are conserved variables for vertical displacements. Due to entrainment processes, $\theta_{\ell,\mathrm{u}}$ and $q_{\mathrm{t},\mathrm{u}}$ are pushed to their grid mean values by lateral mixing as the ascent progresses, as can be seen in Fig. 6.12 where the conditional sampled updraft profiles are shown. Note that within the bulk mass flux formulation, the detrainment process only influences the shape of the mass flux profile, but not the temperature and humidity properties of the updrafts. This is the result of the bulk decomposition, implicitly assuming that the detrainment processes influence the total mass of the updrafts but not its composition.

To illustrate how the steady-state cumulus ensemble can influence the mean state, one has to realise that the tendency due to cumulus convection, in the absence of precipitation, is equal to the turbulent flux divergence

$$\rho \left(\partial_t \overline{\varphi} \right)_{\mathrm{conv}} = -\partial_z \rho \, \overline{w' \varphi'}. \tag{6.65}$$

Applying the mass flux approximation (Eq. (6.53)) to the right-hand side of Eq. (6.65) and substituting the continuity Eq. (6.61b) and the entraining plume equations (Eq. (6.64)) gives

$$\rho \left(\partial_t \overline{\varphi} \right)_{\mathrm{conv}} = M_{\mathrm{u}} \partial_z \overline{\varphi} + D_{\mathrm{u}} \left(\varphi_{\mathrm{u}} - \overline{\varphi} \right). \tag{6.66}$$

The first term on the right hand represents warming and drying due the compensating subsidence in the environment,[9] driven by the updraft mass flux of the steady-state cumulus ensemble. In addition, the second term represents the effect of detrainment of the updraft fields on the environment which will in general moisten and warm the atmosphere near the cloud tops where most of the detrainment is expected to take place.

In order to further develop the updraft Eqs (6.64a)–(6.64c) into a comprehensive mass flux parameterisation, a number of issues needs to be resolved.

- It must be decided whether moist convection will be activated or triggered, for instance, by rising thermals from the boundary layer. Trigger mechanisms will be discussed in Section 6.5.3
- The updraft equations require formulations for the fractional entrainment and detrainment rates ε and δ as well as for the autoconversion rate G. Entrainment end detrainment parameterisations are discussed in Section 6.5.4 and microphysical parameterisations in Section 6.5.6.
- The vertical extent of the updraft equations needs to be determined. Cloud-top height can be defined as the level where the cloud updrafts have exhausted there vertical momentum. Parameterisations for the vertical velocity in cloudy updrafts are derived in Section 6.5.5.
- Estimates of $q_{t,\mathrm{u}}$, $\theta_{\ell,\mathrm{u}}$ and M_{u} at the cloud base are required as a lower boundary condition of the updraft equations. Especially parameterisations of the cloud-base mass flux $M_{b,\mathrm{u}}$ are usually referred to as the *principal closure* of the convection parameterisation as it is $M_{b,\mathrm{u}}$ that determines the overall intensity of convection. Closures will be discussed in Section (6.5.7).

6.5.3   Triggering of Convection

One of the main reasons that the CAPE can accumulate is that free atmosphere is in general *conditionally* unstable rather than absolutely unstable. This accumulation of CAPE can continue until lifted air from the boundary layer starts to condense, is able to reach its LFC and starts to initiate cumulus convection. From that moment onwards, cumulus convection starts to be active, liberating CAPE and turning it into kinetic energy of the updrafts thereby enabling the transport of heat, moisture and momentum. Diagnosing when this occurs is the first task of any convection scheme and is usually referred to as the trigger function. The simplest trigger mechanism is in the case of surface driven cumulus convection in which case initiation of cumulus convection depends on the capability of dry thermals to penetrate through the convective inhibition (CIN) layer and reach the LFC (see Fig. 6.9).

The simplest way of deciding whether surface-based convection can 'trigger' cumulus convection is to mimic a dry thermal by an adiabatic parcel that is released from the lowest model level in the surface layer with the mean thermodynamic properties of that level with an additional small positive temperature perturbation. If this undiluted parcel generates enough kinetic energy to reach its LFC, convection may be triggered. The same undiluted ascent can also be used to obtain the cloud-base values for temperature and humidity that can be used in conjunction with a mass flux closure as a lower boundary condition for the cloud model. A vertical velocity equation for the adiabatic rising parcel (Section 6.5.5) can be used to determine whether the updraft has enough kinetic energy to overcome the CIN layer between the LCL and the LFC.

Moist convection can also be triggered by large-scale convergence or through mesoscale organisation that can initiate lifting air at other levels than near the surface. This is usually parameterised by lifting parcels also from other model levels in the boundary layer. Dynamical triggering for these parcels can be taken into account by making the temperature perturbation dependent on the grid-mean vertical velocity.

6.5.4   Entrainment and Detrainment

6.5.4.1   Single Updrafts

The first and most well-established parameterisation of entrainment originates from laboratory studies of buoyant dry plumes rising in a non-turbulent environment. As known from everyday experience, the horizontal extent of a buoyant plume is growing with height due to mixing

---

[9] This can be understood by realising that $M_{\mathrm{u}} \simeq -(1 - a_{\mathrm{u}})w_e$, because $\overline{w} \ll w_u$.

of the quiescent environmental air with the turbulent air from the plume. In other words, the entrainment process can be viewed as the growth of turbulence by invading the quiescent air. In terms of Eq. (6.59), because $u_b < 0$, lateral mixing is a pure entrainment process in which detrainment plays no role, unless turbulence is eventually diluted to the extent that it is absorbed in the chaos of molecular motion. To obtain an entrainment formulation for a rising plume, it is useful to define A_u as the area of the plume, L_b as the perimeter of the boundary and $\bar{u}_b$ as the averaged velocity at the plume boundary, which allows us to write the left-hand side of the continuity Eq. (6.64c) as

$$\frac{1}{\mathcal{M}_u}\partial_z \mathcal{M}_u = -\frac{L_b}{A_u}\frac{\bar{u}_b}{w_u} \approx \frac{2\eta}{R}. \qquad (6.67)$$

In the last step, scaling results $\bar{u}_b \approx -\eta w_u$ from water tank experiments are used and a circular geometry of the plume with a radius R has been adopted. Comparison of Eq. (6.67) with Eq. (6.64c) leads directly to the famous fractional entrainment relation for plumes

$$\varepsilon = \frac{2\eta}{R} \qquad (6.68)$$

Laboratory experiments suggest typical values for the entrainment constant $\eta \approx 0.1$. The inverse scaling of the fractional entrainment rate with the radius of the plume can be understood by the simple geometrical argument that larger volumes have a relatively smaller surface area and are therefore less prone to entrainment processes at their boundaries.

Unfortunately, clouds are not as simple as dry plumes. Entrainment of unsaturated environmental air into cloud updrafts leads to the evaporation of the cloud condensate. As a result, cloud elements will lose their buoyancy and their vertical momentum so that they are no longer part of the cloud updraft: they are detrained. This naturally requires the inclusion of the detrainment process.

An insightful model of a single steady-state cloud updraft with a *fixed* radius R has been proposed by Asai and Kasahara (1967) and is schematically sketched in Fig. 6.13. By decomposing the boundary flux across the interface (see Eq. (6.57a)) into a mean and a turbulent part

$$\overline{u\varphi}^b = \bar{u}_b\bar{\varphi}_b + \overline{u'\varphi'}^b, \qquad (6.69)$$

a scale separation between a dynamical and a turbulent mixing term is created. This motivates a similar separation of the fractional entrainment and detrainment rates

$$\frac{1}{\mathcal{M}_u}\partial_z \mathcal{M}_u = \varepsilon_{dyn} + \varepsilon_{turb} - \delta_{dyn} - \delta_{turb}. \qquad (6.70)$$

Applying this decomposition in Eq. (6.60), using an ED approach for the turbulent part, and let associate

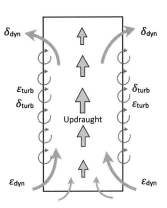

FIGURE 6.13: Schematic illustration of the Asai–Kasahara model of a steady-state cloud updraft model with a constant radius R. The large-scale inflow in the lower part of the cumulus cloud and the large-scale outflow in the upper part are characterised by the dynamical entrainment and detrainment rates while the smaller scale mixing between the updraft and the environment is represented by the turbulent entrainment and detrainment.

convergent plumes contribute to dynamical entrainment and divergent plumes to dynamical detrainment gives

$$\varepsilon_{turb} = \delta_{turb} = \frac{2\eta}{R} \qquad (6.71a)$$

$$\varepsilon_{dyn} = H(-\bar{u}_b)\frac{1}{\rho w_u}\frac{\partial \rho w_u}{\partial z} \qquad (6.71b)$$

$$\delta_{dyn} = -H(\bar{u}_b)\frac{1}{\rho w_u}\frac{\partial \rho w_u}{\partial z}, \qquad (6.71c)$$

where H denotes the Heaviside function. The interpretation of this result is intuitively clear. At the small scales, turbulent mixing occurs with equal turbulent entrainment and detrainment rates. At the larger scale, dynamical entrainment occurs in the case of convergence while detrainment occurs in the case of divergence. Convergence mainly occurs in the lower part of the cloud where the condensational heating feeds the buoyancy, leading to an acceleration of the updraft velocity. This acceleration has a negative feedback since it induces a dynamical entrainment that slows down the updraft, leading to divergence and a dynamical detrainment in the upper part of the cloud. The derived Eqs (6.71b) and (6.71c) of the dynamical entrainment and detrainment rates are a direct result of the assumption that the cloud has a constant radius R and therefore also a constant updraft area fraction a_u. Therefore, if the updraft velocity w_u can be diagnosed by an updraft equation (see Section 6.5.5), ε_{dyn} and δ_{dyn} can be determined and the model can be closed. Therefore, the constant R assumption can be seen as the closure of the dynamical entrainment and detrainment rates in the Asai–Kasahara model. The small-scale turbulent entrainment

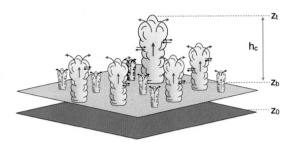

FIGURE 6.14: Schematic view of a cloud ensemble with clouds that have a common cloud-base height z_b but different cloud-top heights. The depth of the cloud layer h_c is determined by the highest cloud of the ensemble

and detrainment processes mimic the small-scale mixing processes that dilute the updraft but they do not affect the total cloud mass.

6.5.4.2   Ensemble of Updrafts

The entrainment and detrainment rates that follow from the Asai–Kasahara model for one cloud updraft cannot be used directly in a bulk cloud updraft model (Eq. (6.64)), where exchange rates representative for the whole cloud ensemble are required. To estimate such bulk entrainment and detrainment rates, information on the cloud updraft size distribution is required. In the case of surface-driven convection, it is a reasonable assumption that all clouds have a common cloud-base height z_b, as sketched in Fig. 6.14. The highest cloud has a vertical extent h_c and defines the top of the cloud layer $z_t = z_b + h_c$. If furthermore a constant aspect ratio γ between the vertical extent h of each individual cloud updraft and its radius R is assumed

$$h(R) \approx \gamma R, \tag{6.72}$$

it is clear that with increasing height there are progressively fewer smaller clouds so that the entrainment rates are more dominated by larger clouds that have lower turbulent entrainment rates.

More quantitatively, a bulk fractional turbulent entrainment for an ensemble of cloud updrafts with different radii R at a height z can be written as

$$\bar{\varepsilon}_{\mathrm{turb}}(z) = \frac{\int_{R_{\min}(z)}^{\infty} N(R)\, \mathcal{M}_{\mathrm{u}}(R)\, \epsilon(R)\, dR}{\int_{R_{\min}(z)}^{\infty} N(R)\, \mathcal{M}_{\mathrm{u}}(R)\, dR}, \tag{6.73}$$

with $N(R)$, the number of cloud updrafts with a radius between R and $R + dR$ and $R_{\min}(z)$, the radius of the smallest cloud updrafts that still can reach a height z. In general, the cloud size number decreases with cloud size. To calculate the bulk fractional entrainment rate,

we assume a number density that decreases exponentially with the radius

$$N(R)\, dR \approx N_0 \exp\left(-R/R_0\right) dR, \tag{6.74}$$

where the radius R_0 sets the typical horizontal scale of the updraft ensemble $\gamma R_0 \simeq \frac{1}{2} h_c$. Evaluating Eq. (6.73), for an ensemble of cloud updrafts with individual entrainment rates given by Eq. (6.71a) and assuming that the mass flux of a plume with a radius R scales as $\mathcal{M}_{\mathrm{u}}(R) \sim R^2$, gives after some algebra

$$\bar{\varepsilon}_{\mathrm{turb}}(z_b) \approx 0.9 \frac{2\eta}{R_0} \qquad \bar{\varepsilon}(z_t)_{\mathrm{turb}} \approx 0.5 \frac{2\eta}{R_0}. \tag{6.75}$$

Two conclusions can be drawn from Eq. (6.75). First, the typical fractional entrainment rate for a cloud ensemble that extends over a cloud layer with a depth of h_c is of the order h_c^{-1}, in reasonable agreement with LES analyses and observations that suggest typical values of the bulk fractional entrainment of $O(10^{-3}\ \mathrm{m}^{-1})$ for shallow cumulus convection with a typical cloud depth of $1\ \mathrm{km}$ and $O(10^{-4}\ \mathrm{m}^{-1})$ at altitudes at $10\ \mathrm{km}$ or more in the presence of deep convection (de Rooy et al., 2013). Second, within a cloud layer the bulk fractional entrainment decreases between cloud base and cloud top with a factor of 2. This is due to the aforementioned limited capability of smaller clouds to grow deep.

Because turbulent detrainment rates are equal to the turbulent entrainment rates, the shape of the mass flux profile is purely determined by the dynamical bulk entrainment and detrainment rates, as can be concluded from inspecting Eq. (6.70). Using the the definition of the mass flux (Eq. (6.53)), the fractional change of the mass flux with height can be divided into a contribution due to the fractional changes of the cloudy updraft fraction a_{u} and fractional changes of the updraft momentum ρw_{u}

$$\frac{1}{\mathcal{M}_{\mathrm{u}}} \partial_z \mathcal{M}_{\mathrm{u}} = \frac{1}{a_{\mathrm{u}}} \partial_z a_{\mathrm{u}} + \frac{1}{\rho w_{\mathrm{u}}} \partial_z \rho w_{\mathrm{u}}. \tag{6.76}$$

The first term on the right-hand side results from the outflows at the top of the individual cloud updrafts and can therefore be interpreted as dynamical detrainment. Because in our simple cumulus updraft model the cloud updraft fraction decreases monotonously, the bulk dynamical entrainment and detrainment rates can be written as

$$\bar{\varepsilon}_{\mathrm{dyn}} = H(-\bar{u}_b) \frac{1}{w_{\mathrm{u}}} \frac{\partial w_{\mathrm{u}}}{\partial z} \tag{6.77a}$$

$$\bar{\delta}_{\mathrm{dyn}} = -H(\bar{u}_b) \frac{1}{w_{\mathrm{u}}} \frac{\partial w_{\mathrm{u}}}{\partial z} - \frac{1}{a_{\mathrm{u}}} \partial_z a_{\mathrm{u}}. \tag{6.77b}$$

LES analyses show that the main contribution to the bulk organised fractional detrainment comes from the second term in Eq. (6.77b) (Böing et al., 2012). Using the exponential cloud updraft number density (Eq. (6.74)) and the constant aspect ratio assumption, this fractional decrease in cloud updraft area scales near cloud base as

$$\frac{1}{a_u}\partial_z a_u \sim -\frac{1}{h_c}\left(\frac{z - z_b}{h_c}\right)^2. \qquad (6.78)$$

From this, the following qualitative picture emerges for the description of the bulk entrainment and detrainment for a cumulus ensemble. At the cloud base, entraining inflow occurs through unstable and moist updrafts from the subcloud layer. These updrafts continue their ride in the cumulus ensemble where their kinetic energy is further fuelled by condensational heating but is slowed down at the same time by the lateral turbulent mixing. The smaller clouds are the first that run out of kinetic energy by entrainment and produce outflows, thereby giving way to dynamical detrainment at relative low altitudes, followed by next class of cloud updrafts with a larger size and a larger vertical extent and so on until finally the largest and most vigorous clouds detrain at the maximum height the cloud ensemble can reach.

One important drawback of the results in Eqs (6.75) and (6.78) is that it they are based once a prescribed shape of the cloud updraft number density as given by Eq. (6.74). In reality, $N(R)$ is dependent on the environmental conditions such as the vertical stability and the environmental relative humidity. One of the few models for entrainment and detrainment parameterisations that incorporate such a dependency is the buoyancy sorting model proposed by Kain and Frisch (1990). This model assumes that the periphery of the cloudy updrafts consists of air parcels that have distinct fractions χ of environmental air and $(1 - \chi)$ of cloud updraft air as a result of lateral mixing. The basic idea of buoyancy sorting is that positive buoyant mixtures will remain part of the updraft and will contribute to the entrainment process while negative buoyant mixtures will be detrained. Fig. 6.15, similar but not equal to Fig. 2.8, shows the density temperature as a function of the fraction of environmental air χ for the case where cloud updraft air is positively buoyant with respect to the environmental air. As explained in Chapter 2, mixtures can become negative buoyant with respect the environment, due to the evaporative cooling. The most negative buoyant parcels are those for which all the liquid water has just been evaporated.

The critical fraction $\chi_{\rm crit}$ is defined as the fraction of environmental air needed to make the mixture just neutrally buoyant. Within the buoyancy sorting framework, all mixtures for which $\chi > \chi_{\rm crit}$ are assumed to detrain. This implies that detrainment rates increase with decreasing values of $\chi_{\rm crit}$ which has been confirmed by LES analyses (Böing et al., 2012). The relationship of $\chi_{\rm crit}$ with the environmental conditions can be further

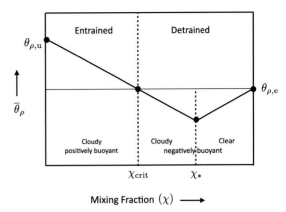

FIGURE 6.15: Density temperature of mixtures with a fraction χ of environmental air and a fraction $(1 - \chi)$ of cloudy air.

understood by expressing $\chi_{\rm crit}$ in terms of the differences between the updraft and the environmental values, that is, $\Delta\theta_\rho \equiv \theta_{\rho,u} - \theta_{\rho,e}$ as

$$\chi_{\rm crit} = \frac{\Delta\theta_\rho}{\beta\Delta\theta_l + (\beta - \alpha)\dfrac{l_v}{c_p\Pi}\Delta q_t}, \qquad (6.79)$$

where Π denotes the exner function and typical values for the coefficients are $\alpha \simeq 0.12$ and $\beta \simeq 0.4$.[10] Inspection of Eq. (6.79) demonstrates that $\chi_{\rm crit}$ increases for a more unstable atmosphere, as measured by $\Delta\theta_\rho$, and for a more humid atmosphere, as measured by Δq_t. Small values of $\chi_{\rm crit}$ correspond to marginally buoyant, often relatively small, clouds rising in a dry environment. Under such conditions, the mass flux will decrease rapidly, consistently with large values of detrainment and a large decrease of the cloud fraction, consistent with Eq. (6.77b). Vice versa, a more humid and unstable atmosphere, described by larger values of $\chi_{\rm crit}$, will support stronger cloud updrafts that have a larger vertical extent with smaller detrainment rates and a smaller decrease of the cloud updraft fraction with height. The buoyancy sorting model also suggests that larger values of $\chi_{\rm crit}$ would promote larger fractional entrainment rates but LES analyses do not support such relationships (Dawe and Austin, 2011; Böing et al., 2012). The physical relationships between entrainment and the mean environmental conditions is therefore still far from being well-established.

Besides dependencies on the mean environmental conditions, entrainment and detrainment conditions may also vary considerably due to the spatial organisation

[10] A derivation of Eq. (6.79) can be found in the appendix of de Rooy and Siebesma (2008).

of convective clouds. For instance, during the transition from shallow clouds to deeper precipitation clouds, cold pools are formed that promote spatial organisation of clouds, yielding wider clouds which entrain less and are therefore capable of reaching higher cloud tops. How to incorporate the effect of spatial organisation into cumulus parameterisation is still a scientific challenge.

6.5.5   Cloud Updraft Vertical Velocity

The bulk mass flux concepts discussed so far are relying on a cloud updraft model for heat and moisture only. There are however good reasons to extend the cloud updraft model with a vertical velocity equation. First of all, a vertical velocity equation allows for a more accurate determination of cloud-top height, as it determines how far a cloud updraft can overshoot its LNB. Second, as discussed in Section 6.5.3, it can be used to estimate when dry thermals can overcome the CIN layer and trigger cumulus convection. Finally, the vertical velocity equation can be used to determine the mass flux directly through its definition, that is, $M \equiv a_u w_u \rho$, provided that an estimate of the cloud updraft a_u is available.

A widely used parameterised form of the vertical updraft velocity for a cloudy updraft ensemble is

$$\frac{1}{2}\partial_z w_u^2 = aB_u - b\,\varepsilon\,w_u^2, \quad B_u \equiv \frac{g}{T_0}\left(\theta_{\rho,u} - \overline{\theta_\rho}\right),$$
(6.80)

where the constants a and b are tunable prefactors that mainly aim to incorporate the pressure gradient term which is not taken into account explicitly. Eq. (6.80) essentially describes the efficiency of the transformation from potential energy, as measured by the buoyancy term B_u, into kinetic energy in the form of the vertical updraft velocity of the updraft ensemble. Although the precise physical basis of Eq. (6.80) is debatable, it is used in many cumulus convection parameterisations. LES analyses of shallow cumulus convection show that there is a whole range of combinations (a, b) that give equally good estimates for determining the same vertical velocity profile (de Roode et al., 2012). The reason for this is that the same LES analyses show that the entrainment term and the buoyancy term scale with each other as

$$\varepsilon\,w_u^2 \simeq \gamma B_u \quad \gamma \approx 0.4\,.$$
(6.81)

Substituting Eq. (6.81) in Eq. (6.80) gives

$$\frac{1}{2}\partial_z w_u^2 = \eta B_u \quad \eta \approx 1/3,$$
(6.82)

where the numerical value η has been derived from LES analyses and can be interpreted as an efficiency factor

that measures the fraction of CAPE that is converted into kinetic energy. Apparently, roughly one-third of the buoyancy production can be utilised to produce kinetic energy for the updrafts which explains why Eq. 2.106 provides such a strong overestimation of the maximum vertical velocity. The relations (6.81) and (6.82) span up a range of combinations (a, b) that provide equal optimal values in Eq. (6.80)

$$a = \gamma b + \eta.$$
(6.83)

Many of the used combination (a, b) used in the literature for parameterisations of Eq. (6.80) fall roughly on the line provided by Eq. (6.83) which explains the wide range of used values of a and b. Note that Eq. (6.82) corresponds to the simplest combination $(a, b) = (1/3, 0)$.

6.5.6   Microphysics in Cumulus Parameterisations

In a bulk mass flux scheme the treatment of cloud microphysics has to be kept very simple, because the mass flux scheme does not provide sufficient information, for example, on updraft velocities of the individual updrafts, for a more detailed treatment of the microphysical processes. Here the elegance of the bulk mass flux scheme, which avoids any explicit assumptions on the individual updraft structures and velocities, leads to a disadvantage in the treatment of microphysical processes. The three main parts of the microphysics in a cumulus parameterisation are the assumption on the glaciation of the clouds, which is important for the latent heat release, the formation of precipitation from the cloud condensate and the evaporation of rain in the downdraft. The latter is crucial not only for the amount of surface precipitation but also for the formation of cold pools (insofar as model has sufficient resolution or the scheme includes a cold pool parameterisation).

6.5.6.1   *Glaciation of Convective Updrafts*

Convective updrafts are to a large extent mixed-phase regions, that is, parts of the cloud where liquid water in the form of cloud and rain drops co-exists with various forms of ice like frozen drops, small crystals, but also rimed snow, graupel and even hail. In a convection parameterisation the complicated mix of cloud particles is described by a simple statistical relation that quantifies the fraction of liquid water of the total condensate written as

$$\alpha_1 = \frac{q_1}{q_1 + q_i} = \begin{cases} 0, & T < T_{ice} \\ \left(\dfrac{T - T_{liq}}{T_{liq} - T_{ice}}\right)^{\xi_1}, & T_{ice} \le T \le T_{liq}\,, \\ 1, & T > T_{liq} \end{cases}$$
(6.84)

where T_{liq} and T_{ice} are temperature values that imply a fully liquid or fully glaciated cloud, that is, a lower bound on T_{ice} is the homogeneous freezing threshold at about 237 K. The exponent ξ_1 is most often set to 1 for a linear relation, but some schemes use a quadratic relation which provides a better fit to observations. This function is then used, for example, in the saturation adjustment to interpolate the saturation pressure between liquid and ice. By construction, such an approach will only capture the average behaviour as it is observed and then expressed by the empirical relation. Actual deep convective updrafts may deviate strongly from this average behaviour.

6.5.6.2 Formation of Precipitation in Convective Updrafts

Convective updrafts contain larger amounts of condensate exceeding 1 g/kg. For such conditions, the formation of precipitation by collision-coalescence processes is quite rapid and rain or graupel form usually at heights of a few kilometres. Nevertheless, cloud droplets can exist in convective updrafts up to 7–8 km height where they freeze homogeneously. Once rain or graupel has formed those particles grow rapidly by accretion or riming, that is, by collecting the small cloud droplet. In convection parameterisation, this complicated chain of microphysical growth mechanisms is described by very simple formulae. Common is a formulation for the precipitation generation G following Sundqvist

$$G = c_0 \frac{M_u q_{l,u}}{\rho w_u} \left[1 - \exp\left(-(q_{l,u}/q_{l,crit})^2 \right) \right], \qquad (6.85)$$

where c_0 is a coefficient of order 10^{-3} s^{-1}. Precipitation is only generated if the threshold $q_{l,crit}$ is exceeded, that is, $q_{l,u} > q_{l,crit}$. This threshold is usually chosen between 0.3 g/kg and 0.5 g/kg and is a quite common tuning parameter in climate models. Alternatively, many schemes apply the even simpler Kessler-type conversion rate

$$G = c_0 (q_{l,u} - q_{l,crit}) \qquad (6.86)$$

with similar values for c_0 and $q_{l,crit}$ as in the Sundqvist-type scheme. The precipitation formation in convective updrafts is a special case of the so-called autoconversion problem, which is discussed later in Section 6.7.5, but due to the simplicity of the convection parameterisations it is less well defined as convection schemes often ignore the accretion or riming stage and everything is shuffled together into a single conversion rate.

6.5.6.3 Evaporation of Convective Rain

The evaporation of rain in the sub-cloud layer or in the downdrafts depends on the size distribution of the rain

drops, that is, on their size, fall speed and so on, but in addition the spatial distribution is important. Does the rain cover only a small fraction of the grid box and the local rain rate is therefore more intense? Is the spatial distribution of moisture expressed by $\mathrm{RH} \equiv q_v/q_s$ homogeneous or does the rain fall into a moister or drier region? This combination of microphysics and sub-grid variability leads to a considerable uncertainty in the formulation of evaporation rates in convection parameterisations. The evaporation rate of convective rain depends on saturation deficit $q_s - q_v$ and the rain rate R

$$E_{conv,rain} = c_{evap} A_{conv} (RH_{crit} q_s - q_v) R^{\xi_2}, \qquad (6.87)$$

with $\xi_2 \approx 0.5$, the area fraction of convective rain A_{conv} and c_{evap} a coefficient that includes mostly microphysical parameters. Most importantly, this evaporation rate is usually only applied if the grid box mean RH is below a certain threshold RH_{crit} (e.g., 90% over sea and 70% over land), because for moist environments it is assumed that the rain-covered fraction of the grid box is already at water saturation. These thresholds related to assumptions about sub-grid variability are often more important than the details of the (microphysical) relation.

6.5.7 Closures of Moist Convection

The cloud updraft model and its parameterisation discussed in the previous subsections provides the shape of the mass flux profile and the values of the updraft fields φ. To complete the cumulus convection parameterisation, the mass flux M_b and the fields φ_b at the cloud base are required. Finding a good estimate for M_b is considered to be the principal closure of convection since it determines the overall strength of the convective fluxes. It therefore requires understanding of the relation between the intensity of convection and the properties of the environment. Two classes of closures can be distinguished: one based on convective quasi-equilibrium and another based on boundary-layer quasi-equilibrium.

6.5.7.1 Convective Quasi-equilibrium

Convective quasi-equilibrium assumes that cumulus convection stabilises the atmosphere by convective mixing at a rate much faster than the destabilisation through radiative cooling and other non-convective processes. Quasi-equilibrium can be phrased most conveniently in terms of the CAPE, defined in Eq. (2.104), as the buoyancy excess of an adiabatic parcel lifted from near the surface and integrated from the LFC to its LNB

$$\mathcal{A} = \beta \int_{z_f}^{z_n} (T_{\rho,p} - \overline{T_\rho}) \, dz, \qquad (6.88)$$

with the buoyancy parameter $\beta \equiv g/T_{\rho,0}$. The time rate of change of $\mathcal{A}$ can formally be decomposed into a contribution due to convective processes and one due to all the other non-convective processes, including large-scale processes

$$\partial_t \mathcal{A} = (\partial_t \mathcal{A})_{\text{conv}} + (\partial_t \mathcal{A})_{\text{ls}}. \quad (6.89)$$

Quasi-equilibrium assumes that the convective response counteracts changes due to $(\partial_t \mathcal{A})_{\text{ls}}$ fast enough, such that

$$|\partial_t \mathcal{A}| \ll |(\partial_t \mathcal{A})_{\text{ls}}|, \quad (6.90)$$

while strict quasi-equilibrium or convective equilibrium would impose $\partial_t \mathcal{A} = 0$, not allowing for any change in $\mathcal{A}$. Quasi-equilibrium suggests that convection consumes the CAPE as

$$(\partial_t \mathcal{A})_{\text{conv}} = -\frac{\mathcal{A}}{\tau_{\text{adj}}}, \quad (6.91)$$

with a convective adjustment time τ_{adj} of several hours, much shorter than the time τ_{ls} it takes by the large-scale processes to produce the CAPE.

The CAPE adjustment closure (Eq. (6.91)) can be used to parameterise the cloud-base mass flux. Eq. (6.66) shows how convection influences the environment by compensating subsidence and detrainment. Assuming that compensating subsidence is the dominant mechanism, this implies for the convective tendency of $\overline{\theta}_\rho$

$$(\partial_t \overline{\theta}_\rho)_{\text{conv}} \simeq \frac{M_u}{\rho} \partial_z \overline{\theta}_\rho, \quad (6.92)$$

which can be rewritten in terms of a convective tendency of CAPE

$$(\partial_t \mathcal{A})_{\text{conv}} \simeq -\beta M_{u,b} \int_{z_b}^{z_n} m \, \partial_z \overline{\theta}_\rho dz, \quad (6.93)$$

where m is a normalized mass flux

$$m(z) \equiv \frac{M_u}{M_{u,b}}. \quad (6.94)$$

Substituting Eq. (6.93) in the CAPE adjustment closure, Eq. (6.91), gives for the cloud-base mass flux

$$M_{u,b} = \frac{\mathcal{A}}{\tau_{\text{adj}}} \left(\beta \int_{z_b}^{z_t} m \, \partial_z \overline{\theta}_\rho dz \right)^{-1}, \quad (6.95)$$

which can be interpreted as the cloud-base mass flux that is needed to exhaust the existing CAPE in an adjustment time τ_{adj} by compensating subsidence (see Fig. 6.16a). More sophisticated versions of the closure, Eq. (6.91), are possible in which diluted CAPE is used based on an entraining parcel and where the effect of detrainment on the environment is taken into account. The basic principle however remains the same.

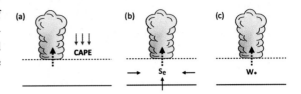

FIGURE 6.16: schematic representation of (a) CAPE closure based on convective quasi-equilibrium, (b) boundary-layer quasi-equilibrium and (c) relaxed boundary layer quasi-equilibrium.

6.5.7.2 Boundary Layer Quasi-equilibrium

Boundary-layer equilibrium closure is based on observations that the subcloud layer, especially over the subtropical oceans, is in an approximate steady state. Such an equilibrium can be enforced by assuming that the venting of heat and moisture out of the boundary layer by cumulus convection is controlled by the moist static energy budget in the subcloud layer. Boundary-layer equilibrium in the subcloud layer requires

$$M_u(s_{e,u} - \overline{s}_e)|_b = \int_0^{z_b} \left(\rho \, \partial_t \overline{s}_{e,\text{ls}} - \partial_z \rho \overline{w's'_e} \right) dz. \quad (6.96)$$

It states the net turbulent flux and horizontal large-scale advection of moist static energy s_e entering the subcloud layer and the radiative cooling is exactly balanced by its convective flux out of the boundary layer through the cloud base (see Fig. 6.16b). Therefore, in the presence of cumulus convection, the vertical partitioning of s_e in the boundary layer can vary but not its integral value. In that respect, equilibrium boundary-layer closure is rather restrictive in that they do not allow for any integral change of moist static energy in the subcloud layer in the presence of cumulus convection. A similar formulation as Eq. (6.96) where the specific humidity q_v is used instead of s_e exists as well and is usually referred to as moist convergence closure.

A more relaxed version of the boundary-layer equilibrium closure relates the mass flux activity at cloud base to the turbulent kinetic energy. This idea is based on observations that cumulus clouds are often rooted deeply in the subcloud layer as dry thermals. Accordingly, the kinetic energy and distribution of the condensated thermals at the cloud base must bear the fingerprint of the dry turbulence convection in the subcloud layer. This links the physics of cumulus mass transport to subcloud-layer dry convection, which forms the conceptual basis for this closure. In its simplest form this can be done by relating the $M_{u,b}$ to the Deardorff convective vertical velocity scale w_* (see also Fig. 6.16c)

$$M_{u,b} = \alpha w_*. \quad (6.97)$$

LES of shallow cumulus convection supports a linear relation between $\mathcal{M}_{u,b}$ and w_* with a proportionality factor $\alpha \simeq 0.03$. When comparing Eq. (6.97) to the definition of the mass flux, it is tempting to interpret α as a typical value for the cloud updraft fraction $a_{u,b}$ at the cloud base. For steady-state shallow cumulus convection typical numbers are indeed $a_{u,b} \simeq 0.03$. In addition, if independent estimates for $a_{u,b}$ and $w_{u,b}$ are available, for instance, by an updraft model in the subcloud layer, this closure can be made even more flexible by using the definition of the mass flux, Eq. (6.53), at the cloud base.

The closure Eq. (6.97) self-regulates the venting of moist static energy out of the boundary layer. To demonstrate this, consider the continuity equation in the absence of large-scale advection. Integration from the surface to the cloud-base height gives

$$\partial_t z_i = w_e - \mathcal{M}_{u,b}. \tag{6.98}$$

In the presence of cumulus convection, $M_{u,b}$ is typically much larger than the dry top entrainment w_e leading to a net venting of air out of the boundary layer so that the inversion height decreases up to a point that z_i is substantially lower than the level of condensation z_b. In that situation, thermals in the boundary layer are not able anymore to reach the condensation level so that cumulus convection will not be triggered and the transport of mass by cumulus convection out of the boundary layer will be inhibited. From that moment on, the boundary layer can subsequently slowly grow again by top entrainment until the inversion height will coincide with the LCL, so that cumulus convection will be triggered again. This self-regulating venting mechanism will even become stronger and more realistic if an interactive updraft cloud fraction $a_{u,b}$ can serve as an extra valve.

6.5.7.3 *Discussion*

Numerous observational studies have demonstrated that convective quasi-equilibrium, defined by Eq. (6.90), does not hold over land on sub-diurnal times. Moreover, observed correlations between CAPE and convective precipitation are often surprisingly low in sharp contrast with the CAPE based closure (Eq. (6.95)). To explore convective quasi-equilibrium in more detail, it is useful to to decompose changes in CAPE (see Eq. (6.88)) into contributions due to changes of the environmental density temperature $\overline{T}_\rho$ and due to changes of the parcel density temperature $T_{\rho,p}$

$$\partial_t \mathcal{A} = \partial_t \mathcal{A}^P + \partial_t \mathcal{A}^e. \tag{6.99}$$

Because $T_{\rho,p}$ depends solely on the near–surface atmospheric conditions where the parcel is released, $\partial_t \mathcal{A}^P$ can

be thought of as the result of changes of T_ρ near the surface. The tendency $\partial_t \mathcal{A}^e$, on the other hand, is due to changes of environmental density temperature $\overline{\theta}_\rho$ in the free troposphere. It is therefore reasonable to refer to the first term of the right-hand side of Eq. (6.99) as a boundary-layer contribution and to the latter as a free tropospheric contribution. Combining Eq. (6.99) with Eq. (6.89), the decomposition into convective and non-convective processes, gives

$$\partial_t \mathcal{A} = \partial_t \mathcal{A}^P_{\text{conv}} + \partial_t \mathcal{A}^P_{\text{ls}} + \partial_t \mathcal{A}^e_{\text{conv}} + \partial_t \mathcal{A}^e_{\text{ls}}. \tag{6.100}$$

The first term $\partial_t \mathcal{A}^P_{\text{conv}}$ refers to changes in the boundary layer by convective processes such as convective downdrafts. The second term $\partial_t \mathcal{A}^P_{\text{ls}}$ refers to changes in the boundary layer by non-convective processes such surface fluxes and advection. The third term $\partial_t \mathcal{A}^e_{\text{conv}}$ refers to changes in the free troposphere by convective processes such as compensating subsidence and detrainment while the last term $\partial_t \mathcal{A}^e_{\text{ls}}$ represents changes in the free troposphere by non-convective processes such as radiative cooling and advection.

Analyses of the three terms in Eq. (6.99) using data of the Tropical Ocean and Global Atmosphere/Coupled Ocean Atmosphere Response Experiment (TOGA COARE) and from Southern Great Plains (SGP) of the Atmospheric Radiation Measurement (ARM) program show that variations in CAPE $\partial_t A$ are primarily controlled by boundary-layer variations $\partial_t A^P$ while changes of CAPE due to free tropospheric processes $\partial_t A^e$ are relatively small (Zhang, 2002, 2003). These analyses suggest that the accumulation of CAPE by boundary-layer warming is so fast that deep convection cannot consume it on sub-diurnal timescales. These observations indicate that, rather than convective quasi-equilibrium (Eq. (6.90)), a *modified* quasi-equilibrium holds

$$\partial_t A \simeq \partial_t A^P \quad \Rightarrow \quad \partial_t A^e = \partial_t A^e_{\text{conv}} + \partial_t A^e_{\text{ls}} \simeq 0 \tag{6.101}$$

This modified quasi-equilibrium indicates that there is a balance between non-convective processes that destabilise the free atmosphere and cumulus convection undoing these instabilities while the boundary layer instabilities by surface fluxes can develop unrestricted on these short timescales. Obeying this modified quasi-equilibrium is probably key to representing the diurnal cycle of cumulus convection.

6.6 Cloud Parameterisation

6.6.1 Introduction

Clouds manifest themselves in a tremendous richness of forms and shapes which is one of the reasons why they

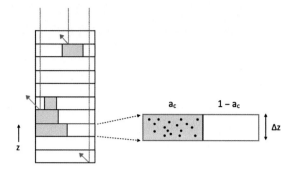

FIGURE 6.17: Schematics of the representation of the cloud macroscopic properties and their geometry in a vertical column of a large-scale model. Each vertical level of the column is characterised by a horizontal area fraction a_c that is considered to be cloudy. For the cloud-radiation interaction, additional assumptions on the cloud geometry are required to determine the overlap of the horizontal area fractions of the various vertical levels in the column to determine the total cloud cover.

keep fascinating us every day. One way to demonstrate this is by considering their global horizontal cloud size distribution. Observations show that this is well described by

$$N(R) \sim R^{-\beta} \exp\left[-(R/R_*)^2\right] \, , \, \beta \simeq 1.66, \quad (6.102)$$

where the cut-off length-scale R_* is approximately 2,000 km and cloud sizes range from 10^2 m to 10^7 m (Wood and Field, 2011). This example not only shows a scale invariance of cloud sizes over more than 5 orders of magnitude but also demonstrates that 50% of the global cloud cover originates from clouds with a horizontal size smaller than 200 km. This implies that a large portion of the global cloud fraction is due to clouds with a spatial scale smaller than the resolution of a typical climate model. Recognising the importance of clouds for radiation, it is clear that large-scale models must consider the sub-grid cloud effects. This is one of the main tasks of cloud parameterisation.

As a minimum requirement, a cloud parameterisation should specify (i) the horizontal fraction a_c of a grid box occupied by clouds, (ii) the associated cloud liquid water $q_{l,c}$ and cloud ice $q_{i,c}$ in the cloudy part and the associated thermodynamic effects and (iii) the overlap of these cloud fractions in a vertical column as schematically visualised in Fig. 6.17. We will refer to these properties as the cloud *macrophysical* properties. Note that at this point already two important approximations have been made. Neither sub-grid variability of the cloud fraction a_c in the vertical part of a grid box nor sub-grid variability of the condensed water q_c within the cloudy part is assumed.

The basic cloud parameterisation framework is discussed in Section 6.6.2, followed by three different cloud

parameterisations that utilise this basic framework at different levels of complexity. Section 6.6.6 describes how the overlap between different vertical levels in a model column can be parameterised and how horizontal cloud inhomogeneity can be taken into account.

In addition, a cloud *microphysical* parameterisation is required that describes the cloud droplet number concentration, the phase of the cloud condensate (liquid, ice, mixed phase) and the autoconversion rate from the condensed phase to precipitation in terms of rain, snow and graupel. Cloud microphysical parameterisations will be discussed in Section 6.7.

6.6.2   Basic Concepts of a Cloud Parameterisation

Before setting up a framework for a cloud parameterisation that addresses all of the tasks described in the introduction, we will start with the simple case of a high-resolution model with a cloud-resolving resolution of $O(100\,\text{m})$. In that case, there is no urgent need to take subgrid variability into account and it is sufficient to use the grid box mean values of specific humidity and temperature. The grid box will be completely cloud free if the total water specific humidity q_t is below its saturation value, or completely cloudy in the case of supersaturation.

For a high-resolution model using moist conserved variables such as q_t and θ_ℓ, a grid box is assumed to be cloudy if supersaturation is diagnosed, that is, $\overline{q}_t > q_s(p, \overline{T})$. In that case, the cloud fraction $a_c = 1$ and the cloud condensate $\overline{q}_c$ may be found by condensing water vapour until the grid box is just saturated. In practice, the testing on supersaturation is done on $q_{sl} \equiv q_s(p, \overline{T}_\ell)$ with $T_\ell \equiv \pi \theta_\ell$, because the temperature T is the variable that still needs to be determined.

Such an 'all-or-nothing' or saturation adjustment scheme can be summarised as[11]

$$\begin{cases} a_c = 0, \\ \overline{q}_c = 0 \end{cases} \text{if} \quad \overline{q}_t - \overline{q}_{sl} < 0$$

$$\begin{cases} a_c = 1, \\ \overline{q}_c = \alpha(\overline{q}_t - \overline{q}_{sl}) \end{cases} \text{if} \quad \overline{q}_t - \overline{q}_{sl} > 0, \quad (6.103)$$

with $\alpha \equiv \left(1 + \left(\ell_v/c_p\right)\beta\right)^{-1}$ and $\beta = \partial_T q_s(\overline{T}_\ell)$.

---

[11] The reduction factor α for the condensed water can be derived by approximating the saturation specific humidity at temperature T by a Taylor expansion around T_ℓ, that is, $q_s \approx q_{sl} + (T - T_\ell)\beta$ and using this in the expression of condensed water in case of saturation adjustment $q_c = q_t - q_s$. The reduction factor corrects for the fact that q_{sl} underestimates the actual saturation specific humidity.

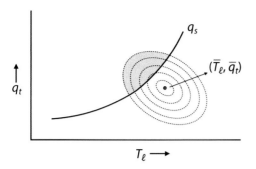

FIGURE 6.18: Schematic view of the joint probability distribution of T_ℓ and q_t along with the saturation curve.

If non-conserved variables (q_v, q_c, T) are used as prognostic variables, the same procedure can be be applied to (q_v, T) to estimate the incremental condensate. The all-or-nothing scheme (Eq. (6.103)) is widely used in LES. It is also used in the cloud updraft models of cumulus parameterisations to calculate the condensational heating rates discussed in Section 6.5.2.

In reality, grid boxes of large-scale models have a typical size of 10–100 km. As a result, both temperature and specific humidity can deviate locally substantially from their grid box mean values. If the mean RH is low, less than typically 0.6, these fluctuations will not cause any local supersaturation and the whole grid box will be cloud free. If the RH is approaching unity there will be regions in the grid box where q_t will exceeds its saturation value, so that a non-zero cloud fraction results, even before the mean RH exceeds unity. As illustrated in Fig. 6.18, the cloud fraction a_c can be interpreted as the part of the joint PDF $P(q_t, T_\ell)$ that is supersaturated. Likewise, the amount of condensed water q_c can be interpreted as the mean supersaturation. Both a_c and $\bar{q}_c$ can be written exclusively in terms of the joint probability distribution $P(q_t, T_\ell)$ as

$$a_c = \int_0^\infty \int_0^\infty H(q_t - q_s)P(q_t, T_\ell)\, dq_t\, dT_\ell$$

$$\bar{q}_c = \int_0^\infty \int_0^\infty (q_t - q_s)H(q_t - q_s)P(q_t, T_\ell)\, dq_t\, dT_\ell,$$

(6.104)

where $q_s = q_s(p, T)$ and H denotes the Heaviside function which is probing the part of the integrand that is supersaturated. Inspection of Fig. 6.18 shows that the degree of supersaturation is equal to the distance to the saturation curve s

$$s = q_t - q_s(p, T),$$

(6.105)

a variable, closely related to the condensed water content via $q_c = H(s)s$. The main advantage of introducing the

saturation deficit s is that it allows us to rewrite Eq. (6.104) into a one-parameter form

$$a_c = \int_0^\infty P(s)\, ds \quad , \quad \bar{q}_c = \int_0^\infty sP(s)\, ds,$$

(6.106)

thereby simplifying the cloud parameterisation problem into a single parameter problem. Many shapes for $P(s)$ have been suggested in the literature, ranging from simple symmetric unimodal shapes such as uniform, triangular or Gaussian, to more complex asymmetric bimodal shapes such as double Gaussian distribution (Tompkins, 2002). Which shape is the most appropriate depends in part on how well one is capable of constraining the shape parameters of the PDF. A double Gaussian distribution can represent the observed PDF of s more realistically than a simpler single Gaussian distribution (see Fig. 6.12), but the drawback is that it also requires parameterisation for three additional shape parameters. So again, the message here is that increased complexity should only be introduced as long as it can be well constrained.

For a single Gaussian distribution, the standard deviation of s is the only free parameter. In that case, it is convenient to non-dimensionalise s by rescaling it by its standard deviation σ_s

$$t \equiv \frac{s}{\sigma_s},$$

(6.107)

so that the cloud fraction is uniquely related to the mean non-dimensional saturation deficit $\bar{t}$ as

$$a_c = 1/2 \left(1 + \text{erf}(\bar{t}/\sqrt{2})\right).$$

(6.108)

If a parameterisation of σ_s can be provided, Eq. (6.108) can be used to estimate the associated cloud fraction. A similar result can be obtained for the cloud condensate $\bar{q}_c$

$$\frac{\bar{q}_c}{\sigma_s} = a_c\bar{t} + \frac{1}{\sqrt{2\pi}} \exp\left(-\bar{t}^2/2\right).$$

(6.109)

Fig. 6.19 shows aircraft observations of cloud fraction versus $\bar{t}$ for a wide range of different boundary-layer cloud types, along with the Gaussian estimates (6.108) and (6.109). Given the fact that a single Gaussian PDF is an overly simplified representation of the observed PDFs, the results are remarkably good.

A number of conclusions can be made on the basis of these results. First, in the limit of zero variance of s, Eqs. (6.108) and (6.109) reduce to the all-or-nothing limit (6.103). This implies that the cloud parameterisation (Eqs (6.108–6.109)) can be made scale-aware by designing resolution dependent formulations that diagnose smaller values of σ_s with increasing resolution. Second, inspection of Fig. 6.19 learns that for $\bar{t} < -1$, the single Gaussian

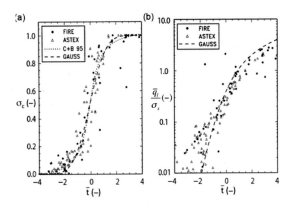

FIGURE 6.19: (a) Cloud fraction a_c and (b) normalized condensed water as a function of the normalised saturation deficit $\bar{t}$ from observations from flight legs of 60 km. The dashed line represents the Gaussian solution as given by Eqs (6.108) and (6.109). Figure from Wood et al. (2002). Copyright © 2002 Elsevier

PDF underestimates cloud fraction and cloud condensate. This can be understood when realising that the low cloud fractions in this regime are due to cumulus clouds for which it has been shown in Section 6.5.1 that the PDF is strongly skewed to large negative values of s which is not captured by a single Gaussian PDF.

The standard deviation σ_s takes into account sub-grid fluctuations of both humidity and temperature and cross-correlations. This can be demonstrated by deconstructing σ_s into contributions of sub-grid variations of T_ℓ and q_t

$$\sigma_s = \alpha \left[\overline{q_t'^2} - 2\beta \overline{q_t'T_\ell'} + \beta^2 \overline{T_\ell'^2} \right]^{\frac{1}{2}}, \qquad (6.110)$$

where the parameters α and β are the same as defined in Eq. (6.103). LES analyses of (Eq. (6.110)) reveal that actually all three terms contribute significantly to σ_s. However, this does not imply that temperature fluctuations are equally important as humidity fluctuations. The main contributions, the second and third term on the right-hand side of Eq. (6.110) to σ_s, are due to fluctuations of q_c in T_ℓ, rather than fluctuations of the temperature T, especially in the tropics where gravity waves remove temperature fluctuations on short timescales. So to first order it is the task of a cloud parameterisation scheme to estimate the humidity fluctuations of q_t. There are numerous ways of doing this and the coming subsections are providing pathways for obtaining such estimates in an increasing order of complexity.

6.6.3 Mean State-Based Diagnostic Cloud Schemes

The mean state variable that correlates best with cloud fraction is the RH, defined as $RH \equiv q_v/q_s$, and many

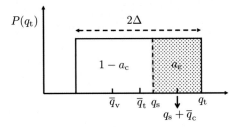

FIGURE 6.20: Simple uniform PDF of total specific humidity q_t characterized by a total width 2Δ.

early cloud schemes are based on relationships between RH and a_c. Even these simple RH schemes are based on an underlying assumed PDF. One of the most used RH schemes uses the relation

$$a_c = 1 - \sqrt{\frac{1 - RH}{1 - RH_{crit}}} \quad RH > RH_{crit}$$

$$a_c = 0 \qquad RH < RH_{crit}, \qquad (6.111)$$

where RH_{crit} has been introduced as the critical RH for which Eq. (6.106) starts to produce a non-zero cloud fraction. The RH scheme (6.111) can be derived from Eq. (6.106) if for the sub-grid variability of q_t a uniform distribution with a width 2Δ is assumed where

$$RH_{crit} = 1 - \frac{\Delta}{q_s}, \qquad (6.112)$$

without any temperature fluctuations as illustrated in Fig. 6.20. Eq. (6.112) shows that a larger sub-grid variability, as indicated by a larger width Δ of the uniform distribution, leads to a smaller RH_{crit} and hence to an earlier onset of cloud formation with lower RH conditions. It also demonstrates that a RH scheme with a fixed prescribed RH_{crit} is equivalent to the assumption that the sub-grid variability of q_t is proportional to q_s.

A fixed prescribed RH_{crit} is too restricted because it assumes that the sub-grid variability only depends on q_s. Many generalisations have been proposed in addition to Eq. (6.111), by making RH_{crit} dependent on height or by adding extra predictors such as the grid mean vertical velocity or the grid mean condensed water content (Quaas, 2012). Another more general pathway, which will be discussed in the next subsection, is to link the sub-grid variability more directly to the physical processes that cause this variability.

6.6.4 Process-Based Diagnostic Cloud Schemes

Vertical transport of moisture is a major source of the subgrid variability. Turbulence and convection transport humid air out of the boundary layer into a dryer troposphere, which is the reason that these vertical mixing processes are a major source of the sub-grid variability of moisture. The Reynolds averaged equations of the second moments, introduced in Chapter 2, establish a formal relationship between transport and sub-grid variability. A simplified version of the variance equation for q_t in which all horizontal transport terms have been neglected can be written as

$$\partial_t \overline{q_t'^2} = -2\overline{w'q_t'}\partial_z\overline{q}_t - \partial_z\overline{w'q_t'^2} - 2\varepsilon_q - (\partial_t\overline{q_t'^2})_{au},$$

(6.113)

where the terms on the right-hand side describe the various sources and sinks of the humidity variability as quantified by its variance. The first term on the right-hand side is a production term due to a turbulent flux in the presence of a humidity gradient, the second term transports variance in the vertical while the third term destroys variance by dissipation. The last term indicates the loss of variance due to autoconversion of q_t to precipitation. Advection of the variance has been omitted for simplicity, but can be added without any problems. The dissipation term is usually parameterised in a similar way as the dissipation term in the TKE Eq. (6.33)

$$\varepsilon_q = c_q \frac{\overline{q_t'^2}}{\tau},$$

(6.114)

where c_q is a dimensionless coefficient. Assuming a simple balance between the production and the parameterised dissipation term (6.114), a diagnostic expression for the variance of q_t follows from Eq. (6.113)

$$\overline{q_t'^2} = \frac{-1}{c_q}\tau\,\overline{w'q_t'}\,\partial_z\overline{q}_t,$$

(6.115)

which can be used to estimate variance production by turbulence and convection. For boundary-layer turbulence, this can be done by making the usual ED approximation for the turbulent flux in Eq. (6.115), so that

$$\left(\overline{q_t'^2}\right)_{bl} = \frac{1}{c_q}K\tau\,(\partial_z\overline{q}_t)^2,$$

(6.116)

for which the usual estimates of the ED K and the dissipation timescale τ in boundary-layer parameterisations may be used.

To estimate the sub-grid variability due to cumulus convection, a similar procedure can be used in which the flux in Eq. (6.113) can be estimated by applying the convective mass flux approximation (6.53) to the flux in Eq. (6.115)

$$\left(\overline{q_t'^2}\right)_{conv} = -\frac{M_u}{\rho c_q}\tau(q_{t,u} - \overline{q}_t)\partial_z\overline{q}_t$$

(6.117)

The dissipation timescale can be estimated as a convective turnover time $\tau \simeq w_{u,*}/h_c$ where h_c is the depth of the cloud layer and an integral cumulus vertical velocity scale $w_{u,*}$ can be estimated as[12]

$$w_{u,*} = \left[\frac{g}{\theta_0}\int_{z_b}^{z_t}\frac{M_b}{\rho_b}(\theta_{\rho,u} - \overline{\theta}_\rho)dz\right]^{1/3}.$$

(6.118)

Both Eqs (6.116) and (6.117) diagnose more variability in regions near the inversion where the humidity gradients are steep thereby making those regions more attractive for cloudiness. Similar relations can be derived for the variance of temperature and for the cross-correlation term. In combination with Eqs (6.110), (6.108) and (6.107) this process-oriented approach offers obviously more flexibility in estimating cloud fraction than the RH schemes discussed in the previous section.

But there are also drawbacks. Most importantly, being a diagnostic relation, Eqs (6.116) and (6.117) only diagnose variance in the presence of turbulence of convection, which leads to strong underestimation of cloud amount during night-time conditions or other situations where turbulence and convection are suppressed. One may include a background variance to mimic memory effects and other contributions due to mesoscale contributions, but the lack of any memory in the variability is a serious shortcoming, certainly in view of the fact that certain cloud types, such as long-lived cirrus fields, may evolve in the absence of turbulence and convection over timescales of hours to days. Therefore, memory effects need to be included. How to do this is the topic of the next section.

Variability in humidity can also result from larger-scale dynamical phenomena, ranging from mesoscale organisation to large-scale convective waves. The contributions of the latter phenomena to the sub-grid variability are difficult to quantify but are obviously dependent on the size of the grid box and expected to decrease for smaller grid boxes, or equivalently, higher resolution.

6.6.5 Prognostic Cloud Schemes

In general, the most natural way to include memory effects into a parameterisation is through introducing an extra prognostic variable. In view of the previous section, it seems a logical choice to use the moisture variance σ_q^2 as a prognostic variable and prognose its value using the variance Eq. (6.113). In this case, the pair $(\overline{q}_t, \sigma_q^2)$ is sufficient to constrain a two-parameter PDF such as a uniform or a Gaussian distribution that can be used to

---

[12] Various different generalisations of the convective velocity scale $w_{u,*}$ in the cumulus layer are possible, as discussed in Grant and Brown (1999).

diagnose the relevant cloud macroscopic variables as the cloud fraction a_c and $\overline{q}_c$.

A more traditional and popular pathway is to choose the pair $(\overline{q}_v, \overline{q}_c)$ as prognostic variables, as already foreshadowed in Eq. (6.2). Under partially cloudy conditions, this pair can equally well be used to constrain a two-parameter PDF of q_t. For instance, in the case of the uniform PDF, the width 2Δ can simply be expressed in terms of q_v and q_c as

$$\Delta = \overline{q}_s + \overline{q}_c - \overline{q}_v$$

so that the cloud fraction a_c and the variance $\overline{q_t'^2}$ can be easily diagnosed, consistent with the assumed uniform PDF.

Using $\overline{q}_c$ as a prognostic variable requires exploration of the sub-grid terms indicated by the right-hand side of Eq. (6.2c)

$$(\partial_t \overline{q}_c)_{\text{sub-grid}} = -\frac{1}{\rho}\partial_z \overline{\rho w' q_c'} + (c - e) - G. \qquad (6.119)$$

The first term of the right-hand side is usually determined by the convection scheme. It includes moistening by detrainment of condensed water and tendencies due to compensating subsidence as expressed in Eq. (6.66) and calculated by a convection scheme. The remaining two terms of Eq. (6.119) are determined by the cloud scheme. The large-scale condensation and evaporation processes of q_c are described in terms of how $q_s(p, T)$ is changing as a result of cooling and warming processes due to advection, radiation and turbulence. In addition, an evaporation term is added that represents the erosion of clouds through mixing of cloudy air with environmental air and is usually chosen to be proportional to the cloud fraction and the saturation deficit in the environment

$$e_{\text{erosion}} = a_c \frac{\overline{q}_s - \overline{q}_v}{\tau_{\text{erosion}}}, \qquad (6.120)$$

where τ_{erosion} is a typical timescale required to dilute the clouds through lateral mixing. The erosion term plays a similar role as the dissipation term in the variance equation of q_t. The parameterisation of the autoconversion term G finally will be further discussed in Section 6.7.

Simultaneous use of the prognostic variance of the total water specific humidity $\overline{q_t'^2}$ and the condensed water q_c may lead to an overspecification of the probability distribution and needs to be avoided. Therefore, a choice needs to be made which of the two variables should be used as a prognostic variable. Unfortunately, either choice comes with a number of pros and cons.

One main advantage of using a prognostic humidity variance equation is that it also provides information on the subgrid variability for cloud-free conditions, so that the onset of cloud formation can be diagnosed. In contrast, if q_c is used as a prognostic variable, no information is available of the sub-grid variability under cloud-free conditions. Therefore, in cloud-free conditions, additional diagnostic assumptions of the sub-grid variability are needed such as a critical RH.

Another advantage is that a prognostic humidity variance can be well combined with turbulence and convection schemes that use moist conserved variables such as q_t and θ_ℓ. In that case, a prognostic humidity variance-based cloud scheme can be regarded as a physical consistent way of disentangling (q_t, θ_ℓ) into (q_v, q_c, θ). Finally, if the variance for humidity vanishes for sufficient high resolution, variance-based cloud schemes simplify automatically to an all-or-nothing scheme.

One major disadvantage of using a prognostic variance is that it is difficult to incorporate realistic cloud microphysics. First, there is the question how to link the autoconversion from condensed water to rain in terms of the sink term in the variance Eq. (6.113). Formally, this sink term can be written as

$$(\partial_t \overline{q_t'^2})_{\text{au}} \equiv S_{\text{au}} = \overline{G' q_t'}. \qquad (6.121)$$

Only for a simple PDF, such as a top-hat distribution, and assuming that autoconversion does not affect cloud fraction, it is possible to derive a simple expression for Eq. (6.121)

$$S_{\text{au}} = a_c(1 - a_c)(2q_{t,c} - q_{t,e})G. \qquad (6.122)$$

However, for more realistic PDFs it is impossible to derive a closed expression for S_{au}. An even more fundamental problem relates to the fact that deriving condensed water and cloud fraction from Eq. (6.104) assumes thermodynamic equilibrium, which allows diagnosis of the condensed water simply as the amount that exceeds the saturation value. While this is an excellent assumption for warm cloud microphysics, it cannot always be used for mixed-phase and glaciated clouds where high supersaturation values of 100% before ice nucleation sets in are not uncommon. If a prognostic variance is used, the amount of condensed water can only be estimated for thermodynamic equilibrium conditions and the further breakdown into a liquid and a ice component can only be done diagnostically as a function of temperature such as Eq. (6.84).

Most of the disadvantages of using a prognostic $\overline{q_t'^2}$ are advantages for using prognostic q_c and vice versa. The main advantage for using prognostic q_c is that it is easier to include microphysical processes. Autoconversion rates can be directly implemented as a sink term in Eq. (6.119) and ice microphysics can be taken into account more easily, especially if separate prognostic equations for liquid and the ice phase are used.

6.6.6   Cloud Overlap and Inhomogeneity

Even in the case of a perfect cloud scheme that diagnoses realistic horizontal cloud fractions a_c at each model level, assumptions are still necessary to decide how the cloud fractions need to be positioned geometrically at the various vertical levels in a column of a large-scale model. Different overlap assumptions can lead to significantly different values of the total (projected) cloud cover a_p which is defined as the fractional area of all clouds projected on to the ground. Obviously a_p is a crucial parameter for the radiative transfer. Fig. 6.21 shows three common options for the overlap assumptions. In Fig. 6.21a the *maximum* overlap assumption is displayed, where all cloudy areas are assumed to overlap to a maximum extent so that the projected cloud cover is equal to the level with the largest cloud fraction. This assumption has the most efficient overlap and consequently the lowest possible projected cloud cover. The *random* overlap assumption (Fig. 6.21c) assumes that clouds at all levels are randomly overlapping with respect to each other. Most models use the *maximum-random* overlap assumption which implies that consecutive levels with a non-zero cloud fraction are assumed to have maximum overlap, while levels that are separated by cloud-free levels are randomly overlapping.

Cloud radar observations indicate that the 'true' overlap between adjacent levels is in general in between the two extremes of random and maximum overlap. To find a quantitative estimate of the true projected cloud cover a_p^{true} of two adjacent levels with a non-zero cloud fraction $a_{c,1}$ and $a_{c,2}$, it is useful to introduce an overlap parameter γ that interpolates between the maximum and the random overlap

$$a_p^{\mathrm{true}} = \gamma a_p^{\mathrm{max}} + (1 - \gamma)\, a_p^{\mathrm{rand}} \qquad (6.123)$$

where a_p^{max} is the maximum overlap limit

$$a_p^{\mathrm{max}} = \max(a_{c,1}, a_{c,2}) \qquad (6.124)$$

and a_p^{rand} is the random overlap limit

$$a_p^{\mathrm{rand}} = a_{c,1} + a_{c,2} - a_{c,1}a_{c,2} \qquad (6.125)$$

Cloud radar observations of a_p indicate that the overlap parameter γ is a function of the separation length Δz between the two layers and can be well fitted as

$$\gamma = \exp\left(-\frac{\Delta z}{\Delta z_0}\right) \qquad (6.126)$$

with Δz_0 being a decorrelation length describing the decay of the overlap parameter from maximum to random overlap. Reported values of Δz_0 range from 1 to 3 km depending on spatial and temporal discretization. Since the used radar observations have a (vertical) resolution of around 300 m, the results do not take into account overlap effects on a scale smaller than a few hundred metres (Hogan and Illingworth, 2000).

Another source of a biased error is the simplified assumption that the liquid water is uniformly distributed over the cloudy part in the grid box as sketched in Fig. 6.22a. In reality, the liquid water can vary substantially within the cloudy area. A simple way to quantify this variability is through the ratio of the standard deviation and the mean of the in-cloud liquid water

$$f_q \equiv \frac{\overline{q_l'^2}^c}{q_{l,c}}, \qquad (6.127)$$

where $q_{l,c}$ denotes the average condensed water amount in the cloudy region and where the variance is determined over the cloudy part only. Typical values of f_q depend on

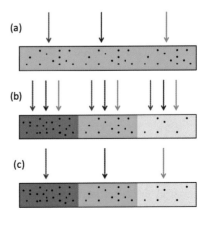

(a)

(b)

(c)

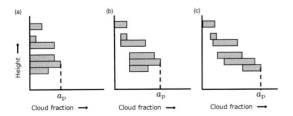

FIGURE 6.21: Schematic representation of three different cloud overlap assumption: (a) maximum cloud overlap, (b) maximum-random cloud overlap and (c) random cloud overlap.

FIGURE 6.22: Schematic illustration of the Monte Carlo independent column approximation (MCICA). In panel (a) Radiative transfer on a homogenised condensed water field. (b) The independent column approximation. (c) The principle of the MCICA.

the cloud type under consideration and the grid box size, but are generally in the range $[0.4, 1]$. Neglecting the fact that clouds are inhomogeneous introduces a systematic error, also known as the cloud albedo bias. This bias is the result of the fact that the albedo α_{alb} is a non-linear and concave function of the liquid water path (LWP) in the atmospheric column so that

$$\overline{\alpha_{\mathrm{alb}}(LWP)} \leq \alpha_{\mathrm{alb}}(\overline{LWP}). \tag{6.128}$$

This implies that neglecting the cloud inhomogeneity leads to a systematic overestimation of the albedo at the top of the atmosphere, and consequently to an underestimation of the downward short-wave radiation at the Earth's surface.

The traditional way of correcting for this cloud albedo bias is to make clouds more transparant by reducing the cloud optical depth that is used in the calculations for the radiative transfer by a factor χ that is often referred to as the *cloud inhomogeneity factor*. In that case, the radiation scheme uses a *reduced* optical depth τ_{red} defined as

$$\chi_{\mathrm{inh}} = \frac{\tau_{\mathrm{red}}}{\tau}. \tag{6.129}$$

Values for χ as small as 0.7 are used in large-scale models, although the general consensus now is that this value is too small, even for large-scale models with a coarse resolution.

Recently, a more elegant and physically sound method has been proposed to take the effect of the cloud inhomogeneity into account. The grid box is subdivided in N smaller sub-grid boxes. Each sub-grid box is stochastically filled with condensed water amounts drawn from a probability distribution in such a way that the average value of q_c over all sub-grid boxes remains invariant. The variance of this PDF can be determined through a prescribed value of f_q. If such a sub-grid division of the condensed water is realised, one can perform a radiative transfer calculation that requires a double summation over all N_v spectral wavebands and over all N_x sub-grid columns

$$F_{\mathrm{rad}}(z) = \sum_i^{N_x} \sum_j^{N_v} F_{\mathrm{rad},i,j}(z), \tag{6.130}$$

where $F_{\mathrm{rad},i,j}$ is the radiative flux for sub-grid column i and waveband j. This method is referred to as the independent column approximation (ICA) and does take into account most of the cloud inhomogeneity effects (see Fig. 6.22b). Unfortunately, this method makes the already expensive radiative transfer calculation a factor N_x more expensive. A simple but intelligent way to reduce the computational costs is to perform the radiative transfer calculations for each subcolumn i monochromatically through

a calculation of only one band j (see Fig. 6.22c). In the case of $N_x = N_v$, the computational cost is equal as for the original case where no inhomogeneity was introduced. This approximation which is referred to as the Monte Carlo ICA (MCICA) reduces thereby a systematic error into a random error, the latter being the result that for all the subcolumns radiative calculations are only calculated monochromatically.

6.7   Cloud Microphysics Parameterisation

The parameterisation of cloud microphysics aims at providing a closed set of equations to calculate, for example, the formation of rain in a cloud and the resulting surface precipitation. So far in this chapter this was largely summarized in the so-called autoconversion to precipitation G. The details are of course much more complicated. As described in Chapter 3, the formation of rain requires many microphysical processes to act in concert to form particles that are large enough to fall out and reach the ground. The whole sequence of processes from activation/nucleation through growth by condensation/deposition and collision processes to sedimentation and evaporation/melting is by far not well understood. Therefore, the challenge for microphysical parameterisation is at least two-fold. The original equations have to be simplified to make them computationally tractable, and the uncertain assumptions have to be constrained by observations or a posteriori testing of the model.

6.7.1   Parametric versus Non-parametric Description

To simulate the evolution of clouds and precipitation in a numerical model it would be desirable to use a discretized form of the particle size distribution $n(D)$ over a size range from $1\,\mu\mathrm{m}$ to $10\,\mathrm{mm}$. Unfortunately, the number of bins (or variables) required is too large for this approach to be computationally feasible, especially in climate models. Microphysical parameterisations in large-scale models usually predict few moments

$$M_k = \int_0^\infty D^k n(D)\, \mathrm{d}D \tag{6.131}$$

of the particle size distribution $n(D)$ defined as the number of particles per unit volume in the size range $[D, D + \mathrm{d}D]$. Such parameterisations are known as bulk microphysical schemes. A third alternative are Lagrangian particle methods that predict the evolution of individual particles in a cloud. This is of course the most expensive, but also the most natural way to simulate clouds and precipitation. Reducing the number of simulated particles and taking each of the particles as being representative for a large number of actual cloud particles with similar properties and growth histories leads to the super-droplet method which is becoming increasingly popular in cloud microphysical research.

6.7.2 Single- versus Double- versus Triple-Moment

For bulk microphysical schemes, a choice has to be made on how many moments are predicted and which. Most models predict only one moment and the natural choice is then the mass mixing ratio. For droplets this is simply

$$q_1 = \frac{\pi \rho_1}{6\rho} \int_0^\infty D^3 n(D) \, dD = \frac{\pi \rho_1}{6\rho} M_3, \qquad (6.132)$$

where ρ_1 is the density of water. The mass mixing ratio is chosen because mass is the most important conserved variable in the system, and because for many applications we are interested in the rain rate, which is the vertical flux of the q_1. Parameterising the cloud properties in terms of moments corresponds to the assumption of a certain statistical distribution of $n(D)$, for example, a Gamma distribution

$$n(D) = N_0 D^\mu e^{-\lambda D}, \qquad (6.133)$$

where N_0 is sometimes called the intercept parameter (although this terminology makes most sense only for the exponential distribution, i.e., $\mu = 0$), μ is the shape parameter and λ the slope of the distribution. Using a one-moment scheme corresponds to the assumption that two of these three parameters are either constant or can be expressed as function of other variables like temperature, and the predicted moment q_1. For example, for snow there exist quite reliable empirical relations to parameterize N_0 as a function of temperature (or a combination of T and q_s). This temperature dependency does correspond to an increase in the size of snowflakes at warmer temperatures (or lower heights above ground), which is the typical behaviour in frontal clouds and other stratiform clouds due to aggregation and size sorting.

In general, however, no such empirical relations exist and it is therefore very much desirable to predict the average size of the particles independently from the mass mixing ratio. To do so, a second moment of the size distribution is needed. A good and also the most popular choice for the double-moment scheme is to predict, besides the mass mixing ratio q_1, also the droplet number $N \equiv M_0$, thereby providing the mean mass as $\bar{m} = q_1/N$ as explicit size information. The number density is

conserved during many microphysical processes like condensational/depositional growth or melting, which simplifies the formulation of the prognostic equations. Also for many collection processes the resulting numbers are a priori known. A third argument to choose the number density as prognostic variable is the importance of or interest in aerosol effects, which primarily modulate the number of particles. Double-moment schemes have many advantages due to the explicit size information they carry. Maybe most important is the improved representation of the life cycle of deep convective clouds from the formation of the first small particles in the updraft core to the formation of the downdraft, the cold pool and the convective stratiform region. Double-moment schemes are therefore often used in storm-resolving and LESs.

Using a double-moment scheme we can determine two parameters of the distribution, for example, N_0 and λ, but the shape parameter μ would still be constant or diagnosed from the two predicted moments. This sounds like a minor issue, but it is not. Some microphysical processes do change the shape of the distribution, most prominently gravitational sorting, that is, the fact that large particles fall faster than small particles. Gravitational sorting makes the particle distribution narrower and would if it is the dominant process, for example, increase the shape parameter in the cloud because large particles leave the cloud more quickly. With a fixed shape parameter, these particles are artificially recreated in the cloud layer, leading to a unphysical spectral flux in the size distribution, and only the effect on the mean size is modelled correctly. This leads to an overestimation of the rain rate, especially when μ is set to small values, that is, a broad distribution. The best way to avoid this would be to use a triple-moment scheme and predict also μ. But such schemes have other issues, besides being computationally expensive, starting from the fact that there is no natural choice for a third moment. A reasonable choice is the sixth-order diameter or second mass moment which is proportional to radar reflectivity, simply because it can be measured.

6.7.3 Diagnostic versus Prognostic

The budget equation for the specific masses of the microphysical species is

$$\frac{\partial q_x}{\partial t} + \mathbf{v} \cdot \nabla q_x - \frac{1}{\rho} \frac{\partial}{\partial z} \left(\rho q_x u_{x,\infty} \right) = S_x, \qquad (6.134)$$

where the terms on the left-hand side are the time rate of change, the advection by the three-dimensional air velocity and the divergence of the vertical flux of q_x with the sedimentation velocity $u_{x,\infty}$. The right-hand side are the microphysical sources and sinks, that is, processes like rain formation, condensation/evaporation and so on. The sedimentation velocity is the weighted terminal fall velocity for this moment of the particle size distribution, that is, for the mass mixing ratio of droplets

$$u_{x,\infty} = \left(\frac{\rho_0}{\rho}\right)^{\xi_T} \frac{\int_0^\infty D^3 n(D) u_\infty(D) \, \mathrm{d}D}{\int_0^\infty D^3 n(D) \, \mathrm{d}D}, \qquad (6.135)$$

where $u_\infty(D)$ is the terminal fall velocity of droplets at the reference density ρ_0 which is only a function of the drop diameter D. The dependency on air density is explicitly taken into account by the pre-factor with an exponent ξ_T between 0.3 and 0.5 for precipitation particles. Here it is important to know that the sedimentation velocity of rain is about $1 \, \mathrm{m \, s^{-1}}$ for small drizzle drops, but reaches almost $10 \, \mathrm{m \, s^{-1}}$ for large rain drops. Small ice particles in the form of hexagonal plates or snowflakes have fall velocities of the order of $1 \, \mathrm{m \, s^{-1}}$, while large high-density particles like hail can reach fall speeds up to $30 \, \mathrm{m \, s^{-1}}$ and higher. With these numbers we can easily estimate whether we actually need the horizontal advection term in the budget equation. For rain that forms at 5 km height and grows rapidly to large enough sizes to reach $10 \, \mathrm{m \, s^{-1}}$ and assuming a horizontal wind speed of $10 \, \mathrm{m \, s^{-1}}$ we find that the rain is advected over a distance of 5 km, that is, only in models with a horizontal grid spacing below 5 km we would actually need to take this into account. For snow that forms higher up, for example, at 8 km, and falls much slower at $1 \, \mathrm{m \, s^{-1}}$, we find a distance of 80 km. At such scales also the vertical advection can be neglected, because the vertical velocities are small. In climate models with coarse grids of 100 km grid spacing, the precipitation particles are therefore often treated by a diagnostic scheme in which the vertical flux $P_x = \rho q_x u_{x,\infty}$ is directly calculated from the sources and sinks, that is,

$$-\frac{1}{\rho}\frac{\partial}{\partial z}\left(\rho q_x u_{x,\infty}\right) = -\frac{1}{\rho}\frac{\partial P_x}{\partial z} = S_x \qquad (6.136)$$

which is also know as column-equilibrium. By neglecting advection and the time derivative we assume that the precipitation flux in the column is in equilibrium with the sources and sinks in that column. Although this approximation is reasonable at large scales, one has to keep in mind that microphysical processes are highly non-linear and simplifications like the column-equilibrium can therefore lead to unwanted side effects. By neglecting the time derivative, for example, we force the system into stationarity and this has consequences for the balance of the source and sinks, for example, autoconversion becomes artificially large while accretion is reduced compared to a system that can show intermittency. Such shifts in the

processes can then affect the sensitivity of the system, for example, the response to aerosol perturbations. Therefore, it can be a better choice or even necessary to at least retain the time derivative, that is, using a prognostic scheme and neglecting only the advection terms can give different results than the column-equilibrium model.

6.7.4 Ice Particles and Non-equilibrium Ice Clouds

Most clouds do contain ice and in mid latitudes most precipitation is due to ice or mixed-phase processes. The formation of ice in the atmosphere is not yet understood in all details. At cold temperatures below $-38\,°\mathrm{C}$, ice can form homogeneously by freezing of liquid cloud droplets or liquid aerosol particles. At higher temperatures, solid aerosol particles are needed that serve as ice nuclei (IN; see Chapter 3). A main difference between liquid clouds and ice clouds is that ice nuclei IN are much more rare than cloud condensation nuclei (CCN). With the exception of extremely clean maritime environments, the supersaturation over liquid water that is reached in clouds is therefore very small, usually below 1%, because already at such low supersaturations enough droplets form and the supersaturation is consumed by condensational growth. Therefore, most liquid clouds are very close to thermodynamic equilibrium between water vapour and liquid water. This is also the fundamental assumption made in most diagnostic and prognostic cloud schemes (Section 6.6.2). Ice clouds behave fundamentally different. The number of IN depends strongly on temperature and supersaturation over ice, and at warm temperature only a few IN per litre are activated (compared to at least 1 CCN per cubic centimetre). Such few ice particles are not able to deplete the supersaturation efficiently and ice clouds are therefore often far away from equilibrium, that is, the supersaturation over ice in (or outside of) clouds can be as large as 50%. It can take several hours for an air parcel inside an ice cloud to form enough ice particles and reach equilibrium. This difference between liquid and ice clouds has important consequences for cloud parameterisations. While the condensational growth is usually a diagnostic calculation, that is, the amount of condensed liquid water can be estimated directly from the thermodynamic variables and the assumptions about sub-grid variability, the depositional growth of ice particles has to be calculated explicitly, taking into account the number, size and geometry of the ice particles which are present in a given volume. This is possible in high-resolution models in which sophisticated microphysical schemes can be applied and the dynamical forcing is reasonably well resolved to estimate the ice particle number densities. In coarse models and especially climate models the non-equilibrium nature of ice clouds is more challenging as a consistent parameterisation of sub-grid and in-cloud variability combined with non-equilibrium microphysical growth equations is still not available.

6.7.5 The Autoconversion Problem

Rain formation via the liquid phase is an important process, especially in the tropics and subtropics. The starting point for any parameterisation of warm rain collision-coalescence processes is the stochastic collection equation, also known as Smoluchowski equation or kinetic equation (see also Eq. (3.24)).

$$\frac{\partial n(x,t)}{\partial t} = \frac{1}{2} \int_0^x n(x-x',t)\,n(x',t)\,K(x-x',x')\,\mathrm{d}x'$$
$$- \int_0^\infty n(x,t)\,n(x',t)\,K(x,x')\,\mathrm{d}x', \qquad (6.137)$$

where x and x' are the drop masses and $n(x,t)$ is the drop size distribution as a function of drop mass. The kinetic equation describes the formation of larger drops by collision and coalescence of smaller droplets, a process which is also known as colloidal instability. If we neglect effects of turbulence, the collection kernel K can be specified by the gravitational kernel (cf. Chapter 3)

$$K(x,y) = \pi [r_x + r_y]^2 E(x,y)|u_x - u_y|, \qquad (6.138)$$

where r_x, r_y are the drop radii, u_x, u_y their terminal fall velocities and E is the collection efficiency. With realistic assumptions for the terminal velocity and collection efficiency, the collection kernel is non-linear in x and y, especially for small droplets, and for such a kernel the kinetic equation is very difficult to solve analytically. Known analytic solutions exist only for a few special cases including constant and linear kernels. For parameterisation we therefore rely to a large extent on numerical solutions of the kinetic equation. Such simulations show that the mass distribution of drops $g(x,t) = xn(x,t)$ becomes bimodal after some time, that is, starting from a unimodal distribution of small droplets it develops a second maximum at larger sizes. This typical behaviour of the colloidal instability suggests separation of the drop size distribution into two parts, small droplets which we call cloud droplets and larger drops which we identify as rain drops. The minimum in the bimodal drop size distribution is usually at sizes of 40–50 μm radius, that is, this value can be used as a separation radius r^* separating cloud droplet from

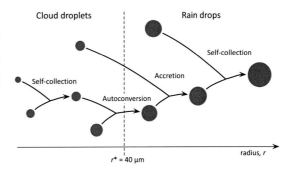

FIGURE 6.23: Illustration of the warm rain processes autoconversion, accretion and selfcollection of cloud droplets and rain drops.

rain drops. In bulk microphysical parameterisations we consequently aim to describe the evolution of the partial moments of the drop size distribution

$$M_{k,c} = \int_0^{x^*} x^k n(x)\,\mathrm{d}x = \left(\frac{\pi \rho_w}{6}\right)^k \int_0^{D^*} D^{3k} n(D)\,\mathrm{d}D \qquad (6.139)$$

$$M_{k,r} = \int_{x^*}^\infty x^k n(x)\,\mathrm{d}x = \left(\frac{\pi \rho_w}{6}\right)^k \int_{D^*}^\infty D^{3k} n(D)\,\mathrm{d}D. \qquad (6.140)$$

For $k = 0$ these are the number densities of cloud droplets and rain drops, and $k = 1$ is the corresponding mass density. Fig. 6.23 depicts collection processes that occur in a bulk system with a separation of cloud droplet and rain drops. Autoconversion results from the collision of two cloud droplets that are large enough to form a rain drop. By this process, the number of rain drops is increased, for each new rain drop, two cloud droplets are lost, and mass is transfered from the cloud water to rain water. Accretion is the collection of cloud droplets by rain drops, and this also results in a mass transfer from cloud to rain, but does not create any new rain drops, it only reduces the number of cloud droplets. In additon there are so-called self-collection processes which describe the collision within each category. Obviously, self-collection does not change the mass of cloud water or rain water, but only reduces the number densities. More specifically we can define these processes as integrals of the size distribution and the collection kernel, for example, the change in number densities due to the self-collection processes and the mass transfer due to accretion and autoconversion can be defined as

$$\partial_t N_c \Big|_{sc} = -\frac{1}{2} \int_{x'=0}^{x^*} \int_{x''=0}^{x^*} n(x')\,n(x'')\,K(x',x'')\,\mathrm{d}x'\mathrm{d}x'' \qquad (6.141)$$

$$\partial_t N_r \Big|_{sc} = -\frac{1}{2} \int_{x'=x^*}^{\infty} \int_{x''=x^*}^{\infty} n(x') \, n(x'') \, K(x',x'') \, \mathrm{d}x' \mathrm{d}x'' \tag{6.142}$$

$$\partial_t \rho q_c \Big|_{ac} = -\int_{x'=0}^{x^*} \int_{x''=x^*}^{\infty} n(x') \, n(x'') \, K(x',x'') \, x' \, \mathrm{d}x' \mathrm{d}x'' \tag{6.143}$$

$$\partial_t \rho q_c \Big|_{au} = -\int_{x'=0}^{x^*} \int_{x''=x^*-x'}^{x^*}$$
$$\times n(x') \, n(x'') \, K(x',x'') \, x' \, \mathrm{d}x' \mathrm{d}x''. \tag{6.144}$$

With some approximations, for example, using the Long kernel (Eq. (3.35)), the first three integrals are relatively straightforward to evaluate, but the last one, the autoconversion, is challenging. First, because the integral limit of the inner integral depends on x', making the evaluation cumbersome. Second, in contrast to self-collection or accretion, it is not sufficient to replace $n(x)$ in the autoconversion integral by a simple analytic distribution. It is the very nature of the colloidal instability that the size distribution changes with time and ignoring this fact in the autoconversion integral leads to errors in the timing and amount of rain water. Currently, there exist various different appoaches to arrive at a parameterisation of the autoconversion rate.

Threshold-based parameterisations approximate the autoconversion rate by a threshold process, for example, the Kessler-type formulation

$$\partial_t q_c \Big|_{au} = \begin{cases} k_{au}(q_c - q_0), & \text{for } q_c > q_0 \\ 0, & \text{else.} \end{cases} \tag{6.145}$$

Such a formulation has little theoretical basis, as the kinetic equation itself does not describe a threshold process, in fact, all cloudy air parcels are colloidally unstable and would develop rain, if allowed to exist long enough. Intuitively, the threshold ansatz makes sense because small and thin clouds do not rain, but this mixes cloud dynamics and cloud microphysics in a way that makes it very difficult to derive an estimate for the threshold value q_0 or relate it to other microphysical properties of the cloud, like number density or particle size. The latter would of course be necessary for example, modelling aerosol effects on clouds.

A pragmatic approach uses numerical simulations based on the kinetic equation itself, either from simple parcel models or more sophisticated LESs, and then approximates this data using a power law ansatz, that is, for a two-moment scheme

$$\partial_t q_c \Big|_{au} = \kappa_{au} \, q_c^n N_c^m \tag{6.146}$$

with a coefficient κ_{au} and exponents n and m that are estimated from the simulation results, for example, by a least-square fit to the data. This has the advantage that the dependency on the number density of cloud droplets is included, but the values of the exponents differ substantially between different schemes that originate from different simulation results casting doubt on the modelled sensitivities.

Instead of using a simple power-law ansatz, one can try to extract more information from the kinetic equation itself before fitting the resulting equation to numerical solutions. For example, the Long kernel provides a good starting point and allows the derivation of analytical estimates of the autoconversion rate in some asymptotic limits. Another valuable property of the kinetic equation is its invariance under transformations of the form $t \to t/c$ and $f \to cn$ with a constant c. Therefore, solutions of the kinetic equation have the similarity property $\tilde{n}(x,t) = cn(x,ct)$. Any parameterisation of autoconversion should, if possible, preserve this invariance and the similarity properties of the solutions. Using such properties of the original equation allows to make a more educated guess for the form of the autoconversion rate.

Maybe the most natural way to attack the autoconversion problem is to make assumptions about the analytic form of the drop size distribution and use this to calculate the process rate based on the integral definitions already given. As already mentioned, this is hindered by the fact that the temporal evolution needs to be represented in quite some detail to capture the correct behaviour of the colloidal instability. Such efforts have been largely unsuccessful for the simple two-class decomposition in cloud and rain categories, but when introducing a third intermediate category of large cloud droplets or, alternatively, drizzle drops, the problem becomes more tractable. Unfortunately, the resulting parameterisations schemes become quite expensive due to the evaluation of the integrals which has to be performed each time step.

6.8 Challenges

So far, this chapter has described current parameterisations of cloud-related processes in present-day state-of-the-art large-scale models. The discussed parameterisations are deterministic and are formulated in terms of the grid box-averaged fields and their impact of these are directly communicated with the grid box mean state.

Many of the discussed parameterisation developments evolve around finding the proper sub-grid variability, ideally in terms of a joint probability density function

$P(w, \theta_\ell, q_t)$. This sub-grid information is required for turbulent and convective fluxes since the vertical velocity is strongly correlated with the thermodynamical variables at sub-grid scales. Similarly, cloud variables such as cloud fraction and cloud amount can be well estimated if the joint PDF of temperature and humidity is known. Neglecting this sub-grid variability leads to strong systematic errors. A canonical example is the parameterisation of cloud fraction discussed in Section 6.6. The cloud fraction in a grid box can be written as

$$a_c = \overline{H(q_t - q_s(p, T))} \geq H\left(\overline{q_t} - q_s(\overline{p}, \overline{T})\right), \qquad (6.147)$$

demonstrating that calculating the cloud fraction on the basis of the grid mean values introduces a systematic underestimation, due to the non-linear character of the Heaviside function H. These sub-grid processes have in common that (i) they can be resolved, provided that sufficient information on the sub-grid variability is available and (ii) they are alleviated by increasing model resolution to a degree that the inequality in Eq. (6.147) becomes an equality at large eddy resolving resolutions. This demonstrates the usefulness of LES as numerical atmospheric laboratories providing reliable benchmarks for the sub-grid variability. New developments for this class of parameterisations concentrate on representing the parameterisation of turbulent and convective transport and cloud macrophysics in a more consistent and unifying way, based on a common joint probability distribution and are discussed in Section 6.8.1.

A second challenge relates to sub-grid physical processes for which we do not have reliable descriptive equations. Prime examples are the many microphysical processes in the mixed cloud phase and ice clouds such as described in Section 6.7.4. For such processes, further experimental research is required and, in addition, new theoretical concepts. Clearly, these are errors that do not alleviate with increasing resolution and cannot be resolved by LESs.

A final challenge is associated with the common quasi-equilibrium assumption of a deterministic relationship between the sub-grid process and the resolved state. Although quasi-equilibrium is defendable for resolutions much coarser than typical size of the sub-grid process under consideration, it breaks down if the size of the grid box becomes comparable to the size of the sub-grid process of interest. For such cases, the resolution is still too coarse to fully resolve the process but too fine to allow for a traditional deterministic statistical approach. In Section 6.8.2, it will be argued that in such cases a stochastic approach is more appropriate.

6.8.1 Unification and Consistency

Traditionally, parameterisation packages of large-scale models describing the atmospheric sub-grid processes were subdivided into separate parameterisation schemes for radiative transfer, boundary-layer processes, moist convection and clouds as schematically illustrated in Fig. 6.2. The first-generation large-scale models in the sixties and early seventies of the last century all had some form of radiative transfer and used moist convective adjustment to mimic the effect of moist convection. Because of the coarse vertical resolutions used by these models, the lowest model level was often well above the boundary-layer height, so that early boundary-layer parameterisations were merely bulk surface flux formulations such as Eq. (6.21). Clouds were mostly climatologically prescribed for the purpose of radiative transfer calculations. These simple parameterisations only interacted with the large scale mean state.

These archetypes of parameterisations have evolved over the last forty years into the more complex and adequate parameterisations described in this chapter. This development is in part the result of the increased understanding of the underlying physical processes but also due to increased computer resources allowing for a finer vertical resolution that can accompany more detailed descriptions. The first-generation GCMs only used a few vertical layers for the whole troposphere while nowadays large-scale models can accommodate more than 100 vertical levels supporting a vertical resolution of tens of meters.

In comparison with the simple parameterisations used in the first-generation large-scale models, Fig. 6.24 demonstrates two changes in present-day parameterisation packages. First, similar physical processes are now shared by different parameterisations. Cloud microphysical processes are represented by the cloud scheme, the convection scheme and sometimes even in boundary-layer schemes while turbulent and convective transport is shared between the convection and the boundary-layer scheme. Second, present-day parameterisation schemes do not only interact with the mean state but also between themselves. The cloud scheme requires information of the sub-grid variability of moisture and temperature from the convection and the boundary-layer turbulence scheme. Boundary-layer turbulence schemes that use a moist TKE formulation require knowledge of the cloud fraction from the cloud scheme. The strength of the updrafts in the convection schemes depends on the surface fluxes and the resulting thermals in the boundary layer while the boundary layer can be modified by saturated downdrafts that can trigger the formation of cold pools.

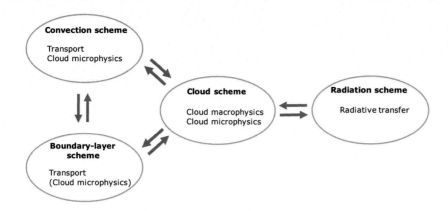

FIGURE 6.24: Tasks and interactions of the parameterisation schemes.

The intimate coupling between the boundary layer, convection and cloud processes requires a redesign of parameterisation schemes in which the physical sub-grid processes are represented in a more consistent manner. Two interesting unifying pathways that have been developed over the last decade are the ED mass-flux (EDMF) approach and an assumed PDF approach, which is essentially a higher-order closure using an assumed shape of the joint probability density distribution $P(w, \theta_\ell, q_t)$. Both methods will be briefly discussed.

6.8.1.1 Higher-Order Closures Using an Assumed PDF

This approach makes use of higher-order closures in conjunction with an assumed functional shape of the joint PDF of vertical velocity, temperature and humidity. The schematics of this approach is illustrated in Fig. 6.25. It makes use of prognostic equations of vertical velocity w, the total water specific humidity q_t and the liquid-water potential temperature θ_ℓ up to second order: the means, the variances and the correlations. In addition, also a prognostic equation for the third moment of the vertical velocity $\overline{w'^3}$ is included so that level of skewness of the velocity field can be predicted.

The novelty of this approach is that an assumed functional form of the joint PDF of these three variables is used as a closure. The prognostic equations for the first- and second-order moments are used to determine the values of the free parameters of the assumed joint PDF. Once these free parameters are determined, the resulting joint PDF can be used to analytically determine the third-order moments that are needed to close the second-order prognostic equations such as the variances and the cross-correlations. This way, a joint PDF can be determined iteratively that consistently describes all the higher-order moments and in addition all the relevant cloud macrophysics such as cloud fraction and condensed water.

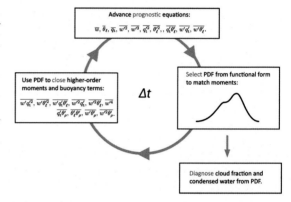

FIGURE 6.25: Schematic illustration of a higher-order closure scheme using an assumed PDF.

To some degree this assumed PDF approach can be viewed as an extension of the PDF-based cloud schemes that were discussed in Section 6.6. For these cloud parameterisations, the objective is to estimate the cloud fraction and cloud condensate from a PDF $P(\theta_\ell, q_t)$. Within this unified approach, the dynamics is included by adding w to the probability distribution so that the turbulent fluxes can be estimated consistently with the same joint PDF.

Different functional shapes of the joint PDF can be used. First pioneering attempts (Randall et al., 1992; Lappen and Randall, 2001) used a double delta function probability density function for of up- and down-drafts which can be regarded as a higher-order mass flux scheme. More recently, Golaz et al. (2002) developed a higher-order approach method, cloud layers unified by binormals, where a double Gaussian PDF is used which has a larger flexibility to describe the observed PDFs. The price of this flexibility is that the double Gaussian joint PDF has a large number of free parameters (fifteen) that need to be constrained, larger than the number of prognostic equations (ten), used to constrain these parameters, so that extra assumptions are needed. These assumed PDF approaches are well capable of parameterising boundary-layer cloud regimes. Extension of this approach to deep convection remains challenging, in part due to the difficulty of incorporating ice and mixed-phase cloud microphysics into this approach.

6.8.1.2 EDMF Approaches

The EDMF approach is inspired on the notion that cumulus convection is usually rooted in the sub-cloud layer in which dry rising thermals feed moist buoyant air from the surface layer into the cumulus clouds aloft. It is therefore natural to extend the updraft calculations of the mass flux scheme all the way down into the surface layer. This implies that the turbulent transport in the boundary layer can be shared between a mass flux contribution describing the organised thermal updrafts and a remaining part that is described by the traditional ED approach.

In its simplest form, EDMF can be formulated in terms of a mass flux contribution describing the organised updrafts and a diffusion term describing the smaller-scale turbulence

$$\overline{w'\varphi'} \simeq -K\partial_z\overline{\varphi} + M_u\left(\varphi_u - \overline{\varphi}\right). \qquad (6.148)$$

In essence, EDMF lowers the initialisation of the updrafts from cloud base into the sub-cloud layer. The updraft is initiated in the surface layer by choosing initial excess values of temperature, humidity proportional to the associated surface fluxes. Next, updraft equations, like the ones derived in Sections 6.5.2 and 6.5.5 can be used to calculate temperature, humidity and vertical velocity of the updraft. If the updrafts become supersaturated, condensational effects are taken into account. This procedure continues as long as the vertical velocity remains positive.

The EDMF approach has a number of advantages over the more traditional approaches described in the previous sections. First, it provides a unification between turbulent transport between the sub-cloud layer and the moist convective transport in the cloud layer aloft. The non-local transport in the convective boundary layer and in the cloud layer is described by the same set of updraft equations, thereby making the parameterised convective transport more consistent. Second, no explicit trigger mechanism for the initiation of cumulus convection is required anymore. If the updraft extends beyond the LCL, the scheme will parameterise both dry and moist convective transport by the updrafts. If an updraft does not reach the LCL, the mass flux term will still contribute to non-local transport by thermals in the dry convective boundary layer. This way the mass flux term in Eq. (6.148) takes into account the counter-gradient transport discussed in Section 6.4.6

in a more natural and consistent way (Siebesma et al., 2007).

As a minimum, one updraft, representative for the average of all updrafts, is required, but the extension to multiple updrafts is straightforward

$$\overline{w'\varphi'} \simeq -K\partial_z\overline{\varphi} + \sum_{i=1}^{n} M_i^u\left(\varphi_i^u - \overline{\varphi}\right), \qquad (6.149)$$

In Eq. (6.149), updraft type i is characterised by its horizontal radius R_i and by the number of updrafts $N(R_i)$ of type i. Each updraft type is described with the same entraining plume Eqs (6.64a), (6.64b) and (6.80), but differ in entrainment rates, for which an inverse relationship with the radius R is assumed. As a result, the larger plumes will have also a larger vertical extent.

If the initial updraft size distribution $N(R_i)$ is given, the mass flux contribution to the turbulent flux can be determined. An advantage of this method is that no separate equation for the mass flux is required, because the mass flux follows naturally from the multi-plume equations that determine both the vertical velocity and the updraft fractions as a function of height. Therefore, a separate explicit parameterisation for detrainment is no longer required. Promising results with this ED(MF)^n have been reported in Neggers (2015).

There are also disadvantages with the EDMF approach. Since it is essentially a first-order scheme, the partitioning between the non-local transport by the updrafts and the local ED driven transport has to be imposed, in terms of the updraft size distribution $N(R)$. This is especially an issue when this partitioning is changing over time, like in transitions from stratocumulus into cumulus.

6.8.2 Resolution, Grey Zone and Stochastic Parameterisations

At present, climate models operate at a 10–100 km resolution routinely for decadal runs while global numerical weather prediction (NWP) models are using resolutions of 10 km or less for ten-day forecasts. Given the ever increasing computational capability, it is therefore reasonable to expect that in the next decade, also decadal climate runs with resolutions in the range 1–10 km will become possible.

This will bring global models in a resolution regime, already encountered by operational limited-area mesoscale model, where convective overturning will be partially resolved. This has the benefit that mesoscale structures such as squall lines and cold pools

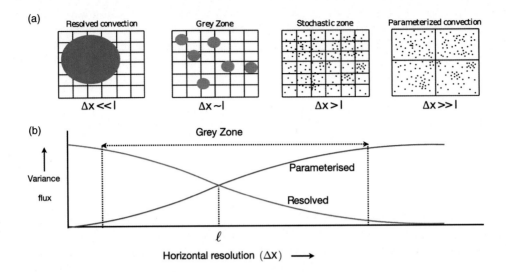

FIGURE 6.26: (a) Schematic illustration of the Grey Zone for convective updrafts. See text for further explanation. (b) Idealised representation of the resolved and parameterised turbulent fluxes (or variances) across the Grey Zone, ranging from resolutions much finer than the horizontal size ℓ of a typical convective updraft to resolutions much coarser than the size of convective updrafts.

will be partly resolved. But it will also require new demands for parameterisations, as will be discussed in this section. Cumulus convection will be partly resolved so that parameterisations should become aware of the used resolution. Moreover, assumptions such as quasi-equilibrium will not hold anymore and will require a more stochastic relationship between the resolved state and the parameterised sub-grid processes.

6.8.2.1 The Grey Zone

Historically, convection parameterisations have been developed under the assumption of quasi-equilibrium. If the grid size Δx is much larger than the individual size of the convective updrafts ℓ, this implies that the number of convective updrafts N in a grid box will be large, that is, $N \gg 1$. In that case, a deterministic statistical treatment of the parameterisation of the cumulus cloud ensemble is allowed. If, furthermore, the rate of change of the large-scale state is slow enough to allow the cloud ensemble to adjust quickly, quasi-equilibrium can be assumed. Under such conditions, a cumulus parameterisation does not aim to predict the behaviour of individual updrafts but rather the overall effect of the cloud ensemble in a statistical sense.

More generally, only in the case of quasi-equilibrium, it is reasonable to expect a deterministic one-to-one correspondence between the resolved processes and the parameterised response of the sub-grid processes. This situation is schematically sketched in the right panel of Fig. 6.26a where $\Delta x \gg \ell$ so that $N \gg 1$. Because the horizontal

size of deep convective updrafts ℓ are in the range of 1–10 km, this implies a grid sizes of at least several hundreds of kilometres for deep convection.

At the other end of the spectrum, when $\Delta x \ll \ell$, we have the situation where the resolution is fine enough that simulations become turbulent resolving and where no explicit parameterisation of the updrafts is required anymore. This is the resolution at which LESs operate.

In between these two extremes, when $\Delta x \simeq \ell$, effects of cumulus convection are partially resolved but additional parameterisation is still required. Resolutions around this value are referred to as the Grey Zone. Within the Grey Zone, the relative contribution of the parameterisation depends on the used resolution. It is for this reason that in this Grey Zone, parameterisations need to be scale-aware.

This notion can be made by considering the turbulent flux of an arbitrary field $\overline{w'\varphi'}^L$ on a domain size L and with a grid size Δx_0. If the grid size is now coarse grained to a size $\Delta x_0 < \ell < L$, the total turbulent flux can be formally decomposed into a resolved and a sub-grid contribution as

$$\overline{w'\varphi'}^L = \langle \overline{w''\varphi''}^\ell \rangle + \langle \overline{w}^\ell \overline{\varphi}^\ell - \overline{w}^L \overline{\varphi}^L \rangle$$
$$= \overline{w'\varphi'}_{\text{sub-grid}} + \overline{w'\varphi'}_{\text{resolved}}, \qquad (6.150)$$

where single primes denote deviations from the average over the whole domain of size L. The double primes denote deviations from the subdomain averages of size ℓ. The chevrons indicate averaging over the $(L/\ell)^2$ subdomains of size ℓ. The first term on the right-hand side of Eq. (6.150) represents the average of $(L/\ell)^2$ fluxes, each determined on a submain of size ℓ and therefore represents the sub-grid flux as if a resolution of $\Delta x = \ell$ was used. The second term on the right-hand side of Eq. (6.150) determines the flux based on the deviations of the coarse-grained fields at a resolution $\Delta x = \ell$ with respect to the averages of the whole domain of size L. This term therefore represents the resolved flux at the resolution $\Delta x = \ell$.

The variation of the two terms as a function of ℓ is shown schematically in Fig. 6.26. By construction, the total turbulent flux is constant with resolution, the sub-grid contribution monotonically increases for larger ℓ while the resolved contribution decreases for larger ℓ. The regime where both terms provide a significant contribution is in general referred to as the Grey Zone. These are the resolutions where a scale-aware parameterisation is required.

For cumulus convection, the Grey Zone extends from several hundred metres which is the minimum resolution required to resolve cumulus updrafts to ~10 km, which is the resolution at which convective overturning starts to become resolved. An interesting scale-aware parameterisation approach for cumulus convection has been proposed by Arakawa and Wu (2013). The starting point of their approach is that the turbulent flux of φ can be written within a top-hat approximation as (see also Eq. (6.52))

$$\overline{w'\varphi'} = a_u(1 - a_u)(w_u - w_e)(\varphi_u - \varphi_e)$$
$$= a_u(1 - a_u)\Delta w \Delta \varphi, \qquad (6.151)$$

which can be approximated for coarse resolutions $\Delta x \gg \ell$ in the quasi-equilibrium limit where $a_u \ll 1$ as

$$\overline{w'\varphi'}_{QE} = a_u \Delta w \Delta \varphi. \qquad (6.152)$$

This approximate form implies for the updraft fraction a_u in the quasi-equilibrium limit

$$a_u = \frac{\overline{w'\varphi'}_{QE}}{\Delta w \Delta \varphi} \ll 1. \qquad (6.153)$$

This requirement becomes too restricted in the Grey Zone, where a_u can take finite values and approaches 1 for cloud-resolving resolutions (see Fig. 6.26). Therefore, Arakawa and Wu (2013) proposed a more relaxed version of Eq. (6.152)

$$a_u = \frac{\overline{w'\varphi'}_{QE}}{\Delta w \Delta \varphi + \overline{w'\varphi'}_{QE}}. \qquad (6.154)$$

The motivation for this ansatz is that it satisfies $0 \le a_u \le 1$, while it reduces to Eq. (6.152) under the condition (6.153). Eliminating $\Delta w \Delta \varphi$, using Eq. (6.151) gives

$$\overline{w'\varphi'} = (1 - a_u)^2 \, \overline{w'\varphi'}_{QE}, \qquad (6.155)$$

which is the relevant result for parameterisations. In practice, the parameterised quasi-equilibrium flux $\overline{w'\varphi'}_{QE}$, valid outside the Grey Zone can be used and has to be reduced by the factor $(1 - a_u)^2$ for finite values of a_u within the Grey Zone. For resolved convection, in the limit $a_u \rightarrow 1$, the parameterised flux finally vanishes. Pathways for estimating a_u in the Grey Zone are discussed in Arakawa and Wu (2013).

For turbulent transport in the boundary layer, the Grey Zone extends from the Kolmogorov scale to the depth of the boundary layer, that is, from the millimetre to the kilometre scale. ED parameterisations in LES that use a Smagorinsky–Lilly length scale ℓ_{smag} proportional to the grid resolution (see Eq. (6.49)), are perfect examples of a scale-adaptive parameterisation of turbulent transport. For boundary-layer turbulence parameterisations, the current challenge is how to bridge such length-scale formulations, valid for resolutions of 100 m or less, to the traditional length-scale formulations ℓ_{bl}, such as Eq. (6.34), used in large-scale models and suitable for resolutions of several kilometres or more. A pragmatic way of bridging these two different length-scale methods is through introducing a blending length scale (Boutle et al., 2014)

$$\ell_{blend} = W\ell_{bl} + (1 - W)\ell_{smag}, \qquad (6.156)$$

which is essentially a weighted average of a boundary-layer length scale ℓ_{bl} and a Smagorinsky–Lilly length scale ℓ_{smag}. The weighting function W interpolates between these length-scale formulations as

$$W = 1 - \tanh\left(\beta \frac{z_t}{\Delta x}\right) \qquad (6.157)$$

where z_t is the relevant turbulent scale and β a scaling parameter that determines the width of the Grey Zone for turbulence.

Recognising the scale invariance of the cloud size distribution, as displayed by Eq. (6.102), on might argue that the Grey Zone of cloud macrophysics ranges from 100 m to the planetary scale. Indeed, measurements show that the moisture variance scales with the linear size L over which it is determined as L^α with a scaling exponent α close to 2/3, up to scales of hundreds of kilometres (Kahn et al., 2011). This implies that cloud schemes that utilise a moisture variance to estimate cloud fraction and cloud amount should use a scale-aware determination of the sub-grid variance decreases. How to do this is in practice is still an open question.

6.8.2.2 Stochastic Parameterisations

The intermittent and random character of moist convection vanishes for the case $\Delta x \gg \ell$ when only statistical effects of a whole ensemble are relevant. For resolutions $\Delta x \geq \ell$, close or in the Grey Zone as depicted in the second panel from the right in Fig. 6.26a, it can be seen that the number of updrafts present in a model grid box varies significantly from grid box to grid box, even when the grid boxes are subjected to the same large-scale forcing. Therefore, for these resolutions, stochastic parameterisations have more potential to adequately represent convection, because a deterministic convection parameterisation will provide a too strict and too hard-wired relationship with the corresponding large-scale state.

Inspection of Fig. 6.26 suggests that the level of stochasticity should increase with finer resolution and should be manifested in the updraft area fraction. At coarse resolutions, a constant updraft area fraction $a_u \ll 1$ is sufficient and can be absorbed in the mass flux formulation, as explained in the previous section. At finer resolutions, not only an explicit parameterisation of a_u is required where it can take any value between 0 and 1, but in addition, this explicit parameterisation of a_u should also be stochastic. Along with an appropriate parameterisation for the updraft velocity w_u, this could replace the necessity to have a separate parameterisation of the mass flux, as expressed by Eq. (6.64c). This will also have the advantage that an explicit parameterisation of detrainment will become obsolete.

Originally, stochastic parameterisations were introduced in NWP models to represent model uncertainty and to correct for the under-dispersive character of ensemble weather prediction. The first of these type of parameterisations is the stochastically perturbed parameterisation tendencies scheme (SPPT), whereby a spatially and temporally correlated random number field is used to perturb the parametrisation tendencies from the models physics package in a multiplicative way (Buizza et al. 2007). The imposed spatial and temporal correlation scales have no direct physical basis, although coarse graining studies have been used to retrospectively justify the multiplicative nature of the noise and its magnitude. Despite the ad hoc choices in the formulation of the scheme, SPPT is surprisingly effective at improving ensemble reliability, reducing forecast error and improving biases in the mean and variability of the models climate.

More recently, more physically based models for stochastic convection parameterisations are being developed that take into account the variability of the parameterised convective response. One class of stochastic convection schemes are stochastic multi-plume models (Plant and Craig, 2008). The variance of the convective mass flux scales inversely with the number of convective clouds in the ensemble, and this variability about the

mean becomes increasingly important as the grid box size is reduced. This way, the level of stochasticity scales with increasing resolution of the model. Sakradzija et al. (2015) used this framework to parameterise shallow convection stochastically, while Sušelj et al. (2013) introduced a stochastic EDMF approach. In all these papers it has been shown that the natural variability of convection can be better captured with a stochastic rather than with a deterministic parameterisation. Finally, stochastic multicloud models (SMCM) form an interesting framework (Khouider et al., 2010). It allows for the evolution of a cloud population on a microgrid within each grid box of a large-scale model. This population can consist of different cloud types such as shallow cumulus, cumulus congestus, deep convective cumulus and anvil clouds. Each lattice site from the microgrid is characterised by one of these cloud types and can change from cloud type to cloud type according to probabilistic rules, conditioned on the large-scale state. The transition probabilities can be estimated from observational data or from LES data. The resulting fractional areas can be combined with a multi-plume model that can represent the vertical mixing of all these different cloud types. The stochastic multicloud model approach has a number of advantages over traditional cumulus parameterisations. It allows for different cloud types at the same time within a grid box. It introduces a memory for the different cloud types as well as the possibility that the different cloud types can evolve through different stages as dictated by the transition probabilities. And finally it allows for incorporating spatio-temporal cloud organisation on the microgrid at scales smaller than the grid size of the large-scale model. Recent implementations of SMCMs in large-scale models show encouraging results (Dorrestijn et al., 2016; Peters et al., 2017).

Further Reading

The only book exclusively dedicated to common parameterisation schemes in weather and climate models is *Parameterization Schemes* by David J. Stensrud (2007). Besides parameterisations of clouds, convection and turbulence, which is the topic of this chapter, it also describes current radiation, land surface and soil vegetation parameterisations. *Parameterization of Atmospheric Convection*, edited by Plant and Yano (2015) provides a broad and extensive overview of parameterization approaches of cumulus convection. The book *Storm and Cloud Dynamics* by William R. Cotton et al. (2010) provides an excellent overview of the dynamics of clouds and of precipitating mesoscale meteorological systems

TABLE 6.1: Symbols and key variables used in this chapter

Symbol	Description	Units (SI)
Generic fields		
s_d	Dry Static Energy	$J\,kg^{-1}$
s_e	Moist Static Energy	$J\,kg^{-1}$
s_ℓ	Liquid-Water Static Energy	$J\,kg^{-1}$
s	Saturation deficit	–
RH	Relative humidity	–
φ	Generic variable	$[\varphi]$
σ_φ	Standard deviation of φ	$[\varphi]$
$\bar{e}$	Turbulent Kinetic Energy	$m\,s^2$
ω	Vertical pressure velocity	$Pa\,s^{-1}$
Generic rates		
c	Condensation rate	s^{-1}
e	Evaporation rate	s^{-1}
G	Autoconversion rate	s^{-1}
P_s	Surface Precipitation rate, Eq. (6.3)	$kg\,m^{-2}\,s^{-1}$
Surface-layer variables		
u_*	Friction velocity	$m\,s^{-1}$
θ_*	Temperature scale for the surface layer	K
$q_{v,*}$	Humidity scale for the surface layer	–
u_*	Friction velocity	$m\,s^{-1}$
z_0	Roughness length	m
L	Obukhov length scale	m
C_m	surface drag coefficient	–
C_φ	surface bulk transfer coefficient	–
Φ_m	Universal gradient function for momentum	–
Φ_h	Universal gradient function for heat	–

Symbol	Description	Units (SI)
Boundary-layer variables		
K	ED, Eq. (6.9)	$m^2\,s^{-1}$
w_*	Boundary-layer convective velocity scale	$m\,s^{-1}$
z_i	Inversion height	m
ℓ_t	Turbulent length scale	m
w_t	Turbulent velocity scale	$m\,s^{-1}$
β	buoyancy parameter	$m\,s^{-2}\,K^{-1}$
ϵ_e	Dissipation rate of TKE	$m^2\,s^{-3}$
w_e	Entrainment velocity	$m\,s^{-1}$
Ri	Richardson number	–
P_r	Prandtl number	–
Cloud-related variables		
a_c	Cloud fraction	–
a_u	Updraft fraction	–
w_u	Cloud updraft velocity	$m\,s^{-1}$
$w_{u,*}$	Cumulus vertical velocity scale	$m\,s^{-1}$
R	Horizontal radius of a cloud	m
$N(R)$	Cloud size distribution	m^{-1}
h_c	Vertical extent of a cloud	m
z_b	Cloud-base height	m
$\mathcal{M}$	Mass flux	$kg\,m^{-2}\,s^{-1}$
E	Entrainment rate	$kg\,m^{-3}\,s^{-1}$
D	Detrainment rate	$kg\,m^{-3}\,s^{-1}$
ϵ	Fractional entrainment rate	m^{-1}
δ	Fractional detrainment rate	m^{-1}
ξ	mixing ratio	–
n	Particle size distribution	m^{-1}
D	Cloud particle diameter	m

and their parameterisation. *An Introduction to Boundary Layer Meteorology* by Roland Stull (1988) provides a pedagogical introduction to atmospheric boundary-layer turbulence and its parameterisation. A comprehensive overview of the history of the development of GCMs can be found in the book *General Circulation Model Development: Past, Present, and Future*, edited by David A. Randall (2000) and published in honour of Akio Arakawa's lifelong contribution to GCM development.

Section 6.3 The Design of Parameterisation
The level of complexity in parameterisation schemes is closely related to their number of tunable parameters.

The article 'The art and science of climate model tuning' by Frederic Hourdin et al. (2017) provides an insightful overview of climate model tuning and should be read by any researcher working on parameterisation development.

Section 6.4 Parameterisation of Boundary-Layer Turbulence
Chapter 4 of the book *Clear and Cloudy Boundary Layers* edited by Bert Holtslag and Duynkerke (1998) provides a comprehensive overview of parameterisations in the clear boundary layer. An insightful quasi-steady analysis of the convective boundary layer using ED profiles non-local fluxes can be found in Stevens (2000). Moist

turbulent kinetic energy-based parameterisations for the cloudy boundary layer are those of Cuxart et al. (2000), Lenderink and Holtslag (2000) and Bretherton and Park (2009). A comprehensive review of higher-order closures for the turbulence in the cloudy boundary layer has been published by Mironov (2008).

Section 6.5 Parameterisation of Cumulus Convection
A review paper on the past, present and future of cumulus parameterisation by Akio Arakawa (2004) provides a useful overview of the current challenges of cumulus parameterisations. Michael Tiedtke's bulk version of Arakawa's spectral cumulus parameterisation still forms the basis for many operational cumulus convection schemes and the original paper of Tiedtke (1989) provides a clear and practical description of a bulk mass flux scheme.

Section 6.6 Cloud Parameterisation
Statistical cloud schemes using a Gaussian PDF were first introduced by Sommeria and Deardorff (1977) and Mellor (1977). Extensions to more general PDFs are presented in Tompkins (2002) and more recently in Schemann et al. (2013). The coupling between mass flux schemes and a statistical cloud scheme is discussed in Klein et al. (2005). The link between RH-based cloud schemes and statistical cloud schemes are discussed in Quaas (2012). Tiedtke (1993) provides most of the basics of a prognostic cloud scheme.

Section 6.7 Cloud Microphysics
The standard book on cloud microphysics is *Microphysics of Clouds and Precipitation* written by Pruppacher and Klett (2010). Another interesting book on this topic concentrated more on parameterisation of cloud microphysics is *Cloud and Precipitation Microphysics: Principles and Parameterizations* written by Straka (2009).

Excercises

(1) Typical near-surface atmospheric conditions over subtropical ocean in the trades are characterised by a 10 metre wind of $10\,\mathrm{ms}^{-1}$, a surface heat flux $\rho c_\mathrm{p}\overline{w'\theta'}_\mathrm{srf} = 10\,\mathrm{Wm}^{-2}$, a surface latent heat flux of $\rho \ell_v \overline{w'\theta'}_\mathrm{srf} = 150\,\mathrm{Wm}^{-2}$ and a roughness length $z_0 = 10^{-4}$ m.

   (a) Calculate the friction velocity u_* and the drag coefficient C_m at 10 m. Assume neutral conditions for simplicity.

   (b) Calculate the surface buoyancy flux $\overline{w'\theta'}_{\rho\,\mathrm{srf}}$. What are the relative contributions from the surface latent and sensible heat flux?

   (c) Calculate the Obukhov length scale L. Is the value for the drag coefficient found in (a) providing an under- or over-estimation?

   (d) Assume an inversion jump of $\Delta\theta_\rho$ of 4 K. Give an estimate of the top-entrainment velocity for a clear convective boundary layer.

   (e) If the clear boundary layer is well-mixed and 500 m deep, what is the tendency (in $\mathrm{K\,day}^{-1}$) for θ_ρ?

(2) Consider a non-dimensional K-profile

$$\hat{K} = c\hat{z}\,(1 - \hat{z})^2$$

with $\hat{K} \to K/(w_* z_i)$ and $\hat{z} \to z/z_i$ and a non-dimensional counter-gradient parameter $\hat{\gamma} \to \gamma w_* z_i / Q_0$ where Q_0 is the surface heat flux. The potential temperature can be made non-dimensional through $\hat{\theta} \to \theta/\theta_*$ where $\theta_* = Q_0/w_*$.

   (a) Show that in quasi-steady state, when using the non-local ED parameterisation given by Eq. (6.45), the non-dimensional heat flux can be written as

$$\overline{w'\theta'}/Q_0 = -\hat{K}\left(\partial_z\hat{\theta} - \gamma\right) = 1 - \hat{z} + A\hat{z}$$

where $A = E/Q_0 \approx -0.2$ with E the entrainment flux defined in Eq. (6.44).

   (b) Show that for the given non-dimensional K-profile, the potential temperature profile in quasi-steady state is given by

$$\hat{\theta} = \hat{\gamma}z - \frac{1}{c}\ln\left(\frac{\hat{z}}{1 - \hat{z}}\right) - \frac{A}{c(1 - \hat{z})} + \mathrm{cst.}$$

(3) (a) Show that the bulk fractional entrainment of an ensemble of cloudy updrafts that all share the same cloud base z_b and a common aspect ratio $\gamma R = h$ between their radius R and vertical extent $h = z - z_\mathrm{b}$ is given by Eq. (6.73).

   (b) Derive an analytical expression for the bulk fractional entrainment rate if the cloud updraft size distribution is given by $N(R)dR \approx N_0 \exp(-R/R_0)\,dR$, if the entrainment rates of the individual cloudy updrafts are described by Eq. (6.71a) and the individual mass fluxes of the updrafts scale as $M \sim R^2$. The highest updrafts have a vertical extent of h_c and the updrafts with a vertical extent of $1/2h_\mathrm{c}$ have a radius R_0.

   (c) Verify whether the results of (b) agree with Eq. (6.75).

(4) Derive the continuous growth approximation of the accretion rate for the collection kernel (Eq. (6.138)) with $E \equiv 1$ and $u_\infty(D) = \alpha D^{1/2}$ for the terminal fall velocity, and neglect the contribution of the cloud

droplet in the collision kernel. Assume an exponential distribution $n(D) = N_0 e^{-\lambda D}$ for the raindrop size distribution. Make the approximation $x^* \to 0$ where appropriate and show that

$$\partial_t q_r\Big|_{ac} \cong \frac{15}{32} \alpha \, \pi^{\frac{5}{8}} N_0^{\frac{1}{8}} \rho_w^{\frac{7}{8}} \rho^{\frac{7}{8}} \, q_c \, q_r^{\frac{7}{8}}.$$

(5) Derive the alternative parameterisation for the accretion rate using the Long kernel, Eq. (3.35). Show that

$$\partial_t q_r\Big|_{ac} = k_{ac} \rho \, q_c q_r,$$

with $k_{ac} = 5.78 \times 10^3$ cm^3 g^{-1} s^{-1}.

(6) Numerical models that apply a non-parametric description of clouds are usually based on the logarithmic mass distribution function $g(\ln x) = x \, n(\ln x)$. Use the definition of the (continuous) drop size distribution and prove that

$$g(\ln x) = x^2 \, n(x) = \frac{Dx}{3} \, n(D).$$

(7) Prove that for a uniform distribution of the sub-grid variability of q_t as illustrated in Fig. 6.20 characterised by a total width Δ, the cloud fraction is given by Eqs (6.111) and (6.112).

(8) Consider a stratocumulus field with a cloud fraction $a_c = 1$. The stratocumulus field is spatially inhomogeneous: 50 % is characterised by LWP $= 50$ g m^{-2}, while for the remaining 50 % of the cloud field, a LWP of 150 g m^{-2} is determined. Assume that all cloud droplets have an identical radius of $r = 10$ µm.

(a) Calculate the cloud optical depth $\tau(\overline{\text{LWP}})$ based on $\overline{\text{LWP}}$ of the stratocumulus field.

An empirical relation between the cloud optical depth τ and the albedo α is given by

$$\alpha = \frac{\tau}{\tau + 7}. \qquad (6.158)$$

(b) Calculate the cloud albedo $\alpha(\overline{\text{LWP}})$, again based on the mean LWP.

A more accurate estimate of the cloud albedo of the stratocumulus field is obtained by calculating the mean cloud albedo $\overline{\alpha(\text{LWP})}$, thereby taking into account the inhomogeneity of the stratocumulus field.

(c) Calculate the mean cloud albedo $\overline{\alpha(\text{LWP})}$.

(d) Make a graph of the cloud optical depth τ versus the cloud albedo α. Show graphically that $\overline{\alpha(\text{LWP})} \le \alpha(\overline{\text{LWP}})$.

7 Evaluating Clouds

Christian Jakob and Jean-Louis Dufresne

Large-scale atmospheric models are used in many applications ranging from numerical weather prediction (NWP), seasonal prediction and climate projections to scientific inquiry. Unlike many models in use in other fields, weather and climate models are not inherently statistical models, but they are based on the fundamental laws of physics applied to the Earth system. A common feature of large-scale models is the need to include cloud and precipitation processes by means of parameterisation (Chapter 6). This is true for almost all global models, which will provide the main focus of this chapter. After briefly discussing the purpose and limitations of model evaluation, key techniques of modern model evaluation will be introduced in detail. While its main focus will be the description of the various model evaluation techniques, the chapter will provide illustrative examples of the use of the techniques, which collectively will provide an introductory overview of the current state-of-the-art in simulating clouds and precipitation in contemporary global models. This chapter makes extensive use of the concepts introduced in earlier chapters.

7.1 The Purpose and Limitations of Model Evaluation

7.1.1 What Is Model Evaluation?

Models come in different incarnations. Hence, defining the term *model evaluation* is intricately linked to the model in use. Models can be qualitative (or conceptual) or quantitative. Only quantitative models, that is, those who perform calculations resulting in a set of numbers, will be considered here. Even within that class there exists a variety of modelling approaches and it is important to define where in this wide range large-scale models of weather and climate fall.

A large class of models in science rely on empirical fits to available data and are often dubbed *statistical models*. A simple example for a common statistical model is the linear regression model, in which two data sets (x and y) are assumed to be connected through an equation of the form

$$y = \mathrm{a}x + \mathrm{b}. \tag{7.1}$$

The model describes a linear relationship between x and y with two parameters, a and b, which can be estimated from an existing (or 'observed') collection of concurrent values of x and y. Evaluation of this model is usually carried out by applying it to data not used in the parameter estimation process and by assessing the 'goodness of fit' of the model values y_m to their 'observed' values y_o. The 'goodness of fit' can be measured both qualitatively, for instance, by drawing a scatter diagram of the predicted versus the observed values in the x-y, or quantitatively by calculating error measures, such as those defined in Eq. (7.2).

$$\mathrm{ME} = \frac{1}{N} \sum_{i=1}^{N} \left(y_{m,i} - y_{o,i} \right)$$

$$\mathrm{MAE} = \frac{1}{N} \sum_{i=1}^{N} \left(|y_{m,i} - y_{o,i}| \right) \tag{7.2}$$

$$\mathrm{RMSE} = \sqrt{\frac{1}{N} \sum_{i=1}^{N} \left(y_{m,i} - y_{o,i} \right)^2}$$

$$\mathrm{R}^2 = \frac{\frac{1}{N} \sum_{i=1}^{N} (y_{m,i} - \bar{y}_m)(y_{o,i} - \bar{y}_o)}{\sqrt{\frac{1}{N} \sum_{i=1}^{N} (y_{m,i} - \bar{y}_m)^2} \sqrt{\frac{1}{N} \sum_{i=1}^{N} (y_{o,i} - \bar{y}_o)^2}}$$

Here, ME is the mean error, often also referred to as the bias, MAE is the mean absolute error, RMSE is the root mean square error and R is the correlation coefficient.

N is the number of values in the sample and the subscripts m and o refer to the modelled and observed values, respectively. The over-bar denotes the average. All four measures are commonly used in assessing the departure of a model from observed values including when assessing weather and climate models. In the simple example of the linear regression model, the error measures in Eq. (7.2) can be directly used in finding the best values for the model parameters a and b as those that minimise one or all the previous error measures. Traditionally, a and b are found by minimising the RMSE. While simple, this example already highlights the strong connection of developing and evaluating a model. Choices in the model design, limited here to two parameters, are made in such a way that prediction errors are minimised. An increase in model complexity will complicate the interaction of model design and evaluation – an issue at the heart of what follows.

In contrast to the previous simple model, large-scale weather and climate models *are not* simple statistical models. Instead, they are numerical expressions of the fundamental laws of physics, including momentum, mass and energy conservation, applied to the atmosphere, ocean and land. The time-dependency included in many of these laws (e.g., see Chapter 2) is exploited to formulate prognostic equations for key quantities, such as the wind components and temperature, as well as water vapour and potentially condensed forms of water such as cloud water or ice. Numerical methods are applied to solve these equations as a function of time and space, which ultimately permits the prediction of the future behaviour of weather and climate. In other words, rather than built into statistical relationships such as those described earlier, the behaviour of the atmosphere, ocean and land *emerges* from the model. This has profound consequences for model evaluation, as it allows it to focus on evaluating the closeness of a model's *emergent behaviour* to the behaviour of the true system as revealed by observations, theory and more accurate process models.

As weather and climate models are used in a number of applications, their evaluation is necessarily diverse and is adjusted depending on the purpose of the application. Furthermore, as the models are continuously improved, model evaluation also has to contribute to this *model development* process by revealing not only model shortcomings themselves but also the reasons for their existence. Hence, in its most fundamental form, model evaluation has two main tasks:

(1) determine whether a model is fit for its purpose,
(2) identify the reasons for model shortcomings.

This broad definition of the purpose of model evaluation contrasts the more frequently used one of a model being *right* or *wrong*. All models are approximations of the real world and the key question to be answered by evaluation is whether the model is an adequate tool. This applies to both the predictive and scientific applications of weather and climate models. With this in mind, examples for possible questions posed for model evaluation are:

- *Does the model meet the needs of the application and to what extent?* An NWP model is applied to predict the sea-level pressure distribution five days ahead. By carrying out many predictions of this type for past cases (so-called hindcasts) and comparing the results to observations, it can be ascertained if the model succeeds in such predictions and how close to reality they are. Based on the results of this evaluation, the model can then be judged to be fit-for-purpose or not.
- *What are the key model shortcomings influencing the model application?* The NWP model succeeds in predicting sea-level pressure, but how well does it predict *the actual weather*? An obvious key weather quantity that should be predicted well is rainfall. Other examples are the prediction of low clouds for air traffic management around airports or fog for road traffic conditions. Once it is known which of these quantities are the least well represented in the model predictions, this knowledge can be combined with some judgement on their importance to provide priorities for future model development.
- *What are the causes for the model shortcomings?* Having identified and prioritised model shortcomings, a further step in model evaluation might be to apply diagnostic techniques to gain insight into their causes. Is the rainfall in a region poorly represented because the model does not capture key dynamical processes or because the representation of microphysical processes is incomplete? This is an important task of model evaluation in support of future model development.
- *What are the limits of applicability of the model?* A coupled ocean–atmosphere model is meant to be used to study changes in the El Niño Southern Oscillation (ENSO) phenomenon in a future climate. An obvious task for model evaluation is to determine if the model simulates the key features of ENSO in the current climate. If it does not, it cannot be applied to study its changes. It is not fit-for-purpose.

These examples highlight the close connection of model evaluation and model development. Model evaluation also directly contributes to an improved understanding of the system that is being modelled. It does so by identifying the processes that are crucial for a realistic simulation of the system and hence likely play a key role in the real world.

7.1.2   The Limitations of Model Evaluation

Model evaluation requires an estimate of truth. As weather and climate models aim to represent a very complex system, defining truth for their evaluation is far from easy. In general, there are three estimates of 'truth' that are frequently applied:

- *Observational estimates.* The use of observations is the most common way of evaluating models. Observations, for the purpose of this chapter, are either direct measurements of a quantity of interest, such as a radiance at a certain wavelength, or quantities derived from such measurements, such as the cloud optical thickness (Chapter 4). It is important to recognise that observations of all kinds are subject to errors themselves. Direct measurements, such as radiances, are subject to instrument errors, while derived quantities contain additional errors resulting from the assumptions made in their derivation. As large-scale model results usually represent averages over large areas (the model grid boxes), another important source of observation 'error' is the mismatch of the scale of the observation with that of the model. This is referred to as *representativeness error.* A common way of taking observation uncertainty into account is to use more than one set of observations in the evaluation of the models. Another is the use of instrument simulators (Section 8.2.2).
- *Estimates from more accurate models.* Quantities that are parametrised in large-scale models can sometimes be modelled more accurately using more detailed models. Examples are the simulation of radiative fluxes using line-by-line radiative transfer models, the simulation of cloud systems using large eddy simulation (LES) and storm-resolving models (SRMs), and the direct numerical simulation (DNS) of small-scale turbulence. Simulations with such models are computationally very expensive and, therefore, such models provide an estimate of truth for a very limited set of situations. Nevertheless, using them in the evaluation process for large-scale models has proven very effective.
- *Theoretical insights.* Some of the behaviour of the atmosphere and ocean can be predicted from theoretical arguments. For example, it is known that in the middle to upper troposphere away from the equator the atmospheric wind field is close to geostrophic balance.

We can test if this behaviour emerges from the model equations.

The need for an estimate of truth is a significant limitation of model evaluation. Only quantities for which such estimates exist can be evaluated, limiting the assessment of models to what is observed and known. This usually does not severely limit the evaluation of weather predictions, as many of the weather variables that are of interest are reasonably well observed. However, it is less clear what influence this limitation has on the evaluation of climate simulations. What is clear is that a lack of observations as well as the uncertainties associated with the observations that are available often prevents the identification of the root causes for model shortcomings. This stresses the need to closely link model evaluation efforts with the broader scientific goals to better observe and understand the climate system.

The evaluation of climate models is more difficult than that of NWP models. This is because the purpose of using an NWP model is more easily defined than that of a climate simulation. NWP is an initial value problem and its goal is to predict 'the weather' for the next few days to weeks. Model evaluation therefore reduces to defining 'the weather', waiting for a few days until the predicted weather can be observed and comparing the forecast with the observations. Beginning in the early days of NWP and continuing today, one of the most important aspects evaluated is the representation of the large-scale circulation features that determine local weather. More recently, NWP models are also evaluated in their ability to directly forecast important weather parameters, such as near-surface temperatures, precipitation and clouds.

Fig. 7.1 provides an example for the evaluation of NWP models. It shows a simple scalar error measure (the anomaly correlation) for the simulation of the geopotential height of the 500 hPa pressure surface using the European Centre for Medium-Range Weather Forecasts (ECMWF) NWP model. While relatively simple, the calculation of this measure for several areas, forecast ranges and over thirty or so years enables the assessment of the performance of this particular NWP system. Forecasts at all ranges have improved significantly for both the northern and southern hemispheres. The rate of improvement has been roughly one forecast day per decade and has been slightly larger in the southern hemisphere than in the northern hemisphere, to the extent that the forecast quality is now nearly equal. As a result, five-day forecasts today have the same quality that three-day forecasts had twenty years ago. This example stresses the usefulness of simple scalar performance measures – often referred

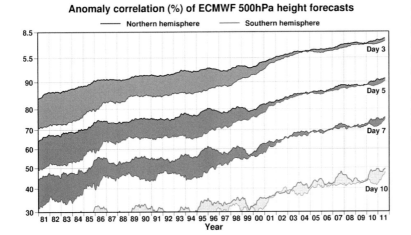

Anomaly correlation (%) of ECMWF 500hPa height forecasts

FIGURE 7.1: The evolution of the performance of the European Centre for Medium-Range Weather Forecasts (ECMWF) NWP model as measured by the accuracy in the simulation of the 500 hPa geopotential height field in the northern (thick lines) and southern (thin lines) hemisphere for three-day, five-day, seven-day and ten-day forecasts. Extended version based on Simmons and Hollingsworth (2002)

to as *metrics* – in measuring success and progress. Not surprisingly, metrics have been central to the evaluation of both weather and climate models.

Climate simulation is a boundary value problem and the purpose of the simulations made with climate models is generally much broader than that of NWP. The main purpose of climate simulation is to understand and predict the behaviour of the climate system under changed forcing including, but not limited to, those from anthropogenic greenhouse gas emissions. A particular interest in applying complex climate models is to try and assess regional changes associated with changes in the global climate. This requires the models to not only skilfully predict global mean features, but to correctly simulate regional circulations and their potential changes.

Given the goal of predicting a future far ahead, a simple fit-for-purpose evaluation for climate models is not possible. The observations of the projected changes lie decades ahead. They likely represent climate states that have not been observed in the past. Consequently, it is not possible to simply assess the performance of climate models by comparing their projected behaviour to that observed. Instead, evaluation of climate models must rely on the behaviour of the models for current and past climates. This raises the question of what aspects of climate the models must simulate well to be deemed reliable in simulating future climate states. Despite significant efforts, simple links between the skilful representation of aspects of current climate and the behaviour of a climate model in projecting future climates have proven difficult to establish.

This is illustrated in Fig. 7.2, which shows the relationship between the performance of climate models in simulating broad aspects of clouds and precipitation (*y*-axis) and the change in global mean temperature for

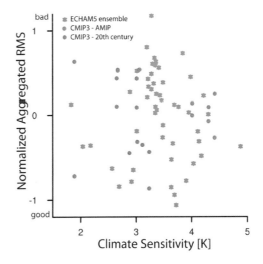

FIGURE 7.2: Relationship between an overall climate model skill measure and the model's climate sensitivity for a single model ensemble (green) and the CMIP3 multi-model ensemble (blue and orange). The skill measure takes into account precipitation, cloud radiative effect and cloud fraction information.
From Klocke et al. (2011). Reprinted with the permission of the American Meteorological Society

a doubling of carbon dioxide, an important quantity also known as climate sensitivity (*x*-axis, see Chapter 13). It is evident that the general performance in simulating clouds and precipitation in a model provides little indication at all as to what its climate sensitivity will be. One may conclude from this result that it is pointless to evaluate climate models in current conditions. In fact, the opposite is more likely to be true.

A faithful representation of current climate as measured for instance by performance metrics provides a necessary condition for model credibility. However, it is far from a sufficient condition for judging a model as fit for purpose. To gain confidence in a climate model, it is necessary to deeply understand if the underpinning processes leading to its overall results are well represented. This is so, because in the absence of simple measures of success, such as those more readily available for NWP, 'trust' in climate model performance must rely on the models' ability to represent such processes well. This additional requirement of 'getting the right answer for the right reason', leads to the need for more insightful model evaluation techniques that are able to probe relationships between climate variables and, importantly, their underlying mechanisms.

7.1.3   Evaluating Large-Scale Models: An Overview

The previous section established model evaluation as a wide-ranging field with often multiple purposes from assessing whether a model is fit-for-purpose to investigating potential causes for model shortcomings. Model evaluation involves comparisons to the 'truth', which is either estimated from observations or from other models.

Usually, the first step in model evaluation is an overall assessment of the model performance in its application (Fig. 7.3). The purpose of this step is to monitor model performance and, importantly, to *identify* model shortcomings. Fig. 7.1 is an example for this evaluation step. *Application-oriented* model evaluation is the most common and wide-spread form of model evaluation. It can be carried out in a *quantitative* fashion (e.g., Fig. 7.1) or more qualitatively, for example, through the visual

comparison of maps or graphs showing observed and modelled quantities.

Fig. 7.4 provides an example for this more qualitative approach to model evaluation. Fig. 7.4a shows the climatology of the observed annual mean precipitation from the Global Precipitation Climatology Project (GPCP). Fig. 7.4b shows the same climatology averaged over

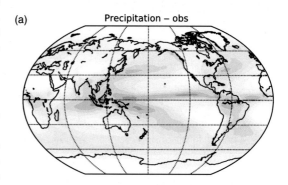

(a)   Precipitation – obs

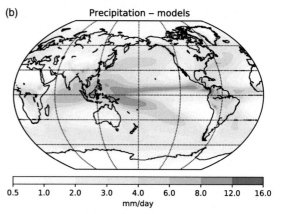

(b)   Precipitation – models

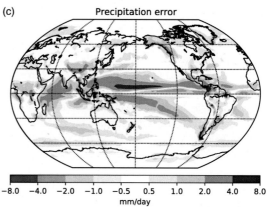

(c)   Precipitation error

FIGURE 7.4: Annual mean precipitation (a) derived from GPCP observations, (b) simulated by models and (c) the model error (*mm/day*) for the average of twenty-nine atmospheric models used in the AMIP experiments of CMIP5. Results are obtained by running the models with historical climate forcings.

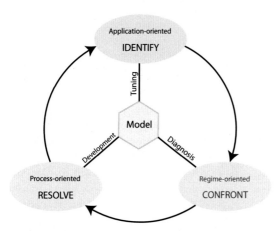

FIGURE 7.3: Schematic of the the model evaluation process including the three components covered in this chapter: application-oriented evaluation; regime-oriented model evaluation and process-oriented model evaluation.

twenty-nine models from the Atmospheric Model Intercomparison Project (AMIP), which is part of the fifth edition of the Climate Model Intercomparison Project (CMIP5). Fig. 7.4c shows the model error. Comparing the multi-model average annual mean precipitation distribution to observations has the goal of highlighting those model errors that are consistent between many models. In this example, it is clear that models have common errors in their precipitation climatology, with an overestimation of precipitation on both sides of the equator in the Pacific Ocean and a more mixed signal in the extratropics. Note again that the details of the model errors can vary with the observational data set chosen for the evaluation, due to uncertainties in our estimates of precipitation, especially over the ocean (Stephens et al., 2012b).

While of obvious interest, this type of model evaluation is of limited use to the model development process, as it usually does not provide much insight into the causes of model shortcomings. An important final step of model development, known as *model tuning*, takes place at this level of model evaluation. The term refers to the adjustment of a small set of poorly known model parameters so that the model adheres to known global constraints of the climate system, such as the top-of-the-atmosphere (TOA) balance between short- and long-wave radiative fluxes in equilibrium. Key approaches to application-oriented model evaluation will be discussed in Section 7.3.1.

Having identified the overall shortcomings of a model, the next step in model evaluation is to gain insight into their causes by trying to identify specific phenomena or regimes that make a large contribution to the overall error. The goal of this step is to *confront* models with data in such a way that the likely causes of errors can be identified. Is a mean error in TOA radiative fluxes due to clouds in ascending or descending motion? Is the radiation error also associated with rainfall errors? Regime-oriented model evaluation requires the development of techniques that aid in the identification of typical recurring patterns and that allow the contribution of each regime to the model error to be quantified. Simple examples for such regime-based approaches are the focus on geographic regions that are dominated by a single cloud type (e.g., the stratocumulus regions in the Eastern parts of the subtropical oceans) or the compositing of cloud structures around typical dynamical features of the atmosphere, such as extratropical cyclones. More sophisticated techniques use pattern-matching algorithms to define dynamically based or cloud-structure-based regimes. The techniques developed for this step of model evaluation can be quite sophisticated and their application is often referred to

as *model diagnosis*. This type of evaluation will be the subject of Section 7.3.2.

Once the key regimes contributing to a model error have been identified, the final step of model evaluation is to identify the processes involved in causing the error. This step is referred to as process-oriented model evaluation. Its main goal is to *resolve* issues in the model formulation to reduce the overall model error. It is therefore very tightly linked to model development. The process-oriented approach requires detailed information on the particular contribution individual processes make to the model errors. Consequently, it tends to focus on analysing the rates of change of cloud-related variables, rather than only the variables themselves. As a result, this approach is often closely connected with highly specialised and intensive observational campaigns, in which many additional observations are taken, allowing the identification of the role of particular processes in an observed phenomenon. Another frequently used approach to process-oriented model evaluation is the concurrent use of field study observations and LES or SRM. Those can be compared to simpler versions of large-scale models, for instance, by extracting a single column of the global model and prescribing the inputs from neighbouring columns either in an idealised fashion or from field observations, an approach known as single-column modelling (SCM). Process-oriented model evaluation will be discussed in Section 7.3.3.

Model evaluation, as encapsulated in the combined use of the application-oriented, regime-oriented and process-oriented approaches is a rapidly expanding and evolving field of research that integrates many communities. Model users are interested in establishing the usefulness of the model for the application they are interested in. The model development community has a keen interest in identifying possible causes of model shortcomings so that their development efforts can be focussed on aspects that are likely to provide model improvements. The extensive use of observations as estimates of truth – both at the large-scale and field study level – and the sophisticated application of these observations in model evaluation is also increasingly being integrated into model evaluation activities. The use of LES and SRMs to provide process-level understanding to the model evaluation process engages yet another community.

7.2 Evaluating Clouds and Precipitation

7.2.1 Model versus Observed Clouds

Clouds have several profound effects on weather and climate. Many of these effects relate to their role in Earth's energy and water cycles. Note that some clouds, such as deep convective clouds and organised convective systems, can also strongly affect the local momentum budget. The main heating effects of clouds are associated with their interaction with radiation (Chapter 4) as well as the release and consumption of latent heat associated with phase changes (Chapter 3). As the sole agents for generating condensed forms of water in the atmosphere, clouds also play an important role in the water cycle, as they produce the observed distribution of precipitation. Many of the processes in the energy and water exchanges in clouds critically depend on the macrophsyical (e.g., cloud shape, depth and size) and microphysical (e.g., phase, cloud particle shape and size) structure of the clouds. It is that connection that requires not only to evaluate the effects of clouds on radiation and precipitation, but the cloud properties themselves. This is necessary to (i) ensure that where the cloud effects appear realistic, they are so for the right reasons and (ii) identify shortcomings in the cloud representation where they lead to unrealistic effects on the energy and water cycles.

Very few clouds are subject to direct, *in situ* observation, as this requires aircraft equipped with a wide range of instruments to penetrate clouds. As a consequence, most cloud observations rely on remote-sensing techniques which are often assessed and calibrated using the above aircraft observations. Chapter 4 has discussed the issues that arise when attempting to observe clouds by remote sensing. In particular, it was shown that the definition of a cloud in such observations strongly depends on the instrument used to observe it. This is so because clouds interact with radiation of different wavelengths in different ways. Passive instruments, such as the eye or radiometers in the visible or infrared part of the electromagnetic spectrum, tend to be most sensitive to the the parts of the clouds nearest to them, that is, the cloud base when looking from the ground and the cloud top when looking from above. Furthermore, passive instruments frequently miss clouds that are optically thin at the measurement wavelength. The use of active remote sensing techniques, such as lidars and radars, from both the ground and from space can alleviate some of these problems, but it comes with its own set of measurement issues, such as small instrument footprints and sensitivity to particular parts of the droplet size spectrum. A more complete discussion of these issues can be found in Chapter 4.

The sensitivity of cloud observations to the instrument used in measuring them poses a fundamental problem for model evaluation: what is a cloud? In large-scale models, clouds are represented by a small set of macrophysical (e.g., cloud fraction, cloud overlap) and microphysical

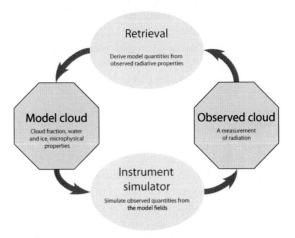

FIGURE 7.5: The two main approaches to comparing model and observed clouds.

(e.g., cloud particle size) properties. From the previous discussion it is evident that the definition of 'cloud' fraction will depend on the instrument taking the measurement. Given that, is the difference between a model and an observation the result of model error or simply a reflection of the difference in definition? The two main approaches currently in use to overcome this important conundrum are summarised in Fig. 7.5. In short, they are the *observation-to-model* approach, in which it is attempted to convert observed quantities into those produced by models; and the *model-to-observation* approach, in which model quantities are converted into observables.

The first, and still most common, approach is to derive the modelled cloud properties from observations, a technique referred to as *retrieval*. The obvious advantage of this approach is that once completed the retrieved variables apparently match the model ones and the evaluation becomes a 'simple' comparison, either quantitative or qualitative, between the 'observations' and the model. However, in the process of retrieving the cloud characteristics, assumptions had to be made to convert the observed radiative measurements into cloud properties. These assumptions are often hidden to the users of the data set. This can lead to inconsistencies between the model and retrieved variables often referred to as a comparison of 'apples to oranges'. This problem is particularly acute when the sensitivity of the measurement to the existence of condensed matter (a cloud) differs significantly from assumptions made in the model as to what constitutes a cloud. It therefore strongly affects measurements of cloud existence – or cloud fraction – and any subsequent retrievals dependent on this existence in the observations. Where cloud existence is less of an issue, for instance, in the retrieval of bulk liquid-water path (LWP) from microwave measurement, the retrieval method remains a powerful tool in model evaluation. However, care must always be taken in interpreting comparisons between model clouds and retrievals of cloud properties. Sometimes, but not always, it is possible to estimate the observation uncertainty by applying several different measurements and techniques to retrieving the same quantities, as illustrated in Fig. 7.12.

An alternative to retrievals is the application of instrument simulators. Here the model's cloud and precipitation fields are converted into observable quantities. There is a wide range of cloud- and precipitation-related observations ranging from radiances at particular frequencies to radar reflectivity to lidar backscatter and polarisation information. Hence, for the simulator approach to be applied successfully requires the development of a range of different instrument simulators, one for each of the measurements of interest. Some details of the most common simulators will be discussed in the following subsections. While necessarily varying in detail, all instrument simulators contain two basic ingredients:

(1) **An algorithm to match the scale of the observations with that in the model.** This step is necessary as most observations of cloud- and precipitation-related quantities are taken at much smaller scales than that of large-scale model grid boxes. For instance, geostationary and polar orbiting passive satellite instruments have pixel sizes of the order of 1–10 km^2, while the footprints of many of the active CloudSat and Cloud-Aerosol Lidar and Infrared Pathfinder Satellite Observation (CALIPSO) measurements are even smaller. This part of the simulator often applies the model's cloud overlap assumption (Chapter 6) to first distribute the model clouds at various levels and then generates 'model pixels' of comparable size to the observations.

(2) **An algorithm to calculate the actually observed quantity.** This part of the simulator turns the model information – usually cloud and precipitation water and ice content as well as some information on particle size, phase and shape – into the observable. Actual algorithms to do so vary in complexity from simple parametric assumptions to complete radiative transfer calculations (Chapter 4).

With the basic construction principles for instrument simulators in mind, we will further illustrate the simulator idea using one of the most widely used simulators to date, the International Satellite Cloud Climatology Project

(ISCCP) simulator. A broad overview over other existing simulator tools will follow that discussion.

7.2.2 An Instrument Simulator Example: The ISCCP Simulator

A commonly used and yet simple instrument simulator is that developed to evaluate model results against observations collected and analysed by the ISCCP (Klein and Jakob, 1999; Webb et al., 2001). ISCCP provides a range of cloud products derived from visible and infrared measurements from geostationary and polar-orbiting satellites since 1983. One particular product frequently used in model evaluation is a two-dimensional histogram of the frequency of occurrence of optical thickness (τ) and cloud-top pressure (CTP) at pixel level in a (280×280 km^2) grid box. The former is derived from visible radiance measurements while the latter relies on infrared cloud-top temperature measurements that are converted into pressure using additional information on the vertical temperature structure of the atmosphere. The histograms are available globally every 3 hours for the sunlit hours of the day. A schematic of the construction of these histograms as well as an example for them are shown in Fig. 7.6.

The histograms are constructed by determining both τ and CTP for every individual pixel in the grid box (note that some subsampling is applied). After doing so, each pixel is assigned to one of forty-two pre-determined classes of joint CTP–τ values in six optical thickness and seven CTP bins. The number of pixels in each class is divided by the total number of pixels sampled, including clear-sky pixels, to yield the relative frequency of occurrence of pixels in each CTP–τ class. The relative frequency of occurrence, either expressed as a fraction of one or in per cent, is then displayed. An example of an ISCCP histogram is shown in the right panel of Fig. 7.6. It shows the mean histogram for tropical clouds that have been found to be associated with large and often organised convective systems. The occurrence of different cloud types can easily be determined from this diagram. In this instance, there is a prevalence of clouds with high tops and medium to large optical thickness as is typical for the convective and stratiform parts of large convective systems. The total cloud cover is simply the sum over all frequencies of occurrence, in this example it is close to 95%.

There are several issues in using passive satellite instruments in determining cloud properties. They are usually related to the instrument sensitivity and viewing properties (Stubenrauch et al., 2013). Without giving

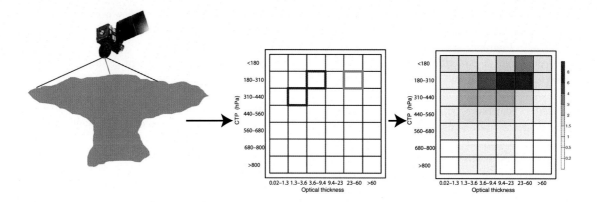

FIGURE 7.6: Schematic for the construction of the ISCCP histogram.

an exhaustive list, these issues include the blocking of low clouds in the presence of high clouds, the potential misinterpretation of CTP for optically thin clouds and the potential misinterpretation of τ for partially cloudy pixels. As models provide a much more complete picture of their clouds, some of these problems do not exist for model clouds. This can lead to misleading comparisons between the model and the observations. Perhaps the simplest example is the comparison of low cloud cover (say clouds with a CTP > 800 hPa). In the model, this can simply be calculated using the vertical distribution of clouds below 800 hPa, but in the observations, some of these clouds might be obscured by clouds above them, leading to a potential misinterpretation of model behaviour. For this and other reasons, it would be desirable to 'simulate' what the satellite constellation used by ISCCP would see if it observed the model clouds. This led to the development of the ISCCP simulator.

When designing an observation or instrument simulator, there are still several choices to be made as to what level of complexity to apply. These choices will affect our expectations of how well models and observations can agree after the application of the simulator. Trade-offs between high complexity – often involving high computational cost and difficulties in implementing the tool in the model – and capturing the key features of the observational techniques need to be made. As a result, not all observation artefacts can necessarily be removed. The trade-offs are not only informed by cost and complexity, but also by the overall expected quality of the model simulations. As clouds remain among the most poorly simulated aspects in large-scale models, it is often sufficient to capture the main observational features in a simulator. The ISCCP simulator aims to alleviate the two largest issues in the observations, namely the CTP determination in optically thin clouds as well as

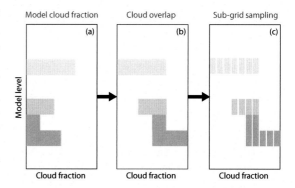

FIGURE 7.7: An example for the construction of the sub-grid distribution of model clouds for use in instrument simulators.

the blocking of the satellite view of lower-level clouds in the presence of high clouds. At the same time, it aims to provide the possibility of evaluating the entire sub-grid distribution of the model cloud field in the context of the CTP-τ diagrams introduced earlier. As discussed earlier, achieving this goal requires the simulator to implement two basic steps: (i) match the scale of the observations by determining the sub-grid scale distribution of the model clouds and (ii) perform some radiative calculations to adjust for observational artefacts, such as the ambiguous CTP determination.

The first step in the ISCCP simulator is to divide the model cloud field into 'pixels' for which the CTP and optical thickness can then be determined. To do so, the model grid-box is divided into a fixed number of sub-boxes, K, where K varies between a few tens and a hundred depending on application and model resolution. The model's cloud overlap assumption determines how clouds in different vertical levels overlap with each other (Chapter 6). Fig. 7.7a shows a hypothetical model cloud fraction configuration as a function of height. The shading of the boxes indicates the layer cloud optical thickness, with lighter shades representing optically thinner clouds. In Fig. 7.7b, the model's cloud overlap is applied to place the clouds in each model level horizontally. Finally, the grid box is subdivided into several 'pixels' (ten in this case), yielding ten cloud configurations for each of which the CTP and τ can now be determined.

Calculating the optical thickness is relatively straightforward, as all models perform this calculation within their radiative transfer codes. This requires knowledge of the liquid-water and/or ice content of the cloud as well as an assumption about the cloud particle shape, size and/or number embedded in the radiation code. Note that the pixel liquid-water and ice content can in principle depend on an assumption of the water/ice horizontal distribution across the grid box. In most models, this distribution is assumed to be uniform, making the water/ice content at each model level the same in each sub-grid box. Some models do assume an inhomogeneous distribution of condensate, which will then require translation into the simulator as varying condensate contents across the sub-grid boxes. Note that even with a homogeneous condensate distribution in each level, in a vertically integrated sense the liquid-water and ice-water path (IWP) will vary significantly due to the model's cloud overlap assumption (see Fig. 7.7).

The calculation of CTP for each model pixel is less straightforward. Simply using the physical CTP (the pressure of the top layer of clouds) would successfully account for the blocking effects of underlying cloud layers. However, it would not capture the fact that the radiative CTP can be significantly larger if the upper cloud layers are optically thin and hence partially transparent to radiation emanating form lower levels in the atmosphere. In this case, the ISCCP-'observed' CTP would lie somewhere between the highest cloud top and the surface, depending on the cloud constellation. It is an obvious goal for the simulator to at least mimic this behaviour of the observations. To do so, the radiance at 11 µm for each sub-column is calculated, taking into account the emissivity and transmissivity of each model layer including its clouds (note that in the sub-boxes cloud cover in each layer is either zero or one). After this, the ISCCP approach of calculating CTP is broadly followed. It assumes that clouds form a single layer in the atmosphere – in other words, it only assesses the column-integrated optical thickness of clouds regardless of the number of true cloud layers in the column. The emissivity of this 'bulk' cloud is calculated from its optical thickness. The total radiance at the top of the column is then assumed to be the sum of the radiation emitted from this bulk cloud and the radiation transmitted from the surface. The cloud emission temperature is varied until the so-calculated radiance matches that of the full calculation. This emission temperature is taken as the cloud-top temperature and is easily converted to CTP using the model's vertical temperature profile.

The results of this procedure for the hypothetical cloud configuration in Fig. 7.7 is shown in Fig. 7.8. By construction, the physical CTP increases from the left to the right, a direct result of the application of the model's cloud overlap assumption. However, for the first seven sub-columns, the radiative CTP is higher than the physical one. This is due to the fact that the high-level clouds in the schematic are assumed to be optically thin. As a result, in the left-most columns, the 'measured' radiation at the top of the atmosphere is a mixture of that from the surface and the layers below the cloud transmitted through the cloud layer and that from the cloud layer itself, placing the apparent cloud top at a much lower level. As the presence of lower-level optically thicker clouds increases when moving to the right, the inferred CTP decreases. However, it is only the last three sub-columns, which do not contain any high level thin cloud, for which the physical and radiative CTP are identical.

The final step in the ISCCP simulator is to collect the 'pixel'-values of CTP and τ in the two-dimensional histograms introduced earlier, which can then be compared to those derived from observations.

7.2.3 A Simulator Toolbox: The Cloud Feedback Model Intercomparison Project Observation Simulator Package

The success of the ISCCP simulator combined with a need to more thoroughly evaluate model clouds and precipitation has led to the development of instrument simulators for various cloud and precipitation observing satellites. These simulators have been combined into a single tool named the Cloud Feedback Model Intercomparison Project (CFMIP) Observation Simulator Package (COSP) (Bodas-Salcedo et al., 2011). Fig. 7.9 shows the current basic ingredients of COSP. Starting from the model profiles of cloud fraction and condensate, the first step for all simulators is to design sub-grid-scale information for the cloud and precipitation fields. The sub-grid information is then passed to various instrument simulators. These currently include a radar simulator for the CloudSat radar, a lidar simulator for CALIPSO measurements, the ISCCP simulator, simulators for both the moderate resolution imaging spectroradiometer (MODIS) and multiangle imaging spectro-radiometer (MISR) as well as a fast

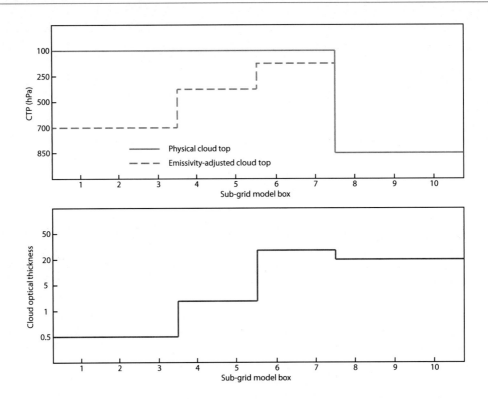

FIGURE 7.8: CTP and τ determined by the ISCCP simulator for the idealised example in Fig. 7.7.

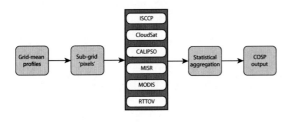

FIGURE 7.9: The basic structure of COSP.
Adapted from Bodas-Salcedo et al. (2011)

radiative transfer model (RTTOV), which can be used to simulate radiances at a variety of wavelengths. The package is designed in a modular fashion, so that one can execute any subset of the simulators. The results of the simulator are then statistically aggregated (i.e., into CTP–τ histograms in the case of ISCCP).

The simultaneous use of several instrument simulators allows a more comprehensive assessment of model clouds and to some extent precipitation, as all comparisons are carried out more consistently with the observational assumptions made. An example for such an assessment is presented in Fig. 7.10. A version of the UK MetOffice global atmospheric model is compared to observations from five instruments over the Southern Ocean for December, January and February. The ISCCP, MISR and MODIS results all indicate a lack of mid-level clouds compensated by a slight overestimate of thick high-top clouds. This is confirmed by the CALIPSO results, which show a lack of mid-level clouds with high scattering ratios, typically found in water clouds. The CloudSat observations add additional detail, showing a distinct bimodality in radar reflectivity at low levels, indicative of non-precipitating clouds (small reflectivity) and clouds with relatively large precipitating particles (high reflectivity). In contrast, the model produces a maximum of occurrence between −10 and 0 dBZ, indicative of an almost constant presence of drizzle. While only illustrative, this example highlights the potential for the combined use of instrument simulators to better understand shortcomings in model simulations of clouds.

7.2.4   Observation Simulators or Retrievals?

The introduction of simulators to the evaluation of clouds enables a better matching of cloud observations with model clouds. Simulators provide a complementary approach and enhancement, rather than a replacement, of the use of retrievals. Both have advantages and limitations (Pincus et al., 2012).

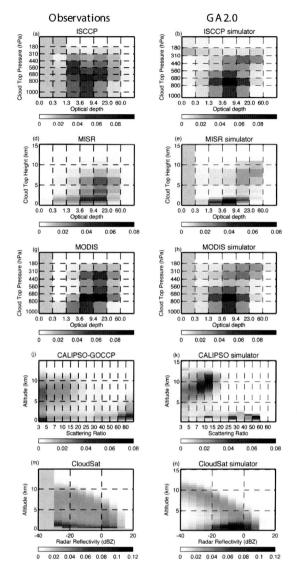

Observations　　　　　GA2.0

FIGURE 7.10: An example of the application of the COSP group of simulators using the ISCCP, MODIS, MISR, CALIPSO and CloudSat simulators.
From Bodas-Salcedo et al. (2012). Reprinted with the permission of the American Meteorological Society

• One of the most obvious limitations of simulators is that by their very design they introduce observational artefacts into the model results. This can make the interpretation of the results more difficult and in some cases even obscure the real model error. A simple example would be the absence of low CTP (i.e., high-level), medium τ clouds in a model-simulated ISCCP histogram, when they are observed. Instead, the model shows mid-level top clouds. Interpreting this error from the simulator output alone is difficult, as it could be realised in more

than one way. The model might truly have no high clouds and mid-level clouds instead, or the high clouds may exist but be too optically thin. As the observations cannot tell the difference between those two situations, nor can the simulator results. In this example, it is simple to overcome this problem by just analysing the model output without applying a simulator. This illustrates that the complementary use of simulator and raw model output is often a necessity.

• There are observational artefacts that are very difficult to implement in simulators without significantly increasing their complexity. An example for the ISCCP simulator discussed in Section 7.2.2 is that no account is being taken of the viewing angle an instrument makes with the clouds. For instance, geostationary satellites observe clouds at a wide range of angles. The photon path taken through a given clouds is much shorter at nadir than with a limb view, leading to (i) an easier detection of thin clouds away from the nadir view and (ii) an increase in the estimated cloud optical thickness. The viewing angle is not considered at all in the current ISCCP simulator and doing so would lead to a major increase in its complexity.

• While simulators attempt to match the observational pixel scale, this can of course not be done perfectly, leading to expected differences between the simulator results and observations. Perhaps more importantly, for many observations, and hence simulators, assumptions still need to be made about the cloud distribution at sub-pixel scale. Currently, those assumptions can be different between simulators and observations. For example, the ISCCP simulator assumes 100% cloud cover in each sub-grid pixel, while the observations make no such assumption. This can lead to differences in the calculated pixel optical thickness.

• Both the observations and simulators rely on ancillary data for their calculations. For instance, radiative transfer calculations require temperature and humidity profiles in the atmosphere. Another example is the need to know the surface albedo in calculations in the shortwave spectrum. In simulators, those inputs are taken from the model, while for the observations, a variety of sources are used depending on the satellite product. Differences in the ancillary data can therefore lead to differences in the results, even if the model clouds themselves are perfectly simulated.

This discussion leads to the obvious question that if tasked with evaluating clouds in models, which of the approaches should one use? Is there an obvious reason to not use a particular one? There is no simple answer to that question. One could argue that careful model evaluation will include all the different approaches, but resources are often limited to make this a reality. Obviously, much depends on the purpose of the evaluation (see also Section 7.1.3). Ultimately, using a number of different approaches combined with a detailed knowledge of each of their strengths and weaknesses is likely to provide the best outcome for model evaluation. Combining both the simulator and retrieval techniques with regime- and process-oriented techniques is another obvious strategy in making sense of the performance of large-scale models in simulating clouds and precipitation.

7.3 Approaches to Model Evaluation

The previous section provided an overview of some techniques to evaluate clouds in models through the use of both retrievals applied to observed data and observation simulators applied to model output. The following sections will discuss each of the main classes of model evaluation (cf. Fig. 7.3) in some more detail.

7.3.1 Application-Oriented Model Evaluation

As discussed in Section 7.1, many large-scale models are used in 'operational' contexts to make weather predictions or climate projections. Clouds and precipitation are often at the centre of these predictions irrespective of the timescale of interest. In NWP, they represent variables of interest per se, as customers are interested in weather forecasts of clouds and precipitation. In climate projections, cloud and precipitation processes are at the heart of important and poorly understood feedback processes as they strongly interact with the circulation (Chapter 13). It is therefore natural to evaluate the overall performance of cloud and precipitation predictions and projections, without specifically aiming to understand the root of any shortcomings discovered in that way.

7.3.1.1 Qualitative Evaluation

Application-oriented model evaluation can be conducted both in a 'qualitative' and 'quantitative' fashion (cf. Section 7.1.3). Fig. 7.4 provides an example of a qualitative evaluation. Note that the term 'qualitative' refers only to the final step of the evaluation, namely the qualitative assessment of maps or equivalent information. Naturally, several quantitative steps are usually involved in producing this information, such as averaging over models and/or time (e.g., months or seasons). Figs. 7.11 and 7.12 provide

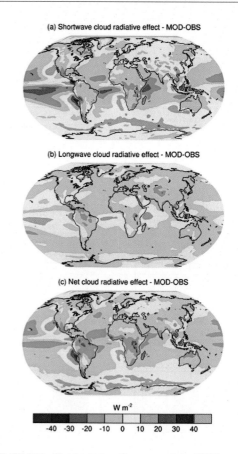

FIGURE 7.11: Cloud radiative effect errors of the CMIP5 multi-model mean compared to Clouds and Earth's Radiant Energy System Energy Balanced and Filled (EBAF) observations version 2.6 observations.
From Flato et al. (2013)

two more examples of qualitative model evaluation aimed at assessing the performance of cloud simulations in climate models.

Fig. 7.11 depicts the error of the CMIP5 multi-model mean cloud radiative effect (CRE) in the short-wave (top) and longwave (middle) part of the electromagnetic spectrum as well as the resulting error in the net CRE (bottom). The full fields for the observed net CRE and its components can be found in Fig. 4.8. It is evident that the models have significant positive and negative errors in all three quantities. The short-wave CRE is underestimated (note that the absolute values are negative) in the stratocumulus cloud regions off the west coasts of the subtropical continents, over the Southern Ocean, and to a lesser extent over Central Eurasia. This lack of reflectivity can result from too few clouds being simulated, that is, too low a cloud fraction, or from an underestimation of the cloud optical thickness. Regions of too strong a cloud effect in the short wave include the trade-wind regions as well as

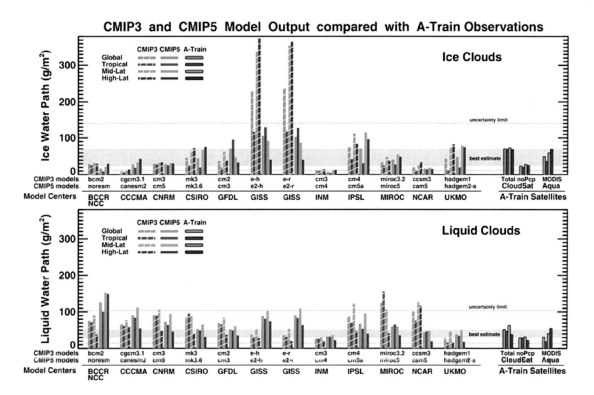

FIGURE 7.12: LWP and IWP of selected CMIP3 and CMIP5 models compared to three observational estimates. From Jiang et al. (2012). Reprinted with the permission of the American Geophysical Union

regions of deep tropical convection most strongly over oceanic regions. In the long-wave part of the spectrum, the largest errors appear in the tropical regions, where the CRE is again overestimated (note that the absolute values of long-wave CRE are positive). This likely indicates an overabundance and/or overestimated optical thickness of high-reaching tropical clouds associated with tropical convection and/or cirrus clouds resulting from it. Elsewhere, the longwave CRE errors are relatively small. The net CRE errors are dominated by the short-wave effects, except in the tropics, where the too strong reflection of short-wave radiation by clouds is compensated by their too strong greenhouse effect. This indicates that many of the net CRE errors are likely dominated by low-level clouds, which can exert large short-wave effects on the atmosphere, without affecting the long-wave radiation in a major way due to their cloud tops being close to the surface (see Chapter 4).

Fig. 7.12 shows the vertical integral of cloud ice (top) and liquid water (bottom) – known as the IWP and LWP, respectively – for several CMIP3 and CMIP5 models and three retrievals from observations. This figure once again highlights the large observational uncertainty, as indicated by the dotted lines. The model results cover a wide range of values of IWP and LWP, with many of them lying within the observational uncertainty but not always near the best estimates. Overall, models perform better for LWP than IWP. The figure also reveals significant improvements in some models in simulating both quantities from the earlier generation CMIP3 models to their more recent CMIP5 counterparts.

The strength of evaluating models using a subjective judgement of graphs and maps, as exemplified earlier, is that it allows for an easy integration of prior knowledge and experience. For instance, the large errors in the short-wave CRE off the subtropical west coasts of the major continents can be connected with the knowledge that those regions are dominated by stratus and stratocumulus clouds, providing a focus for future model development on this cloud type.

7.3.1.2 Quantitative Evaluation
Performance Metrics
Another common approach to evaluate models in their application is to calculate and display quantitative measures of success (Gleckler et al., 2008; Pincus et al., 2008). These measures are usually scalar and are calculated by measuring the 'distance' of the model fields from

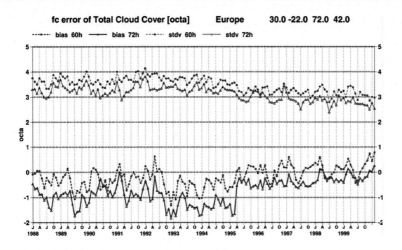

FIGURE 7.13: Mean error (solid lines) and standard deviation (dashed lines) of cloud cover (in octa) predicted by the ECMWF forecast system over Europe from 1983 to 1999. Red, 72-hour forecasts (local daytime); blue, 60-hour forecasts (local night-time).

observations. Collectively they are frequently referred to as *performance metrics*. Examples of performance metrics commonly used in cloud and precipitation evaluation are those given in Eq. (7.2). Note that the summation in those measures can be carried out in different ways, the most common of which are:

- in time, for example, by comparing an annual or monthly average field to an observed equivalent;
- in space, for example, by comparing the time evolution of an area average field to an area average from observations;
- both in space and time, for example, by calculating the measure for a sequence of monthly average global maps against observations. In this example, one would measure both the closeness of the model spatial distribution as well as its evolution over an annual cycle.

Just like the qualitative evaluation of models described earlier, the use of performance metrics usually does not allow for an easy discovery of the reasons for model errors. However, there is a significant advantage in their use in that they allow for an objective comparison of the performance across a number of different models as well as for the tracking of the evolution of model errors in time as the models are developed and improved. Examples for the use of simple performance metrics to measure progress in model development are shown in Figs. 7.13 and 7.14.

Fig. 7.13 shows two measures of forecast errors of clouds from the ECMWF NWP system over Europe averaged over individual months from 1988 to 1999. The lower lines indicate the model mean error (or bias) and the top lines the RMSE (Eq. (7.2)). Red lines represent 72-hour forecasts valid around local noon, and the blue lines a 60-hour forecasts valid around midnight. The mean errors tend to be smaller at night with an underestimation of

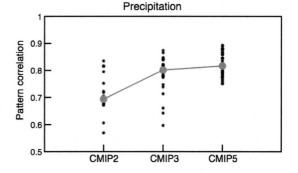

FIGURE 7.14: Evolution of the model errors in annual mean precipitation – measured as the pattern correlation with observations – for three generations of climate models. Stars depict individual models while the dots represent the multi-model mean. From Flato et al. (2013)

daytime cloudiness, in particular in the early years. In the RMSE, the night-time errors are larger than those during the day, indicating a possible compensation of errors in different parts of the European domain. The forecast behaviour can change abruptly on some occasions. Most notably, the errors decreased in early 1995. This can be directly traced to a change in the representation of clouds in the ECMWF model, indicating the potential strength of performance metrics in identifying changes in the forecast system.

Fig. 7.14 shows the pattern correlation (Eq. (7.2)) of the annual mean geographic distribution of precipitation against observations for three generations of climate models. The CMIP2 models were the state-of-the-art climate models in the early 2000s. CMIP3 represents models around 2007 and CMIP5 those around 2013. Only models that participated in all three projects are shown. Each star represents an individual model with the dots representing the multi-model average of the individual correlation coefficients. From this graph, it is possible to deduce that models have improved in representing the precipitation patterns from one generation to the next. From CMIP2 to CMIP3, there was a large improvement in the model behaviour, driven by significant improvements in several individual models. The average correlation does not improve as much between CMIP3 and CMIP5, however, both the best and worst models show improvements in their behaviour. Again, the graph highlights in a very simple way the usefulness of performance metrics in tracking model performance in time.

Like all model evaluation techniques, the use of performance metrics has limitations, which need to be recognised to avoid a misinterpretation of their results. As already mentioned earlier, performance metrics are not usually designed to give insight into the causes of model errors, but merely to document them. Using a single performance metric in isolation can often lead to misleading conclusions. For instance, it is tempting to conclude from Fig. 7.13 that the ECMWF model has generally improved its cloud prediction in 1995. However, the figure only shows two measures (bias and standard deviation) for one variable (cloud fraction) in one region (Europe). As a result, general conclusions about overall cloud prediction changes should not be drawn from it. Likewise, Fig. 7.14 does not allow the conclusions that overall the CMIP5 climate models are better than those in CMIP3, as only one particular metric applied to one particular variable was investigated. It is entirely conceivable that other error measures have not improved or have even deteriorated. Only an assessment of many performance metrics together might provide a more balanced picture of the overall model performance.

Finally it is worth noting that the use of performance metrics in NWP applications is generally easier than those for climate models. This is so as the desired outcome of an NWP forecast, say a good cloud fraction forecast over Europe, can be directly assessed several days later, as done in Fig. 7.13. In contrast, the desired outcome of a climate simulation, a reliable projection of future climates given changes in the forcing, can only be directly measured in decades to come, while performance metrics measure the model behaviour in simulating the current climate. As shown in Fig. 7.2, it remains difficult to establish the link between the two.

The Taylor Diagram

Performance metrics can be calculated in a large variety of ways. The four examples given in Eq. (7.2) merely scratch the surface of the multitude of measures that exist today. Furthermore, there are many different model variables that can be evaluated. Finally, there are many community-wide modelling experiments, such as the CMIP family, in which twenty or more models participate together and for which comparisons of model quality are desirable. This can quickly lead to an overwhelming number of graphs needing to be produced and assessed. It is therefore desirable to design ways of quickly providing an overview over model performance both across different model variables and potentially across models. One of the most common ways of doing this is the so-called Taylor diagram (Taylor, 2001).

The diagram is based on a simple decomposition of the RMSE (Eq. (7.2)) into two terms as

$$RMSE^2 = \overline{E}^2 + E'^2, \tag{7.3}$$

with the two components representing a contribution from the mean error

$$\overline{E} = \overline{y}_m - \overline{y}_o, \tag{7.4}$$

where the overbar denotes the mean; and a contribution from anomalies from the mean

$$E' = \sqrt{\frac{1}{N} \sum_{i=1}^{N} \left[(y_{m,i} - \overline{y}_m)(y_{o,i} - \overline{y}_o) \right]^2}. \tag{7.5}$$

The expression in Eq. (7.5) can be further modified to finally yield

$$E' = \sigma_m^2 + \sigma_o^2 - 2\sigma_m \sigma_o R, \tag{7.6}$$

where σ_m and σ_o represent the standard deviations of the model and observational estimates and R is the linear correlation coefficient (see Eq. (7.2)) between the model and observations. The expression in Eq. (7.6) resembles the law of cosines and inspires the Taylor diagram.

Fig. 7.15 provides a hypothetical example for a Taylor diagram that compares four models (A–D) to observations. The diagram adopts polar coordinates to display two variables. The correlation between the observations and model is displayed as the angle each point makes with the abscissa. Furthermore, the ratio of the model standard deviation to that of the observations is shown as the distance from the origin. Recall from the discussion earlier, that the correlations and standard deviations can be calculated in space or time, depending on the purpose of the evaluation.

Point P in Fig. 7.15 depicts perfect agreement between the model and the observations. It lies on the abscissa, indicating a perfect correlation of one. It also lies on the quarter-circle that depicts the ratio of the model to

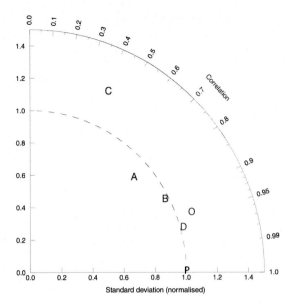

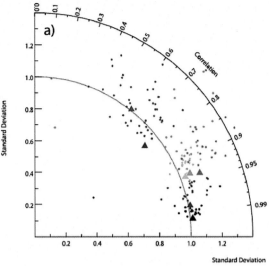

FIGURE 7.15: An example for a Taylor diagram comparing models A, B, C and D to observations. Point P indicates the theoretical perfect result. Point O denotes a comparison to an alternative observational data set, providing a simple estimate of observational uncertainty.

FIGURE 7.16: Taylor diagram of the performance of the CMIP3 family of climate models in the northern hemisphere for the annual mean geographic distribution of seven different variables: 850 hPa air temperature (black), 850 hPa zonal wind (dark green), sea-level pressure (green), precipitation (red), total cloud cover (purple), outgoing long-wave radiation (brown) and top-of-the-atmosphere reflected radiationa (light blue).
From Gleckler et al. (2008). Reprinted with the permission of the American Geophysical Union

observed standard deviation of one. It is important to remember that the Taylor diagram measures 'perfection' only for spatial and/or temporal variability without any reference to mean errors, which have been removed by construction (see Eq. (7.3)). In other words, a perfect model in a Taylor diagram might be far from perfect in its depiction of the overall mean of the quantity evaluated. Furthermore, the observations themselves are imperfect. As discussed earlier, a rough estimate of observational uncertainty can be achieved by comparing two different observational data sets, as indicated by the letter O in Fig. 7.15.

With this in mind, the performance of the three models in the diagram can be assessed. Model C is the 'poorest' model with a weak correlation of about 0.4 and an overestimation of the observed standard deviation by about 20%. Model A underestimates the observed variability but has a better correlation of about 0.7. Model B shows a near perfect standard deviation and a correlation close to 0.9. All three models have errors larger than the observational uncertainty, while Model D lies within that uncertainty and can therefore be considered as adequately reproducing the observed behaviour.

As the Taylor diagram only displays the ratio of standard deviations and the correlation coefficient, its entries are both dimensionless and normalised. This allows for

displaying a variety of different variables in the same diagram. Fig. 7.16 shows an example for the application of the Taylor diagram to the evaluation of the performance of a large number of climate models in simulating the spatial and temporal distribution of seven different variables over the climatological annual cycle (see figure caption for the individual variables). Each model is represented by a dot, while the triangles denote the results if the models are averaged into a multi-model mean before calculating the evaluation measures. Several important conclusions can be drawn from this diagram. Models perform best in simulating air temperature, wind, radiation and sea-level pressure, although many models overestimate the standard deviation. Despite a relatively good simulation of radiation, models simulate cloud cover and precipitation poorly. This indicates the existence of compensating errors, as clouds directly and strongly contribute to the distribution of radiative fluxes.

As is evident from this example, the Taylor diagram is a useful tool to summarise model evaluation results succinctly. Its use squarely falls into the application-oriented class of model evaluation approaches, as its main goal is to identify errors in predictions and climate simulations, rather than to diagnose their potential causes. To achieve the latter requires more insightful evaluation techniques, which will be discussed in the following sections.

7.3.2 Regime-Oriented Model Evaluation

Application-oriented model evaluation aims at *identifying* major areas of success of a model as well as areas that likely require improvement. It provides the starting point for further evaluation (Fig. 7.3) by providing model users and developers with a broad overview of the strengths and weaknesses of the model predictions. However, it is difficult from this approach alone to identify where exactly the model improvement effort should be directed. Are the cirrus clouds in the tropics too optically thin because the model's convection transports too little ice into them or is there something wrong with the microphysical assumptions? If it is the former, does convection occur too infrequently or is it simply too weak? How is all this related to the coexistence of an overestimation of precipitation in the same region? These are potential questions that might follow from the completion of an application-oriented set of model evaluation studies. A common next step to disentangle the model errors further is the application of *regime-oriented* techniques to model evaluation with the goal of *confronting* the model errors in more insightful ways. The basic principle of these techniques follows the common scientific approach to 'divide and conquer'. If the world can be divided into physically meaningful but separate states – or regimes – it might be possible to identify which of those regimes contributes the most to a model error of interest. Further in-depth evaluation can then be focused on the regime with the largest contribution to the model error. Doing so will make it more likely that any resulting model improvements will have a positive impact on the overall performance of the model in its applications. This immediately poses a new question: how can physically meaningful regimes relevant to cloud and precipitation processes be defined? There are three common approaches to this question:

- the identification of geographic regions that are dominated by one cloud system type, such as the subtropical stratocumulus regions that are prominent along the west coasts of major continents;
- the separation of cloud behaviour by dynamical categories, such as ascending or descending branches of major circulation systems;
- the separation by cloud characteristics themselves, such as separating high-top thick clouds from low-top thin clouds.

All of these approaches are used in the evaluation of clouds and each of them will be discussed in turn in the following subsections.

7.3.2.1 Geographic Regimes

Certain regions of the Earth are dominated by a single cloud type. The most prominent examples are the stratocumulus regions that exist over the cold upwelling regions on the eastern boundaries of the major subtropical ocean basins as well as the shallow cumulus clouds that dominate cloud fields of the trade-wind regions of the subtropics. Both these cloud regimes exist in large regions of persistent subsidence associated with subtropical high-pressure systems (see Chapters 4 and 8). As those circulation systems are relatively steady, with variations mostly associated with the seasonal cycle, so are the cloud fields embedded in them. Nature's kindness therefore assists in studying these particular cloud systems simply by isolating them geographically.

A common approach to evaluating geographically isolated subtropical cloud regimes is to draw a cross-section along a region of steady regimes, such as that shown in Fig. 7.17. In this case, the cross-section emanates in the stratocumulus region off the coast of California and extends in a south-westerly direction, roughly following the north-easterly trades over the subtropical North Pacific. It roughly follows an imaginary air column as it moves within the trades towards the equator and thereby forms part of the Hadley Circulation (see Chapter 8 and Fig. 1.15). Along this trajectory, the column moves from lower to higher sea-surface temperature and from higher to lower low-level static stability. The right panel of Fig. 7.18 shows the observed frequency distribution of cloud cover as a function of location along the cross-section from the north-east (top) to the south-west (bottom). The observed cloud field starts off as a high-cloud cover solid stratocumulus deck, as indicated by a very high frequency

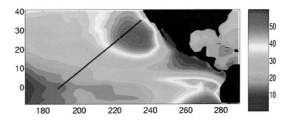

FIGURE 7.17: Low cloud cover (in %) for the months of June, July and August over the subtropical North Pacific and definition of the model evaluation cross-section. See text for details.
From Teixeira et al. (2011). Reprinted with the permission of the American Meteorological Society

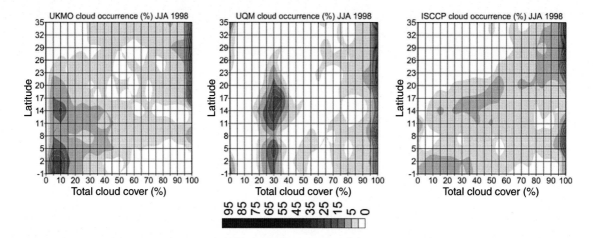

FIGURE 7.18: Histograms of the frequency of occurrence of instantaneous total cloud cover along the cross-section defined in Fig. 7.17 for two models and the ISCCP observational estimates. UKMO denotes the United Kingdom Meteorological Office, UQM the University of Quebec at Montreal and JJA indicates the results are taken for June/July/August.
Adapted from Teixeira et al. (2011). Reprinted with the permission of the American Meteorological Society

of large cloud cover at the most northern latitudes. The frequency of smaller cloud covers increases southward with values of 30–40% that are typical for broken cumulus cloud fields becoming the most common around 15–20° latitude. Further into the tropics deep convective clouds and their associated anvil cirrus systems lead to a higher frequency of large cloud cover.

Similar histograms of instantaneous cloud cover along the cross-section can be constructed from models and two examples are shown in Fig. 7.18. The models capture the general sense of large cloud cover present in the stratocumulus regions but fail to reproduce the transition to the smaller cloud cover regimes in the trade-wind regions. Rather than a smooth transition, both models have preferred values of cloud fraction in the trade cumulus regime, indicating that the physical processes involved in the stratocumulus to cumulus transition are not well captured. The low subtropical cloud regions are regions of large errors in CREs (Fig. 7.11) and they have been shown to crucially contribute to uncertainties in the climate sensitivity of the current generation of climate models (Chapter 13). While simple, the geographical separation of cloud regimes was able to provide guidance for the prioritisation of model development activities in highlighting difficulties in the transition from stratocumulus to cumulus in the models.

7.3.2.2   Dynamic Regimes

The previous subsection has indicated that it can be useful to divide the observed cloud field into subcategories to gain further insight into model capabilities in simulating each of the categories reliably. Where there is a geographic separation of cloud regimes, such as over the subtropical oceans, separating regimes is simple. Unfortunately, nature is not always that kind. Both at tropical and extratropical latitudes, the dynamical systems that are responsible for much of the observed cloud distribution are not steady and at any given location large variations in the dynamical state, and consequently the cloud state, do occur. Well-known examples are the passage of low-pressure systems with their associated fronts in the extratropics (Chapter 9) as well as the strong organisation of the cloud field in the Madden-Julian Oscillation (Chapter 8) at tropical latitudes. Given this strong connection of clouds to the circulation, an obvious approach to separating different cloud regimes is to identify key dynamical and thermodynamic variables and to distribute clouds into classes of those fields. Many such attempts have been made and used in cloud evaluation. Due to its strong influence on the heat and water budget as well as the stability of the atmosphere, vertical motion has proven to be by far the most potent dynamical parameter to separate cloud fields.

A significant advantage of separating clouds into distinct regimes is that it allows for a decomposition of model errors into contributions from model errors in the frequency of occurrence of the chosen regimes and contributions from errors in the representation of clouds within each regime. It also allows us to connect errors in the geographic distribution of clouds with errors in cloud processes. Let Y be a model variable of interest, such as the CRE or cloud fraction. Furthermore, assume that there

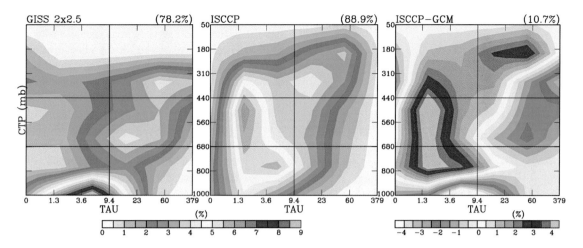

FIGURE 7.19: ISCCP histograms sorted by upward vertical motion at 500 hPa for a model (left) and the observations (middle) and the difference between them (right). The numbers in the top right give the total cloud cover in %. Adapted from Tselioudis and Jakob (2002). The model used is the Goddard Institute for Space Studies (GISS) general circulation model (GCM). Reprinted with the permission of the American Geophysical Union

is an objective classification method that divides the state of the atmosphere into N regimes, each of which occurs with a relative frequency of occurrence f_n. The average of Y in space and/or time can then be expressed as

$$\overline{Y} = \sum_{n=1}^{N} f_n \hat{Y}_n, \qquad (7.7)$$

where $\hat{Y}_n$ denotes the average of the variable Y taken for all cases in which the atmosphere is in regime n. If both Y and the regime-decomposition are carefully chosen, then Eq. (7.7) can be written for both the observations and the model. The mean model error can then be expressed as

$$\Delta Y = \overline{Y}_m - \overline{Y}_o = \sum_{n=1}^{N} f_{n,m} \hat{Y}_{n,m} - \sum_{n=1}^{N} f_{n,o} \hat{Y}_{n,o}, \qquad (7.8)$$

where the subscripts m and o denote the model and observations as before. This expression can be further modified to yield

$$\Delta Y = \sum_{n=1}^{N} f_{n,o} \Delta \hat{Y}_n + \sum_{n=1}^{N} \Delta f_n \hat{Y}_{n,o} + \sum_{n=1}^{N} \Delta f_n \Delta \hat{Y}_n. \qquad (7.9)$$

The first term in Eq. (7.9) represents the contribution to the overall model error from errors in variable Y in each of the N regimes, the second term is the contribution from errors in the frequency of occurrence of each regime, while the last term is a cross-term between the two error contributions, which is usually small. Decomposing the overall model error in this fashion allows for the identification of the largest contributing regimes as well as a first insight into the source of the error as frequency or within-regime cloud errors. This information can be vital in setting priorities for further investigation as well as model development. The following subsections will discuss the application of this method using some common regime-identification methods.

Vertical Velocity Sorting of Cloud Fields

Vertical velocity at all scales strongly and directly modifies the cloud field. It therefore makes for an obvious choice for the separation of different regimes of cloudiness. However, there are no direct observations of vertical air motions at the scales of interest and most certainly not globally. Consequently, diagnostic techniques that aim to separate clouds by vertical motion must make use of analyses of vertical motion performed by NWP centres as part of their data assimilation procedures. Currently, for a number of reasons, the quality of instantaneous NWP analysis estimates of vertical motion is deemed very good at extratropical latitudes but not so at tropical latitudes, where only averages on timescales of weeks or longer have been shown to be reliable.

In the extratropics, a simple sorting of cloud fields by instantaneous vertical motion has proven a very successful technique to identify model shortcomings. Fig. 7.19 shows an example for the application of this technique. ISCCP histograms from a model (left) and the observations (middle) are collected over the extratropical northern hemisphere (30–60°N) ocean areas for all occurrences of upward mid-tropospheric motion. The figure reveals that in upward motion clouds are often composed of deep systems of medium to high optical thickness. The model

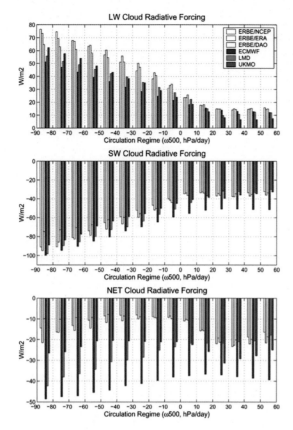

FIGURE 7.20: Tropical CRE in observations and three models as a function of monthly mean vertical motion at 500 hPa. From Bony et al. (2004)

underestimates the overall cloud fraction and the model clouds are too optically thick.

Fig. 7.20 gives an example for the utility of the vertical velocity sorting method in the tropics. It shows the longwave, short-wave and net CRE in the tropics from three different global models and observations sorted by monthly mean vertical motion. Three different NWP analysis data sets are used to determine the monthly mean vertical motion to provide an estimate of the observational uncertainty associated with the choice of analysis. It is evident from the figure that all models underestimate the long-wave CRE in ascending regimes, with the error increasing with the strength of ascent. One of the models also overestimates the short-wave CRE, in particular in the subsiding regimes. The net result is that models overall overestimate the net CRE as a combination of the two effects.

Sorting by Multiple Large-Scale Characteristics

Sorting cloud characteristics by vertical motion provides a useful first-order separation of cloud and radiative

properties. However, it is evident in Fig. 7.20 that this separation works most strongly in ascending conditions, while it is less discriminating in subsiding dynamical regimes. This is so because the dominant cloud types in those regimes are low clouds, whose properties are not strongly controlled by vertical motion alone. It can therefore be useful to extend the regime definition beyond a single large-scale variable. As subtropical low clouds are strongly affected by static stability, it is common to add the lower tropospheric stability (LTS, defined as the potential temperature difference between the 700 hPa and 1,000 hPa levels) to the regime definition. This leads to a classification of the large-scale state in two dimensions.

Fig. 7.21 provides an example for applying a two-dimensional regime classification to model evaluation. The observations show the very strong dependence of low cloud cover to LTS, while precipitation is much more strongly related to vertical motion. The model is underestimating the LTS for strong downward motion and is missing the rare low-stability regime associated with weak downward motion. The model also underestimates the overall low cloud cover, but captures the general increase of cloud cover with increasing LTS. The overall rainfall distribution is well captured, but the values at strong upward motion are overestimated.

7.3.2.3 Cloud Regimes

The previous section has highlighted the large potential of using regimes to study model shortcomings in the simulation of clouds. So far, the regime definitions have been made using a priori knowledge of either the geographic location of particular cloud systems or the relationship of the cloud field to dynamic and thermodynamic variables. Another option of identifying relevant regimes is to look for them in the cloud fields themselves. Such states could be identified subjectively or by searching for recurring states in objective descriptions of the observed cloud field from satellite observations. The latter method has been applied in many model evaluation studies over the past decade and will be described here.

There are a large number of cloud observations available that can be used to define recurring cloud states. The most wide-spread methods make use of the ISCCP CTP–τ histograms introduced in Section 7.2.2. Recall that the histograms are a two-dimensional representation of the statistical distribution of clouds in terms of CTP and optical thickness in an area of roughly (280 × 280 km^2). They are available three-hourly across the globe during the sunlit hours of the day. As the histograms themselves represent patterns in CTP–τ space, pattern recognition algorithms can be used to divide the plethora of

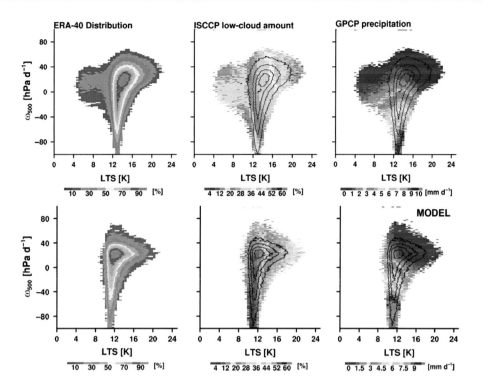

FIGURE 7.21: Frequency of occurrence (%), low cloud amount (%) and precipitation (mm/day) as a function of vertical motion and LTS in observations (top) and a model (bottom). The observed large-scale fields are taken for the ECMWF 40-year Reanalysis (ERA40). From Medeiros and Stevens (2011)

histograms available into recurring cloud states (Jakob and Tselioudis, 2003). The simplest and most frequently used algorithm to do so is that of cluster analysis. The basic idea of cluster analysis to quantitatively divide a data set into groups (or clusters) using a distance measure so that the members of each group are closer to the average of their own group than to that of any other group. When applied to the ISCCP histograms, the distance between the different histograms is determined by treating each histogram as a forty-two-dimensional vector, $\vec{c}$, and calculating the Euclidean distance between two histograms $\vec{c}_1$ and $\vec{c}_2$ as

$$d = \sqrt{\sum_{i=1}^{42}(c_{1,i} - c_{2,i})^2}. \qquad (7.10)$$

Fig. 7.22 shows the results of such an analysis applied to the ISCCP histograms over the southern hemisphere (30–65°S). Each of the states (S1–S8) is characterised by a different distribution of clouds in CTP–τ space. The relative frequency of occurrence and total cloud cover of each of the cloud states are indicated at the top of each panel. Low cloud states dominate the frequency of occurrence, with the S1 state, characterised by mostly

optically thin clouds and low total cloud cover, being the most prominent. Optically thick, high top clouds, as one would associate with warm and cold fronts, are relatively rare and are represented by the S7 cloud state.

Applying the cloud state idea to model evaluation once again allows us to pose questions beyond the simple one of 'How well are clouds simulated?'. Instead, one can ask: 'Do models reproduce the same eight CTP–τ patterns as observed? If so, do they reproduce them with the correct frequency of occurrence? Is it possible identify regimes that contribute particularly strongly to a model error say in radiation or precipitation?' If the latter question can be answered in the affirmative, the regimes provide a powerful tool to guide further investigations and, ultimately, model development.

A priori, there are several approaches to measure a model's ability to reproduce the observed cloud states. The first step in all of them is to apply the ISCCP simulator (Section 7.2.2) to the model cloud fields to produce model CTP–τ histograms. Having done so, the first option is to measure the distances of each model histogram to all the *observed* mean cloud state histograms and assign the model to the closest observed. This has the advantage that the 'base states' correspond to those of the

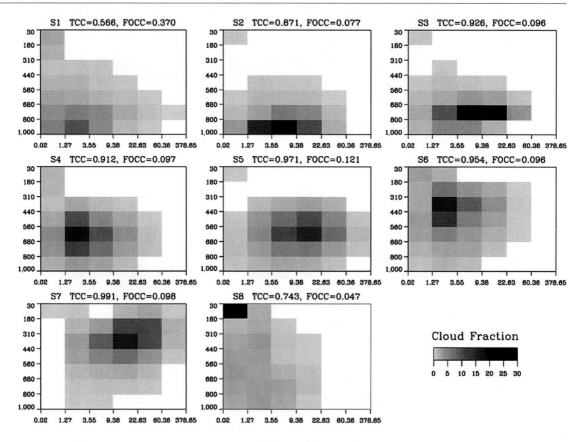

FIGURE 7.22: The eight cloud states identified by applying a cluster analysis to ISCCP histograms over the southern hemisphere. TCC stands for total cloud cover and FOCC for frequency of occurrence.
From Haynes et al. (2011). Reprinted with the permission of the American Meteorological Society

observations and therefore represent observed cloud behaviour. Another advantage is that multiple models, each with their own set of model errors, can be compared to observations using a single set of base states. However, the disadvantage of this approach is that one can envisage model cloud states that do not resemble any of the observed ones very well. Because of the nature of the cluster analysis method, those will still be assigned to the closest of the observed states, potentially distorting the analysis. A second approach is to carry out a cluster analysis on the model histograms themselves. The advantage here is that the model cloud states are determined independently of the observed. However, if the model has difficulties in representing some cloud states, the states so derived might not resemble the observed ones, making the interpretation of the results more difficult and prohibiting the use of a common set of 'base states' in further analysis. This makes the application of this method to several models at the same time difficult, but provides deeper insight into the model behaviour for a single model.

A third, hybrid, approach is to carry out a combined state analysis using both the observed and model histograms in the cluster analysis. Again, since the cluster analysis involves the model states, this method is best applied to a single model. The advantage of this approach is that if the model produces cloud states close to the observed, then those will be identified together with the already known observed states, while states that only occur in the model will be identified easily at the same time. In fact, the result of a perfect model simulation would be indicated by finding the same eight states as in the observation-only analysis, with both the model and the observations producing each state with the same frequency and geographic distribution. If the model was unable to reproduce an observed cloud state, this state will still appear in the hybrid approach, but it will only contain observed histograms. Likewise, if the model produced a cloud state that is not observed, it would appear as a state populated by model results only.

Fig. 7.23 shows the results of applying this hybrid approach to a climate model over the southern

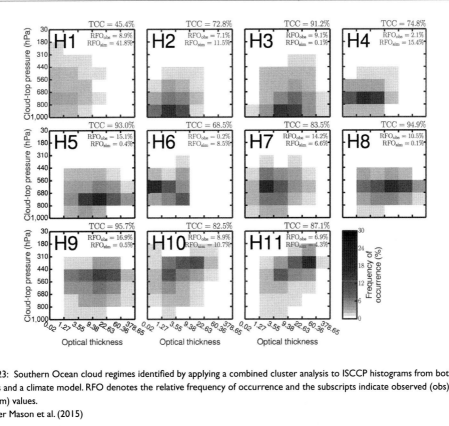

FIGURE 7.23: Southern Ocean cloud regimes identified by applying a combined cluster analysis to ISCCP histograms from both observations and a climate model. RFO denotes the relative frequency of occurrence and the subscripts indicate observed (obs) and model simulated (sim) values.
Redrawn after Mason et al. (2015)

hemisphere. The figure indicates that the combined observation and model cluster analysis reveals eleven unique states, named H1–H11. Those can be directly compared with the eight observation-only states in Fig. 7.22, which were the result of analysing observations only. A careful analysis of the results provides deeper insight into the model behaviour. The overall conclusions that can be drawn from this analysis are:

- The model frequently underestimates the optical thickness of the observed cloud states. Clear examples for this behaviour are the cloud state pairs of H2 and H3, and H4 and H5, as well as H6 and H7. In each case, the mostly model-populated histograms constitute an optically thinner version of similar cloud fields in the observations
- The model fails to produce relatively optically thick mid-level top clouds (H8 and H9).
- The model significantly overpredicts a very thin low cloud cover regime (H1).

While interesting in itself, the strength of the regime-separation is that it provides more insights into the model errors identified in the application-oriented model evaluation step. Fig. 7.11 showed that the CMIP5 family of climate models underestimates the short-wave CRE over the Southern Ocean. This error is also prominent in the

model used to construct Fig. 7.23. Is it possible to use the decomposition in cloud states to better understand the sources of the CRE error? To do so, Eq. (7.9) is applied to the hybrid regimes derived earlier. Here, $N = 11$, the number of states, f_n is the frequency of occurrence of each cloud state and $\hat{Y}_n$ is set to the short-wave CRE of each regime.

Fig. 7.24 shows this error decomposition applied to the Southern Ocean area (50–65°S) for the same model simulation used to construct the hybrid cloud states for the months of December, January and February. Contributions by errors in the frequency of occurrence of the regimes are shown in the dark-grey shade, those due to errors in within-regime shortwave CRE in the second shade and the cross-terms in the lightest shade. The mean short-wave CRE error over the region is an underestimate (positive value) of nearly 48 W m^{-2}. There are several insightful conclusions that can be drawn from this analysis.

- Most, but not all, error contributions are dominated by errors in the frequency of occurrence of the regimes. This indicates that alleviating this error is likely going to have the largest effect on model improvement.
- There are large compensating errors resulting from different regimes. The pairwise cloud states for which the

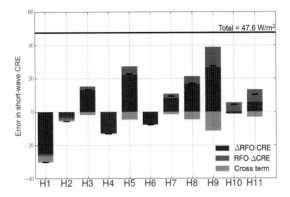

FIGURE 7.24: Contribution of each of the cloud states defined in Fig. 7.23 to the error in short-wave CRE over the southern hemisphere (30–65°S).
Redrawn after Mason et al. (2015)

model underestimates the optical thickness (H2 and H3, H4 and H5, H6 and H7) contribute pairs of positive and negative errors, with the positive errors larger than the compensating negative ones. This indicates an overall small but non-negligible contribution of these regimes to the total error.

- The strongest contribution to the overall error results from the optically thick mid-level cloud regimes (H8 and H9), which the model fails to produce.
- A strong compensating error results from the strong overestimation of the very thin low cloud cover regime (H1). While the short-wave CRE for this regime is relatively small, the large overestimation of its frequency of occurrence leads to a large radiative error compensation.

Analyses as that presented in Fig. 7.24 are at the very heart of using regime-oriented approaches, as they can act as a magnifying glass on the overall model error. The previous example shows that to alleviate the positive CRE error over the Southern Ocean first and foremost requires to address the lack of mid-level clouds in the model (cloud state H9). Furthermore, the optical thickness of low clouds is too low. It also became evident that alleviating these errors without addressing the compensating error of having too many instances of thin, low cloud cover situations would likely lead to an overall error of opposite sign. This highlights that the application of regime-oriented approaches even to regions of no mean error might still be very insightful, as compensating errors from different states might still be identified.

7.3.2.4 *Spatial Compositing of Cloud Fields*

Separating the cloud field into regimes has already been shown to be a simple and yet powerful tool to further develop insights into model errors that have been

identified in the application-oriented model evaluation. It was shown that separating regimes geographically works in some regions, while for other regions, a separation by dynamical or cloud state was a more useful approach. In addition to those, there exists a technique that combines geographical and dynamical separation by compositing the cloud field around major dynamical features of the atmosphere, such as extratropical or tropical cyclones. Baroclinic cyclones are perhaps the signature dynamical feature of the northern and southern midlatitudes and it has been known for a long time that clouds are strongly organised by the flow around those systems and their associated fronts (Chapters 1 and 9). Cyclones are easily identified in the surface pressure field, and to this day this provides the foundation for cyclone compositing techniques. In simple terms, low-pressure centres are objectively identified and the centre of each cyclone becomes the origin of a new coordinate system. Properties of the cyclone, including its cloud and dynamical fields are then averaged in this new 'cyclone-relative' coordinate system and analysed. The technique can easily be applied to both observations and models and the results compared. Like many of the earlier techniques, this provides further insight into not only the correct simulation of the cloud properties but also tests their correct association with the dynamical features of the system. Once again, the aim of the technique is to answer the question: are the clouds right for the right reason?

Fig. 7.25 shows an example for the application of the compositing technique to model evaluation. The figure shows a composite of cloud fraction for Southern Ocean cyclones from a climate model at three different heights together with the differences with two observational estimates derived from the CloudSat and CALIPSO data sets (Chapter 4). The two observational estimates form an upper and lower bound of the observed cloud fraction and are a crude measure of observation uncertainty. Overlaid on the cloud fraction are key dynamical parameters, namely mean sea-level pressure at the 1.8 km height, vertical velocity (ω) at 4.5 km and relative humidity (RH) with respect to water at 9.7 km. Several conclusions about the model cloud field and its connection to the cyclone dynamics can be drawn. At low levels, the model severely underestimates cloud fraction, in particular in the north-west sector of the cyclone, which coincides with the strongest subsidence in the system, creating what is known as the 'dry slot'. Clouds at mid-levels are simulated within the observational uncertainty, while upper-level clouds are slightly overestimated in the warm frontal region. Perhaps the most interesting result is that all dynamic and thermodynamic variables indicate that the composite

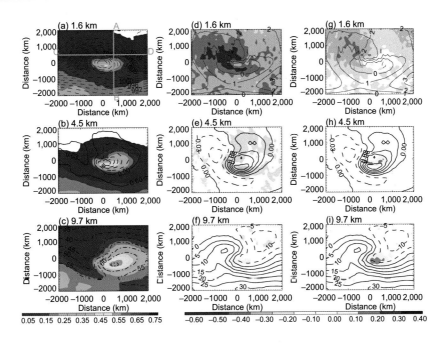

FIGURE 7.25: Composite of a Southern Ocean cyclone in a climate model with cloud fraction shown at three heights (1.6 km, 4.5 km and 9.7 km from top to bottom) as well as differences from two observational estimates of cloud fraction from CloudSat/CALIPSO. From Govekar et al. (2014). Reprinted with the permission of the American Geophysical Union

cyclone is on average too weak. Sea-level pressure values are higher than observed, both ascent and descent at mid-levels are underestimated in strength and RH is too low in the region of its maximum and too high elsewhere. Given the too weak circulation through the cyclone it is perhaps surprising how well the clouds are simulated, with the exception of low clouds behind the system, and one might ask if the underlying cloud processes are correctly simulated or if the result is an expression of compensating errors. Highlighting such issues is the very purpose of studying model cloud properties in regimes, of which cyclone composites are but one example.

In concluding the section on regime-oriented model evaluation, it is worth noting that, while powerful, these techniques by themselves are not usually sufficient to identify causes for model errors. However, they provide a vital link from the application-oriented approaches to model evaluation to the process-oriented evaluation techniques discussed in the next section.

7.3.3 Process-Oriented Model Evaluation

The application-oriented approach to model evaluation provides information on the main errors encountered in the simulation of clouds and precipitation. Applying an appropriate regime-oriented evaluation technique will

shed light on the broad contribution to those errors by different cloud or dynamical regimes. A final remaining step in model evaluation is to assist in the identification of the errors in the representation of particular processes that are involved in producing the model errors. This constitutes the most difficult step in model evaluation. It is also the most important step in applying model evaluation techniques in aid of model development. Approaches that fall into this category are often characterised as *process-oriented* model evaluation. By their very nature they are strongly intertwined with model development.

Process-oriented model evaluation relies on the combined application of observations, process models, such as LES and SRM, and 'simplified' versions of the large-scale models used in the previous two classes of model evaluation (Randall et al., 2003). Process-oriented model evaluation studies often focus on the parametrised processes in large-scale models (Chapter 6). The basic idea behind many process-oriented approaches to cloud and precipitation evaluation is to

(1) choose a relevant case informed by application- and/or regime-oriented model evaluation and the availability of observations;
(2) execute the model in such a way that the dynamic and thermodynamic state of the atmosphere are held close to the observed state;

(3) if possible, perform simulations with more comprehensive process models for the same case;

(4) perform an in-depth comparison of the model's clouds and precipitation behaviour against the available observations and process-model results.

There are two main approaches to ensure the large-scale state is close to observed when performing process studies:

- Isolate a single model grid cell and prescribe the dynamic terms usually provided from the full three-dimensional model equations using observations. This is usually referred to as the *single-column model* approach.
- Run the full three-dimensional global model either initialised with dynamic and thermodynamic states that include information from observations through data assimilation or by relaxing the model state to observations as the model is executed.

The general philosophy of both these approaches will be illustrated in the following sections.

7.3.3.1 *Field Studies and Single-Column Models*

As the parametrised processes in current climate models are calculated for each model column in isolation (see Chapter 6), it is possible to study their behaviour by extracting a single column from the full model. As the column is isolated from its neighbours, it is necessary to prescribe the dynamic terms of the model equations, such as advection and convergence. This is why many meteorological field studies include measurements that allow the derivation of these terms from observations. When those are combined with detailed observations of the cloud and precipitation structures within the experiment domain, often chosen to be of the size of a climate model grid box, these field studies become excellent focal points for model evaluation studies. Their utility is further enhanced by the ability to use process models, such as LES and SRM, with the same set of large-scale input, evaluate them and then extract information that cannot be gleaned from the observations themselves from the process models. Examples for field studies that have been extensively used in process-oriented model evaluation include the Barbados Oceanographic and Meteorological EXperiment (BOMEX), the Atlantic Stratocumulus Transition Experiment (ASTEX), the Tropical Ocean Global Atmosphere Coupled Ocean Atmosphere Response Experiment (TOGA COARE), the Global Atmospheric Research Program's Atlantic Tropical Experiment (GATE) and

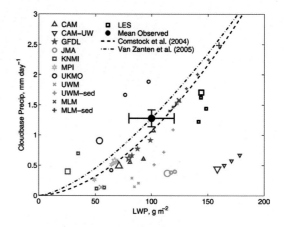

FIGURE 7.26: Mean LWP versus cloud-base precipitation. The symbols represent different SCM, a simple conceptual model (mixed-layer model) and LES. Each symbol represents a single hour of simulation. The observed mean value is plotted with uncertainty bars. The dotted and dashed lines represent empirical relations. Each symbol represents a single hour of simulation with a particular model indicated by its acronym.
For details see Wyant et al. (2007). Reprinted with the permission of the American Geophysical Union

more recently the Dynamics and Chemistry of Marine Stratocumulus field study (DYCOMS), Mixed-Phase Arctic Cloud Experiment (M-PACE) and the Tropical Warm Pool International Cloud Experiment (TWP-ICE).

There are a number of approaches to exploiting the results of model simulations applied to field studies for model evaluation and development. The most obvious is to carry out a comprehensive intercomparison of the SCM results with those from the process models and observations for a process of interest. Fig. 7.26 provides an example for such an intercomparison from the DYCOMS experiment. The figure shows the relationship between the LWP of a drizzling stratocumulus cloud and the precipitation flux at the cloud base. It shows that the parameterisations used in the various SCMs produce a wide range of estimates of this relationship, from models that produce very little drizzle even for large LWP (Community Atmospheric Model – University of Washington [CAM-UW]) to some that drizzle heavily with small LWP (UKMO). The graph also demonstrates that the performance of the process models (LES) cannot automatically be assumed to be perfect, an important lesson to keep in mind in their use for model evaluation. In this study, the LES overestimate the LWP of the cloud, indicating the likely very important role of microphysical processes, which are still parametrised in LES models. Despite these limitations, important conclusions for future development can be drawn for most models, in this particular case,

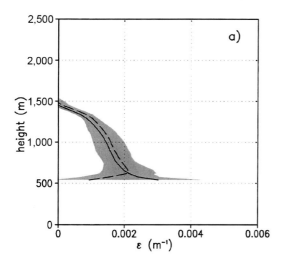

FIGURE 7.27: Entrainment rate as a function of height derived from LES models for the BOMEX field study. The two lines use two different variables to estimate entrainment and the shaded area marks the uncertainty estimated by using several LES models. From Siebesma et al. (2003). Reprinted with the permission of the American Meteorological Society

highlighting the likely need for revision in the microphysical treatment of drizzle.

A second way of exploiting field studies is to first evaluate the process models applied to them, and, where they succeed, to extract information that is critical to parameterisations but difficult or impossible to observe. Fig. 7.27 gives an example for this technique. It exploits a number of LES model applied to the BOMEX field study. After evaluating the LES models overall performance in simulating the observed thermodynamic and cloud structures, we can make use of the knowledge of all variables at all LES grid-points and separately sample the cloud and environment as defined in the mass-flux approximation used in parameterisations (Chapters 5 and 6). Having done so, the entrainment rate, a key parameter that cannot easily be observed, can be estimated from the model states. In this case, it was found that the actual entrainment rates for shallow cumulus clouds are an order of magnitude larger than what is traditionally assumed in many parameterisation schemes.

7.3.3.2   Process Studies Using Global Models

A basic idea of process-oriented model evaluation for clouds and precipitation is to isolate the physical parameterisations and to ensure that the dynamic and thermodynamic state they operate in is as close to observations as possible. The SCM approach achieves this by directly prescribing those states. While SCMs are an important

tool, their application is limited to a small number of field studies at a few locations for which sufficient information to drive and evaluate the model has been collected. So the question arises if it is possible to use the large-scale models in their entirety to study parameterisation errors while keeping the atmospheric state close to observations.

Two techniques to do so have emerged. The first is the use of short NWP-style hindcasts, that is, predictions of past events, in the evaluation of clouds and precipitation. The rationale behind this approach is similar to that in SCMs, namely, to isolate the behaviour of the parametrised processes from feedbacks that occur with dynamical processes and to therefore more easily ascribe model shortcomings to poor representation of cloud processes. Unlike in SCMs, interactions between parametrised and resolved processes can occur, but by keeping the evaluation focused to the first few days of the hindcasts, the large-scale dynamical features of the atmosphere remain close to observations, again making it easier to ascribe any shortcomings in clouds directly to their representation in the model. An additional advantage of this technique is that the explicit forecasting of the conditions for a specific time and location allows the direct comparison of the forecast results with observations at that time and location. This complements the more statistical approach necessary in the evaluation of climate simulations, in which individual model states do not represent particular observed states.

The general philosophy of the use of initialised hindcasts is illustrated in Fig. 7.28, which investigates the common solar radiation biases over the Southern Ocean discussed earlier. The example is taken from the work of the Transpose AMIP (T-AMIP) (Williams et al., 2013), which has coordinated the application of hindcasts to climate models over the past decade. The top left panel shows the solar radiation error of a twenty-year simulation with climate model composited around cyclones over the Southern Ocean, similar to Fig. 7.25. The red areas around the cyclone indicate an underestimation of reflected solar radiation. The bottom left figure shows the same error from a set of two-day predictions made with the same model. Comparing the two figures demonstrates that over much of the cyclone, the errors in reflected solar radiation are very similar in short-range forecasts to those in the climate simulation, especially in the southern half of the composite cyclone. Having established the similarity of key errors between the climate simulation and the hindcasts, we can now select an individual two-day prediction (middle panel) to establish the existence of the same error for just one case. Here, a strong lack of

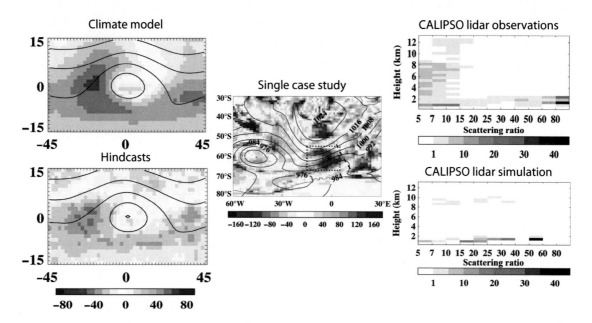

FIGURE 7.28: Illustration of the use of initialised hindcasts in model evaluation. See text for details. Adapted from Williams et al. (2013)

reflected solar radiation is visible behind a low-pressure trough (dashed box). A detailed analysis of observations of back scattering ratio from the CALIPSO satellite lidar in this area (top right panel) in the form of altitude-scattering ratio histograms reveals a high frequency of low clouds with large scattering ratios, indicating highly reflective clouds. A smaller amount of much less reflective clouds is found throughout the troposphere. Applying the CALIPSO instrument simulator to the model results (bottom right panel) shows that while the model produces significant amounts of low clouds, their reflectivity is lower than observed. Further investigations into the treatment of the boundary layer in the model showed that its boundary layers are too shallow, producing clouds that are physically too thin. This example illustrates the utility of the hindcast approach, especially when combined with the regime and compositing techniques discussed earlier.

Initialising climate models is far from trivial as unlike their NWP counterparts, they usually are not equipped with the data assimilation capabilities necessary to derive initial conditions. An alternative to initialised predictions is the so-called relaxation technique. Here, a restoring term is added to the equations for the large-scale variables, preventing them from drifting too far away from the observed large-scale atmospheric state represented by NWP analyses or reanalyses. As with short-range predictions, this allows a more direct comparison of the model with observations as the large-scale circulation

and the characteristics of synoptic phenomena such as cyclones and anticyclones are at the same place at the same time both in the model and in the observations. The variables that are nudged and the strength of the nudging depend on the objective of the evaluation. For instance, nudging the horizontal winds drives the large-scale circulation without a direct modification of the thermodynamic and cloud state of the atmosphere. Hence, the physical parameterisations are allowed to freely evolve within a realistic dynamic field. As stated earlier, this enables a more straightforward attribution of model errors to parameterisation errors as feedbacks on the circulation are suppressed. Just as the hindcasts, the evaluation technique is an intermediate between SCM studies, where the large-scale dynamics are prescribed, and freely evolving three-dimensional simulations. As the two techniques are equivalent in that goal, the evaluation of models using the relaxation technique is very similar to that of initialised hindcasts.

7.4 The Future of Model Evaluation

Model evaluation is a crucial activity that serves multiple purposes. It measures the quality of predictions. It assesses a model's ability to represent observed behaviour. It helps identify the causes for model shortcomings and as a result it supports model development. Many techniques for model evaluation have been developed and applied, a selection of which are summarised in the previous sections of this chapter. Many new data sets for the evaluation of clouds and precipitation have become available (Chapter 4). Process models have improved significantly and modern computational resources allow for their increased use at higher and higher resolution and for larger and larger areas including the entire globe. As a result, models have never been scrutinised more thoroughly. The opportunities for model evaluation in aiding model improvements have never been better than they are today. And yet, key aspects of model uncertainty, such as the wide range of climate sensitivities of modern models (Chapter 13), have been shown to result from shortcomings in the representation of clouds and precipitation in large-scale models that have been present for several decades.

The alleviation of these shortcomings will require continuing advances in the field of model evaluation. Application-oriented model evaluation approaches need to be enhanced for clouds and precipitation in particular, to better identify which of the errors are key to the application. Modern NWP system are evolving away from simply predicting atmospheric flow patterns to predicting the weather, as represented by clouds and precipitation. This provides both a challenge and an opportunity for enhancing the routine and quasi-real-time evaluation of the weather predictions, for instance through the innovative use of radar observations as well as current and future satellite observations. Regime-oriented model evaluation requires significant developments to better identify crucial contributors to the model errors. Here, the interaction of cloud and precipitation processes with circulation features provides both a challenge and an opportunity for the development of new diagnostic techniques. Details of many tropical circulation systems and their connection to cloud systems remain poorly understood and poorly analysed in modern NWP data assimilation systems. This not only hinders tropical weather prediction but also limits the degree to which NWP analyses can be used to better understand the tight connection between clouds and dynamical processes in the tropics (Chapter 8). Progress in process-oriented model evaluation will require the better exploitation of existing field observations, the collection of new relevant data sets informed by application- and regime-oriented evaluation and the better integration of short-range NWP techniques, including data assimilation, in the evaluation process. This naturally requires overcoming the often strict separation of the weather and climate modelling communities and the application of a more unified approach to model development.

Finally, and perhaps most importantly, the various approaches of model evaluation, often represented by different research communities, need to be brought together more closely. Regime-oriented approaches must be more carefully aimed at explaining some of the long-standing model errors in large-scale models, such as the errors in the radiative effects of subtropical and polar clouds as well as the poor representation of tropical precipitation. More rigorous connections of the process-oriented approaches to the findings of regime-oriented studies need to be established, so that the comprehensive process studies are carried out for relevant rather than convenient cases. While widely accepted as useful, the iterative evaluation process displayed in Fig. 7.3 has not been applied comprehensively and repeatedly to any of the major problems identified in the simulation of clouds and precipitation, in part because of the wide-range of tools and community interactions required. None of the three main approaches to model evaluation discussed here is superior to the other. Nor are individual techniques developed within each of the three categories. Instead, it is the thoughtful combination of a diverse set of approaches and techniques that will progress the representation and evaluation of clouds and precipitation in large-scale models.

Further Reading

There are no textbooks dedicated to model evaluation. Many of the statistical methods used in model evaluation and in particular in the verification of numerical weather forecasts can be found in the textbook *Statistical Methods in the Atmospheric Sciences* by Daniel Wilks (2011).

Exercises

General

(1) Based on what you know about their construction, describe some expected key artefacts in the ISCCP data sets. Which of those artefacts are addressed by the ISCCP simulator and which ones are not?

(2) Compare Figs. 7.4, 7.11 and 7.12. Some areas of model error of CRE, LWP and precipitation overlap, while others do not. Using your knowledge about the global distribution of clouds, explain these commonalities and differences.

(3) Class discussion: How can model evaluation best serve model development? Which of the approaches should be used? What are there advantages and disadvantages in supporting model development?

Data analysis and programming

This chapter provides a collection of model output and observations for practical use in model evaluation exercises. Before starting the exercises below, please download the necessary data files from www.cambridge.org/siebesma. The data consists of netCDF files of twelve monthly climatologies (1980–2005) of atmosphere-only (AMIP) simulations from twenty-nine CMIP5 models as well as two observational data sets for each variable used. The observations are identical to those used by Pincus et al. (2008). The files are in the common netCDF format, so you will need a set of tools that can read this format. The variables included can be found in Table 7.1.

Many different exercises can be constructed from these and many different software packages used in the analysis. We leave it to the students and teachers using this book to choose those appropriate to them, but provide some suggestions for exercises below.

(4) Qualitative versus quantitative evaluation
   (a) Calculate the annual mean precipitation climatology and plot a global map for each model and the primary observation data set. Using these maps, rank the models in their performance and record your ranking.
   (b) Plot difference maps between each model and the primary observation data set. Using these maps, rank the models and record your ranking.
   (c) Calculate the root mean square difference between each model and the primary observations. Record the ranking of the models using this measure.
   (d) Compare the three rankings you have made. Record differences and provide an explanation why they may have occurred.
   (e) Repeat the exercise with the secondary set of observations and discuss the differences you find.
   (f) If you want, you can repeat this exercise for any of the other variables contained in the data sets.

(5) Evaluation of variability
   (a) Construct a time-versus-latitude diagram to display the seasonal cycle of zonal mean rainfall for the models and the primary observations.
   (b) Construct diagrams of the differences between the models and the observations.

TABLE 7.1: Variables available in the exercise data files

Acronym	Variable name	Units
pr	Precipitation	mm/day
clt	total cloud fraction	%
rlut	TOA outgoing long-wave radiation	W/m^2
rsdt	TOA incident short-wave radiation	W/m^2
rsnt	TOA new short-wave radiation	W/m^2
rsut	TOA outgoing short-wave radiation	W/m^2
rsutcs	TOA outgoing clear-sky short-wave radiation	W/m^2
rlutcs	TOA outgoing clear-sky long-wave radiation	W/m^2
lwcre	TOA long-wave CRE	W/m^2
swcre	TOA short-wave CRE	W/m^2
netcre	TOA net CRE	W/m^2
tas	Near-surface air temperature	K

   (c) Calculate the RMSE between the models and observations.
   (d) Based on all the results, rank the models in their performance of the seasonal cycle. How does this ranking compare to that of the annual mean precipitation maps?

(6) Multi-variate evaluation
   (a) Calculate the annual mean climatologies of TOA outgoing long-wave radiation, TOA net short-wave radiation, TOA long-wave CRE, TOA short-wave CRE and total cloud fraction for all models and the primary observations. Construct difference maps to the observations for all models and all variables. Sort your results by model.
   (b) Assess each model's representation of TOA radiative fluxes both qualitatively and quantitatively (choose any measure(s) you see fit). Now compare the TOA radiation errors to those in CRE and in total cloud fraction. Which of the radiation errors are cloud-related and which ones are not? What are possible shortcomings in the cloud field that explain the model behaviour?
   (c) Bonus: Construct a Taylor diagram of all models and the secondary observations that includes all five variables. See Fig. 7.16 for an example.

Part III   Clouds and Circulation

8 Tropical and Subtropical Cloud Systems

Gilles Bellon and Sandrine Bony

Tropical and subtropical clouds are characterised by a wide variety of spatial organisations from the mesoscale to the planetary scale. Mesoscale convective systems (MCSs) regroup different types of convective and high stratiform clouds and last up to a few days; at the synoptic scale, tropical cyclones form circular cloud structures with a lifetime of one to two weeks. At larger scales, more cloud systems occur in some regions than others, forming cloud-system envelopes that include numerous cloud systems and clear-sky regions in between. Some of these envelopes propagate, such as the convectively coupled equatorial gravity waves and the Madden–Julian oscillation (MJO), with lifecycles that last from a couple of days to a couple of months. Others, such as the tropical convergence zones (CZs), are more stationary, but they are still modulated on seasonal timescales, even significantly displaced in the cases of monsoons, and on inter-annual scales.

Tropical regions differ from higher latitudes by the smallness of the Coriolis parameter. Rotation is therefore relatively unimportant for tropical circulations, and as a result gravity waves easily disperse temperature anomalies and minimise horizontal temperature gradients in the free troposphere. This means in turn that the energetic balance of the free troposphere is essentially between vertical transport and diabatic terms. This balance is characteristic of the tropics, similarly to the geostrophic balance in the extratropics; it implies a strong interaction between divergent circulation and diabatic heating. Because the tropics absorb a larger amount of solar radiation than the higher latitudes, the diabatic terms are very large. In this context, clouds are a very active part of the energy cycle in the tropics. First, most of the solar energy is absorbed at the surface and a large fraction of it is transmitted to the atmosphere through surface turbulent heat fluxes that warm and moisten the boundary layer; convective clouds redistribute this energy throughout the troposphere by extracting humidity from the boundary layer and car-

rying it upwards where the latent heat is released by condensation and freezing. For most convective clouds (e.g., congestus and cumulonimbus clouds), this is their dominant diabatic contribution. Second, clouds interact with radiation in both the solar and terrestrial ranges. Some of this radiative effect can be local, cooling or warming the atmosphere, and some of this radiative effect acts remotely by changing radiative fluxes and in particular modulating the surface radiative budget. For some clouds such as cirrus and stratocumulus clouds, their radiative effect is their main diabatic contribution. Rather than causing a uniform heating, as would be expected from a random spatial and temporal occurrence of clouds, clouds create gradients of diabatic heating and drive circulations on many spatial scales because of their multi-scale spatial organisation.

On the other hand, cloud occurrence and type are sensitive to the circulation, in particular to the vertical motion of the atmosphere, and to the surface conditions. The two-way interactions between dynamics and clouds and between the surface and clouds underlie feedback mechanisms that are at the heart of many modes of variability of the tropical climate. However, many aspects of these interactions remain poorly understood.

This chapter aims to present the tropical and subtropical cloud systems and their main modes of spatial organisation and variability in light of their interactions with the large-scale tropical circulations and with the surface.

8.1 From Tropical Clouds to Cloud Systems

The tropics exhibit a large diversity of clouds, from stratiform thin clouds to deep convective clouds (DCCs). Cloud types can be classified according to their cloud-top pressure and cloud optical thickness (see Chapter 1). Fig. 8.1

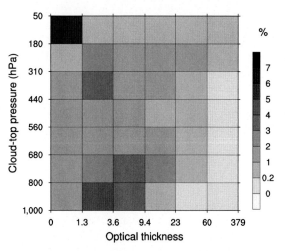

FIGURE 8.1: Cloud frequency of occurrence in the tropics stratified in bins of cloud-top pressure and cloud optical thickness (International Satellite Cloud Climatology Project data, 30°S–30°N, see Fig. 1.8 for the global distribution).

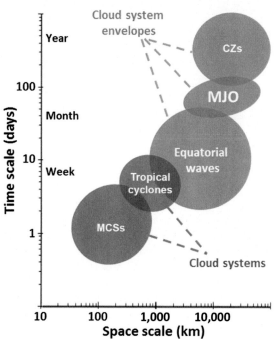

FIGURE 8.2: Types and scales of organisation of tropical convective clouds.

shows the two-dimensional distribution of the frequency of cloud occurrence in the tropics as a function of these properties. High and thin clouds known as cirrus clouds (Ci) are ubiquitous in the tropics, generated by local stratiform condensation or by dissipating convective systems. Low clouds such as the optically thick stratocumuli (Sc) and optically thinner, shallow-convective cumuli (Cu) are frequent as well. Stratocumuli populate the eastern subtropical basins while cumuli are ubiquitous in the subtropical trade-wind regions as well as the deep tropics. In convective regions, the cumuli cohabit with deeper congestus clouds (Cg) as well as DCCs and their extension of stratiform precipitating clouds (SPCs) and non-precipitating anvils (An). In these regions, the distribution of convective cloud-top altitudes is actually trimodal (Johnson et al., 1999): there is an absolute maximum of shallow cumuli (visible in Fig. 8.1) and two secondary maxima for the cumulus congestus and for the DCCs and associated stratiform anvils. These three categories of clouds extend approximately from the lifting condensation level up to one of three characteristic layers with enhanced stability. The cumuli reach their maximum heights at the trade-wind inversion, a strongly stratified layer at the top of the atmospheric boundary layer between 1 and 3 km of altitude. The congestus clouds develop up to the freezing level around 6 km, where the solid–liquid phase change creates a more stable layer, and the DCCs extend all the way to the tropopause (at about 17 km of altitude).

The distribution of tropical clouds is rarely random. On the contrary, it is often associated with different forms

of organisation of the atmosphere over a wide range of scales. For example, stratocumuli form decks over the cool subtropical waters off the western coast of continents, while convective clouds organise in systems from the mesoscale to the planetary scale that are illustrated in Fig. 8.2. MCSs[1] of tens to hundreds of kilometres such as squall lines, cloud clusters and superclusters, feature a transition from shallow to deep convection, to stratiform rain, as illustrated in Fig. 8.3, with a lifecycle of up to a few days. At the synoptic scale (hundreds to a thousand kilometres), tropical cyclones form circles of convective and high stratiform clouds that last up to two weeks and their extreme versions, the hurricanes, organise deep convection and circulation in spiralling structures around their circular eye wall. At larger scales, these cloud systems can be embedded in envelopes of high cloud activity with other regions witnessing fewer cloud systems. These envelopes are often associated to propagating modes of variability such as the convectively coupled equatorial gravity waves (on time scales from the synoptic to the intraseasonal) and the thirty-to-sixty day MJO, which is the most prominent large-scale

[1] Here we use a broad definition of MCSs, including all types of mesoscale organisation. This term is often used with a narrower definition in the literature.

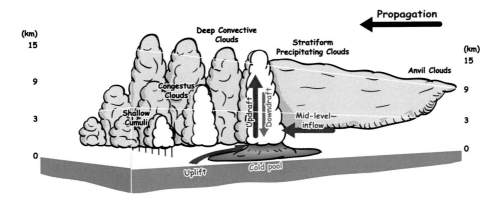

FIGURE 8.3: Schematic of an MCS. Courtesy of Norbert Noreiks

intraseasonal mode of tropical variability. At the planetary scale (thousands to tens of thousands kilometres), convection is spatially organised in CZs, where the column-integrated divergence of humidity is negative and latent heat release is large (see Fig. 8.13). These CZs are quasi-steady at the seasonal timescale. The best-known CZ is the Inter Tropical CZ (ITCZ), a rainband that extends zonally along the equator over most of the globe. Monsoons can be seen as a latitudinal shift of the ITCZ (in West Africa), a strengthening of a CZ (the South Pacific CZ or SPCZ, over Australia and the south-western tropical Pacific) or the emergence of an additional CZ (the Monsoon CZ around 20°N in South Asia).

While some of the basic mechanisms of these cloud systems are well understood, there is still significant uncertainty on what controls the lifecycle and propagation of MCSs and tropical cyclones, and what determines the location of the CZs, in particular in the case of monsoons. These uncertain properties are determined by the interaction of these cloud systems or cloud envelopes with the larger-scale environment, and in particular with the atmospheric circulation and the surface.

The spatial organisation of clouds is often associated with dynamical patterns that can favour the development of the clouds and/or can be a response to the heating inside the clouds. This association is observed at all scales. For example, at the mesoscale, MCSs can be embedded in easterly waves which propagate westward along monsoon troughs (particularly in Africa) and the ITCZ. An MCS can also create in its wake a mid-level mesoscale vortex that can develop into a tropical storm or hurricane, which are associated with strong cyclonic circulations in their mature phase. At a larger scale, transient organised convection is often associated with planetary waves.

An example is the MJO, which exhibits a characteristic dynamical pattern known as Gill circulation (Gill, 1980) which is the equilibrium dynamical response to a steady mid-tropospheric equatorial heating. On the planetary scale, CZs are regions of large-scale convergence in the boundary layer, ascent in the mid troposphere and divergence in the upper troposphere. For example, the ITCZ is embedded in the ascending branch of the Hadley circulation, and the rising branches of the Walker circulation are collocated with CZs.

Cloud systems are coupled with atmospheric circulations through a two-way interaction that is multi-faceted, not perfectly understood and poorly represented in climate models. On the one hand, the advection associated with a given circulation can influence the temperature and humidity stratifications in an atmospheric column, conditioning the type of mixing and convection that occurs in this column and therefore the type of cloudiness. The vertical velocity exerts a particularly strong control on convection, promoting low clouds in the subsiding regions and deep, high clouds in the ascending regions. Horizontal transport can also play a very significant role by carrying away the free-tropospheric moisture, for example, in the case of strong vertical shear. On the other hand, clouds can influence dynamics through diabatic heating that changes the temperature stratification and hence the pressure gradients. Most of this diabatic heating is latent, due to the condensation and freezing during the cloud formation and development, or melting and re-evaporation of cloud or precipitating hydrometeors. The clouds also modify the radiative properties of the atmosphere, modulating both the solar and terrestrial radiative heating.

Clouds also exert radiative effects on the surface, and this yields coupled surface–atmosphere interactions.

Clouds shield the surface from incoming solar radiation, and the resulting surface cooling decreases the turbulent heat fluxes. The decrease in evaporation dries the boundary layer and this reduces most types of cloudiness (from Sc to DCC). But the decrease in sensible heat flux cools the boundary layer and increases the lower tropospheric stability (LTS), a condition that is unfavourable for convective clouds but favourable for stratocumuli. Over continents, competing mechanisms resulting from the characteristics of the land surfaces give rise to complex feedbacks (see Chapter 12).

In this chapter, we will illustrate how the many features of the tropical climate and its variability result from mechanisms involving clouds. We will first present how clouds can organise spatially either spontaneously or forced by the solar insolation or land–sea contrasts. Second, we will present how clouds interact with the atmospheric circulation, and what mechanisms result from this interaction. Third, we will focus on how the clouds interact with the surface.

8.2 The Spatial Organisation of Clouds

Most of the large-scale features of the tropical cloud distribution (Sc decks, CZs) and their variability (monsoons, El Niño-Southern Oscillation [ENSO]) result from a forcing external to the atmosphere (e.g., the solar forcing and its seasonal cycle), from the surface heterogeneities (land–sea contrasts, sea-surface temperature [SST] gradients) or from coupled ocean–atmosphere mechanisms. At the mesoscale, surface heterogeneities can also organise convection along coasts through the diurnal cycle of land and sea breezes (see Chapter 12 for more details). But spontaneous spatial organisation or self-organisation, defined as resulting exclusively from atmospheric processes, can be observed frequently as well. At small scales, spontaneous organisation can be observed in open and closed cells in stratocumulus decks and cloud lines. At the mesoscale, organised cloud systems are routinely observed in CZs. At the synoptic scale, tropical storms and hurricanes are self-maintaining organised cyclonic systems. At large scales, convectively coupled equatorial gravity waves and MJO events are forms of spontaneous organisation of clouds.

8.2.1 Mesoscale Convective Organisation

Cloud clusters and superclusters are of particular importance in the tropical climate variability because most propagating larger-scale tropical modes of variability present a similar cloud structure. These MCSs exhibit

a transition from convectively suppressed conditions to shallow convection, then to deep convection and to stratiform precipitation before a return to dry conditions. This transition though convective regimes corresponds to a transition from cumulus to congestus, to DCCs, to stratiform precipitation and cirriform anvils, as illustrated in Fig. 8.3 (Houze, 2004). The lifecycle of an MCS lasts from a few hours to a few days. First documented in the tropical Atlantic during the GATE campaign (Global Atmospheric Research Program – Atlantic Tropical Experiment) in 1974, tropical MCSs were further documented in the Western Pacific during TOGA-COARE (Tropical Ocean Global Atmosphere Program – Coupled Ocean–Atmosphere Response Experiment) in 1992–1993. Recently, satellites such as the Tropical Rainfall Measurement Mission (TRMM) provide a lot of information of the clouds and precipitation in MCSs.

The transition from shallow to deep convection to stratiform anvil is accompanied by complex cloud-scale circulations. The main features of this circulation are the deep convective updrafts and downdrafts represented in Fig. 8.3. The updrafts are composed of positively buoyant parcels from the boundary layer; upward throughout the DCC, the moisture in these parcels condenses and rains out, resulting in an enhancement of the parcels' buoyancy. The downdrafts include liquid-water-laden, negatively buoyant parcels that are cooled on their way down by rain evaporation. These downdrafts create a thin layer of cool air at the surface under the DCCs, called a cold pool. Extending around the deep cloud, the cold pool lifts the pre-existing, moist and warm surface layer, facilitating the onset of convection ahead of the system. Cold pools also increase the spatial variability of low-level moisture at the MCS scale, and further favour convection around the cold pools. Smaller cold pools occur under shallower precipitating clouds (Cg and the deepest Cu), and they are active in the transition to deep convection. The MCS is further powered by mid-level inflow from the back of the system. This inflow responds to the pressure gradients created by heating in the deep-convective core and the evaporative cooling by stratiform precipitation.

8.2.2 Organisation at Synoptic Scales

Although some cloud superclusters are large enough to be considered synoptic, the main mode of organisation of clouds in this range of scales is the tropical cyclones, ranging from tropical depressions, beneficial for water resources, to severe tropical cyclones (also called hurricanes or typhoons), with dire impacts on human societies.

The main characteristic of these cloud ensembles is the cyclonic circulation around its centre. The Earth's rotation plays a significant role in tropical cyclones and they tend to develop away from the equator and share some characteristics with extratropical storms (many tropical cyclones actually travel to the extratropics and transition to extratropical storms). DCCs organise in spiralling rainbands around the centre of the cyclones, and cirrus clouds develop on top. The cyclonic circulation is accompanied by convergence in the lower levels and divergence in the upper troposphere. Tropical cyclones usually travel westwards along the trade winds and slightly polewards due to a mechanisms of self-advection known as β-drift.

In a severe tropical cyclone, a circular region called the eye develops at the centre where the winds are weak, deep convection is inhibited, and shallow cumulus or stratocumulus clouds may form. The spiralling rainbands converge to a circular ring around this eye, which is called the eye wall. The most intense deep convection and cyclonic winds occur in the eyewall. Detrainment from this intense ring of deep convection creates a thick layer of cirrus clouds that extend outwards above the spiralling rainbands. Fast ascending motion in the eye wall is compensated by the combination of very intense subsidence in the eye as well as subsidence between the spiralling rainbands and away from the cyclone.

Often, tropical cyclones develop from perturbations of vorticity resulting from easterly waves or in the monsoon troughs. Their intensification relies on large turbulent surface heat fluxes, which explains why the most intense cyclones are observed over the warm waters of the western tropical oceans. This sensitivity to SST raises concerns in the perspective of a warming world. It is expected that future tropical cyclones will be more intense than past ones but, due to other sensitivities (in particular to vertical shear that limits the development of tropical cyclones), they should also be less frequent. Quantifying these changes is a very active area of research.

8.2.3 Organisation at Larger Scales

At large scales, multiple modes of variability can be observed in the tropical deep convection. Most of these modes propagate zonally along the equator, and they can be seen in the Hovmöller diagram of deep-convective activity in Fig. 8.4. A spatial and temporal filter is used to extract the anomalies corresponding to ranges of wave number and frequency of particular interest, and these filtered anomalies are represented by contours in Fig. 8.4.

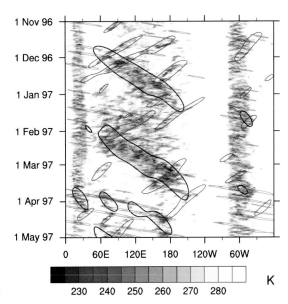

FIGURE 8.4: Example of a Hovmöller diagram of the equatorial brightness temperature anomalies (shading) and filtered anomalies (contours) for Kelvin waves (green contours), equatorial Rossby waves (blue contours) and the MJO (black). Low brightness temperatures correspond to deep clouds. Data from the Cloud Archive User Service (CLAUS).
Courtesy of Stéphanie Leroux

Green contours correspond to Kelvin waves that propagate eastward at a speed of about 15–20 m s^{-1}, and blue contours correspond to equatorial Rossby waves that propagate westward at a speed of 5–10 m s^{-1}. Kelvin and equatorial Rossby waves are known solutions of the momentum and dry energy equations of the tropical atmosphere (waves are solutions to the linearised equations). Analytic expressions of these waves' structure and characteristics can be derived from the shallow-water equations on an equatorial beta plane (Matsuno, 1966). These equations that describe the homogeneous momentum and depth of an incompressible fluid can also describe the first-baroclinic circulation of the dry equatorial atmosphere. The dynamical structure of Kelvin waves is non-rotational, symmetric with respect to the equator, with alternating regions of ascent and descent along the equator resulting from the convergence or divergence of the zonal wind. The Rossby waves feature alternating pairs of low-level gyres straddling the equator. Another prominent type of waves in the observations, the mixed Rossby-gravity waves, is not visible in Fig. 8.4 because their signal is antisymmetric about the equator. All these waves (Kelvin, equatorial Rossby and mixed Rossby-gravity waves) are observed in the tropical atmosphere. They can be associated with deep convection; they are then called convectively coupled equatorial gravity waves and propagate more slowly than the dry waves. The influence of moist convection on the wave dynamics can be accounted for by a simple change of parameter to take into account the impact of convection on the mean thermal structure of the atmosphere. The propagation speed is reduced because of this parameter, but the wave structure is not significantly modified by this coupling. This means that moist processes do not play an active part in determining the structure of the waves, although they do feed back on the dynamics and can be instrumental in exciting and sustaining the waves. Despite our theoretical understanding of convectively coupled gravity waves, many climate models are still very poor at simulating them.

The black contours in Fig. 8.4 correspond to the MJO. This large-scale convective disturbance propagates eastwards along the equator from the Indian Ocean to the Central Pacific with a propagation speed of about 5 m s^{-1}. Fig. 8.5 shows the pertubation of the outgoing long-wave radiation[2] and the low-level wind associated to this phenomenon, with its propagation divided in eight phases following Wheeler and Hendon (2004). The MJO has a characteristic dynamical pattern with a pair of off-equatorial, low-level cyclonic gyres straddling the equator west of the convective disturbance, a strong surface westerly jet between these gyres, surface easterlies over an extended region east of the disturbance and the opposite circulation in the upper troposphere. There is mid-tropospheric ascent around the convective region and subsidence elsewhere. This pattern is similar to Gill's circulation in equilibrium with the diabatic heating from the MJO clouds. The dynamical signal regularly propagates around the globe from the East Pacific back to the Western Indian Ocean with a faster propagation speed of about 20 m s^{-1}, resulting in a period of about forty days. This extension of the dynamical propagation might involve a conversion of the disturbance into a dry Kelvin wave.

First documented in the 1970s (Madden and Julian, 1972), the MJO still eludes our theoretical understanding. It is thought to be a moisture mode, that is, moisture, and in particular moisture transport, is crucial to its development and propagation (and it cannot be described by the equations for a dry atmosphere). But no simple model of the MJO has been widely accepted yet. The fundamental mechanism of the MJO results from the interaction between clouds and circulation, but which aspects are

essential remains debated. Ocean–atmosphere coupling has also be shown to strengthen the MJO in models.

Simulation of the MJO by general circulation models has been improving, but it is still poor in most climate models. At the large scale, these models are expected to represent the dynamics accurately, so the failure to simulate the MJO is attributed to the parameterisations of moist physics (and to the feedbacks between dynamics and moist physics). Two flaws of these parameterisations are hypothesised to play a role in this failure:

- The parameterisations do not produce the correct profile of diabatic heating associated with the MJO. This yields biased pressure gradients and the resulting dynamical response does not provide the appropriate transports of moisture and energy necessary for the development and propagation of the MJO.
- The parameterisations are unable to simulate the spontaneous, upscale organisation of convection from the subgrid scale to the large scale, which prevents the development of a large-scale dynamical response which would then be able to further develop and propagate the MJO disturbance.

In particular, parameterisations of deep convection are suspected to underestimate the sensitivity of convection to free-tropospheric humidity, a factor that could underlie both flaws.

As mentioned earlier, the large-scale modes of atmospheric variability such as the convectively coupled equatorial gravity waves or the MJO have a cloud signature similar to the MCSs (Riley et al., 2011), even though their spatial scales are much larger and their temporal scales are better measured in days or weeks rather than in hours. Fig. 8.6 shows the composite cloud signatures of the MJO and Kelvin wave. These cloud patterns are indeed similar and exhibit a transition from shallow cumuli to DCCS to stratiform anvils similarly to MCSs.

The convectively active, ascending phases (phases 4 and 5) of the MJO and Kelvin wave are associated with maxima in all types of clouds (Cu, Cg, DCCs, SPCs, An). Shallow convection starts deepening in the subsiding free troposphere two to three phases beforehand, and the occurrence of congestus increases. At the same time, cloudiness starts increasing at the tropopause thanks to very intense, isolated DCCs. Organised DCCs and SPCs develop one to two phases before the MJO peak, and remain for one phase afterwards, with lowering cloud tops. Anvil clouds linger for another phase afterwards during phases of subsidence.

This similarity of the cloud patterns of MCSs and larger-scale modes of variability (such as Kelvin waves

---

[2] Minima of outgoing long-wave radiation correspond to maxima of convective disturbance, because of the additional greenhouse effect due to the convective and high clouds

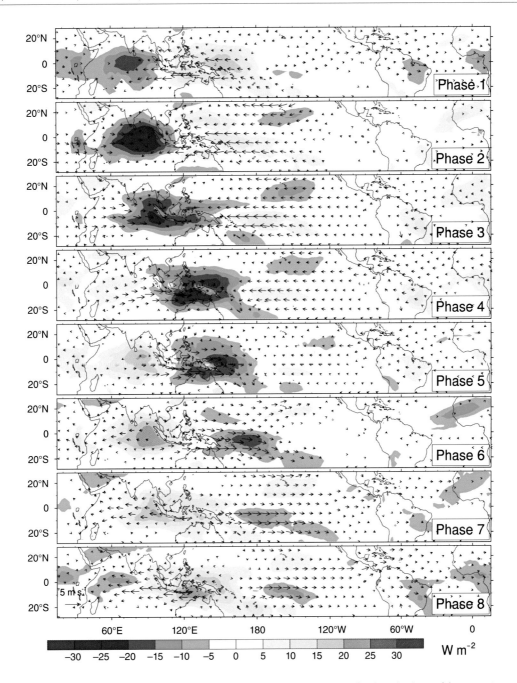

FIGURE 8.5: Anomalies of outgoing long-wave radiation (shading) and 850-hPa wind vectors for the eight phases of the composite austral-summer MJO event.
Data from the National Oceanic and Atmospheric Administration and ERA-Interim.

and MJO) is quite intriguing, especially since the large-scale envelopes of MJO events or Kelvin waves often contain full lifecycles of MCSs (as can be seen in Fig. 8.4, e.g., around 60°E on 22 Nov. 1997). This raises the question whether cloud variability follows a self-similarity across scales due to some fundamental cloud mechanism.

The 'stretched building block' hypothesis aims to explain this similarity (Mapes et al., 2006), suggesting that depending on the phase of the large-scale wave, the MCS lifecycles (which are the building blocks) are modified, with one phase of these lifecycles lasting longer or covering more surface (stretched) than in an average

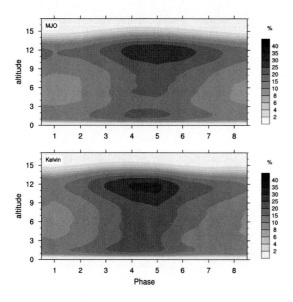

FIGURE 8.6: Cloudiness as a function of the phase of the MJO and convectively coupled Kelvin wave (CloudSat, 15°S–15°N). The x-axis can be considered as a time axis, with a typical timescale between two phases of about 5 days for the MJO and 1 day for the Kelvin wave. Phases 4 and 5 are the active phases, with increased deep convection and ascending motion, while Phases 1 and 8 are suppressed phases, with subsidence and inhibited deep convection. Courtesy of Emily Riley

lifecycle. From this viewpoint, MCSs during the pre-convective phase of the large-scale wave have a longer or larger than average shallow-convective phase in their lifecycle, while MCSs during the mature phase have a larger than average DCC fraction and MCSs during the decaying phase exhibit a larger stratiform anvil. These variations in the duration or size of the different phases of the MCS lifecycle are likely to result from their sensitivity to the large-scale environment that is modulated by the large-scale wave. In particular, the sensitivity to the large-scale circulation is expected to play a major role: convergence of the horizontal wind favours deep convection, ascent can sustain a stratiform anvil and subsidence damps deep convection, limiting convection activity to the boundary layer. Still, this hypothesis has not been validated. How the internal mechanisms of MCSs (updraft, cold-pool uplift) react to larger-scale variability, and how MCSs give rise to, or feed back onto, larger-scale variability is yet to be understood.

8.2.4   Meridional Organisation Due to Insolation

The solar heating of the atmosphere and surface induces cloud spatial structures at the planetary scale and, despite its weak annual cycle in the tropics, forces a very significant seasonal variability.

The average insolation is maximum at the equator. As a result, the sun heats the equatorial regions more than the subtropics. Part of this meridional gradient of heating is due to direct atmospheric absorption of solar radiation, but most of it is actually mediated by the surface, through terrestrial radiation and turbulent heat fluxes. Indeed, the SST is warmer at the equator (or nearby if there is upwelling of deep waters at the equator) than in the subtropics. In response to this heating gradient, a meridional circulation arises with ascent around the equator, poleward motion in the upper troposphere, subsidence in the subtropics and equatorward motion in the lower troposphere. It is known as the Hadley circulation. The Hadley circulation can be modelled by very simple equations of a dry, zonally symmetric atmosphere, such as the shallow-water equations on an equatorial beta plane. Simple models predict the near-surface easterly trade winds and the subtropical upper-tropospheric jets based on angular momentum conservation (Held and Hou, 1980), although it has been shown that eddies due to the instability of these jets are instrumental in controlling the poleward momentum transport and driving the Hadley circulation (Walker and Schneider, 2006). Indeed, moist processes are essential to the real-world Hadley circulation, since they control how the surface heating gradients are transmitted to the free troposphere (or not) and since they create a feedback between circulation and heating (Fang and Tung, 1996).

The cloud patterns embedded in the Hadley circulation are on the planetary scale, clearly visible from space. Fig. 8.7 shows a typical latitude-altitude view of the Hadley circulation with the associated cloud patterns,[3] here for the boreal summer over the Atlantic Ocean. In the subtropical, descending branch of the Hadley circulation, deep clouds are scarce, boundary-layer clouds are ubiquitous over the oceans, and continents tend to be desertic. Shallow cumuli are abundant in most of the subtropical oceans except for the eastern coastal regions where the

---

[3]   Note that in the GOCCP data used here and in the following figures, a systematic minimum on the vertical profile of cloudiness in convective regions can be seen between 3 and 5 km of altitude. This feature might be more prominent in the GOCCP data than in the real world because this data set is based on lidar observations from the satellite Cloud-Aerosol Lidar and Infrared Pathfinder Satellite Observations (CALIPSO), and it is difficult for space lidars to detect clouds shielded by other thick clouds. But this minimum has also been noted in satellite infrared observations and surface observations. It results primarily from the fact that this layer of the troposphere is less stable than the inversion below and the layer around the freezing level above.

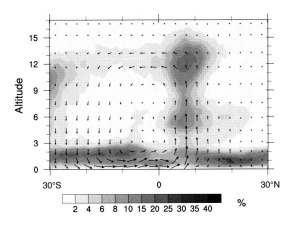

FIGURE 8.7: Average boreal-summer cloudiness (shading, GCM-Oriented CALIPSO Cloud Product [GOCCP]) and circulation (vectors, ERA-Interim) over the Atlantic ocean (40°W–20°W).

SST is cool and stratocumuli form decks. The shallow convection deepens equatorwards and congestus clouds appear. In the ascending branch of the Hadley circulation lies the ITCZ, that extends along the equator (between 5°N and 10°N over most of the tropical oceans). The ITCZ is populated by all types of convective and high clouds (Cu, Cg, DCC, SPC, An).

Because the insolation does not vary seasonally as much in the tropics as in the extratropics, the seasonal cycle in the tropical belt is generally small over the ocean. But in regions with subtropical continents, the interaction between ocean, land and atmosphere creates the most prominent features of the seasonal cycle in the tropics: the monsoons. Depending on the criteria for the seasonality of precipitation and surface winds used to define a monsoon, three[4] or five monsoons can be observed on Earth: the Asian and West African monsoons (in boreal summer) and the Australian monsoon (in austral summer) are the most prominent monsoon systems. The North American and South American monsoons are smaller, less-pronounced monsoon systems that are observed, respectively, over the Southwestern North America in boreal summer and over Amazonia in austral summer.

Fig. 8.8 shows the cloudiness in the West African monsoon region in winter and summer. The West African

---

[4] Using Ramage's (1971) criteria:

- Prevailing wind direction shifts by at least 120° between January and July.
- Prevailing wind direction persists for at least 40% of the time in January and July.
- Mean wind exceeds 3 m s^{-1} in either month.
- Fewer than one cyclone–anticyclone alternation occurs every two years in either month in a 5° latitude–longitude square.

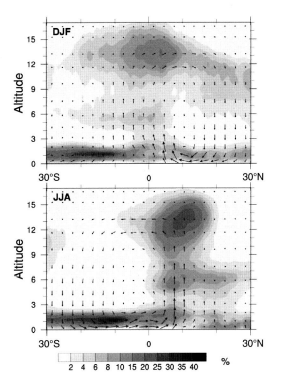

FIGURE 8.8: Average boreal winter and summer cloudiness (shading, GOCCP) and circulation (vectors, ERA-Interim) over West Africa (20°W–20°E). DJF, December–January–February; JJA, June–July–August.

monsoon is characterised by a northward shift of the ITCZ from its winter equatorial location to its summer monsoonal location around 12°N, accompanied by an enhancement of the ITCZ cloudiness. In the austral tropics, increased subsidence during the monsoon favours an increase in low-cloud cover. The Hadley circulation intensifies during the monsoon, with an enhancement of the asymmetry between hemispheres, the development of a strong cross-equatorial flow and a reversal of the trade winds in the summer hemisphere (not shown). This is the canonical pattern of a monsoon, except for the clouds in the middle troposphere over the northern subtropics, that sit atop the Saharan low and are specific to the West African monsoon.

Other monsoons can exhibit features that differ from this canonical pattern. The Asian monsoon is characterised by two CZs in the Indian sector, one over the equator and one around 20°N, and the Australian monsoon is associated with an intensification of the SPCZ.

Monsoons can be thought of as the Hadley circulation resulting from off-equatorial warm surface temperatures. These warm temperatures generate large turbulent surface fluxes that create convective instability, favouring the development of convective clouds that in turn anchor an off-equatorial CZ over the warm surface. Typically, for the solar annual cycle to produce such warm surface temperatures, the surface heat capacity has to be small, as in the case of land surfaces, so that the surface temperatures adjust very fast to the solar forcing. But, in the case of continents, a high surface albedo can reduce the solar warming of the surface and limit the poleward extension of monsoons. This is what happens over subtropical deserts.

Although the Hadley circulation can be understood using simple equations of the dry atmosphere, understanding and simulating the observed circulation and clouds has proved somewhat unsuccessful. There is no consensus understanding of what controls the ITCZ regime: why is the oceanic ITCZ north of the equator? Why are monsoons characterised by distinct ITCZ patterns? This is particularly vexing because the simulation of these features by general circulation models is unsatisfactory and some theoretical understanding would help improve them. These models are still challenged by the simulation of monsoons that are often underestimated. They tend to simulate two ITCZs straddling the equator rather than one ITCZ north of the equator, a systematic bias known as the 'double ITCZ syndrome' (e.g., Oueslati and Bellon, 2015).

8.2.5   Zonal Organisation Due to Surface Heterogeneity

Fig. 8.9 shows the climatological cloudiness and circulation along the equator (10°S–10°N). From the Eastern Indian Ocean to the Western Pacific, as well as over Amazonia and Central Africa, ascending motion can be observed over the equator. In between, regions of subsidence can be observed with mean easterlies at the surface, except in the western equatorial Indian oceans where the surface winds are westerlies, converging into the regions of ascent. This circulation is known as the Walker circulation. In the ascending branches of the Walker circulation, all types of convective and precipitating clouds (Cu, Cg, DCCs, SPCs, An) occur frequently, while the descending branches are populated by cirrus and low clouds. In these regions of reduced elevated cloudiness at the equator, a well-defined ITCZ is located off the equator (as in Fig. 8.7). Fig. 8.9 makes it clear that the continents and their orography play a role in the emergence of the Walker circulation, since rising branches are over continents and islands. In particular, the orography blocks the easterly trade winds, which are prevalent in the Pacific and Atlantic Oceans (they are in fact part of the Hadley circulation), and this mechanically forces ascending motions over the

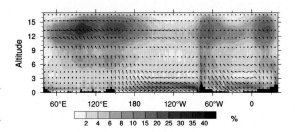

FIGURE 8.9: Climatological cloudiness (shading, GOCCP) and circulation (vectors, ERA-Interim) over the equator (10°S–10°N). Equatorial orography is filled in black.

Maritime Continent and Amazonia. Other mechanisms such as sea-breeze systems and land–atmosphere feedbacks contribute to the locking of CZs over continents and islands (see Chapter 12). In the equatorial Pacific, and to some extent in the Atlantic as well, the trade winds also force upwelling of deep, cold water in the eastern part of the basin and push warm equatorial waters to the western part of the basin, creating an east–west SST gradient. This favours deep convection in the western part, particularly in the equatorial West Pacific CZ (WPCZ).

The main mode of inter-annual variability in the planetary climate is essentially a variation of this mechanism involving surface temperature and wind. It is the ENSO, whose warm phase, El Niño, is characterised by a warming of the Central and/or East Pacific equatorial surface waters, collocated with an increase in precipitation and associated with a weakening of the equatorial easterly winds and a weakening of the east–west SST gradient. The cold phase, La Niña, consists of an enhancement of the east–west SST and precipitation gradients in the equatorial Pacific, with a reinforcement of equatorial easterly winds. El Niño can be seen as an eastward shift and a weakening of the Walker circulation, as well as a strengthening of the Hadley circulation: the WPCZ moves eastwards and east–west contrasts of deep convection and vertical motion are reduced. Conversely, La Niña is associated with an enhancement of the Walker circulation and a weakening of the Hadley circulation.

Fig. 8.10 shows the anomalies of cloudiness associated with both phases of ENSO. During El Niño, convection in the WPCZ weakens and shifts eastwards, resulting in a marked increase in high- and mid-level cloudiness over the equatorial Central Pacific, extending over the equatorward margin of the ITCZ in the East Pacific. Convection and the associated middle and high clouds also intensify in the Western Indian Ocean, and weaken in Amazonia as a result of the shift of the Walker circulation. In the subtropical Pacific, the enhanced subsidence in the descending branch of the Hadley circulation inhibits convection and

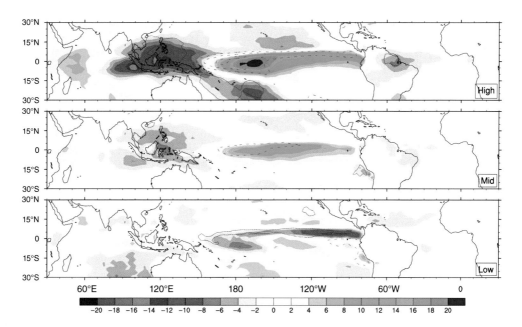

FIGURE 8.10: High, middle and low cloudiness anomalies (ISCCP) for the different phases of ENSO: El Niño (shading) and La Niña (contours: interval is 4%, solid for positive anomalies, dashed for negative anomalies, and the zero contour is omitted).

results in a decrease in middle and high clouds. In general, El Niño low-cloud patterns are opposite to the high-cloud patterns in the Pacific Ocean: deep convection replaces shallow convection (and low clouds occurring below high clouds are not detected by the satellites used to produce the ISCCP data) in the Central and East equatorial Pacific, and the opposite change occurs in the subtropical Pacific and over the equatorial Maritime Continent. But low clouds also decrease in the subtropics around the Maritime Continent and in the Eastern Indian Ocean, and increase in the Western Indian Ocean independently from the deep-cloud patterns, probably due to the reorganisation of the Hadley/Walker circulation.

During La Niña, the cloud anomalies are generally opposite to the El Niño anomalies, but they are smaller and only the main patterns are visible.

8.2.6 Convective Self-aggregation

Large-domain cloud-resolving models and general circulation models have shown that under certain conditions, the horizontally homogeneous radiative-convective equilibrium (RCE)[5] can become unstable (Wing et al.,

2017). This instability manifests itself by the spontaneous tendency of the atmosphere to organise into dry and moist areas associated with large-scale descent and ascent, respectively. The process generally starts with the formation of a dry patch that expands, eventually forcing the convection to aggregate into narrow areas, in the form of a few clusters, at the edges of or in-between the dry patches. This phenomenon is referred to as 'convective self-aggregation', and could play a significant role in the development of MCSs. The physical mechanisms responsible for RCE instability and self-aggregation are still a matter of research. However, the role of interactions between atmospheric water (in the form of water vapour and clouds), radiative cooling and circulation, between surface wind anomalies and surface turbulent fluxes, and between moisture and convection have all been shown to play an important role in the initiation and development of self-aggregation. On the other hand, the maintenance of self-aggregation seems to rely primarily on the interaction between DCCs and radiation, and in particular the reduction of the atmospheric radiative cooling by clouds. Both numerical and observational studies show that situations associated with more aggregated convection are generally drier and less cloudy in the upper troposphere, which tends to increase the emission of long-wave radiation to space and to reduce the planetary albedo. This impact suggests that the spatial organisation of convection might have a climate impact. Furthermore, models suggest that the

---

[5] RCE is an idealised model configuration without rotation or atmospheric transport into or out of the domain, and with uniform vertical boundary conditions and/or forcings (see Chapter 5).

phenomenon of self-aggregation is temperature dependent, and that it is favoured at high surface temperatures (some studies suggest that it can also be favoured at very low temperatures). Given the impact that convective aggregation has on the large-scale atmospheric state and the radiation budget, the question thus arises whether this temperature dependence might affect climate feedbacks (see Chapter 13). Although a few studies suggest that the MJO might be a large-scale manifestation of convective self-aggregation, the relationship between this phenomenon and the large-scale organisation of convection is still an open question.

8.3   Cloud-Stratification Interaction

We have shown in the previous section that clouds exhibit a vast diversity of organisation, on a large range of scales; this has been known since the 1970s thanks to field studies and satellite observations. What is now increasingly recognised is the fact that clouds play an active role in these different forms of organisation. It stems from the different ways through which clouds interact with their environment. Clouds are sensitive to the environmental atmospheric stratification, and affect their environment in turn through moistening and heating.

Temperature stratification has a strong influence on clouds, because it constrains the static stability of the atmosphere and thus the type of clouds that develop. For large LTS (LTS = θ(700 hPa) $-\theta$(surface)), convection is limited to the boundary layer whose depth decreases with increasing LTS. The occurrence of stratocumulus clouds exhibits a strong positive correlation with the LTS. For lower LTS, shallow convection and shallow cumuli occur. DCCs are associated with very low LTS. This is due to the fact that convective mixing participates to the control of the LTS by relaxing the lower-tropospheric temperature profile towards a moist adiabat.

Clouds are also sensitive to humidity: boundary-layer moisture determines, together with temperature, whether and which boundary-layer clouds develop, and is the reservoir for convective moistening of the free troposphere by shallow and deep convection. Free-tropospheric humidity has also been increasingly recognised as a key variable for cloud development at all scales. In subsiding regions, low-level clouds are sensitive to lower-tropospheric humidity because it regulates the ventilation of the boundary layer by subsidence and the buoyancy of mixed parcels at the top of the boundary layer. In general, the moister the free troposphere, the larger the low-cloud cover. Convective clouds are also sensitive to the free-tropospheric

environmental humidity because of the lateral entrainment of environmental air into convective plumes. This mixing dries (and cools) the updrafts, reducing their buoyancy, and dampens or even inhibits deep convection. Regions with a moist free troposphere are therefore favourable for deep convection.

In turn, convective clouds influence the large-scale humidity profile via convective mixing and detrainment, which redistributes boundary-layer humidity within the atmospheric column, and via precipitation, which dries the troposphere. This interaction between clouds and humidity introduces *moisture-convection feedbacks* that are likely critical for many phenomena, from convective self-aggregation to the location of the CZs, and in particular in the lifecycle of organised convection from the mesoscale to the intraseasonal scale.

Interactions between clouds and the thermodynamic vertical structure of the atmosphere are essential to the life cycle of the MCS. Deep convection is sensitive to free-tropospheric humidity because of convective entrainment. Through convective mixing, cumulus and congestus clouds moisten the free troposphere and facilitate the onset of organised deep convection, although this effect might not be as large as the effect of cold pools resulting from congestus convection (see Section 8.2.1). The water detrained from DCCs moistens the free troposphere and this moistening helps sustain deep convection, creating a moisture-convection feedback. On the other hand, DCCs and SPCs warm the upper troposphere and cool the lower troposphere. This thermal restratification of the troposphere dampens convection; it is instrumental in the demise of the MCS. Precipitating clouds also desiccate the atmospheric column, which dampens convection as well. How these different mechanisms compete or cooperate to control the lifecycle of MCSs remains an open area of research.

The same combination of effects found in MCSs is also active in larger-scale variability such as in Kelvin waves and the MJO. It contributes to the transition from shallow to deep convection, from convective to stratiform precipitation and anvil clouds observed in such large-scale modes. In particular, lower free-tropospheric humidity increases ahead of the convective maximum of the MJO, due to processes that are still debated but moistening by shallow cumulus, congestus and isolated DCCs is thought to contribute.

Understanding and quantifying how the different sensitivities and contributions play out to determine moisture-convection feedbacks is still a matter of research. Representing the interaction between clouds and large-scale environment is essentially the goal of moist-physics parameterisations, and our lack of a complete understanding of these processes is one of the reasons why parameterising clouds and convection is still challenging. Chapter 7 on parameterisations details these processes and in the following we will focus on indirect effects, due to processes that change the large-scale thermodynamic stratification and that clouds in turn influence. The most important of these is the atmospheric circulation.

8.4   Cloud–Circulation Interaction

In the mid-latitudes, where the Coriolis parameter is larger than in the tropics and the potential latent heat release is smaller because of the limited amount of moisture in the atmosphere, Earth's rotation is more important for atmospheric dynamics than diabatic processes, although the latter have recently been found to be significant. Conversely, diabatic processes are at the core of tropical circulations, and a large part of these diabatic processes occur within clouds. Heating by clouds creates horizontal gradients of temperature between the cloudy regions and the surrounding clear-sky regions, and the circulation responds to the resulting pressure gradients. The circulation in return affects the cloudiness because the associated energy and moisture transport can change the environmental stratification, and hence the types of cloud that will develop or decay.

The interaction between clouds and circulation is essential in almost every aspect of the tropical climate and its variability. Its most prominent facet is the interaction between deep convection and ascent. Both processes taken independently are essentially self-attenuating (except for dry waves that are free solutions of the momentum and dry energy equations): deep convection is damped by the thermal restratification and boundary-layer drying associated with deep clouds, and ascent of dry air is damped by the pressure gradients created by the adiabatic cooling it causes. But the interaction between these two processes can maintain deep-convective patterns associated with large-scale ascending motion on longer timescales. Historically, as early as the 1960s, the interaction between large-scale circulation and deep convection was thought to be crucial to the atmospheric variability in the tropics. This led to the development of the theory of Convective Instability of the Second Kind that describes the growth of all tropical disturbances as a result of a strong positive feedback between low-level convergence caused by mid-tropospheric latent heating and precipitation caused by the convergence of

humidity. This theory inspired closures on the column-integrated convergence or moisture convergence in the parameterisation of deep convection. It was shown later that this theory and the related closures exaggerate the role of the circulation in controlling convective clouds and can even suggest a wrong causality relationship between moisture and circulation.

In terms of amplitude, latent heat release in convective clouds is the dominant contribution to diabatic heating in the tropical atmosphere. As a result, most tropical modes of variability involve a convective component. By converting latent energy into internal energy, this contribution does not change the energy content of atmospheric columns, but it can create strong gradients of temperature and pressure.

Another important way through which all clouds interact with circulations is through radiation. DCCs radiatively warm the troposphere while boundary-layer clouds radiatively cool the atmosphere (see Chapter 4). Low-cloud radiative effects (CREs) strengthen the winds at the surface of tropical oceans and amplify the atmospheric overturning circulation. DCCs exert an even stronger impact on circulations. First, they warm the atmosphere, which increases the static stability of the free troposphere, shrinks the anvil cloud amount and enhances the horizontal gradients of atmospheric radiative cooling between convective and non-convective areas. These effects were found to impact the organisation of convection over a large range of scales and the structure of large-scale CZs. Second, they enhance the large-scale meridional gradients in upper-tropospheric temperature, which affects the baroclinicity of the extratropical atmosphere and the mid-latitude jets. Finally, as deep clouds radiatively warm the atmosphere during the rising phase of large-scale tropical disturbances and low clouds radiatively cool the atmosphere during episodes of large-scale subsidence, CREs reduce the effective stratification felt by propagating waves and slow down their propagation. Consistent with these different influences, CREs have been found to influence tropical intra seasonal oscillations, ENSO variability, the intensity and width of the ITCZ, the position and strength of extratropical storm tracks, and the structure and intensity of tropical cyclones.

8.4.1   Influence of Circulation on Cloudiness

Vertical motion can influence the environmental stratification very efficiently. In the free troposphere where humidity decreases with altitude and dry static energy

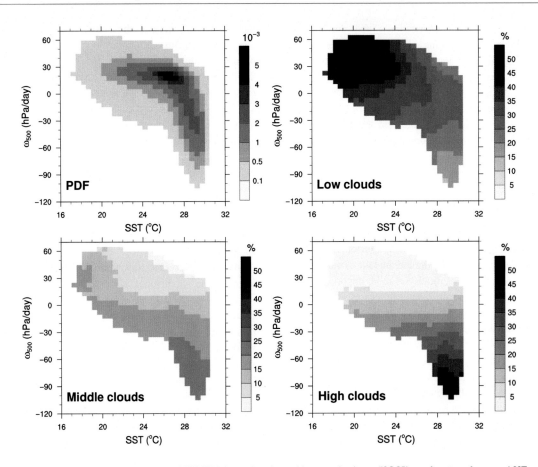

FIGURE 8.11: PDF of monthly mean ω_{500} and SST (ERA-Interim), and monthly mean cloudiness (ISCCP) as a function of ω_{500} and SST, over tropical and subtropical oceans (30°S–30°N).

increases, ascent moistens and cools while subsidence dries and warms. Within the boundary layer where the vertical gradients of conserved variables are small, vertical advection is very small. As a result, the lower-tropospheric temperature and humidity stratification are affected by vertical motion. Note that large-scale ascent refers to an average value that conflates updrafts and downdrafts within convective clouds with slow, mostly subsiding motion in the environment (which means that vertical transport is actually more complex than pictured here, although the overall contribution follows the same principles), while large-scale subsidence averages more horizontally uniform values.

By decreasing tropospheric stability and moistening the free troposphere, ascent favours deep convection: the development of all types of middle and high clouds are facilitated by large-scale ascent. Conversely, subsidence warms and dries the free troposphere, thus increasing the LTS and limiting convection and clouds to the boundary layer. Once an inversion forms at the top of the boundary

layer, subsidence and cloud-top entrainment also transport warm and dry air into the boundary layer, a process known as ventilation which tends to reduce cloudiness there. But a large LTS also keeps the boundary layer shallow and moist, favourable to stratocumulus formation characterised by a large cloud cover.

Fig. 8.11 illustrates these sensitivities. It shows the typical high-, middle- and low-cloud fractions over the tropical and subtropical oceans, as a function of the SST and mid-tropospheric large-scale vertical velocity ω_{500} at 500 hPa (in pressure coordinates so that the negative values correspond to ascent and the positive values to subsidence). It also shows the joint probability density function (PDF) of SST and ω_{500}. The high–cloud fraction exhibits little dependence on the SST, and it is almost linearly related to ω_{500}, with large high-cloud fraction for strong ascent. Still, from the joint PDF, it appears that strong monthly-mean ascent ($\omega_{500} < -30$ hPa/day) only occurs over warm SSTs, pointing to a role of SST gradients in maintaining deep convection and ascent.

Precipitation exhibits the same sensitivity as the high-cloud fraction (not shown). Mid-tropospheric clouds exhibit roughly the same sensitivity as high clouds, except over cool SSTs, where mid-level clouds associated with extratropical intrusions can be observed.

Low clouds have the opposite sensitivity; low-cloud fraction decreases with increasing ascent: shallow convection is replaced by deep convection and high clouds mask the low-level clouds below them. Low-cloud fraction also increases with increasing subsidence, which shows that the effect of subsidence on LTS dominates that of ventilation. Unlike in ascending regions, low-cloud fraction is very sensitive to SST in subsiding regions.

The relationship between large-scale vertical motions and cloudiness shown in Fig. 8.11 for monthly means is valid at shorter timescales in reanalyses and field-campaign measurements.

Although vertical advection is the predominant contribution, horizontal advection can be significant. For example, (i) cooling by horizontal advection in the eastern subtropical basins facilitates saturation and contributes to maintaining stratocumulus decks. (ii) Vertical shear can damp convection by transporting free-tropospheric moisture anomalies away from their convective source and cancelling the moisture-convection feedback. (iii) Horizontal moisture transport is also thought to contribute to the propagation of the MJO.

8.4.2 How Clouds Influence the Circulation

The large-scale circulation of the atmosphere is well described by the quasi-static equations that result from applying the anelastic and hydrostatic approximations to the Navier–Stokes equations. Here, we will use these equations in pressure coordinates, as described in Chapter 2: the continuity Eq. (2.141) and the horizontal momentum Eq. (2.142):

$$\nabla_p \cdot v_p + \partial_p \omega = 0, \tag{8.1}$$

$$D_t v_p + \mathbf{f} \times v_p = -\nabla_p \phi, \tag{8.2}$$

with v_p the horizontal wind in pressure coordinates, ω the pressure velocity, ϕ the geopotential and $\mathbf{f}$ the Coriolis parameter vector. Eq. (8.2) reveals that the geopotential gradient is the primary forcing of the horizontal flow. The other terms of the momentum flux (transport and Coriolis acceleration) are functions of the flow. And Eq. (8.1) shows that, to ensure mass conservation, the vertical motion is the vertically integrated convergence of the horizontal flow.

So the key to the action of clouds on circulation is the modulation of the geopotential gradient by cloud processes. This modulation results from diabatic processes. In the tropics and subtropics, the hydrostatic approximation (2.136) and equation of state (2.7) can be combined as follows:

$$\partial_p \phi = -\frac{R_d T_\rho}{p}, \tag{8.3}$$

where $T_\rho = T(1 + \epsilon q_v - q_c)$ is the density (or virtual) temperature, with q_v the specific humidity, q_c the specific condensed water and $\epsilon = R_v/R_d - 1$, with R_v the gas constant for water vapour and R_d the gas constant for dry air.

Let's consider a cloud with its base at p_b and its top at p_t. Eq. (8.3) can be differentiated on an isobaric surface and integrated in the vertical to obtain an expression of the difference in horizontal geopotential gradient between the top and bottom of the cloud:

$$\nabla_p \phi_t - \nabla_p \phi_b = R_d \int_{p_t}^{p_b} \nabla_p T_\rho \frac{dp}{p}, \tag{8.4}$$

which shows that the change of the horizontal geopotential gradient across the depth of the cloud is proportional to a vertical integral of the density temperature gradient (it is the same equation that leads to the thermal wind shear relationship in the extratropics).

Note that the contribution of the density temperature gradient at a given pressure level is inversely proportional to the pressure. Now let's consider an atmosphere with uniform density temperature stratification except for this cloud which is warmer than the rest of the atmosphere. This corresponds to anomalies of geopotential smaller at the bottom of the cloud than at its top. Let's first assume that the geopotential anomaly is zero at the top of the cloud; in this case, we have a depression at the bottom of the cloud. In the absence of rotation, the geopotential gradients would essentially drive a converging horizontal flow below the cloud. With the Coriolis acceleration, the circulation is cyclonic and the convergent circulation is weaker (and with a spatially varying Coriolis parameter, the circulation response is not symmetric with respect to the geopotential perturbation[6]), but convergence occurs nonetheless. By continuity, convergence below the clouds corresponds to ascending motion within the cloud and, since this type of circulation does not penetrate significantly in the stratosphere, this implies divergence above the cloud. This is in contradiction to our assumption that the geopotential anomaly is zero above the clouds. Continuity in fact dictates that a weighted average of the

---

[6] See Exercise 3 on the role of rotation and β effect.

geopotential anomalies at the top and at the bottom of the clouds is zero, so that the two anomalies have opposite signs. A warm cloud is associated with a geopotential minimum below and a geopotential maximum above, and therefore with convergence below the cloud and divergence above.

Rather than controlling temperature *per se*, as in the simple example considered earlier, clouds contribute to diabatic heating via phase changes, radiative processes and in-cloud mixing. The partial time derivative of Eq. (8.4) yields the same integral, linking the change in geopotential gradient to the tendency of density temperature gradient, and we can isolate the contribution of clouds:

$$\partial_t^{cl} \nabla_p \phi_t - \partial_t^{cl} \nabla_p \phi_b = R_d \int_{p_t}^{p_b} \nabla_p \left(\partial_t^{cl} T_\rho \right) \frac{dp}{p}, \quad (8.5)$$

where ∂_t^{cl} indicates the contribution of the clouds to the local derivative. Strictly speaking, $\partial_t^{cl} T_\rho$ depends on the cloud heating and moistening, but the main contribution is the heating ($\partial_t^{cl} T$) and in the following discussion, we will focus on this term. Eq. (8.5) describes how cloud diabatic heating can change the horizontal gradient of geopotential across the depth of a cloud. Our reasoning with a simple warm cloud above still holds, although with some delay effects. If a cloud warms the troposphere, the geopotential difference across the cloud increases and outward horizontal geopotential gradients are created below the cloud while inward geopotential gradients are created above the clouds. This forces convergence below the clouds and divergence above. This clearly creates a positive *convection-circulation feedback*: warming by a deep cloud results in ascent which is favourable to deep clouds, as shown in the previous section. As noted earlier, Coriolis acceleration limits convergence, so that this type of feedback is expected to be stronger in the deep tropics than in the subtropics. The latitudinal variation of the Coriolis parameter also breaks the symmetry of the circulation response with respect to the cloud diabatic pattern, and this could intervene in the propagation of cloud systems and their envelopes. Because clouds can vary on a wide range of spatial and temporal scales, the geopotential gradients created by the cloud diabatic effect can feed back on a large range of transient and steady features of the tropical climate.

As pointed out earlier, the contribution of the heating gradient to the geopotential gradient is inversely proportional to the pressure. Therefore, in order to represent properly the effect of clouds on the circulation, it is not enough to simulate properly the vertically integrated heating; it is also necessary to simulate correctly the details of the vertical profile of cloud heating. Errors in the

profiles of diabatic heating are thought to be an important source of biases in general circulation models. At the same time, it is very difficult to extract reliable diabatic heating profiles from available observations and to discriminate between the different contributions to this heating. The lack of an observational basis limits our ability to constrain this aspect of model parameterisations. The vertically integrated heating is better constrained since the column latent heating is essentially proportional to the surface precipitation and the column radiative heating is given by the surface and top-of-the-atmosphere budgets.

8.4.3 Cloud Diabatic Effect

Clouds have a strong diabatic effect on the atmosphere through three main processes: phase change, radiation and turbulent mixing due to cloud-scale circulations. Cloud formation releases latent heat by condensation and freezing. Melting and re-evaporation of precipitation cause latent cooling. Hydrometors also interact with solar and terrestrial radiation: they scatter and absorb incoming radiation, and emit in the terrestrial range. A lot of mixing occurs inside convective clouds, due to updrafts in the cloudy convective plumes and downdrafts in precipitating convective clouds as well as in the vicinity of convective clouds.

The total diabatic heating can be estimated from field-campaign observations (e.g., GATE, TOGA-COARE) as a residual term from the dry static energy budget because storage and transport can be computed from the temperature and wind observations (Yanai et al., 1973). Cloud-resolving simulations of periods during these campaigns can provide an estimate of the latent, radiative and mixing contributions to this total heating. Some modern satellites, such as TRMM, provide radar observations of rainfall, which can be used to estimate the latent heating, often using input from cloud-resolving models. The various existing methods yield estimates of vertical profiles that differ significantly (Hagos et al., 2010). Estimates of radiative heating can be obtained combining satellite measurements and radiative transfer models (see Chapter 4): the satellites provide cloud properties (e.g., ISCCP products) and the radiative budget of the Earth (provided by the Earth Radiation Budget Experiment – ERBE – and more recently by Clouds and Earth's Radiant Energy System – CERES – instruments) are used to constrain the numerical simulations. Extracting the cloud contribution to the radiative heating can be done by numerical experiments or by subtracting composites of clear-sky observations from the all-sky observations. There is still some uncertainty on the vertically integrated estimates, and a lot of uncertainty on the details of profiles, and better documenting and understanding the vertical profile of mixing, latent and radiative heating will continue to be a challenge in the years to come. Nevertheless, we have rough estimates of what type of heating profile is associated with the major cloud types.

Fig. 8.12 shows typical tropospheric profiles of diabatic heating due to the three processes (phase change, radiation and mixing) for different cloud types (first row). Typical surface precipitation rates are also indicated, and they are proportional to the vertically integrated latent heating. This figure is based on a compilation of data from observation campaigns, satellite retrievals (some using cloud-resolving models) and weather-station soundings; it should be considered as illustrative rather than exact: while the vertical profiles of heating for convective clouds are fairly robust, the altitude of sign change and amplitude of the latent heating for SPCs are more variable, and the radiative profiles are sensitive to a variety of factors, including aerosols. Fig. 8.12 also shows the profiles of solar and terrestrial contributions to the radiative heating (second row) and their vertical integral that corresponds to the radiative flux divergence across the atmospheric column (third row). Radiative effects are given for an average daily solar insolation, so depending on the time of day during the cloud occurrence, its radiative effect in the solar range can be significantly different. The main cloud type that is not represented in Fig. 8.12 is the cirrus type; cirrus clouds have the same heating properties as anvil clouds except over a thinner layer in the upper troposphere.

We also chose to represent the profiles typical of non-precipitating stratocumuli and shallow cumuli (typical of the Barbados Oceanographic and Meteorological Experiment – BOMEX, 1969) even though stratocumuli can drizzle and a small percentage of cumuli, the deepest ones, do precipitate, as observed during the Rain in Cumulus over Ocean (RICO) campaign in the subtropical western Atlantic in 2004. The typical latent heating profile of these cumuli is in between that of a non-precipitating cumulus and that of a congestus.

From the first row in Fig. 8.12, it is evident that the cloud latent effect is dominant for all precipitating clouds (Cg, DCCs and SPCs) by one order of magnitude. Cloud mixing and latent heating have the same order of magnitude for the boundary-layer clouds, and CREs have the same order of magnitude as the two other effects in stratocumuli. In non-precipitating stratocumuli and cumuli, condensation occurs in the lower part of the cloud and evaporation occurs in the upper part. Precipitating convective clouds (Cg and DCCs) release latent heat with a maximum in the middle of the cloud. SPCs release latent heat in the cloud, and part of the precipitation is re-evaporated between the cloud base and the surface, so that there is latent cooling from the surface to the cloud base. The non-precipitating anvil clouds are fairly passive, with little condensation.

Heating tendencies due to mixing are typically negative in the lower part and very top of the cloud, and positive in the upper part. The heat released by condensation in the lower part of the clouds is transported upwards by cloud turbulence and warms the upper part of the cloud (which is cooled by evaporation). At the very top of the cloud, the turbulent overshoots tend to cool the layer, particularly when a stable layer, such as an inversion, caps the cloud. The respective magnitudes of these patterns depend on the cloud type. The cooling by mixing at the base of the cloud is clear for all types of boundary-layer and convective clouds (Sc, Cu, Cg, and DCCs), but the other features are significant only in boundary-layer clouds (Sc and Cu).

The atmospheric CRE is small compared to the other terms except in the case of the stratocumuli, at the top of which radiative cooling is a major term in the energy budget. The second and third rows of Fig. 8.12 further detail the solar CRE and the terrestrial CRE. The short-wave CRE is due to the absorption of incoming solar radiation by clouds (possibly after multiple scattering), and its vertically averaged atmospheric contribution is generally positive (i.e., a warming), roughly increasing with the optical depth of the cloud. Most of the absorption occurs in the upper part of the cloud. Below this absorption layer, the short-wave CRE can be negative because the downward short-wave radiative flux is reduced by the absorption and scattering above, so that the local absorption, which is proportional to this incoming flux, is reduced. The long-wave CRE results from a competition between increased emission (i.e., a function of temperature) and increased absorption (i.e., proportional to the incoming fluxes) due to the presence of cloud hydrometeors. It is positive in the lower part of the cloud where absorption dominates, and negative in the upper part of the cloud where emission dominates. The net vertically integrated contribution depends strongly on the cloud type, mostly because of the temperature at the cloud top: moderate cooling for stratocumuli, small cooling for cumuli, small warming for congestus clouds and a large warming for clouds with high tops (DCCs, SPCs, An).

The variety of heating profiles associated with different cloud types clearly shows the challenge that parameterisation developers face. Parameterisations need to simulate the effect of a population of clouds with

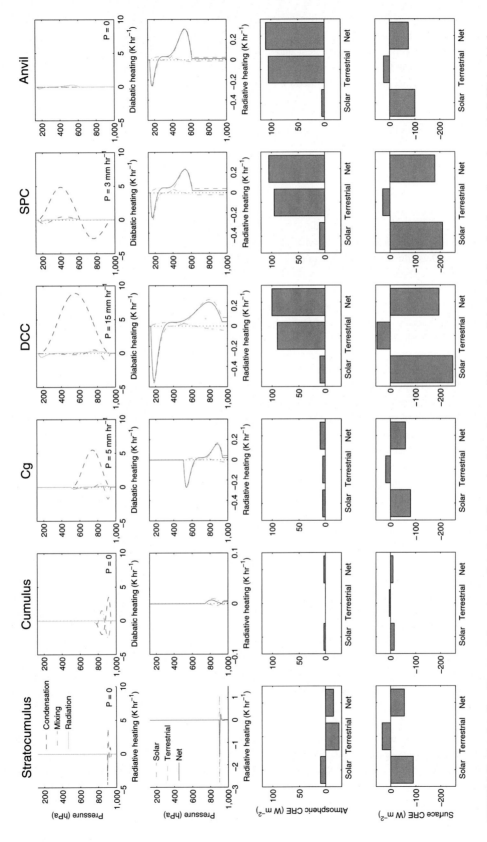

FIGURE 8.12: Typical profiles of latent, turbulent and net radiative heating (first row), and the solar, terrestrial and net radiative heating (second row), as well as total atmospheric (third row) and surface (fourth row) radiative heating derived from observation campaigns and satellite retrievals, associated with different types of clouds: stratocumulus, shallow cumulus, congestus DCCs, SPCs and anvil clouds. Typical precipitation rates are indicated in each panel on the first row.

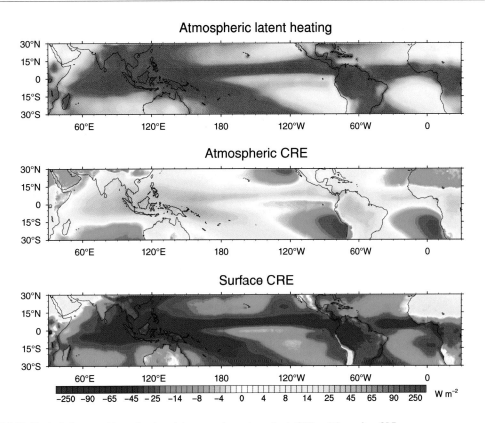

FIGURE 8.13: Vertically integrated latent heating of the atmosphere, atmospheric CRE, and the surface CRE. Data from NASA's Surface Radiation Budget (SRB) Project for the period 1998–2007

distinct characteristics in order to obtain the correct cloud-driven circulation. As an example, we can compare the circulation caused by the latent heating from a DCC versus that caused by an SPC. The DCC warms the troposphere with a maximum in the mid troposphere, creating density temperature gradients between the cloud and the environment; this results in outward geopotential gradients with increasing amplitude downwards. Horizontal convergence with a maximum at the surface and ascent in the cloud will ensue. The SPC warms the upper troposphere and cools the lower troposphere; because the vertically integrated heating is positive (i.e., the surface precipitation is positive) and because the elevated heating is more efficient at creating geopotential gradients due to the weight $1/p$ in the integral, the SPC diabatic heating will create horizontal convergence over the whole troposphere like the DCCs, even at the surface. But unlike in the case of the DCCs, the maximum convergence will occur in the mid troposphere, where the SPC heating changes sign. Furthermore, for a given precipitation rate and cloud area, the overturning circulation resulting from SPCs will be much larger than the circulation resulting from

DCCs.[7] Simulating the correct proportion of convective and stratiform rain is currently a challenging issue in general circulation models.

It would certainly be interesting to show a figure similar to Fig. 8.12 for cloud-moistening profiles. Cloud moistening is crucial for its direct effect on moisture stratification rather than for its effect on circulation through density temperature gradients. Unfortunately, there is even more uncertainty on these moistening profiles than on heating profiles.

Fig. 8.13 shows the climatology of the CRE vertically integrated over the atmosphere, as well as the column-integrated latent heating (estimated by multiplying the surface precipitation by a constant latent heat of vaporisation L_V). The climatologies of the CRE are related to the values in Fig. 8.12 (last two rows) by the climatological frequency of occurrence of the different cloud types.

In CZs (ITCZ, SPCZ, WPCZ), the column-integrated diabatic heating is overwhelmingly dominated by latent heat release in precipitating clouds (typically 250 W/m^2).

[7] Exercise 2 illustrates these points in the case of the Walker circulation.

Initially, the precipitation was thought to occur mostly under DCCs, but campaigns such as TOGA-COARE showed that the congestus clouds account for about a quarter of the precipitation in CZs, and satellite data (TRMM) have shown that stratiform clouds account for almost half the surface precipitation. This is surprising considering that the rain rates associated with each cloud type (Cg, DCC and SPC typical rain rates: 5 mm/hr, 15 mm/hr and 3 mm/hr, see Fig. 8.12) suggest that the DCCs precipitate much more. In fact, DCCs are very infrequent compared to congestus and SPCs. For example, within an MCS, the area covered by stratiform clouds is five- to ten-times larger than that covered by DCCs.

In convective regions, the net atmospheric cloud radiative heating is about five-times smaller than the latent heating (about 50 W/m^2). The net CRE in these regions is dominated by the large terrestrial warming associated with deep clouds. Over the ocean, the solar CRE is a small warming, but over continents and islands, it can be a small cooling, because of the reduced absorption in the lower troposphere. In the trade-wind, shallow-convective regions, small cumuli exert a moderate radiative cooling on the atmospheric column, which has roughly the same magnitude as the latent heating resulting from occasional deep convective events or the deepest shallow cumuli, resulting in near-zero net diabatic heating. In coastal regions west of the subtropical continents, stratocumuli have a significant cooling effect (about -30 W m^{-2}) resulting from strong terrestrial cooling (about -40 W m^{-2}) at the top of the cloud that is hardly modulated by a small solar warming. The column-integrated latent heating is negligible in these regions (very little drizzle reaches the surface).

Overall, the climatological CRE enhances the diabatic heating contrasts between precipitating and dry regions; the CRE warms the atmosphere in the CZs and cools the atmosphere in convectively suppressed regions populated by boundary-layer clouds (Sc and Cu).

8.4.4 Idealised Frameworks

Eq. (8.5) is a diagnostic tool and it cannot be used to study the convection-circulation feedback loop. On the other hand, the quasi-static equations are a closed mathematical system so complex that it is difficult to use them to grasp the fundamental interplay between moist physics and dynamics. Consequently, a number of approaches have been developed to simplify the problem and study interactions and feedbacks in idealised frameworks. The most basic of these approaches focuses on the interaction

between vertical motion and moist physics in an atmospheric column and essentially parameterise the large-scale vertical velocity.

The first of such approaches is called the weak temperature gradient (WTG) approximation (see Chapter 5). It relies on the observation that gravity waves rapidly dampen the free-tropospheric horizontal gradients of temperature, and as a result the subsequent horizontal transport of energy is negligible (Sobel and Bretherton, 2000). This approximation is used to simplify the thermodynamic Eq. (2.143) in the free troposphere, which can be written as follows in steady state:

$$\omega \partial_p (c_{p_e} T + \phi) = \nabla \cdot \boldsymbol{F}_h + Q_{\mathrm{cnd}}. \tag{8.6}$$

The right-hand side of this equation is the diabatic heating, with $\boldsymbol{F}_h$ the total heat flux (due to radiation and mixing) and Q_{cnd} the net condensation heating. If the free-tropospheric profile of temperature is known, Eq. (8.6) provides a parameterisation of the pressure velocity ω that sets it proportional to the diabatic heating: a warming causes collocated ascent, and a cooling causes subsidence. The pressure velocity is then used to compute the vertical transport of the other variables. The horizontal transport of variables other than temperature has to be imposed or parameterised. A relaxed version of this approximation has also been proposed: the large-scale vertical motion is assumed to relax the free-tropospheric temperature profile towards a reference profile. This formulation is particularly useful in cloud-resolving models, in which Eq. (8.6) cannot be implemented. These WTGs and relaxed WTG approximations have proven useful to understand the very basic interaction between moisture transport and cloud physics. Still, the collocation of diabatic heating and vertical motion stipulated in this approach might be too local in the light of Eq. (8.5) that establishes the dependence of geopotential on heating. Horizontal convergence thus depends on the vertical profile of heating and vertical motion and, through the continuity equation, has a non-local relationship to the heating.

Another idealised approach does take into account this non-local relationship. It is called the damped gravity wave (DGW) approximation because it uses a steady wave solution to the damped, quasi-static momentum equation with a motionless basic state and no rotation (Kuang, 2008). This simplified momentum equation can be written:

$$0 = -\nabla_p \phi - \eta \boldsymbol{v}_p, \tag{8.7}$$

which is essentially a linearised, steady Eq. (8.2) without Coriolis acceleration and with an additional damping term; η is the damping rate. Wave solutions with wavevector $k: (v_p, \omega, \phi, T_\rho) = (\hat{v}_p, \hat{\omega}, \hat{\phi}, \hat{T}_\rho) \, e^{ik \cdot x}$, with x the horizontal space vector, to Eq. (8.7) together with the continuity and hydrostatic Eqs (8.1) and (8.3), verify:[8]

$$\partial_p(\eta \, \partial_p \hat{\omega}) = \frac{R_d}{p} k^2 \hat{T}_\rho. \tag{8.8}$$

Using this equation, we can compute the pressure velocity ω that results from a perturbation of T_ρ from a reference profile, using $\omega = 0$ as boundary condition at the (flat) surface and at the tropopause. This pressure velocity is then used to compute the vertical transport of thermodynamic variables (including T), and the horizontal transport has to be imposed or parameterised. Compared to the WTG approximation, the DGW approximation reintroduces a complex relationship between the perturbation of the temperature profile (resulting from diabatic heating) and the profile of vertical motion, as shown by the double derivative in Eq. (8.8). The main simplification in this approach is that the horizontal momentum is exactly proportional to the geopotential gradients. This implies that the horizontal convergence is collocated with the Laplacian of the geopotential. In wave solutions, the Laplacian is proportional to the variable, so that the horizontal convergence is collocated with the geopotential perturbation in the DGW approximation. A more complex approximation named the weak pressure gradient approximation follows a similar approach, but it uses the general solution of linearised, time-dependent Boussinesq equations.

The potential and limits of these idealised approaches are still being investigated. They already provide an opportunity to study the interaction between moist physics and dynamics in an atmospheric column, providing a more tractable framework than the full tri-dimensional equations over a large horizontal domain. This type of approach can be used in realistic or idealised case studies, implemented in cloud-resolving or single-column models. They have not been widely used so far for model improvement, but the comparison between single-column models and cloud-resolving models using this type of representation of the large-scale circulation is seen as a promising avenue.

8.4.5 Cloud–Circulation Interaction and Climate

The *convection-circulation feedback* is active in all the major features of the tropical climate and their variability. It is one of the crucial mechanisms that produce cloud systems and their envelopes: as deep convection concentrates in cloud clusters, a circulation develops in

between the heated convective regions and the radiatively cooled dry regions. This circulation transports moisture from the dry-region boundary layer to the convective clusters, fuelling convection in these regions, and drying the non-convective free troposphere via downward transport in the dry regions, inhibiting deep convection there. The same mechanism is active in tropical cyclones in which the warm core, heated by latent heat release (and subsidence in the eye of severe tropical cyclones) creates pressure gradients between the centre and the periphery that drive the cyclonic circulation, convergence in the lower troposphere and divergence in the upper troposphere; nevertheless, most of the latent heat provided to a severe tropical cyclone does not result from atmospheric transport but from evaporation, through the wind-flux mechanism explained in Section 8.5.3.

The cloud diabatic heating in the ITCZ intensifies the Hadley circulation, which would exist in the absence of cloud processes just from the differential solar heating between the equator and the subtropics. This intensification is particularly strong when the ITCZ is away from the equator according to the angular momentum theory of the Hadley circulation; this is one of the mechanisms that maintains the ITCZ north of the equator rather than on the equator where the solar forcing is maximum (see Fig. 8.14). Some model experiments even suggest that the interaction between clouds and circulation is enough to break the inter-hemispheric symmetry and locate the ITCZ away from the equator. The cloud diabatic effect in monsoon CZs is even further from the equator than in the case of the ITCZ, and this contributes to the strength of monsoon circulations. Because turbulent heat fluxes increase with the surface wind, they feed back on the circulation: a more intense circulation means larger surface winds and larger surface latent and sensible heat flux; this additional energy is transported by the circulation to the ITCZ region and intensifies deep convection there.

The convection-circulation feedback also reinforces the Walker circulation, which is primarily forced by orography and land–sea contrasts, and its main form of inter-annual variability (ENSO), which is a coupled ocean–atmosphere mode. The cloud diabatic heating in the ascending branch of this circulation creates zonal gradients of geopotential that sustain convergence into these ascending branches. When the Walker circulation is perturbed during an El Niño event, the cloud diabatic effect causes ascent in the central equatorial Pacific and strengthens the eastward shift of WPCZ (see Fig. 8.15).

The issue of the vertical profile of cloud diabatic heating, through the factor $1/p$ in Eq. (8.5), was raised as early as the 1980s in the context of the simulation of the

---

[8] Details of the derivation are the subject of Exercise 1.

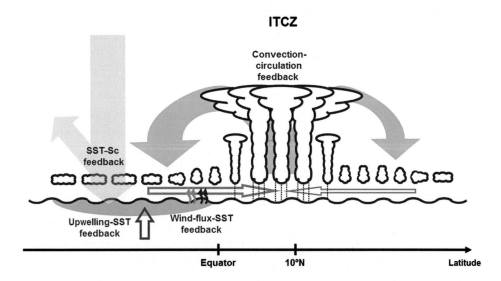

FIGURE 8.14: Schematic of the ITCZ and the associated cloud mechanisms: (i) the convection circulation feedback enhances the ITCZ in its off-equatorial location; (ii) SST-stratocumulus feedback cools the SST south of the equator; (iii) the near-surface circulation is enhanced by the cross-equatorial SST gradient; (iv) the surface winds also enhance surface turbulent heat fluxes and upwelling south of the equator, resulting in additional cooling of the ocean surface there.

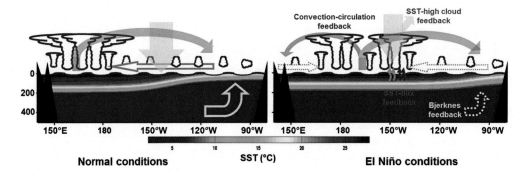

FIGURE 8.15: Schematic of the normal and El Niño conditions and the associated cloud mechanisms: (i) the positive Bjerknes feedback tends to maintain the El Niño (or normal, or La Niña) conditions through the interaction of the SST gradients, surface winds and upwelling; (ii) the SST-high cloud and the SST-flux negative feedbacks are responsible for the return to normal conditions.

Walker circulation: it was shown that a top-heavy vertical profile of diabatic heating (associated with the presence of clouds with a significant fraction of stratiform precipitation) induces a stronger Walker circulation in better agreement with observations than a profile that exhibits a maximum heating in the mid troposphere (as that induced by DCCs).[9] Later, TRMM observations confirmed this deduction by showing that almost half the surface precipitation originates from SPCs. More recently, observations in the East Pacific and Atlantic have revealed a shallow meridional overturning circulation with outflow in the lower free troposphere (associated with a bottom-heavy

profile of diabatic heating) additional to the troposphere-deep canonical Hadley circulation. These observations point to a complex latitudinal distribution of diabatic-heating profiles in these regions.

The circulation associated with the MJO is similar to the quasi-steady dynamical response following the cloud diabatic forcing. In this circulation pattern, known as the Gill circulation, large-scale ascent is almost co-located with the diabatic heating, and the resulting atmospheric transport is a strong positive feedback on the MJO convective disturbance. This convection-circulation feedback is thought to be essential to the development of an MJO event. Also, more ascent occurs east of the maximum of diabatic heating than west of it. This should moisten the

[9] This is illustrated in Exercise 2.

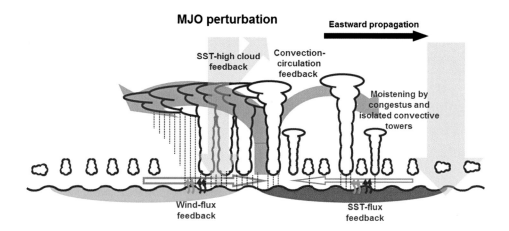

FIGURE 8.16: Schematic of an MJO perturbation and the associated cloud mechanisms: (i) the convection-circulation positive feedback is crucial to the development of the MJO event; (ii) the strong westerlies under the convective disturbance cause large turbulent surface heat fluxes that are also instrumental in maintaining the disturbance; (iii) together with this wind-flux feedback, the SST-high cloud negative feedback creates an SST gradient across the convective disturbance that enhances its eastward propagation through the SST-flux effect and the free-tropospheric moistening east of the disturbance.

lower free troposphere and favour the eastward propagation of deep convection (see Fig. 8.16). As pointed out earlier, the leading hypotheses to explain the difficulties encountered in modelling the MJO are errors in the profile of diabatic heating, or the lack of large-scale organisation of DCCs. Both shortcomings likely yield a wrong dynamical response and prevent the development and propagation of an MJO event.

8.5 Cloud–Surface Interaction

Clouds are sensitive to surface conditions because these constrain the surface turbulent fluxes of energy and humidity, which affect the boundary-layer temperature and humidity and thus the stratification of the atmospheric column. Also, the surface turbulent buoyancy flux conditions the intensity of turbulence in the boundary layer and therefore its deepening or thinning and the boundary-layer cloud type. Clouds are thus sensitive to the SST over the oceans and they are sensitive to multiple factors over land: the vegetation type, land characteristics, soil temperature and moisture (see Chapter 12).

Reciprocally, by interacting with radiation, clouds change the downwelling flux at the surface and generally cool the surface. Clouds also affect the surface winds locally through convective gustiness and at larger scales by the change of circulation discussed in the previous section. Since turbulent surface fluxes increase with surface wind speed, these changes in wind can affect the surface conditions. Precipitating clouds also change the soil moisture content over the continents, which can

also change vegetation type and associated characteristics (albedo, exchange coefficients) over long timescales. Here, we will focus on the oceanic surfaces; discussion on the land surfaces can be found in Chapter 12.

8.5.1 Influence of SST on Clouds

Both the surface sensible heat flux and the latent heat flux increase with increasing SST. This warms and moistens the boundary layer and increases the conditional instability of the atmospheric column. It also increases the turbulence in the boundary layer and deepens it. As a result, we expect cloudiness to be sensitive to SST.

Fig. 8.11 shows the sensitivity of cloudiness to the underlying SST. As noted in Section 8.4.1, high clouds and precipitation do not exhibit a strong intrinsic sensitivity to SST, except that intense precipitation occurs mostly in regions of ascent over warm SSTs, and the occurrence of large-scale ascent exhibits a strong increase for SSTs exceeding 27 °C. However, this threshold should not be considered as an absolute threshold, as its value depends on the tropical-mean SST. Middle-cloud fraction exhibits roughly the same sensitivity, except some mid-level clouds associated with extratropical intrusions, over low SSTs in neutral and subsiding regions.

Low clouds are more sensitive to SST than higher clouds, with large cloud fractions up to 50% associated with stratocumuli over low SSTs in subsiding regions, and small cloud fractions associated with cumuli over warmer SSTs. A large part of this sensitivity is directly related to the lower-tropospheric stability: the cooler the

SST, the larger the LTS, and the thinner and moister the boundary layer. In a very stable lower troposphere, shallow convection is inhibited and stratocumuli develop. As the SST increases, the lower troposphere becomes less stable and the surface buoyancy flux increases; both factors are conducive to shallow convection.

SSTs can influence cloudiness indirectly through the effect of their gradient on circulation: SST gradients are communicated to the air temperature in the boundary layer by the sensible heat flux, and geopotential gradients result from air temperature gradients following Eq. (8.4) and force atmospheric circulations from the regions of cool SST to regions of warm SST. These circulations influence cloudiness as described in Section 8.4.1. At the seasonal scale and on longer timescales, direct and indirect effects of the SST on cloudiness have similar effects: ascent and warm SSTs are generally collocated and both favour deep clouds, while subsidence and cool SSTs are also collocated and favour boundary-layer clouds.

8.5.2   Clouds and the Surface Energy Budget

If clouds are sensitive to the surface temperature, they also have a strong effect on the surface energy budget. Clouds are responsible for a large part of the global albedo because of backward scattering on cloud hydrometeors, and this is particularly felt at the surface (see Chapter 4). The last row of Fig. 8.12 shows the CRE on the surface, for a surface albedo typical of the ocean. Conversely to the atmospheric CRE (third row), the surface CRE is dominated by the solar contribution, and it is a cooling effect. The terrestrial CRE is a small warming due to the clouds' additional greenhouse effect. The magnitude of both the solar and terrestrial contributions are essentially a function of their optical depth: they are small for the optically thin cumuli, moderate for stratocumulus, congestus and anvil clouds, and large for deep clouds (DCCs, SPCs). The magnitude of the net surface CRE is sufficient to influence the surface temperature significantly. This cooling effect opens the way for a negative feedback between convection and SST and a positive feedback between low clouds and SST (all clouds cool the SST, which is unfavourable for deep, high clouds and favourable for low-cloud cover).

The bottom panel of Fig. 8.13 shows the climatology of CREs at the surface. In the CZs, the net radiative effect is a cooling of a similar magnitude as the radiative warming of the atmospheric column above (about 50 W/m^2). This net cooling is due to the strong solar contribution (up to 100 W/m^2) associated with the additional albedo of high, deep clouds, partly compensated by the terrestrial contribution due to the additional cloud greenhouse effect.

In the oceanic, shallow-convective regions, the net radiative effect due to optically thin cumuli is close to zero because the solar cooling and terrestrial warming compensate (both are about 30 W m^{-2}). Under the optically thick stratocumuli, the net cloud effect at the surface is a moderate cooling (about 30 W/m^2) that results from strong solar cooling (up to 100 W m^{-2}), modulated by significant terrestrial warming.

Over subtropical deserts, in particular the Sahara, the net CRE at the surface is a warming due to the terrestrial contribution. Because the albedo of the desert surfaces is large, the surface absorbs a limited fraction of the incoming solar radiation and the impact of clouds on incoming radiation is not felt as strongly as over regions of smaller surface albedo.

Clouds also affect the surface heat budget indirectly through their influence on surface turbulent fluxes because these fluxes increase with surface wind and sea-air temperature difference. By ventilating and drying the subcloud layer, shallow and deep convection increase the evaporation and cool the surface. Convective downdrafts and cold pools also cool the subcloud layer and this cooling is transmitted to the surface through an increase in sensible heat flux. Radiative cooling of the stratocumulus-capped boundary layer also increases turbulence and turbulent surface fluxes. Finally, changes in large-scale surface wind resulting from remote cloud diabatic effects can also modulate the SST via changes in these fluxes.

Clouds have multiple other indirect effects on the surface temperature through changes in oceanic mixing and circulation: precipitation on the other hand can locally restratify the surface and reduce mixing by injecting fresh water with low density at the surface. On the other hand, turbulent surface heat and momentum fluxes power the turbulence in the oceanic mixed layer, and therefore the entrainment of subsurface waters into this layer, so changes in surface wind and sea–air difference due to the clouds impact oceanic mixing. The momentum flux also forces the large-scale oceanic circulation, which results in oceanic energy transport that influences the SST.

8.5.3   Cloud–Ocean Interaction and Climate

Interactions between clouds and the oceanic surface are at play in many features of the tropical climate and its variability. The off-equatorial location of the ITCZ is partly explained by its proximity to the latitudinal maximum of SST north of the equator: warm SST permits the development of high clouds and precipitation, while cool SSTs south of the equator in the Eastern Pacific and Atlantic oceans inhibit deep convection and maintain stratocumulus decks off the coast of Peru and Namibia. The off-equatorial maximum of SST and precipitation results from the interaction between clouds, circulation and SST. The *SST-stratocumulus feedback* illustrated in Fig. 8.14 is the direct positive feedback between SST and low clouds mentioned earlier: the large cloud fraction associated with stratocumulus decks in subtropical eastern basins yields a strong cooling effect on the surface ocean. As shown in Section 8.5.1, cloud fraction increases with decreasing SST and the resulting positive feedback between stratocumuli and SST prevents the development of any deep convection south of the equator. Another indirect feedback involving circulation has been documented. Cloud diabatic heating in the ITCZ and the south–north SST gradient create a southward low-level geopotential gradient which forces a monsoon-like anomalous circulation: enhancement of the easterly trade winds south of the equator, northward cross-equatorial wind and weakening of the trade winds north of the equator. Because the surface turbulent heat fluxes depend on the surface wind, they are reduced north of the equator and increased south of the equator, which reinforces the north–south SST contrast: this is called the *wind-flux-SST feedback*. These feedbacks might be enough to break the inter-hemispheric symmetry and locate the ITCZ in the northern tropics by themselves; in the real world, the geography of the western coasts of the Americas and Africa play a role via coastal upwelling, and current orbital parameters (set by the phase of the precession of the equinoxes) might also have an influence.

The negative feedback between high clouds and SST is active in the ENSO, as well as an indirect feedback between SST, atmospheric circulation and oceanic circulation (see Fig. 8.15). After a westerly wind burst in the equatorial West Pacific triggers an El Niño, the reduced east–west SST gradient and anomalous cloud atmospheric heating gradient weaken the equatorial easterly winds, which reduces the surface easterly current and the upwelling in the eastern Pacific. This in turn reduces the east–west SST gradient and the gradient of the thermocline depth and constitutes a positive feedback on the El Niño conditions called the *Bjerknes feedback* (Bjerknes, 1969). Note that this feedback is also active in maintaining the zonally asymmetric normal conditions. The return to normal conditions is due to oceanic equatorial waves and to surface fluxes in the Central and East Pacific, with two contributions: an increase in surface turbulent heat fluxes resulting from the increased SST constitutes a negative *SST-flux feedback*, and a decrease of incoming solar radiation at the surface

due to the high-CRE at the surface constitutes the negative *SST-high cloud feedback*. The contribution of the latter is dominant in the return to normal conditions.

The negative *SST-high cloud feedback* is also active in the MJO: the convective disturbance cools the surface as well, but with a different outcome because of the propagative nature of this mode. As shown in Fig. 8.16, the SST cools below the convective perturbation, so that along the propagation the SST is warm east of the convective perturbation and cool west of it. This tends to increase surface sensible and latent heat fluxes east of the convective perturbation (and decrease fluxes to the west) and favour the eastward propagation of convection, by an *SST-flux mechanism*. An indirect effect of the clouds on the surface, through circulation, is thought to be important for the MJO as well: the Gill circulation associated with the MJO is characterised by a strong westerly jet just west of the maximum of deep convection. This jet increases surface turbulent heat fluxes that fuel the MJO convective disturbance by a *wind-flux mechanism*, which also cools the surface and enhances the SST contrast across the convective disturbance.

This wind-flux mechanism is also crucial to the intensification of tropical depressions into severe tropical cyclones: the cyclonic surface winds around the centre of the cyclone extract latent and sensible heat from the ocean that fuel deep convection in the spiralling rainbands (and in the eye wall during the mature stage); in turn, latent heat release in the clouds warms the core of the cyclone, increases the pressure gradients between its center and its periphery, accelerates the cyclonic circulation and increases the turbulent surface heat fluxes.

8.6 Concluding Remarks

Through their interaction with atmospheric circulations and the surface, clouds are essential contributors to the climatology and the climate variability in the tropics. While clouds have long been seen as passive expressions of tropical variability, we now increasingly recognise their active role in controlling this variability. Consistently, while clouds have long been characterised through their morphology or physical properties, we now realise that they should also be considered as 'energetic bodies' in the climate system.

Understanding the physical mechanisms through which tropical clouds interact with the large-scale climate remains a challenge. One of the key mechanisms is the interaction between large-scale circulation and convective clouds. Better understanding the dependence of cloudiness on large-scale conditions, in particular on free-tropospheric humidity, is necessary to parameterise convection and understand the effect of large-scale transport on the cloudiness. Reciprocally, better documenting the profile of cloud diabatic effects is crucial to understand the influence of clouds on the circulation. Another key interaction is that between clouds and the surface at the large scale. In particular, studying interaction between clouds, circulation and a heterogeneous surface is a challenging topic in need of creative approaches and frameworks.

This understanding is all the more necessary since clouds are still poorly represented in many state-of-the-art models. Climate models often exhibit some systematic flaws in the simulation of clouds, such as the double ITCZ syndrome, a weak MJO or an underestimation of stratocumulus cover in the eastern subtropical basins (see Chapter 7). In terms of climate projections, the models do not agree on the sign of the predicted precipitation change during the twenty-first century over most of the tropical continents, and tropical low clouds are a major contributor to the spread in predicted global-mean temperature change. A better understanding of cloud–climate interactions in the tropics will help improve the parameterisations of moist physics and to remediate these biases and uncertainties.

Fortunately, new opportunities arise to advance this understanding. New observations are becoming available (e.g., cloud latent heating and CRE estimates from satellite observations). Large-domain high-resolution modelling (with cloud-resolving models and large-eddy simulations) is now computationally affordable and opens perspectives in the study of the interaction of clouds and the large-scale environment (e.g., the large-scale organisation of convection). Finally, the scientific communities working on fine-scale cloud processes and large-scale climate studies are increasingly collaborating, opening new scientific perspectives.

Further Reading*

The review articles on 'Mesoscale convective systems' (Houze, 2004) and 'Tropical cyclones' (Emanuel, 2003) are references. On convectively coupled equatorial waves, another review article is very detailed (Kiladis et al., 2009). Most aspects of the tropical intraseasonal variability are explained in the book *Intraseasonal*

* The authors are grateful to Dominique Bouniol and Françoise Guichard for inspiring discussions, and to David Saint-Martin for his multiple comments. Thanks to Cathy Hohenegger, Christopher Holloway, and Brian Mapes for their reviews.

Variability in the Atmosphere–Ocean Climate System (Lau and Waliser, 2011); see also the review article on the MJO by Zhang (2005). General elements on tropical waves and dynamics can be found in chapter 11 of *Atmosphere–Ocean Dynamics* (Gill, 1982). An extensive overview of the Hadley circulation is given in the book *The Hadley Circulation: Present, Past, and Future* (Diaz and Bradley, 2004) and a clear explanation of theoretical aspects can be found in chapter 11 of *Atmospheric and Ocean Fluid Dynamics: Fundamentals and Large-Scale Circulation* (Vallis, 2006). The book *An Introduction to the Dynamics of El Niño and the Southern Oscillation* (Clarke, 2008) will give you the basics on tropical inter-annual variability. Chapters 6–11 of *The Global Circulation of the Atmosphere* (Schneider and Sobel, 2007) address theoretical aspects of the tropical dynamics and its interaction with precipitation. An article by Li et al. (2015) sums up the influence of CREs on atmospheric circulation.

Exercises

(1) **DGW approximation:**
Derive Eq. (8.8).

(2) **Walker circulation:**
We can build a 'simple', linear model of the Walker circulation responding to latent and surface heating. We use the DGW approximation to describe the Walker circulation as a wave extending across the Pacific ocean, with the zonal wavenumber $k = \pi/L_x$, with L_x the width of the equatorial Pacific. We also neglect the contribution of humidity to the density temperature anomaly, so that $\hat{T}_\rho = \hat{T}$. Since damping is larger at the surface than above, we will opportunistically set the damping rate proportional to the pressure: $\eta = \nu p$. We will also consider that the surface and the tropopause are isobaric surfaces at $p_s = 1,000$ hPa and $p_t = 150$ hPa.
Together with Eq. (8.8), we use the linearised thermodynamic Eq. (2.100) at equilibrium, in which we consider that horizontal advection and clear-sky radiation act as a damping on the wave:

$$\hat{\omega}\Gamma + \hat{Q}_c + \hat{Q}_r + \hat{Q}_t - \mu\hat{T} = 0, \qquad (8.9)$$

in which $\Gamma = -\partial_p(c_{p_e}T_{\text{ref}} + \phi_{\text{ref}}) > 0$ is the opposite of the basic-state vertical gradient of static energy and $\nabla \cdot \boldsymbol{q}$ has been divided into three terms: $\hat{Q}_r$ is the cloud radiative heating, $\hat{Q}_t$ is the turbulent heating, and μ is the damping rate due to clear-sky radiation and horizontal advection. Eq. (8.9) relates the perturbation

TABLE 8.1: Parameter values

k	ν	μ	Γ	α
$2.5 \times 10^{-7}\,\mathrm{m\,s^{-1}}$	$2 \times 10^{-10}\,\mathrm{s^{-1}\,Pa^{-1}}$	$5 \times 10^{-3}\,\mathrm{J\,s^{-1}\,K^{-1}}$	$0.4\,\mathrm{J\,Pa^{-1}}$	$18\,\mathrm{kg\,m^{-3}\,s^{-1}}$

to a sum of forcings and, since the model is linear, the total circulation due to the sum of forcings equals the sum of circulations due to each forcing.

(a) **Walker circulation forced by latent heating:** we set $\hat{Q}_r$ and $\hat{Q}_t$ to zero.

(i) Establish the relationship between the pressure velocity $\hat{\omega}$ and the latent heating $\hat{Q}_c$. We will use $\alpha = R_d k^2 / \nu / \mu$ to simplify the expressions.

(ii) What are the solutions to the corresponding homogeneous equation? (A change of variable makes it easy.)

(iii) Assuming that the heating profile is described by a polynomial $\hat{Q}_c = \sum_{n=0}^{N} Q_n p^n$, look for a polynomial particular solution $\hat{\omega} = \sum_{n=0}^{N} \Omega_n p^n$ to the non-homogeneous equation established in question 1(i).

(iv) Using the boundary conditions of no vertical velocity at the tropopause ($p = p_t$) and surface ($p = p_s$), write the solution to the equation established in question 1(i) for a polynomial heating $\hat{Q}_c$.

(v) Let's assume that convective and stratiform clouds produce heating profiles described by the following normalised profiles $\zeta_c(p)$ and $\zeta_s(p)$:

$$\zeta_c(p) = 6 \frac{(p - p_t)(p_s - p)}{(p_s - p_t)^3}$$

and

$$\zeta_s(p) = 60 \frac{(p - p_t)(p_m - p)(p_s - p)}{(p_s - p_t)^4},$$

with $p_m = (3 p_s + 2 p_t)/5 = 660$ hPa the pressure where the stratiform heating changes sign. The amplitude of the vertically integrated latent heating in the WPCZ is about 100 W/m^2 (corresponding to an east–west contrast of 200 W/m^2 or 7 mm/day). What would be the profile of the pressure velocity if all the precipitation was convective? What if all the precipitation was stratiform? What if half the precipitation is stratiform, as observed in the WPCZ? Plot the profiles of latent heating $\hat{Q}_c$ and the

corresponding profiles of $\hat{\omega}$, $\hat{\phi}$ and $\hat{T}$. What error on the Walker circulation is made if no stratiform precipitation is simulated? Use the constants $R_d = 288\ \mathrm{J\,kg^{-1}\,K^{-1}}$, $g = 10\ \mathrm{m\,s^{-2}}$ and the parameter values given in Table 8.1.

(vi) In the solution found in question 1(vi), we have both the response of the circulation to the latent heating (term in $\hat{Q}_c$ in the original differential equation from question 1(i)), and the feedback of the advection (term in $\hat{\omega}\Gamma$). We can investigate the role of each contribution. For the realistic case with a stratiform to convective ratio of one, plot the temperature perturbation $\hat{T}_d$ due to the diabatic heating and the perturbation $\hat{T}_\omega$ due to the circulation feedback. Derive and plot the corresponding geopotential and pressure velocity perturbations $\hat{\phi}_d$, $\hat{\phi}_\omega$, $\hat{\omega}_d$ and $\hat{\omega}_\omega$ (a change of variable might help again). What is the effect of the circulation feedback on the dynamics? What would be its effect on precipitation if precipitation was interactive?

(b) **Walker circulation forced by the SST gradient:** we set $\hat{Q}_r$ and $\hat{Q}_c$ to zero and we assume that $\hat{Q}_t$ communicates the SST gradient very efficiently up to the top of the boundary layer at $p_i = 800$ hPa, so that the air temperature anomaly is equal to the SST anomaly $\hat{T}_s$ up to p_i: $\hat{T} = \hat{T}_s$ for $p > p_i$.

(i) Establish the differential equations for $\hat{\omega}$ below and above p_i.

(ii) Solve for $\hat{\omega}$ using the boundary conditions of zero vertical velocity at the surface and tropopause, as well as the continuity of $\hat{\omega}$ and $\partial_p \hat{\omega}$ at p_i required by mass conservation and the continuity of the geopotential.

(iii) Plot the profiles of $\hat{\omega}$, $\hat{\phi}$ and $\hat{T}$ for a perturbation $\hat{T}_s = 5$ K (corresponding to an east–west SST contrast of 10 K).

(iv) Plot the profiles of the perturbations of vertical circulation, temperature and geopotential due to the 'realistic' latent heating and the

SST gradient. What is the contribution of each term?

(c) **Walker circulation forced by CRE:**
On the basis of the CREs depicted in Fig. 8.12, what contribution to the Walker circulation do you expect from the clouds in the WPCZ?

(3) **Horizontal circulation: effect of rotation and β effect:**
This simple exercise aims at illustrating how the dynamical response of the tropical atmospheric circulation to a geopotential perturbation does not conserve the symmetries of the geopotential field, which can be instrumental in the propagation of the disturbance. We use the linearised primitive equation for momentum (Eq. (8.2)) with a motionless basic state and a damping term that account for vertical diffusion and surface friction, that we rewrite for the zonal and meridional winds u and v:

$$\partial_t u - fv = -\partial_x \phi - \eta u,$$

$$\partial_t v + fu = -\partial_y \phi - \eta v.$$

We will use $\eta = 2\ 10^{-5}\ \text{s}^{-1}$ typical of the atmospheric boundary layer.

(a) Establish the equation for the steady horizontal wind (u, v) and divergence $\delta = \partial_x u + \partial_y v$ for a steady geopotential perturbation.

(b) Let's consider a Gaussian geopotential perturbation, that could result from cloud diabatic heating above the level under consideration:

$$\phi = -\phi_0 \exp\left(-\frac{x^2 + y^2}{2L^2}\right),$$

with $\phi_0 = 500\ \text{m}^2\ \text{s}^{-2}$ the perturbation amplitude and $L = 100$ km the horizontal scale of the perturbation. Plot the wind and divergence response in the case without rotation, in the case of the equatorial beta plane ($f = \beta y$ with $\beta = 2.27\ 10^{-11}\ \text{m}^{-1}\text{s}^{-1}$), for a subtropical f-plane ($f = f_0 = 1.27\ 10^{-5}\ \text{s}^{-1}$) around 15°N, and for the corresponding beta plane ($f = f_0 + \beta y$ with $f_0 = 1.27\ 10^{-5}\ \text{s}^{-1}$ and $\beta = 2.20\ 10^{-11}\ \text{m}^{-1}\text{s}^{-1}$). Does the divergence field conserve the symmetry of the geopotential field in the case without rotation? On an f-plane? On a beta plane?

(c) How does the circulation feed back on such a geopotential perturbation in the boundary layer? What can we expect from the energy advection and the moisture advection?

9 Midlatitude Cloud Systems

George Tselioudis and Kevin Grise

In contrast to the tropics and subtropics, the middle latitudes are characterised by large meridional temperature gradients, created as a consequence of differential radiative heating between high and low latitudes. These meridional temperature gradients often concentrate in relatively narrow baroclinic zones that become unstable to wave-like perturbations called baroclinic eddies, or more commonly baroclinic storms or mid-latitude cyclones. Baroclinic storms constitute the primary source of poleward energy transport at mid latitudes, which is accomplished through contrasting transports of warm air masses polewards (warm fronts) and cold air masses equatorwards (cold fronts). Baroclinic storms also flux momentum into mid-latitude regions, driving a region of enhanced westerly winds from the surface to the upper troposphere called the eddy-driven jet stream.

Baroclinic storms are the primary source of mid-latitude cloud formation, and thus play an important role in determining the radiative and hydrologic budgets in mid-latitude regions. High clouds of various vertical extents tend to form in the uplift regimes of the warm and cold fronts associated with the baroclinic storms, while low cloud decks often form in the subsidence regimes of the cold air masses that follow a cold frontal passage. As a result, the mid-latitude cloud field encompasses almost the full range of cloud types, and the appearance of particular cloud types can be viewed as a tracer of mid-latitude dynamic regimes. Because the clouds in the vicinity of the warm and cold fronts tend to be optically thick and often have tops in upper tropospheric layers, they produce substantial short- and long-wave radiative signatures. Section 1.2.1 and Chapter 4 provide a detailed review of the short- and long-wave radiative signatures of clouds. Briefly, the large optical depth of the clouds in the vicinity of the fronts promotes the reflection of incident solar radiation (a short-wave cooling effect), whereas the cold cloud-top temperatures promote the reduction of outgoing long-wave radiation to space (a long-wave warming effect). In contrast, the low cloud decks in mid-latitude cold air regimes have primarily a short-wave cooling effect, as their cloud-top temperature differs little from the underlying surface temperature.

These short-wave cooling and long-wave warming effects make mid-latitude clouds a key contributor to the global radiative budget and thus a potential source of significant radiative feedbacks in climate change situations (Chapter 13). Changes in the climate system, such as a climate warming, could affect meridional temperature gradients as well as the moisture availability of the atmosphere, which, through latent heat release, constitutes an additional energy source for baroclinic storms. Consequently, significant changes in the track and strength of baroclinic storms could occur with climate warming, and these changes would alter the mid-latitude cloud field and produce radiative and hydrologic climate feedbacks. At the same time, altered cloud fields through their radiative effects and latent heat release have the ability to change temperature gradient patterns which in return can alter the characteristics of the mid-latitude atmospheric circulation. The examination of mid-latitude cloud processes and feedbacks, therefore, requires a detailed understanding of the relationships between the dynamical features of baroclinic storms and the properties of the clouds that they produce.

This chapter will examine in detail the relationships between cloud properties and atmospheric dynamics in mid-latitude regions. Cloud structures and formation mechanisms in baroclinic storms will be examined first, with an emphasis on the latest satellite retrievals of cloud properties. Next, the climatologies of mid-latitude clouds and their radiative properties will be discussed. Then, the interactions of the mid-latitude atmospheric circulation with clouds and their radiative properties will be examined, including an examination of the effects of clouds on the mid-latitude circulation. Finally, the chapter will address how mid-latitude clouds may be affected in as well as affect a changing climate.

9.1 Mid-Latitude Cloud Structures

9.1.1 Cloud Structures in Baroclinic Storms

Baroclinic storm clouds have been extensively observed from both ground sites and space-based platforms, as they are integrally linked to weather forecasting at mid-latitudes. Since the early days of weather observations, attempts have been made to explain the circulation features in a baroclinic storm and to map the cloud and precipitation structures that are formed in the different components of those circulation features. In this section, a brief description will be provided of early theories that explain the dynamical circulations and the resulting cloud formations in mid-latitude storms, and then more recent results will be presented that use satellite retrievals, meteorological observations and field-campaign results to quantitatively characterise the properties of the mid-latitude cloud field and their relationship with mid-latitude atmospheric dynamics.

9.1.1.1 Norwegian Cyclone Model

As early as the late nineteenth century, it was recognised that mid-latitude baroclinic storms had a typical pattern in their cloudiness, such that systematic cloud observations might provide clues as to the development of weather systems. The first *International Cloud Atlas* published in 1896 allowed for the standardisation of remote cloud observations (Chapter 1). As a result of these early observations, the first conceptual model to depict the circulation and cloud features in a mid-latitude cyclone, the Norwegian cyclone model, was formulated in the late 1910s and early 1920s by Jacob Bjerknes.[1] The model was constructed through the accumulation of surface observations of a multitude of storm passages, which typically include the appearance first of the warm frontal sector with southerly winds and warm temperatures followed by the passage of the cold front that turns the winds to northerly directions and drops the temperatures

---

[1] Jacob Aall Bonnevie Bjerknes (1897–1975) was a Norwegian meteorologist, who is widely considered to be one of the pioneers of modern meteorology. His contributions range from his early work on the Norwegian cyclone model at the Geophysical Institute in Bergen, Norway, to his later work on the general circulation and the El Niño-Southern Oscillation as a professor at the University of California at Los Angeles.

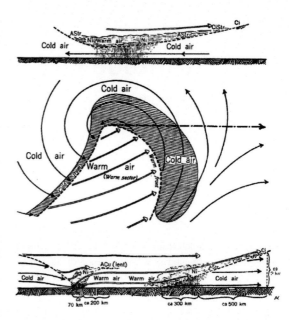

FIGURE 9.1: A schematic illustration of frontal clouds and precipitation constructed based on the Norwegian cyclone model, as conceptualised in the work of Bjerknes and Solberg (1922). The middle panel shows the warm and cold fronts and the movement of air around the storm centre, and the top and bottom plots show schematics of two west–east cross-sections through the storm with indications of the locations of different cloud types, precipitation and air movements. Adapted from Posselt et al. (2008).
Copyright © 2008 American Meteorological Society. Used with permission

significantly. A schematic of the model is shown in Fig. 9.1, where the warm and cold air circulation regimes that form the warm and cold frontal regions around a northern hemisphere low-pressure centre are shown with arrows, and the typical cloud and precipitation types that occur in each regime are drawn schematically and noted using cloud classification definitions. The classification of cloud types is based on the definitions used by surface weather observers that have changed very little over time (see Chapter 1). Those types identified by the weather observers as the primary clouds in mid-latitude storms, included thin cirrus (Ci) and cirrostratus (Cs) clouds occurring upstream of the storm centre, nimbus clouds (Ni) together with rain or snow along the warm and cold frontal zones, and altostratus clouds (As) in the cold air mass behind the cold front.

9.1.1.2 Quasi-geostrophic Theory

In the 1940s and 1950s, quasi-geostrophic theory was introduced, which helped to resolve with a high degree of accuracy the synoptic atmospheric motions associated with midlatitude baroclinic storms. A detailed explanation of quasi-geostrophic theory is provided in many atmospheric dynamics textbooks (see Further Reading). A brief summary is provided here.

The quasi-static (primitive) equations describe the dynamics of large-scale flows in the atmosphere. The origin of these equations is discussed in detail in Section 2.4.2. Using pressure as a vertical coordinate, the horizontal momentum equations can be written concisely as:

$$D_t v_p = -f \times v - \nabla_p \phi, \qquad (9.1)$$

where ϕ is the geopotential and $D_t = \partial_t + v_p \cdot \nabla_p$. In mid-latitude synoptic-scale weather systems, a scale analysis of the terms in Eq. (9.1) reveals an approximate balance between the Coriolis force and pressure gradient force terms, which is referred to as geostrophic balance:

$$0 \approx -f \times v - \nabla_p \phi. \qquad (9.2)$$

Likewise, the horizontal wind that reflects an exact balance between the Coriolis and pressure gradient forces is referred to as the geostrophic wind $v_{\mathbf{g}}$:

$$v_{\mathbf{g}} = \{u_g, v_g, 0\} = (1/f) \times \nabla_p \phi. \qquad (9.3)$$

To first order, the winds in mid-latitude weather systems are dominated by their geostrophic component. The ageostrophic component of the wind, $v_{\mathbf{a}} = v - v_{\mathbf{g}}$, is comparatively much smaller, usually by an order of magnitude. Using this knowledge, the primitive equations can be greatly simplified by assuming that mid-latitude synoptic-scale flows are approximately (but not exactly) in geostrophic balance ($v \approx v_{\mathbf{g}} >> v_{\mathbf{a}}$). This *quasi-geostrophic approximation* simplifies the horizontal momentum equations (9.1) as follows:

$$\frac{d_g v_{\mathbf{g}}}{dt} = -f_0 \mathbf{k} \times v_{\mathbf{a}} - (\beta y) \mathbf{k} \times v_{\mathbf{g}}. \qquad (9.4)$$

Here, note that, in the quasi-geostrophic approximation, the total derivative D_t in Eq. (9.1) is replaced by $\frac{d_g}{dt} = \frac{\partial}{\partial t} + u_g \frac{\partial}{\partial x} + v_g \frac{\partial}{\partial y}$. In other words, advection is approximated by the horizontal advection by the geostrophic wind; any vertical advection is neglected. Note also that the Coriolis parameter $f = \{0, 0, f\}$ has been replaced by $\{0, 0, f_0 + \beta y\}$ using the so-called beta-plane approximation where $f_0 >> \beta y$.

Similarly, using pressure as a vertical coordinate, the thermodynamic equation (2.143) can be written concisely as:

$$D_t T + \frac{\omega}{c_p} \partial_p \phi = \frac{Q}{c_p}. \qquad (9.5)$$

Here, we assume a dry atmosphere, such that the right-hand side of Eq. (9.5) equals the diabatic heating rate Q

divided by the specific heat of dry air at constant pressure ($c_p = 1{,}004 \text{ J kg}^{-1} \text{ K}^{-1}$).

We can rewrite the thermodynamic equation (9.5) in terms of the dry potential temperature θ (2.46):

$$\frac{\partial T}{\partial t} + u\frac{\partial T}{\partial x} + v\frac{\partial T}{\partial y} - S_p \omega = \frac{Q}{c_p}, \qquad (9.6)$$

where $S_p = -\frac{T}{\theta}\frac{\partial \theta}{\partial p}$ is a measure of the static stability of the atmosphere in pressure coordinates. Under the quasi-geostrophic approximation, Eq. (9.6) simplifies to:

$$\frac{d_g T}{dt} - S_{p_0}\omega = \frac{Q}{c_p}, \qquad (9.7)$$

where the subscript zero indicates a reference state temperature profile that varies only in the vertical direction ($S_{p_0} = -\frac{T_0}{\theta_0}\frac{\partial \theta_0}{\partial p}$).

It is often convenient to express the quasi-geostrophic horizontal momentum equations (Eq. (9.4)) in terms of the quasi-geostrophic relative vorticity $\zeta_g = \frac{\partial v_g}{\partial x} - \frac{\partial u_g}{\partial y}$:

$$\frac{d_g \zeta_g}{dt} = -f_0\left(\frac{\partial u_a}{\partial x} + \frac{\partial v_a}{\partial y}\right) - \beta v_g. \qquad (9.8)$$

Now, the quasi-geostrophic thermodynamic equation (Eq. (9.7)) and vorticity equation (Eq. (9.8)) can be combined to yield one diagnostic equation for vertical motions in mid-latitude weather systems, the quasi-geostrophic omega equation (9.9):

$$\left(\nabla_p^2 + \frac{pf_0^2}{R_d S_{p_0}}\frac{\partial^2}{\partial p^2}\right)\omega = -\frac{1}{S_{p_0}}\nabla_p^2(-v_{\mathbf{g}} \cdot \nabla_p T)$$
$$- \frac{pf_0}{R_d S_{p_0}}\frac{\partial}{\partial p}[-v_{\mathbf{g}} \cdot \nabla_p(\zeta_g + f)] - \frac{1}{S_{p_0}}\nabla_p^2\frac{Q}{c_p}. \qquad (9.9)$$

Assuming that the atmosphere is statically stable ($S_{p_0} > 0$), the term on the left-hand side of Eq. (9.9) behaves like the Laplacian of the vertical velocity (omega, ω). Recalling that the Laplacian operator can be approximated as a negative sign (particularly near minima or maxima in a scalar field), the left-hand side of Eq. (9.9) can be qualitatively interpreted as $-\omega$, or rising motion. The first term on the right-hand side of Eq. (9.9) is related to the Laplacian of temperature advection by the geostrophic wind on constant pressure surfaces. This term implies that warm air advection is associated with $-\omega$ (rising motion) and cold air advection is associated with $+\omega$ (sinking motion). The second term on the right-hand side of Eq. (9.9) is related to the differential absolute vorticity ($\zeta_g + f$) advection by the geostrophic wind on constant pressure surfaces. This term implies that cyclonic vorticity advection increasing with height (decreasing with pressure, $\frac{\partial}{\partial p}$) is associated with $-\omega$ (rising motion) and anticyclonic vorticity advection

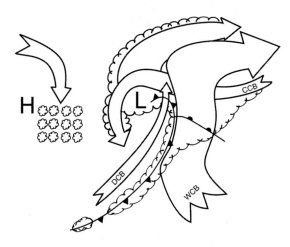

FIGURE 9.2: A schematic view of the conveyor belt model for the air circulation in mid-latitude storm systems. The figure shows the positions of the fronts, the low-pressure centre (labelled as L), a high-pressure centre (labelled as H) and the three major conveyor belt circulations (warm conveyor belt, WCB; cold conveyor belt, CCB; and dry conveyor belt, DCB). The sense of circulation assumes $f > 0$ and would be reversed for $f < 0$, as in the southern hemisphere.

increasing with height is associated with $+\omega$ (sinking motion). Finally, the third term on the right-hand side of Eq. (9.9) is related to the diabatic heating rate, including effects from radiation and the latent heat of condensation. This term implies that diabatic heating is associated with $-\omega$ (rising motion) and diabatic cooling is associated with $+\omega$ (sinking motion). The diabatic heating term is often neglected, if air motions are assumed to be adiabatic.

9.1.1.3   The 'Conveyor Belt' Model

Using knowledge from quasi-geostrophic theory, the Norwegian cyclone model can be extended into the so-called conveyor belt model, which is a full-scale model of the circulation, cloud and precipitation components of baroclinic storms. The conveyor belt model is summarised in Fig. 9.2 and includes three major circulation features ('conveyor belts') surrounding the low-pressure centre of a baroclinic storm. Note that the latitudinal transports shown in Fig. 9.2 and discussed later apply to the northern hemisphere ($f > 0$) and must be reversed for southern hemisphere ($f < 0$) baroclinic storms.

The first feature of the conveyor belt model is a warm conveyor belt that originates at low levels in the southeastern quadrant of the storm and carries warm, moist air towards the north. This warm, moist air mass is often convectively unstable and therefore produces convective clouds and precipitating systems, including squall lines and strong thunderstorms. Even in the absence of convection, however, the large scale lifting of the warm, moist air

by the approaching cold front produces thick clouds with high tops that often precipitate. As the warm conveyor belt moves northwards, this air motion is associated with warm air advection, and thus consistent with Eq. (9.9), the air rises as it flows northwards and lifts over the warm front. This large-scale rising motion often produces large regions of stratiform cloud cover and precipitation ahead of the warm front. As the warm conveyor belt ascends, it is deflected to the east by the strong westerly winds in the mid-latitude upper troposphere. As the warm conveyor belt moves to the east and north-east of the storm centre, the lifted moist air forms cirrostratus clouds that thin and change to cirrus as the air stream moves further eastwards. Therefore, the appearance of cirrus clouds moving into a mid-latitude region from the west is considered a harbinger of a storm passage.

The second circulation feature of the conveyor belt model is a cold conveyor belt that originates to the north-east of the low-pressure centre and brings cold air westwards, as it passes underneath the warm conveyor belt and parallels the warm front along its northern edge. As the cold conveyor belt passes underneath the warm conveyor belt and warm frontal precipitation, it becomes significantly moistened and begins to rise slowly as it nears the low-pressure centre. The rising motion near the surface low is consistent with cyclonic vorticity advection increasing with height (from Eq. (9.9)), as an intensifying baroclinic weather system tilts westwards with height such that the cyclonic vorticity advection associated with a mid-tropospheric trough (cyclonic vorticity anomaly) is typically positioned above the surface low. Once in the vicinity of the surface low, the cold conveyor belt splits into two branches: (1) an anticyclonic branch that is deflected eastwards by the mid-tropospheric westerly winds and follows at lower levels the warm conveyor belt and (2) a cyclonic branch that wraps around the surface low, producing a distinctive comma-shaped cloud feature to the west of the surface low. Throughout its track, the low-level cold conveyor belt favours the formation of stratus cloud decks and fogs, which are often enhanced by evaporating precipitation falling from the warm conveyor belt above. Further to the west of the low-pressure centre, shallow cumulus cloud decks often form over ocean, as the cold air advection from the cyclonic circulation of air passes over warmer ocean waters.

The final circulation feature of the conveyor belt model is a dry conveyor belt. The dry conveyor belt originates in the upper troposphere and descends cyclonically to the south of the surface low, creating a dry tongue, or dry slot, at the western edge of the cold frontal circulation. The descending motion of the dry conveyor belt behind the cold front is consistent with the cold air advection occurring in this region (from Eq. (9.9)).

9.1.1.4 Satellite Retrievals of Storm Cloud Properties

Surface weather observers provided these first, qualitative descriptions of the cloud structures in mid-latitude baroclinic storms. Since the early 1980s, retrievals of cloud, radiation and precipitation properties from satellite instruments have started to provide more quantitative information about the radiative and hydrologic structures of the global cloud field. Satellite retrievals provide detailed properties of mid-latitude clouds and therefore make it possible to quantitatively relate those properties to the properties of the baroclinic storms that constitute the primary mechanism for the formation of these clouds.

A composite of satellite and surface retrievals of cloud and precipitation properties in a baroclinic storm is shown in Fig. 9.3. The satellite cloud property retrievals come from the International Satellite Cloud Climatology Project (ISCCP), while the cloud type and weather type observations come from surface weather observers (see Chapter 1 for a detailed discussion of cloud observing systems). The figure shows a composite for cyclones over the North Atlantic, for the ISCCP cloud type retrievals (Fig. 9.3a), the surface observers' cloud types (Fig. 9.3b) and the surface observers' weather types (Fig. 9.3c). The satellite definition of cloud height is based on a satellite measurement of the thermal emission from the top of the cloud, which indicates the temperature and therefore the vertical location of the cloud radiative top. The definition of cloud optical thickness is based on the cloud reflectance of solar radiation, which is indicative of the vertical thickness of the cloud and the density of the water and/or ice in it (Chapter 3). It is remarkable that the classical view of the cloud distribution around baroclinic storms derived first by surface observers as early as the late nineteenth century (Fig. 9.1) is verified with high accuracy by the global satellite retrievals and the more recent surface observer data; high-top thick and high-top medium or nimbostratus and altostratus clouds along with heavy precipitation dominate the warm and cold frontal regions of the storm, while low-top thick and low-top thin or stratus and cumulus/stratocumulus clouds along with snow are found in the cold air outbreak region behind the storm. One major difference between the surface observers and the satellite cloud types relates to the fact that the surface observers find large amounts of stratus cloud in the frontal sectors where the satellite retrieves higher top thick clouds. This is due to the bottom-up view of the surface observers versus the top-down view of the satellite and the fact that cloud

top is not readily retrievable from the surface when the sky is fully covered by cloud.

The cloud structures derived by surface observers and by the first satellite cloud retrievals provide an illustration of the average structure of cloud type distribution in a typical mid-latitude storm. However, cloud cover and cloud type distributions tend to vary widely both with changes in the strength of the baroclinic storms as well as with changes in atmospheric conditions that the storm is embedded in, such as the moisture availability of the atmosphere. In recent years, the global, multi-year nature of satellite observations along with reconstructions of atmospheric conditions by reanalysis data sets, allow us to analyse large ensembles of baroclinic storms and to derive meaningful classifications that make it possible to examine how the cloud structures and properties change with changing dynamic and thermodynamic conditions. Fig. 9.4 illustrates the change in high clouds in mid-latitude storms with changing storm strength (horizontal axis) and moisture availability of the atmospheric column (vertical axis). Here the strength of the storm is measured by the pressure value of the storm's low pressure centre. It can be seen that the amount of high cloud in a storm is strongly dependent on the storm strength since, for the same atmospheric moisture conditions, high clouds become more abundant and cover a larger region along the cold and warm frontal zones as the storm strength increases. The dependence of high cloud amount on atmospheric moisture availability is much weaker than the dependence on storm strength.

The first generation of satellite observations included cloud imagers that were able to retrieve cloud properties such as cloud-top temperature and optical depth from radiances and irradiances originating from the cloud top. These measurements provided information on the column-mean cloud properties but did not provide information on the vertical variability of the cloud layers. The advent of active remote sensing measurements from space with the launch of the CloudSat radar and the lidar, Cloud Aerosol Lidar with Orthogonal Polarisation, in 2006 made it possible to examine the vertical distribution of cloud layers in the vicinity of baroclinic storms. Fig. 9.5 shows a cross-section of the vertical profile of cloud layers in a baroclinic storm over the North Atlantic, derived from CloudSat radar retrievals. It can be seen that there is again a remarkable similarity of vertical structure with the cross-section of the cloud vertical profiles derived by the Norwegian cyclone model some eighty years earlier and shown in Fig. 9.1. The cloud type progression during the storm passage, with the cirrus and cirrostratus clouds ahead of the storm, the nimbostratus clouds along the

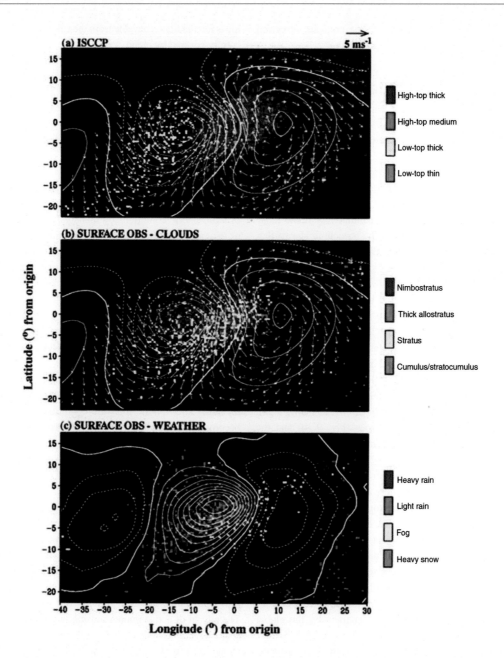

FIGURE 9.3: Patterns of the number of selected cloud types based on (a) ISCCP data and (b) surface observations, and (c) frequency of occurrence of selected surface weather types, for a composite of mid-latitude storms over the North Atlantic during the cool season (October–March). The colours on the plots indicate the relative abundance of a cloud type or weather regime as they are listed on the right of each plot, while the arrows (see scaling at top right) and contours (interval: 10 m) in (a) and (b) indicate the composite anomalous 1,000 hPa wind and geopotential height fields from European Centre for Medium-Range Weather Forecasts (ECMWF) analyses, respectively. The composite anomaly pattern for the vertical velocity ω (with positive values indicating upward motion) from the ECMWF data is shown as contours (interval: 2×10^{-2} Pa s^{-1}) in (c). Solid (dashed) contours indicate positive (negative) values. Adapted from Lau and Crane (1997). Copyright © 1997 American Meteorological Society. Used with permission

High cloud fraction

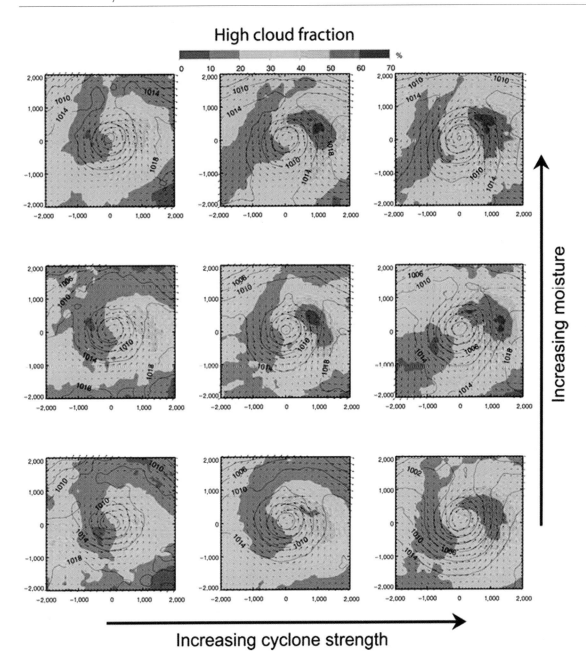

FIGURE 9.4: Composites of the amount of high cloud from mid-latitude storms (cyclones) in all the major storm tracks, classified by the cyclone strength (increasing left to right) and the atmospheric moisture (increasing bottom to top). Cyclone strength is defined as the mean wind speed within 2,000 km of the low-pressure centre and the atmospheric moisture as the mean precipitable water within 2,000 km of the low-pressure centre. The composite mean surface pressure contours (in hPa) and surface wind vectors are also shown. Adapted from Field and Wood (2007). Copyright © 2007 American Meteorological Society. Used with permission

frontal zones, and the low-top post-frontal clouds, is now verified through the use of radar retrievals. However, the use of active remote sensing also makes clear the existence of multi-layered clouds in the storm, a detail that cannot be readily resolved either from passive remote sensing or from surface observations. The pre-frontal cirrus are accompanied by low thin cloud decks, while part of the cold frontal clouds are formed by two distinct thick layers

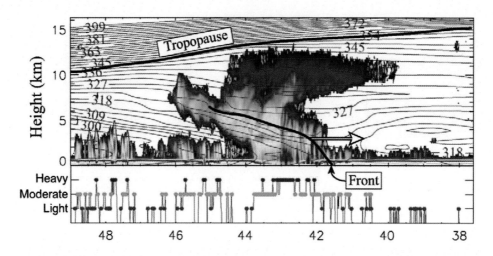

FIGURE 9.5: Cross-section along a North Atlantic baroclinic storm of CloudSat observed radar reflectivity (dBZ, colour shaded), overlaid with ECMWF-analysed equivalent potential temperature (K, solid red lines). The positions of the cold front and tropopause are marked with heavy black lines, and the direction of movement of the front is indicated with a white arrow. CloudSat-estimated precipitation rate is depicted directly below the plot of observed reflectivity. Adapted from Posselt et al. (2008). Copyright © 2008 American Meteorological Society. Used with permission

of low and high cloud tops. This explains the difference in the view of the frontal clouds between satellite and surface observations shown in Fig. 9.3.

9.1.2 Regional Patterns of Mid-Latitude Cloud Organisation

The observational analyses presented so far show that, at synoptic scales, baroclinic storm systems constitute a major source of the variety of cloud types, and that cloud cover and type varies significantly depending on location within the baroclinic storm domain. High clouds dominate the warm conveyor belt circulation while low clouds occur in the cold conveyor belt. When averaged to form a climatology, however, the mid-latitude cloud field does not show the pronounced regional patterns found in other climate regimes, other than some distinct cloud cover differences between land and ocean in the northern hemisphere. This relative uniformity of the mid-latitude cloud field is due to the transient nature of baroclinic storms, which tend to distribute cloud amounts in a some-what uniform manner throughout the three major oceanic storm tracks, located across the North Atlantic, North Pacific and Southern Oceans.

The mid-latitude storm tracks have been thoroughly documented and studied for well over a century, as they constitute the major weather makers for many of the world's populated regions. Mid-latitude storms are commonly tracked by following the movement of the

low-pressure centre that constitutes the centre of the storm over the duration of a storm's lifecycle (from cyclogenesis to cyclolysis). Storm tracking can either be performed manually (by hand) or through the use of an automated computer algorithm, which applies propagation and direction criteria to follow a low-pressure centre. Fig. 9.6 shows two maps of the North Atlantic storm track, Fig. 9.6a from the climatology atlas published by Wladimir Köppen[2] in 1882 and Fig. 9.6b from the application of a storm-tracking algorithm on sea-level pressure (SPL) data over the years 2000–2015. It is remarkable to note the similarities between the two maps, given the sparse observational network available near the end of the nineteenth century. Several centres of large storm activity are present in both maps, including the maximum between Greenland and Iceland (the Icelandic Low) and the secondary storm track in the Mediterranean region. The location of the storm track is not necessarily stationary in time. There is some evidence that suggests that the North Atlantic storm track shifted polewards over the latter half of the twentieth century. The potential effects of this storm track shift on the cloud field and its radiative effects are discussed in Section 9.3.1.

---

[2] Wladimir Köppen (1846–1940) was a Russian-born climatologist, who spent much of his professional career at the German Naval Observatory in Hamburg. He is best known for establishing the Köppen classification of the world's climate zones, but was also an author on the first International Cloud Atlas (see Section 1.1.1).

Some of the highest cloud amounts on the planet are observed in the mid-latitude storm tracks (see Fig. 4.6). In the three major oceanic storm tracks, annual mean total cloud cover is almost everywhere higher than 70%, and shows only weak regional patterns over the northern hemisphere oceans. In those basins, cloud cover tends to be higher in the western parts, maintaining values above 80%, especially in the winter season when the storm tracks are generally most active. These larger cloud covers are caused by larger amounts of high-top clouds in the western basins, while low cloud amounts are somewhat higher in the eastern parts of the northern hemisphere ocean basins. The western parts of the basins are the cyclogenesis areas of the oceanic storm tracks (Hoskins and Hodges, 2002), and storms tend to produce larger amounts of high clouds in their formation stages than in their decay stages. Low cloud decks, on the other hand, occur more frequently in the eastern parts of the basins, because (a) the sub-tropical and lower mid-latitude parts of the Eastern North Atlantic are dominated by the semi-permanent Azores high-pressure system, which tends to deflect storm tracks to the north (Fig. 9.6) and favours the formation of stable boundary layers and stratocumulus cloud decks, and (b) in the eastern North Pacific, cold surface waters favour the formation of fogs and stratocumulus clouds through mechanisms explained in more detail in Chapter 8.

Mid-latitude continental total cloud cover ranges in the annual mean between 40% and 70% and is, therefore, significantly lower than the mid-latitude oceanic cloud cover. This is true for all cloud types with the exception of high thin cirrus clouds. The overall lower continental cloud covers are due both to the smaller moisture supply of the land compared to the ocean surface and to the overall weaker storm systems occurring over land. The continental storm tracks also tend to have smaller frequencies of storm systems than the oceanic storm tracks, partly because of shorter-lived storms over land. These factors contribute to the lower cloud cover over mid-latitude continents.

The large amounts of mid-latitude continental cirrus clouds are primarily orographic in nature. Orographic cloud formation occurs preferably on the leeward side of mountain ranges, as eastward flowing air is lifted by the mountains and cools forming cloud layers throughout the extent of the resulting gravity wave. Orographic cirrus coverage can reach values of about 30%, thus constituting a major contributor to the total cloud cover in regions downwind from major mountain ranges, such as the Rocky Mountains in North America, the Andes Mountains in South America and the Himalayas in Asia.

9.2 Effects of Mid-Latitude Clouds on the Atmospheric Radiation Budget

In the climatology, mid-latitude clouds exert stronger effects on the Earth's top-of-the-atmosphere (TOA) radiative budget than clouds in any other latitude band. Annual mean shortwave cloud radiative cooling at TOA in the mid-latitude regions ranges between 30 and 90 W m^{-2}, and these values are some of the largest observed on the planet, comparable only to the values found in the tropical Western Pacific warm pool region, Inter Tropical Convergence Zone (ITCZ), and the stratocumulus regions of the eastern subtropical ocean basins (Fig. 4.8). Mid-latitude long-wave cloud radiative warming ranges between 20 and 50 W m^{-2}, lower only than the values observed in the ITCZ, the tropical Western Pacific and the deep convective regions in the Amazon and equatorial Africa (Fig. 4.8). As a result, net cloud radiative cooling in the midlatitudes ranges between 10 and 40 W m^{-2}, only rivalled by the net cloud radiative cooling in the stratocumulus regimes in the eastern subtropical ocean basins (Figs. 4.8 and 4.9). The large radiative cooling effect of mid-latitude clouds makes them a significant contributor to the equator-to-pole temperature difference and hence acts to affect the large-scale atmospheric circulation (see Section 9.3.2).

As mentioned in the previous section, the transient nature of baroclinic storms distributes clouds, and therefore their radiative effects, rather uniformly along the major storm tracks. Large regional differences occur mainly between northern hemisphere land and ocean regions. Shortwave cloud radiative cooling over land ranges between 30 and 60 W m^{-2} while over ocean it ranges between 40 and 90 W m^{-2}. Short-wave cooling maxima occur in the cyclogenesis regions in the western parts of northern hemisphere ocean basins, and relative minima in short-wave cooling occur in the cyclolysis regions in the eastern parts of the ocean basins (Fig. 4.8). Cloud long-wave warming over land ranges between 20 and 40 W m^{-2} with the maxima occurring in the cirrus prone regions downwind from the Andes, the Rockies and the Himalayas. Cloud long-wave warming over ocean ranges between 30 and 50 W m^{-2} with the maxima occurring in the high-cloud prone cyclogenesis regions (Fig. 4.8). The values listed here are annual mean values. However, in the mid-latitudes, cloud radiative effects vary significantly with season, both because of seasonal storm-track variability and the significantly lower solar insolation during the winter season. The lower insolation during winter reduces the wintertime short-wave cloud radiative effect, as by definition, a cloud of a certain

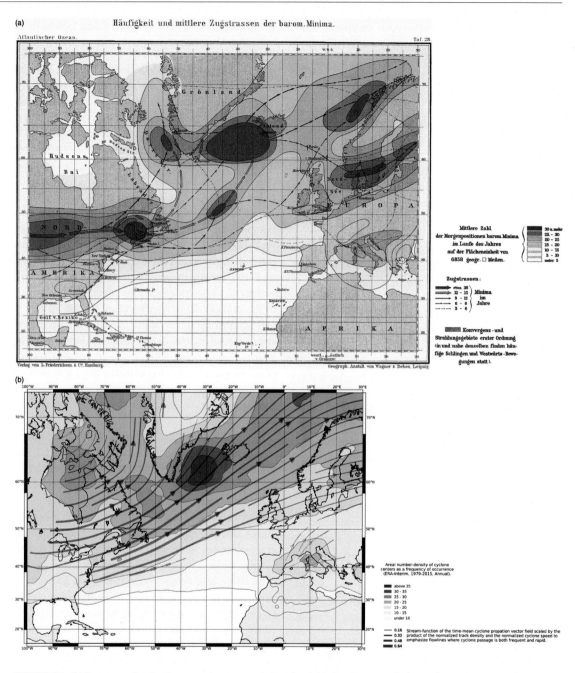

FIGURE 9.6: (a) Frequency of surface lower pressure centres (colours) and density of centre paths (arrows) over the course of a year from the 1882 Köppen climatology atlas. (b) Areal number density of cyclone centres (colours) and stream function of mean cyclone propagation from a fifteen-year climatology (2000–2015), derived using the storm-tracking analysis of Bauer et al. (2016).

optical depth will reflect less solar radiation in the winter than in the summer.

It is important to note here that the two components of the cloud radiative effect, the short-wave cooling and the longwave warming, manifest themselves at differ-ent heights in the atmospheric column. Short-wave cool-ing manifests itself at the surface and, therefore, affects surface–atmosphere energy exchanges and temperature gradients in the boundary layer. The long-wave warming, on the other hand, manifests itself in the atmospheric column with emphasis on the upper troposphere. As a result, long-wave warming, along with latent heat release, plays a significant role in the vertical distribution of energy exchanges and in determining temperature gradients in the upper troposphere. The customary addition of the short- and long-wave cloud radiative effects to yield a net TOA cloud radiative effect does not properly resolve the different nature of the two contrasting effects, and is only useful in determining the net cloud contribution to the overall atmospheric energy budget. A net zero TOA cloud radiative effect that arises from the cancellation of contrasting short- and long-wave cloud radiative effects implies vertical heating gradients that will be balanced by atmospheric circulation changes, which in turn can alter the cloud field and affect the radiative budget. Therefore, a net zero TOA cloud radiative effect does not necessarily correspond to a state of radiative equilibrium.

At the synoptic scale, large cloud radiative effect differences occur among the different circulation regimes of baroclinic storms. The existence of clouds with different cloud-top heights and optical depths in the different storm sectors creates distinct radiative signatures. One way to understand these signatures is to separate the cloud radiative effects into instances when the SLP is higher and lower than the climatological value. This separation captures the climatological signatures of high- and low-pressure regimes, respectively. Combined analysis of recent satellite observations and weather data shows that the short-wave flux differences between high and low SLP regimes are significant and can vary with season: in the winter, an excess short-wave cooling of 5–20 W m^{-2} occurs in the low SLP regime while in the summer this excess cooling increases to 20–50 W m^{-2}, mostly due to the higher amounts of solar insolation (Tselioudis et al., 2000). This excess cooling manifests itself at the surface and would act to stabilise the surface horizontal temperature gradients in the storm, as it occurs primarily in the warm sector of the storm. The long-wave differences between high and low SLP regimes are uniform across seasons and show an excess long-wave warming of 5–35 W m^{-2} in the low SLP regime. This excess warming is manifested mostly as an additional warming of the atmospheric column and would act to increase the upper tropospheric temperature gradients of the storm, as it occurs again in the warm sector of the storm.

9.3 Interaction of Clouds with the Mid-Latitude Circulation

The results presented in the previous sections show that, at synoptic scales, the amount and type of cloud produced in a storm, and therefore the cloud radiative effect, depend on the strength of the storm and the moisture availability of the atmosphere (Fig. 9.4). This implies that changes in the strength of the mid-latitude circulation or the atmospheric moisture content will initiate changes in the properties of the mid-latitude cloud field that have the potential to produce strong cloud radiative feedbacks. Clouds, on the other hand, can also alter the atmospheric thermal structure through their diabatic effects, which can then impact the strength of the mid-latitude baroclinic circulation. Hence, changes in the atmospheric circulation can modify clouds, and changes in clouds can in return modify the atmospheric circulation.

For a discussion of the interaction of mid-latitude clouds with the atmospheric circulation, it is crucial, therefore, to examine in brief the properties and processes of the mid-latitude atmosphere that determine the strength and character of its circulation. From the theoretical perspective, if one considers a purely zonal jet stream with geostrophic wind speed U_g that only depends on height z, then a suitable measure of the baroclinicity of the flow is the Eady growth rate maximum:

$$\sigma_{BI} = 0.31 \frac{f}{N} \left| \frac{dU_g}{dz} \right|, \tag{9.10}$$

where f is the Coriolis parameter, and N is the Brunt–Väisälä frequency, a measure of the atmospheric static stability (Eq. (2.90); Table 2.7). The coefficient of 0.31 is derived from numerical calculations (Lindzen and Farrell, 1980).

Assuming hydrostatic balance (Eq. (2.136)), the geostrophic wind (Eq. (9.3)) varies with height according to the thermal wind relationship:

$$\frac{\partial v_g}{\partial z} = \frac{g}{f} \times \nabla(\ln T). \tag{9.11}$$

Substituting for $\frac{\delta v_g}{\delta z}$ from Eq. (9.11), the maximum Eady growth rate can be written in terms of the magnitude of the meridional temperature gradient $\frac{\partial T}{\partial y}$:

$$\sigma_{BI} = 0.31 \frac{g}{TN} \left| \frac{\partial T}{\partial y} \right|. \tag{9.12}$$

Eq. (9.12) implies that baroclinic storm generation depends on the existence and magnitude of meridional temperature differences and the static stability of the mid-latitude atmosphere. According to Eq. (9.12), any processes that enhance the static stability of the mid-latitude atmosphere would act to suppress baroclinic storm generation, whereas any processes that enhance the meridional temperature gradient would act to promote baroclinic storm generation. This theoretical framework will be used as the basis for understanding the interactions between clouds and the midlatitude circulation discussed in this section. Results from recent studies of both observations and model simulations will be examined.

9.3.1 Influence of the Mid-Latitude Circulation on Cloud Properties

Baroclinic storms propagate preferentially within the mid-latitude storm track regions of the North Atlantic, North Pacific and Southern Oceans, so at planetary scales, mid-latitude cloud and radiation changes can result either from (a) changes in the location of those storm track regions or (b) from changes in the strength or frequency of the individual baroclinic storms.

9.3.1.1   Changes in Storm Track Position

The positions of the baroclinic storm tracks at both northern and southern hemisphere mid latitudes are not static in time, but rather fluctuate between more poleward and more equatorward positions on the timescale of approximately ten days. These modes of unforced variability in the extratropical atmospheric flow are referred to as the northern and southern annular modes, respectively. The exact dynamics that govern the annular modes are still being researched. In brief, from Eq. (9.12), the maximum growth rates of baroclinic eddies occur where the meridional temperature gradients are largest. The baroclinic eddies formed at the latitude of the maximum meridional temperature gradient act to flux momentum back towards their formation region, driving a region of enhanced westerly winds from the surface to the upper troposphere called the eddy-driven jet stream. Consequently, if the location of the strongest meridional temperature gradient is perturbed polewards or equatorwards, the baroclinic storm tracks and eddy-driven jet stream will closely follow. The momentum fluxes by the baroclinic waves are thought to be crucial in maintaining the storm track and eddy-driven jet stream in a more poleward or equatorward position for one to two weeks (Lorenz and Hartmann, 2001).

The latitudinal shifts in the baroclinic storm tracks and eddy-driven jet stream associated with the annular modes should be accompanied by corresponding cloud property shifts, particularly in the high cloud field, as the intense upward motions in individual baroclinic storms should closely follow the location of the storm tracks and jet stream. In addition, the cloud shifts should be associated with coherent changes in the cloud radiative effects. Poleward storm track shifts are also sometimes linked to increases in mid-latitude storm strength, but this may be due to the lower climatological SLP closer to the poles

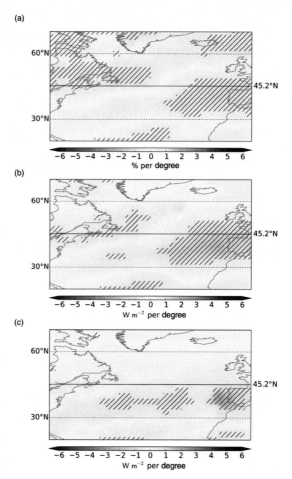

FIGURE 9.7: Changes in (a) high cloud amount, (b) long-wave cloud radiative effect and (c) short-wave cloud radiative effect for a 1° poleward shift of the eddy-driven jet during winter over the North Atlantic ocean. The high cloud changes are shown in percentage units and the radiation changes in $W\,m^{-2}$. Shaded regions denote statistically significant changes and the thicker horizontal line shows the mean position of the eddy-driven jet for that region and season. Adapted from Tselioudis et al. (2016). Copyright © 2016 John Wiley & Sons, Inc

which makes a storm shift appear as an increase in the SLP-defined storm strength even if the storm circulation is not stronger.

Analyses of both observational data gathered over the last thirty-five years and of model simulations can be used to examine the relationships between eddy-driven jet shifts and cloud properties. High clouds, for the most part, tend to shift consistently with the eddy-driven jets in most ocean basins and seasons. The high cloud shifts produce distinct long-wave atmospheric warming in the region of increased high cloud cover (i.e., the latitudes towards which the jet and storm tracks have moved) and cooling in the region of reduced high cloud cover (i.e., the latitudes from which the jet and storm tracks have moved). Changes in the short-wave cloud radiative effects with eddy-driven jet shifts, however, are more complicated because low-top clouds do not respond in a systematic manner to eddy-driven jet shifts. The response of low-top clouds to eddy-driven jet shifts varies widely by ocean basin and often differs between observational data and model simulations (Grise and Medeiros, 2016).

In the North Atlantic, the poleward movement of the eddy-driven jet stream during the winter season produces a shortwave warming effect in areas equatorwards of the jet, caused by a reduction in the total cloud amount in those areas that allows more sunlight to reach the ocean surface. The North Atlantic high cloud and cloud radiative effect changes are illustrated in Fig. 9.7, which shows the changes in high cloud amount (Fig. 9.7a), long-wave cloud radiative effect (Fig. 9.7b) and short-wave cloud radiative effect (Fig. 9.7c) for a 1° poleward shift in the North Atlantic jet during winter. The high cloud amount shows a dipole change, with increases of 1–2% in the northern part of the basin and decreases of similar magnitude in the southern part of the basin, while the long-wave cloud radiative effect change shows a corresponding warming of 1–2 W m^{-2} in the northern part of the basin and a cooling of 1–3 W m^{-2} in the southern part of the basin. The short-wave cloud radiative effect change shows mainly a 1–3 W m^{-2} warming in the southern part of the basin, since in the northern part of the basin, solar insolation is very small during the winter and therefore cloud changes do not produce a significant short-wave radiative signature.

In contrast to the North Atlantic winter season, in the summertime northern hemisphere and in all seasons in the southern hemisphere, a poleward eddy-driven jet shift produces a small short-wave cloud radiative cooling anomaly at the poleward part of the mid-latitude domain (~50–60% latitude) and does not produce any observable short-wave cloud radiative warming anomaly equatorwards of the jet position. Understanding these complicated signatures is an area of active research, as they are not represented well in many present-day global climate models. The short-wave cooling at higher latitudes is likely produced through an increase in the liquid-water path of the storm cloud field. The lack of short-wave warming equatorwards of the jet shift is likely due to the presence of large amounts of low-top clouds in the lower mid-latitude areas (~30–45% latitude) of these ocean basins, and these clouds do not appear to respond in a coherent manner to eddy-driven jet shifts. As a matter of fact, a poleward movement of the eddy-driven jet is often correlated with a poleward expan-sion of the subsiding branch of the Hadley circulation, which creates favourable conditions for the formation of more extensive low cloud decks at the equatorward side of the mid-latitude zones. Such increases in low clouds would negate the radiative effects of any decreases in the high cloud field with the poleward storm track shift. Consequently, variability of the mid-latitude circulation and its effect on the mid-latitude cloud field cannot be examined in isolation from the variability of the large-scale tropical circulation.

9.3.1.2 Changes in Strength and Frequency of Baroclinic Storms

Changes in mid-latitude clouds and their radiative proper-ties are also expected with changes in the strength and fre-quency of individual baroclinic storms. From Eq. (9.12), recall that the growth rates of baroclinic storms are closely related to the strength of the meridional temperature gra-dient and to the atmospheric stability, with more unstable lapse rates promoting greater baroclinic eddy growth. As illustrated in Fig. 9.4, the amount of high cloud in baro-clinic storms depends strongly on storm strength changes and, to a lesser extent, on the moisture availability of the atmospheric column. Table 9.1 quantifies the changes in cloud radiative effect associated with changes in mid-latitude storm strength and frequency, as derived from observations. The values in Table 9.1 represent the short- and long-wave cloud radiative effect responses to a 7% decrease in the overall mid-latitude storm frequency and to a 5% increase in overall storm intensity. These changes in frequency and intensity were used because they represent typical values occurring in climate warming simulations (see Section 9.4 for further discussion of climate change impacts on mid-latitude clouds).

The results in Table 9.1 show that changes in storm strength and frequency have opposite effects on cloud radiative effects. Storm strength increases produce short-wave cooling and long-wave warming through the pro-duction of optically thicker clouds with higher cloud tops. These cloud changes occur in the cold and warm frontal sections of the storms as shown in Fig. 9.4. The short-wave cooling is of order 2–5 W m^{-2}, depending on hemisphere and season, and the long-wave warming is of order 1.5–2.5 W m^{-2}. As a result, a 5% increase in storm strength results in a net cooling of order 1.5 W m^{-2}. It must be noted again, however, that short-wave cooling materialises at Earth's surface while longwave warming materialises in the tropospheric column, and therefore the short- and long-wave effects affect fundamentally different processes in the Earth system.

TABLE 9.1: Net TOA short- and long-wave flux changes in W m^{-2} with storm strength and frequency over mid-latitude bands (30–65°N and 30–65°S) for winter and summer seasons.

		Winter		Summer	
		Short wave	Long wave	Short wave	Long wave
Northern hemisphere	Increasing storm strength	−3.7	+1.5	−1.9	+1.6
	Decreasing storm frequency	+2.6	−1.4	+1.9	−1.0
	Net change	−1.1	+0.1	0.0	+0.6
Southern hemisphere	Increasing storm strength	−4.9	+2.5	−3.7	+1.4
	Decreasing storm frequency	+1.4	−0.3	+1.9	−0.4
	Net change	−3.5	+2.2	−1.8	+1.0

Adapted from Tselioudis and Rossow (2006). Copyright © 2006 John Wiley & Sons, Inc.

Storm frequency decreases, on the other hand, produce short-wave warming and long-wave cooling through an overall decrease in the amount, optical depth and height of the cloud field. The short-wave warming is of order 1.5–2.5 W m^{-2} and the long-wave cooling is of order 0.5–1.5 W m^{-2}. As a result, a 7% decrease in storm frequency results in a net warming of order 1 W m^{-2}. It is clear that the net cloud radiative effect of any combined storm strength and frequency change depends on the relative magnitude of change of the two dynamic components and the hemisphere and season where they occur. For this particular storm change configuration, the overall radiative change is dominated by the short-wave cooling caused by the storm strength increase. Note that the magnitudes of the radiative signatures from these relatively modest changes in storm strength and frequency are of order 1–5 W m^{-2}, implying the potential for strong radiative effects from climate perturbations in the characteristics of baroclinic storms.

9.3.2 Influence of Clouds on the Mid-Latitude Circulation

Unlike the tropics where diabatic processes are the key driver of the atmospheric circulation, the midlatitude circulation is governed more strongly by the presence of large horizontal temperature gradients. As discussed earlier, the location and strength of the eddy-driven jet streams and storm tracks depend on the thermal structure of the mid-latitude atmosphere, and mid-latitude clouds apply thermal forcings to the atmospheric column, both through radiative and latent heating effects. Differential thermal forcing by the clouds can produce horizontal and vertical temperature gradients that in turn could influence the mid-latitude circulation, including the eddy-driven westerly jets. This forcing can happen at different spatial scales ranging from the global and climate scales to the local and synoptic scales. The Eady growth rate equation (9.12) indicates that storm growth responds to both temperature gradients and static stability changes, and both of those processes can be induced by changes in the structure and the properties of the mid-latitude cloud field.

Clouds produce atmospheric thermal gradients either through the release of latent heat via condensation or through their interactions with atmospheric radiation. At the synoptic scale, emphasis has been put on the release of latent heat because the cloud-induced radiative heating is generally smaller by an order of magnitude than latent heating, and because observations of cloud-modified radiative fluxes have been sparse and less reliable. At such synoptic and local scales, cloud diabatic effects can damp or amplify the in-atmosphere energy balance of atmospheric eddies depending on whether the passage of storm clouds tends to warm an otherwise warm atmosphere or to warm an otherwise cold atmosphere. The cloud diabatic effects often depend on latitude and season as well as on the vertical distribution of the diabatic heating.

Modelling and observational studies generally show that latent heat release leads to stronger storms and faster storm development. In the vertically tilted structure of a baroclinic storm, the latent heat release associated with cloudy air ascending the frontal structure of the warm conveyor belt is often accompanied by evaporative cooling from precipitation falling into the dry air intrusion behind the cold front (Fig. 9.2). This creates a heating dipole along the front that increases baroclinicity, which could strengthen the storm circulation. Despite their relatively smaller magnitude, radiative heating terms could also play a role in affecting baroclinic storm development. As discussed in Section 9.2, short-wave cloud radiative cooling effects tend to dampen the surface horizontal temperature gradients in low pressure systems, while long-wave cloud warming effects tend to strengthen the upper tropospheric horizontal temperature gradients. In addition, long-wave cooling that peaks above the cloud top combined with long-wave warming below can create vertical diabatic heating gradients especially near the tropopause, which can induce circulations that strengthen the amplitude of the baroclinic storm. It is difficult to observe diabatic heating rates with sufficient spatial and temporal resolutions to resolve interactions between clouds and synoptic-scale dynamics. However, modelling analyses indicate that latent heat release is an important contributor to the storm energy budget at the early stages of high latitude winter storms (Booth et al., 2013), while radiative effects could play an important role in the development of high latitude summertime storms when short-wave effects are the strongest. It is fair to say that we are still at the beginning stages of understanding the interactions between cloud-scale thermodynamic processes and synoptic-scale mid-latitude weather systems.

On global climate scales, the radiative effects of mid-latitude clouds, particularly the effects on short-wave radiation, are generally considered to be the most important diabatic cloud effects on the mid-latitude circulation. Short-wave cloud radiative effects cool the surface and therefore modulate horizontal temperature gradients and surface baroclinicity, particularly during the summer season when the solar insolation at mid-latitudes is maximised. The existence of bands of strong short-wave cloud radiative forcing over the mid-latitude storm tracks introduces an additional cooling component that sharpens the meridional temperature gradient on the equatorward side of the mid-latitude storm tracks, which affects the baroclinicity (see Eq. (9.12)) and could therefore subsequently feed back upon the strength and location of the eddy-driven jets and storm tracks. The strength of this shortwave cloud radiative forcing depends on the cloud amount and the optical depth of the mid-latitude cloud field. It is important to also note, however, that the existence of short-wave cloud radiative cooling at mid latitudes decreases meridional temperature gradients on the poleward side of mid-latitude ocean basins and therefore can decrease the baroclinicity in higher latitude regions. Therefore, the short-wave radiative effects of mid-latitude clouds depend strongly on the mean position of the storm tracks.

Studies of the radiative effects of mid-latitude clouds on the large-scale midlatitude circulation are done mostly in the context of atmospheric models, often run in idealised frameworks such as the aqua-planet framework in which the Earth's surface is entirely ocean. Such modelling studies suggest that cloud changes that favour increases in the gradient of absorbed short-wave radiation between high and low latitudes increase mid-latitude baroclinicity and produce a poleward shift and a strengthening of the eddy-driven jet streams (Ceppi and Hartmann, 2016). Model simulations also suggest that cloud radiative effects within the atmosphere, achieved mostly through long-wave absorption in the atmospheric column, exhibit important influences on the large-scale circulation. Model sensitivity experiments demonstrate that these in-atmosphere radiative effects strengthen the eddy driven jet stream and increase eddy kinetic energy by as much as 30% (Li et al., 2015). This effect is achieved through an increase of the in-atmosphere meridional temperature gradient and decreases in static stability in the mid-latitude upper troposphere.

The modelling results point to a strong dependence of the mid-latitude circulation on cloud radiative effects. It must be noted, however, that atmospheric models include in their simulations strong biases in their mid-latitude radiative budgets, manifested in most models through a strong positive bias in absorbed short-wave radiation at mid and high latitudes (Trenberth and Fasullo, 2010). This is due to the fact that most climate models simulate a mid-latitude cloud field that includes too few yet too optically thick clouds. Models simulate high, optically thick cloud decks along the warm conveyor belt circulations of mid-latitude storms. However, for the most part, they fail to simulate the extensive middle and low cloud decks that occur in the cold air outbreak regimes that follow the passage of the frontal structures (Bodas-Salcedo et al., 2014). These cold air outbreak regimes are associated with large-scale subsidence, so the failure of models to accurately simulate low cloud decks here is consistent with their similar struggles in the subsiding regimes of the tropics and subtropics (Chapter 8). As a result, most climate models allow excess amounts of short-wave radiation to reach the mid-latitude surface and at the same time misrepresent the horizontal radiative gradients that exist between the different sectors of baroclinic storms, which are important factors in the storms' development. The existence of these cloud errors and radiative biases make the results of model studies of mid-latitude cloud/dynamics interactions difficult to interpret and motivates the need for stricter observational constraints.

9.4 Mid-Latitude Clouds in a Changing Climate

The mid-latitude cloud systems discussed in this chapter are likely to be significantly altered over the coming century (see also Chapter 13). With increasing greenhouse gas concentrations, the troposphere is expected to warm and the stratosphere is expected to cool, which will raise the height of the global tropopause. A rising tropopause height will allow high-topped clouds to extend to higher altitudes at mid-latitudes, increasing their long-wave radiative warming capacity. However, because many mid-latitude clouds are composed, in part, of ice, a warming troposphere will also promote the existence of more in-cloud liquid, increasing the optical depth of these clouds and consequently their short-wave radiative cooling capacity. Furthermore, as the troposphere warms, the saturation vapour pressure will increase exponentially with temperature via the Clausius–Clapeyron relationship (Eq. (2.37); Fig. 2.1), and thus there will be increased moisture availability for mid-latitude clouds and the energetics of mid-latitude storms. Absolute humidity gradients are also projected to increase, which would make mixing more effective in stabilising saturated regions.

The changing climate will also likely alter mid-latitude circulation patterns. Global climate models suggest that, with increasing greenhouse gas concentrations, the storm tracks and eddy-driven jet streams will shift polewards, particularly in the southern hemisphere. The mechanisms for these poleward jet and storm track shifts are not well understood. Several leading hypotheses rely on the fact that, in a warming climate, upper tropospheric temperatures will increase substantially more than surface temperatures, particularly in the tropics and subtropics. The large warming of the tropical upper troposphere (compared to that at the surface) is due to the fact that the tropical atmospheric lapse rate is approximately moist adiabatic. Greater warming in the subtropical and mid-latitude upper troposphere would enhance the static stability on the equatorward side of the present-day storm track, acting to suppress baroclinic eddy growth there (see Eq. (9.12)) and thus promoting a poleward shift of the storm track (Vallis et al., 2015). Because warming of the tropical upper troposphere strengthens the meridional temperature gradient in the upper troposphere-lower stratosphere, the enhanced meridional temperature gradient in the upper troposphere-lower stratosphere might also be a critical component in driving the poleward shift in the storm track and eddy-driven jet. A complicating factor is that, as the climate warms, enhanced warming of the Arctic surface ('polar amplification') will act to reduce the surface pole-to-equator temperature gradient in the northern hemisphere, which has been shown to promote an equatorward shift in the storm tracks and eddy-driven jet stream in model experiments (Butler et al., 2010).

Regardless of the mechanism involved, systematic poleward jet and storm track shifts are likely to have fundamental impacts on mid-latitude cloud fields (as noted in Section 9.3.1.1). Some recent satellite observations suggest that mid-latitude cloud fields have already begun to shift polewards (Bender et al., 2012). Recent observed poleward jet and storm track shifts are more pronounced in the southern hemisphere and can be attributed, at least in part, to the existence of the Antarctic ozone hole, which is thought to have contributed to a pronounced poleward jet and storm track shift in the southern hemisphere summer months since the early 1980s. The ozone hole is associated with substantial cooling in the southern hemisphere polar lower stratosphere during spring and summer months, which enhances the meridional temperature gradient in the upper troposphere-lower stratosphere. Antarctic polar stratospheric ozone levels are expected to recover by the mid to late twenty-first century, so the southern hemisphere summertime storm tracks would be expected to return to a more equatorward position by the mid-twenty-first century in the absence of increasing greenhouse gases.

The strength and frequency of individual mid-latitude storms is also likely to be impacted by the changing climate. Enhanced warming in the upper troposphere compared to that at the surface would increase mid-latitude static stability, which would have a weakening effect on storm intensity via Eq. (9.12). Likewise, enhanced warming of the Arctic surface would act to reduce the surface pole-to-equator temperature gradient in the northern hemisphere, reducing the surface baroclinicity available for the growth of baroclinic storms. However, the predicted increase in atmospheric moisture content and the upper tropospheric-lower stratospheric temperature gradient might be expected to increase the strength of baroclinic storms. There is little consensus about whether baroclinic storms will increase or decrease in strength in a warming climate, while there is some agreement among climate models that baroclinic storms will reduce in frequency, particularly in the northern hemisphere (Chang et al., 2012). As noted in the introduction to this chapter, the purpose of baroclinic storms is to transport energy from equator to pole. If the equator-to-pole temperature gradient is reduced, the amount of required energy transport would be reduced. Similarly, if the amount of moisture in baroclinic storms increases, weaker storms could attain the same required energy transport. From these simple arguments, fewer or weaker storms might be expected in a warmer climate. Systematic changes in storm magnitude and frequency will have fundamental impacts on mid-latitude cloud fields and their radiative impacts, as noted in Section 9.3.1.2.

Mid-latitude cloud changes induced directly by warming temperatures or indirectly through mid-latitude circulation changes can potentially feed back on the position and strength of the atmospheric circulation via changes in horizontal and vertical temperature gradients in diabatic heating (as discussed in Section 9.3.2). It remains an area of active research as to how much mid latitude clouds and their responses to a changing climate will impact future changes in baroclinic storms and the large-scale mid-latitude circulation.

Further Reading

This chapter provided an overview of a wide range of extratropical cloud issues, related to their properties, their formation mechanisms, and their interaction with radiation and with atmospheric dynamics. Each of those issues is examined in more detail in the resources listed below.

Section 9.1 Mid-Latitude Cloud Structures

T. N. Carlson, *Mid-Latitude Weather Systems* (Carlson, 1998) provides a detailed discussion of the observed structures and dynamics in extratropical cyclones from a synoptic meteorology perspective. Topics covered include the Norwegian cyclone model, baroclinic storm development, and quasi-geostrophic theory. W. R. Cotton, G. Bryan and van den Heever, *Storm and Cloud Dynamics* (Cotton et al., 2010) provides extensive coverage of the relationships between cloud formations and dynamical circulations across a range of mesoscale and synoptic-scale weather systems. Their chapter 10 includes further detail on the relationships between clouds and dynamics in extratropical cyclones. J. R. Holton and G. J. Hakim, *An Introduction to Dynamic Meteorology* (Holton and Hakim, 2013) provides a detailed introduction to atmospheric dynamics, ranging from fundamental concepts such as geostrophic and thermal wind balance to more advanced topics on the atmospheric general circulation. Their chapters 6 and 7 provide further detail

on quasi-geostrophic theory and baroclinic instability, respectively.

Section 9.3 Interaction of Clouds with the Mid-Latitude Circulation

P. Ceppi and D. L. Hartmann, 'Connections between clouds, radiation, and midlatitude dynamics: a review' (Ceppi and Hartmann, 2015) provides an overview of the recent scientific literature linking variability in mid-latitude dynamics with clouds and their associated radiative feedbacks.

Section 9.4 Mid-Latitude Clouds in a Changing Climate

T. A. Shaw, M. Baldwin, E. A. Barnes et al., 'Storm track processes and the opposing influences of climate change' (Shaw et al., 2016) provides a detailed summary of the recent scientific literature on the dynamics of baroclinic storm tracks, their linkages with moist processes and processes relevant for explaining their changes in a warming climate.

Exercises

(1) Using $v_{\mathbf{g}} = (1/f_0) \times \nabla_p \phi$, show that the right-hand side of Eq. (9.1) is equivalent to the right-hand side of Eq. (9.4). Make the quasi-geostrophic beta-plane approximation.

(2) Show that the thermodynamic equation for a dry atmosphere (Eq. (9.5)) can be rewritten in terms of the dry potential temperature θ (Eq. (2.46)) as Eq. (9.6).

(3) Show that the quasi-geostrophic vorticity equation (Eq. (9.8)) can be derived from the quasi-geostrophic horizontal momentum equations (Eq. (9.4)).

(4) Show that the quasi-geostrophic omega equation (Eq. (9.9)) can be derived from the quasi-geostrophic thermodynamic equation (Eq. (9.7)) and vorticity equation (Eq. (9.8)).

(5) Explain why the Laplacian operator can be approximated as a negative sign in the interpretation of the quasi-geostrophic omega equation (Eq. (9.9)).

(6) Using your knowledge of the quasi-geostrophic omega equation (Eq. (9.9)), draw a sketch of a typical mid-latitude cyclone (as in Figs. 9.1 and 9.2). Label where the thermal advection, differential vorticity advection and diabatic heating terms are likely to contribute to upward or downward motion.

(7) The average daily solar insolation reaching the top of Earth's atmosphere is given by the following formula (see Hartmann (2016) for details):

$$\frac{S_0}{\pi}\left(\frac{\bar{d}}{d}\right)^2 [h_0 \sin\phi \sin\delta + \cos\phi \cos\delta \sin h_0], \quad (9.13)$$

where $S_0 = 1{,}360\,\mathrm{W\,m^{-2}}$ is the solar constant at Earth, $\bar{d} = 1{,}496 \times 10^{11}\,\mathrm{m} = 1\,\mathrm{AU}$ is the mean Earth–sun distance, d is the Earth–sun distance, ϕ is latitude and δ is the solar declination angle (the latitude at which the sun is directly overhead at noon, which varies from $-23.45°$ to $23.45°$). h_0 is the hour angle at sunrise and sunset (in radians), which is defined by $\cos h_0 = -\tan\phi \tan\delta$.

Using this formula, estimate the variation in solar insolation between the summer and winter solstices at 45°N and 45°S. The Earth–sun distance on 21 December is 0.98376 AU, and the Earth–sun distance on 20 June is 1.01618 AU.

(8) Assuming a hydrostatic atmosphere (Eq. (2.136)), derive the thermal wind relationship (Eq. (9.11)) from the definition of the geostrophic wind (9.3). *Hint: You can assume that $v_{\mathbf{g}}\frac{d(\ln T)}{dz}$ is small and can be neglected.*

(9) Using reanalysis data, plot the annual-mean climatology of the maximum Eady growth rate in the latitude-height plane. Explain the observed features within the context of the climatologies of the meridional temperature gradient and static stability.

(10) Speculate on how increasing high and low cloud cover at mid-latitudes may affect the mid-latitude circulation. Justify your answer using the thermal wind equation (Eq. (9.11)) and the Eady growth-rate equation (Eq. (9.12)).

10 Arctic Cloud Systems

Gunilla Svensson and Thorsten Mauritsen

The Arctic is a cloudy region. The hostile environment has long made the North Pole, situated in the Arctic Ocean, as well as the Antarctic region highly desired targets of exploration. Initial modern-time exploration was by whale hunters following their prey into the highly nutritious waters near the sea ice edge. But in the nineteenth century more of a scientific exploration began, partly motivated by a race to reach to the two poles, by famous explorers such as Fritjof Nansen, Otto Sverdrup, Roald Amundsen, Robert Peary, Robert Scott, Ernest Shackleton and Umberto Nobile, to mention just a few, several of whom paid for fame and discovery with their lives. These early explorers knew that to advance in their endeavour they had better understand the weather, and many of them made detailed observations of the atmosphere and ocean that inspired scientific breakthroughs.

Today we understand that the meteorology of the Arctic is driven by a net energy loss to space of about 100 W m^{-2}, when averaged over the year and 70–90°N. The atmosphere and ocean circulations respond by transporting energy into the region from lower latitudes to compensate the loss. If we, as a first approximation, assume that the Arctic radiates to space as a blackbody, this means the temperature is more than 30 K warmer than it would have been without the imported energy. In the zonal mean, this results in a direct thermal cell, denoted the polar cell, with rising motion around 60°N and subsidence on the poleward side (Fig. 10.1), somewhat resembling the tropical Hadley cell. However, the picture is deceiving in that most of the transport of energy is in the form of internal energy, that is, air that intrudes the Arctic has a higher temperature than air that leaves the region at a given altitude. This advective transport could be caused by synoptic-scale storms or other larger-scale motion that moves air in the meridional direction. Subsidence and latent heat release, as well as ocean energy transport, are of secondary importance for closing the energy budget.

They do, however, play important roles in forming the mean state and causing variability.

The combination of long-wave cooling and atmospheric circulation leads to the formation of the Arctic inversion which is practically always present (Fig. 10.1). Because tropospheric absolute humidity is much lower than at lower latitudes, and so the atmospheric emissivity is relatively low, most of the long-wave cooling to space occurs either at the Arctic surface, in the initially humid boundary layer or at the cloud top. Therefore, warm air intruding into the region is first cooled at lower levels, forming the Arctic temperature inversion. The presence of the Arctic inversion is a key ingredient in maintaining the persistent low-level clouds, much like subtropical stratocumulus. In addition, because the warmer temperature with height is caused mainly by horizontal moist and warm air advection, unlike the sub-tropical inversion which is caused by subsidence, the specific humidity often increases across the inversion. This means that cloud-top entrainment of air from aloft moistens the boundary layer, thereby contributing to cloud formation.

There is no strict definition of the Arctic. The polar front, which separates cold air masses from warmer air at mid latitudes, is on average at around 60°N, but further south in winter and more polewards in summer. Some studies apply a fixed latitude boundary such as 60°N, 70°N or the polar circle at 66.6°N, others might use a zonally varying definition to distinguish between the relatively mild climates of the North Atlantic and the North Pacific as well as the western parts of the continents. The colder eastern parts of the continents are more influenced by polar air masses.

Regions polewards of the polar circle experience periods when the sun is continuously below the horizon, known as the polar night, and periods when the sun does not set, known as the polar day. This makes for a substantial contrast between summer and winter. On the other

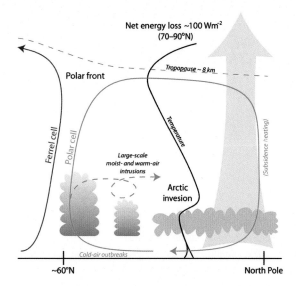

Net energy loss ~100 Wm⁻²
(70–90°N)

FIGURE 10.1: Sketched zonal mean polar cell circulation and associated clouds. The polar regions lose energy to space which is compensated by transport from lower latitudes through a direct thermal cell. Most of the transfer is through internal energy, whereas latent heat and potential energy play secondary roles.

This chapter provides an overview of cloud formation in the Arctic region with some references to Antarctic conditions. The ubiquitous low-level clouds which often contain both liquid droplets and ice crystals are in focus. These clouds are formed when warm and moist air is cooled while it is advected into the high Arctic. The clouds dissipates as the cooling continues and an Arctic airmass is formed. Eventually, this cold and dry arimass is transported southwards again in cold-air outbreaks with vigorous convection and cloud formation. Some of the challenges involved in observing and understanding Arctic clouds are discussed and how Arctic clouds may feed back on climate change are presented. By necessity, the presentation is somewhat descriptive and phenomenological, and ideas are sketchy, sometimes speculative, as less is known about the role of the clouds in the Arctic compared to elsewhere on the globe.

10.1   Observations of Arctic Clouds

The Arctic is a remote region with few habitants and the sea ice-covered ocean makes it challenging to take observations, even in summer. Land-based synoptic weather stations are located where people live along parts of the coast of the Arctic Ocean. Direct observations in the interior are only accessible using icebreakers. While such expeditions provide valuable information about the clouds and local processes that control the lower atmosphere conditions over the pack ice, the expeditions are short and most take place during summer. The year-long Surface Heat and Energy Budget of the Arctic (SHEBA) programme is an exception. In October 1997, an icebreaker was frozen into the ice north of Alaska with a group of scientists that made continuous observations using ground-based and remote sensing instruments. SHEBA observations expanded the knowledge about Arctic processes and clouds substantially.

hand the diurnal cycle is much weaker than elsewhere, and in principle on the pole the diurnal cycle is non-existent. Combined with a highly reflective surface, consisting mostly of snow and ice, and a low thermal inertia, it means that the surface temperature is mostly controlled by long-wave radiation and freezing and melting processes. In summer, the near-surface temperature is typically locked in between 0 and −2°C which are the freezing temperatures of fresh and saline waters, respectively. During winter, when the surface is frozen and usually covered with insulating snow, the surface temperature is mostly controlled by the presence or absence of clouds: if an optically thick low-level cloud moves away and the skies clear, then the temperature can easily drop 10–20°C in a matter of hours due to long-wave cooling. An example of such a rapid drop will be shown in Section 10.1.2.

Most of the mechanisms and phenomena outlined in the this chapter carry over to the southern polar region. However, the contrasting geography with a continent on the pole covered by a thick ice sheet surrounded by an ocean, makes for a more zonal atmospheric circulation, relative to the northern hemisphere, with fewer warm- and moist-air intrusions. Over the Antarctic plateau, clouds are therefore less abundant than over the Arctic Ocean. Observations at Dome C, located at an altitude of about 3,200 m above sea level more than 1,000 km inland, reveal that it is one of the least cloudy and coldest places on Earth, with an annual mean temperature of −54°C.

Icebreaker expeditions are heading for the poles every summer and some support large atmospheric programmes. During the Arctic Summer Cloud Ocean Study (ASCOS) in summer 2008, cloud observations were obtained during about a month beginning on 2 August when the Swedish icebreaker Oden left Svalbard. In transit in to and out of the pack ice, Oden stopped four times for more intense measurements. An ice camp located at 87°N was established on 12 August when a mast was erected for observations of boundary-layer processes. A suite of remote sensing instruments onboard the icebreaker observed cloud properties during the entire expedition along with regular atmospheric soundings and

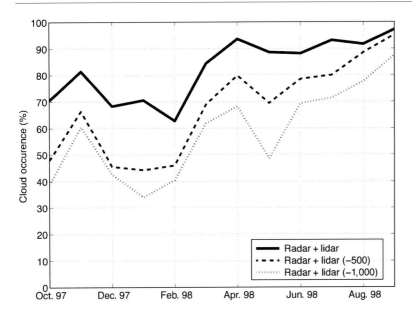

FIGURE 10.2: Monthly mean cloud occurrence during the SHEBA campaign obtained by combining cloud detection from both the ground-based radar and lidar instruments and measuring the fraction of time a cloud is detected somewhere in the vertical (solid). Also shown is the occurrence one obtains by omitting clouds detected in the lower-most 500 m (dashed) and 1,000 m (dotted), respectively. The figure is modified from Zygmuntowska et al. (2012) and based on data obtained from Janet M. Intrieri, NOAA/ESRL.

observations of many other environmental parameters such as aerosols and ocean processes.

Surface meteorological observations that cover the Arctic Ocean year around are provided by automated buoys. These buoys are deployed in the ocean, freeze into the ice and drift across the pole while automatically transferring data via satellite. Since the buoys drift with the ocean currents and sea-ice movements, their coverage vary with time. Thus far, the information has been limited to ocean and surface parameters, but the next-generation buoys will carry remote sensing instruments that provide information on the vertical structure in the atmosphere, including cloud information.

Polar-orbiting satellites pass over the Arctic more frequently than any other location on Earth. There are thus plenty of opportunities for observations, but there are substantial challenges to interpreting them. Problems are related to the weak contrast in reflectivity and temperature between the surface of snow or sea ice and the clouds. The very dry environment, due to the cold temperatures, result in clouds with little water content which makes them more difficult to detect. The variation between seasons with polar night and day and large zenith angles when the sun is present furthermore complicates the analysis of satellite-based observations. Passive sensors such as the Advanced Very High Resolution Radiometer (AVHRR) instrument are sensitive to these issues. With careful calibration they can be useful for forming climatologies, and possibly even for calculating trends, since AVHRR have long records. Active sensor satellites, such as the instruments onboard Cloud-Aerosol Lidar and Infrared

Pathfinder Satellite Observations (CALIPSO), have a better chance to detect clouds in the Arctic, but the records are shorter. Internal variability is inherently large in the region and climate change signals are therefore difficult to separate out.

10.1.1 Cloud Cover

During the SHEBA campaign, clouds were observed using ground-based radar and lidar instruments. Fig. 10.2 shows the best estimates of a combination of these two instruments. The total cloud occurrence exhibits two regimes over the year with more clouds during summer than winter with about 90% and 70%, respectively. The largest values are found in late summer during August and September. During ASCOS, the total cloud fraction was about 90%.

Multi-year observations using similar remote sensing techniques at the coast of Alaska and northern Canada show annual cloud fractions that are slightly lower than during the SHEBA year. At Barrow in Alaska, an annual cycle similar to that at the SHEBA site is observed, whereas at Eureka in Canada a slightly delayed cycle is observed with a minimum in spring/early summer and maximum during the autumn. Observations at the summit of the Greenland ice sheet show the least presence of clouds (58%) among these Arctic stations, followed by observations at Svalbard (61%). The average of stations that have both cloud radar and lidar observational records of five to twelve years is 72%. The combined estimates are considered the most accurate techniques to observe clouds, but the average value corresponds surprisingly well with other historical climatologies (manual observations) that estimate the annual cloud fraction to 65–70%.

Many of the clouds at Arctic locations are observed in the lowest part of the troposphere, with a base lower than 1,200 m present 40–55% of the time. These low-level clouds are often quite persistent, about 25% of them last for ten hours or longer. The most persistent liquid clouds were observed continuously for more than seventy hours in autumn and more than thirty hours in winter. Mid-level clouds are more related to synoptic variability and are therefore more variable in space and time. These clouds are more frequent during early spring and autumn.

The difference between the solid and dashed (dotted) lines in Fig. 10.2 gives the occurrence of clouds above 500 (1,000) m for the SHEBA record. Deliberately omitting the low-level clouds in the record gives an underestimation of the total cloud occurrence of about 20% in winter and 10% in summer. This analysis illustrates the magnitude of the error in total cloud cover for instruments that have problems detecting low-level clouds. For example, the satellite instrument CloudSat does not detect any clouds below 500 m and have problems detecting them in layers up to 1,000 m above the surface. Many of the clouds in the region are also very thin and have low water content and are therefore sometimes not detected at all. Instruments observing the atmosphere from satellite have a harder time detecting them than when they are observed from the ground. These detection limitations can explain some of the differences that are seen between data sets.

Annual total cloud fraction observed from space on CALIPSO is shown in Fig. 10.3. The climatology is dominated by clouds over the oceans where some regions are almost always overcast. The mid-latitude storm-track regions influence the western part of the continents, and on the Atlantic side it reaches well into the Arctic. Further east over the continents, the cloud fraction decreases.

There are three main pathways for warm and moist air to enter the Arctic, shown schematically by black arrows in Fig. 10.3, and their locations are related to maxima in-cloud cover: there are two regions in the Atlantic, to the west and east of Greenland, and one in the Pacific near Bering Strait. There is more air-mass exchange on the Atlantic sector where the mid-latitude atlantic storm tracks extend into the Arctic. Clouds are also frequently present over the open ocean during cold-air outbreaks (see Section 10.2.3). Furthermore, the sea ice edge is more dynamic in the Atlantic sector, specially in the Barent Sea between Svalbard and Novaya Zemlya, and Kara Sea east

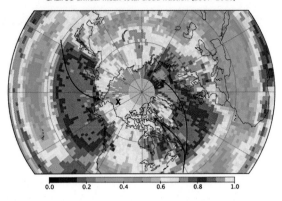

CALIPSO annual mean total cloud fraction (2007–2010)

FIGURE 10.3: Total cloud fraction as observed with the Cloud-Aerosol Lidar with Orthogonal Polarisation (CALIOP) instrument flying aboard the CALIPSO satellite. The data (Obs4MIPs CALIOP L3C) is gridded and averaged over the years 2007–2010. The black cross marks the approximate position of the SHEBA field experiment, for comparing with surface-based observations see Fig. 10.2. Most of the ASCOS observations were taken within the grey area that is not seen by the lidar. The black arrows illustrate the three primary pathways of warm air intrusions into the Arctic.

thereof. This region has longer periods that are ice free than other locations north of the Arctic circle and this influences the cloud formation.

At the SHEBA location, we can compare observations from space and the ground. The CALIPSO climatology presented in Fig. 10.3 (the SHEBA location marked by the black cross) gives a value of about 70% annual total cloud fraction. This is lower than for the SHEBA ground-based observations that give about 80% (Fig. 10.2). This difference could be due to the difference in observation length (four CALIPSO years and one SHEBA year) with no time overlap or to the changing sea-ice conditions. However, these are likely not the main reason for the difference, rather the ability to detect low and thin clouds, which is a known problem for the satellite-based instruments, as discussed earlier. This is an issue of particular importance in the Arctic because even though these clouds are optically thin, they do interact with the long-wave part of the spectrum and, therefore, as we shall see next, play an important role for the surface climate.

10.1.2   Cloud Phase and Impacts on Surface

Observations taken during the SHEBA campaign (Fig. 10.4) can be used to illustrate the cloud impact on the radiation and consequently on the surface temperature. The figure shows the signal retrieved for a period in November 1997, where the low-level clouds were

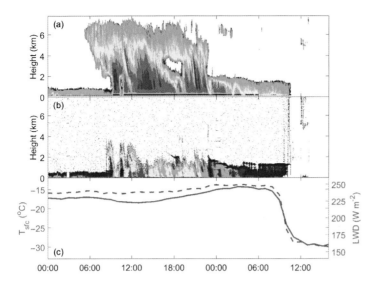

FIGURE 10.4: Time–height cross-section and time series from the SHEBA site on 23–24 November 1997 of (a) cloud radar reflectivity, indicating cloud presence, (b) lidar depolarisation ratio, the darkest blue show values below 0.11 which indicate predominantly liquid-water droplets (LWDs), whereas green, yellow and red indicate increasing amounts of ice phase, and (c) surface (blue solid) temperature (T_{sfc}), and incoming long-wave radiation at the surface (red dashed). Data provided by NCAR/EOL under sponsorship of the US National Science Foundation

interrupted by a frontal system. Also shown in the figure are surface observations of temperature and the downward component of the long-wave radiation. The cloud radar (Fig. 10.4a) shows a cloud top of about 1 km that then extends to 7 km during the frontal passage. After the frontal system has passed, the low-level clouds return, now with a cloud top of 2 km followed by a sudden clearing ten hours later. The depolarisation ratio observed by the lidar is shown in Fig. 10.4b. A value lower than 0.11 (darkest blue) indicates presence of predominantly liquid droplets, whereas higher values of depolarisation indicate the presence of a mixed or ice phase. The lidar signal is attenuated by cloud meteors, then only the lower part of the frontal clouds is detected.

The impact of the clouds, both the frontal and the thin low-level clouds, is clearly seen in the downwelling long-wave radiation at the surface (Fig. 10.4c). The surface is receiving about 240 W m^{-2} when overcast, regardless of whether the clouds are frontal or confined to low levels which indicates that the cloud radiative temperatures are very similar. Although this is in mid-winter and the ground temperature is below −15 °C, the low-level clouds contain liquid and are thus able to absorb and re-emit long-wave radiation efficiently. The surface temperature is a few degrees higher after the frontal passage (Fig. 10.4c).

As soon as the clouds disappear around 10 UTC on 24 November, the downwelling long-wave radiation quickly drops by almost 100 W m^{-2} and as a consequence the surface temperature decreases almost 15°C over the course of a few hours. Thus, there is a large difference between the cloudy and the clear states in terms of the radiative fluxes at the surface, and the snow-covered ice surface

temperature responds quickly to the changes in the energy budget. A shorter period when an ice cloud appears at about 5 km height occurs some hours after the disappearance of the low-level clouds. This cloud is not clearly visible in the downwelling long-wave radiation signal as it is an ice cloud (depolarisation ratio larger than 0.11). From this short period, we can identify two distinct states: the cloudy and the radiatively clear states. In the second state, ice clouds can be present since they have little influence on the downwelling long-wave radiation at the surface in these very cold and dry conditions. We shall return to discussing the bi-modality of Arctic low-level clouds and surface state in Section 10.2.

The presence and importance of liquid and mixed-phase clouds for understanding the Arctic surface heat budget and thus for the surface temperature climate is well recognized. Local observations help understand how and when these conditions occur but space-borne lidar observations are needed for the area coverage. CALIPSO observations show liquid-containing cloud fraction over the Arctic Ocean during all seasons both for low- and higher-level clouds. The liquid clouds are observed at almost all temperatures down to homogeneous freezing at about −35°C, during all seasons and up to heights of about 4 km in winter and all throughout the troposphere (tropopause at about 8 km) during summer, although they are most frequent in the lowest 2 km during the melting season. Autumn months has the largest fraction of liquid-containing clouds, but even during the deep winter, more than 20% of the clouds over the Arctic Ocean contain liquid. The areas with the highest fraction of liquid-containing clouds coincide with the areas of maximum cloudiness in Fig. 10.3.

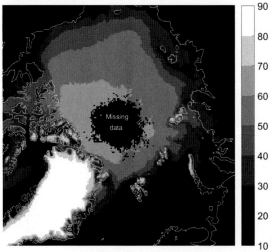

FIGURE 10.5: Summer (June-July-August) annual mean surface albedo (%) based on AVHRR Global Area Coverage data from the European Organisation for the Exploitation of Meteorological Satellites Application Facility on Climate Monitoring (CM SAF). The dark area around the North Pole is due to lack of data.

10.1.3   Arctic Cloud Radiative Effects

The previous section discusses the cloud systems and cloud types that are found in the Arctic region that affect the radiative fluxes. The strong seasonal cycle means that there is no short-wave component acting during the polar night. In summer, the sun is present continuously, but over a surface with mostly high albedo that reflects most of the radiation. The summer average (June–August) surface albedo is observed to be about 60% for most of the Arctic Ocean (Fig. 10.5). Clouds over a high albedo surface do not alter the top-of-the-atmosphere (TOA) short-wave flux much, and so the difference between the clear-sky and all-sky short-wave irradiance are small relative to other regions (Eq. (4.19)). The long-wave component of the CRE_{TOA} (cloud radiative effect at the top-of-the-atmosphere, see Section 4.3.1) determines the total effect in winter while it is present during all seasons. But the temperatures of the cloud tops do not differ much from the temperature of the surface, and therefore the long-wave component of the CRE_{TOA} is also typically small. When taken together, it means that the annual net CRE_{TOA} is only slightly negative in the Arctic (Fig. 10.6.a), despite clouds being plentiful (Figs. 10.2 and 10.3).

The presence of clouds has a profound effect on the radiative fluxes at the surface, as discussed earlier. So clouds are important for the surface energy balance and for the temperature climate in the Arctic. Their impact can be

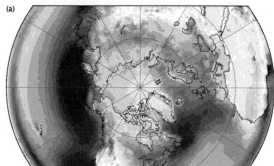

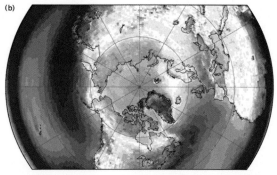

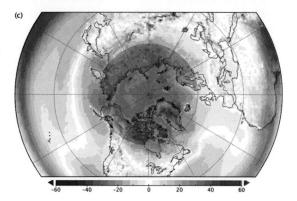

FIGURE 10.6:   Cloud radiative effect at (a) TOA, (b) at the surface and (c) for the atmosphere. The retrievals are from Clouds and the Earth's Radiant Energy System Energy Balance and Filled (CERES-EBAF) Ed2.8 and annually averaged over 2001–2014. Note that these estimates depend critically on our ability to detect the presence of clouds, which is problematic in the Arctic region, as discussed in the text. However, we believe the sign of the cloud radiative effect is probably correct in most places.

understood by inspecting the surface cloud radiative effect CRE_{surf} which can be approximated as:

$$\begin{aligned} CRE_{\mathrm{surf}} \approx\ & (SW_{\mathrm{all\text{-}sky}} - SW_{\mathrm{clear\text{-}sky}})(1 - \alpha) \\ & + (LW_{\mathrm{all\text{-}sky}} - LW_{\mathrm{clear\text{-}sky}}), \end{aligned} \quad (10.1)$$

where SW and LW are the downwelling short- and long-wave fluxes at the surface under all-sky and clear-sky conditions, and α is the surface albedo. Here we ignore multiple reflections between the cloud and surface, and it is assumed that the reflectivity of long-wave radiation at the surface is zero, which is the case in the limit of a perfect blackbody with an emissivity of one, a reasonable approximation for snow and ice.

The global CRE_{surf} is on average negative because the short-wave effect of clouds or the cloud albedo effect is dominating over the long-wave effect (Fig. 4.9). In the polar regions, the positive long-wave CRE_{surf} dominates the annual mean (Fig. 10.6b). In fact, it is only a short period during the Arctic summer that the net CRE_{surf} is negative. Most of the summer, the clouds have a warming effect because of the attenuating albedo effect on the short-wave CRE_{surf}. Following Eq. (10.1), the short-wave effect is small if the surface albedo is high as the short-wave part of CRE_{surf} increases linearly with a decreasing albedo. Hence, the radiative effect of the clouds increases during the summer season in the Arctic when snow and sea ice is melting and the surface albedo decreases.

It is also interesting to calculate the cloud radiative effect on the atmosphere as $CRE_{\mathrm{atm}} = CRE_{\mathrm{TOA}} - CRE_{\mathrm{surf}}$ (Fig. 10.6c). This measure is dominated by the long-wave component because clouds hardly cause atmospheric absorption in the short wave. The emerging picture is that clouds cool the atmosphere in the Arctic by transferring long-wave radiation to the surface and to space, and that at lower latitudes high-level clouds heat the atmosphere by trapping long-wave within the atmosphere. The resulting pattern alone would act to enhance the atmospheric temperature contrast between low and high latitudes and therefore would require an increase of the meridional atmospheric energy transport.

The radiative effects of clouds can also be important for the evolution in the transitional seasons. For instance, longwave surface warming by enhanced cloudiness at the end of the winter season can help to pre-condition an earlier onset of the melt season, and a subsequent deeper summer sea-ice minimum. This effect is amplified by the positive surface albedo feedback. Likewise, in autumn, a clearing in the clouds can initiate the freeze-up of the surface, and once started, the resulting brighter surface is less likely to melt again.

10.2 Arctic Cloud Processes

The cloudy Arctic region offers a variety of clouds, including surreal polar stratospheric mother of pearl clouds, dramatic mountain wave lenticular and rotor clouds in the lee of mountains, or intriguing Kelvin–Helmholtz

instabilities forming billow clouds in the Arctic inversion. Here, however, the focus is on the commonly occurring low-level clouds. We shall trace an air mass as it enters the Arctic, bringing heat and moisture, and transforms to a cold dry air mass until it eventually leaves the region in the form of convectively unstable cold-air outbreaks. We shall further discuss how synoptic-scale storms steer this flow, mention the violent polar lows which bear similarities with tropical cyclones, and briefly discuss the role of aerosol particles.

10.2.1 Low-Level Arctic Stratocumulus

As warm and moist air intrudes in over the Arctic pack ice, it cools off, horizontally near-uniform condensation occurs and thereby a highly persistent low-level stratocumulus cloud deck forms. A typical summer cloud scene is displayed in Fig. 10.7. The base of Arctic stratocumulus may reside at altitudes close to the surface, in which case it may be classified as fog, and up to 1 km or more above the surface, whereas the cloud top is usually at around or less than 1–2 km in height, as was discussed in Section 10.1. The Arctic stratocumulus usually consists of mixed-phase or super-cooled liquid, as the upper cloud temperature is typically on the order of 10 K cooler than the surface.

A summary of processes related to marine stratocumulus is presented in Fig. 10.8. Stratocumulus is sustained by convective mixing driven by cloud-top long-wave cooling. Cooling of air parcels near cloud-top leads to negative buoyancy and as a result they sink, ensuring that condensation occurs in a thicker mixed layer. The resulting latent

FIGURE 10.7: A photo of a cloud scene on 26 August 2008 at 87°N during the ASCOS campaign. The typical optically thick low-level stratocumulus cloud dominates with a rare clearing on the horizon. Weak horizontal inhomogeneities in the cloud thickness can be observed which are indicative of convective mixing processes. Photo by Thorsten Mauritsen

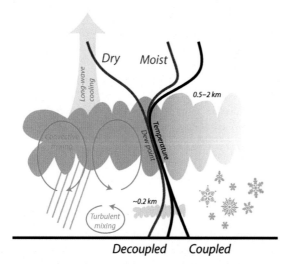

FIGURE 10.8: Processes maintaining Arctic stratocumulus clouds. The black and pink lines show temperature and dew point temperature profiles, respectively. Cloud-top radiative cooling (pink arrow) drives convective mixing that extends below the cloud base in coupled cases all the way to the surface. In decoupled cases, a low-level inversion inhibits turbulent exchange between the surface and the cloud. Horizontal moisture advection into the Arctic may give an increase in specific humidity across the inversion and the top of Arctic stratocumulus clouds is frequently observed to extend into the temperature inversion.

heat release counteracts some of the long-wave cooling of the mixed layer. These are the exact same processes acting in subtropical stratocumulus, although the latent heat release is smaller in the Arctic due to the generally colder conditions. The latent heat flux at the surface is not nearly strong enough to balance radiative cooling at the top of the cloud in the Arctic. To sustain a cloud, an approximate balance is instead attained through a second source of moisture.

Because there is on average subsidence in the central Arctic – a consequence of the polar cell circulation discussed earlier (Fig. 10.1) – turbulent cloud-top entrainment of air from aloft must occur in order to sustain a vertically stationary cloud-top level. In the subtropical stratocumulus, cloud-top entrainment leads to reduced cloudiness because the air that is entrained from above the cloud is usually warmer and much drier. This is because of the dominating large-scale subsidence in the Hadley cell. However, the Arctic inversion is frequently observed to be moist; the specific humidity increases with altitude, and in addition the top of the stratocumulus cloud is observed to extend well into the inversion. This increase of specific humidity with height means that cloud-top entrainment is a source of moisture to the cloud layer. The cause of this increase of moisture across the inversion

is primarily due to the horizontal warm- and moist-air advection from lower latitudes that is largest in the lowest 1–2 km of the troposphere.

Below the Arctic stratocumulus there is frequently a small temperature inversion and so the cloud-generated turbulence does not always extend to the surface – the turbulent cloud layer is de-coupled from the surface (Fig. 10.8). This lower layer is a weakly turbulent stably stratified boundary layer driven by near-surface wind shear. The presence of this layer is important because it slows the turbulent exchange between the surface and the cloud layer. Occasionally, a secondary cloud can form at the top of the near-surface inversion layer.

Decoupling (see Section 5.4.4.5) also occurs underneath subtropical stratocumulus and is usually attributed to evaporating drizzle. However, because most of the Arctic stratocumulus precipitation is in solid form, which does not evaporate much, this is less likely to be the cause. Another possibility is that there is simply not enough energy at the surface to sustain a dry adiabatic lapse rate from the surface up to the cloud base against long-wave cooling of the surface. Think of the surface and cloud as two planar blackbodies facing each other; then, equilibrium is reached once they have the same temperature. Now, add some energy to the surface in the form of solar absorption, latent heat of fusion or conduction through the ice, and the temperature at the surface can be larger than that of the cloud. However, under Arctic summertime conditions, this surplus seldom amounts to more than 5–15 W m^{-2} as long as the surface is covered with insulating and reflective snow, and so the temperature of the surface quickly adjusts to be about 1–3 K warmer than the cloud-base radiative temperature, that is, close to thermal equilibrium. If the cloud base is relatively high, it is therefore necessary that an inversion forms below the cloud base as dry adiabatic mixing would have otherwise lead to a large temperature difference between the surface and cloud base. If, however, the cloud base is low (1–300 m), or for some reason more energy is available at the surface, for example, due to a darkening of the surface, then the surface and cloud mixed layer can couple and the whole layer becomes well mixed.

In summary, the Arctic stratocumulus is maintained by cloud-top long-wave cooling and bears similarities with the subtropical stratocumulus (Section 5.4.4), but there are important distinctions. Foremost, there is the presence of a moist inversion and the frequent decoupling of the cloud layer from the surface. In addition, the presence of mixed-phase condensate in the cloud plays a central role in understanding the Arctic stratocumulus, as we shall see next.

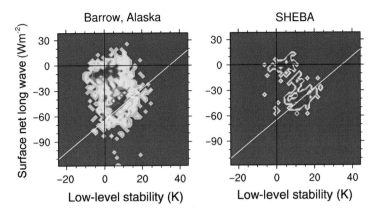

FIGURE 10.9: Bi-variate probability of wintertime (November–February) low-level stability and surface long-wave cooling for Barrow, Alaska (2000–2009) and the SHEBA field experiment (1997/1998). Low-level stability is the difference between 850 hPa and near-surface temperatures. Modified from Pithan et al. (2014)

10.2.2 The Radiatively Clear State

During the SHEBA campaign, it was noted how the wintertime surface conditions would vary between two preferred states (Section 10.1.2): a cloudy state with fairly little surface long-wave cooling, corresponding to that described in the previous section, and a radiatively clear state with strong surface long-wave cooling. In the midst of winter the difference is substantial. A cloudy state would typically have a surface temperature of $-20°C$ and a radiatively clear state of perhaps $-40°C$. The cloudy state exhibits neutral to weakly stable stratification underneath a cloud, whereas the radiatively clear state exhibits a strong surface-based temperature inversion (see Fig. 10.9). The radiatively clear state needs not be absent of cloud, in fact, often it is not, but if a cloud is present, it is sufficiently thin that the infrared emissivity is too small to cool the atmosphere effectively. Such clouds are optically thin and thus difficult to see by the naked eye and most instruments, except perhaps a sensitive surface-based lidar, will have difficulty detecting them.

We take a Lagrangian perspective and let time progress, such that the air mass transformation can be examined (Fig. 10.10). Air that enters the Arctic from the open ocean is well mixed and humid (a) it is then gradually cooled and a self-sustained low-level stratocumulus cloud forms. Due to the cloud-top cooling, a new mixed-layer forms beneath an elevated inversion (b). As the cooling and resulting depletion of total water continues, at some point the cloud becomes radiatively so thin that rather than cooling at the cloud top, radiative cooling occurs near the surface, and, with that, a surface-based inversion forms (c).

As we alluded to earlier, the long-wave cooling at the stratocumulus cloud top is counteracted by latent heat release through condensation in the cloud. However, the thereby consumed water vapour is not replenished at a

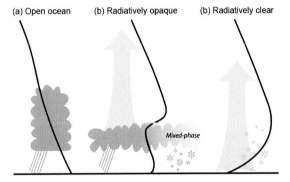

FIGURE 10.10: A schematic of Arctic air mass formation in a Lagrangian perspective. Three states are shown (a) the initial state over open ocean, (b) the cloudy state over ice underneath an elevated inversion and (c) the final cold state with no optically thick clouds and a surface-based inversion. The black lines show temperature structure and the red arrows radiative cooling.

sufficient rate by surface evaporation and the moisture in the inversion is only a finite reservoir for a given air mass, so that the vapour in the lower troposphere eventually depletes. This, though, does not alone explain the bi-modality of the Arctic low-level cloud state (Fig. 10.9). Here we go through a few additional ingredients necessary to understand the observed behaviour.

First, long-wave emissivity of the cloud layer is strongly dependent on the amount of condensate in the cloud. The emissivity (ϵ) is a measure of the clouds ability to absorb and emit long-wave radiation. The emitted radiation scales with emissivity as $R = \epsilon \sigma T^4$ and a blackbody has $\epsilon = 1$. Clouds consisting of ice particles have emissivities that are less than that of a cloud consisting of liquid droplets, for a given amount of condensate. This is because ice particles are larger but fewer than cloud droplets and so the effective absorption cross-section is smaller. Typical liquid or mixed-phase marine stratocumulus in the subtropics and Arctic have liquid-water paths (LWPs) of 50–100 g m^{-2} and an effective cloud droplet radius of 10–15 μm since any larger droplets drizzle out rapidly. These clouds act practically as blackbodies. If the water content is decreased to LWP = 20 g m^{-2}, ϵ still is about 0.9 and thus alters the long-wave radiation substantially. For lower LWP values, however, ϵ drops rapidly. Reducing the LWP to 2 g m^{-2} gives ϵ = 0.2. Thus, there is substantial non-linearity in the response of long-wave radiation to cloud phase and thickness which plays a role in interpreting the role of Arctic clouds.

Second, the efficiency with which cloud condensate is precipitated out depends on the phase. Cloud droplets are smaller than typical ice crystals and therefore sediment more slowly. For the transition from supercooled liquid to ice clouds it is important to appreciate the Wegener–Bergeron–Findeisen process (WBF, see Section 3.3.4.1), which may however not be very effective in Arctic clouds. At freezing temperatures, the saturation vapour pressure over ice is lower than that over a supercooled liquid. Therefore, if both ice and liquid hydrometeors are present, water vapour will deposit on the ice crystals and eventually evaporate the liquid droplets. This is in principle a fast process that could explain a bi-modality. At a larger scale, however, the WBF process is not very effective because often liquid and ice are separated spatially. In Arctic stratocumulus, the small supercooled liquid droplets are observed to preferentially reside in the upper part of the cloud whereas the larger ice crystals sediment to the lower parts or fall out as snow (see Fig. 10.4). On top of that, aerosol particles that can act as ice nuclei (IN) are scarce in the Arctic, at least in the lower part of the troposphere, such that it is difficult to form ice in the first place. Sources and compositions of IN are not known but they may enter the cloud through top entrainment where they are quickly activated, and after rapid growth they are removed as falling ice. At decreasing temperatures in the cloud, more aerosol particles may also act as IN and with that the likelihood of transitioning to ice through the WBF process increases. Glaciation will eventually happen in any case when homogeneous freezing dominate at temperatures below about −35°C.

Third, liquid clouds may also dissipate without going to the ice phase if the cloud condensation nuclei (CCN) aerosol particles are depleted. Under typical summertime conditions, there are some 50–100 cm^{-3} CCN in remote Arctic air. Thus, if each CCN would activate one cloud droplet, they result in droplets much smaller than the 10–15 μm that in practice marks the onset of drizzle. Drizzle can still form through collection by collisions,

FIGURE 10.11: A tenuous cloud during the ASCOS campaign on 31 August 2008. Although the skies appear clear, the boundary layer was supersaturated with respect to water and housed a sub-visible cloud that revealed its presence by the double fog-bow. The fog-bow is caused by Mie scattering of sunlight, and the double bow occurs for droplets of 20–50 μm effective radius which is large enough to sediment as drizzle. At this time, the cloud condensation nuclei count with respect to 0.2% supersaturation was less than one particle per cubic centimetre, and the surface temperature dropped from about −4 to −12 °C. From Mauritsen et al. (2011)

although this is a relatively slow process for droplets of this size. However, if there are less than about 10 cm^{-3} CCN available, then each droplet can grow to drizzle sizes without having to collide with other droplets. In this case, the cloud will begin to loose its liquid through sedimentation which in turn leads to a further depletion of the CCN as they are deposited at the surface with the precipitation. This constitutes a positive feedback loop which can eventually lead to very low CCN counts. One such tenuous cloud is shown and explained in Fig. 10.11.

Regardless of the microphysical pathway, it is clear that Arctic stratocumulus clouds will eventually transition from the radiatively opaque to the radiatively clear state (Fig. 10.10b and 10.10c). One can think of the former state as a quasi-stationary state and the latter as a stable state; once in the radiatively clear state, the system stays there until the air mass moves over a warmer surface. The timescale for the transition from the radiatively opaque to the clear state is dependent on a number of factors, but is likely on the order of ten days. This is long enough that not all air masses will spend sufficient time over snow and ice in the Arctic to make the full transition. The time to transition to the clear state is likely shorter in winter when the mixed-phase processes are more effective than they are in summer, and one could speculate that this is the reason that clouds are more abundant in summer than during winter (Fig. 10.2).

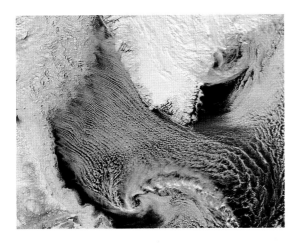

FIGURE 10.12: A cold-air outbreak over the Labrador Sea between Canada and Greenland on 16 March 2014. As the cold and dry Arctic air mass moves from the ice-covered to the open ocean in the north-westerly flow, large cloud streets form which are aligned with a slight angle (10–20°) to the mean wind direction. Further downstream, the cloud streets break up and convective cell structures dominate. A synoptic-scale low is forming in the lower part of the image. For reference, the distance from Newfoundland and southern Greenland is about 1,000 km. Image credit to Jeff Schmaltz, MODIS Land Rapid Response Team, NASA

10.2.3   Cold-Air Outbreaks

After the air has spent substantial time over the pack ice or snow-covered land and transformed to a cold and dry Arctic air mass, it is eventually advected out over the open ocean. The coldest air is formed during the late winter-early spring when the sea ice extent is largest. The cold-air outbreaks occur on a daily basis at various positions in the North Atlantic or North Pacific, depending on the governing flow. The temperature difference between the air and the ocean surface leads to very strong turbulent heat fluxes of heat and moisture of several hundreds of W m^{-2} from the surface to the atmosphere and in turn lead to rapid heating and moistening of the boundary layer and convective clouds form.

Fig. 10.12 shows a case from the Labrador Sea between Canada and Greenland where arctic air is advected out over the ocean. Close to the ice edge, the cloud streets are narrow, a few kilometres across. The horizontal and vertical scales are initially determined by the depth of the convective boundary layer in these horizontal rolls. As the air moves downstream these narrow streets combine to form fewer and broader streets. In the middle of the Labrador Sea, they are 10–20 km across. Further downstream, south of Greenland, the air moves over even warmer Atlantic waters. The convective instability increases further and the

cloud streets begin to break up and more of a convective cell-like structure is formed, eventually organized on a scale of about 100 km. One can think of this as an upscale organization of the convection from an initially shear-dominated boundary layer to a mostly convectively driven state.

Cold-air outbreaks mark the end of the lifecycle of an Arctic air mass; once heated and moistened over the warm oceans, the process can start over again. Cold-air outbreaks, such as the case in Fig. 10.12, are instrumental in cooling surface ocean waters in the region of deep water formation in the North Atlantic. In winter and spring when the polar front is the furthest south, cold-air outbreaks may penetrate to, for instance, central Europe and bring about fresh chilly winds, clear blue skies and fair weather afternoon cumuli. Further north, along the Norwegian coast are instead hit by many intensive snow showers and possibly development of more organized convective systems called polar lows.

10.3   Synoptic Storms and Polar Lows

Mid-latitude synoptic systems (see Chapter 9) usually develop over ocean and travel east or north-east and together they make up the storm tracks. Most of these systems stay in the mid-latitudes but a small fraction do enter into the polar regions. Occasionally, systems are formed in the Arctic Ocean. The ones found in the polar areas are generally weaker and have smaller horizontal extent due to the increasing Coriolis parameter with latitude. The change of preferred size of a system due to Earth's rotation can be understood by diagnosing the Rossby radius of deformation, λ_R:

$$\lambda_R = \frac{NH}{f}, \tag{10.2}$$

where H is the depth of the system, N is the Brunt–Väisälä frequency and f the Coriolis parameter. Thus, mid-latitude storms that reach high latitudes will decrease in size and locally formed systems have smaller initial radii.

Cyclones occur in the Arctic in all seasons but are more common in summer than winter both because the polar front is located further north and as there is more local baroclinicity and cyclogenesis within the region at this time of year. The intensities in summer are, however, generally weaker and the fronts are also less well defined near the surface as the surface temperature is constrained to stay close to the melting point as long as sea ice is present.

These synoptic systems bring sensible heat and moisture to the polar regions and this eddy transport is important for the climate in the region. The largest number and most energetic cyclones enter the Arctic in the Atlantic sector. Some are also entering from the Pacific side, but they have less of an impact. The large-scale mid-latitude circulation determines the regions of cyclogenesis and storm-track positions discussed in connection to Fig. 10.3. Some storms do enter the Arctic but at other times the large-scale circulation favours warm-air advection to the Arctic without a low-pressure system entering the Arctic. This occurs preferentially in the Atlantic sector when an atmospheric high-pressure centre is situated over the European continent and is blocking the flow. Such an atmospheric blocking event steers the marine air northwards. Very few mid-latitude cyclones are able to reach the high-elevation central Antarctic but they are not uncommon in the ice-shelf areas around the continent.

Returning to Fig. 10.4, already discussed in Section 10.1.2, the cloud structure is shown during a period in 1997 when a mid-latitude cyclone entered the Arctic north of Alaska and reached the SHEBA camp. The system has a clear double-front structure with deep clouds up to the tropopause. The tropopause is generally lower in the polar regions and on this day it is found at a height of about 8 km. Precipitation is recorded as light or moderate snowfall during the frontal passage and the near-surface temperature is well below freezing. The total water content in polar frontal clouds is lower than at lower latitudes because the air is colder. During the SHEBA winter, about ten such synoptical systems passed the ice camp location.

A polar low is a small-scale cyclonic system that exists in the polar regions in both hemispheres. Polar lows form polewards of the major baroclinic zone and cause heavy precipitation and strong winds (>15 m s^{-1}). They have a horizontal length scale of roughly 200–1,000 km, which is small compared to the typical 1,500–5,000 km for a mid-latitude storm. The polar lows develop not too far from the ice edge and can intensify rapidly and have shorter lifetimes than synoptic systems, at most a couple of days. In Fig. 10.12, a polar low is seen developing downstream in the cold-air outbreak.

It is not well established under which conditions polar lows are formed but some ingredients that favour their development are known. They tend to form during cold-air outbreaks when the large-scale environment provides low-level baroclinicity and/or upper-level potential vorticity anomalies. The large turbulent heat fluxes of heat and moisture from the ocean to the atmosphere during the outbreaks provide latent energy that can be released within the developing polar low. In some cases, structures similar to hurricanes have been observed, with a clear central eye and remarkably axisymmetric cloud structure. Not all polar lows have this structure and it is suggested that there is an influence of the diabatic heating by convection on the development and that it can interact non-linearly with the baroclinic dynamics. The development of these small-scale intensive systems, that involves interaction between the large-scale environment and local boundary layer and cloud processes, also creates strong atmosphere-ocean coupling that possibly plays a climatological role in both mixing the atmosphere and cooling of the ocean surface. Their small scale and rapid development means that they are particularly challenging to forecast and their effects are not captured well in climate models.

10.4 Climate Change Cloud Feedbacks

Understanding Arctic cloud feedbacks to a changing climate is challenging – to no small extent due to the large annual cycle with no short-wave radiation part of the year and a dramatically changing cloud environment through, for example, melting snow and sea ice. For example, an increase in cloudiness warms the surface in winter but usually acts to cool it during summer. But this view is complicated by the fact that the cloud net surface radiative effect changes from positive to slightly negative as snow and sea ice melts during the annual cycle in today's climate, and so further surface melting will make it negative during a longer period. The latter is, strictly speaking, not a cloud feedback because the cloud itself need not necessarily change in order to induce a shift in cloud radiative effect (see Eq. (10.1)). When accounting for such environmental changes, it is found that climate models on average exhibit small positive cloud feedback over the Arctic, but about a third of the current models have negative values. However, evaluation of climate models displays their inability to correctly simulate the properties of the clouds in the present-day climate which could potentially lead to an erroneous change when forcing is applied.

One type of cloud feedback, which is active mainly at mid to high latitudes, is related to cloud phase changes. Liquid clouds consisting of numerous small droplets are far more reflective than ice clouds which consist of larger but fewer ice crystals. As discussed in Section 10.2, an ice cloud also sediments faster, resulting in a shorter cloud lifetime. Assuming that the transitions from liquid to ice phase is primarily a function of absolute temperature leads to a shift from ice to liquid phase in a warmer climate. It follows that clouds will be more reflective and possibly become more abundant in regions where clouds frequently glaciate today.

One common way to think about this mixed-phase cloud feedback, which is most applicable at mid latitudes, is that in a warmer climate the tropopause – and, with it, clouds – extend upwards such as to maintain an approximately constant temperature. Provided that the cloud bases do not move appreciably, then the layer with liquid-phase clouds has thickened. In the Arctic, however, most clouds reside in or just above the boundary layer. The depth of this layer is perceivably not directly related to the tropopause height. Here it is more helpful to think of an air parcel undergoing an Arctic air-mass transition (Section 10.2). If parcels start out being warmer and more humid, then all other things being equal, it will take more cooling and thus a longer time to undergo the transition to the radiatively clear state. Therefore, one can expect a shift to more frequent radiatively opaque cases at the expense of radiatively clear cases.

It is not straightforward to figure out how this mixed-phase cloud feedback impacts the Arctic in a changing climate because of the competition between short-wave cooling and long-wave warming associated with an increase in cloudiness. However, if we consider certain limiting cases, we may determine the sign. On the one hand, in the summer over the open ocean or on snow-free land, the high insolation and low surface albedo will cause a negative cloud feedback with an increasing LWP. On the other hand, during the Arctic winter, when the sun is below the horizon, more frequent radiatively opaque cases will exert a positive cloud feedback at the surface and, thereby, alone, contribute to exacerbate the warming.

It should be noted that the full Arctic climate change cloud feedback is more complex than outlined here, in particular, if the ocean and atmospheric circulations change or if one considers regions becoming snow or ice free.

10.5 Modelling Challenges

Both polar regions are very remote and explain why there are only a few observational sites, specially over the Arctic Ocean or the interior of Antarctica. Thus, there is a lack of both climatological and process-level observations in environments that are quite different from the rest of the world. The range of cloud-related processes active in the polar regions that may contribute to the Arctic amplification are thus quite challenging to fully understand and thereby also difficult to model. The process descriptions included in climate models are based on studies using observations at lower latitudes and are sometimes not well suited for polar conditions. The shallow layers with weak turbulence combined with strong coupling and interaction

with the readily changing surface makes it even more challenging. Climate models, for example, have problems in maintaining the low-level liquid phase in the clouds with large biases in the radiation fluxes and surface temperature as a consequence. This bias also limits the model's ability to correctly capture the feedback related to the phase of the clouds in a warmer climate. Likewise, the representation of stably stratified turbulent mixing, for example, near the cloud top and during the radiatively clear state, and the coupling to the surface snow and ice can severely impact modelled climate.

It is well known that in climate models, the North Atlantic storm tracks often exhibit a too zonal propagation towards Europe with the consequence that the air exchange between the mid latitudes and the polar region is not captured correctly. Atmospheric blockings increase the meridional transport that bring warm, moist and cloudy air to the polar regions. The large-scale circulation determines the time scales of the air mass exchange and, thus, the properties of the Arctic air since the longer it resides, the colder and dryer it gets. Thus, biases can occur even with perfect local process descriptions.

To advance knowledge on the processes, a range of models can be used in combination with observations. Large eddy simulations are used to study the delicate balance between the different small-scale processes that control the evolution of mixed-phase low-level Arctic clouds. It is a demanding problem since the microphysical processes must handle both the liquid and ice phases as well as turbulence and interaction with radiation. The evolution of a mixed-phase stratocumulus show sensitivity to ice nucleation processes, cloud-top entrainment and surface-driven turbulence. Although the precipitation rate is less than elsewhere, it is still important for the water budget of the cloud and, thus, assumptions on crystal shape through its impact on the fall velocity becomes important for the evolution of the modelled cloud. Finally, the clouds have the ability to change the temperature and albedo of the surface that, in turn, feeds back to the cloud properties. Properly accounting for the interaction of the near-surface cloudy layer with the surface, sea ice and ocean below is therefore desirable.

Further Reading

This chapter provided an overview of the dominant Arctic cloud types and how they follow and to some extent interact with the high-latitude circulation. The main idea is that relatively warm and moist air masses are transported into the Arctic, cooled radiatively, and thereby highly persistent low-level stratocumulus clouds form. The long-wave radiative cooling of the lower troposphere eventually leads to the formation of dry and cold polar air masses – an idea that dates back to Wexler (1936), refined by Curry (1983) and revisited in terms of the representation in general circulation models by Pithan et al. (2014). In this regard, Arctic clouds and circulation are intimately related.

Observations are the backbone for understanding Arctic cloud processes. In this chapter, we have used observations from two campaigns, SHEBA (Perovich et al., 1999; Uttal et al., 2002) and ASCOS (Tjernström et al., 2014). A recent analysis linking transport and local climate processes, including clouds, during the SHEBA winter can be found in Persson et al. (2016). More information about the CM-SAF data can be found in Schulz et al. (2009), and Karlsson and Dybbroe (2010) present a comparison between different satellite-based products and discuss some of the issues with the instruments. Cesana et al. (2012) provide satellite-based observations of cloud phase for the Arctic.

The Arctic low-level stratocumulus bear similarities to their subtropical counterparts with important differences; specific humidity often increases across the inversion such that cloud-top entrainment is a source of cloud water, the presence of mixed-cloud phases and the frequent decoupling due to low-level inversions below the cloud base. Morrison et al. (2012) explain the particular challenges involved in understanding why in the Arctic mixed-phase clouds can be as persistent as they are at temperatures well below freezing, whereas Shupe et al. (2013) provide an overview of the interactions between clouds and turbulent mixing in the boundary layer. All relevant small-scale physical processes that contribute to the low-level structure, including clouds, and surface budget in the Arctic is reviewed in Vihma et al. (2014). Many more useful references for continued reading on the subject of Arctic clouds and climate can be found in Kay et al. (2016).

Exercises

(1) Make a back-of-the-envelope estimate of the latent and sensible heat convergence caused by Arctic air-mass transitions per unit area. *Hint: estimate the moist static energy of the lower troposphere before and after the transformation to an Arctic air mass and estimate the time it takes.*

(2) Estimate the Rossby radius of deformation for a tropical cyclone and a polar low.

(3) Discuss the similarities and differences between marine low-level clouds found over the subtropical oceans and the Arctic ocean. Estimate the cloud radiative effect at the surface and TOA for the two cases using Eqs (10.1) and (4.19).

Part IV   Cloud Perturbations

11 Clouds and Aerosols

Johannes Quaas and Ulrike Lohmann

The discipline 'meteorology' got its name from the mete-ors ('*meteoros*' is Greek for the 'things high up') or particles that are suspended in the air. Many of those consist mostly of water (hydrometeors), namely the cloud and precipitation particles. All other particles, of both liquid and solid phases, are subsumed under the term 'aerosol particles'. Literally 'aerosol' is the ensemble of the particles and the air surrounding them, from Latin '*aer*' (air) and '*solutio*' (solution). In the following, we also adopt the term 'aerosol' when more precisely 'aerosol particles' are meant, since in atmospheric science the particles are always surrounded by air, so the distinction is unnecessary. This usage is in line with most of the relevant literature. Aerosol, cloud and precipitation particles interact in many ways – most prominently, because cloud particles form on aerosol particles and because aerosol particles can be scavenged by precipitation. In certain regions, a considerable fraction of today's atmospheric aerosol is anthropogenic. This exerts a radiative forcing of the climate system, considered the second largest one in anthropogenic climate change, after the anthropogenic greenhouse effect, but of much larger uncertainty and of opposite – negative – sign. This chapter explains the fundamentals of aerosols and aerosol–cloud–precipitation interactions (see Fig. 11.1 for a schematic explanation of the different hypotheses, details are provided later), the established and more hypothetical interaction processes, and means to observe and understand these interactions and their consequences for climate.

Doing so is largely based on the established theory and concepts such as the cloud microphysical processes presented in Chapter 3 and the various impacts of clouds on radiation that are discussed in Chapter 4. The chapter's special focus is on the extent anthropogenic aerosol and aerosol precursor emissions affect climate. It first discusses the aerosol, its chemical and physical characteristics, its sources, transformations and sinks.

This is followed by a description of the interaction of aerosols with radiation and, more extensively, with clouds and the implied alteration of the radiation budget. Subsequently, the role of aerosols and their radiative forcing for large-scale climate phenomena, for air quality and climate change are discussed. The last part introduces the challenges in observing the perturbations in clouds and radiation introduced by anthropogenic aerosol, and the challenges in modelling aerosol–cloud–radiation interactions.

11.1 Aerosol

To fully describe liquid cloud and precipitation particles, it is usually sufficient to know the size distribution. In addition, aerosol particles may vary in density and consist of various chemical compounds (see Section 11.1.3). These diverse physical and chemical aspects are relevant for their lifecycle and their interaction with radiation, clouds and precipitation. Atmospheric aerosols may stem from natural or anthropogenic emissions. Their lifecycle is governed by emissions, chemical transformation, microphysical interaction of particles, and sedimentation or wet scavenging.

11.1.1 Chemical Characterisation and Emission

There is a large variety of chemical species forming aerosols in the atmosphere. The most common constituents are mineral dust, sea salt (NaCl), sulphate (SO_4), black carbon, organic carbon and nitrate (NO_3). The chemical differences lead to differences in physical properties of the particles. For example, sea salt, sulphate, nitrate and organic carbon are more hydrophilic than dust or black carbon; the latter, in turn, absorb solar radiation which the other particles do only to a much lesser extent.

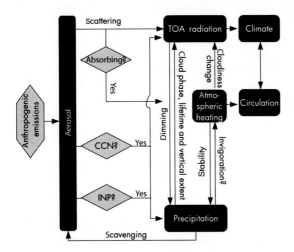

FIGURE 11.1: Interactions in the climate system induced by anthropogenic aerosol and aerosol precursor emissions. CCN, cloud condensation nuclei; INP, ice nucleating particles; TOA, top-of-the-atmosphere radiation.

The particles also differ in their size. Sulphate and nitrate particles are typically much smaller than most sea salt and dust particles.

Some aerosols are emitted directly as particles; these are called primary particles. Of these, mineral dust is lifted from soils by wind erosion, and sea salt is dispersed from the sea surface. Black carbon is formed by incomplete combustion of fossil fuels, biofuels and wild fires. Organic carbon is emitted by a variety of processes, including combustion of fossil fuels and biofuels, also in wild fires, and emission from the biosphere such as pollen, bacteria and vegetation debris.

Other aerosols are formed by nucleation from precursors emitted in the gas phase; these are called secondary aerosols. Volatile gases transform into compounds that may cluster and thus can nucleate to particles in the atmosphere. Nitrate forms from precursor gases emitted from fertilisers and combustion, but also from oxidation of molecular nitrogen by lightning. Some of the organic carbon is formed as secondary organic carbon from precursor gases emitted, for example, by the vegetation.

In the remainder of this section, some aspects of sulphur chemistry relevant for the atmosphere are explained, in order to schematically illustrate the relevance of atmospheric chemistry for aerosol science, and to show how the in-cloud production of sulphate particles is an important mechanism of aerosol–cloud interactions. Sulphate aerosols originate from gaseous sulphur dioxide (SO_2) emitted by the combustion of fossil fuels and biofuels and – to a lesser extent – by biomass burning and

volcanoes, and from oxidation products originating from dimethylsulfide emitted by ocean plankton.

SO_2 is oxidised in the atmosphere in the presence of another, arbitrary, molecule M that takes up excess energy, by the hydroxyl radical, OH, to form $HOSO_2$,

$$SO_2 + OH + M \rightarrow HOSO_2 + M, \tag{11.1}$$

which further oxidises to HO_2 and SO_3

$$HOSO_2 + O_2 \rightarrow HO_2 + SO_3. \tag{11.2}$$

This then consumes water to form gaseous-phase H_2SO_4, which can nucleate to form liquid sulphuric acid

$$SO_3 + H_2O + M \rightarrow H_2SO_4 + M. \tag{11.3}$$

There is an alternative, much more efficient pathway of oxidation of sulphur dioxide, namely within cloud droplets. The concentration of dissolved sulphur dioxide in the aqueous phase is determined from its partial pressure in the gas phase according to Henry's law that describes the equilibrium between the gas phase and dissolved molecules:

$$SO_2(g) + H_2O \rightleftharpoons SO_2 \cdot H_2O. \tag{11.4}$$

SO_2 is oxidised to hydrogen sulphite:

$$SO_2 \cdot H_2O \rightleftharpoons H^+ + HSO_3^- \tag{11.5}$$

$$HSO_3 \rightleftharpoons H^+ + SO_3^- \tag{11.6}$$

and then SO_3 further oxidises consuming ozone

$$SO_3 + O_3 \rightarrow SO_4 + O_2. \tag{11.7}$$

(Similarly, hydrogen peroxide, H_2O_2 may react with SO_3 to form SO_4 and water.)

In either way, much of the SO_4 is dissolved in cloud water, and, subsequently, in rain water. At high levels of sulphur emissions, the rain thus becomes acid, a problem that has received much attention in Europe and North America in the 1980s due to the acidification of soils and detrimental impacts on trees and plants. Besides the aim of improving air quality, acid rain was a main driver towards a legislation to reduce sulphur emissions. In present-day climate, more than two-thirds of the sulphate aerosol is formed by this aqueous pathway. One of the pathways of aerosol–cloud interactions thus is the efficient production of aerosol particles by nucleation from the gas phase via aqueous chemistry within clouds.

11.1.2 Anthropogenic Emissions

Some of the primary aerosols and aerosol precursor gases are emitted by anthropogenic sources or at least influenced

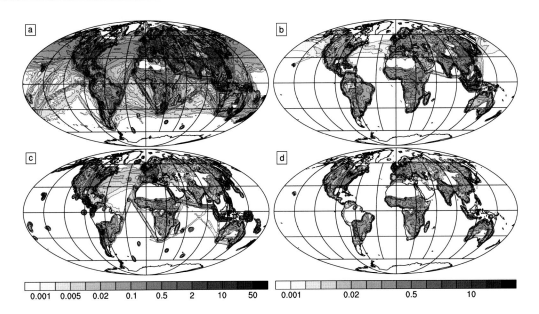

FIGURE 11.2: (a, c) Present-day (year 2005) and (c, d) pre-industrial (year 1850) distributions of the emissions of sulphur dioxide (a, c; in g(S) m^{-2} year^{-1}) and of black carbon (b, d; in g(C) m^{-2} year^{-1}) as imposed in Phase 5 of the Coupled Model Intercomparison Project (Lamarque et al., 2010), multi-model average, ten-year average.

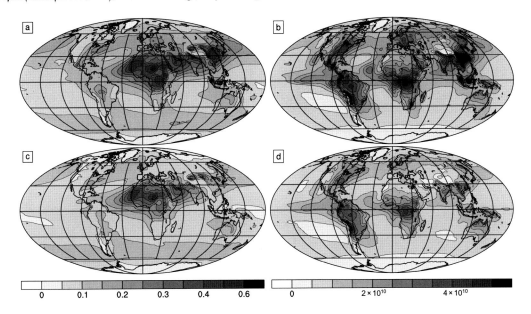

FIGURE 11.3: As Fig. 11.2, but for (a, c) aerosol optical depth and (b, d) column-integrated cloud droplet number concentration [m^{-2}].

by them (see Fig. 11.2; for a distribution of emissions for two aerosol species). It is estimated that about 40 % of today's aerosols are of anthropogenic origin (Bellouin et al., 2013). Particles forming from combustion (sulphate, black carbon, organic carbon) are to a large part anthropogenic, namely due to the combustion of fossil fuels and biofuels (with a minor contribution from wild fires).

Nitrate from fertilisers and formed from nitrogen oxide emissions is also anthropogenic. Part of the dust is anthropogenic since it is blown from roads and agricultural areas, and from deserts existing due to anthropogenic activities. The model-simulated distribution of aerosol optical depth, τ_a – the column-integral of the extinction due to the aerosol, see Chapter 4 – is shown in Fig. 11.3, for

TABLE 11.1: Size ranges for typical aerosol modes or size classes. 'PM' stands for 'particulate matter' and the number thereafter for the upper limit in diameter (in μm) considered for the specific mass. All particles smaller than 1 μm in diameter are sometimes summarised as fine mode particles.

Nucleation mode	Aitken mode	Accumulation mode	Coarse mode	PM2.5	PM10
Radii	<0.5 nm	5–50 nm	50–500 nm	<1.25 μm	<5 μm

pre-industrial and present-day aerosols. While large aerosol optical depths due to natural emissions of dust and sea salt are similar (especially at low latitudes and in the southern hemisphere, comparing the remote oceanic regions as well as the desert regions in Figs. 11.3a and 11.3c), substantial differences are simulated especially for the mid latitudes of the northern hemisphere.

The relatively short lifetime of aerosols (see Section 11.1.5) that allows for transport over horizontal distances of less than a few thousand kilometres, leads to a very heterogeneous distribution of the aerosol, with large concentrations mainly close to the main source areas (Fig. 11.3).

11.1.3 Physical Characterisation and Particle Interaction

Individual aerosol particles may consist of single chemical compounds. A population of different aerosols is then described as externally mixed; in this approximation, an air parcel contains different particles each of which has a pure chemical composition. A more likely case is that individual particles consist of several constituents. The mixing state of the aerosol is then called internally mixed. Internal mixing can happen by co-emission of different compounds and collision-coalescence of different aerosols (see Section 3.2.5, for a description on the physics of this process, which essentially is the same as for cloud particles). It can also happen as low-volatility gases such as H_2SO_4 condense onto a pre-existing aerosol (cf. condensational growth of cloud droplets as described in Section 3.2.3). Also, water vapour may condense or be deposited onto aerosol particles (see later). Such processes occur regularly in the atmosphere. They are called 'ageing' of particles in the atmosphere and occur on timescales of hours to days and change the state from a pure chemical composition to an internally mixed one. Both condensation of vapours (including water vapour) onto particles and collision-coalescence processes lead to the growth of particles towards larger sizes.

As for cloud particles, a fundamental physical characterisation of the aerosol particles within a volume of air is their size distribution. Observations show that for aerosols, log-normal size distributions often are a good approximation. Depending on the application, a number, surface or volume size distribution might be considered. The log-normal shape of the distribution implies that the distributions are very different when considering zeroth (number), second (surface) or third (mass) moments of the size distributions. Small particles usually heavily contribute to the number and little to the mass, and the opposite is true for large particles. Production of new particles from the gas phase occurs at very large number concentrations of particles but the nucleation mode contributes little to the overall mass. Collision-coalescence then reduces the particle number concentrations and leads to larger particles. Often, a superposition of two or more modes is observed. In the Aitken mode, particle radii typically range from 5 to 50 nm, in the accumulation mode, from 50 to 500 nm and particles with radii larger than about 5 μm are in the coarse mode (Stier et al., 2005). Sometimes, a simplified size categorisation is applied with a fine mode for particles with radii less than 0.5 μm and a 'coarse mode' for larger particles. For air quality applications, often PM2.5 and PM10 are measured, as the mass of particles with diameters smaller than 2.5 and 10 μm, respectively. A summary of the names assigned to typical modes is provided in Table 11.1.

11.1.4 Aerosol Sinks

Two ways of deposition of aerosol to the Earth's surface can be distinguished, namely dry and wet deposition. Aerosols settle in the air due to gravitation. This sedimentation is described analogously to the fall velocity of spherical cloud droplets in the Stokes regime (valid for air densities within the troposphere and particle sizes smaller than 30 μm in radius; Section 3.2.4). Thus, the terminal fall velocity, u_∞ (Eq. (3.28)) is proportional to the square of the particle radius, r_a

$$u_\infty \sim r_a^2.$$ (11.8)

Since aerosol particles typically are small compared to cloud and precipitation particles, so are the typical fall velocities. When particles reach the vicinity of the Earth's surface, they are intercepted by direct impact or by diffusion. Aerosols are taken up by cloud particles when serving as cloud condensation and ice-forming nuclei (Sections 11.3 and 11.4), or by collision and coalescence of cloud particles with the aerosol particles within the cloudy air surrounding the cloud particles, which are called interstitial aerosol particles. When precipitation is formed and reaches the surface, these aerosol particles are deposited. Also, precipitation particles collide and coalesce with aerosols, scavenging the aerosol along their way through the atmosphere. The washout of aerosols by precipitation by these two means is called wet deposition. Where present, wet scavenging is much more efficient than dry deposition due to the much higher fall velocity of precipitation particles. This is one of the most important pathways of aerosol–cloud–precipitation interactions.

Aerosols inside cloud and precipitation particles may be released back to the atmosphere in case these evaporate or sublimate before reaching the ground. Then, a relatively large, internally mixed aerosol particle is released which makes it particularly suitable as a CCN. This process is a sink to the aerosol number concentration, but leaves the mass concentration unchanged.

11.1.5 Aerosol Lifetime and Distribution

Typical lifetimes of aerosol mass, the time between emission (primary) or formation (secondary particles) and dry or wet deposition, are on the order of 0.5–7 days in the troposphere. For a typical horizontal wind speed in the lower troposphere of about $10 \, \mathrm{m \, s^{-1}}$, this translates to distances of 400–6,000 km over which the aerosols are transported.

Consequently, aerosols are often concentrated rather close to their sources and distributed heterogeneously (see also Fig. 11.3). Considering that aerosol sources are mostly at the surface, usually a vertical profile is observed which is well mixed in the atmospheric boundary layer and decreases exponentially above it unless mixing by deep convection, or nucleation of aerosol in the upper troposphere, is present in which case concentrations may even increase with height. Very low aerosol concentrations are found above the troposphere except after strong volcanic eruptions. In such cases, aerosols can penetrate into the stratosphere and have lifetimes of up to several years since wet deposition is negligible.

Aerosol emissions typically are larger over continents than over oceans (Fig. 11.2, see also Sections 11.1.1 and 11.1.2). Particularly large aerosol emissions are found over deserts (dust) and over polluted industrialised regions. For this reason, aerosol concentrations are particularly large in the northern hemisphere mid latitudes over the continents or downwind, close to them.

11.2 Aerosol–Radiation Interactions

Aerosol particles affect radiation in a similar, albeit more complex due to their more complex composition and shapes, way as cloud particles (cf. Chapter 4). They scatter sunlight, an effect that depends on the surface size distribution and aerosol number concentration (Fig. 11.1). Absorption of solar radiation by aerosols can be much larger than is the case for cloud particles of the same size. The single scattering albedo in the visible part of the spectrum (c.f. Eq. (4.2)), for example, of black carbon can be as low as 0.4. Usually, absorption and emission of terrestrial radiation by aerosol particles is relatively small, but non-negligible for large dust particles (Fig. 11.4). The overall direct effect of aerosols on radiation is typically a net cooling at the top of the atmosphere, a warming within the atmosphere, and a stronger cooling at the surface resulting from the difference of the cooling at the top of

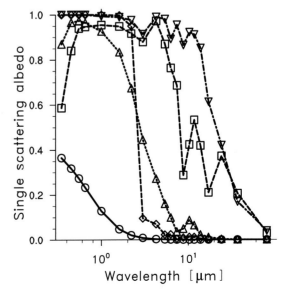

FIGURE 11.4: Spectrum of single scattering albedo for five aerosol species (circles/straight line, black carbon; triangles up/dotted, organic carbon; diamonds/dot-dot-dashed, sulphate; squares/dashed, dust; triangles down/dot-dashed, sea salt). Data from Kinne et al. (2013)

the atmosphere and the warming in the atmosphere. The warming of the atmosphere by absorption of sunlight may have consequences for atmospheric stability, potentially affecting cloud occurrence. Because the predominant heat balance in the atmosphere is between net condensational heating and the radiative flux divergence, a reduction in the latter associated with aerosol absorption can influence the globally averaged precipitation, not just its distribution (Ming et al., 2010). The reduction in solar radiation at the surface for larger aerosol concentrations and thus more scattering and absorption of sunlight is called solar dimming (Wild et al., 2005). This can lead to a reduced surface evaporation rate (Liepert et al., 2004) which in turn reduces precipitation rates (see Fig. 11.1). Reductions in aerosol emissions, in turn, lead to the opposite trend, a solar brightening.

11.3 Aerosols acting Acting as Cloud Condensation Nuclei

Water vapour condenses onto aerosol particles. In thermodynamic equilibrium, that is, assuming that chemical potential and temperatures are equal in the gas and liquid phases and that the pressure difference between the cloud particle and the vapour is balanced by the surface tension, the ratio of the saturation water vapour pressure above a solution droplet e and the saturation vapour pressure over a flat surface e_1 is obtained by the Köhler equation (Eq. (3.14); see also Fig. 3.4):

$$S = \frac{e}{e_1} = \exp\left(\frac{A(T)}{r} - \frac{B_s}{r^3}\right). \tag{11.9}$$

The left-hand side is called the saturation ratio, S. It needs to be compared with the relative humidity of an air parcel: for a certain relative humidity, one can infer e as a function of $A(T)$ and B_s at which a solution droplet is in a thermodynamic equilibrium. The first term on the right-hand side is called curvature term or Kelvin term and accounts for the fact that for small solution droplets, a large relative humidity of several hundred per cent is necessary for the formation of stable cloud droplets. The Kelvin term shows that the saturation ratio increases exponentially as r decreases. $A(T)$ depends mainly on the temperature T. The second term is called solution term or Raoult term. It takes into account the aerosol mass and chemical composition. The term B_s can be specified in terms of empirical parameters and depends on the effectiveness of an aerosol to be humidified, and on the aerosol mass. The denominator r^3 shows that the term decreases as the solution droplet mass increases. The

dependency of B_s on aerosol mass (Eq. (3.13)), or the cube of its dry radius, shows that the size is particularly important. Solubility or suitability for humidification also is important, but some humidification is possible for most internally mixed aerosols.

Evaluation of Eq. (11.9) shows that for small r, the saturation ratio is smaller than 1. This means that at relative humidities below saturation, humidified aerosols stably exist in thermodynamic equilibrium, and the size grows strongly as relative humidity increases (see also Fig. 3.3). This effect becomes easily visible when watching a hazy atmosphere in the vicinity of clouds: the haze gets thicker towards the clouds, where relative humidity is larger. This swelling of aerosols can be so large that it is difficult to distinguish haze and cloud. One manifestation of this swelling is that statistics show aerosol optical depth increasing strongly as a function of cloud fraction or even precipitation intensity, because all these quantities are strongly dependent on humidity. The very large spatial variability of relative humidity (discussed in the context of cloud parameterisations in Section 6.6.2) thus is reflected in the spatial variability of haze. While not actually a process of aerosol cloud-precipitation interactions, the influence of relative humidity is an important effect for aerosols, clouds and precipitation.

One can compute the maximum from Eq. (11.9) for a given temperature, aerosol mass and chemical composition to obtain the 'critical' radius (see Chapter 3 for more details, and note that no cluster formation or phase change is involved, in contrast to the nucleation process) r_c and the critical saturation ratio above which a solution droplet is not in thermodynamic equilibrium anymore but freely grows by condensation of water vapour up to cloud droplet size. Such an aerosol is called 'activated'.

In the atmosphere, typical saturation ratios in liquid-water clouds are on the order of 0.1–1% above water saturation. Whether the critical saturation ratio for a given aerosol particle is reached depends, in adiabatic ascent, on the cooling rate that is generated by the updraught, considering the depletion of water vapour by condensation. For a given updraft, all aerosols that are suitable chemically and that have dry sizes large enough will activate into cloud droplets. It follows that in a perturbed atmosphere with more suitable condensation nuclei available, the peak supersaturation is reduced, in turn limiting the amount of activated particles. It also implies that in vigorous convective clouds with updraught speeds increasing above cloud base, a considerable fraction of the aerosol is activated not at the cloud base, but aloft inside the cloud.

Note that growth by vapour deposition is proportional to the square root of the particle size (Section 3.2.3), and thus larger particles need a longer time to grow than the smaller particles. Thus, during short supersaturation peaks, the larger particles may not have time to grow beyond r_c. The condensation rate is larger when more aerosol particles activate, reducing peak supersaturation. This competition effect limits the number of cloud condensation nuclei (CCN) that are activated. Still, the more suitable CCN are available, the larger the cloud droplet number concentration (N_d). This relationship can be approximated by a logarithmic increase in N_d with increasing CCN.

Fig. 11.3 shows the model-simulated response of column-integrated cloud droplet number concentration to anthropogenic aerosol emissions, which largely increases in regions in which anthropogenic emissions lead to larger aerosol concentrations. This is the most obvious effect of aerosols on clouds and thus an important aspect of the aerosol–cloud–precipitation interactions.

11.4 Aerosol Effects on Ice Crystal Formation and Mixed-Phase Clouds

Cloud droplets freeze readily only once a certain temperature threshold below $0\,°C$ is reached because the crystalline structure has to be built. Below about $-45\,°C$ in theory, and about $-38\,°C$ in the real atmosphere, this homogeneous freezing can occur. The term homogeneous freezing refers to the fact that no foreign particles are involved in the process. At such low temperatures, water vapour may also directly deposit onto suitable dry aerosol particles to form ice crystals (deposition nucleation). Haze droplets may freeze homogeneously as well, at temperatures below the homogeneous freezing threshold. If aerosol concentrations are perturbed, ice crystal perturbations can also be expected in case of homogeneous freezing, although the sensitivity is found not to be very large.

At temperatures between $0\,°C$ and $-38\,°C$, the ice formation in the atmosphere occurs via heterogeneous nucleation (Section 3.3.2) on ice nucleating particles (INPs), which are solid or crystalline aerosol particles. Mineral dust particles have been identified as good INPs that probably dominate ice formation in the atmosphere. While the ice nucleating ability of soot and organics remains controversial, bio-aerosols such as pollen, fungal spores and bacteria have been found to activate ice at the highest temperatures. To date, only a few studies comparing the freezing abilities of different pollen grains are available and it is not yet clear if the number concentrations of bio-aerosols is sufficiently large to noticeably affect ice nucleation.

In special situations in mixed-phase clouds, ice multiplication may occur, producing dozens of small ice splinters per milligram of ice. This is called the Hallett–Mossop ice multiplication process and when it occurs, it dominates the ice crystal number concentration. It happens in a temperature range of about $-8\,°C$ to $-3\,°C$, for situations where riming occurs, in this case the collision coalescence of ice particles (crystals or snow flakes) with numerous cloud droplets or drizzle drops. Graupel particles generated this way may break up into splinters.

Because the saturation vapour pressure over ice is lower than over liquid water, ice crystals in mixed-phase clouds can grow at the expense of cloud droplets if the actual vapour pressure is at or below water saturation. This process is called Wegener–Bergeron–Findeisen process and can cause a supercooled or mixed-phase cloud to rapidly glaciate and form precipitation (Chapter 3). Therefore, an increase in INP, for instance from anthropogenic activity, can lead to a glaciation of clouds at a higher temperature, more precipitation, a reduced cloud cover and hence less reflection of solar radiation from these clouds (Lohmann, 2002). An opposite effect is obtained for situations where soluble material condenses onto suitable INPs and limits their ice nucleating capabilities (Kärcher and Lohmann, 2003) – something referred to as the 'deactivation' of INPs. In this case, the temperature at which freezing is initiated is shifted to lower temperatures.

11.5 Aerosol Effects on Collision and Coalescence

The fundamentally positive relationship between aerosol concentrations and cloud particle concentrations at least for liquid-water clouds has been discussed earlier (Section 11.3). An effective process by which cloud particles grow is by collision and coalescence. In an air parcel, the larger particles of the size spectrum fall faster than the smaller ones, with the terminal fall velocity proportional to the square of the particle size for small particles ($\lesssim 30\,\mu m$; Stokes regime) and proportional to the square root of the size for large particles ($\gtrsim 300\,\mu m$; turbulent regime; see Chapter 3). This velocity difference enables collisions and potentially coalescence (Section 3.2.5). The collision rate further depends quadratically on the particle size since this determines the surface area on which a collision can occur. So, the overall coalescence rate is a strongly non-linear function of the particle size, with an exponent of about five for cloud droplets. A shift in the cloud droplet size spectrum thus strongly alters the collision and coalescence efficiency, reducing it if the entire cloud droplet size distribution shifts to smaller sizes or becomes narrower. This could happen due to anthropogenic aerosol particles. In this case, the transformation of cloud to precipitation water is slowed down, and cloud lifetime, geometric thickness and water path may be enhanced in the absence of compensating effects (Fig. 11.1). Please note that depending on cloud type, precipitation may not be the dominating sink, but the cloud dissolves due to mixing (addressed in the following section).

In this context, it is important to note that on a global average, the water budget is determined by the balance between evaporation and precipitation at the surface. Thus, as long as evaporation does not change, accumulated precipitation does not change either. The potential effects of aerosols on collision and coalescence and thus precipitation formation rate thus may alter precipitation characteristics (frequency of occurrence, timing and intensity), and importantly also cloud properties, but not the overall large-scale precipitation amount unless via a cooling of the surface, which in fact might be the dominant climate mechanism of aerosol particles.

11.6 Aerosol Effects on Cloud Dynamics

A shift in the droplet size distribution can feed back onto the cloud dynamics. In this context, we consider air motions as 'dynamics' (Fig. 11.1). The evaporation rate of cloud droplets is dependent on their surface area (Chapter 3). If by additional aerosol particles, cloud droplets are smaller, the evaporation rate thus is disproportionally larger. The cooling by evaporation at cloud edges enhances turbulent mixing. As a result, cloud-top and lateral entrainment may be enhanced, diluting the cloud and further reducing its liquid-water path (Ackerman et al., 2004).

In convective clouds which grow beyond the freezing level, an 'invigoration' may take place (Rosenfeld et al., 2008). If by additional aerosol particles, cloud droplets are smaller, the freezing is delayed since it depends on the droplet size. Thus, the smaller cloud droplets can continue to grow, accumulating more liquid water that freezes. Also, if the precipitation rate is reduced in the liquid phase, even more water freezes. In consequence, the water mass that freezes is enhanced, and more latent heat of freezing is released. A substantial part of this additional buoyancy due to freezing is offset by less buoyancy due to more lofted liquid water. The net effect may lead to an increase in positive buoyancy of the air parcel. The resulting precipitation may ultimately be more intense than in an unperturbed case. Note, however, that possible aerosol effects on convective clouds are a topic of active research (Tao et al., 2012), and that the large-scale precipitation amount is governed by the radiation budget, as noted earlier.

11.7 Radiative Effects of Altered Cloud Droplet Size

Cloud albedo is a strictly monotonic function of cloud optical depth (τ_c, Chapter 4). For liquid clouds, τ_c can be written in terms of cloud liquid-water path W (Eq. (3.5)), and cloud droplet effective radius r_{eff}, for a cloud layer in which the droplet size distribution is vertically and horizontally constant (see Eqs (3.3) and (3.4)):

$$\tau_c = \frac{3}{2} \frac{W}{\rho_w r_{eff}}, \tag{11.10}$$

with ρ_w the density of liquid water. Depending on the dispersion of the droplet size distribution, the effective radius often simply scales with the volume-mean radius, $r_{eff} = \beta r_{vol}$. From aircraft observations, the scaling factor β is determined at approximately $\beta = 1.1$–1.5 (Martin et al., 1994; Pawlowska et al., 2006; Freud and Rosenfeld, 2012), depending on the cloud regime. With the cloud droplet number concentration N_d the effective radius is then inversely proportional to the cube root of N_d:

$$r_{eff} = \beta \sqrt[3]{\frac{3}{4\pi} \frac{q_l}{\rho_w} \frac{1}{N_d}}. \tag{11.11}$$

In turn, τ_c is (at given W) proportional to the cube root of the droplet number concentration,

$$\tau_c \sim \sqrt[3]{N_d} \tag{11.12}$$

In a two-stream approximation, a relative change in cloud albedo is proportional to a relative change in optical depth

$$\Delta \ln \alpha_c = (1 - \alpha_c) \Delta \ln \tau_c. \tag{11.13}$$

For a constant droplet spectrum shape and a constant liquid-water path, cloud albedo is thus strictly monotonically increasing as the droplet number concentration increases:

$$\frac{\partial \ln \alpha_c}{\partial \ln N_d} = \frac{1 - \alpha_c}{3} \tag{11.14}$$

This shows that an increase in cloud droplet concentrations leads to an increase in cloud albedo for constant or increasing W, unless α_c already is 1. Cloud droplet number concentration is perturbed, as described in Section 11.3, as the number concentration of aerosols suitable as CCN is perturbed. Ice crystal number concentration also may change, but in a less clear manner (Section 11.4). Such changes of cloud particle concentrations due to anthropogenic aerosols, which subsequently alter cloud albedo, lead to the radiative forcing due to aerosol–cloud interactions (RF$_{ACI}$). They previously have been termed 'first aerosol indirect effect' in the literature (or 'Twomey' effect; Twomey, 1974), or also 'cloud albedo effect'. Changes in microphysical (Section 11.5) and dynamical (Section 11.6) processes that in turn alter cloud water paths and cloud cover are microphysical adjustments to the RF$_{ACI}$. Along with the microphysical adjustments, thermodynamic profiles are also adjusted. The sum of the RF$_{ACI}$ and the microphysical, dynamical and thermodynamic adjustments is called the effective radiative forcing due to aerosol–cloud interactions (ERF$_{ACI}$). Previously, the microphysical adjustments have been called 'second aerosol indirect effect' (and sometimes individual hypotheses are also referred to, such as a 'cloud lifetime effect' or an 'invigoration effect').

11.8 Effective Radiative Forcing Due to Aerosol–Cloud Interactions

An external perturbation to the energy balance of the climate system is called a 'radiative forcing' (see also Chapter 13). Anthropogenically emitted aerosols and aerosol precursor gases can be considered external to the climate system and thus, modifications of the energy balance of the Earth system, that is, of the top-of-the-atmosphere net radiation, caused by aerosols can be considered a radiative forcing, or rather an 'effective' radiative forcing since rapid adjustments, or feedbacks that modify the initial top-of-the-atmosphere radiation budget change, also take place. The effective radiative forcing due to aerosol–cloud interactions ΔF_{aci} can be defined in terms of the net top-of-the-atmosphere radiation R, the cloud particle number concentrations N_c (which is true for both liquid and ice particles so a generic 'N_c' is used here), and a measure of the CCN and INP concentrations in terms of aerosol α_a as well as the anthropogenic perturbation of α_a, $\Delta \alpha_{a,ant}$:

$$\Delta F_{aci} = \frac{dR}{d \ln N_c} \frac{d \ln N_c}{d \ln \alpha_a} \Delta \ln \alpha_{a,ant}. \qquad (11.15)$$

The total change in R with relative change in N_c can be expanded by its contributions due to changes in N_c in a partial derivative sense, and via changes in cloud liquid- and ice-water path W, cloud cover f and cloud-top temperature T_{top}:

$$\frac{dR}{d \ln N_c} = \frac{\partial R}{\partial \ln N_c} + \frac{\partial R}{\partial f} \frac{df}{d \ln N_c} + \frac{\partial R}{\partial W} \frac{dW}{d \ln N_c}$$
$$+ \frac{\partial R}{\partial T_{top}} \frac{dT_{top}}{d \ln N_c}. \qquad (11.16)$$

In such a notation, the radiative forcing due to aerosol–cloud interactions refers to the contribution of the first term on the right-hand side of Eq. (11.16) to the forcing (Eq. (11.15)). Changes in the droplet size spectrum dispersion are included in this term as well. The physics underlying this first term are described in Section 11.3 for liquid, and in Section 11.4 for ice clouds. The adjustment effects, in turn, are composed of the contributions by the other three terms to the forcing in Eq. (11.16). Specifically, changes in cloud cover, cloud water path and cloud-top temperature occur due to alterations of cloud microphysical processes (Section 11.5), but also in response to alterations of cloud dynamics (Section 11.6).

The top-of-the-atmosphere net radiation usually is further decomposed (Chapter 4) into solar and terrestrial spectral ranges. The largest effect by the anthropogenic aerosol affects the solar spectrum via an alteration of planetary albedo due to clouds. However, the cloud greenhouse effect in the terrestrial spectrum may also change, and the change in cloud-top temperature just affects the terrestrial part of the spectrum.

To date, the radiative forcing by aerosol–cloud interactions, let alone the effect by the adjustments, is not well constrained. Model simulation results span a very large range from zero to more than offsetting the current anthropogenic greenhouse effect. A current estimate (Boucher et al., 2013) is that the overall effective radiative forcing due to anthropogenic aerosol offsets about one third of the 2011 versus 1750 radiative forcing due to anthropogenic well-mixed greenhouse gases of 3 Wm^{-2}. Climate models suggest that the radiative forcing by aerosol–cloud interactions is approximately doubled by rapid adjustments (Lohmann and Feichter, 2005), although in observations the adjustments are not statistically significantly detectable given the large cloud variability (Quaas, 2015). This is, besides the uncertainty of the contribution of cloud-climate feedbacks to climate sensitivity, another large uncertainty for the quantification of anthropogenic climate change. One can relate ΔF_{aci} and climate sensitivity given the observed global warming over the last century, so that a reliable quantification of either quantity also constrains the other (Schwartz, 2012). Such a relation, however, also has to account for the ocean heat uptake and its variability. A good quantification of climate sensitivity thus allows us to infer at a global scale the total aerosol radiative forcing, and, vice versa, a quantification of the transient aerosol forcing to quantify transient climate sensitivity. In addition, qualitative constraints that consider the fact that the signs of total forcing and temperature change need to agree on global and also on hemispheric scales, allow to infer clues about the aerosol forcing (Stevens, 2015).

11.9 Aerosol-Cloud Climate Effects in the Context of Global Climate Change

11.9.1 Impacts of Aerosol-Cloud Effects on Large-Scale Climate Dynamics

As discussed in Chapters 8, 9 and 13, global warming due to the anthropogenic greenhouse effect impacts the general circulation. An even more complex impact of the aerosol forcing on many aspects of global climate dynamics is expected due to the heterogeneous spatial distribution of the forcing.

Based on the results of sensitivity studies with general circulation models, it has been hypothesised that the aerosol indirect radiative forcing above the North Atlantic Ocean might have had pronounced impacts on the general circulation. Specifically, it has been proposed that the declining precipitation trend over the Sahel region during the second half of the twentieth century was caused by this feature (Rotstayn and Lohmann, 2002). The aerosol indirect forcing above the North Atlantic Ocean, according to other model-based studies, might also have driven variability of the North Atlantic Oscillation (Booth et al., 2012), and potentially also the variability of hurricane activity (Dunstone et al., 2013), over much of the past century.

Since anthropogenic aerosol loads are particularly high over East Asia, variability and trends in monsoon precipitation have been linked to the rising aerosol forcing in this region. It has been proposed that absorption of sunlight above the Tibetan plateau might have enhanced the monsoon circulation. Satellite-based quantification, however, does not support this hypothesis (Kuhlmann and Quaas, 2010). Also, the cooling of the surface, and even the cloud microphysical effects (Section 11.5) might have affected long-term trends in precipitation. The strong hemispheric contrast in particular in the Tropical Indian ocean might have contributed to circulation changes impacting the Indian monsoon (Bollasina et al., 2011).

These effects are, however, still debated in the literature. They underline the potentially large impact of anthropogenic aerosols on climate.

11.9.2 Climate Feedbacks Involving Aerosols

A climate feedback is usually defined as a process that changes as the (global mean) surface temperature changes and that itself affects the surface temperature via an alteration of the Earth's energy budget (see Chapter 13). As seen earlier (Sections 11.2 and 11.7), aerosols do affect radiation. Some aerosol and aerosol precursor emissions are temperature- and wind-speed-dependent, namely, sea salt and dimethyl-sulfide emitted by ocean plankton and precursor gas to sulphate, and volatile organic compounds, precursor gases to organic aerosols (Charlson et al., 1987; Kulmala et al., 2004). However, current quantitative understanding suggests these feedbacks are small compared to the 'classical' water vapour, surface albedo and cloud feedbacks. On longer timescales such as interglacial transitions, aerosol-climate feedbacks might be more important. Since precipitation co-varies with temperature, so does aerosol wet scavenging. In a colder climate, aerosol lifetime is prolonged and aerosol radiative effects may contribute to a further cooling.

11.9.3 Geo-engineering Exploiting Aerosol-Cloud Effects?

'Geo-engineering' or 'climate engineering' is the consideration to artificially cool the global climate (Royal Society, 2009). Some means are proposed by which carbon dioxide is removed from the atmosphere and stored (carbon capture and sequestration), other proposals suggest increasing emitted terrestrial radiation (e.g., by aiming at a reduction in cirrus cloudiness). A relatively cheap and effective way might seem to be solar radiation management. Since some of the radiative effects of aerosol–cloud interactions reduce solar radiation, an intended increase in tropospheric or stratospheric aerosol concentrations would likely counterbalance part of the anthropogenic greenhouse effect. It thus has been proposed as one mechanism[1] that involves aerosol–cloud interactions, to seed marine boundary layer clouds with sea salt. However, it is unclear to date whether such a 'geo-engineering' could be effective and what the climatic as well as socio-economic side effects of such a perturbation would be. Also, only part of anthropogenic climate change could be balanced since geo-engineering via solar radiation management, such as the seeding of marine boundary-layer clouds, is only effective during

---

[1] Other mechanisms have been proposed, too, such as the seeding of the stratosphere with aerosols, e.g., sulphate.

daytime, and since precipitation, for example, is affected differently by solar aerosol and terrestrial greenhouse gas forcings.

As explained in Sections 11.9.1 and 11.9.2, various complex and large-scale consequences may occur in response to intentional aerosol perturbations that need to be understood before geo-engineering might be considered.

11.10 Reducing Short-Lived Pollutants and Improving Air Quality?

Some aerosols are detrimental to human health. In particular, fine-mode particles enter human lungs and can cause cardiovascular diseases and possibly cancer. Clean air with low aerosol concentrations is thus considered healthier and of higher air quality. This is a main reason for legislation that led to reductions in anthropogenic aerosol emissions. Sunlight-absorbing anthropogenic aerosols (see Section 11.2) exert a positive direct radiative forcing, notably black carbon, but also less absorbing organic carbon. Implementing policies that reduce black carbon might thus be effective at short timescales to reduce the anthropogenic global warming. At the same time, due to the detrimental health effects of soot, there might be a co-benefit for air quality as well. However, aged black carbon aerosol, and also organic carbon may serve as CCN and thus it is unclear whether the net effect on climate of reducing black carbon is substantial. Also, black carbon is usually co-emitted with other species so that the net effect of emission reduction policies lead to a positive change in the radiative budget (Stohl et al., 2015), because the radiative effect of the co-emitted species prevails.

11.11 Attempts to Observe the Anthropogenic Aerosol Indirect Effect

The principles of aerosol–cloud–precipitation interactions as outlined earlier mostly rely on theoretical considerations or statistical analysis from observational data and model results. There have been attempts, nevertheless, to directly observe the anthropogenic aerosol indirect effect. In the following, the attempts and the results obtained so far are outlined.

11.11.1 Ship Tracks and Contrails

'Ship tracks', the clouds behind ships as seen occasionally in satellite pictures (e.g., Fig. 11.5) when meteorolog-ical conditions are favourable, are evidence for clouds formed on or altered by pollution aerosols. The white lines are recent ship routes that are embedded either in a background stratus cloud deck or occur in otherwise very thin clouds or (rarely so) cloud-free air (Conover, 1966). Because the marine boundary layer has a high relative humidity, the additional water vapour from the ships can be discarded as the source for these clouds. Ship tracks can thus be regarded as evidence for cloudiness altered by anthropogenic aerosol particles via their role as CCN (Christensen et al., 2014). The high number concentrations of aerosol particles causes the ship track clouds to have higher cloud droplet number concentrations than their neighbouring clouds (case studies show about a factor of 5–10 increase). As a result the ship track clouds have a higher optical depth and thus cloud albedo than their neighbouring clouds (Goren and Rosenfeld, 2014). However, only about 1 % of all ships cause ship tracks because the marine boundary layer needs to meet certain characteristics for ship tracks to form: it needs to be shallow enough that the emissions from the ship are mixed within the cloud layer, it needs to be thermodynamically stable which in practice implies that ship tracks form mainly over cold ocean currents when the pre-existing cloud is thin enough for the ship tracks to become visible.

Similar to ship tracks, condensation trails or contrails are seen behind aircraft in favourable situations. However, in contrast to ship tracks, it is believed that the additional source of water vapour from the aircraft exhaust is mostly responsible for this phenomenon. Altering pre-existing thin cirrus, the aerosols from aircraft emissions lead to changes in ice crystal number concentrations.

Both ship tracks and contrails themselves have a negligible effect on climate due to the very small horizontal extent. Only if the trail spreads out in space and time, a relevant radiative forcing could be expected. When analysing cloud and radiation properties from satellite data along shipping routes in an otherwise pristine environment, and comparing these between the windward and leeward sides of the track, however, no statistically significant difference has been found.

11.11.2 Hemispheric Contrast

Comparing satellite data and also model simulations (e.g., Fig. 11.3) in the two hemispheres, a clear contrast is found in terms of aerosol loading, as manifest in the aerosol optical depth from satellite retrievals even above oceans, with much larger concentrations in the northern hemisphere

FIGURE 11.5: Ship tracks over the Atlantic ocean among stratus and stratocumulus clouds as well as fog off the coast of Portugal. A composite of three channels in the visible spectrum and and the high-resolution visible channel is shown for a range of 2–20% reflectivity (copyright (2017) EUMETSAT).

(Feng and Ramanathan, 2010). This is also true for the fine mode, less influenced by the dust loading which is also larger in the northern hemisphere where more of the deserts are located. The hemispherical difference reflects the fact that most aerosol sources, and specifically most anthropogenic aerosol sources, are located in the northern hemisphere. Also, for the cloud droplet effective radius retrieved from satellite data, a difference is found with on average smaller radii in the northern hemisphere (due to the mechanisms explained in Section 11.3). However, cloud optical depth is not larger in the northern hemisphere, when comparing the oceanic clouds, implying a larger cloud water path on average in the southern hemisphere. Thus, the differences between the hemispheres in clouds due to other reasons seem to be larger than the differences in aerosol particle effects on clouds.

11.11.3 Trends Following Air Pollution Mitigation Policies

Large trends in aerosol emissions and subsequently aerosol concentrations have been observed in various regions of the globe. Increasing aerosol loads have been accompanied by decreasing trends in surface solar radiation, an effect called 'solar dimming'. Over Europe, since about 1990, a decline in aerosols is observed, following air-quality-improvement legislation in Western Europe as well as cleaner energy and industrial production following the political changes in Eastern Europe (Cherian et al., 2014). Similar trends are found over North America. Subsequently, a 'solar brightening' is observed, but is statistically significant only for clear skies. Trends in cloudy skies are unclear. Aerosol trends in other parts of the worlds are currently either still increasing (e.g., over East Asia) or less pronounced.

11.11.4 Weekly Cycles

When analysing aerosol temporal variability over Europe and North America, a clear weekly cycle is found with less aerosol during the weekends and early in the week, and a buildup during the week with a maximum towards its end (Sanchez-Lorenzo et al., 2012). Satellite data also show a consistent weekly cycle in cloud droplet number concentrations. A seven-day period is hardly attributable to natural variability and thus an unambiguous weekly cycle is evidence for an anthropogenic perturbation. Other meteorological quantities, however, such as radiation, temperature or precipitation, show inconclusive cycles – some publications do hint at weekly cycles, others either do not support these results or call the statistical methods into questions. For quantities for which one distinct maximum and one distinct minimum is found, modelling shows that even this is difficult to attribute to an aerosol effect. Rather, a random maximum and minimum is occurring as a result from the natural variability. Since only seven instances are to be distinguished in the analysis, it is likely that one distinct maximum and minimum, respectively, occur and the significance of it must be tested thoroughly.

11.11.5 Low Signal-to-Noise Ratio

The difficulties in observing large-scale aerosol–cloud interaction effects are perhaps not surprising considering the large natural variability of clouds and their effect on radiation. Indeed, the many studies aiming at weather modification particularly in the twentieth century did not yield statistically significant results either. Also, the effect to be distinguished is small: the global mean solar cloud radiative effect is about $-50\,\mathrm{W\,m^{-2}}$, whereas the aerosol radiative forcing is on the order of $-1\,\mathrm{W\,m^{-2}}$, that is, only up to 2% of the average value.

11.12 Modelling of Aerosol–Cloud–Precipitation Interactions

Given the potentially large importance of aerosol–cloud–precipitation interactions in the context of global climate change (see Section 11.8), and provided that these interactions might be relevant also for various aspects of numerical weather prediction, it is increasingly desirable to include the relevant processes in climate and weather prediction models. In this section, we first outline how aerosol processes may be represented in atmospheric models, followed by a description of the cloud processes that need to be considered in models in order to characterise the effects of aerosols on clouds and precipitation. In the third subsection, we discuss how such parameterisations may be evaluated using observational data.

11.12.1 Aerosol Modelling

Depending on how much of the relevant aerosol characteristics are to be represented in the model, given computational constraints, a more or less comprehensive scheme

will be applied. Typically, a number of chemical constituents is represented, together with a representation of their mixing state. The next most important quantity is the aerosol particle size distribution. In parameterisations called 'bulk' schemes, only the specific mass concentrations of the individual components, often sulphate, sea salt, dust, black and organic carbon (Section 11.1.1) are simulated by mass continuity equations. Modal schemes simulate the size distribution by the representation of a superposition of individual modes. Note that similar concepts apply to the modelling of both cloud and aerosol particles. Usually, the shape of the mode distribution is assumed, such as a log-normal distribution. The standard deviation may also be fixed and then only the median radius of each mode needs to be calculated. Besides the advection term, the continuity equations include sources and sinks. The principal source of the aerosol is the emission, usually at the surface or at low height above it. For secondary aerosols, some atmospheric chemistry needs representation to quantify the source of aerosol from the emission and chemical transformation in the atmosphere, either as monthly mean oxidant fields or by coupling the aerosol module to an atmospheric chemistry module (see Section 11.1.1 for a description of the mechanisms). Interaction of aerosols among each other and between aerosol and water vapour as well as clouds (Section 11.3) leads to changes in the chemical composition of individual particles and in size distributions. Sinks are dry and wet deposition (Section 11.1.4).

When considering the modes, usually at least the 'soluble' aerosol, capable of serving as CCN, and the insoluble aerosol need to be distinguished. In medium-complexity aerosol modules (e.g., Stier et al., 2005), six chemical species are considered (see Section 11.1.1), and the four modes (nucleation, Aitken, accumulation and coarse modes, see Table 11.1) are resolved for both soluble and insoluble compounds. In the nucleation mode, only soluble aerosols are generated from the vapour phase. This yields seven modes times six species. For each of these modes, at least the specific masses of each chemical compound are predicted, taking into account sources, sinks and microphysical as well as chemical transformations, yielding forty-two equations. Assumptions such as the presence of only sulphate in the nucleation mode, or that some species are always hydrophilic, lead to a substantial reduction in equations. Seven additional equations are necessary for the mode-median particle number concentrations. Additional equations may become necessary if the – even simplified – air chemistry for secondary aerosols for sulphate (cf. Section 11.1.1), and possibly for nitrate and secondary organic aerosols, is considered.

In order to assess the influence by anthropogenic aerosols, it is usually necessary to conduct two model simulations, with and without anthropogenic emissions. In this regard, it is necessary to properly define the reference atmosphere. Usually, it is defined as equivalent to 1750 or 1850, but in terms of aerosols, there have already been sizeable anthropogenic emissions by this time. Besides, also the 'background' level of natural emissions is highly uncertain. Given that the radiative forcing typically is about logarithmic with the aerosol perturbation (Section 11.7), this is an important difference. From model simulations it has been inferred that for the aerosol–radiation interactions, about a third of the anthropogenic radiative forcing was already realised in 1850 (Bellouin et al., 2008).

11.12.2 Cloud Microphysical Modelling

For many aspects of cloud–climate interactions, it is sufficient to use a bulk cloud model that just considers cloud liquid and ice mass mixing ratios, and treats precipitation diagnostically, reaching the surface, within one model time step but allowing for evaporation of precipitation below the cloud base (see Chapter 6). Typically, a certain threshold in size is considered to separate cloud liquid and ice water from precipitation (rain and snow). For more details, please see Sections 6.5.6.1 (Eq. (6.84)) and 6.7.5. If aerosol–cloud interactions are to be taken into account, at least the cloud particle number concentrations need to be considered. Typically, this is done by introducing equations for these quantities and thus applying a 'two-moment' cloud microphysical scheme, which uses the zeroth and third moments of the number size distributions, namely total number and mass concentrations as the two physically most meaningful quantities.

In high-resolution models used, for example, for numerical weather prediction and increasingly also for climate projections, precipitation needs prognostic treatment, because of the small time steps where diagnostic precipitation is not justified anymore. Since densities, fall speeds and weather impact are very different for drizzle, rain, snow, graupel and hail, sometimes these species are treated separately. Comprehensive cloud microphysical schemes thus treat up to seven classes of hydrometeors, some with two moments. The even more sophisticated alternative is to resolve the size distributions of the cloud particles. This can be done by representing the size spectrum as the superposition of different modes with specified shape, with some of the shape parameters prognostic; or by discretising the size spectrum in discrete size bins with prognostic mass and possibly number of cloud particles in each size bin. In these schemes, each modal parameter or each bin of the spectrum needs one model equation. The interactions of the particles then are represented in the form of sink terms to one cloud particle (e.g., collision-coalescence would deplete cloud droplets) and a corresponding source to another hydrometeor (e.g., collision-coalescence would generate drizzle and rain drops). Aerosol–cloud interactions come into play as source terms for the cloud droplet and ice crystal number concentrations. In addition to the availability of CCN and INPs, the updraught velocities play a role (Section 11.3 for cloud droplets and Section 11.4 for ice crystals).

11.12.3 Evaluation of Parameterisations Using Observational Data

The parameterisations of the individual source and sink processes for aerosols and cloud microphysical quantities are usually derived from theory where possible (e.g., Köhler theory for the cloud droplet activation, Section 11.3). Where this is not possible because the principles are too complex or not well-enough understood, empirical formulations, derived from laboratory or *in situ* observations are applied (Lohmann et al., 2007). Lacking adequate observations, process-resolving model results may be used (Chapter 6). However, all such formulations are valid only at the process scale. While interactions of individual particles are relatively easily described, the behaviour and interactions of the entire ensemble of particles in the turbulent flow at all scales that are not resolved by the numerical model is very difficult to parameterise. It is thus necessary to further evaluate the validity of the parameterisations at the actual scale of the model resolution, which, in case of general circulation models, may be on the order of 100 km. The large subgrid-scale variability and non-linearity of processes may prohibit the simple extrapolation of process-scale formulations. Metrics, derived from large-scale data sets, are then useful for evaluation (cf. Chapter 7). Relevant metrics for the aerosol-cloud interactions for liquid-water clouds are, for example, the relative change in droplet number concentration with a relative change in column aerosol concentration (Quaas et al., 2009), $\Delta \ln N_d / \Delta \ln \alpha_a$. Often, satellite retrievals are the most useful data sources for such evaluation. In many cases, it is recommendable to apply satellite simulators for the comparison to such data (please see Sections 7.2.2 and 7.2.3 for more details).

To assess the adjustment effects, it further has proven very useful to investigate the frequency of occurrence of precipitation, or probability of precipitation, as a function of cloud optical depth or cloud water path (Suzuki et al., 2011). The quantities are provided by cloud or precipitation radar and co-located optical retrievals (cf. Chapter 4). This then further may be separated into polluted and pristine scenes and compared to the equivalent diagnostics from models.

As discussed in Section 11.3, it is difficult to use the retrieved aerosol optical depth when trying to interpret statistical relationships between cloud and aerosol quantities, since it is strongly influenced by relative humidity and its small-scale fluctuations (Quaas et al., 2010). Furthermore, the aerosol optical depth is representative often just for aerosol layers in the boundary layer, and it is proportional to the total aerosol surface whereas for the CCN concentration the number of particles is relevant (Stier, 2016). A somewhat preferable alternative might be the aerosol index, defined as the product between aerosol optical depth and the Ångström exponent[2] since it emphasises the contribution of smaller particles. However, the problem of availability of just a column-integral quantity remains. Re-analyses of aerosols in numerical weather prediction models and the application of appropriate satellite simulators in the models to be evaluated may help overcome this problem.

11.13 Outstanding Challenges

There remain very important open questions about aerosol–cloud–precipitation interactions and their impact on the Earth's energy budget and, subsequently, on climate. Provided the large differences of aerosol–cloud interactions expected for different aerosol–cloud regimes, the most promising way forwards is to disentangle regimes. While extensive marine stratocumulus clouds might be most important in terms of the radiative perturbations, perturbations of deep convective clouds are likely most relevant for impacts on atmospheric circulation and severe weather. The most important challenges include

- A reliable, ideally observations-based quantification of the aerosol radiative forcing by anthropogenic aerosol. More specifically, what is needed is: (i) a quantification of the effect of anthropogenic aerosols on the albedo of

liquid-water clouds that also accounts for the cooling rate variability; (ii) a qualitative and quantitative assessment of the modification of cloud liquid-water path and cloud cover due to anthropogenic aerosols (adjustment effects); (iii) a qualitative and quantitative assessment of the radiative effect of anthropogenic aerosols on ice- and mixed-phase clouds; (iv) a quantitative characterisation of the anthropogenic and natural contributions to present-day aerosol load.

- An understanding how aerosols acting as CCN and/or INPs affect precipitation characteristics in terms of frequency of occurrence, intensity and geographical distribution. Potentially this can be relevant to weather prediction as well.
- A better understanding of how large-scale circulations react to aerosol perturbations.
- An assessment how aerosol-climate feedbacks operate and which of these might be of relevance at different time scales.

The methods to advance these outstanding challenges are likely the exploitation of the wealth of satellite observations in combination with airborne and ground-based reference observations. Laboratory studies are fundamental at achieving an understanding of the details of processes. Modelling at process-resolving scales including interactive aerosols is essential to interpret statistical results. Modelling necessarily involves a thorough use of data in order to constrain the process representations. It is, in turn, needed to corroborate the significance and interpretation of observational studies. The exploitation of observational simulators (Chapter 7) in models for the comparability of retrievals and model diagnostics is indispensable for a reliable comparison in order to evaluate model simulations. The analysis should focus on specific aerosol–cloud regimes, distinguishing notably liquid from mixed-phase and ice clouds, and distinguishing shallow- and deep-convective clouds from stratiform ones. In terms of aerosol, a distinction of those well-suited to serve as CCN from those serving as INPs and those absorbing sunlight and affecting local radiative heating rates is necessary.

Further Reading

Besides the literature already referenced earlier, two textbooks are recommended that cover in depth the basics, as well as recent developments in the field of clouds and aerosols and their interaction.

---

[2] The Ångström exponent defines the spectral dependency of τ_a as $-\log \frac{\tau_{a,\lambda_1}}{\tau_{a,\lambda_2}} / \log \frac{\lambda_1}{\lambda_2}$ with two distinct wavelengths λ_1 and λ_2.

Aerosols

O. Boucher, *Atmospheric Aerosols: Properties and Climate Impacts* (Boucher, 2015) is a textbook that covers an in-depth and broad introduction into the topic of atmospheric aerosols. It introduces their physical, chemical and optical properties; describes in-depth modelling approaches, as well as *in situ* and remote sensing observations; and explains the interaction of aerosols with radiation and clouds, and their subsequent climate impacts.

Clouds

U. Lohmann, F. Lüönd and F. Mahrt, *An Introduction to Clouds from the Microscale to Climate* (Lohmann et al., 2016) is a textbook that starts from fundamental thermodynamics to introduce the scientific basis for the formation of clouds. It describes how at the microphysical level cloud particles form and introduces in depth microphysical processes within clouds. On the basis of this microscale understanding, the book explains the role of clouds in large-scale climate.

Exercises

(1) What is the relative change in aerosol optical depth for an otherwise unchanged layer of small aerosol particles when relative humidity (RH) increases from 50% to 90% due to moistening?

(2) At what albedo are clouds the most susceptible to droplet number concentration perturbations?

(3) A polar-orbiting satellite observes a scene once a day at 10.30 a.m. local time. Over Europe, the average retrieved solar cloud radiative effect at this time of the day is about $-100\,\mathrm{W\,m^{-2}}$, and the temporal standard deviation, $50\,\mathrm{W\,m^{-2}}$. How long does an observation time series need to be to statistically confirm the hypothesis that on Sundays at 10.30 a.m. the solar cloud radiative effect is $2\,\mathrm{W\,m^{-2}}$ smaller than on Wednesdays at the same time of the day (assuming normal distributions and applying a t-test at 95 % confidence)?

(4) A perturbation in CCN may affect the droplet size distribution dispersion. To illustrate this, consider a simple bi-modal size distribution in a cloud layer with one delta peak at $r = 10\,\mu\mathrm{m}$ and a second one at $20\,\mu\mathrm{m}$. A narrowed size distribution with the same mean radius has one peak at $r = 12\,\mu\mathrm{m}$ and the second one at $18\,\mu\mathrm{m}$. The vertical and mass-integral over the two modes yields liquid-water paths of $W_1 = W_2 = W = 100\,\mathrm{g\,m^{-2}}$. These are the same for the perturbed and unperturbed spectra. How does the average optical depth differ? What is the radiative forcing at $\alpha_c = 0.5$ for an incoming solar radiation of $R = 340\,\mathrm{W\,m^{-2}}$?

12   Clouds and Land

Cathy Hohenegger and Christoph Schär

Looking from space to the Earth often reveals intriguing cloud features. From the typical summer day shown in Fig. 12.1, one thing becomes evident: clouds are not randomly distributed, and the presence of land surfaces matters. Whereas shallow cumuli abound over the land areas in Fig. 12.1, they systematically avoid water and sea surfaces. The English Channel, the Ysselmeer (Y on Fig. 12.1), the smaller sea intrusions south-west of Rotterdam (Ro) or the Rhine River (R) – all remain cloud free. Even over land, distinct spatial structures emerge. The boundaries between clear and cloudy sky, on both sides of the English Channel, are perfect redrawings of the coastlines shifted by a few kilometres inland. Likewise, clouds seem to have been pushed away from the Ysselmeer. Moreover, the rougher terrain around Rotterdam, as well as the gentle forest-covered hills (F) south of Amsterdam, stand out through their denser cloud population. The emergence of such spatial (and temporal) patterns in the cloud field constitutes a visual expression of the effects of the land surface on clouds. The fact that the land surface strongly modulates the cloud distribution and in general the climate has long been known, at least since sailing times (Halley, 1686). Knowing the modification of the wind pattern by the presence of continents, in particular in the Arabian Sea where the winds reverse seasonally due to the heating of the Indian subcontinent, was of vital importance for the commerce of spices. Through human alterations of the land cover as well as the possible role of land–atmosphere interactions in modulating climate variability and climate change, studies on land–atmosphere interactions have recently flourished.

Understanding and modelling land–atmosphere interactions is a difficult task due to the involvement of various processes acting on a wide range of spatial and temporal scales. Clearing an area of a forest to plant crops, for example, alters the heat and moisture input from the surface to the atmosphere and introduces a new, partly human controlled, seasonal cycle in such quantities. It also induces spatial heterogeneity in surface fluxes, favouring the development of diurnal circulations at the crop–forest boundaries that may lead to cloud formation and even deep convection. In case of widespread clearing, the regional atmospheric circulation and the resultant precipitation distribution can ultimately be altered. As another example, consider the multiple possible interactions between the temperature at the Earth's surface, soil moisture and clouds on a summer day in the mid latitudes. As will be described in more detail in this chapter, the surface temperature depends upon the soil moisture content, because the soil moisture content affects the partition between sensible and latent heat flux. The surface temperature also depends upon the cloud field given the effects of clouds on the radiation budget. But clouds that precipitate will replenish the soil moisture, hence, feeding back on the partitioning between sensible and latent heat flux, and thus on the temperature. Moreover, under certain synoptic conditions, the development of clouds is actually directly related to the soil moisture content, making for a fully coupled land–atmosphere system.

Besides their variety, many of the land–atmosphere interactions involve biophysical components and are poorly understood. There is no equation based on first principles describing the growth of a tree, neither its transpiration. The prediction of the behaviour of a tree as the climate warms, and its effect upon the local climate, thus depends upon the chosen vegetation model and its underlying sensitivities. Many of the processes also happen on scales that are too small to be explicitly resolved in current global or regional climate models. Convection is a good example, being parameterised in climate models, but at the heart of land–atmosphere interactions. Although observations give us clear examples of the effects of the land surface on clouds, their interpretation is ambiguous. A farmer may happen to water his crop field a couple of hours before a severe thunderstorm forms: is the thunderstorm a result of the

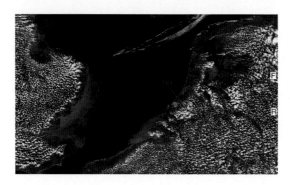

FIGURE 12.1: Satellite view on the Southern Bight (North Sea) with adjacent land areas (mainly Great Britain, the Netherlands and Belgium) taken on 6 July 2004 with swath data at 12:15 UTC, 13:50 UTC and 13:55 UTC. Note peculiar distribution of clouds. For explanation of symbols see main text. From NASA Earth data: https://wvs.earthdata.nasa.gov/. We acknowledge the use of Rapid Response imagery from the Land, Atmosphere Near real-time Capability for EOS (LANCE) system operated by the NASA/GSFC/Earth Science Data and Information System (ESDIS) with funding provided by NASA/HQ

TABLE 12.1: Symbols and key variables used in this chapter

Symbol	Description	Units (MKS)
S	Terrestrial water storage	$m^3\,m^{-2}$
P	Precipitation	$m\,s^{-1}$
R	Run-off	$m^3\,m^{-2}\,s^{-1}$
E	Evapotranspiration	$m\,s^{-1}$
E_{pot}	Potential evaporation	$m\,s^{-1}$
G	Ground heat flux	$W\,m^{-2}$
$Q_{\mathrm{net}}^{\mathrm{sw}}$	Net surface short-wave radiation	$W\,m^{-2}$
$Q_{\mathrm{net}}^{\mathrm{lw}}$	Net surface long-wave radiation	$W\,m^{-2}$
H	Sensible heat flux	$W\,m^{-2}$
T_{skin}	Temperature skin layer	K
C_{skin}	Heat capacity skin layer	$J\,K^{-1}\,m^{-2}$
ϕ	Soil moisture	$m^3\,m^{-3}$
n	Porosity	$m^3\,m^{-3}$
ϕ_{fc}	Field capacity	$m^3\,m^{-3}$
ϕ_{pwp}	Permanent wilting point	$m^3\,m^{-3}$
K	Soil hydraulic conductivity	$m\,s^{-1}$
k	Soil thermal conductivity	$W\,K^{-1}\,m^{-1}$
h_{w}	Hydraulic head	m
ψ	Pressure head	m
r_{a}	Atmospheric resistance	$s\,m^{-1}$
r_{s}	Surface resistance	$s\,m^{-1}$
α	Albedo	–
ρ_{w}	Density of water	$kg\,m^{-3}$
ρ	Density of air	$kg\,m^{-3}$

watering? Or is it rather due to the prevailing atmospheric environment (i.e., synoptic-scale conditions) and the watering a pure coincidence? Did the watering maybe make the thunderstorm more severe than it would have been otherwise? Disentangling such effects from each other is a thorny challenge.

This chapter dwells into this world of interactions. It presents the underlying land surface processes in Section 12.1, disentangles the different possible feedbacks between those land surface processes and the atmosphere in Section 12.2 and applies the developed conceptual framework to the fully coupled land–atmosphere system in Section 12.3 to highlight the potential role of the land surface in the climate record. It is the aim of this chapter that a student can understand when and why the land surface is important, how a change in the land surface might influence clouds, precipitation and the climate, and what the uncertainties are.

12.1   Basic Land Surface Processes

We start our discussion by recalling the basic balances underlying the coupled land–atmosphere system, followed by an overview of the main land surface processes. The task is to identify those factors that control land surface processes to an extent that may end up driving an atmospheric (cloud and precipitation) response. A list of the key variables introduced in this section can be found in Table 12.1 with their meaning and units.

12.1.1   Water Balance and Surface Energy Budget

Both energy and water fulfill conservation equations. The land–water balance (Fig. 12.2) may be expressed as

$$\partial_t S = P - E - R. \tag{12.1}$$

Here, the left-hand side represents the rate of change of terrestrial water storage over some area. This term embraces all water components that are stored below the soil–atmosphere interface, including interception water, snow, soil moisture and ground water content. Changes in terrestrial water storage are balanced by precipitation P, evapotranspiration E and run-off R. Run-off denotes the water that is removed from the soil storage through surface and subsurface channels. With some delay, it will in general manifest itself as river streamflow. Evapotranspiration embraces all water vapour fluxes across the soil–atmosphere interface. This includes evaporation of water from the soil and vegetation canopy, transpiration by plants and sublimation of snow, as well as dewfall (with a negative sign).

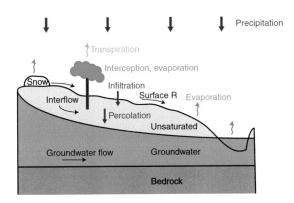

FIGURE 12.2: Main components of the hydrological cycle over land. Run-off is the sum of surface run-off, interflow and groundwater flow. Evapotranspiration contains contribution from soil, snow, water bodies and vegetation (transpiration by plants and evaporation of intercepted water).

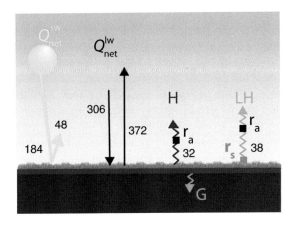

FIGURE 12.3: Components of the surface energy budget over land. The numbers ($\mathrm{W\,m^{-2}}$) represent best estimates of their annual mean at the Earth's surface averaged over land from a combination of direct observations and modelling results. Numbers taken from Wild et al. (2015)

The energy balance (Fig. 12.3) expresses the conservation of energy at the Earth's surface:

$$Q_{\mathrm{net}}^{\mathrm{sw}} + Q_{\mathrm{net}}^{\mathrm{lw}} = H + \ell_{\mathrm{v}}\rho_{\mathrm{w}}E + G. \qquad (12.2)$$

Here $Q_{\mathrm{net}}^{\mathrm{sw}}$ and $Q_{\mathrm{net}}^{\mathrm{lw}}$ stand for net short-wave and net long-wave radiation, respectively, H denotes the sensible heat flux, $\ell_{\mathrm{v}}\rho_{\mathrm{w}}E$ the latent heat flux and G the ground heat flux. In physical terms, the sensible and latent heat fluxes are related to the exchange of heat and moisture across the soil interface, respectively, while the ground heat flux is related to heat conduction and heat transport in the soil. At daytime, the net short-wave (solar) radiation is positive (i.e., it heats the soil), while net long-wave (terrestrial) radiation is usually negative during day and night. The

other terms are defined to be positive when the flux is away from the interface.

Eq. (12.2) and Fig. 12.3 state that the amount of radiative energy absorbed at the Earth's surface, that is, $Q_{\mathrm{net}}^{\mathrm{sw}} + Q_{\mathrm{net}}^{\mathrm{lw}}$, must be balanced by some other energy fluxes. Eq. (12.2) is only valid at the soil–atmosphere interface of infinitesimal thickness with zero heat capacity or under equilibrium conditions (no surface warming). The full and more general form of Eq. (12.2) reads

$$C_{\mathrm{skin}}\partial_t T_{\mathrm{skin}} = Q_{\mathrm{net}}^{\mathrm{sw}} + Q_{\mathrm{net}}^{\mathrm{lw}} - H - \ell_{\mathrm{v}}\rho_{\mathrm{w}}E - G. \qquad (12.3)$$

T_{skin} is the temperature of the surface layer of the Earth, called skin temperature, and C_{skin} the heat capacity of the skin layer. This equation is often employed to predict the evolution of T_{skin} in bulk models of the land–atmosphere system (see Chapter 5), or to predict the evolution of the sea-surface temperature in mixed-layer ocean models. C_{skin} varies with the properties of the skin layer, namely, density, specific heat capacity and thickness. It is obviously smaller over land than over ocean. Many atmospheric models actually neglect any heat storage over land, thus putting C_{skin} to zero, whereas a slab ocean of 50 m depth would exhibit a heat capacity of $2.047 \times 10^8\,\mathrm{J\ K^{-1}\ m^{-2}}$. If a finite soil layer is considered, then the freezing of water in the soil needs to be considered in Eqs (12.2) and (12.3), since this may involve significant energy transformations (latent heat of melt). Also, Eqs (12.2) and (12.3) neglect the metabolic heat used in photosynthetic functions, which is minute and uncertain.

We can directly see from Eqs (12.1) and (12.2) that E couples the surface energy with the hydrological cycle, so that a perturbation in any of the terms listed in Eqs (12.1) and (12.2) has the potential to impact the other terms in both equations. The land surface itself affects the energy and hydrological cycle primarily through its control on $Q_{\mathrm{net}}^{\mathrm{sw}}$, G, E and R. Hence, the two basic tasks of a land surface model are (i) to partition the net radiative energy into H, E and G and (ii) to partition P into E and R. The partitioning between sensible heat flux and latent heat flux is called the Bowen ratio ($= H/(\ell_{\mathrm{v}}\rho_{\mathrm{w}}E)$). As clouds can also affect the energy and hydrological cycle, either through their direct control on P and radiation or through their indirect control on H and E, this allows for multiple interactions between the land surface and the cloudy atmosphere, interactions which are detailed in Section 12.2.

12.1.2 Soil Moisture

Before explaining in more detail the surface control on $Q_{\mathrm{net}}^{\mathrm{sw}}$, G, E and R, a few words about soil moisture are needed. Soils are composed of a mixture made of water, air and solid particles. Soil moisture characterises the water content in the soil and is often expressed as volumetric soil moisture (or water) content ϕ, which is formally defined as the fraction of soil volume that is filled with water, that is,

$$\phi = \frac{V_{\mathrm{w}}}{V_{\mathrm{t}}}, \qquad (12.4)$$

with V_{w} volume of water and V_{t} total soil volume. The role of ϕ has not been mentioned very explicitly yet; ϕ is implicit in most of the terms of Eqs (12.1) and (12.2). Especially important is the control of ϕ on E (explained in Section 12.1.5) and the control of ϕ on R (explained in Section 12.1.6).

The solid part of a soil (the soil matrix) is made of different grain size distributions. Clay are soil particles with diameters below 0.002 mm (USDA classification), silt particles have diameters between 0.002 and 0.05 mm, and sand particles are above 0.05 mm. The relative amount of clay, silt and sand characterises the texture of the soil and gives a soil its name. Loam, for instance, contains around 40% sand, 40% silt and 20% clay. The ability of a given soil to retain water depends upon the size of its particles, the form of its particle and the degree of compaction (arrangement and distribution of the soil particles).

In general, the soil moisture content varies within a soil column. Many soils (see Fig. 12.2) exhibit a three-floor structure with unsaturated conditions near the surface, saturated (groundwater) conditions underneath and bedrock with small porosity and water content at the bottom of the soil column. Rain water infiltrates into the soil and, through the action of gravity, moves downwards, modulated by run-off generation.

If gravity would be the only acting force, soils would be devoid of water. Hence, further forces exist in the soil, forces that are strong enough to retain water against the action of gravity. The most important of these forces are capillary forces. Capillarity occurs when the adhesive forces between water and soil particles are larger than the cohesive forces between water molecules. A typical example of capillarity is water rising in a small tube. The associated pressure difference between the water and air phase is called matrix suction. Capillary forces depend upon surface tension, viscosity of water and geometrical effects. Capillary forces are stronger with smaller pores. A consequence of this is that with the same suction, more water is available in fine-textured (loam) than in coarse-grained soil (sand), making fine-textured soils better for plant growth, as long as the plants manage to develop the necessary suction.

The balance between capillary and gravity forces mainly determines the water movement in the soil and thus soil moisture. This has first been recognised by Darcy[1] for saturated soils but has been shown to be applicable for unsaturated conditions as well, as long as the flow remains laminar. The general form of Darcy's law for vertical water transport reads:

$$v = -K(\phi)\frac{\partial h_{\mathrm{w}}}{\partial z} \qquad (12.5)$$

where

$$h_{\mathrm{w}} = -z + \psi \qquad (12.6)$$

denotes the hydraulic head measured in metres, v the velocity (m s^{-1}) and K the hydraulic conductivity (m s^{-1}). z is the elevation head due to gravity, measured as a depth (in metres). ψ is he pressure head and is primarily due to capillarity. The hydraulic head h_{w} is a potential, it measures the energy difference between different states of water in units of height. Free water at the height $z = 0$ m has the reference potential of $h_{\mathrm{w}} = 0$ m, corresponding to negligible capillary forces. As energy is needed to extract water from soils, ψ is negative. For instance, a value of $\psi = -10$ m means that extracting the water from the soil requires the same energy as lifting it against gravity by 10 m.

According to Eq. (12.5), water moves in the direction of decreasing hydraulic head (or potential); similar relationships apply in horizontal directions. K is a function of the physical properties of the soil constituents and of water, of the soil layering, as well as of the soil moisture content. K increases with ϕ, and water movement is thus more rapid in moist soils, other things being equal. Eq. (12.5) implies that water tries to move from higher to lower levels, from moist to dry soils, and from coarse to fine soils.

Applying water conservation and assuming a horizontally homogeneous soil, Eq. (12.5) can be transformed into the following equation

$$\frac{\partial \phi}{\partial t} = \frac{\partial}{\partial z}\left[K(\phi)\frac{\partial h_{\mathrm{w}}}{\partial z}\right] + S_{\phi}. \qquad (12.7)$$

The term S_{ϕ} represents sources and sinks of soil moisture, that is, precipitation, evapotranspiration and run-off. Eq. (12.7) is a key component of land surface models to predict the evolution of soil moisture in such models. Without the S_{ϕ} term, it is called the Richards equation.[2]

---

[1] Henry Philibert Gaspard Darcy (1803–1858), a French engineer.
[2] From Lorenzo Adolph Richards who formulated the equation in 1931.

The groundwater table forms the boundary between unsaturated (above) and saturated (below) soil conditions. It represents the physical surface at which $\psi = 0$. Groundwater is made of rain water that has percolated through the soil. The occurrence of groundwater strongly depends on geographical, geological and climatological factors. It occurs in permeable geological formation, typically granular sedimentary deposit.

Several quantities have proven useful to characterise the hydraulic properties of a given soil. They all depend upon the soil texture. These quantities are:

- Porosity n: the volume of void space per total soil volume. ϕ lies between 0 and n, where for most soils n is in the range of 0.2–0.6. If $\phi = n$, then the soil is saturated.
- Field capacity ϕ_{fc}: the maximum volume of water that a soil can hold through its capillary forces against gravity. ϕ_{fc} is smaller than n and generally lies between 0.1 and 0.4.
- Permanent wilting point ϕ_{pwp}: the minimum water content below which plants begin to wilt. ϕ_{pwp} varies between 0.05 and 0.3. This evidently is an approximate concept, as wilting depends upon the plant type considered. A more objective definition is feasible based on the pressure head. More specifically, the wilting point is defined as the volumetric water content at which $\psi = -15$ m. At this head, extracting water from the soil requires the same energy as lifting it by 15 m.

This discussion only highlights the effects of water, air and solid inorganic particles on the soil properties. Real soils contain organic material, mainly resulting from the breakdown of dead organic matter, being plants or animals. Such effects are not considered in land surface models except for applications that attempt to explicitly simulate the full cycle of organic matter. A higher level of organic matter increases the stability of the soil which allows for the presence of larger pores. This manifests itself in an increase of the soil porosity. The presence of organic matter also increases the water holding capacity of the soil with, in particular, an increase in ϕ_{fc}.

Real soils not only contain organic matter but are rarely homogeneous. The aforementioned soil properties like K, n, ϕ_{fc} and ϕ_{pwp} can objectively be defined for homogeneous soil samples, but their application to heterogeneous soils becomes more difficult. The hydraulic conductivity K varies by many orders of magnitude depending on the soil type and water content. The presence of a macropore in an otherwise homogeneous soil would thus drastically affect the hydraulic conductivity of the soil sample. It

follows that a heterogeneous soil may behave very differently than its most prevalent component. Representing heterogeneous soils in land surface models remains a tremendous challenge.

12.1.3 Albedo

As a first surface control on the water and energy cycles, we consider the effect of the land surface on $Q_{net}^{sw} = Q_{in}^{sw} - Q_{out}^{sw}$. Q_{net}^{sw} directly depends upon the reflection of incoming solar radiation at the Earth's surface as $Q_{out}^{sw} = \alpha Q_{in}^{sw}$. The reflection coefficient α is referred to as albedo. It depends upon the incident radiation, see Chapter 4 for a comprehensive discussion of such effects, as well as upon the physical characteristics of the soil surface, the plant canopy and the water storage. It is typically lower over forests (0.12–0.16) than over grassland (0.19–0.2) or sand desert (0.28) (e.g., Hagemann, 2002), leaving more energy for E, all other fluxes remaining equal. It is also larger with smaller soil moisture (drier soils).

12.1.4 Ground Heat Flux

The ground heat flux G is generally an order of magnitude smaller than the other terms of the surface energy budget and thus often neglected, especially when averaging over longer time periods (days/years). G may nevertheless peak up to $200\,\mathrm{W\,m^{-2}}$ over bare soils during the course of a day. As visible from Eq. (12.2), G is ultimately driven by the available net radiative energy. The main mechanism that transports heat into the soil is heat conduction along temperature gradients. During the daytime, the Earth's surface is warmer than the underlying soil layers, which leads to conduction of heat into the soil:

$$G = -k \frac{\partial T_{soil}}{\partial z}. \tag{12.8}$$

z denotes the depth into the soil so that a positive G means that the flux goes into the soil. The thermal conductivity k depends upon the thermal conductivity of the individual soil components (solid particles, air and water) as well as of their spatial arrangement. Air is a good insulator whereas water conducts heat well. As such, the thermal conductivity of a given soil primarily varies with its soil moisture. Eq. (12.8) not only allows to compute G but can also be used to predict the evolution of the temperature inside the soil T_{soil}. Besides conduction, infiltration of water into the soil and phase changes of water transport heat into and through the soil, although these contributions are generally ignored in land surface models.

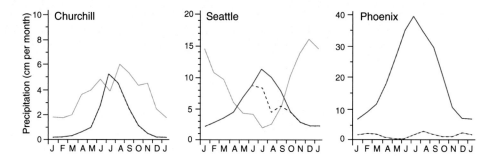

FIGURE 12.4: Yearly cycles in precipitation (blue), potential evaporation (red) and evaporation (dashed black) for three cities located in different climatic regimes (redrawn from Hartmann (2016)). P is from observations whereas E_{pot} and E are computed with formulae like Eqs (12.9) and (12.12), fed with observed values of wind, temperature and humidity. The soil moisture, needed to compute E, is obtained from Eq. (12.1), using the bucket model to compute R (see Section 12.1.6) and assuming $r_{\mathrm{s,min}} = 0$.

12.1.5 Evapotranspiration

The term evapotranspiration E includes evaporation from interception and bare soil as well as transpiration from vegetation (Fig. 12.2). E depends upon the availability of energy and water (see Eqs (12.1) and (12.2)), and thus implicitly upon the soil type, the vegetation cover and type, as well as upon the atmospheric conditions. To illustrate those different controls, we start with the simple case of bare soil evaporation, thus neglecting the effect of vegetation. The three controls on E are in that case soil moisture, the energy input and the atmospheric conditions. To isolate the effect of soil moisture on E, it is first useful to define a quantity called potential evaporation E_{pot}. The latter corresponds to the maximum possible evaporation as obtained over a water surface for given atmospheric conditions and energy input. Fig. 12.4 displays yearly cycles of precipitation P, evaporation E and potential evaporation E_{pot} in Churchill, Seattle and Phoenix, three locations characterised by different regimes of evaporation due to different soil moisture regimes.

In Churchill, there is always enough soil moisture available and E equals E_{pot}. Soil moisture is unimportant, this is the wet regime and E is limited by the energy input and atmospheric conditions (see later). At the other extreme, in Phoenix, E is much smaller than the potential evaporation, indicating that E is limited by the soil moisture availability, and the remaining controls (atmospheric conditions or energy input) are unimportant to first order. An increase in E in Phoenix primarily requires an increase in soil moisture. Phoenix belongs to the soil moisture-limited regime. This regime is called a dry regime, when ϕ lies below the wilting point and there is no E, and a transition regime for $\phi_{\mathrm{pwp}} < \phi < \phi_{\mathrm{crit}}$. Above ϕ_{crit}, soil moisture is unimportant for E and we are in the wet regime. Seattle lies in-between the two extreme cases of

Churchill and Phoenix. There is ample soil moisture in winter, whereas, in summer, the warmer temperatures, the increased insolation and the lack of precipitation conspire to decrease the soil moisture reservoir and limit E. Temperate climates often experience a shift from a wet to a soil moisture-limited regime during summer. Semi-arid regions, which lack sufficient precipitation to replenish soil moisture, are more often under water-stressed conditions than temperate climates. From those simple considerations, it is clear that soil moisture is critical in determining the climate over particular regions, but not everywhere. The terminology used to distinguish different climatic regimes, like arid, semi-arid or temperate, often reflects those distinct E regimes and the distinct importance of soil moisture on climate.

What then determines the strength of E in the wet regime? Remember that in this case, E happens at its potential rate and is controlled by the energy input and by the atmospheric conditions. To formally disentangle those two controls we start from the following equation, which is generally used to compute E_{pot}:

$$E_{\mathrm{pot}} = \frac{\rho}{\rho_{\mathrm{w}}} \frac{q_{\mathrm{s}}^{\mathrm{surf}} - q_{\mathrm{v}}^{\mathrm{atm}}}{r_{\mathrm{a}}}. \tag{12.9}$$

$q_{\mathrm{s}}^{\mathrm{surf}}$ denotes the saturation specific humidity at the surface and $q_{\mathrm{v}}^{\mathrm{atm}}$ the specific humidity of the atmosphere near the surface, typically taken at 2 m. Note that $q_{\mathrm{s}}^{\mathrm{surf}} = q_{\mathrm{v}}^{\mathrm{surf}}$ over a fully saturated surface. The equation states that E_{pot} is proportional to the vertical gradient in specific humidity, and the proportionality factor is referred to as aerodynamic resistance r_{a}. It expresses the resistance of air to the turbulent transport of water vapour. An expression for r_{a} can be derived using Monin–Obukhov theory and similarity functions. In essence, r_{a} increases with decreasing wind speed and with increasing atmospheric stability. r_{a} also varies with the roughness of the surface, where rougher surface conditions yield more mixing and thus smaller r_a. The aerodynamic resistance can alternatively be expressed as $r_a = 1/(C_H \bar{u})$, the product of a turbulent transfer coefficient for heat C_H and mean wind speed $\bar{u}$.

Eq. (12.9) can be further transformed to isolate the atmospheric contribution to E_{pot}, given by the terms r_a, ρ and q_v^{atm}, and the energy contribution, hidden in the term q_s^{surf} through its dependency on skin temperature. Realising that by linearising q_s^{surf} the implicit dependence on the temperature can be eliminated from Eq. (12.9) and using Eq. (12.2) yield the famous Penman[3] formula:

$$E_{pot} = \frac{\Gamma}{\ell_v \rho_w}(Q_{net}^{sw} + Q_{net}^{lw} - G)$$
$$+ (1 - \Gamma)\frac{\rho}{\rho_w}\frac{q_s^{atm} - q_v^{atm}}{r_a} \qquad (12.10)$$

with

$$\Gamma = \frac{\partial q_s}{\partial T}\bigg/\left(\frac{\partial q_s}{\partial T} + \frac{c_p}{\ell_v}\right) = 1\bigg/\left(1 + \frac{c_p}{\ell_v}\left(\frac{\partial q_s}{\partial T}\right)^{-1}\right). \qquad (12.11)$$

The Penman formula clearly identifies the two main controls on E_{pot}: the first term is the energy component and the second term the aerodynamic (or atmospheric) component. The weight Γ determines the relative contribution of the two terms. In warm climates (high temperatures), $\frac{\partial q_s}{\partial T}$ is large, Γ tends towards 1 and E_{pot} is mostly dependent on the available energy. As the temperature decreases, $\frac{\partial q_s}{\partial T}$ becomes smaller, Γ decreases and E_{pot} becomes more dependent upon the aerodynamic term, that is, the capacity of the air to dry the soil. At temperature near or below freezing, E_{pot} is primarily determined by the second term of Eq. (12.10).

In the dry and transition regimes, E can be limited by the soil moisture availability, as already exemplified in Fig. 12.4. The resistance of the surface against the evaporation process must be taken into account. This can easily be achieved by adding a surface resistance r_s to the aerodynamic resistance r_a, like it is done in electricity. The general formulation for E thus reads:

$$E = \frac{\rho}{\rho_w}\frac{q_s^{surf} - q_v^{atm}}{r_a + r_s}. \qquad (12.12)$$

Eq. (12.9) is recovered by setting r_s to zero, as expected. On the one hand, r_s encapsulates the effect of soil moisture on E. Over bare soils, r_s is a sole function of soil moisture. Soil suction holds water in the void space: the smaller the void space, the larger the soil suction

and the harder it is to evaporate water from the soil. Hence, as the soil dries, the resistance increases, limiting evaporation. On the other hand, r_s can also be used to represent the control of the vegetation on the transpiration process. Over a vegetated surface, r_s is then primarily a function of the vegetation type and environmental conditions. Forest and grass have, for instance, very different strategies concerning their handling of water. Forests possess deeper roots than grass and can thus extract more water from the soil. Forests have larger leaves, implying more surface for transpiration. These and other factors (e.g., insolation, atmospheric carbon dioxide content, temperature, extent of the interception reservoir, roughness length) ultimately determine the transpiration of a vegetated surface and must be parameterised in land surface models (Arora, 2002). These relationships are generally based on laboratory measurements made on single leaves and plants, and thus difficult to generalise to the vegetation canopy.

E_{pot} is a useful theoretical concept to highlight and isolate the contribution from the energy and from the aerodynamic component to the evaporation process. In practice, it has limited use as the definition of a water surface is ambiguous over a vegetated surface and even a fully saturated soil always exerts a minimum resistance $r_{s,min}$ on the water molecules. It is thus more useful in practice to compare E to a reference value, E_{ref}, than to E_{pot}. Eqs (12.9) and (12.10) can be easily modified by adding $r_{s,min}$ to r_a. In this case, the Penman formula is called the Penman–Monteith[4] combination formula. The Food and Agricultural Organization (FAO) of the United Nations defines what are the surface characteristics of the reference case. For instance, E_{ref} over crop is defined as evapotranspiration for a theoretical grass with a height of 0.12 m, an albedo of 0.23 and a constant surface resistance of $70\,\text{s}\,\text{m}^{-1}$.

12.1.6 Run-off

Run-off refers to the export of liquid water from the surface and the soil column, ultimately into a river, lake or sea. The run-off formation term R in Eq. (12.1) is defined as the respective transfer of water per unit area and time, and has the unit $\text{m}^3\,\text{m}^{-2}\,\text{s}^{-1}$. The discharge of the water takes place through a complex multiscale network that includes sub-surface flow paths and aquifers in the soil,

---

[3] Howard Penman (1909–1984), a British meteorologist. He derived his famous equation in 1948.

[4] John Lennox Monteith (1929–2012) worked with Howard Penman on evapotranspiration. He was strongly interested by the micro-climatology of field crops.

as well as streams and rivers at the surface. Once the run-off has been concentrated into a river, it can be measured using a stream gauge. The level of detail in describing the run-off process differs from model to model. Hydrological models usually contain a detailed description of the river network and the run-off concentration process, whereas simpler representations are used in climate models.

Although much of the discussion in Sections 12.2 and 12.3 will relate to the surface control on E, we know from observations that the run-off ratio R/P is an important characteristic affecting the mean and variability of the climate (temperature, precipitation). In the long-term mean, precipitation is balanced by run-off and evapotranspiration, $R = P - E$.

Considering a vertical slab of soil integrated over a catchment area (see Fig. 12.2), one distinguishes several contributions to run-off formation R: surface run-off at the Earth's surface (also called overland flow), interflow in the unsaturated soil and groundwater flow (or baseflow) in the saturated soil. Surface run-off occurs under two conditions. First, if the rainfall rate exceeds the infiltration ability of the soil, part of the water cannot enter the soil and flows away as surface run-off. Second, if the soil is saturated, or frozen, water will also be unable to infiltrate and will flow away as surface run-off. It is thus evident that surface run-off depends upon different factors. On the one hand, it depends upon the soil characteristics and the soil moisture content. Coarse-grained soils (e.g., sand) have a higher infiltration rate than fine-grained soils (e.g., clay). On the other hand, surface run-off also depends upon the rainfall characteristics, especially intensity, duration and rainfall area. In addition, surface run-off is also influenced by topography and vegetation cover. Under conditions of light to intermediate precipitation, surface run-off is rare over fully vegetated soils, as the rain is first intercepted by the vegetation cover.

Interflow primarily occurs when lateral differences in soil hydraulic conductivity are larger than vertical differences (see Eq. (12.5)). If rain infiltrates faster in the upper soil layers than it can enter the lower layers, some fraction of the rain will flow laterally. Interflow is thus favoured in soils containing impermeable soil layers. A second-generation mechanism of interflow is flow through macropores. Holes and channels of various scales and of irregular pattern/distribution that exist in the soil due to cracks, soil inhomogeneities, roots, mice, worms and other animals can constitute a preferred pathway for water. Measuring, modelling and predicting this component is very difficult, even impossible. However, the contribution of the interflow to run-off formation is small on the scales considered in climate models, and can safely be neglected in large-scale and mesoscale models.

Groundwater run-off consists of saturated lateral flow in aquifers and porous regions of the soil. Aquifers are extended groundwater systems in water-bearing soils, and are often connected to surface rivers. The primary driver of groundwater run-off is gravity. Groundwater will continuously flow following the hydraulic head (see Eq. (12.5)), albeit at a slow pace. Groundwater will emerge at the surface when the water table intersects with the surface due to change in orography, as displayed in Fig. 12.2. Groundwater flow is strongly linked to variations of the groundwater table, and can become unusually large with an abnormally high water table. Groundwater flow represents the long-term component of run-off and is especially important during dry spells. Groundwater run-off is not as well temporally correlated with precipitation as surface run-off, as it takes time for water to percolate through the soil.

The earliest method used to compute R in land surface models is the so-called bucket model. As with a bucket under a water tap, the water fills the bucket up to the point when the bucket gets full and the water spills over, that is, precipitation fills the soil as long as it remains unsaturated, and forms run-off when saturation is reached. The bucket method ignores vertical and horizontal subsurface fluxes. More recent land surface models try to parameterise run-off by means of relationships to relate rainfall intensity to infiltration to lateral water movement for given land surface and soil hydraulic properties. Such methods especially ignore flow through macropores and remain fairly simple, in contrast to the at times complicated formula employed to compute r_s and E. Paradoxically and in opposition to E, catchment averaged R is actually a relatively easy to measure quantity, and has been measured for many decades for major river basins.

12.1.7 Snow Cover

The accumulation of snow on the land surface affects the water and energy cycles and thus needs to be mentioned here. Snow exhibits a higher albedo than other surface types. The typical albedo of dry fresh snow is 0.8, versus 0.07 for ocean, 0.28 for desert and around 0.17 for vegetation. As more radiation is reflected, the skin temperature of a snow-covered surface is colder than the skin temperature of a snow-free surface. The snow insulates the land surface. Moreover, when the snow starts to melt, a fraction of the available net radiation is used to melt the snow rather than to warm the Earth's surface and the atmosphere. Snow may even sublimate. Eq. (12.2) does not include sublimation and melting. Those terms need to be taken into account in the presence of snow. Frozen ground plays a similar role of thermal storage as snow. This leads to a dampened but extended cold season given that some of the energy is used to freeze or thaw the ground.

Snow also keeps water away from the deeper soil layers. Precipitation falling in solid form does not penetrate into the soil but accumulates at the Earth's surface. When the snow melts, part of the water infiltrates the soil and replenishes soil moisture and ground water. The rest flows away as run-off, making a substantial contribution to spring and summer river flows in regions with long-lasting snow cover, like mountain areas.

In essence, the snow stores heat and water that are released when the snow melts. This introduces a memory component to the system (see Section 12.1.8). Snow can also introduce a strong variability into the system, both temporally with short-lived snowfalls and spatially. Spatial variations in snow cover are important for the melting process. A correct representation of the melting process is an issue in many land surface models.

12.1.8   Land Surface Memory

A very important effect of the land surface on the weather/climate is that it introduces a memory component to the system. A first example of memory was given in Section 12.1.7 when speaking of snow cover. Likewise, vegetation and soil moisture introduce a memory component to the system. If the vegetation dies or if the soil dries out, it may take many months for the land surface to recover, thus possibly influencing the atmospheric conditions over an extended period. Likewise, if soil moisture and groundwater become saturated, for instance, following a flood, the implied anomalous run-off and evapotranspiration regime may last over a long time period. Moreover, the deeper rooting of forests increases the soil moisture memory to several months by basically increasing the size of the soil moisture reservoir that can be accessed by the atmosphere.

One way to quantify soil moisture memory consists of computing time lag correlations of soil moisture. The soil moisture values of all years at time t (e.g., June first) are correlated with the soil moisture values of all years at time $t + $ lag. Expanding the correlation function in its various terms and using the water balance (Eq. (12.1)) allow isolating the different processes determining the soil moisture memory (Koster and Suarez, 2001). Examples of

soil moisture memory values are 1.2 months as averaged over the equatorial region and subtropics (4°S–31°N), 1.9 months over mid latitudes (31–54°N) and 2.5 months at high latitude (54–76°N) with values varying between 0.5 and 10 months locally. Those values are decay timescales obtained by fitting a theoretical red-noise spectrum to the soil moisture output of a global climate model at each grid point (Delworth and Manabe, 1988). They in essence characterise the time lag after which the autocorrelation function is reduced to 1/e. Applying similar methods to observations of soil moisture leads to similar values. The 'long' land surface memory is in strong contrast to temperature, evapotranspiration or other atmospheric variables, which rapidly loose any memory. The memory associated with soil moisture and snow cover is crucial for monthly and seasonal prediction.

12.2   Interactions between Processes

Section 12.1 has introduced the basic land surface processes acting in the climate system by describing those processes in isolation. Here we discuss how a sequence of processes can induce a feedback loop. We have, for instance, seen that the surface net radiation plays an important role for evapotranspiration. This surface net radiation can also be interpreted as one element of a feedback loop, for instance, in the case when the evapotranspiration affects boundary-layer cloud formation, which in turn feeds back on the surface net radiation. Feedback loops may be characterised as positive or negative. A negative feedback loop moderates the response of the system to an imposed perturbation. Positive feedback loops, in contrary, amplify the response of the system. A myriad of feedbacks exist between the surface and the atmosphere. However, most feedbacks follow a few characteristic pathways that will be described in this section.

It happens that the response of the atmosphere to a spatially homogeneous perturbation of the land state can be very different than the response of the atmosphere to a spatially localised perturbation. This is illustrated in Fig. 12.5. In the case of a homogeneous perturbation (Fig. 12.5a), we may take a column-view, where a land-surface anomaly affects the atmospheric column by vertical fluxes of heat and moisture, which in turn will affect the atmospheric profiles and ensuing cloud formation. In the case of heterogeneous perturbations (Fig. 12.5b), the situation is more complex. Gradients in surface conditions may drive three-dimensional mesoscale circulations that will non-linearly influence the horizontal

(a) Column view   **(b) 3D view**

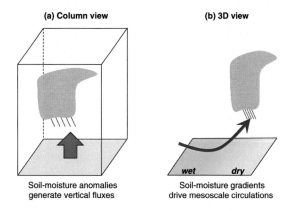

Soil-moisture anomalies   Soil-moisture gradients
generate vertical fluxes   drive mesoscale circulations

FIGURE 12.5: Schematic illustration of the two categories of land–atmosphere feedback processes. (a) In the column view, the feedbacks are relying on vertical fluxes. (b) In the three-dimensional view, gradients in surface conditions may drive mesoscale circulations.

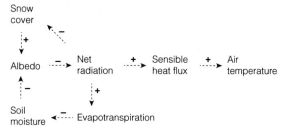

FIGURE 12.6: Possible interactions involving changes in surface albedo. Only the interactions discussed in Section 12.2.1 are displayed. A + sign indicates that the two variables are positively correlated: an increase (decrease) in the first variable yields an increase (decrease) in the second one. A − sign indicates negative correlation. Absence of − signs along a feedback loop, or an even number of − signs, implies a positive feedback. Otherwise, the feedback is negative.

transport between different surface patches. In discussing the most important feedback processes, we will distinguish between these two viewpoints. Sections 12.2.1–12.2.3 first consider the homogeneous case and neglect any effect that arises from induced or imposed surface heterogeneity. Effects of surface heterogeneity are presented in Sections 12.2.4 and 12.2.5. Section 12.2.6 summarises those different views based on results from observations and model simulations.

As will become evident in the following, changes in the surface state can sustain feedbacks of opposite signs, depending on the prevailing atmospheric conditions, as well as on the time and spatial scales involved. The following subsections only describe those feedbacks in isolation. They constitute the building blocks that, put together, will determine how the climate is shaped by land surface processes under past, current and future conditions. This final aspect is addressed in Section 12.3.

12.2.1   Surface Albedo Feedbacks

We start by considering the feedback loop induced by a change in surface albedo (see Fig. 12.6). Assume a homogeneous surface and assume that the surface state changes so that the surface albedo decreases, everything else remaining the same. How does the system respond? A decrease in albedo yields an increase in net surface radiation and hence an increase in H, an increase in E and an increase in air temperature. If the decrease in surface albedo is caused by a decrease in snow cover, the resulting larger net surface radiation favours the melting of snow, further reducing the snow cover: a positive feedback

between snow cover and the net surface radiation via albedo changes is generated. This is the well-known snow-albedo feedback, which is of particular relevance for cold climates, as already mentioned in Chapter 10.

Besides snow cover, changes in soil moisture or vegetation properties also directly affect the surface albedo. An increase in soil moisture leads to a decrease in surface albedo. In this case, the resulting higher E depletes the initial soil moisture anomaly and a negative feedback emerges (see Fig. 12.6). This simple response assumes no change in the Bowen ratio, which is often not the case as the soil moisture can also directly affect E via r_s. This can lead to a very distinct atmospheric response, in particular in terms of air temperature, as detailed in Section 12.2.3.1.

The simple interactions displayed in Fig. 12.6 remain valid as long as changes in cloud cover or precipitation can be neglected. Consider, for instance, the change in the plant phenology at the beginning of the rainy season in the tropics. Precipitation boosts the plant development, the plants turn green and this significantly reduces the surface albedo. This reduction in surface albedo then increases $Q_\mathrm{net}^\mathrm{sw}$ and leads to warmer air temperatures, which, up to a certain threshold, is beneficial for plant growth. More importantly, the increase in E following the albedo change can induce a precipitation response that, if positive, would maintain a positive feedback between vegetation and precipitation. This leads us to one of the key question in the field of land–atmosphere interactions, namely: how do clouds and precipitation respond to a change in E and, as its counterpart, to a change in H? Does it rain more over a wet or over a dry surface? Quite a simple question with a quite complicated answer, which will occupy ourselves for most of the rest of Section 12.2.

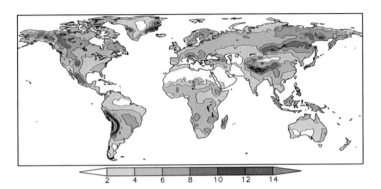

FIGURE 12.7: Climatological (25-year) annual mean recycling ratio (%) at a representative spatial scale of $1.0 \times 10^5 km^2$, taken from Dirmeyer and Brubaker (2007). The analysis methodology employs a computation of backward trajectory based on a combination of data from reanalysis and observations. Note that estimates of recycling ratios depend upon the data sets used as well as the method and assumptions made to compute the recycling ratios. Copyright ©2007 American Meteorological Society (AMS)

12.2.2 Moisture Recycling

We next turn our attention to feedbacks involving changes in r_s (assuming no change in surface albedo) and include the response of clouds and precipitation in the feedback loop. We take again a column view of the atmosphere, neglect any horizontal variations and assume a perturbation that leads to a decrease in r_s, for instance, through an increase of soil moisture, and hence an increase in E. The direct effect of an increase in E is a moistening of the atmosphere. If more water is put in the atmosphere, then one expects an increase in relative humidity and consequently an increase in cloud cover and precipitation. If precipitation increases, r_s decreases. This is evident if the initial change in r_s was due to an increase in soil moisture. It is also generally true if the initial change in r_s was related to a change in vegetation properties in the sense that vegetation likes precipitation. As such, a positive feedback loop is sustained between the land state and precipitation. It is one of the simplest feedback loops of the land–atmosphere system and is generally referred to as moisture recycling.

To assess the relevance of moisture recycling in reality, recycling ratios can be computed from the output of climate model simulations. The recycling ratio is defined as the fraction of precipitation in a chosen analysis domain that originates from evapotranspiration from within the same domain. The computation of the recycling ratio requires choosing a budget analysis domain. If we take the full Earth and wait for equilibrium conditions, the recycling ratio amounts to 1, that is, all the water falling as precipitation on the Earth's surface previously evaporated from the Earth's surface. On a point measurement, the recycling ratio is generally 0, given that the wind rapidly transports away the evaporated water vapour.

Fig. 12.7 shows the annual mean recycling ratio based on analysis domains of size $1.0 \times 10^5 km^2$. The analysis methodology employs the computation of backward trajectories based on a combination of reanalysis data and observations. The large values over the mountain areas visible in Fig. 12.7 are likely artefacts resulting from the analysis methodology. Except over some continental regions (China), the recycling ratios are generally low. This indicates that over most regions, a direct feedback between the underlying surface and the overlying atmosphere is not obvious, at least at a $10.0 \times 10^5 km^2$ scale. Water vapour that falls as precipitation over a given region tends to originate from upstream of that region. Similar conclusions have also been reached using isotopic tracers. In addition, this is also compatible with the recognition that the average residence time of water molecules in the atmosphere amounts to about 7 days. This implies that evapotranspiration in general takes place far upstream (by several thousand kilometres) of the location of precipitation.

Instead of asking the question as to how much precipitation over a particular region originates from evapotranspiration from within the same region, a more general question is as follows: how much of the precipitation falling at a point originates from moisture evaporated from the land surface? Answering this second question does not require any assumption about the chosen analysis domain. Computation of such continental recycling ratios based on reanalysis data (see Fig. 12.8) reveals that, on average, 40% of the precipitation falling on a land point is of terrestrial origin. In regions like China, Central Asia, the western part of Africa and central South America, the percentage climbs up to as much as 80%. In these regions, most of the falling precipitation is thus of terrestrial origin. Continental recycling ratios also vary seasonally and are far less dominant in winter than in summer.

12.2.3 Local Land–Atmosphere Coupling

Even though the simple moisture recycling mechanism described in Section 12.2.2 often does not contribute significantly to the precipitation falling over a region, the

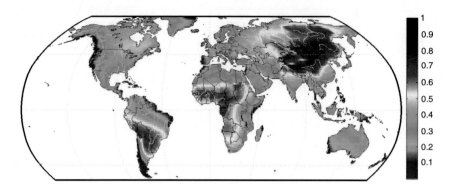

FIGURE 12.8: Climatological (1999–2008) annual mean continental recycling ratio taken from van der Ent et al. (2010). The computation of the continental recycling ratio uses the atmospheric water balance equation and data from reanalysis.

land state can affect clouds and precipitation formation via indirect mechanisms. Again we take a column view of the atmosphere, neglect any horizontal effect and assume a perturbation in the land state leads to a decrease in r_S and hence an increase in E. The underlying idea is that the induced change in the Bowen ratio affects the ability of the atmosphere to transform advected water vapour into clouds and precipitation by acting on the atmospheric profile. Such mechanisms are generally referred to as local coupling or, in the case of the feedback between soil moisture and precipitation, as indirect soil moisture–precipitation feedback.

We disentangle the involved interactions, which are summarised in Fig. 12.9, by first considering the cloud-free case in Section 12.2.3.1. We then investigate the convective case, analysing the response of shallow and deep convection to a change in surface fluxes (Section 12.2.3.2) before considering the impact of convection on the initially imposed perturbation (Section 12.2.3.3), thus completing the feedback loop.

12.2.3.1   Cloud Free: Temperature and Humidity Response

For a cloud-free planetary boundary layer (PBL), we know from Chapter 5 that the state (i.e., height, temperature and humidity) of the PBL depends on the one hand upon the buoyancy at the surface and the jump in buoyancy at the top of the PBL. The latter jump determines the amount of dry free-tropospheric air that is entrained into the PBL. On the other hand, the PBL state feeds back on latent and sensible heat fluxes, as well as on the jump in buoyancy at the PBL top. This makes a coupled system. The underlying rule is that the PBL adjusts and drives the surface fluxes towards a state in which the atmospheric demand for water becomes constant, providing the land surface can meet up with the atmospheric demand.

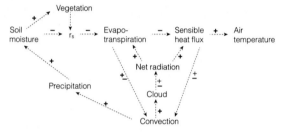

FIGURE 12.9: Possible interactions between processes involving changes in r_S. Only the interactions discussed in Section 12.2.3 are displayed. In cases where both positive and negative correlations are possible, the bolder sign indicates the most frequent occurrence.

Let's start with our thought experiment and decrease r_S (see Fig. 12.9). A decrease in r_S leads to an increase in latent heat flux. Assuming unchanged net radiation, this means that less energy is available for the sensible heat flux and the Bowen ratio decreases. The direct consequence is a moistening and cooling of the PBL. A decrease in sensible heat flux also leads to a shallower PBL that entrains less warm and dry free-tropospheric air. Comparing this response to Fig. 12.6 indicates that an increase in soil moisture leads to a cooling of the PBL through its direct effect on r_S but to a warming of the PBL through albedo change.

12.2.3.2   Response of Shallow and Deep Convection

The just-described response of the PBL state to a change in the Bowen ratio often impacts cloud and precipitation formation. In cases of shallow convection, a decrease of r_S can lead both to an increase or to a decrease in cloud cover depending on the initial atmospheric profile. In cases characterised by weak stratification above the PBL and as long as the atmosphere is not too dry, cloud cover actually decreases with an increase in E (reduced r_S).

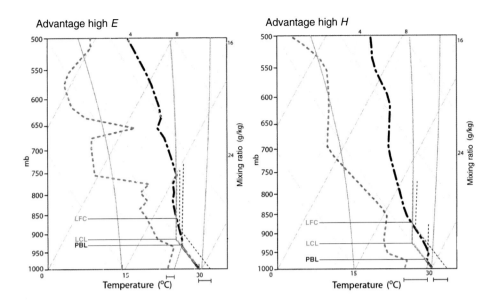

FIGURE 12.10: Thermodynamic diagrams illustrating the response of deep convection to a change in surface fluxes under different atmospheric conditions, adapted from Findell and Eltahir (2003). The green dashed line shows the ascent of a parcel lifted from the surface, which cools dry adiabatically until it reaches its LCL and moist adiabatically afterwards. The red solid line on the x-axis shows the required warming to trigger convection with the resulting parcel ascent (red dashed line) assuming unchanged specific humidity. The blue solid line on the x-axis shows the required moistening to trigger convection with the resulting parcel ascent (blue dashed line) assuming no change in the parcel's starting temperature. The light dashed blue line indicates the corresponding dew point temperature in the PBL. Copyright original Figure ©2003 American Meteorological Society (AMS)

This unexpected result can be understood by analysing the tendency equation of the relative humidity at the top of the PBL (Ek and Mahrt, 1994). Under these circumstances, the warming of the temperature at the top of the PBL due to a decreased boundary layer depth is able to over-compensate the cooling and moistening expected from an increase in E. At the top of a PBL and for a given potential temperature, the temperature is higher in the case of a shallow (in comparison to deep) PBL due to its higher top pressure. The consequence of a warmer temperature is a decrease in relative humidity, thus explaining the decrease in cloud cover with increasing E. In general, however, most atmospheric profiles support an increase in cloud cover with an increase in E (reduced r_s), due to colder and wetter PBLs (more clouds over wetter surfaces).

The response of deep convection to a change in the surface state can be understood using related arguments. One nevertheless has to realise that the conditions for triggering deep convection are more stringent than those for triggering shallow convection. Adopting a parcel theory point of view, forced shallow convection first occurs when the PBL meets the lifting condensation level (LCL). Deep convection in addition requires that the level of free convection (LFC) meets the LCL, although this may not

say much about the actual depth of the clouds and the resulting precipitation amounts (see later).

Taking a bulk energy perspective as first argument, a decrease in the Bowen ratio due to a decreased r_s, even without a change in the total input of energy into the PBL, yields a decrease in the PBL height. This implies that more energy is available per PBL unit of air, favouring deep convection. Moreover, the entrainment of dry air at the PBL top is reduced, increasing the energy. In this view, a decrease in r_s enhances deep convection: more rain is expected to fall over a wetter surface.

As a second argument, parcel theory can be adopted, see Fig. 12.10. As described earlier, a minimum requirement for the triggering of deep convection is that the LCL, LFC and PBL top match. To first order, this may be achieved in two ways. First, through a moistening of the PBL, the LFC and LCL can be brought down to the PBL top. This favours deep convection on surfaces with high E (e.g., wetter soils). Second, through a drying of the PBL, the LCL and PBL top can be brought up to the LFC. This favours deep convection under high H (e.g., drier soils) and deep PBLs. As shown in Fig. 12.10, one or the other mechanism may be more efficient depending upon the initial atmospheric profile. In Fig. 12.10a, the moistening required to trigger deep convection is smaller than the required warming, giving an advantage for surfaces exhibiting high E. The opposite is true in Fig. 12.10b.

Based on such arguments, different measures have been proposed to distinguish between atmospheric profiles that may favour deep convection depending on the surface state (Findell and Eltahir, 2003; Tawfik and Dirmeyer, 2014). They all neglect some aspects of the flow that can affect the development of convection, an obvious one being the large-scale advection of moisture (see Section 12.2.5). The indexes merely reveal the propensity of deep convection to be triggered over a particular surface, without saying much about the actual convective development and precipitation amounts. If convection manages to develop both over wet (or low r_s) and dry surfaces, the wet surface will produce more precipitation. Only in those cases where the convection develops on the dry and not on the wet surface, the rain amounts will increase with increasing r_s and decreasing E.

12.2.3.3 *Modification of the Response by Convection*
The development of convection, through the appearance of clouds and precipitation, affects the local coupling between the land and the atmosphere and closes the feedback loop (see Fig. 12.9).

The formation of clouds modifies the radiation budget at the surface. With an increase in cloud cover, Q_{net}^{sw} decreases, the absolute value of Q_{net}^{lw} decreases so that the net radiation may decrease or increase. Consequently, the surface fluxes may decrease or increase. Based on observations, the short-wave effect normally dominates (see also Chapter 4), even though this is not always true in atmospheric models. On a diurnal timescale, the short-wave cloud radiative effect is nevertheless not strong enough to break the direct and potentially positive feedback between the surface and precipitation.

The effect of precipitation is to replenish soil moisture, thus decreasing r_s. In the cases where a decrease in r_s enhances deep convection, a positive feedback is sustained: deep convection is enhanced, more precipitation is produced, the soil moisture increases and r_s further decreases. In the cases where a decrease in r_s reduces deep convection, a negative feedback emerges and the initial perturbation in r_s is dampened. The surface state is less important for convection and the soil-moisture memory is much shorter.

Precipitation not only replenishes soil moisture but also generates spatial variability, both in the PBL and at the land surface. Precipitation falling on a homogeneous

soil generates wet and dry patches. As explained in the upcoming Section 12.2.4, such heterogeneity can induce a thermally driven circulation that will affect the spatial distribution of precipitation.

12.2.4 Mesoscale Circulations Due to Land Surface Heterogeneities

One characteristic of the land surface is that it is rarely homogeneous. Horizontal variations in surface fluxes affect the development of convection and give rise to a very different precipitation response from the case of a homogeneous surface. Horizontal variations in surface fluxes lead to the formation of thermally driven mesoscale circulations. Depending on the source of the heterogeneity, such circulations are called: mountain–valley circulation (warming/cooling of mountain slopes), land–sea breeze (contrast land versus sea), lake breeze (contrast land versus lake), river breeze (contrast land versus river), vegetation breeze (contrast between different vegetation types or vegetation versus bare soil) and soil moisture-induced circulation (different soil wettnesses). The generation and properties of such circulations are described in Section 12.2.4.1. Section 12.2.4.2 investigates the impact of such circulations on the development of convection, whereas Section 12.2.4.3 considers the feedback of convection on the circulation.

12.2.4.1 *Cloud Free: Thermally Driven Circulations*
Fig. 12.11 shows an example of a thermally driven circulation, in the form of the sea-breeze circulation. Land surface heterogeneities induce corresponding spatial variations in E, Q_{net}^{sw} (albedo difference), and H. The ensuing horizontal temperature gradient generates pressure anomalies. Over the warm surface (land), the warmer air expands, such that above some elevated level more air is now present. The pressure, which is proportional to the weight of air aloft, consequently increases (see H_{igh} in Fig. 12.11). Likewise, the cooler air over the cold surface (ocean) leads to the formation of an elevated low-pressure anomaly (L_{ow} in Fig. 12.11). This pressure difference moves air from H_{igh} to L_{ow}. At the Earth's surface, this leads to the formation of a low-pressure anomaly on the warm surface and of a high-pressure anomaly on the cold surface. The pressure difference at the surface drives the cool air towards the warm surface. The cool air collides with the warm air on the warm surface and the warm air is forced to rise, closing the circulation loop.

In mathematical terms, the circulation C along a closed contour is defined as the line integral of the component of

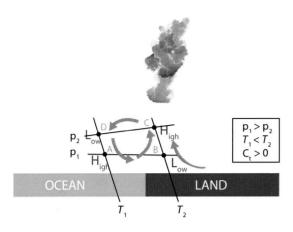

FIGURE 12.11: Schematic representation of the land–sea breeze circulation and on the resulting preferred location of convective triggering in the absence of background wind and without taking into account the modification of the sea-breeze circulation by the developing convection. See text for symbol meaning.

the velocity vector **v** that is tangent to the contour. The acceleration of the circulation C is then given by:

$$\frac{dC}{dt} = \frac{d}{dt}\oint \mathbf{v} \cdot d\mathbf{l} = \oint \frac{d\mathbf{v}}{dt} \cdot d\mathbf{l}. \tag{12.13}$$

Neglecting Coriolis and frictional forces for simplicity yields

$$\frac{dC}{dt} = -\oint \frac{1}{\rho}\nabla p \cdot d\mathbf{l} = -\oint \frac{1}{\rho}dp = -R_d \oint \frac{T}{p}dp, \tag{12.14}$$

where we have used the ideal gas law, the momentum equation as well as the fact that the line integral of gravity around a closed surface is zero. Integrating over the solenoid ABCDA enclosed by the isotherms T_1 and T_2 and the isobars p_1 and p_2 (see Fig. 12.11) finally gives

$$\frac{dC}{dt} = -R_d(\ln(p_2) - \ln(p_1))(T_2 - T_1). \tag{12.15}$$

Eq. (12.15) predicts in the case of Fig. 12.11 a thermally direct circulation, with the near-surface flow from cold to warm anomalies, as expected and observed. It is evident from this equation that lines of constant pressure and constant temperature intersect to yield the circulation, and this is referred to as a solenoid configuration. Thermally driven circulations are shallow circulations extending at best up to the depth of the PBL.

Inserting realistic numbers for pressure and temperature differences in Eq. (12.15) yields an acceleration of the circulation that is too rapid compared to observed values. This mainly results from neglecting friction, as friction acts to slow down the acceleration of the circulation. Furthermore, the advection of the cool air over the

warm surface reduces with time the initial temperature difference between the two surfaces.

The propagation of the cool air over the warm surface can alternatively be explained in terms of the propagation of a dense fluid into a lighter fluid. Theory about the propagation of density currents in the simplest case of a dense fluid of virtual potential temperature $\theta_{v,1}$ released into a lighter fluid of potential temperature $\theta_{v,2}$ predicts a propagation velocity c that has to be proportional to

$$c \sim \sqrt{gh\frac{\theta_{v,2} - \theta_{v,1}}{\theta_{v,2}}}, \tag{12.16}$$

with h the depth of the cool fluid. Comparison with observed propagation velocities of thermally driven circulations reveals a surprisingly good agreement, with a proportionality factor around 0.5.

Eqs (12.15) and (12.16) as well as any other theoretical models for sea breeze-like circulations highlight the fact that thermally driven mesoscale circulations merely owe their existence to the production of a sufficiently strong pressure gradient force for a given atmospheric state. The produced pressure gradient force typically allows sea breezes to propagate inland with a speed of a few m s^{-1}. Hence, the presence of stronger background wind, depending on its direction, can obviously destroy the circulation through the production of strong turbulence, homogenisation of the temperature field at the coast and dilution of the pressure gradient force. As a rule of thumb, thermally driven circulations are generally not observed past a mean wind of 5 m s^{-1}, except in cases of mountain wind circulations, where much larger velocities are feasible. Likewise, strong turbulence, as on a very unstable day, will tend to vertically dilute the temperature gradients and thus prevents the circulation to develop. In opposition and as evident from Eq. (12.15), an increase in the temperature gradient is beneficial for the circulation and results both in an increase of its strength (vertical extent) and propagation velocity.

Another important factor influencing the circulation characteristics is the Coriolis parameter f. The latter yields a clockwise rotation of the onshore wind in northern hemisphere mid latitudes as the day progresses. The effect of the Earth's rotation is only visible for circulations with a significant horizontal scale and in the absence of other factors influencing the pressure gradient force, for instance, changes in the synoptic pressure pattern.

A last class of factors to consider are geometrical effects. The type of the transition between sea (or cold patch) and land (warm patch), being sharp or smooth, is irrelevant. Spatial variations of scale of 100 km, which corresponds to the value of the local Rossby radius of deformation for the PBL in mid latitudes, have theoretically the largest influence on the generation of thermally driven circulation. At the other scale end, a minimum patch size of 5–10 km is required to be able to sustain a circulation. Of interest is the fact that real surfaces exhibit heterogeneities on a wide range of spatial scales. The atmosphere acts here as a kind of medium-pass filter to the various heterogeneity length scales and favours particular wavelengths depending on the synoptic conditions.

12.2.4.2 Response of Shallow and Deep Convection

The presence of a thermally driven circulation strongly modulates the convective lifecycle. Compared to homogeneous surface conditions experiencing same domain mean surface fluxes, the initiation of convection is more rapid over heterogeneous surfaces. This follows from lower convective inhibition over the warmer patches due to a warmer and more rapidly growing PBL. Also, the vertical velocity associated with the solenoid circulation helps the triggering of convection.

The development of shallow or deep clouds is further locked on preferred locations and generates typical spatial patterns (see Figs. 12.1 or 12.11). Cloud generation is favoured at the convergence zone of the induced circulation on the warm patch, as long as the atmosphere is not too dry. Depending on the direction of the background wind, this results in clouds aligned on the upstream side of warm patches. The mechanisms favouring the development of convection at the convergence zone are two-fold. On the one hand, the convergence provides additional energy for vertical lifting, which expresses itself in increased vertical velocity and favoured cloud development. This is a purely dynamical effect. On the other hand, the atmospheric conditions at the convergence zone are more favourable to the development of convection, especially due to higher equivalent potential temperature. The relative importance of these two mechanisms depends upon the overall synoptic situation. When the convective inhibition is strong, strong additional lifting is needed to break through the inhibition barrier, that is, the dynamical effect dominates.

As the temperature gradient driving the circulation does not stay constant in time, the location of the convergence front and of the associated clouds and precipitation exhibits corresponding temporal variations. Taking again our sea-breeze circulation as an example, the sea-breeze front will typically move from the sea to the interior of the continent in the morning and early afternoon hours. As the sun disappears and the land begins to cool more effectively

than the sea, the land breeze front will propagate back from the land towards the sea. Over islands, this results in a typical diurnal pattern, with preferred propagation of convective systems from the sea to the land during the day, and vice versa during night.

Finally, the circulation often impacts cloud cover and precipitation amounts. The response of the precipitation depends non-linearly on the combination of atmospheric profile and scale of the surface heterogeneity, making an a priori prediction difficult. As a general rule and for same domain mean flux, precipitation is enhanced over heterogeneous surfaces. Changes in cloud cover depend non-linearly on the initial atmospheric profile, the scale of the surface heterogeneity and the amount of precipitation. Heterogeneous surfaces tend to exhibit higher cloud cover than homogeneous surfaces for the same domain mean flux, although the production of stronger precipitation can actually lead to a decrease in cloud cover.

12.2.4.3 Modification of Thermally Driven Circulations by Convection

Thermally driven circulations obviously affect convection, but the development of convection also feeds back on the characteristics of the circulation. Convective clouds generate their own circulation (see Chapter 9), which superimposes on the background circulation. This alters the pressure gradient across the front and the circulation characteristics (see Eq. (12.15)). One typical alteration is that the return current of the initially shallow thermally driven circulation merges with the cloud outflow at the cloud top.

Clouds also reflect solar radiation. The induced cooling on the warm patch acts to reduce the temperature gradient between warm and cold patches. This theoretically means a slow down of the circulation, although this effect is not easily observed in large eddy simulations of convection interacting with a thermally induced circulation.

More importantly, the formation of precipitation drastically modifies the structure of the PBL. Melting and evaporation of hydrometeors lead to the formation of dense downdrafts which spread as cold pools when they encounter the surface. Cold pools are density currents propagating at the Earth's surface. Their propagation speed is proportional to the buoyancy difference between downdraft and environmental air (see Eq. (12.16)). The latter is different and generally larger than the thermal difference between warm and cold patches. In a case of a well-defined line of convection sitting at the edge of the sea breeze, for instance, the formation of cold pools thus accelerates the inland propagation of the front but still keeps the convective cells aligned on the front. In a more complex situation with the presence of multiple background thermally driven circulations propagating from different directions, cold pools from different convective systems can either collide with each other or cold pools from one convective system can collide with the front from a thermally induced circulation propagating from another direction. This leads to the formation of secondary convective cells away from their original location. Islands or narrow peninsulas, like the Californian Peninsula, are the places where this complex spatial pattern of interactions between sea breezes, cold pools and convection can be frequently observed.

Adding a last degree of complexity and, as already mentioned in Section 12.2.3.3, precipitation replenishes soil moisture and generates wet and dry patches at the Earth's surface, which can lead to the development of soil-moisture induced circulations. Starting from an initially homogeneous surface and in the absence of background wind, this leads to a random distribution of precipitation with time. Convection is triggered over a dry patch due to the presence of the soil-moisture induced circulation, the ensuing precipitation moistens the dry patch, alters the soil moisture gradient, and the next convective cell is triggered at another location. In the presence of background wind, persistent spatial rainfall and soil moisture anomalies can develop. Convection is still triggered over the dry patch, but, due to flow advection, the convective cells are transported away. Convection is enhanced while travelling over the wet patches and preferentially precipitates over the wet patches. Starting from an initially heterogeneous surface instead of a homogeneous surface, where, for instance, the heterogeneity results from the presence of two different vegetation types, the vegetation breeze forces the development of convection on the warm and dry patch. Precipitation falls on the dry patch, increases the soil moisture of the dry patch, enhances E and thus decreases the initial spatial variations in surface fluxes. In most cases, this effect is nevertheless not strong enough to mask the initial surface heterogeneity, even after many days.

12.2.5 Land Surface Influence on Large-Scale Circulation

As seen in the previous section, pressure gradients at the Earth's surface due to temperature gradients generate circulations. This type of thermally driven circulation is nevertheless not confined to the mesoscale. If the perturbation in the surface state is of large scale, then the surface can alter the development of convection through alteration of the large-scale circulation. A large-scale increase in H following a large-scale and long-lasting perturbation in the surface state induces a corresponding warming, which can produce a heat low. The mechanism is similar to the one described in Section 12.2.4.1 and on Fig. 12.11 to explain the formation of a sea breeze. A heat low manifests itself in divergent flow conditions aloft and convergence at the lower levels. The altered flow direction modifies the advection of moisture into the considered region and thus the precipitation distribution. Depending on the location of the region and prevailing large-scale circulation, precipitation may increase or decrease.

Examples of large-scale perturbation of the land surface are given in Section 12.3 with continent-wide deforestation (Section 12.3.1.1) or continent-wide drying during heat waves (Section 12.3.3.3). Another example would be snow cover anomalies during wintertime. The best-known example of land surface influence on the large-scale circulation is nevertheless the monsoon and in particular the Indian monsoon. The term monsoon refers to the seasonal reversal of the wind, blowing towards the continent during summer and towards the ocean during winter. Even though new or updated theories on the formation of monsoons have emerged (see Chapter 8 for more details), the basic ingredient of the monsoon, namely the differential heating between a large land mass and the ocean, was already uncovered in 1686 by Halley, before Hadley could actually explain the mean large-scale atmospheric circulation in the tropics now known as the Hadley circulation. Large-scale land surface effects can not only directly modify the large-scale circulation but can generate remote effects on the climate of other regions due to the triggering of planetary waves.

The potential modification of the large-scale circulation by the land surface is an efficient way to alter the precipitation distribution, as we have learned in Section 12.2.2, as the advection of moisture by the large-scale circulation provides the main moisture source for precipitation. Along somewhat similar lines, a local change in the precipitation regime over a region due to a local perturbation in the surface state can impact the precipitation regime far downstream due to corresponding alterations in the advected moisture content.

12.2.6 What Do Observations and General Climate Models Tell Us

Up to now we have learned different pathways by which the land surface and the atmosphere can interact. Some of them lead to positive feedbacks, some of them lead to negative feedbacks and all of them may happen concurrently. Here we apply the developed conceptual frameworks to observations and outputs from global climate model simulations to assess their relevance.

Many features of the precipitation climatology are very direct and visible manifestations of the effect of an underlying spatially heterogeneous land surface and can be understood from the considerations given in Section 12.2.4. Those include the precipitation maxima over the mountain ranges, some of the precipitation features associated with the monsoon circulation as well as the precipitation distribution over islands and coastal areas. Land–sea breeze, mountain–valley wind circulations and lake breeze are indubitable, whereas observations of vegetation breezes are less common. Earlier studies have conveyed the view that vegetation breezes are rather a local and infrequent phenomenon and that the effect of such breezes in model simulations was overestimated due to the use of idealised heterogeneity patterns and idealised synoptic conditions. The more recent view is that vegetation breezes can actually be observed over a wider range of synoptic conditions than previously assumed. This has been especially documented for the Great Plains in North America, the Amazon and Sahel region.

Like for vegetation breezes, little direct evidence of soil moisture-induced circulations has first been found in observations. With the development of soil moisture observations from space and through recent dedicated measurement campaigns (such as the African Monsoon Multidisciplinary Analyses, AMMA, over West Africa), widespread indication of soil moisture-induced circulation has been detected in semi-arid regions.

The strength and sign of the coupling between the land surface, in particular soil moisture, and the atmosphere, in particular precipitation, has received significant recent attention. It has been assessed using global or regional climate models, where the involved processes are parameterised, convection-permitting simulations with grid spacings in the kilometre range, where convective processes are explicitly resolved, and observations. Those different approaches have led to conflicting results.

In the model world, the strength and sign of the coupling is diagnosed by conducting sensitivity experiments with surface conditions perturbed at the initial time and comparing the cloud, precipitation and atmospheric response to a control experiment starting from an unperturbed surface state. This especially allows the filtering out of the effect of synoptic variability on the results, which, in observations, is a thorny challenge. From such types of experiments, see Fig. 12.12b, the usual response of regional or global climate models is a positive feedback: an increase in E through an increase in soil moisture (reduced r_s) enhances convection, yields an increase in precipitation, an increase in soil moisture and hence an increase in E.

Performing the same experiments with a storm-resolving model, as displayed in Fig. 12.12a, nevertheless yields the opposite response. An increase in E through an increase in soil moisture yields a decrease in precipitation, which reduces the soil moisture and dampens the initial anomaly. The feedback is negative. The convection-permitting model thus simulates more precipitation over drier soils, whereas the regional climate model simulates more precipitation over wetter soils!

And what do observations tell us? On a local scale, point measurements that have been taken at Cabauw (the Netherlands) and over the Sahel during the AMMA field campaign confirm the existence of atmospheric regimes where cloud formation is enhanced either over wet or over dry soils. Applying indexes to sounding data to a priori infer the sign of the soil moisture–precipitation feedback (see Section 12.2.3.2) reveals a predominantly positive feedback, with the exception of a few regions. In contrast, directly diagnosing the soil moisture–precipitation feedback from global observations of soil moisture and of precipitation taken from satellites reveals a predominantly negative feedback.

Those confusing results between models, observations and different studies can be partly understood. First, there is no guarantee that parameterisations, as used in atmospheric models, can properly reproduce the coupling between the surface and the atmosphere. This is quite strikingly demonstrated by Figs. 12.12c, 12.12d and 12.12e. Altering or replacing the default convective parameterisation of the regional climate model affects the strength of the simulated soil moisture–precipitation feedback, up to producing a negative feedback when the Kain–Fritsch–Bechtold scheme is used in place of the Tiedtke scheme. Note that there is no particular reason that speaks for using one convection scheme over another one. Also, given the coarse grid spacings of regional or global climate models, thermally driven circulations are generally not well (if at all) represented. They are resolved in higher-resolution simulations but the use of a convective parameterisation can distort the representation of the interactions between convection and thermally driven circulations.

Second, observations suffer from several drawbacks: the effect of synoptic variability that is difficult to filter out, the large uncertainties characterising global soil moisture observations from satellites, or the good quality but insufficient spatial coverage of point measurements.

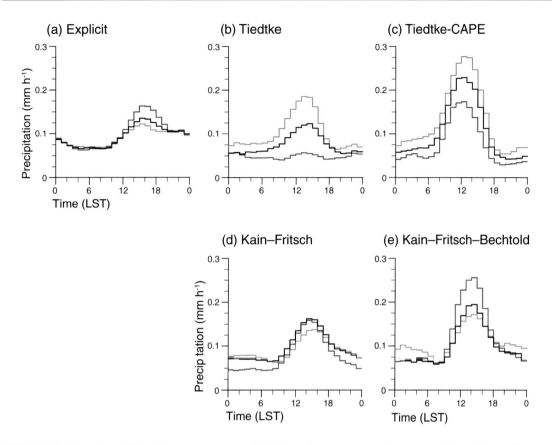

FIGURE 12.12: Dependence of the soil moisture–precipitation feedback on the representation of convection. (a) shows the monthly mean (July 2006) diurnal cycle of precipitation (mm h^{-1}) averaged over the Alpine region for three convection-permitting model integrations (resolution 2.2 km), where convection is explicit. Black is the control run, blue starts from wetter soil moisture conditions (control soil moisture + 30%) and red from drier soil moisture conditions (control soil moisture −30%). (b–e) represent the results of similar simulations but performed at 25 km resolution with parameterised convection: (b) original Tiedtke scheme, (c) Tiedtke scheme with convective available potential energy (CAPE) closure, (d) Kain–Fritsch scheme and (e) Kain–Fritsch–Bechtold scheme. Figure adapted from Hohenegger et al. (2009). Copyright original Figure ©2009 American Meteorological Society (AMS)

Third and maybe most importantly, different studies have used different methodologies to quantify the strength and sign of the coupling between the land surface and the atmosphere. The chosen analysis methodology, depending on its design, emphasises one or the other feedback described in Sections 12.2.1–12.2.5. Surface heterogeneity (Section 12.2.4) favours rain over dry soil. This is a spatial effect due to a gradient in the surface conditions (e.g., Fig. 12.5). For the temporal variations, the feedbacks mentioned in Sections 12.2.2 and 12.2.3 are of relevance, where a positive feedback seems to be more likely. Hence, the emerging idea at the time of writing is that the feedback between soil moisture and precipitation is negative due to the generation of thermally driven circulations between adjacent wet and dry surfaces. Soil moisture spatially homogenises precipitation. But one expects stronger precipitation (over the dry patch) if the

soil moisture of the full region (dry and wet patches) is higher than usual.

12.3 The Role of the Land Surface in the Climate Record

Sections 12.1 and 12.2 have isolated the basic processes and building blocks that allow the land surface and atmosphere to interact. In Section 12.2.6, we already highlighted typical footprints of the land surface on present-day climate. Such footprints arose from natural variations in the land surface state, for instance, because a forest exists next to a lake; or because precipitation continuously modifies soil moisture. Here we follow up upon this discussion with the goal of understanding the response of the fully coupled land–atmosphere system to externally forced and long-lasting land surface changes. Our discussion not only includes anthropogenic alterations of the land surface, which constitutes the obvious example of externally forced land surface changes, but also the response and possible climatic effects of changes in the land surface triggered by climate changes, both past and future.

Due to the absence of adequate observations, much of this discussion relies on results from global climate simulations, where the involved interactions are at best parameterised (see Section 12.2.6) and where the representation of several land surface processes remains uncertain. This concerns the representation of vegetation dynamics or the fact that the groundwater response is drastically simplified or neglected by most global climate models. The Phase 5 Coupled Model Intercomparison Project multimodel ensemble, whose models are typically used to assess effects of the land surface on the climate system and changes thereof, exhibits systematic biases in the representation of temperature, precipitation and E in the present-day climate. Well-known biases are the dry (too little E, too little precipitation) and warm bias over the mid-latitude land area during summertime, the dry bias over the Amazon region and underestimated land-to-ocean precipitation ratios. What follows therefore needs to be interpreted keeping these shortcomings in mind.

12.3.1 Anthropogenic Effect on the Land Surface

Humans live on land and as such have continuously reshaped it according to their needs. Food, goods and energy production, settlements or consumption of natural resources, all these activities leave their fingerprint at the Earth's surface. The fingerprint can be long lasting and at a large scale, as with continental-scale deforestation. Depending on the type of alterations and the climate of a region, they have the potential to feed back on the local, regional and large-scale atmospheric conditions on diurnal to centennial timescales.

12.3.1.1 Deforestation

Most parts of the world have experienced periods of strong deforestation during their history. Europe and China have been deforested in the ancient times with the first developing civilisation and were largely free of trees already ~2,500 years ago. Deforestation in North America started with the European colonisation and persisted into the twentieth century. Some areas of Europe, China or North America have seen their forests regrowing. In contrast, deforestation in the tropics started in the mid-twentieth century and is still ongoing.

Replacing forests by open land pastures has four main and direct effects on the land surface. Assuming no change in the atmospheric conditions, forests have a lower albedo, higher roughness length, they transpire more and they intercept more water than open land. Tropical forests also have very deep roots, more than 10 m in the Amazon, which give them access to months of water storage and hence allow them to maintain high E and high carbon uptake through dry season spells.

In terms of temperature, lower albedo leads to a warming (Fig. 12.6), whereas the higher roughness length and the higher transpiration lead to a cooling (Fig. 12.9). Whether an overall warming or cooling results, depends upon the prevailing atmospheric conditions and hence geographical location. Tropical forests cool the atmosphere due to the large availability of water vapour, whereas boreal forests warm due to weak E. The response of temperate forests is debated as different climate models exhibit responses of different signs.

Changes in precipitation due to deforestation are also a matter of debate. Many studies have looked at Amazon deforestation. A full deforestation, as simulated with global climate models, yields a spatial reorganisation of the precipitation pattern characterised by a dipole structure: the central and western Amazon get drier, whereas the eastern Amazon sees its precipitation amounts increase. Averaged over the full Amazon region, precipitation decreases. Those changes are the result of changes in the large-scale circulation (Section 12.2.5). In opposition, a partial deforestation yields an increase in precipitation due to the development of vegetation breezes (Section 12.2.4). In observations, no trend can be detected as the signal is masked by synoptic variability, especially the strength of El Niño events.

12.3.1.2 Urbanisation

Almost 55% of the world population lives in urban areas. Cities create their own microclimate that is different from surrounding land areas. The most well-documented effect of urbanisation is the urban heat island effect. It is beyond doubt that air temperatures are higher within a city than in its rural surrounding. Urban heating is especially pronounced during night-time, resulting in a smaller diurnal temperature range. It is also more pronounced on clear days and under light wind conditions. The presence of wind and clouds affects the surface energy balance and the interaction between the surface and the overlying air, and can offset the urban heat island effect. The strength of the urban heat island effect in general depends upon meteorological conditions, geographical location, city size and city design. For a city of 1 million people, the annual mean air temperature may be 1 °C–3 °C warmer than the surrounding rural countryside.

Several factors contribute to the urban heat island effect. The albedo characteristics of cities are different from rural areas due to different material properties and geometrical effects. The end result is, during daytime, a larger absorption of solar radiation in cities, which warms the air. At night, urban areas loose less heat than their surroundings for various reasons: the presence of tall buildings leads to an obscuration of the sky, preventing the heat escaping from the city; the higher thermal emissivity and higher heat capacity of building material reduce the night-time cooling within cities; and the presence of pollutants enhances the downward longwave radiation. Besides such thermal effects, the reduced E (increased H) in the city and the production of waste energy from human activity also contribute to the urban heat island effect, although the contribution of waste energy is not thought to be dominant. Planting vegetation, changing the colour or material properties of roofs and buildings are techniques that have been proposed/are used to mitigate the effect of urbanisation on the local climate. The localised higher temperatures in urban area compared to their surrounding can support the formation of a thermally driven circulation, as described in Section 12.2.4.

Urbanisation not only alters the air temperature. The strong drag of building reduces the mean wind except where channelling of wind through a street canyon occurs. The reduction of wind speed supports the urban heat island effect. Problems of air quality in urban areas are a well-known issue. The higher concentration of aerosols over cities may impact the formation of clouds and precipitation (see Chapter 11) and supports a downstream enhancement of precipitation. From the arguments presented in Section 12.2, some impact on convective precipitation can be expected, as cities represent islands of high H and low E. Systematic urban effects on precipitation are nevertheless difficult to find in observations due to the strong influence of the synoptic conditions on the precipitation amounts. When precipitation occurs, run-off is nevertheless increased in urban areas, leading to less groundwater formation compared to the neighbouring countryside.

12.3.2 Paleoclimates

Even without anthropogenic alterations of the land surface, changes in the land surface state due to externally forced climate changes may feed back on the climate state. The basic principle is that interactions between the land surface and climate, as described in Section 12.2,

may induce non-linearities in the system and modify the response of the climate system to a change in external forcing. We present two often mentioned examples from the past record where such effects have likely been at play.

12.3.2.1 Sahara Greening

Paleorecords reveal that northern Africa experienced several humid periods in its history. The African humid period denotes the last of such periods. It started around 15,000 years BP. What is now the Sahara desert was a green landscape of grasslands and trees with numerous lakes and rivers, providing food for various mammals and with records of human activity. The African humid period terminated about 5,000 years BP. By this time, the verdant conditions had been replaced by arid, sandy conditions, and human activity had moved to more fertile areas, like the Nile delta.

Differences in Earth's obliquity and especially precession between the early to mid-Holocene and present-day conditions primarily explain the changes in vegetation cover and in the northern African climate. The mid-Holocene orbital parameters allowed for a larger summer insolation in the northern hemisphere. This strengthened the West African monsoon, allowing a more northwards extension of the monsoon compared to today. The state of the meridional overturning circulation, of the Indian monsoon and of the ice sheets during the mid-Holocene, as well as resulting ocean–atmosphere interactions further contributed to the greening of the Sahara.

Although the basic mechanism linking orbital changes and monsoonal climate response and hence Sahara greening is well known, the details of the termination of the African humid period are subject to debate. A key question is whether the termination was abrupt (on the order of centuries) or not with a slow decline in vegetation cover, on the order of millennia. Paleorecords, due to their sparseness, and climate model simulations, due to their uncertainties, are inconclusive and often in disagreement with each other.

Orbital forcing alone would lead to a slow decline in vegetation. Vegetation–atmosphere interactions have been postulated as a non-linear mechanism able to amplify the changes externally triggered by the orbital forcing and supporting an abrupt termination of the African humid period. One of the first arguments for such a behaviour was expressed by Joseph Otterman and shortly after by Jules Charney. The idea is that deserts make themselves: a reduction in vegetation cover increases the albedo and, all other things remaining equal, the temperature at the Earth's surface cools. This has already been described in

Section 12.2.1. Joseph Otterman argued that under cooler temperatures, the air would have less tendency to rise, suppressing the formation of deep convection, whereas Jules Charney argued that the change in the energy balance of the atmosphere resulting from the change in surface albedo would lead to stronger subsidence; the subsiding air would begin to cap the weakly ascending air, suppressing deep convection and maintaining drier conditions.

In reality, all other things do not remain equal. Reproducing mid-Holocene conditions with global climate models coupled to dynamic vegetation models, where the vegetation may shift or die depending on plant types and climatic conditions, revealed various outcomes: slow decline, multiple stable states or abrupt transition. In the latter case, the vegetation starts decreasing in response to the orbital forcing, the decrease of the vegetation increases the albedo and increases r_s, E decreases (see Figs. 12.6 and 12.9), precipitation decreases and the drying of the atmosphere is accelerated. Below an ecological threshold, the plants collapse, leading to an abrupt climate change. The occurrence of such tipping points presupposes a positive feedback between vegetation and precipitation. The existence of tipping points also depends upon the initial perturbation, the atmospheric conditions, the ecosystem characteristics and the model used. For instance, some plants are more resilient to drought than others: mixing different plant types affects the response of the system, whereby plant diversity tends to attenuate the instability of the system.

Although results of mid-Holocene simulations should be interpreted carefully given the intrinsic and large uncertainties, they do reveal the possible existence of metastable vegetation states. Through vegetation–climate interactions, those states can flip between one and the other in response to an external forcing. A less dramatic expression for metastable states is that a specific climate may sustain two different vegetation populations starting from the exact same meteorological conditions. Here a sparsely vegetated land surface may also maintain itself by its drying effect on the climate, whereas a densely vegetated surface may as so much maintain itself by its moistening effect on the climate. To what extent such effects have played a role in Earth's climate history and are still playing a role remains an open question.

12.3.2.2 Glacial–Interglacial Cycles in the Northern Latitudes

During the Quaternary, the Earth went through several episodes of warm and cold climates. It is now beyond doubt that such a succession of glacial and interglacial phases is caused by changes in the orbital parameters of the Earth, the well-known Milankovitch cycles. The slow forcing of the orbital parameters on Earth's climate may have been amplified by land–atmosphere interactions, as already suggested in our discussion of the Sahara greening in the previous subsection. The presence of forests in the northern latitudes is another mechanism, by which the climate response during glacial–interglacial cycles may have been amplified.

The presence of forests (taiga) in the northern latitudes, and especially its interaction with snow, is important in determining the air temperature. Snow falling on open ground, like tundra, leads to cool air temperature due to the high albedo of snow (Fig. 12.6). If forests are present, snow mostly falls through the forest canopy and accumulates on the ground. The effect of snow on the incoming solar radiation is thus partly hidden by the overlying forest canopy. The consequence is that, for the same depth of snow, a region covered with forest exhibits a lower albedo than a region covered by grass. Albedos for snow-covered forest are in the range of 0.2–0.4, whereas snow-covered grass exhibits albedo in the order of 0.75. What are the implications for glacial–interglacial cycles in the northern latitudes?

Let us start from a warm period. Given that the climate is warm, the trees extend further north than during a cold period. Now the orbital parameters slowly move the Earth's climate towards a cold phase. As the climate cools, the tree line retreats southwards and taiga is replaced by tundra in the northern latitudes. The higher albedo of snow-covered tundra compared to snow-covered taiga means colder temperatures, amplifying the initial perturbation (positive feedback) and preventing tree growth. This presupposes that all other impacts of a replacement of taiga by tundra, especially those involving changes in E, does not counteract the direct albedo's effect.

Now that we understand the basic idea let us look at a warm period to illustrate some important subtleties. Some 9,000–6,000 years ago, the Earth's climate was warmer than today (but colder than it is projected to be in 100 years), both during winter and summer, in Europe and in the northern latitudes. The Earth's orbital parameters for that time imply a stronger insolation during summer but a weaker insolation during winter compared to today. The winter response is thus inconsistent with the expected change from a pure orbital perspective. The unexpected warming in winter results from a combination of ice albedo feedback and of forest-snow albedo feedback. Warmer summers due to the prevailing orbital parameters lead to a melting of ice and snow, which amplifies the warming/reduces the winter cooling. Likewise, warmer summers support the development of forests which further amplifies the warming. Taking both feedbacks separately, each of them is not strong enough to counteract the expected winter cooling. But taken together, they are able to offset the winter cooling. This illustrates the potential importance of the synergy between feedbacks.

12.3.3 Future Climate

If land–atmosphere interactions have been important in modulating past climate variability, they may continue to do so in the future. An extensive discussion about aspects of climate change, especially involving clouds, can be found in Chapter 13. Here we focus on aspects more closely related to the land surface and to the role of land–atmosphere interactions in modulating the climate change signal.

12.3.3.1 Land–Atmosphere Contribution to Change in the Mean Climate

For many areas, global climate models predict a decrease in soil moisture with increasing carbon dioxide as a result of the surface warming and decrease in precipitation. A method frequently used to uncover the impact of such soil moisture changes on the climate change signal consists of reintegrating a global climate model by prescribing its soil moisture to a reference climatology and comparing the result of this experiment to a control experiment where the soil moisture is allowed to vary freely.

Results of such experiments indicate a potential contribution of soil moisture to the climate change signal over specific regions. This should not come as a surprise because soil moisture-limited regimes tend to be found either over arid/semi-arid regions or during summertime in temperate climates (remember Fig. 12.4 and Section 12.1.5). The feedback between the soil moisture and the temperature leads to an additional warming of 1–1.5 K for the late twenty-first-century climate (2071–2100) (Seneviratne et al., 2013). The change seems to primarily result from a change in the Bowen ratio (see Fig. 12.9), whereas the total surface flux remains unaffected. Effects of land–atmosphere interaction on the changes in mean precipitation are small and not robust across different models.

Under global warming, climatic zones are also expected to shift. The semi-arid regions gain terrain polewards, whereas the higher latitudes experience a more temperate climate. Such climatic shifts affect the native vegetation of a region: some species might die out or get threatened by new invasive species. Phenological observations have already revealed alterations of the native vegetation distribution and of the characteristics of the growing season, with, for example, a lengthening of the growing season in temperate climates. Again, changes in vegetation types and/or growing season characteristics imply changes in transpiration (as well as in albedo and run-off) which feed back on the climate of a region and, depending on the significance and extent of such changes, may induce large-scale circulation changes.

An even clearer and stronger effect of the land surface on the temperature change signal can be found in the polar region and is known as polar amplification. Melting of sea ice and snow cover is thought to be the primary reason. It leads to a stronger warming at the pole than at lower latitude through albedo feedbacks (see Section 12.2.1). More details on polar amplification are given in Chapter 10.

As a last, often-discussed, but still highly uncertain effect of the land surface on the mean climate, we mention the potential release of methane from permafrost. Carbon was deposited in the permafrost during the last ice age, when organic material (plants, animals) froze into the ground. Large deposits of carbon have been found in Siberia. The total amount of carbon stored in permafrost is estimated to be larger than the amount of carbon currently present in the atmosphere. As the climate warms, soil temperatures increase and the permafrost thaws. The stored organic material (carbon) may decay. Depending upon circumstances, carbon dioxide or methane (if no oxygen is present) are produced. On the one hand, this carbon dioxide or methane may be released to the atmosphere, increasing the greenhouse effect. On the other hand, warmer temperatures in the northern latitudes also imply a longer plant growing season, more transpiration and thus more carbon dioxide removal from the atmosphere. Whether the Arctic may end up as a sink or a source for carbon dioxide, how much carbon dioxide and methane will be released, and how much carbon is exactly stored in permafrost are all open questions.

12.3.3.2 Land–Atmosphere Contribution to Change in Variability

Not only the mean climate is changing in response to carbon dioxide, but its variability on daily to interannual timescales may change as well. Changes in variability are of high interest, since they are linked to changes in extremes. For instance, if temperature variability increases, the width of the statistical distribution will increase, implying increases in the frequency of cold and hot extremes. This type of mechanism in combination with land–atmosphere feedback processes has been proposed to explain the European summer heat wave of 2003

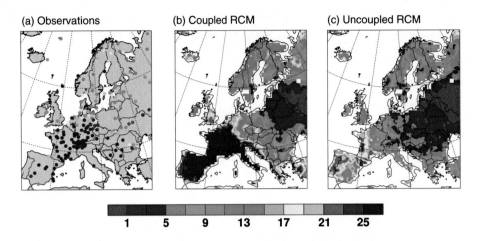

FIGURE 12.13: Contribution of soil moisture to the European 2003 heat wave, taken from Fischer et al. (2007). The panels show the number of hot days derived from (a) observations, (b) a regional climate model simulation, (c) the same regional climate model simulation but with uncoupled land surface. In the latter case, the soil moisture is prescribed at each time step by its climatology (represented by the average annual cycle from the coupled simulation). Copyright ©2007 American Meteorological Society (AMS)

(see Section 12.3.3.3). Climate models project increases in interannual and daily temperature variability, and some observational studies have detected statistically significant variability increases during the last decades over parts of Europe.

Climate models have been used to quantify the contribution of land–atmosphere interactions to the predicted changes in variability. From such studies it appears that the strength of the land surface contribution is highly model dependent, as it depends upon the strength of the coupling between the land and the atmosphere, which is itself highly model dependent. Favourable conditions for a broadening of the temperature distribution by land–atmosphere interactions are regions (or models) that experience a shift from a previously energy limited to a soil moisture-limited regime as the climate warms.

12.3.3.3 Heat Waves

Heat waves usually receive tremendous attention from the public and the media. This is due to their far-reaching consequences that may involve increased mortality, forest fires, water scarcity, crop failure and other impacts. Examples of extraordinary heat waves in the extratropics include the Chicago heat wave (1995), the European heat wave (2003), the Russian heat wave (2010) and the Australian heat wave (2012/13). For Europe and some other regions, model simulations project increases in temperature variability, and heat waves are expected to become more frequent in the future. But do land–atmosphere interactions play any role in their generation?

Extratropical heat waves first and foremost require a specific large-scale constellation, especially a strong and

persistent high-pressure system, to develop. Although soil moisture does not directly cause heat waves (in the sense that it is unable to invert the signal associated with the large-scale atmospheric flow), it can affect the amplitude, extent and duration of such events. As the surface warms, the soil dries out. This increases r_s, reduces E, thus further enhances H, and ultimately increases the air temperature, as long as precipitation changes remain negligible (see Fig. 12.9). A decrease in soil moisture can further generate a heat low at the surface and a high-pressure system at higher levels (see Section 12.2.5). This supports the overall large-scale pattern and helps maintain the heat wave.

Fig. 12.13 shows an estimate of the contribution of soil moisture to the number of hot days during the summer 2003 heat wave. It is based on regional climate model simulations driven by observed lateral boundary conditions. Figs. 12.13a and 12.13b demonstrate that the control integration is credibly reproducing the observed heat wave. Uncoupling soil moisture (i.e., prescribing the evolution of the soil moisture according to climatology) halves the number of hot days over Western and Southern Europe. This type of feedback has also been supported using research-model seasonal forecasting systems, by demonstrating that the appropriate representation of the dry soils in the initial conditions in early summer increases the probability of a strong heat wave to occur.

The apparent existence of a positive feedback between soil moisture and heat waves does not constitute a special feature of the 2003 event. It has been confirmed for other past heat waves. The feedback strength may nevertheless vary from event to event and also depends upon the employed climate model. This is not surprising. We have seen in Section 12.2.6 that the interactions between soil moisture and the atmosphere, being parameterised in regional or global climate models, are uncertain. Different models couple the atmosphere differently to the land surface, yielding stronger or weaker contributions of soil moisture to heat waves. Moreover, almost all climate models support a positive soil moisture precipitation feedback. This can bias the temperature response. A model with a strong positive feedback rapidly shuts down precipitation and generally experiences a rapid drying and strong warming, whereas a model with a negative feedback can maintain a moderate warming.

12.3.3.4 Droughts

Besides heat waves, floods and droughts are very serious threats to society with potentially disastrous ecological, political and socio-economic consequences. In severe drought conditions, the reduction in water availability and associated crop failure can lead to a shortage in food supply and hence malnutrition, famine and abnormal mortality. Droughts are also believed to be important drivers of mass migration, political tensions and armed conflicts. Earth's history has experienced many droughts, but their effects vary from region to region, and strongly depend upon socio-economic circumstances. In wetter regions (e.g., Europe, North America), seasonal droughts can nowadays be managed by using the water supply systems. Dry regions (e.g., Sahel), in contrast, have little adaptation possibility given an already stressed water supply system.

Drought generally refers to a period of abnormally dry weather that lasts long enough to cause a serious hydrological imbalance. Different types of drought are distinguished: meteorological droughts, which result from a deficit in precipitation; soil moisture droughts, which are associated with a deficit in soil moisture; and hydrological droughts, which are due to negative anomalies in stream-flow, lake and/or groundwater levels. We focus here on soil moisture drought given the topic of this chapter.

Soil moisture droughts result either from missing precipitation or from too much evapotranspiration (see Eq. (12.1)). The lack of precipitation is more often considered as the main driver of soil moisture droughts, although many atmospheric forcings act towards both decreasing P and increasing E. For instance, large-scale subsidence in subtropical to extratropical regions will suppress convective precipitation and dry out the lower troposphere as a result of adiabatic warming and drying of the subsiding airmass. Persistence and pre-conditioning are further important aspects that determine the extent and severity of drought. Persistence is not only determined

by the persistence of atmospheric circulation anomalies (e.g., El Niño) but also from the persistence arising from the memory associated with soil moisture (see Section 12.1.8). Preconditioning relates to the state of the land surface, being soil moisture or vegetation, before the event. This means that the same precipitation deficit will affect regions differently or affect the same region differently at different times.

One difficulty in quantifying soil moisture droughts is that there are few direct observations of soil moisture. Various drought indices have thus been proposed to quantify soil moisture droughts. Some are based solely on precipitation, like the standardised precipitation index or the consecutive dry days index. Other indices include both precipitation and estimates of actual or potential evapotranspiration. As these indices emphasise different aspects of drought, they may disagree with each other and exhibit opposite trends. In contrast, soil moisture is directly available from model simulations, but most models have a very simplified representation of subsurface processes and usually lack a description of the major aquifers that are critical for drought. The temporal evolution of soil moisture conditions thus remains uncertain due to limitations in the representation of land surface processes and associated feedbacks between the land surface and the atmosphere (see Section 12.2.6).

Keeping in mind these difficulties in quantifying drought trend, droughts are projected to intensify, becoming longer and/or more frequent in many areas by the end of the twenty-first century. Particularly strong signals are projected for the summer climates of Southern and Central Europe, central and southern North America, Central America, north-east Brazil and southern Africa. In general, spatially coherent shifts in drought regimes are expected with changing global circulation patterns. A key role is played by variations of the Hadley circulation. The Hadley circulation is projected to expand polewards with climate change, thus the region affected by large-scale subsidence is expanding.

12.3.3.5 Floods

In contrast to drought, flood corresponds to the flooding by a stream or by another body of water of an area that is not normally submerged by water. Many types of flood exist, for example, river floods, flash floods, urban floods or glacial lake outbursts. The most frequent cause of flood is intense and/or long-lasting precipitation. Other causes include snow/ice melt, dam break or a clogged stream. The severity of the flood depends upon the characteristics of precipitation, like amount, duration and type, the drainage basin conditions and the characteristics of the land surface.

The latter includes the presence of snow and ice as well as anthropogenic alterations of the land surface (urbanisation, dikes, dams, etc.).

The intensity and frequency of flood is also expected to change under global warming. There are at least three reasons why climate change will lead to increases in floods. First, models project a general increase in global mean precipitation by about 2% per degree global warming. The signal is most pronounced in the high latitudes. The projected increase in precipitation is expected to lead to more frequent pluvial floods. Second, supported by theoretical, modelling and observational studies, there is evidence that the frequency of heavy precipitation events will increase much stronger than increases in precipitation amounts, by up to about 6% per degree warming. This increase is determined by the Clausius–Clapeyron relationship that reflects the moistening of the atmosphere in response to the warming. This process affects also areas with a projected decrease in mean precipitation, such as Southern Europe during the summer season. Many of the associated events are of small-scale convective nature and may have implications on erosion. Third, in regions experiencing snow fall, the increase in mean temperature implies a shift from solid to liquid precipitation, which is likely to affect the intensity and frequency of floods in such regions. In general, the storage of precipitation in the form of snow delays run-off formation and moderates the hydrological response.

Overall, observations of precipitation support this reasoning. Many studies have found an increase of mean precipitation in higher latitudes and also increases in the frequency of heavy events in mid latitudes during the last decades. However, despite the hydrological and meteorological arguments mentioned earlier, river gauge measurements reveal no widespread trend in floods over the last decades. This may be related to the scarcity of long-term homogeneous observations. Moreover, during the last 150 years, many river basins and lakes have undergone substantial anthropogenic influence. In response to urbanisation and agricultural pressures, authorities have sought to decrease the flood risks by engineering of streams, building of dams, regulating of lakes and so on. The river network in most countries is indeed highly regulated, and floods can be at times prevented. Finally, there is a high interannual and interdecadal variability in floods, as flood magnitude and frequency is highly sensitive to modest alterations in atmospheric circulation patterns. As a result of the large natural variability, the emergence of climate change-induced precipitation and hydrological trends from the natural background will take longer than changes in the large-scale warming.

Summary

Through its control on the water and energy budgets, the land surface has the potential to affect the properties of the boundary layer and hence the formation of clouds and precipitation. The effect of the land surface on the temperature is straightforward, with both a decrease in the surface resistance and an increase in surface albedo acting to cool the boundary layer. In contrast, the effect of the land surface on cloud and precipitation is complex. Taking a column view, a decrease in the surface resistance tends to favour cloud and precipitation via changes in the atmospheric profile: rain is favoured over wetter soils. In the presence of surface heterogeneity and in the absence of a strong background wind, though, the generation of thermally induced mesoscale circulations between surfaces of different properties forces the development of convection on the warmer surface: rain falls over drier soils. Beyond the potential existence of contrasting precipitation responses to land surface changes, isolating the resulting dominant response remains highly uncertain: in observations, the effect of land–atmosphere interactions is blurred out by the strong synoptic variability; in climate models, the magnitude of such interactions is affected by the employed parameterisations. With the advancement of computer technologies, it now becomes feasible to conduct climate simulations with convection-permitting resolutions, where at least part of the feedback loops between the land surface and convection can be explicitly resolved. This opens new perspectives to reassess the dominant feedbacks at play and, therefore, the role of vegetation and the land surface in the climate record including its susceptibility to change.

Further Reading

Section 12.1: Basic Land Surface Processes:
Rafael L. Bras, *Hydrology: An Introduction to Hydrologic Science* (Bras, 1990), as the title indicates, this book provides a good introduction to hydrologic science.

Section 12.2: Interactions between Processes:
Jordi Vila-Guerau de Arellano, Chiel C. van Heerwaarden, Bart J. H. van Stratum and Kees van den Dries, *Atmospheric Boundary Layer: Integrating Air Chemistry and Land Interactions* (Vilà-Guerau de Arellano et al., 2015) provides a software package and comprehensive exercises that allow students to investigate aspects of boundary-layer dynamics. The software is made of a bulk model of the PBL coupled to a simple land surface scheme. It

can be used to investigate the response of the dry PBL and of shallow clouds to changes in land surface conditions. Erik T. Crosman and John D. Horel, *Sea and Lake Breezes* (Crosman and Horel, 2010) provides a review of the factors that control the characteristics of sea and lake breezes, and hence factors that generally apply for mesoscale circulations due to land surface heterogeneities. The relationships are nevertheless only valid for a dry atmosphere.

Section 12.3: The Role of the Land Surface in the Climate Record:

Rezaul Mahmood, R. A. Pielke Sr, K. C. Hubbard et al., *Land Cover Changes and Their Biograophysical Effects on Climate* (Mahmood et al., 2014) provides an up-to-date review of the effects of land cover change on climate. Jonathan Adams, *Vegetation–Climate Interactions* (Adams, 2010) is one of the very few books presenting the effect of vegetation on the climate. Martin Claussen, A. Dallmeyer and J. Bader, 'Theory and modeling of the African Humid Period and the Green Sahara' (Claussen et al., 2017) is an encyclopaedia article that gives a detailed explanation of the factors that shaped the climate during the African Humid Period. Christoph Schär, P. L. Vidale, D. Lüthi et al., '*The role of increasing temperature variability in European summer heat waves*' (Schär et al., 2004) is the first article that introduced the idea that temperature variability has to increase as the climate warms to explain the observed record-breaking heat waves. R. Garcia-Herrera, J. Diaz, R. M. Trigo, J. Luterbacher and E. M. Fischer '*A review of the European summer heatwave 2003*' (Garcia-Herrera et al., 2010) is an article that gives a nice overview of the various studies that have looked at the European 2003 summer heat wave. Sonia I. Seneviratne, N. Nicholls, D. Easterling et al., 'Change in Climate Extremes and Their Impacts on the Natural Physical Environment' (Seneviratne et al., 2012) is a chapter of the Intergovernmental Panel on Climate Change report that deals with extremes. It provides an overview of the mechanisms behind extremes and expected changes.

Exercises

(1) Derive the Penman–Monteith combination formula.

(2) What is the recycling ratio of continental precipitation? Assume that the averaged net available energy, Bowen ratio and run-off over land are 65 W m^{-2}, 0.625 and 0.6 mm day^{-1}.

TABLE 12.2: Pressure, temperature and relative humidity from the observed atmospheric sounding

P (hPa)	T ([°C])	RH (%)
975.0	28.8	66.1
950.0	27.0	65.7
925.0	25.2	65.3
917.3	24.3	68.5
885.9	22.1	77.2
873.0	21.0	82.0
855.3	19.5	70.6
850.0	19.1	67.0
825.6	18.0	62.0
823.0	17.0	65.2
796.5	16.2	55.3
768.4	14.7	47.8
741.2	13.7	39.8
700.0	10.6	32.3
664.5	8.1	25.2
641.0	6.4	20.9

(3) We consider a soil of type sand of 1 m depth. The characteristics of the soil are: $\phi_{fc} = 0.196$, $\phi_{pwp} = 0.042$, $n = 0.364$, the heat capacity is 1.28×10^6 J K^{-1} m^{-3} and the resistance r_s is given by: $r_s = \frac{\phi_{fc} - \phi_{pwp}}{50(\phi - \phi_{pwp})}$. The soil is fully saturated, the net radiation amounts to 150 W m^{-2} and we neglect run-off and ground heat flux.

   (a) How long does it take for the soil to dry out?

   (b) The soil has now reached a soil moisture of 0.045. How strong is the sensible heat flux? Take a value of $r_a = 60$ s m^{-1} and assume that the atmospheric state does not change.

   (c) Imagine all this heat would be used to warm the soil. By how much would the soil warm? Take one day as a timescale.

   (d) Now imagine all the initial water in the soil would freeze and this energy would be used to warm the soil. By how much would the soil warm?

(4) We consider an heterogeneous surface constituted by a patch of forest and a patch of cropland. The forest has an albedo of 0.15 and a Bowen ratio of 0.5, whereas the cropland has an albedo of 0.25 and a Bowen ratio of 0.0625. Where does convection develop? Why?

(5) Table 12.2 contains observed values of pressure, temperature and relative humidity for a particular case. Is convection favoured over wet or dry soils? *Hint: apply the framework of Findell and Eltahir (2003).*

13 Clouds and Warming

Sandrine Bony and Bjorn Stevens

Clouds are among the most fascinating manifestations of weather, the most beautiful expression of atmospheric motions and energy transfers through the Earth system, and our vital source of rainfall. In addition to this, it has become well understood that in a context where anthropogenic activities increasingly perturb the radiation balance of our planet, clouds are likely to play a significant role in the fate of our climate over the next decades and centuries. For our generation, understanding and anticipating the behaviour of clouds in a changing climate has thus become a scientific challenge associated with a societal dimension.

This chapter addresses the issues of why and how clouds respond to global warming, and how it affects global mean surface temperature and precipitation patterns. After a discussion of some of the reasons why these issues remain a long-standing scientific challenge (Section 13.1), it reviews some of the concepts and frameworks that help understand the impact of clouds on global mean surface temperature (Section 13.2), shows how the problem may be decomposed into sub-problems and reviews ways through which these concepts can be applied to climate model simulations or observations (Section 13.3). In Section 13.4, it discusses some of the main physical processes that underly the response of clouds to climate change, and in Section 13.5 it assesses our understanding of the behaviour of precipitation in a changing climate. Some concluding remarks are given in Section 13.6, and suggestions for further reading are given at the end of the chapter.

13.1   A Long-standing Scientific Challenge

For more than thirty years, scientific assessments synthesising our physical understanding of the climate system and assessing the risks of future climate change have all identified clouds as a key source of uncertainty. The primary question is whether or not, and how, clouds depend on the state of the climate system, and thereby damp or amplify the climate response to an external perturbation. This is the question of cloud feedbacks. The most glaring expression of the uncertainty in cloud feedbacks is the imprecision in estimates of climate sensitivity, that is, the magnitude of the global mean surface temperature rise associated with a doubling of the atmospheric concentration in carbon dioxide. The fact that most of the projected changes in global and regional climate scale with the global mean surface temperature change, and that this latter strongly depends on climate sensitivity, confers to the assessment of cloud feedbacks a great importance.

Clouds, however, matter for much more than just the climate sensitivity. For instance, and as discussed in Part III of this book, through their role in the general circulation of the atmosphere, clouds matter for the regional changes in precipitation which are projected to occur by the end of this century. By controlling so directly and fundamentally the temperature and precipitation changes associated with climate change, cloud processes also influence a large number of biogeochemical physical feedbacks involved in the climate response to anthropogenic forcings. Examples include the efficiency of carbon sinks over land, which depends on soil moisture, and aerosol burdens in the atmosphere, whose abundance and distribution depend on rainfall.

Why is it so challenging to assess the behaviour of clouds in climate change?

A first challenge is the absence of clear observational constraints on the long-term behaviour of clouds. The longest records (fifty years of ship-based observations, thirty years of satellite data) exhibit a lot of variability in cloud signals. With the exception of a few recent studies, the extraction of robust trends emerging from natural climate variability that are unlikely to result from observational artefacts (related for instance to changes in instrument calibration or satellite drifts) remains difficult. Over the last decade, cloud observations have become much more numerous, comprehensive and accurate. However, we have learned that climate variations occurring on short timescales cannot always be considered as an analogue of long-term climate change. These difficulties have prevented so far a direct evaluation of climate change cloud feedbacks.

Another challenge is that clouds are complex bodies. They result from processes occurring over a wide range of scales (from the microphysical to the planetary-scale), that appear under a large diversity of shapes and heights, and that exert very contrasted effects on the Earth radiation budget (Chapter 4). In the absence of overarching constraints, such as a conservation law or thermodynamic bound, on their collective effect on the Earth radiation budget we are mostly left to understand the role of clouds in climate in terms of the specific processes that regulate their formation and longevity, something that is called a bottom-up approach. Given the myriad of processes having the potential to contribute to cloud feedbacks, it may easily become a no end approach.

Finally, the trivial fact that there are no observations of the future climate implies that our anticipation of climate change primarily relies on model predictions. Unfortunately, like in any discipline, model results are affected by approximations, simplifications and errors. Moreover, climate models exhibit little consensus regarding certain aspects of the behaviour of clouds in climate change (e.g., changes in the low-cloud amount). The reliability and credibility of model results may thus not be taken for granted.

In view of these challenges, what research strategies may help circumvent difficulties and foster progress? One strategy for progress consists in decomposing complex, intractable problems into a series of sub-problems which are individually more tractable. It often requires conceptualising the problem so as to articulate it in terms of a limited number of key processes or components. This conceptualisation process may operate through analysis methods, and/or through the simplification of models. But it may also operate through the consideration of multiple model configurations designed to isolate specific physical processes, examples include the use of aqua-planet, single-column or radiative-convective versions of general circulation models (GCMs) using the same representation of physical processes as the native/original GCM, but applied to a different physical setting. Likewise, although traditional climate models will always be hindered by their necessarily crude representation of clouds, sometimes looking across models applied to the same problem

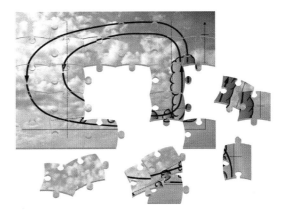

FIGURE 13.1: Our understanding of the role of clouds in climate is like a jigsaw puzzle where each piece represents robust physical understanding of some particular process (e.g., some aspects of the behaviour of anvil clouds, as discussed later), and where its position within the jigsaw is such that it connects properly to the other pieces of established knowledge while ensuring the consistency of the other overall pattern (e.g., global water and energy conservation laws).

can provide insights into larger-scale constraints on cloud-controlling factors. Different perspectives on cloud-controlling factors can also be achieved by looking at similar situations, but by using models that are better grounded in the laws of fluid dynamics, tools like direct numerical simulation (DNS), large eddy simulation (LES), storm-resolving models (SRMs) or hybrid approaches (Chapter 2). Nor is the power of very simple, conceptual models to be underestimated, especially in so far as they encapsulate physical constraints. Our ability to connect robust behaviours across a spectrum of relevant models, and to connect these behaviours to basic physical principles and to observations, is what builds confidence in our understanding.

As patterns in the behaviour of models emerge and an ability to physically interpret them advances, our understanding of climate change and of the role of clouds in climate in particular, does not look like a house of cards for which one missing piece of knowledge is enough to destroy the entire edifice. On the contrary, it looks more like a puzzle for which the position of each new piece is supported by a range of robust evidences and the consistency with the surrounding pieces and the big picture (Fig. 13.1).

13.2   Conceptual Frameworks

In this section, important concepts and methodologies that have proven useful to understand what controls the climate sensitivity and the cloud and precipitation responses to climate change are introduced. We begin by introducing the basic concepts of forcing and feedback through a simpler, and more generic, example. Subsequently we extend these generic concepts to the problem of climate, and discuss particular issues that arise in their application.

13.2.1   Forcing and Feedback: An Illustrative Example

To help fix terminology, we first consider a simple example, that illustrates some of the basic concepts behind the idea of forcing and feedback. Imagine a system that warms up when it is forced. For instance, a pot of water on a gas stove. In this example, the stove provides the forcing to the system, which is just our pot of water. If the water isn't near the boiling point, then the temperature of the water is determined by the balance between how much energy it gets from the stove's burner and how much it loses to its environment. If the heat loss is proportional to the temperature difference between the pot and its environment, then the temperature of the water will increase as the burner is turned up. If we adopt the perspective of explaining how the system changes its equilibrium, then the increment in the burner, rather than its absolute position, can be associated with a forcing F. The response of the system is measured by its change in temperature ΔT which implicitly identifies the temperature of the water T as the state variable. For small perturbations we might expect that $\Delta T \propto F$. If we choose $-1/\lambda_b$ to denote the constant of proportionality, with $\lambda_b < 0$, we can write

$$\Delta T = \Delta T_b = -F/\lambda_b. \tag{13.1}$$

In this system there is no feedback because the change in the burner is assumed to be independent of the warming, and λ_b can be thought of as a simple response parameter.

Now suppose that as the pot warms it cools more efficiently, for instance, because the heating causes the lid to rattle which allows steam to escape, increasingly so as the pot warms. This would reduce the temperature rise compared to the situation where the lid's rattling was not present. The effect of the rattling can be quantified by decomposing the temperature response[1] into a part that depends on the burner, and a part that depends on how well sealed the lid is, such that

$$\Delta T = \Delta T_b + \Delta T_l, \tag{13.2}$$

[1] This approach is based on the article by Dufresne and Saint-Lu (2016).

where subscript "l" has been introduced to denote the effect of the lid. Now how well the lid is sealed depends on the temperature of the pot, and hence ΔT itself. Assuming that they are linearly proportional, so that $\Delta T_l = g_l \Delta T$, then Eq. (13.2) can be rearranged to show that

$$\Delta T = \frac{1}{1 - g_l} \Delta T_b. \tag{13.3}$$

In our example, the proportionality constant g_l denotes the 'gain' in temperature associated with the rattling of the lid. A positive gain increases the overall temperature response. In our example, however, the gain is negative because the temperature change due to the lid rattling more at higher temperatures acts to reduce the overall temperature change from that which otherwise would have been expected. One could also imagine a situation whereby the system loses heat less effectively as it warms, for instance, because the environment of the pot also warms. In such a case, a positive gain would be introduced and the overall temperature rise would be greater than would have been expected. Here we emphasise that the gain measures the difference between the actual response of the system as a result of some process being incorporated, from how it would have been expected to respond were that process not incorporated.

Instead of talking about the gain of the system, one can also express the response of the system in terms of a contribution λ_l from the rattling of the lid to the overall response parameter. In this case we can write:

$$F = -\lambda \Delta T = -(\lambda_b + \lambda_l)\Delta T, \tag{13.4}$$

where in our example we expect λ_b and λ_l to both be negative. Here λ_l can be identified with a feedback parameter. In doing so, we are effectively describing the response of the system in terms of the response of a system without a rattling lid to a forcing composed of the original forcing, and a contribution from a feedback process – in this case, the warmer pot causing the lid to rattle and heat to escape more efficiently from the system. The rattling of the lid can be thought of as a feedback in the literal sense of 'feeding back' on the forcing, if the forcing is thought of in a proximate sense, that is, as the net heating of the pot, rather than in terms of its ultimate origin, as measured by the position of the throttle on the burner.

The relationship between the gain parameter and the feedback parameter can be derived by substituting Eq. (13.1) into Eq. (13.3), such that

$$g_l = -\frac{\lambda_l}{\lambda_b}. \tag{13.5}$$

Hence the gain measures the relative strength of the feedback of the heating on the net heating of the system.

In our example, $g_1 < 0$ which implies that λ_l is negative. This approach can be readily generalised to account for additional processes, each associated with its own gain, or feedback parameter. In so far, as the net gain of the system is less than unity, the system will be stable, but as the net gain approaches unity, the response of the system to an initial forcing becomes unbounded.

Additional material providing a more thorough discussion of the concept for forcing, feedback and gain, and in particular their relation to physical systems, is outlined in the Further Reading section.

13.2.2 Global Energy Budget Analysis

To a very great extent, the only form of energy exchange between the Earth and space is radiation: the Earth receives energy from the sun, reflects a fraction of the incoming solar irradiance back to space and absorbs what remains, a quantity referred to as ASR for absorbed short-wave radiation. At the same time as Earth receives radiant energy from the sun, it radiates energy to space. As discussed in Chapter 4, at Earth-like temperatures, this terrestrial irradiance is predominantly concentrated in the wavelengths of the thermal infrared and when measured at the top-of-the-atmosphere (TOA), is often referred to as outgoing long-wave radiation, or OLR. At equilibrium and on average over the whole planet, the Earth's radiation budget R at the TOA, defined as

$$R = \text{ASR} - \text{OLR}, \tag{13.6}$$

is zero.

What happens if, for some reason, for example, a change in insolation at TOA, or a change in the content of carbon dioxide within the atmosphere, the radiation budget comes out of balance, such that R is no longer close to zero? Any body that accumulates energy, or discards energy, will find its energetic content increased or reduced, respectively. This will affect its temperature and hence the rate at which it accumulates or discards energy, eventually restoring the energetic balance between incoming and outgoing fluxes of energy. The Earth restores this energy balance mainly by increasing its surface and atmospheric temperatures, and through changes in other variables which respond directly, or indirectly, to temperature changes. These changes occur until the global mean TOA net radiation, affected both by the perturbation itself and by all the subsequent climate changes it induces, returns to zero. A new balance (or equilibrium) is then reached.

Changes in temperature constitute such a basic and fundamental response mechanism of the perturbed global energy balance that when considering the climate response to an external perturbation, one may want to isolate the component of the response that depends on surface temperature changes from the rest of the response. Going even further, and *assuming* that the globally averaged surface temperature describes the state of the system sufficiently to determine its radiative response, allows us to write

$$R = R(\varphi, T_s), \tag{13.7}$$

where φ is included to describe factors (or boundary conditions) external to the system such as the solar constant or the anthropogenic contribution to the carbon dioxide concentration that may also influence the radiant energy balance. Given Eq. (13.7), it then follows that,

$$\delta R = \left(\frac{\partial R}{\partial \varphi}\right)_{T_s} \delta\varphi + \left(\frac{\partial R}{\partial T_s}\right)_{\varphi} \delta T_s. \tag{13.8}$$

The second term on the right-hand side represents the part of the radiative response that is mediated by temperature changes and independent of changes in φ, the first term represents the part which is induced by some external perturbation $\delta\varphi$ but independent of changes in T_s.

By analogy with the concepts of forcing and feedback introduced in the previous section, we introduce the symbols

$$F = \left(\frac{\partial R}{\partial \varphi}\right)_{T_s} \delta\varphi \quad \text{and} \quad \lambda = \left(\frac{\partial R}{\partial T_s}\right)_{\varphi} \tag{13.9}$$

to represent what we call the radiative forcing (in W m^{-2}) and the feedback parameter (in $\text{W m}^{-2}\,\text{K}^{-1}$), respectively.[2] In some literature, this same parameter is called the *climate feedback parameter*, the negative of its inverse is often called the *climate sensitivity parameter*. With this notation, Eq. (13.8) takes the simple form:

$$\delta R = F + \lambda\,\delta T_s. \tag{13.10}$$

The quantity δR measures the imbalance between the rate at which the Earth receives versus loses energy, where our sign convention is such that $\delta R > 0$ implies an accumulation of energy within the system. The accumulation of energy that occurs in this situation must be stored within the system (heat uptake), and for the most part this heating is, at least initially, thought to be concentrated in the upper part (700 m) of the ocean. It leads to changes in ocean temperatures, and to T_s, until the energy budget of the

---

[2] Note that by simplicity, λ is often assumed to be time invariant and independent of the spatial pattern of surface warming. However, as will be discussed later, these assumptions are not always valid.

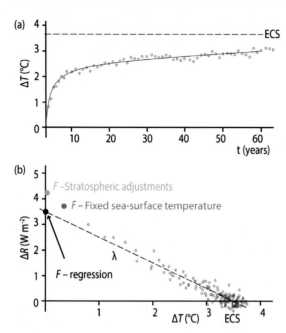

FIGURE 13.2: (a) Time series of the first sixty-five years of global mean surface temperature change predicted by coupled ocean–atmosphere models after an instantaneous doubling of carbon dioxide concentration in the atmosphere. The asymptote (dashed line) represents the long-term global warming reached by the climate system at equilibrium, referred to as the equilibrium climate sensitivity (ECS). Each point represents a yearly multi-model mean anomaly compared to pre-industrial (unperturbed) conditions. (b) Relationship between the global mean anomalies of the surface temperature δT_s and of the net radiation flux at the TOA δR for the full simulation. This is commonly referred to as a Gregory plot. The δR intercept at $\delta T_s = 0$ corresponds to the radiative forcing, the δT_s intercept at $\delta R = 0$ to the ECS, and the slope λ of the relationship to the feedback parameter. Also reported are estimates of the radiative forcing diagnosed from $2\times CO_2$ atmosphere-only experiments with fixed sea-surface temperatures (dark grey marker, the slight global warming associated with it is due to the land surface warming), and radiative forcing diagnostics derived from radiative calculations after allowing for the adjustment of stratospheric temperatures (light grey marker).

system approaches a new balance ($\delta R = 0$, Fig. 13.2, Gregory et al. (2004)). At equilibrium:

$$\delta T_s^{eq} = -\frac{F}{\lambda}. \tag{13.11}$$

In the case where F is the radiative forcing associated with a doubling of the carbon dioxide atmospheric concentration (F_{2CO2}), δT_s^{eq} is referred to as the equilibrium climate sensitivity (ECS).

Due to the long response timescales of the deep ocean, a new climate equilibrium would only be fully attained a few millennia after an applied forcing. A related concept, transient climate response (TCR), takes this lag

into account and is thought to be a better index of the warming that will occur within a century or two (see its precise definition in Section 13.3.4.2). Nevertheless, the ECS is usually taken as the starting point for studies of climate change. There are several reasons for this: (i) the ECS is conceptually similar; (ii) numerical experiments show that many aspects of the climate response to forcing, including the TCR, scale with the ECS; and (iii) many of the processes that control the ECS are central for a broader understanding of the climate system. For these reasons, ECS estimates are also crucial to assessments of the economic cost of climate change and hence for the development of strategies for mitigating and adapting to climate change.

Eq. (13.11) illustrates how, if the assumptions leading to Eq. (13.7) hold the ECS is determined by the ratio of two quantities: the radiative forcing F and the response parameter λ. Hence, to understand the ECS requires an understanding of processes that contribute to the feedbacks on the one hand, and how exogenous perturbations translate into radiative forcing on the other hand. These ideas are developed further in the following sections.

13.2.3   Feedbacks

The concepts of forcing and feedbacks can be applied to the Earth system by exploring perturbations to the system in terms of its radiative response R. To do so, we assume that the OLR depends on the surface temperature as OLR $= \gamma \sigma T_s^4$, where σ is the Stefan–Boltzmann constant and γ is a quantity that accounts for all the factors that, for a given global mean T_s, affect the absorption and emission of long-wave radiation from the surface to TOA: long-lived greenhouse gases (GHGs), water vapour, clouds, temperature lapse rate, horizontal variations of surface temperature and so on.[3] By introducing S_o, the insolation at TOA, and α, the planetary albedo – which depends on surface properties such as vegetation type, snow or sea ice and on atmospheric variables such as clouds or aerosols – then R can be expressed as a function of four climate variables, S_o, α, γ and T_s as follows

$$R = \frac{S_o}{4}(1 - \alpha) - \gamma \sigma T_s^4. \tag{13.12}$$

---

[3]   Note that in this simple model, γ depends on the temperature lapse rate but not on the absolute atmospheric temperature (i.e., a uniform atmospheric warming does not change γ). In the present climate, global mean values of $240\,\mathrm{W\,m^{-2}}$ and $288\,\mathrm{K}$ for the OLR and T_s, respectively, suggest a global mean value of γ close to 0.61.

This equation is identical to that introduced in Eq. (5.31), whereby the effective temperature T_e in that equation is here accounted for by the factor γ, so that $T_e^4 = \gamma T_s^4$.

If we assume that T_s is the only climate variable that responds to an external perturbation and, following the notation of the previous section, use φ to denote S_o and other possible external parameters, then the sensitivity of R to T_s can be calculated analytically as

$$(\partial R/\partial T_s)_\varphi = -4\,\gamma\,\sigma\,T_s^3. \qquad (13.13)$$

This quantity is named the Planck response parameter λ_P. Present-day values for γ and T_s lead to $\lambda_P \approx -3.3\,\mathrm{W\,m^{-2}\,K^{-1}}$. For a doubling of carbon dioxide, which is associated with a radiative forcing of the about $3.7\,\mathrm{W\,m^{-2}}$, this would lead, after Eq. (13.11), to an equilibrium global warming $\delta T_{s,P} = -F/\lambda_P$ (roughly $3.7/3.3 = 1.1\,\mathrm{K}$). This quantity, commonly referred to as the Planck response, corresponds to the ECS of a system devoid of radiative feedbacks.

However, as climate is warming, other variables in addition to T_s are likely to respond: for instance, an increase in the atmospheric water vapour content would increase the atmospheric long-wave opacity and would reduce γ, or an increased melting of snow and ice at the surface would decrease the surface albedo and hence α. Both effects would, compared to the previous case (the Planck response), reduce the degree to which a change in T_s would dampen, or reduce, the initial change in R. In other words, the feedback parameter, while still negative, would have a reduced magnitude, that is, $\lambda_P < \lambda < 0$, implying a larger ECS, such that $\mathrm{ECS} > \delta T_{s,P}$. Alternatively, changes within the climate system that reduce its sensitivity to perturbations (such that $\lambda < \lambda_P < 0$) would produce an $\mathrm{ECS} < \delta T_{s,P}$.

Processes that increase the climate sensitivity are generally conceptualised as positive feedbacks, because as the system warms, they act in a way that increases the radiative imbalance that is responsible for the warming in the first place. Processes that decrease the climate sensitivity are generally thought of as negative feedbacks because they act to reduce the radiative imbalance in response to warming. The Planck response also has the conceptual character of feedback, as it is associated with a change in R, but it is usually taken as the basic response of the system relative to which the other feedbacks are measured. A system can have a number of feedbacks, as, for instance, illustrated in Fig. 13.3a, where the emissivity and albedo feedbacks are also included. Whether or not the system is stable depends on whether the net response

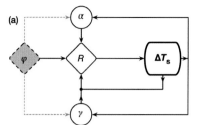

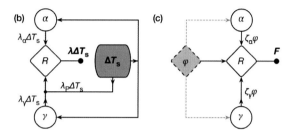

FIGURE 13.3: Diagrammatic illustration of forcing and feedbacks following the system described by Eq. (13.12). (a) In the full system, the feedbacks and adjustments can be diagnosed by relating ΔR to ΔT_s for a given perturbation $\Delta\varphi$ using the regression method. (b) The contributions to feedbacks (λ_P, λ_γ and λ_α), as would be inferred by a prescribing ΔT_s and measuring ΔR, so that $\lambda = \Delta R/\Delta T_s$. (c) likewise illustrates a method for diagnosing the adjustments (ζ_γ and ζ_α) by measuring the response ΔR to the perturbation $\Delta\varphi$ when surface temperature is decoupled from the system, that is, T_s is not allowed to change. In all panels, the grey block designates what is perturbed, and the output is normally measured in terms of the bold variables.

is positive or negative. If the positive feedbacks are too large, so that the overall response parameter becomes positive ($\lambda > 0$), the system will be unstable – more warming leads to a larger radiative imbalance in a fashion that runs away on itself. Therefore, the ECS strongly depends on climate feedbacks.

Many climate variables can respond directly or indirectly to δT_s. Here, only those affecting the TOA radiation budget R are considered as feedbacks. This follows from the defining character of a feedback, that is, something that feeds information from the output (or state) of the system back to the input, for example, Fig. 13.3a and 13.3b. In the case of the conceptual model of the climate system used for interpreting climate change, the state (or output) is the globally averaged surface temperature T_s and the input is the TOA irradiance imbalance R.

To physically understand how different processes contribute to the total feedback, it is useful to introduce auxiliary variables x describing specific processes that respond to warming. Thus, $R = R(\varphi, x(T_s), T_s)$ and the feedback parameter λ, which was defined in Eq. (13.9), becomes (at leading order)

$$\lambda = \left(\frac{\partial R}{\partial T_s}\right)_{\varphi,x} + \sum_x \left(\frac{\partial R}{\partial x}\right)_\varphi \frac{\partial x}{\partial T_s} = \lambda_P + \sum_x \lambda_x. \quad (13.14)$$

In the simple system discussed earlier, $x \in \{\gamma, \alpha\}$.

The Planck response of the system is so basic that it is often taken as the starting point for analysing the effects of the other feedbacks, which is why it was separated from these feedbacks in Eq. (13.14). The global mean surface temperature change occurring when all the climate variables x respond to the change in T_s can then be expressed relative to the Planck response ($\delta T_{s,P} = -F/\lambda_P$) as: $\delta T_s = \frac{\lambda_P}{\lambda} \delta T_{s,P}$. From the definition of λ in Eq. (13.14), δT_s can be expressed as:

$$\delta T_s = \frac{1}{1 - \sum_x g_x} \delta T_{s,P}, \quad (13.15)$$

where, in the analogy to the discussion of the rattling pot (Section 13.2.1), $g_x = -\frac{\lambda_x}{\lambda_P}$ is called the feedback gain for the variable x. As λ_P is negative, the sign of the feedback gain g_x is the sign of the feedback parameter λ_x. Following this convention, a positive feedback parameter thus amplifies the temperature response of the climate system to a prescribed radiative forcing and enhances the climate sensitivity, while a negative feedback damps it.

The feedback gain g_x can be used to partition the temperature response in proportion to the contribution from the different feedback processes. With a few algebraic manipulations, Eq. (13.15) can be rewritten as:

$$\delta T_s = \delta T_{s,P} + \sum_x \delta T_{s,x}, \quad (13.16)$$

where $\delta T_{s,x} = g_x \delta T_s$. Although in Eq. (13.15), the contribution from the different feedbacks to δT_s is not additive, in the formulation (13.16), δT_s and thus the climate sensitivity, can be conveniently decomposed into a Planck response plus a series of contributions associated with the different feedbacks (Dufresne and Bony, 2008).

13.2.4 Radiative Forcing

The radiative forcing F corresponds to the part of the global radiation budget perturbation induced by $\delta\varphi$ that arises independently of global surface temperature changes, that is, from Eq. (13.8):

$$F = \delta R|_{T_s}. \quad (13.17)$$

Generalising the approach that was taken to incorporate feedbacks into the system, for example, in Eq. (13.14), by also allowing x to depend on φ, so that $R = R(\varphi, x(T_s, \varphi), T_s)$, we can expand expression (13.17) as follows,

$$F = \left[\left(\frac{\partial R}{\partial \varphi}\right)_{T_s,x} + \left(\frac{\partial R}{\partial x}\right)_{\varphi,T_s} \frac{\partial x}{\partial \varphi}\right] \delta\varphi. \quad (13.18)$$

It shows that F depends on two terms: a direct response of R to the change φ, and an indirect response of R to changes in φ, both arise independently from changes in T_s. The first, and direct, term is called the instantaneous radiative perturbation. The second, and indirect, term is called the radiative adjustment. Historically, the effect of adjustments on the forcing were not widely recognised, and even when they were recognised, often only stratospheric adjustments were accounted for, and just this subset of contributions to F was called the 'radiative forcing'. As the community became more aware of additional 'adjustments', the term 'effective radiative forcing' was introduced to avoid ambiguity with earlier definitions of radiative forcing. In addition, because the adjustments that are easiest to identify often emerge quite rapidly compared to changes in T_s, the indirect contribution to the radiative forcing, the adjustments, are sometimes called *rapid* adjustments.

13.2.5 Instantaneous Radiative Perturbation and Adjustments

The idea of an instantaneous radiative perturbation is quite simple, as it is the change in R that arises from a change in the state of the system in the absence of any other change. The instantaneous radiative perturbation from a doubling of the atmospheric concentration of carbon dioxide is, for instance, 2.7 W m^{-2}. This is considerably smaller than the value of 3.7 W m^{-2} that is often cited for the value of F. The difference is due to what is called *s*tratospheric adjustment, which adds an additional 1 W m^{-2} to the forcing.

Stratospheric adjustment was historically the first form of adjustment to be recognised, and is conceptually straightforward to understand. Increasing carbon dioxide raises the emission height of the stratosphere, but because of its positive stratification (temperature increases with height), this implies a greater loss of energy, and hence cooling of the stratosphere. Unlike the troposphere, which is thermally coupled to the surface and largely in a state of radiative–convective equilibrium, the stratosphere is thermally decoupled from the troposphere and in a state of radiative equilibrium. Hence, additional cooling of the stratosphere must be balanced by reducing the stratospheric temperature (Fig. 13.4). With an increase in carbon dioxide, the additional emission of thermal radiation by the stratosphere initially offsets part of the reduction of thermal emission by the troposphere, as measured at the TOA. But as the stratosphere cools, so as to reduce its emissions and come back into

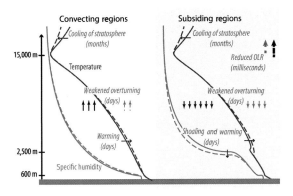

FIGURE 13.4: Illustration of some robust adjustments of the atmosphere to increased carbon dioxide in regimes of large-scale ascent (left) and descent (right) when the surface temperature is fixed. The typical timescale associated with each adjustment is given in parentheses. Adjusted state shown by dashed lines, base state by solid lines.

radiative equilibrium, then this offset is reduced by the aforementioned $1 \, \mathrm{W \, m^{-2}}$.

External perturbations can also induce significant adjustments in the troposphere. Indirect contributions to the forcing, or adjustments, have long been recognised for aerosols, for instance, through their propensity to modify cloud optical properties and possibly other climate properties, independently of T_s changes (Chapter 11). However, GHG perturbations also induce tropospheric adjustments in the absence of T_s changes. An increase of carbon dioxide reduces the radiative cooling and warms the free troposphere (by about 0.5 K in just a few days for a $4 \times CO_2$), hence affecting the tropospheric static stability, the relative humidity profiles and the vertical distribution of clouds. Some of these processes are indicated schematically in Fig. 13.4 and the physical mechanisms which underpin them are discussed further in Section 13.4.2. Going back to expression (13.12), it means that $(\partial \alpha / \partial \varphi)_{T_s}$ and $(\partial \gamma / \partial \varphi)_{T_s}$ are not zero, and thus contribute to δR.

Adjustments in general are often conflated with rapid adjustments because when adjustments are rapid, they are easy to identify. Here rapid implies a change that is quick compared to the timescale over which sea-surface temperature (SST) anomalies develop, that is, the timescale of months or more. Therefore, radiative perturbations that arise on shorter timescales are unlikely to be mediated by δT_s, and readily identified as 'adjustments'. For this reason, the fast radiative response of the perturbed system can be considered as part of the radiative forcing, and is even sometimes considered as a practical definition of it. However, there is no definite timescale to distinguish

the fast response from the slower temperature-mediated response. For example, changes in the land surface properties that arise solely from increased atmospheric concentrations of carbon dioxide, for instance, by favouring different types of vegetation cover, are adjustments, but ones that can be expected to be very slow. For this reason, defining adjustments in terms of their timescale can introduce ambiguity. Nonetheless, to the extent that adjustments arise on timescales much shorter than that of the global warming, this difference in timescales can be used to help separate the effects of adjustments from those of feedbacks experimentally.

Distinguishing between adjustments and feedbacks is important because the atmospheric adjustments are, by definition, sensitive to the details of the radiative heating rate perturbations, and thus to the nature of the forcing. For instance, a GHG forcing will affect the tropospheric heating rate much more than a solar perturbation, and therefore the cloud adjustments will be different in the two cases. It means that perturbations that correspond to similar direct radiative changes can be associated with different radiative forcings. On the other hand, the temperature mediated climate response appears to be less specific to the nature of the forcing, explaining why its pattern, once scaled by δT_s, is similar for very different types of forcing.

13.2.6   Limitations and Caveats

Many of the feedback and forcing concepts that have been developed already are most appropriately applied to a comparison of the equilibrium state of a perturbed versus a control system. But in practice the real system is never in perfect equilibrium and there is a tendency to apply these concepts to time-evolving systems, for instance, in inferring the climate sensitivity from the historical record. This works well if the radiative response varies linearly with the surface temperature, as is approximately the case for the example shown in Fig. 13.2b.

Even in this example, the points tend to lie above the regression line for small δT_s, suggesting that during the initial period of warming, the radiative response changes more strongly with temperature than over later periods. The time series plot also shows evidence of at least two timescales in the warming response, a rapid response whose timescale is a few years, and a more ponderous response that equilibrates the system over hundreds, or even thousands, of years. If the responses emerging on different timescales are associated with different processes, it is easy to imagine them being accompanied by different patterns of warming, which in turn are associated with a different balance of feedback processes, and hence different radiative responses. This violates the assumption used to write down Eq. (13.7) and can be a source of non-linearity in the relationship between δR and δT_s. Physically, this could correspond to a situation where, during transient climate changes, surface temperature changes develop faster over land than over ocean (and thus in the northern hemisphere than in the southern hemisphere). As a result, any feedback depending on the land–sea contrast or more generally on the distribution of temperature changes is likely to appear stronger to begin with, when land–sea contrasts are more pronounced.

Important feedback processes may also operate only on very long timescales (e.g., surface albedo feedbacks associated with millennial timescale changes in continental ice sheets, or changes in the state of the deep ocean). To the extent they are large, the whole concept of an equilibrium approach to understanding climate change becomes questionable. More generally, these slow changes may not even be evident, or considered in experiments designed to reach equilibrium on a timescale of decades. These points highlight the fact that the climate sensitivity is unlikely to be a unique number that applies to all perturbations and all timescales; rather it is some quality that must be conditioned on the feedbacks which are allowed to be operative for a given conceptualisation of the system, ideally chosen as a function of the timescale of interest.

Given the complexity of the climate system, the theoretical framework encapsulated by Eq. (13.10) is more surprising for what it captures, than for what it misses. As such, it has proven to be a useful starting point for the interpretation of climate change experiments, for interpreting the spread of climate sensitivity estimates among models, and for providing rough bounds on the climate sensitivity given observational records. But it is only a starting point, and its success in capturing many aspects of the climate system's response to forcing raises the danger of over-interpretation. For instance, the dependence of climate variables (e.g., clouds) on surface temperature is generally indirect and statistical rather than intrinsic.

Deviations from the simple model resulting in Eq. (13.7) also helps structure ideas about how the climate system works. In particular, the theory raises two important questions: (i) how important is the pattern effect? Alternatively phrased, how much does the radiative response depend on the spatial pattern of the heating anomaly as opposed to its amplitude δT_s averaged globally? (ii) Even if the radiative response R depends only on the global mean surface temperature, is the strength of this response itself dependent on temperature? Pattern effects complicate attempts to infer future changes

from past changes (e.g., some past orbital changes have induced large hemispheric changes in temperature and large changes in climate without any significant change in global mean temperature), or to interchange the contributions from qualitatively different forcings. A strong state dependence complicates interpretations of past changes, and raises the possibility of surprises as the climate system enters uncharted territory.

13.3 Application of the Conceptual Framework to GCMs

As outlined earlier, the analysis of the Earth's energy balance can help to understand and predict how the climate system responds to an external perturbation. But how to relate this framework to the output from actual climate model simulations is not a simple matter. In this section, some important diagnostic approaches and methodologies that have been developed for this purpose are presented.

13.3.1 Regression Approach

One methodology commonly used to diagnose tropospheric adjustments, the ECS and the total feedback parameter is based on a consideration of the evolution of the TOA radiative imbalance R in a transient coupled ocean–atmosphere model experiment in which a radiative perturbation is abruptly imposed to the climate system. An illustration of such an experiment is shown in Fig. 13.2 for an abrupt doubling of carbon dioxide. After the application of the carbon dioxide perturbation, before the global mean T_s has begun to change, but after stratospheric adjustment is complete, the global radiation budget exhibits an imbalance of several $W\,m^{-2}$. This imbalance weakens with time as the climate system responds to the radiative imbalance, and vanishes after a few decades (when a slab ocean model is used to represent ocean–atmosphere couplings) or longer (after centuries to millennia when a full ocean model is used) as the system reaches a new equilibrium.

Plotting the evolution of the TOA net radiation as a function of δT_s (e.g., as is done in Fig. 13.2b) shows that although both δT_s and δR both evolve non-linearly in time, they vary more or less linearly with respect to one another. In so doing, they evolve as would be expected from Eq. (13.10). In this case, a linear regression of δR vs δT_s can be used to infer the radiative forcing F as the y-axis, or δR, intercept of this regression, that is, the point where $\delta T_s = 0$. Similarly, the x-axis, or δT_s, intercept denotes the ECS, as the point where δR vanishes. The global response parameter (λ), follows as the slope of the regression line. Because the values of key quantities are inferred from the regression of R against T_s, this method is often called the regression method, although some authors refer to it as the Gregory method.[4]

The application of the regression method to a large number of coupled ocean–atmosphere climate models yields $F_{CO_2} = 3.44\,\mathrm{W\,m^{-2}}$ (*F*-regression on Fig. 13.2b). This value is about 10 % smaller than what would be estimated ($3.7\,\mathrm{W\,m^{-2}}$) by computing the instantaneous radiative perturbation and stratospheric adjustment. The difference can be attributed to other adjustments, within the troposphere and at the surface. Methodologies for assessing the relative contributions to these tropospheric adjustments – by changes in the thermodynamic structure of the atmosphere on the one hand, or clouds on the other hand – are developed further.

13.3.2 Prescribed SST Approaches

Given the definition of the feedback and the forcing, the most straightforward way to calculate these terms is to disable one or the other, for example, as illustrated in Fig. 13.3b and 13.3c. For instance, the feedbacks can be calculated directly by comparing how the radiative imbalance δR at TOA changes when the globally averaged surface temperature is changed in a simulations with prescribed surface temperatures differing by an amount δT_s. Likewise, by not allowing the surface temperature to change, but measuring how R responds to a doubling of carbon dioxide, provides a natural way of estimating the effective radiative forcing F_{CO_2}. In practice, the highly interactive nature of the land surface, and of sea ice, makes it difficult to perform these types of experiments in a comprehensive climate model. However, the effect of doing so can be approximated by evaluating atmosphere only simulations with specified SSTs and sea-ice concentrations, either as a function of these SSTs (to evaluate feedbacks), or as a function of the atmospheric carbon dioxide (to evaluate forcing).

Prescribed SST perturbation experiments have long been used to assess climate feedbacks. Usually, a globally uniform change in SST (2 K or 4 K warming) is applied, and λ is defined from the ratio of the change in radiative fluxes at the TOA to δT_s (where the latter is usually

slightly larger than the prescribed SST change because in such experiments, the land surface warms more than the imposed ocean warming). The analysis framework developed does not really allow for a change in the pattern of SSTs, however, one can apply a patterned change in SSTs, chosen to mimic the pattern of warming from a coupled simulation. In so far as T_s is the only important control variable, this should make no difference to the estimate of the feedback parameter. In practice, this is the case at first order, as the multi-model spread of the global feedback estimated from a prescribed SST experiment agrees well with that derived from the regression method, wherein the pattern of warming will not in general be uniform. However, there are GCMs for which the strength of climate feedbacks depends more strongly on the SST pattern.

Estimates of F_{CO_2} obtained through the prescribed SST approach are found to be qualitatively consistent with those derived using the regression approach, with quantitative differences that do not exceed a few per cent.

13.3.3 Attributing Responses

The ECS of the different models performing experiments of the type described earlier has been diagnosed using the regression method. The different model estimates exhibit a large spread, ranging from about 2 K to nearly 5 K. Most of these differences can be attributed to differences in the global feedback parameter rather than in the forcing that different models produce in response to a doubling of atmospheric carbon dioxide. To some extent, the contribution to the global feedback parameter from different components of the radiative budget can be estimated, for instance, by comparing the long- or short-wave contributions, or the clear versus all-sky contributions to the regression of R against δT_s. By considering separately the clear-sky and 'all-sky minus clear-sky' (i.e., cloud radiative effect, CRE) anomalies, it is possible to assess the combined non-cloud feedbacks (temperature, lapse rate, water vapour, surface albedo) on the one hand, and an estimate of 'cloud feedbacks' on the other hand. However, the cloud feedback estimates obtained through this method are not consistent with the definition of feedbacks given by Eq (13.14) because: (i) CRE differences do not only depend on changes in cloud properties but also on changes in surface and tropospheric properties (water vapour, surface albedo, carbon dioxide), and (ii) feedbacks are conventionally defined relative to a zero-feedback situation (the Planck response) while the CRE anomalies quantify changes relative to a 'clear-sky Earth'. More formal methods for estimating cloud feedbacks, and for decomposing

---

[4] This in recognition of Jonathan Gregory, a British climate scientist who developed and popularised it, (Gregory et al., 2004).

non-cloud feedbacks into their component contributions are provided later in this chapter. As a rule of thumb, however, comparisons between cloud feedback estimates based on the change in CRE and estimates from more accurate methods show that the former underestimate the cloud feedback parameter by about $-0.4\,\mathrm{W\,m^{-2}\,K^{-1}}$. This results in a diagnosed sign of cloud feedback that may differ from that derived from other methods. Despite these caveats, this method is still widely used to assess, at first order, the range of feedback estimates among models because it is simple and because the inter-model spread of 'cloud feedbacks' derived from this method is close to that obtained using more quantitatively precise methods.

13.3.3.1 Formal Feedback Analysis

To compute the feedback parameters associated with different climate variables more formally, the most natural approach would be to disable a feedback, by prescribing the radiative effect of a particular constituent, and measure the difference in the response. This is sometimes referred to as the feedback locking approach, although in actuality it is not the feedback that is locked, rather it is the radiative effect of the variable that leads to the feedback which is prescribed, or locked.

As an example, the surface albedo feedback can be estimated by comparing three simulations: (i) a control simulation; (ii) a simulation with a forcing from a perturbation (i.e., doubled carbon dioxide); and (iii) another perturbed simulation but with the surface albedo prescribed (perhaps as a time varying field to maintain the correct seasonal correlations) from the control run. The difference between the second and the first simulation yields the temperature response with all feedbacks active, the difference between the third and the first simulation yields the temperature response with the surface albedo feedback disabled. By comparing the two different responses the surface albedo feedback can be estimated. A variant of the same approach would be to replace the third simulation with a second control simulation in which the surface albedo is prescribed from the freely evolving perturbed simulation. This provides an independent estimate of the surface albedo feedback. To the extent that surface albedo changes correlate with other changes in the system, the two estimates may differ. Hence, the feedback is sometimes calculated by performing both 'forward' and 'backward' feedback swaps, and then by averaging the two. This approach can be implemented in turn for all the feedbacks, by alternately locking (or prescribing changes) other relevant climate variables, such as water vapour, clouds and so on, but it requires a great many simulations with extensive

writing to and from files, and thus is computationally demanding, often prohibitively so.

If the perturbations are sufficiently small, the feedbacks measured in these different ways should be identical. However, the signal from small perturbations is difficult to extract from the background noise of the natural variability within a simulation, and thus demands long simulations, increasing the computational demands. For this reason, larger perturbations are often employed, thereby allowing shorter simulations. Larger perturbations, however, introduce a greater potential for correlations among the different changes to dominate the response. When this is the case, estimates of the feedback strength obtained by turning the feedback off in a perturbed run, or incorporating its effect in a control run, may differ substantially. Similarly, when estimating the strength of the feedback using strong perturbations, non-linear changes, which manifest as a state dependence of the response parameter, can be expected to play an increasingly important role.

Some computational savings can be achieved by comparing the radiative effect from changes to different components of the system while the simulation is running, that is, online. For instance, while computing the second, or perturbed, simulation described earlier, the radiative fluxes could be calculated twice, once for the surface albedo as it presently is predicted in the model, and the second time based on values taken from the control simulation, which are indicative of what it would be if the surface albedo was not allowed to change. Then, the feedback can be estimated directly from the changes in the radiative fluxes, rather than the temperature response. This approach is called the partial radiative perturbation method (or PRP). It replaces multiple simulations with multiple calls to the radiative solver. However, because the radiative calculations are often a large fraction of the simulation time, the computational advantages are relatively modest and the demands of extensive reading and writing of files are not avoided.

An alternative to the PRP method is the kernel method. It is less accurate, but by precomputing the average sensitivity of the radiation to a change in a state parameter, it is computationally much more efficient. The kernel method arises from the recognition that the PRP method, in effect, computes the feedback parameters λ_x (e.g., Eq. (13.14)) by brute force, by recomputing $\partial_x R$ for each value of $\partial x/\partial T_s$. However, for small perturbations, in so far as the perturbations sample the state of the system in an unbiased way, the term $\partial_x R$ can be averaged over the state of the system before being multiplied by the changes in the system. Doing so implies estimating

$$\lambda_x = \left\langle \frac{\partial R}{\partial x} \frac{\partial x}{\partial T_s} \right\rangle_{i,j} \approx \left\langle \left\langle \frac{\partial R}{\partial x} \right\rangle_i \left\langle \frac{\partial x}{\partial T_s} \right\rangle_{il} \right\rangle_j, \qquad (13.19)$$

where here the angle brackets denote an average over the state of the system, as indexed by i and j. So, for instance, j could index latitude, and i could index height, longitude and time. In this case, Eq. (13.19) defines an integral equation, and x is a vector, that is, for the water vapour feedback, x is the specific humidity in each grid cell at each time, and for the surface albedo feedback, x denotes the surface albedo at each surface point at each time. The approximation implied by Eq. (13.19) is that the radiative response can be estimated from some aggregation of states, for instance, from zonal averages, given the sensitivity of R to x averaged over the states indexed by i for a given j. This method of estimating feedback factors based on Eq. (13.19) is called the radiative kernel method (Soden et al., 2008).

Although the calculation of the approximate radiative kernel can still be intensive – as it requires one to compute the changes in R to a change in each radiatively active degree of freedom within a model, in a way that adequately samples the natural variability – it only needs to be performed once. Given $\langle \partial_x R \rangle_i$, Eq. (13.19) can then be used to estimate the feedback parameters from output that does not fully sample the state space of the model. This allows the kernel method to be used for the offline computation of the feedbacks from different models using the radiative kernel derived from a single radiation code integrated over the climate response patterns from the different models. Moreover, the robustness of the Kernels can be estimated by comparing kernels calculated from different models, or using different radiation codes, or for different samplings of natural variability.

One difficulty with the kernel method is that when x denotes a cloud property, the assumption of small perturbations is generally not valid; a seemingly small change in cloudiness has a disproportionately large effect on R, thus making the linear estimates of λ_{cld} following Eq. (13.19) less useful. To circumvent this difficulty, one approach has been to compute the cloud effects as the residual between the change in CREs and the non-cloud feedback factors,

$$\lambda_{cld} \delta T_s = (\delta R - \delta R^o) - \delta T_s \sum_{x \neq cld} (\lambda_x - \lambda_x^o), \qquad (13.20)$$

where superscript o denotes the clear-sky perturbation. Thus, Eq. (13.20) endeavours to correct the CRE anomaly (the first term on the right-hand side) for changes attributable to other feedbacks. This correction to the CRE is usually computed from the difference, between all-sky and clear-sky computations, of the feedback parameters associated with temperature, surface albedo and water

vapour. A disadvantage of this approach is that all other contributions to the response, for instance, associated with non-linearities or feedbacks not otherwise accounted for (e.g., changes in surface emissivity) are attributed to cloud feedbacks.

A better approach is to calculate λ_{cld} in a space over which perturbations to clouds are expected to be somewhat smaller, for instance, by calculating the cloud radiative kernels for a more aggregated state of the system, such as its annually and zonally averaged state. By calculating the cloud radiative kernels in the space of the International Satellite Cloud Climatology Project (ISCCP) histogram (Zelinka et al., 2012), specifically in terms of the cloud fraction as a function of cloud-top temperature and optical thickness, it is possible to calculate the cloud radiative kernel directly following Eq. (13.19). The difference between this and the standard approach, is in how the state space is defined. Instead of averaging the effect of a change in cloudiness over the physical space of the system, a radiative transfer model is used to compute δR from cloud fraction perturbations in each bin of the (cloud top pressure, cloud optical thickness) histogram, and these sensitivities are then multiplied by the changes in ISCCP histograms normalised by δT_s (Fig. 13.5). An advantage of this method is that it facilitates the diagnosis of the nature of cloud changes that are responsible for the global cloud feedbacks (the x of Eq. (13.19) can now refer to cloud altitude, cloud amount or cloud optical thickness, and thus to different cloud types) consistently across the different models. By estimating the cloud feedback parameter directly, it is also possible to compute the residual between the total response and that diagnosed from the sum of the individual feedbacks, thereby giving insight into the extent to which the response can be explained in terms of the individual feedbacks.

13.3.3.2 *Formal Forcing Analysis*

The PRP and feedback locking methods are more rigorous methods for assessing the contribution of a particular process or climate system component to the response of the system. But in addition to being computationally expensive to calculate, they don't allow for a natural separation of the contribution of that process to the forcing (through adjustments) and the feedback. This ability to separate the contribution of particular processes to the forcing compared to the feedback is a particular strength of the kernel method. Here it is illustrated by showing how the Kernels can be used to decompose the radiative forcing F into different components.

We begin by noting that the effective radiative forcing resulting from some change in a quantity φ can

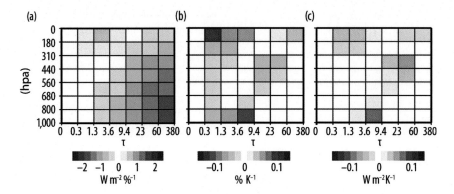

FIGURE 13.5: (a) Global annual mean net radiative kernel (i.e., the sensitivity $\frac{\partial R}{\partial x}$, in $Wm^{-2}\%^{-1}$, of the net TOA fluxes to perturbations in the fraction of different cloud types x characterised by cloud optical thickness and cloud-top pressure); (b) change in ISCCP-like cloud types per unit change in global mean surface temperature ($\frac{\partial x}{\partial T_s}$, in $\%.K^{-1}$) predicted by models in climate change (colours are reported only where the model responses are robust); and (c) cloud feedback parameter ($\frac{\partial R}{\partial x}\cdot\frac{\partial x}{\partial T_s}$, in $W\,m^{-2}\,K^{-1}$). An increased cloud cover at low levels reduces R because the cooling effect of low clouds exceeds their greenhouse effect, while the opposite occurs for high-level clouds. Therefore, the decrease of low-level clouds and the upward shift of upper-level clouds predicted by climate models in response to global warming both contribute to a positive cloud feedback. Adapted from Zelinka et al. (2012)

be decomposed into a direct response to the change in φ and an indirect response or adjustment, such that

$$F = \left(\frac{\partial R}{\partial \varphi}\right)_{T_s,x} \delta\varphi + F_{\text{adj}} \qquad (13.21)$$

where

$$F_{\text{adj}} \approx \sum_x \left\langle\frac{\partial R}{\partial x}\right\rangle_{\varphi,T_s} \left.\frac{\partial x}{\partial \varphi}\right|_{T_s} \delta\varphi, \qquad (13.22)$$

where $\langle\partial_x R\rangle_{\varphi,T_s}$ are given by the kernels and x denotes radiatively active state variables. Relevant are those that result from rapid adjustments (diagnosed, for instance, by comparing SST-fixed experiments runs by atmospheric GCMs in 1×CO$_2$ and 4×CO$_2$ conditions). Like cloud feedbacks, the cloud adjustment term can be inferred from the fast CRE anomaly corrected for the change in non-cloud variables that can alter the change in CRE ($-(F_x - F_x^o) - (G - G^o)$) where $(G - G^o)$ is the difference between the all-sky and clear-sky stratosphere-adjusted radiative change estimated from purely radiative calculations (about $-1.24\,W\,m^{-2}\,K^{-1}$).

A remarkable feature of these diagnostic methods is that they make it possible to reconstitute from radiative calculations and basic model outputs (temperature, water vapour, surface albedo and TOA radiative fluxes in all-sky and clear-sky conditions) estimates of climate sensitivity, radiative forcings and total climate feedback that can otherwise only be inferred from the more direct, top-down regression approach. This provides an opportunity to interpret the spread of climate sensitivity among models, and to relate climate sensitivity, adjusted forcings and feedback estimates to physical processes.

13.3.4 Climate Sensitivity Estimates

13.3.4.1 Spread of Climate Sensitivity

A better understanding of how to calculate the feedback parameter associated with specific processes, or forcing adjustments, makes it possible to quantify the relative roles of the Planck response, of adjusted forcings and of climate feedbacks in the multi-model mean ECS and its spread (Vial et al., 2013). On average, the radiative forcing estimate ($3.4\,W\,m^{-2}$) is only slightly weaker than that given by radiative calculations after allowing for the adjustment of stratospheric temperatures ($3.7\,W\,m^{-2}$). This slight reduction in forcing by adjustments is attributed to the fast warming of land surfaces induced by the carbon dioxide which leads to an increased emission of long-wave radiation to space that is partly compensated by a positive cloud adjustment (dominated by its short-wave component), whose physical mechanism will be discussed later on in this chapter. However, inter-model differences in the radiative forcing estimate remain significant (15–20%).

As shown on Fig. 13.6, the Planck response (including the effect of the fast land surface warming) explains about 1.2 K of the ECS. Tropospheric adjustments (mostly from clouds and water vapour) explain about 0.4 K of the warming. The increased water vapour combined with the weakened temperature lapse rate provide in combination a strong positive feedback that contributes for about 1.4 K to the ECS. The decrease of surface albedo due to snow and sea-ice retreat with warming provide a positive feedback that accounts for 0.2 K of the warming, and cloud feedbacks provide an additional positive feedback that

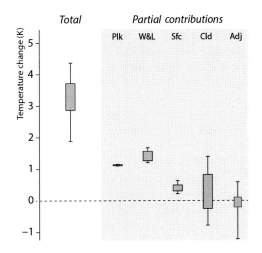

Total   Partial contributions

FIGURE 13.6: Estimates of the ECS from climate models, and their decomposition into partial contributions associated with the Planck response (Plk), the water vapour plus lapse-rate feedback (W&L), the surface albedo feedback (Sfc), cloud feedbacks (Cld) and adjustments (Adj). Boxes indicate the 25–75% spread of model estimates around the multi-model mean value, and whiskers represent the minimum and maximum values. The sum of all partial contributions on the right equals the total on the left.

accounts for an additional 0.4 K on average (but this term is associated with a large spread among models).

The spread of ECS is primarily explained by inter-model differences in cloud feedbacks (Fig. 13.6). These differences arise mainly, but not entirely, from the subsidence regions of the tropics. The spread in the cloud adjustments to the forcing are considerably smaller, but still constitutes a primary source of spread of radiative forcing estimates among models.

13.3.4.2 Observational Constraints

Given the large spread of model estimates of climate sensitivity, one may attempt to estimate climate sensitivity from past climate changes using observations (Forster, 2016). Several methods have been proposed for doing so. Most adopt Eq. (13.10) as their starting point.

From Eq. (13.10), the radiative heat imbalance δR can, to a good approximation, be equated with the rate of change (per unit area) of Earth system enthalpy $\dot{H}$. If most of the change in system enthalpy is assumed to be manifest in changes in ocean temperature, then $\dot{H}$, and hence δR, can be calculated from measurements of ocean warming. Given knowledge of the forcing F, Eq. (13.10) allows for an estimate of the climate feedback parameter λ, and hence the ECS as

$$\text{ECS} = \overbrace{\left(\frac{\delta T_s}{F - \delta R}\right)}^{-1/\lambda} F_{2CO_2}. \tag{13.23}$$

An application of this approach at a point in time, or averaged over time, necessarily assumes that the radiative response parameter λ is constant. This, for the reasons discussed in Section 13.2.6, is not an obviously valid assumption, but it is used nonetheless.

The TCR, is defined as the temperature at the time of carbon dioxide doubling that arises from a system that is forced by an exponential, 1%, increase in carbon dioxide. If one assumes that δR (equivalently the ocean heat uptake) is proportional to δT_s, something that might be a reasonable approximation in a system that is increasingly forced, then by taking $\delta R = \kappa \delta T_s$ where $\kappa > 0$ is a constant proportionality factor, Eq. (13.10) becomes

$$\kappa \delta T_s = F + \lambda \, \delta T_s. \tag{13.24}$$

From this equation, it follows that

$$\kappa - \lambda = F/(\delta T_s).$$

From the definition of TCR, it likewise follows that

$$\kappa - \lambda = F_{2CO_2}/\text{TCR},$$

thereby allowing one to estimate TCR given knowledge of the present temperature change δT_s, the present forcing F and the forcing from a doubling of atmospheric carbon dioxide, as

$$\text{TCR} = \frac{\delta T_s}{F} F_{2CO_2}. \tag{13.25}$$

Eqs (13.23) and (13.25) are often referred to as 'energy budget methods' – as an acknowledgement of their roots in Eq. (13.10), which models the energy balance – and they have found common application in efforts to infer ECS and TCR from measurements. For instance, taking observational estimates of δT_s and δR (0.75 K and 0.65 W m^{-2}, respectively, for the decade 2000–2009 relative to 1860–1879) combined with model estimates of F_{2CO_2} (3.44 W m^{-2}) and of the radiative forcing F (1.95 W m^{-2}) suggest, based on Eq. (13.23), a most likely ECS value of 2 K and, based on Eq. (13.25), a smaller TCR value of about 1.3 K (Otto et al., 2013). These values are smaller than what is currently found by more comprehensive models. Whether these differences are due to deficiencies in the more comprehensive models (i.e., by allowing clouds to change in unphysical ways) or a consequence of inappropriate assumptions, either in the postulation of Eq. (13.10) to begin with or in the additional simplifications required to go from Eq. (13.10) to Eqs (13.23) and (13.25) remains a fertile area of inquiry (Stevens et al., 2016).

Eq. (13.10) also avails itself of other methods to estimate λ and hence ECS. For instance, by direct inspection one might imagine that if the forcing varied slowly, then variations in δR might correlate with variations in δT_s in proportion to $-\lambda$. If so, satellite measurements of year-to-year variability and measurements of T_s could be used to constrain the climate sensitivity. Comprehensive climate models, which explicitly represent δR and δT_s, don't however show a strong relationship between inter-annual fluctuations in these quantities and the climate sensitivity derived from the same model. Assuming that Eq. (13.10) is a good model of the climate system, this could imply that the pattern of the temperature changes with warming are fundamentally different than those associated with inter-annual variability, or that the random component of the system is simply too large to extract a meaningful relationship.

Another 'energy budget' approach to estimating λ using observations is to replace δR in Eq. (13.10) with a term describing the rate of ocean enthalpy increase, that is,

$$C \frac{d(\delta T_s)}{dt} = F + \lambda \, \delta T_s. \qquad (13.26)$$

Here the ocean enthalpy is taken proportional to the mean surface temperature and a heat capacity C. Subjecting a system described by Eq. (13.26), to stochastic white noise forcing results in a 'reddening' of the temperature response, characterised by an auto-correlation timescale given by $-C/\lambda$. Climate models suggest variations in this reddening timescale can largely be explained by λ. Assuming that the models therefore provide a reasonable estimate of C, allows one use observations of the one-year lag correlation in δT_s to estimate λ, and hence ECS as $-\lambda^{-1} F_{2CO_2}$. This procedure suggests values of ECS in the range of 2.2–3.4 K, slightly larger than what is estimated using Eq. (13.23) but still on the lower end of estimates using results from comprehensive climate models (Cox et al., 2018).

In addition to instrumental records, longer-term records of past climate variations (e.g., glacial–interglacial changes) can be used to investigate how the Earth system responds to external perturbations. These too could be referred to as 'energy budget methods' and, though promising, this approach is complicated by the difficulty of inferring global temperature changes from proxy records which have an incomplete spatial coverage. Another difficulty arises from the need to have a good knowledge of solar, volcanic and orbital changes, of atmospheric composition changes (e.g., GHGs, aerosols) and of ice sheets changes, all of which are required to estimate past radiative forcings with a sufficient precision. Given these difficulties, the climate sensitivity estimates inferred from past climate variations such as the Last Glacial Maximum remain very uncertain. These approaches, particularly as applied to changes accompanying glacial cycles agree roughly with the other energy budget methods, in particular, they appear to rule out extreme ECS values (lower than 1 K or larger than 5 K). Interestingly, however, the inter-model spread of climate sensitivity estimates inferred from past climate changes arises primarily from differences in the short-wave cloud feedback, as it is the case for future changes. Given the fairly good correspondence between the inter-model spread of this feedback in past and future climates, there is reason to believe that further direct or indirect constraints on the cloud feedbacks derived from paleo-climatic studies will translate into more effective constraints on ECS.

13.3.4.3 Bottom-up Versus Top-down Approaches

The relationship between uncertainties in cloud feedbacks and in climate sensitivity is so strong that constraining ECS effectively constrains cloud feedbacks, and vice versa. This makes it possible to use 'top-down' approaches, which consist of constraining the climate sensitivity directly so as to determine the cloud feedbacks, in addition to 'bottom-up approaches', which attempt to build up an estimate of cloud feedbacks from its constituent parts and their individual constraints.

Both the top-down and bottom-up approaches have strengths and weaknesses. The top-down approach is tantalising as it does not assume an a priori knowledge of the physics of cloud feedbacks. The drawback, however, is the risk to incorrectly constrain feedbacks by misinterpreting the relative role of cloud feedbacks versus other factors (e.g., heat uptake) in the climate sensitivity estimate, or by assuming (wrongly) that the behaviour of clouds feedbacks is independent on the timescale or the type of climate variation considered, as has been discussed earlier.

In contrast, the bottom-up approach is usually more physically based, and better suited to focused observational analyses (including field experiments) and process studies. Its drawback, however, is that the processes which are truly critical for climate change cloud feedbacks are not necessarily known a priori. Although the laws of physics are likely to be universal, cloud-perturbing factors do not exist in isolation. So cloud responses to perturbations that occur naturally in the present climate can only be confidently taken as indicative of how clouds will change in the future if the perturbations from the present day adequately sampled the perturbations that may occur in the future. For example, if one is interested in how clouds respond to surface warming, studying how clouds respond to warming in the present day will not be indicative of how clouds respond to warming of the climate, if the suite of other cloud controlling factors, like humidity, wind-shear, irradiances and advective tendencies of heat and moisture vary with global warming differently or in a different proportion than they do for the particular case being studied.

One way to circumvent these difficulties is to combine the two approaches: by analysing and prioritising the physical processes that mainly control the large- or global-scale behaviour of the climate system (as can be done by analysing GCM outputs), one can determine those aspects of a simulation that need to be realistic if the simulation's representation of climate change is to engender confidence. Given the complexity of comprehensive climate models, it often helps to work with model configurations that isolate what are believed to be key processes. By reproducing the large-scale behaviour of interest in a suite of model configurations using simpler boundary conditions (e.g., aqua-planets, or even radiative–convective equilibrium) or simplified physics (e.g., with some specific processes or interactions switched off), it can help to pinpoint the critical cloud feedback processes. These processes can then be tested using detailed models (e.g., LES or SRMs) and/or observations, and even motivate dedicated process or observational studies. Constraints on these targeted processes are thus more likely to constitute relevant and efficient constrains on the global cloud feedback.

13.4 Clouds in a Changing Climate

Climate projections have always been associated with large uncertainties, and it is our ability to identify aspects of these projections which are robust, that is, physically understood and supported by multiple lines of evidence, that has allowed us to better assess our relative confidence in various aspects of these projections. These efforts are bearing fruit and uncertainty in estimates of crucial quantities, like cloud feedbacks and the ECS, are beginning to narrow. While much remains to be done, progress over the last decades has added great specificity to the questions that need to be answered to better understand how climate responds to forcing, and clouds respond to warming. We review this progress, and summarise the building blocks of our present understanding.

13.4.1 A Historical Synopsis

Beginning in the 1960s, NASA launched its Television Infrared Observation Satellite (TIROS) (first-generation)

and Nimbus (second-generation) satellites designed for meteorological research, so that by the early 1970s these measurements were providing new insight into Earth's albedo, helping to point to the large impact of clouds on the Earth's radiation balance. In parallel, the first GCMs based on the primitive equations (Chapter 2) were being developed in the United States and in the United Kingdom. By the end of the 1970s, results from these early models and from a hierarchy of simpler models helped inform an influential report led by Jule G. Charney,[5] wherein the impact of increased carbon dioxide on climate was assessed. The Charney report estimated the ECS to be between 1.5 and 4.5 K, about the same as present-day estimates. This report, and other reports at that time, began to identify cloud processes as some of the most challenging in any attempt to model the Earth's atmospheric general circulation and identified cloud feedbacks as a dominant source of uncertainty in estimates of climate sensitivity.

In the late 1970s and early 1980s, formalisms began to be developed to aid the interpretation of model-based estimates of climate sensitivity. These formalisms introduced the ideas of radiative forcing and climate feedbacks. An early model inter-comparison project (Feedback Analysis of GCMs and in Observations, or FANGIO (Cess et al., 1990)) demonstrated that most of the model spread in climate sensitivity estimates from different models could be attributed to inter-model differences in cloud feedbacks. The Earth's Radiation Budget Experiment (ERBE) provided the first measurements of the long- and short-wave components of the radiation balance at TOA in clear-sky and cloudy conditions. These measurements were used to estimate what is now known as the CRE (see Section 4.4.2) and provided the first quantitative estimates of the impact of clouds on the Earth's radiation budget in the present day; on average over the globe, the measurements demonstrated that clouds exert a net cooling effect on climate. For the most part, physical ideas as to why clouds, or the CRE, might change with warming, and hence the physical basis for understanding cloud feedbacks, was lacking.

By the 1990s, many more GCMs had been developed around the world. Analyses of these models showed that a very large number of microphysical and macrophysical processes had the potential to affect the sign and magnitude of cloud feedbacks. Simulators started to be

---

[5] Jule G. Charney (1917–1981) was an American meteorologist at the Massachusetts Institute of Technology who contributed to the development of dynamical meteorology and numerical weather prediction, and promoted the use of a hierarchy of models combined with physical insight to advance understanding of how the atmosphere and climate work.

developed to more closely link satellite observations (Chapter 7), foreshadowing more modern efforts, as groups began to more systematically compare observed and simulated radiative fluxes at TOA and recognise the importance of their magnitude for coupled modelling. At the same time, interest in the possible effect of aerosols on cloud properties developed rapidly as the magnitude of the human perturbation to the atmospheric aerosol during the twentieth century began to be better appreciated.

Refined analysis, aided by the use of simulators, allowed cloud feedbacks in GCMs to be diagnosed in a consistent way. Research in the first years of the twenty-first century began to suggest that the sign of cloud feedbacks was generally positive, but their amplitude was shown to vary a lot across models, and the physical rationale for a positive feedback remained elusive. Model inter-comparisons considering a large number of GCMs (such as carried out in the third phase of the coupled model intercomparison project (CMIP3), or in the Cloud Feedback Model Intercomparison Project (CFMIP)) showed that while some aspects of cloud feedbacks were robust among models (e.g., the positive feedback associated with the rise of high clouds), other aspects were much more variable. The response of low-level clouds to warming was shown to dominate the model uncertainty in cloud feedbacks and climate sensitivity. This result fostered research on low clouds using observations and a hierarchy of numerical models, including SRMs and LES. In parallel, it was shown that the forcing-feedback interpretative framework needed to be revised to account for the presence of tropospheric adjustments (Section 13.2.4) and hence to better assess the role of clouds in climate sensitivity.

Over the past ten years, research on cloud feedbacks has intensified and has produced substantial progress in the observation of clouds and our physical understanding of cloud feedbacks. Some of the first satellite studies positing a response of clouds to warming have been published and lidar and radar instruments on board satellites have provided an increasingly detailed description of the full three-dimensional distribution of clouds in the atmosphere. Cloud-feedback processes began to be routinely studied using GCMs and process models (e.g., LES and SRMs). Several key physical processes involved in cloud feedbacks began to be identified, helping to substantiate the belief that the global cloud feedback was unlikely to be negative, and the strength of the coupling between clouds and convective- and large-scale circulations began to be appreciated. The crucial role that CREs play in the large-scale circulation and climate variability is coming increasingly into focus as their influence on the regional

response to warming is becoming better appreciated. An increasing number of studies have suggested ways to constrain some aspects of cloud feedbacks using observations and steady progress is being made in understanding how the physics of climate models influences cloud feedbacks under climate change.

Thanks to this progress, it is becoming possible to decompose the long-standing climate sensitivity problem into several pieces. As individual processes or mechanisms may be more easily testable using present-day observations than global-scale emergent properties of the climate system, such advances progressively pave the way for future progress in our ability to assess climate sensitivity and cloud feedbacks in a more precise way. We discuss some of the main mechanisms underlying cloud feedbacks and adjustments. This review is selective so as to highlight some fruitful approaches to tackle the climate change problem, and how advances in physical understanding can help improve our assessment of complex climate projections.

13.4.2 Tropospheric Adjustments

The concept of tropospheric adjustments, that is, of climate responses to external perturbations which are fast and not mediated by global surface temperature changes, is reminiscent of the way cloud-mediated effects of aerosols on climate (commonly referred to as indirect and semi-direct effects of aerosols) are considered. The evidence that changing GHG concentrations also result in fast atmospheric adjustments (Section 13.2.4) implies that the traditional concept of radiative forcing and feedbacks needs to be revisited not only for interpreting climate changes of the twentieth century (which are associated with significant aerosol changes) but also for predicting those of the twenty-first century which are associated with a much more dominant radiative forcing from increased carbon dioxide. Hereafter, we will focus on tropospheric adjustments to increased carbon dioxide. Forcing adjustments to aerosols are addressed in Chapter 11.

Although climate models predict adjustments to increased carbon dioxide that exhibit quantitative differences, qualitatively they exhibit a few features which are reproduced by most models, including those which predict cloud processes without using cloud or convective parameterisations. Among these features is the fact that tropospheric adjustments develop on the timescale of days, suggesting a control by very fast physical processes. This short timescale implies that tropospheric adjustments keep track with the evolution of carbon dioxide concentrations instead of lagging behind as it would be the case for temperature-mediated responses. Another robust feature is that tropospheric adjustments appear to be associated with a downward shift of cloud layers at low atmospheric levels, and with contrasted changes in free-tropospheric cloudiness and convective precipitation between land and ocean regions.

These robust changes can largely be understood as the atmospheric response to purely radiative perturbations. An increase of carbon dioxide concentrations reduces the atmospheric radiative cooling. By reducing the radiatively driven turbulent entrainment through the trade inversion, it lowers the depth of the trade inversion; by warming the atmosphere, especially in the lower free troposphere, it increases the strength of the trade inversion and the LTS, and reduces the relative humidity near the top of the planetary boundary layer. These combined effects lead to a shallower atmospheric boundary layer and to a downward displacement of cloud layers (Fig. 13.4).

In addition, as the overall overturning atmospheric circulation is primarily constrained by the amount of radiative destabilisation of the atmosphere, reduced radiative cooling rates weaken the strength of large-scale vertical motions. Over ocean, it is accomplished through the effect of the free-tropospheric warming on (increased) static stability and (reduced) convective instability. Over land, the large-scale subsidence also weakens owing to the enhanced static stability, but the vigour of large-scale ascent strengthens. It is due partly to the increased convective instability associated with the fast warming of the surface relative to the atmosphere aloft, and partly to the mass convergence induced by the change in land–sea surface temperature contrasts.

The net effect of these different changes on the Earth's TOA radiation balance varies substantially among models as it results from the sum of many (and antagonist) influences. Most models produce nevertheless a positive TOA cloud-radiative anomaly (i.e., warming effect), ranging roughly from 0.5 to 1.5 $W m^{-2}$ for a quadrupling of carbon dioxide.

13.4.3 Cloud Feedbacks

The idea that clouds might be sensitive to the mean temperature of the Earth's system and play a role in climate change emerged in the 1960s. Cloud feedbacks were recognised as critical for predictions of climate sensitivity and for other aspects of the large-scale climate as soon as quantitative results from early climate models were compared. With the organisation of model intercomparisons involving ever more models, this recognition only became more widespread. Today, inter-model differences in cloud feedbacks remain the primary source of spread of climate sensitivity estimates (Section 13.3.4.1).

As mentioned in Section 13.3.4.3, the response of clouds to perturbations in the present climate may, by virtue of subtle differences in the perturbations, not be representative of how clouds respond to perturbations accompanying climate change. However, because the basic physics and impact of environmental conditions on clouds should be invariant, an alternative approach consists in decomposing the cloud feedbacks into elementary components associated with different physical processes or cloud-controlling factors, and to assess these different factors individually using observations. For this reason, process-oriented evaluations of cloud variations may be relevant for assessing climate change cloud feedbacks, even if the overall cloud variations inferred from present-day variability may not. Our ability to assess the credibility of model cloud feedbacks – and to interpret inter-model differences – thus depends on our understanding of the physical processes that underly these feedbacks.

When climate models are analysed carefully, some common responses of clouds to warming can be identified. For instance, climate models tend to predict fewer but higher clouds in a warmer climate (Zelinka et al., 2013; Fig. 13.7). The cloud feedback (quantified consistently and relative to the Planck response) associated with these changes is generally positive in the global mean but with substantial inter-model variability. Just because models predict a consistent cloud response, or some aspect of this response, to warming does not imply that the response is correct. But it does encourage the idea that the response can be understood, and its correctness assessed based on this understanding. Some of the basic mechanisms thought to underly the tendency of models to predict optically thicker clouds (at least in particular circumstances), to robustly produce higher clouds, and to some extent fewer clouds with warming are described later. Further mechanisms specific to mid-latitude and polar cloud feedbacks are discussed in Part III of this book, and cloud responses to forcings, such as changing aerosol concentrations, are discussed in Chapter 11.

13.4.3.1 Cloud Optical Depth Feedback

Polewards of about 40° of latitude, climate models tend to predict an increase of the cloud optical thickness, per unit cloud, as climate gets warmer. This appears to be associated with an enhancement in the liquid-water path in

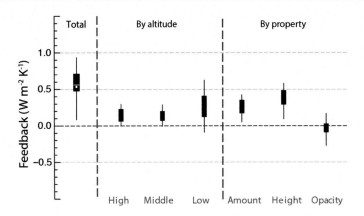

FIGURE 13.7: Partitioning of CMIP5 cloud feedbacks in components associated either with cloud types (low, middle and high clouds) or with cloud property (cloud amount, cloud altitude and cloud optical thickness). The total cloud feedback, and its inter-model spread (characterised by whisker boxes) is represented on the left. Adapted from Zelinka et al. (2013)

these regions. Because the change in the optical thickness primarily reduces the absorbed short-wave radiation by the Earth system, this implies a negative cloud feedback at these latitudes.

This robust behaviour is consistent with a simple thermodynamic argument known as the cloud optical depth feedback (Gordon and Klein, 2014). Consider a cloud forming by adiabatic cooling as air rises. The net condensation that accompanies a given expansion is greater at warmer temperatures, which is why the temperature lapse rate approximately following a saturated isentrope (moist adiabat) increasingly departs from the dry adiabatic temperature lapse rate as the temperature warms (see, for instance, the discussion surrounding Fig. 2.3). This also implies that more condensate will be produced, so that cloud water content is increased. So that for a given rate of saturated expansion, more liquid water (q_l) will be produced at warmer temperatures. The temperature dependance of the condensation rate can be measured by the rate of change of the saturated isentropic lapse rate (moist adiabat) with temperature:

$$\frac{\partial \ln(q_l)}{\partial T} = \frac{\partial \ln(\Gamma_s)}{\partial T}, \tag{13.27}$$

where $\Gamma_s = -(\partial \theta/\partial p)_{\theta_{e,s}}$ is the change in θ with p, the atmospheric pressure, for a saturated isentropic process. As illustrated in Fig. 13.8, the proportional rate of change is positive, and like the Clausius–Clapeyron curve itself, increasingly so at cold temperatures. At the temperatures characteristic of the mid and high latitudes, the relative rate of increase of liquid water following adiabatic ascent increases by about 4 % K^{-1}, roughly twice the temperature dependence found at the warmer temperatures characteristic of the cloud environment in the tropics. Hence, in the colder extra-tropics, clouds are likely to increase their liquid water with warming more readily, and perhaps in a way that overwhelms other feedbacks, thereby

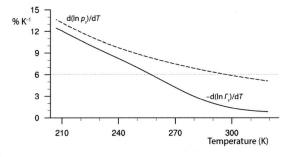

FIGURE 13.8: Sensitivity of the saturation vapour pressure p_s to temperature, and the associated change in the saturated adiabatic lapse rate $-d\ln(\Gamma_s)$ with temperature

contributing to the tendency of models to increase the optical depth of extra-tropical clouds with warming.

Other processes are also likely to contribute to this robust response of models. For example, the systematic change in cloud phase that accompanies warming. As more of the atmosphere is above temperatures at which ice clouds form, the onset of glaciation will occur at higher altitudes and latitudes as the climate warms. This implies that where, in a colder climate, ice crystals formed, liquid water in the form of cloud droplets can continue to exist. Because liquid-water droplets distribute water in smaller particles, the cloud effective radius decreases (Chapter 4), which causes the clouds to back scatter more solar radiation, contributing in the same sense as the negative feedback associated with the increased water content of clouds.

Lastly, a poleward shift of the storm tracks which is expected to accompany warming implies that major storm-track cloud regimes would also be displaced polewards, to areas where the insolation is lower. So doing reduces the amount of sunlight reflected by clouds. Counteracting such effects might be the change in strength and

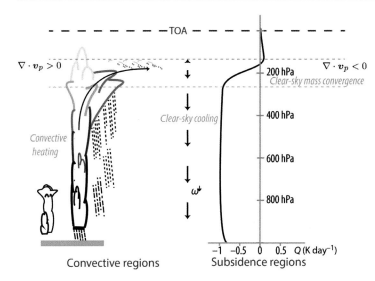

FIGURE 13.9: The fixed anvil temperature (FAT) hypothesis relates the level of convective outflow to the clear-sky driven mass convergence. The radiatively driven mass convergence in the clear sky is determined by the amount of water vapour, which in turn is a function of temperature. In the present atmosphere, at 200 hPa there is sufficient water vapour to maintain a radiatively driven convective layer (Q represents the clear-sky radiative heating rate), and this defines the top of the troposphere. Adapted from Hartmann and Larson (2002)

frequency of storms, which may modulate (and partly oppose) the contribution of these delineated processes to the global feedback (Chapter 9). However, the quantitative balance of the different processes determining model cloud feedbacks remains poorly understood.

If the cloud optical depth feedback is mostly negative in the extra-tropics, it is mostly positive in the tropics. This is primarily due to the fact that in these regions, the change in cloud optical depth is dominated by changes in the cloud amount (which tends to decrease in a warmer climate, as explained later). As a result, the cloud optical depth feedback is only slightly negative at the global scale (Fig. 13.7).

13.4.3.2   Altitude Feedback

Models robustly predict an upward shift of upper cloud layers in a warmer climate. This behaviour, which is particularly marked in the tropics, is not only predicted by GCMs but also by cloud-resolving models simulating radiative–convective equilibrium. The influence of such a shift on the Earth's radiation balance depends on how much the cloud-top temperature varies as the clouds rise.

The cloud-top temperature of deep convective clouds is primarily controlled by the depth of the convection layer, and in radiative–convective equilibrium the latter is determined by the depth of the atmospheric layer destabilised by radiation, loosely speaking the troposphere. Owing to its peculiar spectroscopic properties, water molecules interact very strongly with atmospheric radiation and exert a major control on the radiative cooling rate. Because water vapour pressure decreases as temperature decreases, at some temperature (i.e., altitude), the concentration of water molecules becomes so small that their contribution

to the radiative cooling becomes negligible: this defines the upper limit of the radiatively driven convection layer (Fig. 13.9). In clear skies, the radiative cooling is balanced by adiabatic warming through large-scale subsidence. The strong decline of the radiative cooling with altitude (which in the present atmosphere occurs in a layer around 200 hPa, e.g., Fig. 13.9) must thus be accompanied by a strong lateral convergence of mass at that level. Owing to mass conservation, upper-tropospheric mass convergence in the subsiding areas implies mass divergence in convective areas, that is, by the frequent occurrence of convective anvil clouds at this level (Fig. 13.9). The analysis of GCM outputs, SRM simulations and observations confirms this inference, showing evidence for a maximum cloud amount at the same altitude as the maximum of clear-sky mass convergence in clear-sky regions (i.e., which coincides with the level of maximum convective detrainment). This physical reasoning implies that the altitude of maximum convective detrainment can change in a warmer climate, but the temperature at the detrainment level, and thus the emission temperature of anvil clouds, is largely independent of surface and atmospheric temperature variations.

This mechanism, commonly referred to as the 'fixed anvil temperature' (FAT) mechanism, suggests that cloud-top temperatures remain roughly unchanged as surface temperatures rise (Hartmann and Larson, 2002; Zelinka and Hartmann, 2010). It increases the difference between the temperature at which the deepest clouds emit radiation and the surface temperature, thus increasing their greenhouse effect, and Earth's greenhouse effect overall. Because this implies that the climate system becomes less effective at radiating heat to space, and thus must warm more to balance a positive perturbation in the TOA radiation budget, it constitutes a positive feedback. This robust feedback mechanism, supported by physical understanding, by a spectrum of numerical models and by observations, partly explains why GCMs all predict a positive long-wave cloud feedback in convective areas of the tropics. In contrast, the short-wave cloud feedback primarily depends on changes in cloud amount and cloud optical thickness. The constraints on the upper-level cloud amount imposed by the mass and energy constraints are discussed in the next section.

13.4.3.3 Cloud Amount Feedback

Most climate models predict a decrease of the total cloud amount in a warmer climate, which leads to a negative longwave cloud feedback and a positive short-wave cloud feedback. The net cloud feedback associated with changes in cloud amount appears to be dominated by the short-wave component, consistent with CREs being larger, on average, in the short-wave than in the long-wave part of the spectrum. Further investigation suggests that the disproportionate positive short-wave cloud feedback arises through disproportionate changes in the low-level cloud amount (Bony et al., 2006; Fig. 13.7). The reason why changes in low-cloud amount dominate the net cloud feedback is that low-level clouds are ubiquitous over the global ocean, and that their warm cloud tops prevent them from having a substantial effect on TOA long-wave radiation; that is, their net CRE is more dominated by the short-wave, and hence more negative, than the average cloud. This, combined with the fact that low-cloud feedbacks have been shown to be the primary contributors to the inter-model spread of global cloud feedbacks, has generated intense research efforts towards understanding how the low-level cloud fraction, particularly over the ocean in the tropics and subtropics, might respond to climate change. We discuss our understanding of changes in low-cloud amount with warming, and then we address the case of changes in middle- and upper-level cloud amount.

Low-Cloud Amount

Observations suggest a strong positive relationship between the low-cloud amount and the lower-tropospheric static stability (LTS), as measured for instance by the difference in the potential temperature at 700 hPa relative to the surface. Enhanced stability acts to resist mixing of moist air in the near-surface marine layer with the usually much drier air of the free-troposphere aloft. Hence, more moisture is trapped in the marine layer which enhances cloudiness there (see for example, Chapter 5). Because the LTS is expected to increase in a warmer climate

owing to the dependence of the saturated isentropic lapse rate (moist adiabat) on temperature (Chapter 2), in the context of global warming this would imply more low-level marine cloudiness, and hence a negative cloud feedback due to changes in low-cloud amount. LES, however, when perturbed not just by lower-tropospheric stability changes, but by a range of factors thought to accompany surface warming, predicts a decrease of the low-cloud fraction in response to global warming.

This example, which illustrates how empirical correlations may be misleading when they are naively extrapolated to climate change, suggests that strong theoretical arguments and/or numerical evidence need to be accumulated before correlations derived from present-day climate variability may be used to infer climate change cloud feedbacks. Particularly when present-day relationships are based on dimensional quantities, as such relationships are unlikely to express a basic, or universal, law. In fact, an enhancement in LTS in a warmer climate might be accompanied by other factors, for instance, the amount of energy available for mixing, which could more than offset the effects of increased stability.

To explore how low-level clouds might respond to climate change, an alternative approach consists in investigating the behaviour of clouds in an environment that more completely accounts for the large-scale changes expected to accompany climate change. Focusing only on changes in LTS neglects, for instance, changes in surface fluxes that are expected to accompany warming. These changes can be illustrated directly from a consideration of the bulk formulation of the surface fluxes (e.g., Chapter 5), which expressed in terms of the relative humidity can be written as

$$F_s = V\left[q_s(T_s) - q_{2m}\right] \approx Vq_s(T_s)\left[1 - RH\right], \quad (13.28)$$

where V is the surface wind, q_s the saturation specific humidity, T_s the surface temperature, q_{2m} the near-surface specific humidity and RH the relative humidity at this same height. When clouds are involved, surface moisture fluxes are particularly effective in increasing the buoyancy flux through the depth of the boundary layer (as discussed in Section 2.5.3). This leads to enhanced mixing between the cloud layer and the dry layer above, which mixes more dry and warm air from the lower free-troposphere to the surface and deepens the boundary layer. LES suggest that the enhanced mixing leads to a drying of the planetary boundary layer, and then to a decrease of the shallow cumulus cloud fraction (Rieck et al., 2012; Bretherton, 2015) (Fig. 13.10).

The fact that most GCMs also predict a decrease of the low-cloud cover in warmer climates, especially in shallow

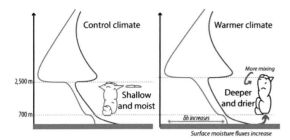

FIGURE 13.10: Representation of the effect of global warming on the thermodynamic stratification of the low troposphere in typical trade-wind conditions. The blue and red curves show respectively the moist static energy h and the saturation moist static energy h_s as observed in the north Atlantic trades near Barbados. The distance between the two curves measures the saturation deficit of the atmosphere. In a warmer climate, surface latent heat fluxes increase, which deepens the layer of cumulus convection and drives more mixing in the lower troposphere. The vertical gradient of moist static energy within the low troposphere also strengthens owing to the Clausius–Clapeyron relationship. The enhanced import of dry, and hence lower, moist static energy air into the boundary layer, and enhanced export of moist, high-moist-static-energy air to the free troposphere lead to a drier cloud layer and fewer clouds.

cumulus regimes but also in other regimes, suggests some robustness in this mechanism. Indeed, shallow convection is common over the entirety of the global ocean. Whereas the moisture transport by deep convection is energetically constrained by the radiative cooling rate of the troposphere and therefore cannot change by more than about $2–3\%$ K^{-1}, the moisture transport by shallow convection is much less energetically constrained (because it is not always associated with precipitation and therefore does not necessarily affect the atmospheric heat budget) and more controlled by the rate of change of humidity with temperature. This simple argument suggests that in a warmer climate, one expects an increasing relative role of shallow convection in exporting humidity out of the planetary boundary layer, with a low-level drying associated with the enhanced shallow convective mixing exceeding the moistening effect by surface fluxes. By depleting the water vapour needed to sustain the low-cloud cover, this mechanism might explain why in global warming simulations most models predict a widespread decrease of the low-cloud fraction which contributes, with FAT, to a positive global cloud feedback (Sherwood et al., 2014).

Middle- and Upper-Cloud Amount

While a number of recent studies suggest a positive cloud amount feedback associated with low-level clouds, the sign of middle- and upper-level cloud amount feedbacks remains more uncertain. Although the vertical shift of

cloud layers is presumably associated with a positive long-wave cloud feedback (Section 13.4.3.2), whether and how the middle- and upper- level cloud fractions might change in a warming climate and thus partly offset or enhance the impact of the altitude feedback remains a more open question.

A change in the strength of the overturning circulation could in principle affect the relative fractions of the Earth covered by convective or subsiding areas, and thus the distribution of cloud types, cloud fraction and TOA radiative fluxes. However, numerical investigations suggest that this effect plays a minor role in cloud feedbacks. A sensitivity to temperature of some microphysical processes within convective clouds (e.g., precipitation efficiency) could affect the amount of water detrained in the upper troposphere and then the cloud amount, but so far observations and models have not provided any compelling evidence for such a sensitivity. On the other hand, recent studies suggest that the anvil cloud amount might be partly controlled by thermodynamic processes (Bony et al., 2016).

GCMs used in a variety of configurations suggest that as the climate warms, the coverage of tropical convective anvil clouds reduces. This behaviour, which is also present in observations at the inter-annual timescale (the tropical anvil clouds exhibit an overall reduction in coverage during warm years compared to cold years) and in idealised radiative–convective equilibrium simulations from SRMs, is likely rooted in the same energetic and thermodynamic mechanisms as those involved in the altitude feedback (Section 13.4.3.2), namely the sensitivity to temperature of the radiatively driven mass convergence in non-convective regions (or mass divergence in convective regions). Consistent with the FAT hypothesis (Fig. 13.9), as the surface temperature increases, the clouds tend to rise and to remain at nearly the same temperature. However, the rise moves the clouds to a lower pressure. Because of the inverse pressure dependence of the static stability

$$S = -\frac{T}{\theta}\frac{\partial \theta}{\partial p} = \left(\frac{R_d}{c_{pd}}\right)\frac{T}{p}\,(1-\gamma), \qquad (13.29)$$

where γ denotes the ratio of the actual over dry adiabatic temperature lapse rates and the pressure dependence of saturation specific humidity and thus γ, the clouds find themselves in a more stable atmosphere (in situations when clouds do not rise much as the surface temperature increases, they find themselves in a warmer and therefore more stable atmosphere as well). As a result, the radiatively driven lateral mass divergence D_r, which depends on the clear-sky radiative cooling Q_r, and the static stability S, as,

$$D_r = \frac{\partial \omega}{\partial p}, \quad \text{whereby} \quad \omega = -\frac{Q_r}{S}, \tag{13.30}$$

decreases as the climate warms. This leads to a decrease of the convective outflow and of the anvil-cloud fraction. This behaviour can be described as a 'stability-iris' effect, whereby more infrared radiation escapes to space as the surface temperature increases.

Finally, recent SRMs, GCMs and observational studies have pointed out that the large-scale relative humidity and upper-level cloud amount were sensitive to the degree of 'clustering' of convective systems (Chapter 8). In models, the decrease of the anvil cloud fraction as convective aggregation increases relates to the same thermodynamic mechanism as the one just explained: the convective cloud-base temperature rises as convection becomes more aggregated. This is due to the fact that convection concentrates in areas where the lower-tropospheric moist static energy is maximum, thus shifting the saturated isentropic lapse rate (moist adiabat) towards higher values, which tends to warm the atmosphere and increase the static stability at the height of anvil clouds, and therefore to decrease the anvil cloud amount. However, whether a warmer climate is associated with changes in the organisation of deep convection and a change in the overall high-cloud (anvil + cirrus) amount remains an interesting but unsettled issue.

Another open issue is how changes in the anvil-cloud amount affect the cloud-radiative feedback. Indeed, their net radiative impact depends on the extent to which changes in the long- and short-wave CREs compensate each other as the cloud amount varies, and on how the balance between these two effects can be affected by changes in the optical thickness of anvil clouds and/or by changes in the low-cloud fraction with surface temperature. At the inter-annual timescale, observations suggest that changes in the anvil cloud amount are associated with a near-neutral radiative feedback. Whether or not this holds in a warmer climate remains unknown.

13.4.3.4 On the Magnitude of Cloud Feedbacks

As discussed earlier, the cloud feedback is actually the sum of a number of cloud feedback mechanisms that operate within and across different cloud regimes. At the moment, the robustness of cloud-feedback mechanisms is assessed primarily from the point of view of our understanding of the sign of the constituent cloud feedback mechanisms. However, the magnitude of each component is often less well constrained and can vary among models. The sign and magnitude of a sum of positive and negative constituent feedbacks makes the sum of these feedbacks more uncertain than that of each individual component,

which helps explain why an assessment of the overall, or global, cloud feedback is progressing more slowly than that of its individual components. Beyond physical mechanisms, one has therefore to understand what controls the magnitude of each component, if not the whole feedback. This type of quantitative understanding remains poorly developed. Nevertheless, a few such controlling factors have already been identified. A few examples related to the size of low-cloud feedbacks are discussed later.

One important factor influencing the strength of cloud feedbacks is thought to be the strength of atmospheric cloud-radiative effects arising from low clouds. In addition to reflecting solar radiation to space, low-level clouds radiatively cool the troposphere in the long wave (cf Chapter 4). By doing so, they tend to enhance the surface latent heat flux, to moisten the boundary layer and thus to contribute to their own maintenance. This positive feedback between cloud atmospheric radiative cooling and low-cloud amount amplifies the response of the low-cloud amount to external perturbations, and thus the size of the cloud amount feedback, whatever the sign of this response is. Because the radiative cooling exerted by low-level clouds is greater when the cloud layers are close to the surface, the magnitude of this positive feedback is stronger the lower the clouds.

A second controlling factor is the strength of the lower-tropospheric mixing in the present-day climate. As explained in Section 13.4.3.3, in a warmer climate, one expects more dry, and hence lower, moist static energy free tropospheric air to be drawn down toward the surface and more moist, and hence higher, moist static energy air to be transported upwards, leading to a drying of the low-cloud layer and to a reduction of the low-cloud amount. In climate models, the rate at which the lower-tropospheric mixing of water increases with temperature scales roughly (through the Clausius–Clapeyron relationship) with the strength of present-day mixing. Schematically we can think of the effect of lower-tropospheric mixing on the drying of the boundary layer as proportional to an exchange velocity and a vertical gradient – similarly to how we think about surface fluxes, that is, Eq. (13.28) – such that

$$\dot{q}|_{\text{ltm}} = M_{\text{ltm}} \partial_z q. \tag{13.31}$$

Here the drying rate is denoted by $\dot{q}|_{\text{ltm}}$, the exchange velocity by M_{ltm} and the vertical humidity gradient by $\partial_z q$. Eq. (13.31) suggests that the amount of drying with warming, and hence the size of the positive low-cloud amount feedback predicted by models, depends on M_{ltm}. Following similar arguments as were used to describe the change in surface fluxes with warming, that is, Eq. (13.28), for a constant relative humidity, one expects the drying rate to scale as

$$\delta \, \dot{q}|_{\text{ltm}} = M_{\text{ltm}} \delta(\partial_z q) \approx -M_{\text{ltm}} \delta q_{\text{s}} (\partial_z \text{RH}). \qquad (13.32)$$

In other words, it depends on how much mixing (M_{ltm}) between the lower and middle troposphere the models produce in the present-day climate through explicit and parameterised processes (Sherwood et al., 2014).

More generally, the strength of the low-cloud feedback depends on two antagonist influences on lower-tropospheric humidity: moistening by surface turbulence and drying by lower-tropospheric mixing. The relative strength of both influences depends on many factors, including the coupling between surface turbulent fluxes, convection and low-level radiative cooling, whose representation differs widely across numerical models. Observational constraints on the strength of these couplings would greatly help constrain the size of the low-level cloud feedbacks.

13.4.3.5 Clouds and Regional Warming Pattern

When the planet warms globally, the changes in surface temperature exhibit important contrasts at the regional scale. Surface temperature change is generally higher over land than over ocean (by about 50% on average), and it is larger near the poles than at the global scale. This latter phenomena is referred to as 'polar amplification', and it is particularly strong in the Arctic. These regional features have been found in reconstructions of past climate changes (e.g., Last Glacial Maximum, 21 ky before present), in the instrumental record (land surfaces are warming faster than the oceans, and the Arctic is warming twice as much as the global average) and emerge from climate projections using numerical models.

The robustness of these regional features can be explained at first order by basic physics. The lower moisture availability over land (relative to ocean) limits the ability of the surface to respond to an external radiative forcing through surface evaporation, which enhances the contribution of sensible heat flux and infrared emission to the restoring of the surface energy budget: a given energy imbalance at the surface thus induces a larger surface temperature change over land than over ocean. The basic reasons for polar amplification is more debated. However, and as discussed in Chapter 10, it seems that in addition to positive short-wave feedbacks from melting of surface snow and ice, two additional factors are important: (i) the non-linear dependence of blackbody emissions on temperature, which implies that a given change in emitted long-wave radiation by the surface requires a larger temperature change at colder background temperatures, and (ii) the different temperature lapse rate at low and high latitudes, which implies that a larger surface temperature change is required to maintain the Earth's effective emission temperature when the warming is confined near the surface.

Clouds can modulate, and respond to, regional temperature changes in various ways. Over most land regions, a local decrease of the cloudiness under global warming tends to amplify the magnitude of surface warming as a result of the reduced shading effect of clouds. Changes in cloudiness coupled to changing patterns of SSTs may lead to different radiative feedbacks from warming, even for the same global temperature change, as discussed in Section 13.2.6. Clouds also play an important role in mediating the heat transfer between the hemispheres, and thus partly determine how the tropical convergence zones may respond to an asymmetric hemispheric forcing.

The role of clouds in polar amplification is less clear owing to the complex influence of clouds on the surface energy budget (Chapter 10). The influence of clouds on polar surface warming can be local and/or remote. For instance, the polar cloudiness is strongly modulated by moisture advection from the mid latitudes, and the frequency and strength of these intrusions depend on the position of the mid-latitude eddy-driven jets which are remotely driven by cloud and temperature changes at low latitudes. In addition, a warming climate can be associated with an increase of the polar cloud amount (e.g., due to the enhanced surface evaporation associated with the melting of sea ice) and/or of the cloud optical thickness (e.g., through the thermodynamic feedback discussed earlier). These changes would tend to increase the amount of long-wave radiation emitted by clouds down to the surface and thus to warm the surface, while it would also tend to reduce the amount of solar radiation reaching the surface and thus to cool the surface. It makes the impact of cloud changes on polar temperatures dependent on season, time (day/night) and surface properties, and this impact is thought to contribute substantially to the polar amplification of warming in winter (Cronin and Tziperman, 2015). Based on these examples, it is clear that beyond their influence on global mean surface temperature, clouds also affect the regional pattern of surface temperature changes in a complex way.

13.5 Precipitation in a Changing Climate

Few aspects of climate change matter as much for humankind, and the biosphere more generally, as changes in precipitation: the welfare of billions of people across the world depends on rainfall, and sustained changes in the frequency or amount of rain can have serious social, environmental and geo-political consequences. Anticipating how rainfall might respond to anthropogenic activities thus seems worth trying to understand.

Since the dawn of climate modelling, regional changes in precipitation predicted by climate models have always been recognised as one of the most uncertain aspects of climate projections. The sensitivity of regional precipitation changes to details of the model's formulation, compounded by the large inter-model differences of precipitation projections at the regional level, may easily give the impression that predicting regional rainfall is an impossible challenge. Yet, when looked at in the right way, main features of the regional pattern of precipitation projections appear to be robust, from one generation of climate models to the next, and for a large range of climate change scenarios. The coexistence of robustness and uncertainties constitutes an invitation, and an opportunity, to understand what controls precipitation in a changing climate.

In the case of the climate sensitivity problem, the previous sections have shown how an analysis framework can be developed to decompose a difficult problem into a set of sub-problems such as forcings, adjustments and feedbacks. Both adjustments, and feedbacks can be further decomposed into components associated with changes in water vapour, thermal stratification, surface properties and clouds, and cloud feedbacks can, in turn, be broken down into components associated with particular cloud regimes. This framework turns out to be useful also to study the hydrological sensitivity, that is, the global mean change in precipitation normalised by the global mean surface temperature change. However, such a framework is much less developed than that which has been used to study clouds, particularly in the case of regional precipitation changes.

It may sound trivial to say that clouds matter for precipitation, as they are the place where water condensate in suspension in the atmosphere, coalesces and eventually precipitates. What is less trivial is the recognition that clouds do not control precipitation only at the process level but also through large-scale constraints related to their coupling to the tropospheric radiation budget and the large-scale atmospheric circulation. These large-scale constraints provide an opportunity to develop an analysis framework that helps interpret robust and uncertain features of the precipitation response to climate change and gives hope to the ambition of designing research strategies to tackle the sources of uncertainty in regional precipitation change estimates.

13.5.1   Global Mean Precipitation

13.5.1.1   Energetic Constraints and Precipitation

Precipitation is a component of the global water cycle, and the mass conservation of water implies a balance between precipitation rate P and surface evaporation (including sublimation) rate E_{sfc} (Mitchell et al., 1987). However, given its large enthalpy of vaporisation l_v, phase changes of water and precipitation do not only matter for the water budget but also for the surface and atmospheric energy budgets: as implied by the budget of dry static energy in the atmosphere (Chapter 2, Fig. 13.11), if it is assumed that the vast majority of precipitation falls in the form of rain, the release of latent heating to the atmosphere is given by $l_v P$. This condensational heating must be balanced by other changes in the atmospheric budget, namely the vertically integrated radiative cooling (defined as the difference between the TOA and surface radiation budgets) and the surface flux of dry static energy (or sensible heat flux S_{sfc}):

$$l_v P = -R_{atm} - S_{sfc} \qquad (13.33)$$

By mass conservation $P = E_{sfc}$, introducing a related constraint wherein the evaporative contribution to the surface energy budget ($l_v E_{sfc}$) must balance the sum of the surface net radiation budget and sensible heat flux. In the extreme case, where the net long-wave irradiance at the surface approaches zero (which can occur for very opaque atmospheres such as found in very warm and wet climates), and to the extent that S_{sfc} is small, the global mean precipitation must balance the solar radiation absorbed at the surface. This defines an approximate upper bound on precipitation.

13.5.1.2   Hydrological Sensitivity

The hydrological sensitivity is usually defined as the sensitivity of rate of precipitation to a change in temperature, $\Delta P / \Delta T_s$, for a given forcing. It can be thought of as an analogue to the climate sensitivity. But, as will be shown later, there are some important differences, as the hydrological sensitivity defined in this manner depends on the type of forcing.

A striking expression of the overwhelming role of water in the climate system as an energetic body is that in climate change, the global mean precipitation is more constrained by the availability of energy than by the availability of moisture: although water vapour increases with temperature at a rate of about 7.5% K^{-1}, in line with the Clausius–Clapeyron equation (Chapter 2, but also Fig. 13.8), precipitation increases with surface temperature at a much slower rate (about 2–3% K^{-1}). Indeed,

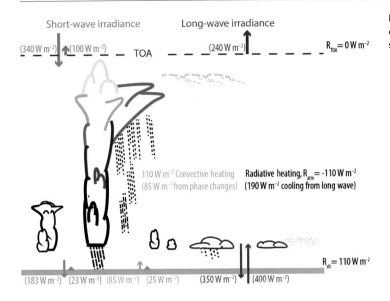

Short-wave irradiance   Long-wave irradiance

(340 W m⁻²)↓ ↑(100 W m⁻²)   TOA   (240 W m⁻²)↑ $R_{TOA} = 0\,W\,m^{-2}$

110 W m⁻² Convective heating   Radiative heating, R_{atm} = -110 W m⁻²
(85 W m⁻² from phase changes)   (190 W m⁻² cooling from long wave)

$R_{sfc} = 110\,W\,m^{-2}$

(183 W m⁻²) (23 W m⁻²) (85 W m⁻²) (25 W m⁻²)   (350 W m⁻²) (400 W m⁻²)

FIGURE 13.11: Illustration of the energy constraints on precipitation (cartoon of the surface and atmospheric energy budgets).

$l_v dP = -dR_{atm} - dS_{sfc}$ and dS_{sfc} is small, which means that the sensitivity of precipitation to climate change is therefore limited by the rate of change of the atmospheric radiative cooling.

The global hydrological sensitivity may thus be understood to a first order by analysing the sensitivity dR_{atm}/dT_s. Like for the TOA radiation budget, radiative kernels can help decompose this sensitivity into different components associated with changes in temperature, water vapour or clouds.

Such an analysis reveals that in response to climate warming, the main contributor to the enhanced radiative cooling of the atmosphere is the increased atmospheric long-wave emission to space (and, to a lesser extent, to the surface) associated with higher atmospheric temperatures, and differences in water vapour, lapse rate and cloud changes across models all contribute to inter-model spread of hydrological sensitivity estimates.

Unlike other GHGs, atmospheric water vapour is associated with competing effects on the tropospheric cooling. Increased absorption of solar radiation by water molecules leads to a radiative heating (mostly within the lower troposphere). At the longer wavelengths of the thermal infrared, water vapour enhances the radiative cooling in the lower troposphere but weakens it in the upper troposphere. Overall, these effects enhance the tropospheric radiative cooling but they are more than compensated by increased absorption in the near infrared, which results in a reduced radiative cooling (the water vapour effect counteracts about 40 % of the temperature effect). This leads to the counter-intuitive result that, on average over the globe, the increase of atmospheric water vapour leads to a

reduced rate of precipitation! This apparent paradox just emphasises again the importance of energy constraints on precipitation and the fundamental energetic role of water in the climate system.

Another contribution to the change in the radiative cooling of the atmosphere comes from CREs. Although clouds strongly modulate the tropospheric radiative cooling at the regional scale, their impact on the global mean radiative cooling of the atmosphere appears to be small in the current climate. It means that the net CRE at the surface is almost the same as the net CRE at the TOA. Yet, the response of clouds to global warming induces changes in R_{atm}. Most importantly, a decrease of the low-cloud amount leads to a radiative warming of the troposphere, as does rising high clouds. Owing to cloud-radiative feedbacks, the hydrological sensitivity is thus weaker than it would be if clouds were transparent to radiation. This suggests that, if for some reason cloud feedbacks were strongly negative, so as to mitigate against future warming associated with increasing carbon dioxide, the hydrological sensitivity would be greater, so that increasing carbon dioxide would be more efficient in inducing changes in precipitation. For instance, if the degree of convective aggregation were sensitive to temperature in ways that are not captured in the present generation of large-scale models, it is conceivable that a strong cloud feedback would result. However, this feedback would imply an additional perturbation of the long-wave atmospheric radiative cooling and a commensurate change (or opposite sign) in the hydrological sensitivity. Whether such effects are realistic remains an open, and much discussed, question.

13.5.1.3 Dependence on Forcing Agents

In addition to the temperature, water vapour and cloud responses that it induces, the radiative forcing can directly affect the tropospheric radiative cooling rate, an effect very much related to the emergence of forcing adjustments as discussed in Section 13.4.2. Because the atmosphere is more transparent to short-wave radiation than to long-wave radiation, changes in R_{atm} are more sensitive to perturbations in GHGs than they are to changes in solar radiation. Consistently, climate models show that the hydrological sensitivity strongly depends on the nature of the radiative forcing applied to the climate system, with sensitivities significantly weaker in the case of a $2 \times CO_2$ forcing than for perturbations of the solar constant or of sulphate aerosols which result in the same amount of surface temperature change (Fig. 13.12). To understand these differences, the budget of R_{atm} can be analysed using the regression method (Section 13.3.1), analogous to what is done to understand how the radiative imbalance at the TOA changes with changing temperature. So doing shows that the component of the hydrological sensitivity related to the temperature-mediated climate response (the slope of the $\Delta P/P - \Delta T$ relationships in Fig. 13.12), something called the hydrological sensitivity parameter, appears to be fairly insensitive to the nature of the forcing

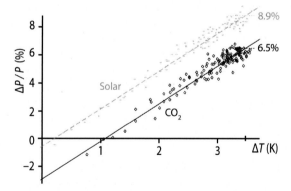

agents, while changes in the hydrological cycle that are independent of the surface temperature change (the y-axis intercept on Fig. 13.12) strongly depends on the forcing agent. This suggests that the differing hydrological sensitivities are related to differences in the nature of the forcing, in a way that is likely related to the different adjustments that arise in response to different forcings.

13.5.2 Regional Patterns

An example of precipitation changes projected by climate models by the end of the twenty-first century for a scenario in which humans continue to use fossil fuels unabated is shown in Fig. 13.13. Overall precipitation increases, but this increase is much less uniform than the change in surface temperatures. Despite an overall globally positive change in precipitation, climate models suggest that many regions will actually experience less precipitation.

It turns out that many aspects of the change in the pattern of precipitation can be understood (Held and Soden, 2006). On average over a season or a year, models predict a regional pattern of precipitation change that features positive anomalies (more precipitation) polewards of 40° of latitude, negative anomalies at subtropical latitudes and positive anomalies in the moist convective zones of the tropics. Qualitatively, such a pattern arises for a large range of models, seasons and climate change scenarios.

Nonetheless, and particularly in physical space, and over land where it matters most, important discrepancies remain to be understood. In any particular region, the magnitude of the changes, and even sometimes their sign, differs among models. The largest uncertainties on the sign arise over tropical continents, especially over South America and over the Sahel. In many regions, especially at the edge of moist and dry regions, the models robustly predict changes which are not statistically significant (Knutti and Sedláček, 2013).

13.5.2.1 Analysis Framework

To interpret the robust features and the uncertainties of regional precipitation patterns, it is useful to extend the analysis framework in the previous section to identify factors that may control precipitation at the regional scale.

On regional scales it proves useful to start with the monthly mean vertically integrated water budget (Chou and Neelin, 2004; Bony et al., 2013a). Following the framework developed in Section 2.5.4, wherein angle brackets denote a mass-weighted vertical average, precipitation can be decomposed as

$$P = E_{sfc} - \left\langle q_v \nabla \cdot v_p \right\rangle - \left\langle v_p \cdot \nabla q_v \right\rangle, \tag{13.34}$$

FIGURE 13.12: Relationship between the change in global mean precipitation and in global mean surface temperature predicted by a climate model after an abrupt increase of the carbon dioxide concentration in the atmosphere or an abrupt increase of the solar constant. The temperature dependence of precipitation changes on temperature (the slope of the relationship, also referred to as the hydrological sensitivity parameter) is similar in both cases, but the rapid adjustment of the precipitation to the carbon dioxide radiative forcing introduces an offset (in the absence of global mean surface temperature change, an increase of atmospheric carbon dioxide leads to a weaker atmospheric radiative cooling and therefore a smaller global mean precipitation), leading to a weaker hydrological sensitivity (computed from initial and final states, not from the slope) in the case of increased carbon dioxide than in the case of increased solar forcing.

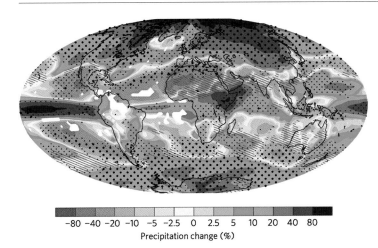

FIGURE 13.13: Multi-model mean relative precipitation change (compared to 1985–2005) projected by climate models (for the December–February season) by the end of the twenty-first century under a scenario in which humans continue to use fossil fuels without mitigation. Stippling indicates regions of robust model results, hatching marks regions of no significant change and white areas regions of inconsistent model responses. From Knutti and Sedláček (2013). Reproduced with permission from Nature Publishing Group

where v_p is the horizontal wind (in pressure coordinates), and q_v is the vertical profile of specific humidity. By introducing the vertical mean pressure vertical velocity $\langle\omega\rangle$, defined in Chapter 2, and by assuming that the vertical velocity profile has everywhere a specified vertical structure $\Omega(p)$ (close to a first baroclinic mode), Eq. (13.34) can be rewritten

$$P = E_{sfc} + \langle\omega\rangle\mathcal{G}(q_v) + H_q, \tag{13.35}$$

where $\mathcal{G}(q_v) = -\langle\Omega\partial_p q_v\rangle$ is the gross moisture lapse rate (< 0) defined in Chapter 2, and $H_q = -\langle v_p \cdot \nabla q_v\rangle$ is the horizontal moisture advection term.

The second term on the right-hand side of Eq. (13.35) represents the contribution of large-scale atmospheric vertical motions to the regional water budget, namely convergence or divergence. Its contribution to the budget is positive in regimes of large-scale rising motion ($\langle\omega\rangle < 0$) and negative in regimes of large-scale subsidence ($\langle\omega\rangle > 0$). The third term on the right-hand side is important mostly in convective regions and is usually negative, consistent with the fact that the precipitating convective regions are usually moister than their surroundings. At first order, $\langle\omega\rangle$ determines the sign and magnitude of $\langle\omega\rangle\mathcal{G}(q_v)$ and thus largely controls the regional variations of precipitation (Fig. 13.14), all the more that surface evaporation and horizontal advection exhibit much smaller variations in comparison. This moisture budget framework can also be applied to climate change variations, in which case it becomes:

$$\delta P = \delta E_{sfc} + \langle\omega\rangle\delta\mathcal{G}(q_v) + \delta H_q + \mathcal{G}(q_v)\delta\langle\omega\rangle. \tag{13.36}$$

In analysing this expression, it proves useful to separate the term involving changes in $\langle\omega\rangle$ from the other terms, such that

$$\delta P = \delta P_{ther} + \delta P_{dyn}, \tag{13.37}$$

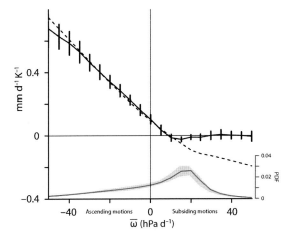

FIGURE 13.14: $\delta P/\delta T_s$ composited as a function of $\langle\omega\rangle$ over tropical oceans. The solid line shows multi-model mean results (vertical bars show the inter-model standard deviation), and the dashed line shows the relationship obtained by assuming that water vapour varies according to the Clausius–Clapeyron relationship (7.5%/K) and that surface evaporation increases everywhere by 2%/K. The relative statistical weight of the different dynamical regimes in the tropics is shown by the probability distribution function at the bottom (light grey, vertical axis on the right). Adapted from Bony et al. (2013a)

with $\delta P_{dyn} = \mathcal{G}(q_v)\delta\langle\omega\rangle$ and δP_{ther} denoting the other terms on the right-hand side of Eq. (13.36). Here the nomenclature is such that the precipitation changes can be thought of as arising from a circulation dependent term, which is called the dynamic term, and a remainder, which is referred to as the thermodynamic term as it mostly summarises thermodynamic contributions to precipitation changes. Similar decompositions have been applied to an understanding of cloud feedbacks, as discussed in Chapter 7. Although the annual mean precipitation

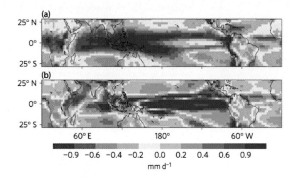

FIGURE 13.15: Multi-model annual precipitation change δP at the end of the twenty-first century in RCP8.5 scenario, decomposed into (a) thermodynamic (δP_{ther}) and (b) dynamic (δP_{dyn}) components. Adapted from Bony et al. (2013a)

changes are dominated by the thermodynamical component at latitudes polewards of 40°, both components contribute significantly in the tropics (Fig. 13.15).

13.5.2.2 Thermodynamical Component

The increase of water vapour with temperature associated with a roughly unchanged relative humidity leads to an increase of $\mathcal{G}(q_{\text{v}})$ by about 7.5% K^{-1}, in line with the Clausius–Clapeyron relationship (Fig. 13.8). With more water vapour in the atmosphere, more water can be transported by a given circulation. This implies that more water converges into wet (or convective) areas, and more water diverges from dry (or subsiding) areas. The associated increase of precipitation in the wet convective regimes and decrease in dry subsidence regimes produces what is referred to as a 'wet-get-wetter, dry-get-drier' response, particularly noticeable on large spatial scales (e.g., zonal average). However, this is not the only effect. For instance, the tendency of surface evaporation to increase with warming further amplifies the 'wet-get-wetter' behaviour, but partly opposes the 'dry-get-drier' response in subsidence regimes. As a result, the precipitation change in dry regions can be positive, negative or close to zero depending on $\langle\omega\rangle$ and on the magnitude of surface evaporation changes (Fig. 13.14).

As expected from the role of the Clausius–Clapeyron relationship in the sensitivity of δP_{ther} to temperature, the regional pattern of δP_{ther} largely scales with the magnitude of global warming. The resultant 'wet-get-wetter, dry-get-drier' pattern partly explains the increase of precipitation found in convergence zones, the drying found in subtropical subsidence areas, and the weak precipitation change found at the edges of convective and subsidence areas where $\langle\omega\rangle$ is close to zero (Fig. 13.13). The fact that

the amplitude of this pattern scales with the magnitude of the globally averaged surface temperature change is one reason why the ECS is so important, as its value helps explain the amplitude of regional changes.

The dependence of $\langle\omega\rangle\delta\mathcal{G}(q_{\text{v}})$ on $\langle\omega\rangle$ indicates that the regional pattern of δP_{ther} is partly controlled by the present-day climatology of the atmospheric circulation and hence precipitation (Fig. 13.15). It implies that biases in the representation of the current climate projects strongly onto the pattern of precipitation projections, and as a matter of fact, the bulk of the inter-model spread in regional precipitation change is not due to differences in ECS, but due to differences in the present-day climatology. Alternatively, it indicates that at least one component of regional precipitation projections may be constrained by observations of the current climate.

13.5.2.3 Dynamical Component

Simulations of global warming often exhibit a poleward shift of the storm tracks (especially in the southern hemisphere) accompanied by a widening of the subtropical belt, that contribute to the reduced precipitation at subtropical latitudes and the increased precipitation at high latitudes. In addition to this, climate change is accompanied by further changes in the strength of the circulations which are particularly noticeable in the tropics and can help explain a large fraction of the precipitation changes.

Given the large contribution of large-scale mass convergence or divergence in the regional water budget (Eq. (13.35)), and its linear dependence on the large-scale vertical velocity $\langle\omega\rangle$, regional precipitation changes are very sensitive to changes in the overturning atmospheric circulation. As explained in Eq. (2.134), the strength of large-scale vertical motions is controlled by the net heat input into the atmospheric column (to a first approximation, the sum of R_{atm} and of surface enthalpy fluxes), and the gross moist static stability (which depends on the vertical profile of moist static energy). In the clear-sky regions of the tropical free troposphere, the balance of dry static energy implies, cf. Eq. (13.30), that:

$$\omega^{\downarrow} = -\dot{Q}_{\text{rad}}S^{-1}, \qquad (13.38)$$

where $\omega^{\downarrow}$ is the pressure vertical velocity of downward motions (in hPa/day), $\dot{Q}_{\text{rad}}$ is the radiative cooling rate and S the stability, which is everywhere positive. Hence, cooling implies subsidence ($\omega^{\downarrow} > 0$).

In response to global warming, the troposphere becomes more stable owing to the temperature dependence of the saturated isentropic lapse rate and therefore $\partial\theta/\partial p$ increases in magnitude. The warmer troposphere also cools more through radiation, but the increased cooling is not sufficient to overcome the effect of the increased stability, so that $\omega^{\downarrow}$ weakens in warmer climates. Provided that the relative areas of rising and sinking motions remain approximately unchanged, the conservation of mass then implies that the large-scale rising motions also weaken. A consistent change is predicted when approximating the global-mean precipitation by $P = Mq$, where M is the global convective mass flux (e.g., Eq. (2.68)) and q the boundary-layer water vapour. With warming, q increases with temperature at a rate close to that implied by Clausius–Clapeyron relationship (which at temperatures characteristic of the tropical boundary layer is about $7.5\% \, K^{-1}$), while the global precipitation increases more slowly (about 2–$3\% \, K^{-1}$) owing to its energetic control (Section 13.5.1.2). It implies that

$$\delta M / M = \delta P / P - \delta q / q < 0 \qquad (13.39)$$

and therefore that the global or tropical mean deep convective mass flux (which composes most of the large-scale rising motions) also weakens. However, the actual rate of change of $\omega^{\uparrow}$ with temperature, which depends on several competing factors, is only loosely constrained by this simple scaling argument, and it is actually quite variable among models.

As discussed in Section 13.4.2, the large-scale circulation can also be directly affected by the radiative forcing itself, even in the absence of global surface temperature change. This can induce a dynamic response when the forcing agents exert a significant impact on the tropospheric radiative cooling and thus affect the atmospheric stratification. In response to increased carbon dioxide, for instance, the radiative cooling of the atmosphere is quickly reduced and the troposphere is warmed within a few days (Fig. 13.4). These changes induce a fast weakening of both downward and upward motions, and of the overall overturning circulation, through a process similar to the dynamical response to precipitation changes discussed earlier.

Even in the absence of significant global mean change in surface temperature, radiative forcings may also induce rapid adjustments of surface temperature patterns. For example, owing to the low thermal inertia of the land surface, radiative forcings generally induce fast changes in land surface temperature and in land–sea contrasts. In response to increased carbon dioxide, it enhances the mass convergence and strengthens the large-scale rising motions over land. Over ocean, the emergence of fast SST pattern changes has also been noticed. This ocean adjustment, which presumably relates to the surface ocean response to the atmospheric circulation adjustments, likely amplifies the atmosphere-alone adjustments of the large-scale circulation in some regions.

Over ocean, rapid adjustments to increased carbon dioxide and global warming affect the large-scale circulation in a qualitatively similar manner and therefore both effects reinforce each other. Over land, on the other hand, they exert opposing influences on the strength of large-scale rising motions. In response to a given change in carbon dioxide, the change in land precipitation may thus be positive or negative depending on the amount of warming that is associated with the carbon dioxide change (i.e., climate sensitivity). This is one of the complications involved in understanding precipitation changes over land, many more are discussed in Chapter 12.

The direct (through adjustments) and indirect (through global surface temperature changes) effects of radiative forcings on the large-scale circulation significantly affect the amount and regional distribution of precipitation on all timescales. Climate models suggest that the precipitation adjustment to increased carbon dioxide primarily stems from dynamical adjustments, while the temperature-mediated response of precipitation results from both the thermodynamical and dynamical components.

13.5.2.4 Sources Uncertainties in Regional Precipitation Projections

In response to an external perturbation, thermodynamical and dynamical processes thus compound to produce large-scale patterns of precipitation change. The δP_{ther} is fairly well understood from basic physical principles: its magnitude scales with the size of global warming through the Clausius–Clapeyron relationship, and its pattern is largely tied to the present-day large-scale circulation pattern. For instance, models that simulate a double Inter Tropical Convergence Zone (ITCZ) pattern in the eastern Pacific in the present day will also exhibit a double-ITCZ pattern in δP_{ther}. This component can be more uncertain over land, however, where the limited availability of moisture introduces an additional degree of freedom in the magnitude of the precipitation response. Another source of uncertainty stems from the uncertainty in climate sensitivity, which directly controls the magnitude of global warming and then of δP_{ther} at the regional scale.

The prediction of the δP_{dyn} pattern is more complex and remains less robust across models, mostly because models predict different responses of the general circulation to climate change. Without any surprise, it is this component that leads to most of the inter-model spread of precipitation projections at the regional scale. Although many processes could contribute to this uncertainty and hence the multi-model spread, including the representation of complex aerosol, chemical or biogeochemical processes and interactions between the atmosphere and the upper ocean, there is strong evidence that it is primarily in the representation of the basic physical atmospheric processes, and perhaps their interaction with differing degrees of moisture availability on the land surface, that the essence of the uncertainty lies. Even in simple and idealised frameworks such as aqua-planets forced by a spatially uniform SST change, few robust responses of the atmospheric circulation to global warming have been pointed out so far. Advancing our understanding of the coupling of cloud and moist processes to the atmospheric circulation, and their response to temperature change, is thus critically needed if one wants to improve our ability to anticipate precipitation changes arising from warming.

13.6 Outlook

Understanding the mechanisms through which climate responds to external perturbations constitutes an extraordinary scientific challenge. In emphasising some of the basic principles and approaches followed in addressing this challenge, this chapter has not been able to touch on every aspect of climate change to which these principles apply. For instance, many of the approaches devoted to an analysis of cloud feedbacks, or precipitation changes, might be fruitfully applied to understanding adjustments associated with aerosol–cloud interactions, perturbations to the biosphere or changes in extremes. And although many important aspects of climate change, such as climate sensitivity or regional precipitation changes, remain difficult to quantify, continuous advances in the development of relevant conceptual frameworks now make it possible to decompose these complex problems into pieces. And these pieces can be investigated using a wider range of tools, from fine-scale simulation to observations. By improving and accumulating knowledge about individual pieces, and their connections to each other, we progressively advance our physical understanding and more and more pieces of the jigsaw puzzle of cloud influences on climate fall into place.

Several important pieces of the puzzle have yet to be connected, however. Among those pieces are those that relate to the role of couplings between clouds and circulation in a changing climate. In addition, the interaction of clouds with humidity, with the diabatic heating of the atmosphere and with the land surface are likely to exert a significant control on atmospheric circulations, but

in ways that we are just beginning to appreciate (Chapters 8 and 9 of this volume). Likewise, the role of convective processes in controlling boundary-layer cloud feedbacks or the short-wave cloud feedback of upper-level clouds remains a critical open issue. Finally, the extent to which clouds and convection might organise and aggregate differently in a warmer climate, and the impact that such changes may have on climate sensitivity or precipitation, remains largely unknown.

These questions are just a few examples of issues which are right at the interface between process studies and climate studies. The more our minds adopt an analytical approach to tackle complex problems, the more we need to ask ourselves how the different pieces connect to each other and to the whole picture, and how overarching constraints might control in turn the behaviour of the individual pieces. In the end, an ability to bridge the process-scale and the large-scale understanding of cloud–climate interactions will be necessary to identify robust features in model projections, to design strategies to evaluate them using observations (which are generally available on scales much smaller and shorter than those typical of the global climate system dynamics) and to better assess future climate changes. Building on the conceptual advances of these last decades, on our improved ability to interpret the results of climate models in terms of physical processes and on the novel possibility of running high-resolution process models over large domains, cloud research has never been more exciting. It gives one the belief that significant progress on long-standing problems might well be in reach.

Further Reading

Section 13.1: Introduction:
As an introduction and a contribution to the historical synopsis of the climate sensitivity issue (Section 13.4.1), it is an inspiring experience to read the Charney report (Charney, 1979) to appreciate the power of physical understanding and of using a hierarchy of models to assess climate change. The topicality of this approach is discussed in a related position paper (Bony et al., 2013a).

Reports from the Intergovernmental Panel on Climate Change (IPCC) always provide a nice up-to-date assessment of climate change (e.g., Boucher et al., 2013).

A discussion of some of the challenges associated with observing and modelling the sensitivity of the Earth's radiation budget and precipitation distribution in a changing climate can be found in Stevens and Bony (2013) and in Bony et al. (2015).

Section 13.2: Conceptual Frameworks:

An excellent introduction to the concepts of feedback and forcing in the atmospheric context is provided by Dufresne and Saint-Lu (2016). The article by Schwartz (2011) links these ideas to their use in circuit analysis, a domain of analysis in which they were first formalised. An introduction to the need to revise the traditional forcing-feedback framework is provided by Sherwood et al. (2015).

Section 13.3: Application of the Framework to GCMs:

The application of the regression method to diagnose radiative forcing and climate sensitivity is discussed in Gregory et al. (2004). An early assessment of climate feedbacks in CMIP3 GCMs was provided in Soden and Held (2006) and in Bony et al. (2006). Climate sensitivity and feedbacks in CMIP5 GCMs were assessed in Andrews et al. (2012) and in Vial et al. (2013).

An excellent discussion of the two-layer model introduced in the exercises is provided by Gregory et al. (2015). This paper also reviews the step model, and discusses differences between models that are linear in forcing and have a radiative response that is linearly related to the surface temperature perturbation. Analysis of CMIP5 models in terms of the two-layer model is given by Geoffroy et al. (2013).

Section 13.4: Clouds in a Changing Climate:

Comprehensive discussions about cloud adjustment and feedback processes may be found in Zelinka and Hartmann (2010), Rieck et al. (2012), and Kamae and Watanabe (2013). An assessment of the different components of cloud feedbacks in climates models is provided in Zelinka et al. (2013). The stability iris effect on anvil clouds is discussed in Bony et al. (2016).

Section 13.5: Precipitation in a Changing Climate:

An illuminating discussion of precipitation changes in a warming climate can be found in Mitchell et al. (1987) and in Held and Soden (2006). A discussion of energetic constraints on precipitation is given in O'Gorman et al. (2012). Further discussions about precipitation changes can be found in Allan et al. (2014) and in Chou and Neelin (2004).

Exercises

(1) Based on the global energy budget of the Earth system, estimate how much the cloud feedback parameter would have to be to make the system unstable to external perturbations.

(2) Starting from Eq. (13.15), show that the global mean temperature change can be decomposed as:

$$\delta T_{\rm s} = \delta T_{\rm s,P} + \sum_x \delta T_{\rm s,x},$$

with $\delta T_{{\rm s},x} = g_x \delta T_{\rm s} = -\frac{\lambda_x}{\lambda_{\rm P}} \delta T_{\rm s}$.

(3) Consider a simple model of the climate system wherein the enthalpy budget of the surface ocean and the deep ocean are explicitly accounted for, such that

$$C \frac{{\rm d}T_{\rm s}}{{\rm d}t} = F + \lambda T_{\rm s} - \gamma \left(T_{\rm s} - T_{\rm d} \right)$$

$$C_{\rm d} \frac{{\rm d}T_{\rm d}}{{\rm d}t} = \gamma \left(T_{\rm s} - T_{\rm d} \right).$$

Hence, the radiative imbalance is given as $\delta R = C\frac{{\rm d}T_{\rm s}}{{\rm d}t} + C_{\rm d}\frac{{\rm d}T_{\rm d}}{{\rm d}t}$ in agreement with Eq. (13.10). Derive the temperature response of this system to an abrupt forcing in the limit when $C_{\rm d} \to \infty$, and for the case when $C \to 0$.

(4) Show that for the case of a step (Heaviside) forcing F that the general solution to the two layer model of the climate system presented earlier has the form

$$T(t) = -\frac{F}{\lambda} \left[a_{\rm f}(1 - e^{-t/\tau_{\rm f}}) + a_{\rm s}(1 - e^{-t/\tau_{\rm s}}) \right]$$

$$T_{\rm d}(t) = -\frac{F}{\lambda} \left[\phi_{\rm f} a_{\rm f}(1 - e^{-t/\tau_{\rm f}}) + \phi_{\rm s} a_{\rm s}(1 - e^{-t/\tau_{\rm s}}) \right],$$

where

$$a_{\rm f} = -\left(\frac{\lambda}{C} - \frac{1}{\tau_{\rm s}} \right)\left(\frac{1}{\tau_{\rm f}} + \frac{1}{\tau_{\rm s}} \right)^{-1} \quad a_{\rm s} = 1 - a_{\rm f}$$

$$\tau_{\rm f} = -\frac{CC_{\rm d}}{2\lambda\gamma}\left(b + \sqrt{\delta} \right), \qquad \tau_{\rm s} = -\frac{CC_{\rm d}}{\lambda\gamma\tau_{\rm f}}$$

$$\phi_{\rm f} = \frac{C}{2\gamma}\left(b^* + \sqrt{\delta} \right), \qquad \phi_{\rm s} = -\frac{C}{C_{\rm d}\phi_{\rm f}}$$

$$b = \frac{(\gamma - \lambda)}{C} + \frac{\gamma}{C_{\rm d}}, \qquad b^* = \frac{(\gamma - \lambda)}{C} - \frac{\gamma}{C_{\rm d}}.$$

(5) For the solution of the two-layer model, plot δR versus $\delta T_{\rm s}$ given the parameters $C_{\rm d} = 10C = 100$ W yr m^{-2} K^{-1}, $\gamma = -(2/3)\lambda = -1.1$ W m^{-2} K^{-1}. What if $\gamma = -1/2\lambda$. Physically interpret your results.

(6) If the response of the climate system is linear in the forcing, then the response $X(t)$ to an arbitrary forcing can be expressed in terms of the response $X_{\rm step}$ to a step forcing $F_{\rm step}$ such that

$$X(t) = \sum_{t'=1}^{t} X_{\rm step}(t - t' + 1) \frac{F(t') - F(t' - 1)}{F_{\rm step}}.$$

$$(13.40)$$

Using the parameters from the previous problem, and the step response of temperature, plot the temperature versus time, assuming that the forcing increases linearly in time for seventy years and then is held constant. Compare this answer to the response of the same system, but with an infinite heat capacity of the deep layer.

(7) Assuming multi-model mean values of the water vapour, lapse rate and surface albedo feedback parameters, estimate how much the cloud feedback parameter should be for the ECS to exceed 2 K.

(8) Estimate the ECS using observations: using $\text{ECS} = \frac{F_{2CO_2}}{F - \delta R} \delta T_s$, quantify how much error on each term of the expression might explain the difference between observation- and model-derived estimates of ECS.

(9) Let the temperature be represented by a discrete variable indexed by three spatial coordinates, i, j, k, denoting the position as a function of latitude, longitude and altitude, respectively. If $k_*(i, j)$ denotes the height of the tropopause, and $k = 1$ denotes the top model level, define an algorithm for calculating the lapse rate feedback using the kernel method. How does the lapse rate feedback defined in this manner depend on the sensitivity of the tropopause height to temperature?

(10) Based on the moist static energy budget of the planetary boundary layer, and by assuming a constant relative humidity and constant cloud-optical properties, explain how the low-cloud fraction is expected to change in response to (i) increased surface turbulent fluxes, (ii) increased atmospheric temperatures and (iii) increased vertical mixing between the boundary layer and the free troposphere.

References

Abercromby, R. 1888. Cloud-land in folk-lore and in science. *The Folk-Lore Journal*, **6**(21), https://books.google.com.au/books?id=FwgNAQA.

Ackerman, A. S., Kirkpatrick, M. P., Stevens, D. E., and Toon, O. B. 2004. The impact of humidity above stratiform clouds on indirect aerosol climate forcing. *Nature*, **432**, 1014–17.

Ackerman, Thomas P., and Stokes, Gerald M. 2003. The atmospheric radiation measurement program. *Physics Today*, **56**(1), 38.

Adams, J. 2010. *Vegetation-Climate Interaction, How Plants Make the Global Environment*. Springer.

Allan, Richard P., Liu, Chunlei, Zahn, Matthias, et al. 2014. Physically consistent responses of the global atmospheric hydrological cycle in models and observations. *Surveys in Geophysics*, **35**(3), 533–52.

Andrews, Timothy, Gregory, Jonathan M., Webb, Mark J., and Taylor, Karl E. 2012. Forcing, feedbacks and climate sensitivity in CMIP5 coupled atmosphere-ocean climate models. *Geophysical Research Letters*, **39**(9), L09712.

Arakawa, Akio. 1969. Parameterization of cumulus convection. Pages 1–6 of: *Proc. WMO/IUGG Symp. Numerical Weather Prediction*.

Arakawa, Akio. 2004. The cumulus parameterization problem: past, present, and future. *Journal of Climate*, **17**(13), 2493–525.

Arakawa, Akio., and Schubert, H. 1974. Interaction of a cumulus cloud ensemble with the large-scale environment. Part I. Theoretical formulation and sensitivity tests. *Journal of the Atmospheric Sciences*, **31**, 674–701.

Arakawa, Akio, and Wu, Chien-Ming. 2013. A unified representation of deep moist convection in numerical modeling of the atmosphere. Part I. *Journal of the Atmospheric Sciences*, **70**(7), 1977–92.

Arora, V. 2002. Modeling vegetation as a dynamic component in soil–vegetation–atmosphere transfer schemes and hydrological models. *Reviews of Geophysics*, **40**, 1–26.

Arrhenius, S. 1896. On the influence of carbonic acid in the air upon the temperature of the ground. *Philosophical Magazine and Journal of Science*, **41**, 237–76.

Asai, T., and Kasahara, A. 1967. A theoretical study of the compensating downward motions associated with cumulus clouds. *Journal of the Atmospheric Sciences*, **24**, 467–96.

Augstein, E., Riehl, Herbert, Ostapoff, F., and Wagner, V. 1973. Mass and energy transports in an undisturbed Atlantic trade-wind flow. *Monthly Weather Review*, **101**(2), 101–11.

Bailey, M., and Hallett, J. 2009. A comprehensive habit diagram for atmospheric ice crystals: confirmation from the laboratory, AIRS II, and other field studies. *Journal of the Atmospheric Sciences*, **66**(9), 2888–9, https://dx.doi.org/10.1175/2009jas2883.1.

Bannon, P. R. 2002. Theoretical foundations for models of moist convection. *Journal of the Atmospheric Sciences*, **59**(12), 1967–82.

Barenblatt, G. I. 1996. *Scaling, Self Similarity, and Intermediate Asymptotics*. Cambridge University Press.

Bauer, M. P., Tselioudis, G., and Rossow, W. B. 2016. A new climatology for investigating storm influences in and on the extratropics. *Journal of Applied Meteorology and Climatology*, **55**, 1287–303.

Beard, K. V. 1977. Terminal velocity adjustment for cloud and precipitation drops aloft. *Journal of the Atmospheric Sciences*, **34**, 1293–8.

Bellon, G., and Sobel, A. H. 2010. Multiple equilibria of the Hadley circulation in an intermediate-complexity axisymmetric model. *Journal of Climate*, **23**, 1760–78.

Bellouin, N., Jones, A., Haywood, J., and Christopher, S. A. 2008. Updated estimate of aerosol direct radiative forcing from satellite observations and comparison against the Hadley Centre climate model. *Journal of Geophysical Research*, **113**, D10205.

Bellouin, N., Quaas, J., Morcrette, J.-J., and Boucher, O. 2013. Estimates of aerosol radiative forcing from the MACC re-analysis. *Atmospheric Chemistry and Physics*, **13**, 2045–62.

Bender, F. A-M., Ramanathan, V., and Tselioudis, G. 2012. Changes in extratropical storm track cloudiness 1983–2008: observational support for a poleward shift. *Climate Dynamics*, **38**, 2037–53.

Betts, A. K., and Ridgway, W. 1988. Coupling of the radiative, convective, and surface fluxes over the Equatorial Pacific. *Journal of the Atmospheric Sciences*, **45**(3), 522–36.

Betts, A. K., and Ridgway, W. 1989. Climatic equilibrium of the atmospheric convective boundary layer over a tropical ocean. *Journal of the Atmospheric Sciences*, **46**(17), 2621–41.

Bjerknes, Jakob. 1969. Atmospheric teleconnections from the equatorial Pacific. *Monthly Weather Review*, **97**(3), 163–72.

Bjerknes, Jakob., and Solberg, H. 1922. Life cycle of cyclones and the polar front theory of atmospheric circulation. *Geophysisks Publikationer*, **3**, 3–18.

Bodas-Salcedo, A., Webb, M. J., Bony, S., et al. 2011. COSP: satellite simulation software for model assessment. *Bulletin of the American Meteorological Society*, **92**(8), 1023–43.

Bodas-Salcedo, A., Williams, K. D., Field, P. R., and Lock, A. P. 2012. The surface downwelling solar radiation surplus over the Southern Ocean in the Met Office model: the role of midlatitude cyclone clouds. *Journal of Climate*, **25**(21), 7467–86.

Bodas-Salcedo, A., Williams, K. D., Ringer, M. A., et al. 2014. Origins of the solar radiation biases over the Southern Ocean in CFMIP2 models. *Journal of Climate*, **27**, 41–56.

Boers, R., de Haij, M. J., Wauben, W. M. F., et al. 2010. Optimized fractional cloudiness determination from five ground-based remote sensing techniques. *Journal of Geophysical Research*, **115**(D24), D24116.

Bohren, Craig F. 1987. Multiple scattering of light and some of its observable consequences. *American Journal of Physics*, **55**(6), 524–33.

Bohren, Craig F., and Clothiaux, Eugene E. 2006. *Fundamentals of Atmospheric Radiation*. John Wiley and Sons.

Böing, Steven J., Siebesma, A. P, Korpershoek, J. D., and Jonker, Harm J. J. 2012. Detrainment in deep convection. *Geophysical Research Letters*, **39**, 20–30.

Böing, Steven J., Jonker, Harm J. J., Nawara, Witek A., and Siebesma, A. Pier. 2014. On the deceiving aspects of mixing diagrams of deep cumulus convection. *Journal of the Atmospheric Sciences*, **71**(1), 56–68.

Bollasina, M., Ming, Y., and Ramaswamy, V. 2011. Anthropogenic aerosols and the weakening of the South Asian summer monsoon. *Science*, **334**, 502–5.

Bolton, David. 1980. The computation of equivalent potential temperature. *Monthly Weather Review*, **108**(7), 1046–53.

Bony, Sandrine, Bellon, Gilles, Klocke, Daniel, et al. 2013a. Robust direct effect of carbon dioxide on tropical circulation and regional precipitation. *Nature Geoscience*, **6**(6), 447–51.

Bony, Sandrine, Colman, Robert, Kattsov, Vladimir M., et al. 2006. How well do we understand and evaluate climate change feedback processes? *Journal of Climate*, **19**(15), 3445–82.

Bony, Sandrine, Dufresne, J.-L., Le Treut, H., Morcrette, J.-J., and Senior, C. 2004. On dynamic and thermodynamic components of cloud changes. *Climate Dynamics*, **22**(2–3), 71–86.

Bony, Sandrine, Stevens, B., Held, I., et al. 2013b. Carbon Dioxide and Climate: Perspectives on a Scientific Assessment. In *Climate Science for Serving Society: Research, Modeling and Prediction Priorities*. Springer. Pages 391–413.

Bony, Sandrine, Stevens, Bjorn, Coppin, David, et al. 2016. Thermodynamic control of anvil cloud amount. *Proceedings of the National Academy of Sciences of the United States of America*, **113**(32), 8927–32.

Bony, Sandrine, Stevens, Bjorn, Frierson, Dargan M. W., et al. 2015. Clouds, circulation and climate sensitivity. *Nature Geoscience*, **8**(4), 261–8.

Booth, B. B. B., Dunstone, N. J., Halloran, P. R., Andrews, T., and Bellouin, N. 2012. Aerosols implicated as a prime driver of twentieth-century North Atlantic climate variability. *Nature*, **484**, 228–32.

Booth, J. F., Wang, S., and Polvani, L. M. 2013. Midlatitude storms in a moister world: lessons from idealized baroclinic life cycle experiments. *Climate Dynamics*, **41**, 787–802.

Boucher, O. 2015. *Atmospheric Aerosols: Properties and Climate Impacts*. Springer.

Boucher, O., Randall, D., Artaxo, P., et al. 2013. Clouds and Aerosols. In *Climate Change 2013: The Physical Science Basis*. Cambridge University Press. Pages 571–657.

Boutle, I. A., Eyre, J. E. J., and Lock, A. P. 2014. Seamless stratocumulus simulation across the turbulent gray zone. *Monthly Weather Review*, **142**(4), 1655–68.

Bras, R. L. 1990. *Hydrology: An Introduction to Hydrologic Science*. Addison-Wesley.

Bretherton, Christopher S. 2015. Insights into low-latitude cloud feedbacks from high-resolution models. *Philosophical Transactions of the Royal Society A*, **373**(2054).

Bretherton, Christopher, S., Blossey, P. N., and Khairoutdinov, M. 2005. An energy-balance analysis of deep convective self-aggregation above uniform SST. *Journal of the Atmospheric Sciences*, **62**, 4273–92.

Bretherton, Christopher, S., Blossey, P. N., and Uchida, J. 2007. Cloud droplet sedimentation, entrainment efficiency, and subtropical stratocumulus albedo. *Geophysical Research Letters*, **34**.

Bretherton, Christopher S., and Park, Sungsu. 2009. A new moist turbulence parameterization in the community atmosphere model. *Journal of Climate*, **22**(12), 3422–48.

Bretherton, Christopher, S., and Wyant, M. C. 1997. Moisture transport, lower-tropospheric stability, and decoupling of cloud-topped boundary layers. *Journal of the Atmospheric Sciences*, **54**, 148–67.

Brient, Florent, and Bony, Sandrine. 2013. Interpretation of the positive low-cloud feedback predicted by a climate model under global warming. *Climate Dynamics*, **40**, 2415–31.

Brooks, C. E. P. 1927. The mean cloudiness over the earth. *Memoirs of the Royal Meteorological Society*, **1**(10), 127–38.

Butler, A. H., Thompson, D. W. J., and Heikes, R. 2010. The steady-state atmospheric circulation response to climate change-like thermal forcings in a simple general circulation model. *Journal of Climate*, **23**, 3274–496.

Buizza, R. and Miller, M. and Palmer, T.N. 2007. Stochastic representation of model uncertainties in the ECMWF ensemble prediction system. *Quarterly Journal of the Royal Meteorological Society*, **125**(560), 2887–2908.

Byers, H. R., and Braham Jr, R. R. 1949. *The Thunderstorm: Final Report of the Thunderstorm Project*. US Government Printing Office.

Caldwell, P., Zhang, Y., and Klein, S. 2013. CMIP3 subtropical stratocumulus cloud feedback interpreted through a mixed-layer model. *Journal of Climate*, **26**, 1607–25.

Callen, Herbert B. 1985. *Thermodynamics, and an Introduction to Thermostatics*. John Wiley & Sons.

Capderou, Michel. 2014. *Handbook of Satellite Orbits: From Kepler to GPS*. Springer.

Carlson, T. N. 1998. *Mid-Latitude Weather Systems*. American Meteorological Society.

Ceppi, Paulo, Brient, Florent, Zelinka, Mark D., and Hartmann, Dennis L. 2017. Cloud feedback mechanisms and their representation in global climate models. *Wiley Interdisciplinary Reviews: Climate Change*, **8**(4), e465.

Ceppi, Paulo, and Hartmann, D. L. 2015. Connections between clouds, radiation, and midlatitude dynamics: a review. *Current Climate Change Reports*, **1**, 94–102.

Ceppi, Paulo, and Hartmann, D. L. 2016. Clouds and the atmospheric circulation response to warming. *Journal of Climate*, **29**, 783–99.

Cesana, G., Kay, J. E., Chepfer, H., English, J. M., and de Boer, G. 2012. Ubiquitous low-level liquid-containing Arctic clouds: new observations and climate model constraints from CALIPSO-GOCCP. *Geophysical Research Letters*, **39**(20), L20804.

Cess, R. D. 1976. Climate change: an appraisal of atmospheric feedback mechanisms employing zonal climatology. *Journal of the Atmospheric Sciences*, **33**(10), 1831–43.

Cess, R. D., Potter, G. L., Blanchet, J. P., et al. 1990. Intercomparison and interpretation of climate feedback processes in 19 atmospheric general circulation models. *Journal of Geophysical Research*, **95**(D10), 16601–15.

Chang, E. K. M., Y., Guo, and X., Xia. 2012. CMIP5 multimodel ensemble projection of storm track change under global warming. *Journal of Geophysical Research: Atmospheres*, **117**, D23118, 10.1029/2012JD018578.

Charlson, R., Lovelock, J., Andreae, M., and Warren, S. 1987. Oceanic phytoplankton, atmospheric sulphur, cloud albedo and climate. *Nature*, **326**, 655–61.

Charney, J. G. 1963. A note on large-scale motions in the tropics. *Journal of the Atmospheric Sciences*, **20**(6), 607–9.

Charney, J. G. 1979. *Carbon Dioxide and Climate: A Scientific Assessment*. National Academy Press.

Chen, Yi-Chun, Christensen, Matthew W., Stephens, Graeme L., and Seinfeld, John H. 2014. Satellite-based estimate of global aerosol-cloud radiative forcing by marine warm clouds. *Nature Geoscience*, **7**(9), 643–6.

Cherian, R., Quaas, J., Salzmann, M., and Wild, M. 2014. Pollution trends over Europe constrain global aerosol forcing as simulated by climate models. *Geophysical Research Letters*, **41**, 2176–81.

Chou, Chia, and Neelin, J. David. 2004. Mechanisms of global warming impacts on regional tropical precipitation. *Journal of Climate*, **17**(13), 2688–701.

Christensen, M. W., Suzuki, K., Zambri, B., and Stephens, G. L. 2014. Ship track observations of a reduced shortwave aerosol indirect effect in mixed-phase clouds. *Geophysical Research Letters*, **41**, 6970–77.

Clarke, Allan J. 2008. *An Introduction to the Dynamics of El Niño and the Southern Oscillation*. Academic Press.

Claussen, M., Dallmeyer, A., and Bader, J. 2017. Theory and modeling of the African Humid Period and the Green Sahara. In *Oxford Research Encyclopedias, Climate Science, Regional and Local Climates, Climate Change*. Oxford University Press.

Connolly, P. J., Möhler, O., Field, P. R., et al. 2009. Studies of heterogeneous freezing by three different desert dust samples. *Atmospheric Chemistry and Physics*, **9**, 2805–24.

Conover, J. H. 1966. Anomalous cloud lines. *Journal of the Atmospheric Sciences*, **23**, 778–85.

Cotton, W. R., Bryan, G., and van den Heever, S. 2010. *Storm and Cloud Dynamics, 2nd Edition*. Vol. 99. Academic Press.

Cox, P. M., Huntingford, C., and Williamson, M. S. 2018. Emergent constraint on equilibrium climate sensitivity from global temperature variability. *Nature*, **553**, 319.

Cronin, T. W., and Jansen, M. F. 2016. Analytic radiative-advective equilibrium as a model for high-latitude climate. *Geophysical Research Letters*, **43**(1), 449–57.

Cronin, T. W., and Tziperman, E. 2015. Low clouds suppress Arctic air formation and amplify high-latitude continental winter warming. *Proceedings of the National Academy of Sciences of the United States of America*, **112**(37), 11490–5.

Crosman, E. T., and Horel, J. D. 2010. Sea and lake breezes: a review of numerical studies. *Boundary-Layer Meteorology*, **137**, 1–29.

Curry, Judith. 1983. On the formation of continental polar air. *Journal of the Atmospheric Sciences*, **40**(9), 2278–92.

Curry, Judith., and Webster, P. J. 1999. *Thermodynamics of Atmospheres & Oceans*. Academic Press.

Cuxart, J., Bougeault, P., and Redelsperger, J.-L. 2000. A turbulence scheme allowing for mesoscale and large-eddy simulations. *Quarterly Journal of the Royal Meteorological Society*, **126**(562), 1–30.

Dal Gesso, S., and Neggers, R. A. J. 2017. Can we use single-column models for understanding the boundary-layer cloud-climate feedback? *Journal of Advances in Modeling Earth Systems*, **10**(2), 245–61.

Dawe, J. T., and Austin, P. H. 2011. Interpolation of LES cloud surfaces for use in direct calculations of entrainment and detrainment. *Monthly Weather Review*, **139**, 444–56.

de Roode, S. R., and Duynkerke, P. G. 1997. Observed Lagrangian transition of stratocumulus into cumulus during ASTEX: mean state and turbulence structure. *Journal of the Atmospheric Sciences*, **54**, 2157–73.

de Roode, S. R., Duynkerke, P. G., and Jonker, H. J. J. 2004. Large eddy simulation: how large is large enough? *Journal of the Atmospheric Sciences*, **61**, 403–21.

de Roode, S. R, Sandu, I., van der Dussen, J. J., et al. 2016. Large eddy simulations of EUCLIPSE/GASS Lagrangian stratocumulus to cumulus transitions: mean state, turbulence, and decoupling. *Journal of the Atmospheric Sciences*, **73**, 2485–508.

de Roode, S. R., Siebesma, A. P., Dal Gesso, S., et al. 2014. A mixed-layer model study of the stratocumulus response to changes in large-scale conditions. *Journal of Advances in Modeling Earth Systems*, **6**(4), 1256–70.

de Roode, S. R., Siebesma, A. P., Jonker, H. J. J., and de Voogd, Y. 2012. Parameterization of the vertical velocity equation for shallow cumulus clouds. *Monthly Weather Review*, **140**, 2424–36.

de Rooy, Wim C., Bechtold, Peter, Fröhlich, Kristina, et al. 2013. Entrainment and detrainment in cumulus convection: an overview. *Quarterly Journal of the Royal Meteorological Society*, **139**(670), 1–19.

de Rooy, Wim C., and Siebesma, A. Pier. 2008. A simple parameterization for detrainment in shallow cumulus. *Monthly Weather Review*, **136**(2), 560–76.

Delworth, T. L., and Manabe, S. 1988. The influence of potential evaporation on the variabilities of simulated soil wetness and climate. *Journal of Climate*, **1**, 523–47.

Diaz, Henry F., and Bradley, Raymond S. 2004. *The Hadley Circulation: Present, Past, and Future*. Springer.

Dirmeyer, P. A., and Brubaker, K. L. 2007. Characterization of the global hydrologic cycle from a back-trajectory analysis of atmospheric water vapor. *Journal of Hydrometeorology*, **8**, 20–37.

Dorrestijn, Jesse, Crommelin, Daan T., Siebesma, A. Pier, Jonker, Harmen J. J., and Selten, Frank. 2016. Stochastic convection parameterization with Markov chains in an intermediate-complexity GCM. *Journal of the Atmospheric Sciences*, **73**(3), 1367–82.

Dufresne, Jean-Louis, and Bony, Sandrine. 2008. An assessment of the primary sources of spread of global warming estimates from coupled atmosphere–ocean models. *Journal of Climate*, **21**(19), 5135–44.

Dufresne, Jean-Louis, and Saint-Lu, Marion. 2016. Positive feedback in climate: stabilization or runaway, illustrated by a simple experiment. *Bulletin of the American Meteorological Society*, **97**(5), 755–65.

Dunstone, N. J., Smith, D. M., Booth, B. B. B., Hermanson, L., and Eade, R. 2013. Anthropogenic aerosol forcing of Atlantic tropical storms. *Nature Geoscience*, **6**, 534–9.

Durant, A. J., and Shaw, A. 2005. Evaporation freezing by contact nucleation inside-out. *Geophysical Research Letters*, **32**.

Duynkerke, P. G., de Roode, S. R., van Zanten, M. C., et al. 2004. Observations and numerical simulations of the diurnal cycle of the EUROCS stratocumulus case. *Quarterly Journal of the Royal Meteorological Society*, **130**, 3269–96.

Duynkerke, P. G., Zhang, H.-Q., and Jonker, P. J. 1995. Microphysical and turbulent structure of nocturnal stratocumulus as observed during ASTEX. *Journal of the Atmospheric Sciences*, **52**, 2763–77.

Ek, M., and Mahrt, L. 1994. Daytime evolution of relative humidity at the boundary layer top. *Monthly Weather Review*, **122**, 2709–21.

Emanuel, Kerry. 1994. *Atmospheric Convection*. Oxford University Press.

Emanuel, Kerry. 2003. Tropical cyclones. *Annual Review of Earth and Planetary Sciences*, **31**(1), 75–104.

Faloona, I., Lenschow, D. H., Campos, T., et al. 2005. Observations of entrainment in eastern Pacific marine stratocumulus using three conserved scalars. *Journal of the Atmospheric Sciences*, **62**, 3268–85.

Fang, Ming, and Tung, Ka Kit. 1996. A simple model of nonlinear Hadley circulation with an ITCZ: analytic and numerical solutions. *Journal of the Atmospheric Sciences*, **53**(9), 1241–61.

Feng, Y., and Ramanathan, V. 2010. Investigation of aerosol-cloud interactions using a chemical transport model constrained by satellite observations. *Tellus*, **62B**, 69–86.

Fermi, E. 1956. *Thermodynamics*. Dover Publications.

Field, P. R., Heymsfield A. J., Bansemer, A. R., and Twohy, C. H. 2008. Determination of the combined ventilation factor and capacitance for ice crystal aggregates from airborne observations in a tropical anvil cloud. *Journal of the Atmospheric Sciences*, **65**, 376–91.

Field, P. R., and Wood, R. 2007. Precipitation and cloud structure in midlatitude cyclones. *Journal of Climate*, **20**, 233–54.

Findell, K. L., and Eltahir, E. A. B. 2003. Atmospheric controls on soil moisture-boundary layer interactions. Part I: framework development. *Journal of Hydrometeorology*, **4**, 552–69.

Fischer, E. M., Seneviratne, S. I., Vidale, P. L., Lüthi, D., and Schär, C. 2007. Soil moisture–atmosphere interactions during the 2003 European summer heat wave. *Journal of Climate*, **20**, 5081–99.

Flato, G. M., Marotzke, J., Abiodun, B., et al. 2013. Evaluation of Climate Models. In *Climate Change 2013: The Physical Science Basis*. On Climate Change, Intergovernmental Panel (ed.) Cambridge University Press. Pages 741–866.

Forster, Piers M. 2016. Inference of climate sensitivity from analysis of Earth's energy budget. *Annual Review of Earth and Planetary Sciences*, **44**(1), 85–106.

Freud, E., and Rosenfeld, D. 2012. Linear relation between convective cloud drop number concentration and depth for rain initiation. *Journal of Geophysical Research*, **117**, D02207.

Frisch, Uriel. 1996. *Turbulence, the Legacy of A. N. Kolmogorov*. Cambridge University Press.

Garcia-Herrera, R., Diaz, J., Trigo, R. M., Luterbacher, J., and Fischer, E. M. 2010. A review of the European summer heat wave of 2003. *Critical Reviews in Environmental Science and Technology*, **40**, 267–306.

Geoffroy, O., Saint-Martin, D., Olivié, D. J. L., et al. 2013. Transient climate response in a two-layer energy-balance model. Part I: analytical solution and parameter calibration using CMIP5 AOGCM experiments. *Journal of Climate*, **26**(6), 1841–57.

Gerst, Alexander. 2017. *Astro Alex*. www.flickr.com/photos/astro_alex/. Accessed: 15/01/2017.

Gill, Adrian E. 1980. Some simple solutions for heat-induced tropical circulation. *Quarterly Journal of the Royal Meteorological Society*, **106**(449), 447–62.

Gill, Adrian E. 1982. *Atmosphere–Ocean Dynamics*. Elsevier.

Gleckler, P. J., Taylor, K. E., and Doutriaux, C. 2008. Performance metrics for climate models. *Journal of Geophysical Research*, **113**(D6), D06104.

Golaz, Jean-Christophe, Larson, Vincent E., and Cotton, William R. 2002. A PDF-based model for boundary layer clouds. Part I: method and model description. *Journal of the Atmospheric Sciences*, **59**(24), 3540–51.

Gordon, Neil D., and Klein, Stephen A. 2014. Low-cloud optical depth feedback in climate models. *Journal of Geophysical Research*, **119**(10), 6052–65.

Goren, T., and Rosenfeld, D. 2014. Decomposing aerosol cloud radiative effects into cloud cover, liquid water path and Twomey components in marine stratocumulus. *Atmospheric Research*, **138**, 378–93.

Govekar, Pallavi D., Jakob, Christian, and Catto, Jennifer. 2014. The relationship between clouds and dynamics in southern hemisphere extratropical cyclones in the real world and a climate model. *Journal of Geophysical Research: Atmospheres*, **119**(11), 6609–28.

Grabowski, Wojciech W., and Smolarkiewicz, Piotr K. 1999. CRCP: a cloud resolving convection parameterization for modeling the tropical convecting atmosphere. *Physica D: Nonlinear Phenomena*, **133**(1), 171–8.

Grabowski, Wojciech W., and Wang, L. P. 2013. Growth of cloud droplets in a turbulent environment. *Annual Review of Fluid Mechanics*, **45**, 293–324.

Grant, A. L. M., and Brown, A. R. 1999. A similarity hypothesis for shallow-cumulus transports. *Quarterly Journal of the Royal Meteorological Society*, **125**(558), 1913–36.

Gregory, Jonathan M., Andrews, Timothy, and Good, Peter. 2015. The inconstancy of the transient climate response parameter under increasing CO_2. *Philosophical Transactions of the Royal Society A*, **373**(2054), pii: 20140417.

Gregory, Jonathan M., Ingram, W. J., Palmer, M. A., et al. 2004. A new method for diagnosing radiative forcing and climate sensitivity. *Geophysical Research Letters*, **31**(3), L03205.

Grise, K. M., and Medeiros, B. 2016. Understanding the varied influence of midlatitude jet position on clouds and cloud radiative effects in observations and global climate models. *Journal of Climate*, **29**, 9005–25.

Gunn, R., and Kinzer, G. D. 1949. The terminal velocity of fall for water droplets in stagnant air. *Journal of the Atmospheric Sciences*, **6**, 243–48.

Hagemann, S. 2002. An improved land surface parameter dataset for global and regional climate models. Max Planck Institute for Meteorology Report No 336.

Hagos, Samson, Zhang, Chidong, Tao, Wei-Kuo, et al. 2010. Estimates of tropical diabatic heating profiles: commonalities and uncertainties. *Journal of Climate*, **23**(3), 542–58.

Hahn, C. J., and Warren, S. G. 2007. *A Gridded Climatology of Clouds over Land (1971–96) and Ocean (1954–97) from Surface Observations Worldwide. Numeric Data Package NDP-026E*. CDIAC, Department of Energy.

Halley, E. 1686. An historical account of the trade winds and monsoons, observable in the seas between and near the tropics, with an attempt to assign the physical cause of the said winds. *Philosophical Transactions of the Royal Society*, **16**, 153–68.

Hartmann, Dennis. 2016. *Global Physical Climatology, 2nd Edition*. Elsevier Science.

Hartmann, Dennis L., and Larson, Kristin. 2002. An important constraint on tropical cloud – climate feedback. *Geophysical Research Letters*, **29**(20), 12-1-12-4.

Haynes, John M., Jakob, Christian, Rossow, William B., Tselioudis, George, and Brown, Josephine. 2011. Major characteristics of southern ocean cloud regimes and their effects on the energy budget. *Journal of Climate*, **24**(19), 5061–80.

Hazeleger, W., Severijns, C., Semmler, T., et al. 2010. EC-Earth, a seamless earth-system prediction approach in action. *Bulletin of the American Meteorological Society*, **91**, 1357–63.

Held, Isaac M., and Hou, Arthur Y. 1980. Nonlinear axially symmetric circulations in a nearly inviscid atmosphere. *Journal of the Atmospheric Sciences*, **37**(3), 515–33.

Held, Isaac M., and Soden, Brian J. 2006. Robust responses of the hydrological cycle to global warming. *Journal of Climate*, **19**(21), 5686–99.

Hogan, Robin J., and Illingworth, Anthony J. 2000. Deriving cloud overlap statistics from radar. *Quarterly Journal of the Royal Meteorological Society*, **126**(569), 2903–9.

Hogstrom, U. 1996. Review of some basic characteristics of the atmospheric surface layer. *Boundary Layer Meteorology*, **78**, 215–46.

Hohenegger, C., Brockhaus, P., Bretherton, C. S., and Schär, C. 2009. The soil moisture-precipitation feedback in simulations with explicit and parameterized convection. *Journal of Climate*, **22**, 5003–20.

Holton, J. R., and Hakim, G. J. 2013. *An Introduction to Dynamic Meteorology, 5th Edition*. Academic Press.

Holtslag, A. A. M., and Duynkerke, P. G. 1998. Clear and cloudy boundary layers. Proceedings of the Colloquium 'Clear and Cloudy Boundary Layers', Amsterdam, 26–9 August 1997. *Verhandelingen der Koninklijke Nederlandse Akademie van Wetenschappen, Afd. Natuurkunde: Eerste reeks*. Royal Netherlands Academy of Arts and Sciences.

Hoskins, B. J., and Hodges, K. I. 2002. New perspectives on the northern hemisphere winter storm tracks. *Journal of the Atmospheric Sciences*, **59**, 1041–61.

Hourdin, Frederic, Mauritsen, Thorsten, Gettelman, Andrew, et al. 2017. The art and science of climate model tuning. *Bulletin of the American Meteorological Society*, **98**(3), 589–602.

Houze, Jr, R. A., and Betts, A. K. 1981. Convection in GATE. *Reviews of Geophysics and Space Physics*, **19**, 541–76.

Houze, Robert A. 1993. *Cloud Dynamics*. Vol. 53. Academic Press.

Houze, Robert A. 2004. Mesoscale convective systems. *Reviews of Geophysics*, **42**(4).

Howard, L. 1865. *Essay on the Modification of Clouds*. https://books.google.com.au/books?id=HvADAAAAQAAJ&printsec=frontcover&source=gbs_ge_summary_r&cad=0#v=onepage&q&f=false. Accessed 14/01/2020.

Howell, Wallace E. 1949. The growth of cloud drops in uniformly cooled air. *Journal of Meteorology*, **6**(2), 134–49.

Hughes, N. A. 1984. Global cloud climatologies: a historical review. *Journal of Applied Meteorology and Climatology*, **23**, 724–51.

Iribarne, J. V., and Godson, W. L. 1981. *Atmospheric Thermodynamics*. D. Reidel.

Jakob, Christian, and Tselioudis, George. 2003. Objective identification of cloud regimes in the Tropical Western Pacific. *Geophysical Research Letters*, **30**(21), 2082.

Jiang, Jonathan H., Su, Hui, Zhai, Chengxing, et al. 2012. Evaluation of cloud and water vapor simulations in CMIP5 climate models using NASA "ATrain" satellite observations. *Journal of Geophysical Research*, **117**(D14), D14105.

Johnson, Richard H., Rickenbach, Thomas M., Rutledge, Steven A., Ciesielski, Paul E., and Schubert, Wayne H. 1999. Trimodal characteristics of tropical convection. *Journal of Climate*, **12**(8), 2397–418.

Kahn, B. H., Teixeira, J., Fetzer, E. J., et al. 2011. Temperature and water vapor variance scaling in global models: comparisons to satellite and aircraft data. *Journal of the Atmospheric Sciences*, **68**(9), 2156–68.

Kain, J. S., and Frisch, J. M. 1990. A one-dimensional entraining/detraining plume model and its application in convective parameterization. *Journal of the Atmospheric Sciences*, **479**, 2784–802.

Kamae, Youichi, and Watanabe, Masahiro. 2013. Tropospheric adjustment to increasing CO_2: its timescale and the role of land–sea contrast. *Climate Dynamics*, **41**(11), 3007–24.

Kärcher, B., and Lohmann, U. 2003. A parameterization of cirrus cloud formation: heterogeneous freezing. *Journal of Geophysical Research*, **108**, 4402.

Karlsson, K.-G., and Dybbroe, A. 2010. Evaluation of Arctic cloud products from the EUMETSAT Climate Monitoring Satellite Application Facility based on CALIPSO-CALIOP observations. *Atmospheric Chemistry and Physics*, **10**(4), 1789–807.

Kay, Jennifer E., L'Ecuyer, Tristan, Chepfer, Helene, et al. 2016. Recent advances in Arctic cloud and climate research. *Current Climate Change Reports*, **2**(4), 159–69.

Kessler, E. 1995. On the continuity and distribution of water substance in atmospheric circulations. *Atmospheric Research*, **38**(1–4), 109–45.

Khouider, B., Biello, J., and Majda, A. J. 2010. A stochastic multicloud model for tropical convection. *Communications in Mathematical Sciences*, **8**(1), 187–216.

Khvorostyanov, V. I., and Curry, J. A. 2006. Aerosol size spectra and CCN activity spectra: reconciling the lognormal, algebraic, and power lows. *Journal of Geophysical Research*, **111**(D12).

Kiladis, George N., Wheeler, Matthew C., Haertel, Patrick T., Straub, Katherine H., and Roundy, Paul E. 2009. Convectively coupled equatorial waves. *Reviews of Geophysics*, **47**(2).

Kinne, S., O'Donnell, D., Stier, P., et al. 2013. MAC-v1: a new global aerosol climatology for climate studies. *Journal of Advances in Modeling Earth Systems*, **5**, 704–40.

Kleidon, A., and Renner, M. 2013. A simple explanation for the sensitivity of the hydrologic cycle to surface temperature and solar radiation and its implications for global climate change. *Earth System Dynamics*, **4**(2), 455–65.

Klein, Rupert. 2010. Scale-dependent models for atmospheric flows. *Annual Review of Fluid Mechanics*, **42**(1), 249–74.

Klein, Stephen A., and Jakob, C. 1999. Validation and sensitivities of frontal clouds simulated by the ECMWF model. *Monthly Weather Review*, **127**, 2514–31.

Klein, Stephen A., Pincus, Robert, Hannay, Cecile, and Xu, Kuan-Man. 2005. How might a statistical cloud scheme be coupled to a mass-flux convection scheme? *Journal of Geophysical Research: Atmospheres*, **110**(D15), 10–15.

Klocke, Daniel, Pincus, Robert, and Quaas, Johannes. 2011. On constraining estimates of climate sensitivity with present-day observations through model weighting. *Journal of Climate*, 24(23), 6092–9.

Knutti, Reto, and Sedláček, Jan. 2013. Robustness and uncertainties in the new CMIP5 climate model projections. *Nature Climate Change*, **369**(3).

Kolmogorov, A. N. 1941. The local structure of turbulence in incompressible viscous fluid for very large Reynolds' numbers. *Doklady Akademiia Nauk SSSR*, **30**, 301–5.

Koster, R. D., and Suarez, M. J. 2001. Soil moisture memory in climate models. *Journal of Hydrometeorology*, **2**, 559–70.

Kuang, Zhiming. 2008. Modeling the interaction between cumulus convection and linear gravity waves using a limited-domain cloud system-resolving model. *Journal of the Atmospheric Sciences*, **65**(2), 576–91.

Kuhlmann, J., and Quaas, J. 2010. How can aerosols affect the Asian summer monsoon? Assessment during three consecutive pre-monsoon seasons from CALIPSO satellite data. *Atmospheric Chemistry and Physics*, **10**, 4673–88.

Kulmala, M., Suni, T., Lehtinen, K. E. J., et al. 2004. A new feedback mechanism linking forests, aerosols, and climate. *Atmospheric Chemistry and Physics*, **4**, 557–62.

Kundu, Pijush K., and Cohen, Ira M. 2002. *Fluid Mechanics. 2nd Edition.* Academic Press.

Lamarque, J.-F., Bond, T. C., Eyring, V., et al. 2010. Historical (1850–2000) gridded anthropogenic and biomass burning emissions of reactive gases and aerosols: methodology and application. *Atmospheric Chemistry and Physics*, **10**, 7017–39.

Lamb, D., and Verlinde, J. 2011. *Physics and Chemistry of Clouds.* Cambridge University Press.

Lappen, Cara-Lyn, and Randall, David A. 2001. Toward a unified parameterization of the boundary layer and moist convection. Part I: a new type of mass-flux model. *Journal of the Atmospheric Sciences*, **58**(15), 2021–36.

Lau, N.-C., and Crane, M. W. 1997. Comparing satellite and surface observations of cloud patterns in synoptic-scale circulation systems. *Monthly Weather Review*, **125**, 3172–89.

Lau, William K.-M., and Waliser, Duane E. 2011. *Intraseasonal Variability in the Atmosphere–Ocean Climate System.* Springer.

L'Ecuyer, Tristan S., Beaudoing, H. K., Rodell, M., et al. 2015. The observed state of the energy budget in the early twenty-first century. *Journal of Climate*, **28**(21), 8319–46.

Lenderink, G., and Holtslag, A. A. M. 2000. Evaluation of the kinetic energy approach for modeling turbulent fluxes in stratocumulus. *Monthly Weather Review*, **128**(1), 244–58.

Lenoble, Jacqueline. 1993. *Atmospheric Radiative Transfer.* Studies in Geophysical Optics and Remote Sensing. A. Deepak Publishing.

Li, Ying, Thompson, David W. J., and Bony, Sandrine. 2015. The influence of atmospheric cloud radiative effects on the largescale atmospheric circulation. *Journal of Climate*, 28(18), 7263–78.

Liepert, B., Feichter, J., Lohmann, U., and Roeckner, E. 2004. Can aerosols spin down the hydrological cycle in a moister and warmer world? *Geophysical Research Letters*, **31**, L06207.

Lindzen, R. S., and Farrell, B. 1980. A simple approximate result for the maximum growth rate of baroclinic instabilities. *Journal of the Atmospheric Sciences*, **37**, 1648–54.

Loeb, N. G., and Wielicki, B. A. 2015. Earth's Radiation Budget. In *Encyclopedia of Atmospheric Sciences. 2nd Edition.* North, Gerald R., Pyle, John, and Zhang, Fuqing (eds). Academic Press. Pages 67–76.

Lohmann, U. 2002. A glaciation indirect aerosol effect caused by soot aerosols. *Geophysical Research Letters*, **29**, 1052.

Lohmann, U., and Feichter, J. 2005. Global indirect aerosol effects: a review. *Atmospheric Chemistry and Physics*, **5**, 715–37.

Lohmann, U., Lüönd, F., and Mahrt, F. 2016. *An Introduction to Clouds: From the Microscale to Climate.* Cambridge University Press.

Lohmann, U., Quaas, J., Kinne, S., and Feichter, J. 2007. Different approaches for constraining global climate models of the anthropogenic indirect aerosol effect. *Bulletin of the American Meteorological Society*, **88**, 243–9.

Long, Alexis B. 1974. Solutions to the droplet collection equation for polynomial kernels. *Journal of the Atmospheric Sciences*, **31**(4), 1040–52.

Lorenz, D. J., and Hartmann, D. L. 2001. Eddy-zonal flow feedback in the southern hemisphere. *Journal of the Atmospheric Sciences*, **58**, 3312–27.

Madden, Roland A., and Julian, Paul R. 1972. Description of global-scale circulation cells in the tropics with a 40–50 day period. *Journal of the Atmospheric Sciences*, **29**(6), 1109–23.

Mahmood, R., Pielke Sr, R. A., Hubbard, K. G., et al. 2014. Land cover changes and their biogeophysical effects on climate. *International Journal of Climatology*, **34**, 929–53.

Malkus, Joanne, and Scorer, R. S. 1955. The erosion of cumulus towers. *Journal of Meteorology*, **12**(1), 43–57.

Malkus, Joanne S., Scorer, R. S., Ludlam, F. H., and Bjorgum, O. 1953. Correspondence – bubble theory of penetrative convection. *Quarterly Journal of the Royal Meteorological Society*, **79**, 288–93.

Manabe, S., Smagorinsky, J., and Strickler, R. F. 1965. Simulated climatology of a general circulation model with a hydrological cycle. *Monthly Weather Review*, **93**, 769–98.

Manabe, S., and Wetherald, R. T. 1967. Thermal equilibrium of the atmosphere with a given distribution of relative humidity. *Journal of the Atmospheric Sciences*, **24**(3), 241–59.

Mapes, Brian, Tulich, Stefan, Lin, Jialin, and Zuidema, Paquita. 2006. The mesoscale convection life cycle: building block or prototype for large-scale tropical waves? *Dynamics of Atmospheres and Oceans*, **42**(1), 3–29.

Marquet, Pascal. 2011. Definition of a moist entropy potential temperature: application to FIRE-I data flights. *Quarterly Journal of the Royal Meteorological Society*, **137**(656), 768–91.

Marquet, Pascal. 2017. A third-law isentropic analysis of a simulated hurricane. *Journal of the Atmospheric Sciences*, **74**(10), 3451–71.

Martin, G. M., Johnson, D. W., and Spice, S. 1994. The measurement and parameterization of effective radius of droplets in warm stratocumulus clouds. *Journal of the Atmospheric Sciences*, 51, 1823–42.

Mason, B. J. 1971. *The Physics of Clouds.* Clarendon Press. Page 671.

Mason, Shannon, Fletcher, Jennifer K., Haynes, John M., et al. 2015. A hybrid cloud regime methodology used to evaluate Southern Ocean cloud and shortwave radiation errors in ACCESS. *Journal of Climate*, **28**(15), 6001–18.

Matsuno, Taroh. 1966. Quasi-geostrophic motions in the equatorial area. *Journal of the Meteorological Society of Japan. Ser. II*, **44**(1), 25–43.

Mauritsen, T., Sedlar, J., Tjernström, M., et al. 2011. An Arctic CCN-limited cloud-aerosol regime. *Atmospheric Chemistry and Physics*, **11**(1), 165–73.

McDonald, James E. 1958. The physics of cloud modification. *Advances in Geophysics*, **5**, 223–303.

McDonald, James E. 1963. Use of the electrostatic analogy in studies of ice crystal growth. *Zeitschrift für Angewandte Mathematik und Physik*, **14**(5), 610–20.

Medeiros, Brian, and Stevens, Bjorn. 2011. Revealing differences in GCM representations of low clouds. *Climate Dynamics*, **36**(1), 385–99.

Mellado, Juan Pedro. 2010. The evaporatively driven cloud-top mixing layer. *Journal of Fluid Mechanics*, **660**(Oct.), 5–36.

Mellado, Juan Pedro. 2017. Cloud-top entrainment in stratocumulus clouds. *Annual Review of Fluid Mechanics*, **49**(Jan.), 145–69.

Mellor, George L. 1977. The Gaussian cloud model relations. *Journal of the Atmospheric Sciences*, **34**(2), 356–58.

Ming, Y., Ramaswamy, V., and Persad, G. 2010. Two opposing effects of absorbing aerosols on global-mean precipitation. *Geophysical Research Letters*, **37**, L13701.

Mironov, Dmitrii V. 2008. *Turbulence in the Lower Troposphere: Second-Order Closure and Mass-Flux Modelling Frameworks*. Springer.

Mishchenko, Michael I., Hovenier, Joop W., and Travis, Larry D. (eds). 2000. *Light Scattering by Nonspherical Particles: Theory, Measurements, and Applications*. Academic Press.

Mitchell, D. L., and Heymsfield, A. J. 2005. Refinements in the treatment of ice particle terminal velocities, highlighting aggregates. *Journal of the Atmospheric Sciences*, **62**, 1637–44.

Mitchell, J. F. B., Wilson, C. A., and Cunnington, W. M. 1987. On CO_2 climate sensitivity and model dependence of results. *Quarterly Journal of the Royal Meteorological Society*, **113**(475), 293–322.

Möhler, O., Field, P. R., Connolly, P., et al. 2006. Efficiency of the deposition mode ice nucleation on mineral dust particles. *Atmospheric Chemistry and Physics*, **6**, 3007–21.

Morrison, Hugh, de Boer, Gijs, Feingold, Graham, et al. 2012. Resilience of persistent Arctic mixed-phase clouds. *Nature Geoscience*, **5**(1), 11–17.

Muller, Caroline J., and Held, Isaac M. 2012. Detailed investigation of the self-aggregation of convection in cloud-resolving simulations. *Journal of the Atmospheric Sciences*, **69**(8), 2551–65.

Murphy, D. M., and Koop, T. 2005. Review of the vapour pressures of ice and supercooled water for atmospheric applications. *Quarterly Journal of the Royal Meteorological Society*, **131**(608), 1539–65.

Neelin, J. David, and Held, Isaac M. 1987. Modeling tropical convergence based on the moist static energy budget. *Monthly Weather Review*, **115**(1), 3–12.

Neelin, J. David, and Zeng, N. 2000. A quasi-equilibrium tropical circulation model-formulation. *Journal of the Atmospheric Sciences*, **57**(11), 1741–66.

Neggers, R. A. J. 2015. Exploring bin-macrophysics models for moist convective transport and clouds. *Journal of Advances in Modeling Earth Systems*, **7**(4), 2079–104.

Neggers, R. A. J., Neelin, J. D., and Stevens, B. 2007. Impact mechanisms of shallow cumulus convection on tropical climate dynamics. *Journal of Climate*, **20**(11), 2623–42.

Nicholls, S., and Turton, J. D. 1986. An observational study of the structure of stratiform cloud sheets: part II. entrainment. *Quarterly Journal of the Royal Meteorological Society*, **112**, 461–80.

Norris, J., and Slingo, A. 2009. Trends in Observed Cloudiness and Earth's Radiation Budget. In *Clouds in the Perturbed Climate System: Their Relationship to Energy Balance, Atmospheric Dynamics, and Precipitation*. Heintzenberg, J., and Charlson, Robert J. (eds). The MIT Press. Pages 17–36.

O'Gorman, Paul A., Allan, Richard P., Byrne, Michael P., and Previdi, Michael. 2012. Energetic constraints on precipitation under climate change. *Surveys in Geophysics*, **33**(3), 585–608.

Otto, A., Otto, F. E. L., Boucher, O., et al. 2013. Energy budget constraints on climate response. *Nature Geoscience*, **6**(6), 415–16.

Oueslati, Boutheina, and Bellon, Gilles. 2015. The double ITCZ bias in CMIP5 models: interaction between SST, large-scale circulation and precipitation. *Climate Dynamics*, **44**(3–4), 585–607.

Park, S., Leovy, C. B., and Rozendaal, M. A. 2004. A new heuristic Lagrangian marine boundary layer cloud model. *Journal of the Atmospheric Sciences*, **61**, 3002–24.

Pauluis, Olivier, and Held, I. M. 2002a. Entropy budget of an atmosphere in radiative-convective equilibrium. Part I: maximum work and frictional dissipation. *Journal of the Atmospheric Sciences*, **59**(2), 125–39.

Pauluis, Olivier, and Held, I. M. 2002b. Entropy budget of an atmosphere in radiative-convective equilibrium. Part II: latent heat transport and moist processes. *Journal of the Atmospheric Sciences*, **59**(2), 140–9.

Pawlowska, H., Grabowski, W. W., and Brenguier, J.-L. 2006. Observations of the width of cloud droplet spectra in stratocumulus. *Geophysical Research Letters*, **33**, L19810.

Perovich, Don K., Andreas, E. L., Curry, J. A., et al. 1999. Year on ice gives climate insights. *Eos, Transactions American Geophysical Union*, **80**(41), 481–6.

Persson, P. Ola G., Shupe, Matthew D., Perovich, Don, and Solomon, Amy. 2016. Linking atmospheric synoptic transport, cloud phase, surface energy fluxes, and sea-ice growth: observations of midwinter SHEBA conditions. *Climate Dynamics*, **49**(4), 1341–64.

Peters, Karsten, Crueger, Traute, Jakob, Christian, and Mobis, Benjamin. 2017. Improved MJO-simulation in ECHAM6.3 by coupling a stochastic multicloud model to the convection scheme. *Journal of Advances in Modeling Earth Systems*, **9**(1), 193–219.

Petters, M. D., and Kreidenweis, S. M. 2007. A single parameter representation of hygroscopic growth and cloud condensation nuclei activity. *Atmospheric Chemistry and Physics*, **7**, 1961–71.

Petty, Grant W. 2006. *A First Course in Atmospheric Radiation. 2nd Edition*. Sundog Publishing.

Pierrehumbert, R. T. 1995. Thermostats, radiator fins, and the local runaway greenhouse. *Journal of Atmospheric Sciences*, **52**(10), 1784–1806.

Pierrehumbert, R. T., and Yang, H. 1993. Global chaotic mixing on isentropic surfaces. *Journal of Atmospheric Sciences*, **50**(15), 2462–80.

Pincus, Robert, Batstone, Crispian P., Hofmann, Robert J. Patrick, Taylor, Karl E., and Glecker, Peter J. 2008. Evaluating the present-day simulation of clouds, precipitation, and radiation in climate models. *Journal of Geophysical Research*, **113**(D14), D14209.

Pincus, Robert, Platnick, Steven, Ackerman, Steven A., Hemler, Richard S., and Patrick Hofmann, Robert J. 2012. Reconciling simulated and observed views of clouds: MODIS, ISCCP, and the limits of instrument simulators. *Journal of Climate*, **25**(13), 4699–720.

Pino, D., Vilà-Guerau de Arellano, J., and Duynkerke, P. G. 2003. The contribution of shear to the evolution of a convective boundary layer. *Journal of Atmospheric Sciences*, **60**(16), 1913–26.

Pithan, Felix, Medeiros, Brian, and Mauritsen, Thorsten. 2014. Mixed-phase clouds cause climate model biases in Arctic wintertime temperature inversions. *Climate Dynamics*, **43**(1), 289–303.

Plant, Robert S., and Yano, Jun-Ichi. 2015. *Parameterization of Atmospheric Convection*. Imperial College Press.

Plant, R. S., and Craig, George C. 2008. A stochastic parameterization for deep convection based on equilibrium statistics. *Journal of Atmospheric Sciences*, **65**(1), 87–105.

Pope, Stephen B. 2000. *Turbulent Flows*. Cambridge University Press.

Posselt, D. J., Stephens, G. L., and Miller, M. 2008. CloudSat: adding a new dimension to a classical view of extratropical cyclones. *Bulletin of the American Meteorological Society*, **89**, 599–609.

Pruppacher, H. R., and Klett, J. D. 2010. *Microphysics of Clouds and Precipitation*. Atmospheric and Oceanographic Sciences Library. Springer.

Quaas, Johannes. 2012. Evaluating the "critical relative humidity" as a measure of subgrid-scale variability of humidity in general circulation model cloud cover parameterizations using satellite data. *Journal of Geophysical Research: Atmospheres*, **117**(D9).

Quaas, Johannes. 2015. Approaches to observe effects of anthropogenic aerosols on clouds and radiation. *Current Climate Change Reports*, **1**, 297–304.

Quaas, Johannes., Bony, S., Collins, W. D., et al. 2009. *Current Understanding and Quantification of Clouds in the Changing Climate System and Strategies for Reducing Critical Uncertainties*. MIT Press. Pages 557–73.

Quaas, Johannes., Stevens, B., Lohmann, U., and Stier, P. 2010. Interpreting the cloud cover – aerosol optical depth relationship found in satellite data using a general circulation model. *Atmospheric Chemistry and Physics*, **10**, 6129–135.

Ramage, Colin S. 1971. *Monsoon Meteorology*. International Geophysics Series. Vol. 15. Academic Press.

Ramanathan, V., Cess, R. D., Harrison, E. F., et al. 1989. Cloud-radiative forcing and climate: results from the Earth radiation budget experiment. *Science*, **243**(4887), 57–63.

Randall, David A. 1980. Conditional instability of the first kind upside downocument. *Journal of Atmospheric Sciences*, **37**, 125–30.

Randall, David. A. 2000. *General Circulation Model Development: Past, Present, and Future*. International Geophysics. Elsevier Science.

David Randall. 2012. *Atmosphere, Clouds and Climate*. Princeton University Press.

Randall, David, DeMott, Charlotte, Stan, Cristiana, et al. 2016. Simulations of the tropical general circulation with a multiscale global model. *Meteorological Monographs*, **56**, 15.1–15.15.

Randall, David A., Krueger, S. K., Bretherton, C. S., et al. 2003. Confronting models with data – The GEWEX Cloud Systems Study. *Bulletin of the American Meteorological Society*, **84**, 455–69.

Randall, David A., Shao, Qingqiu, and Moeng, Chin-Hoh. 1992. A second-order bulk boundary-layer model. *Journal of the Atmospheric Sciences*, **49**(20), 1903–23.

Raymond, David J., Sessions, Sharon L., Sobel, Adam H., and Fuchs, Željka. 2009. The mechanics of gross moist stability. *Journal of Advances in Modeling Earth Systems*, **1**(3).

Rieck, M., Nuijens, L., and Stevens, B. 2012. Marine boundary layer cloud feedbacks in a constant relative humidity atmosphere. *Journal of Atmospheric Sciences*, **69**(Aug.), 2538–50.

Riehl, H. 1954. *Tropical Meteorology*. McGraw-Hill.

Riehl, H., and Malkus, J. S. 1957. On the heat balance and maintenance of circulation in the trades. *Quarterly Journal of the Royal Meteorological Society*, **83**, 21–9.

Riehl, H., and Malkus, J. S. 1958. On the heat balance in the equatorial trough zone. *Geophysica*, **6**(3–4), 503–37.

Riley, Emily M., Mapes, Brian E., and Tulich, Stefan N. 2011. Clouds associated with the Madden-Julian oscillation: a new perspective from CloudSat. *Journal of the Atmospheric Sciences*, **68**(12), 3032–51.

Rogers, R. R., and Yau, M. K. 1996. *A Short Course in Cloud Physics*. Butterworth Heinemann. Page 290.

Romps, D. 2010. A direct measurement of entrainment. *Journal of Atmospheric Sciences*, **67**, 1908–27.

Rosenfeld, D., Lohmann, U., Raga, G. B., et al. 2008. Flood or drought: how do aerosols affect precipitation? *Science*, **321**, 1309–13.

Rossow, W. B., and Schiffer, R. A. 1999. Advances in understanding clouds from ISCCP. *Bulletin of the American Meteorological Society*, **80**(11), 2261–87.

Rotstayn, L. D., and Lohmann, U. 2002. Tropical rainfall trends and the indirect aerosol effect. *Journal of Climate*, **15**, 2103–16.

Royal Society. 2009. *Geoengineering the Climate – Science, Governance and Uncertainty*. Royal Society Policy document.

Sakradzija, Mirjana, Seifert, Axel, and Heus, Thijs. 2015. Fluctuations in a quasi-stationary shallow cumulus cloud ensemble. *Nonlinear Processes in Geophysics*, **22**(1), 65–85.

Sanchez-Lorenzo, A., Laux, P., Hendricks-Franssen, H.-J., et al. 2012. Assessing large-scale weekly cycles in meteorological variables: a review. *Atmospheric Chemistry and Physics*, **12**, 5755–71.

SCEP Study of Critical Environmental Problems. 1970. *Man's Impact on the Global Environment*. The MIT Press.

Schalkwijk, J., Jonker, H. J. J., and Siebesma, A. P. 2013. Simple solutions to steady-state cumulus regimes in the convective boundary layer. *Journal of Atmospheric Sciences*, **70**, 3656–72.

Schalkwijk, J., Jonker, H. J. J., Siebesma, A. P., and Van Meijgaard, E. 2015. Weather forecasting using GPU-based large-eddy simulations. *Bulletin of the American Meteorological Society*, **96**, 715–23.

Schär, C., Vidale, P. L., Lüthi, D., et al. 2004. The role of increasing temperature variability in European summer heatwaves. *Nature*, **427**, 332–6.

Schemann, Vera, Stevens, Bjorn, Grutzun, Verena, and Quaas, Johannes. 2013. Scale dependency of total water variance and its implication for cloud parameterizations. *Journal of the Atmospheric Sciences*, **70**(11), 3615–30.

Schneider, S. H. 1972. Cloudiness as a global climate feedback mechanism: the effects on the radiation balance and surface temperature of variations in cloudiness. *Journal of Atmospheric Sciences*, **29**, 1413–22.

Schneider, Tapio, and Sobel, Adam H. 2007. *The Global Circulation of the Atmosphere*. Princeton University Press.

Schulz, J., Albert, P., Behr, H.-D., et al. 2009. Operational climate monitoring from space: the EUMETSAT Satellite Application Facility on Climate Monitoring (CM-SAF). *Atmospheric Chemistry and Physics*, **9**(5), 1687–709.

Schumacher, Courtney, Zhang, Minghua H., and Ciesielski, Paul E. 2007. Heating structures of the TRMM field campaigns. *Journal of the Atmospheric Sciences*, **64**(7), 2593–610.

Schwartz, Stephen E. 2011. Feedback and sensitivity in an electrical circuit: an analog for climate models. *Climatic Change*, **106**(2), 315–26.

Schwartz, Stephen E. 2012. Determination of Earth's transient and equilibrium climate sensitivities from observations over the twentieth century: strong dependence on assumed forcing. *Surveys in Geophysics*, **33**, 745–77.

Seneviratne, S. I., Nicholls, N., Easterling, D., et al. 2012. Changes in Climate Extremes and Their Impacts on the Natural Physical Environment. In *Managing the Risks of Extreme Events and Disasters to Advance Climate Change Adaptation*. Field, C. B., Barros, V., Stocker, T. F., et al. (eds). A Special Report of Working Groups I and II of the Intergovernmental Panel on Climate Change (IPCC). Cambridge University Press. Pages 109–230.

Seneviratne, S. I., Wilhelm, M., Stanelle, T., et al. 2013. Impact of soil moisture-climate feedbacks on CMIP5 projections: first results from the GLACE-CMIP5 experiment. *Geophysical Research Letters*, **40**, 5212–17.

Shaw, T. A., Baldwin, M., Barnes, E. A., et al. 2016. Storm track processes and the opposing influences of climate change. *Nature Geoscience*, **9**, 656–64.

Sherwood, Steven C., Bony, Sandrine, Boucher, Olivier, et al. 2015. Adjustments in the forcing-feedback framework for understanding climate change. *Bulletin of the American Meteorological Society*, **96**(2), 217–28.

Sherwood, Steven C., Bony, Sandrine, and Dufresne, Jean-Louis. 2014. Spread in model climate sensitivity traced to atmospheric convective mixing. *Nature*, **505**, 37–42.

Shupe, M. D., Persson, P. O. G., Brooks, I. M., et al. 2013. Cloud and boundary layer interactions over the Arctic sea ice in late summer. *Atmospheric Chemistry and Physics*, **13**(18), 9379–99.

Siebesma, A. Pier, Bretherton, C. S., Brown, A., et al. 2003. A large eddy simulation intercomparison study of shallow cumulus convection. *Journal of the Atmospheric Sciences*, **60**(10), 1201–19.

Siebesma, A. Pier, and Cuijpers, J. W. M. 1995. Evaluation of parametric assumptions for shallow cumulus convection. *Journal of Atmospheric Sciences*, **52**, 650–66.

Siebesma, A. Pier, Soares, Pedro M. M., and Teixeira, Joao. 2007. A combined eddy-diffusivity mass-flux approach for the convective boundary layer. *Journal of the Atmospheric Sciences*, **64**(4), 1230–48.

Simmons, A. J., and Hollingsworth, A. 2002. Some aspects of the improvement in skill of numerical weather prediction. *Quarterly Journal of the Royal Meteorological Society*, **128**(580), 647–77.

Simpson, J. 1973. The Global Energy Budget and the Role of Cumulus Clouds. NOAA Technical Memorandum ERL WMPO-8, Boulder, Colorado. ftp://ftp.library.noaa.gov/noaa_documents.lib/OAR/ERL_WMPO/TM_ERL_WMPO_8.pdf. Accessed 14/01/2020.

Smythe, W. R. 1962. Charged right circular cylinder. Journal of Applied Physics, **33**(10), 2966.

Sobel, Adam H., and Bretherton, Christopher S. 2000. Modeling tropical precipitation in a single column. *Journal of Climate*, **13**(24), 4378–92.

Soden, B. J., and Held, I. M. 2006. An assessment of climate feedbacks in coupled ocean-atmosphere models. *Journal of Climate*, **19**(14), 3354–60.

Soden, Brian J., Held, Isaac M., Colman, Robert, et al. 2008. Quantifying Climate Feedbacks Using Radiative Kernels. *Journal of Climate*, **21**(14), 3504–20.

Sommeria, G., and Deardorff, J. W. 1977. Subgrid-scale condensation in models of nonprecipitating clouds. *Journal of the Atmospheric Sciences*, **34**(2), 344–55.

Stensrud, D. J. 2007. *Parameterization Schemes: Keys to Understanding Numerical Weather Prediction Models*. Cambridge University Press.

Stephens, Graeme L. 1994. *Remote Sensing of the Lower Atmosphere: An Introduction*. Oxford University Press.

Stephens, Graeme L. 2005. Cloud feedbacks in the climate system: a critical review. *Journal of Climate*, **18**(2), 237–73.

Stephens, Graeme L., Brien, Denis O., Webster, Peter J., et al. 2015. The albedo of Earth. *Reviews of Geophysics*, **53**(1), 141–63.

Stephens, Graeme L., Li, J., Wild, M., et al. 2012a. An update on Earth's energy balance in light of the latest global observations. *Nature Geoscience*, **5**, 691–6.

Stevens, B. 2005. Atmospheric moist convection. *Annual Review of Earth and Planetary Sciences*, **33**(1), 605–43.

Stephens, Graeme L., L'Ecuyer, Tristan, Forbes, Richard, et al. 2010. Dreary state of precipitation in global models. *Journal of Geophysical Research*, **115**(D24), D24211.

Stephens, Graeme L., Li, Juilin, Wild, Martin, et al. 2012b. An update on Earth's energy balance in light of the latest global observations. *Nature Geoscience*, **5**(10), 691–6.

Stevens, Bjorn. 2000. Quasi-steady analysis of a PBL model with an eddy-diffusivity profile and nonlocal fluxes. *Monthly Weather Review*, **128**(03), 824–36.

Stevens, Bjorn. 2006. Bulk boundary-layer concepts for simplified models of tropical dynamics. *Theoretical and Computational Fluid Dynamics*, **20**(5–6), 279–304.

Stevens, Bjorn. 2007. On the growth of layers of nonprecipitating cumulus convection. *Journal of Atmospheric Sciences*, **64**, 2916–31.

Stevens, Bjorn. 2015. Rethinking the lower bound on aerosol radiative forcing. *Journal of Climate*, **28**, 4794–819.

Stevens, Bjorn, and Bony, Sandrine. 2013. Water in the atmosphere. *Physics Today*, **66**(6), 29, https://doi.org/10.1063/PT.3.2009.

Stevens, Bjorn, and Feingold, G. 2009. Untangling Aerosol Effects on Clouds and Precipitation in a Buffered System. *Nature*, **461**, 607–13.

Stevens, Bjorn, Moeng, C.-H., Ackerman, A. S., et al. 2005. Evaluation of large-eddy simulations via observations of nocturnal marine stratocumulus. *Monthly Weather Review*, **133**, 1443–62.

Stevens, Bjorn, and Schwartz, Stephen E. 2012. Observing and modeling Earth's energy flows. *Surveys in Geophysics*, **33**(3–4), 779–816.

Stevens, Bjorn, Sherwood, Steven C., Bony, Sandrine, and Webb, Mark J. 2016. Prospects for narrowing bounds on Earth's equilibrium climate sensitivity. *Earth's Future*, **4**(11), 512–22.

Stier, P. 2016. Limitations of passive remote sensing to constrain global cloud condensation nuclei. *Atmospheric Chemistry and Physics*, **16**, 6595–607.

Stier, P., Feichter, J., Kinne, S., et al. 2005. The aerosol-climate model ECHAM5-HAM. *Atmospheric Chemistry and Physics*, **5**, 1125–56.

Stohl, A., Aamaas, B., Amann, M., et al. 2015. Evaluating the climate and air quality impacts of short-lived pollutants. *Atmospheric Chemistry and Physics*, **15**, 10529–66.

Straka, J. M. 2009. *Cloud and Precipitation Microphysics: Principles and Parameterizations*. Cambridge University Press.

Stubenrauch, C. J., Rossow, W. B., Kinne, S., et al. 2013. Assessment of global cloud datasets from satellites: project and database initiated by the GEWEX Radiation Panel. *Bulletin of the American Meteorological Society*, **94**(7), 1031–49.

Stull, R. B. 1988. *An Introduction to Boundary Layer Meteorology*. Kluwer Academic Publishers.

Sušelj, Kay, Teixeira, João, and Chung, Daniel. 2013. A unified model for moist convective boundary layers based on a stochastic eddy-diffusivity/mass-flux parameterization. *Journal of Atmospheric Sciences*, **70**(7), 1929–53.

Suzuki, K., Stephens, G., van den Heever, S., and Nakajima, T. 2011. Diagnosis of the warm rain process in cloud-resolving models using joint CloudSat and MODIS observations. *Journal of Atmospheric Sciences*, **68**, 2655–70.

Tao, W.-K., and Adler, R. 2013. *Cloud Systems, Hurricanes, and the Tropical Rainfall Measuring Mission (TRMM): A Tribute to Joanne Simpson.* Meteorological Monographs. Vol. 29. American Meteorological Society.

Tao, W.-K., Chen, J.-P., Li, Z., Wang, C., and Zhang, C. 2012. Impact of aerosols on convective clouds and precipitation. *Reviews of Geophysics,* **50**, RG2001.

Tawfik, A. B., and Dirmeyer, P. A. 2014. A process-based framework for quantifying the atmospheric preconditioning of surface-triggered convection. *Geophysical Research Letters,* **41**, 173–8.

Taylor, K. E. 2001. Summarizing multiple aspects of model performance in a single diagram. *Journal of Geophysical Research,* **106**, 7183–92.

Teixeira, J., Cardoso, S., Bonazzola, M., et al. 2011. Tropical and subtropical cloud transitions in weather and climate prediction models: the GCSS/WGNE Pacific Cross-Section Intercomparison (GPCI). *Journal of Climate,* **24**(20), 5223–56.

Tennekes, H. 1973. A model for the dynamics of the inversion above a convective layer. *Journal of Atmospheric Sciences,* **30**, 558–67.

Thorsen, Tyler J., Fu, Qiang, and Comstock, Jennifer M. 2013. Cloud effects on radiative heating rate profiles over Darwin using ARM and A-train radar/lidar observations. *Journal of Geophysical Research: Atmospheres,* **118**(11), 5637–54.

Tiedtke, M. 1989. A comprehensive mass flux scheme for cumulus parameterization in large-scale models. *Monthly Weather Review,* **117**(8), 1779–1800.

Tiedtke, M. 1993. Representation of clouds in large-scale models. *Monthly Weather Review,* **121**(11), 3040–61.

Tjernström, M., Leck, C., Birch, C. E., et al. 2014. The Arctic Summer Cloud Ocean Study (ASCOS): overview and experimental design. *Atmospheric Chemistry and Physics,* **14**(6), 2823–69.

Tomita, H., Miura, H., Iga, S., Nasuno, T., and Satoh, M. 2005. A global cloud-resolving simulation: preliminary results from an aqua planet experiment. *Geophysical Research Letters,* **32**(8), L08805.

Tompkins, A. M. 2002. A prognostic parameterization for the subgrid-scale variability of water vapor and clouds in largescale models and its use to diagnose cloud cover. *Journal of Atmospheric Sciences,* **59**, 1917–42.

Trenberth, K. E., and Fasullo, J. T. 2010. Simulation of present-day and twenty-first century energy budgets of the Southern Ocean. *Journal of Climate,* **23**, 440–54.

Trenberth, Kevin E., Fasullo, John T., and Kiehl, Jeffrey. 2009. Earth's global energy budget. *Bulletin of the American Meteorological Society,* **90**(3), 311–23.

Troen, I. B., and Mahrt, L. 1986. A simple model of the atmospheric boundary layer; sensitivity to surface evaporation. *Boundary-Layer Meteorology,* **37**(1), 129–48.

Tselioudis, George, and Jakob, C. 2002. Evaluation of midlatitude cloud properties in a weather and a climate model: dependence on dynamic regime and spatial resolution. *Journal of Geophysical Research,* **107**(D24), 4781.

Tselioudis, George, Lipat, B. R., Konsta, D., Grise, K. M., and Polvani, L. M. 2016. Midlatitude cloud shifts, their primary link to the Hadley cell, and their diverse radiative effects. *Geophysical Research Letters,* **43**, 4594–601.

Tselioudis, George, and Rossow, W. B. 2006. Climate feedback implied by observed radiation and precipitation changes with midlatitude storm strength and frequency. *Geophysical Research Letters,* **33**, L02704.

Tselioudis, George, Rossow, William, Zhang, Yuanchong, and Konsta, Dimitra. 2013. Global weather states and their properties from passive and active satellite cloud retrievals. *Journal of Climate,* **26**(19), 7734–46.

Tselioudis, George, Zhang, Y., and Rossow, W. B. 2000. Cloud and radiation variations associated with northern midlatitude low and high sea level pressure regimes. *Journal of Climate,* **13**, 312–27.

Twomey, S. 1974. Pollution and the planetary albdeo. *Atmospheric Environment,* **8**, 1251–8.

Twomey, S. 1977. The influence of pollution on the shortwave albedo of clouds. *Journal of Atmospheric Sciences,* **34**, 1149–52.

Uttal, Taneil, Curry, Judith A., Mcphee, Miles G., et al. 2002. Surface heat budget of the Arctic Ocean. *Bulletin of the American Meteorological Society,* **83**(2), 255–75.

Vallis, Geoffrey K. 2006. *Atmospheric and Ocean Fluid Dynamics: Fundamentals and Large-Scale Circulation.* Cambridge University Press.

Vallis, Geoffrey K., Zurita-Gotor, P., Cairns, C., and Kidston, J. 2015. Response of the large-scale structure of the atmosphere to global warming. *Quarterly Journal of the Royal Meteorological Society,* **141**, 1479–501.

van de Hulst, H. C. 1980. *Light Scattering by Small Particles.* Dover Publications.

van der Ent, R. J., Savenije, H. H. G., Schaefli, B., and Steele-Dunne, S. C. 2010. Origin and fate of atmospheric moisture over continents. *Water Resources Research,* **46**(9).

van Heerwaarden, C. C., and Vilà Guerau de Arellano, J. 2008. Relative humidity as an indicator for cloud formation over heterogeneous land surfaces. *Journal of Atmospheric Sciences,* **65**, 3263–77.

vanZanten, M. C., Duynkerke, P. G., and Cuijpers, J. W. M. 1999. Entrainment parameterization in convective boundary layers. *Journal of Atmospheric Sciences,* **56**, 813–28.

Vial, Jessica, Dufresne, Jean-Louis, and Bony, Sandrine. 2013. On the interpretation of inter-model spread in CMIP5 climate sensitivity estimates. *Climate Dynamics,* **41**(11–12), 3339–62.

Vihma, T., Pirazzini, R., Fer, I., et al. 2014. Advances in understanding and parameterization of small-scale physical processes in the marine Arctic climate system: a review. *Atmospheric Chemistry and Physics,* **14**(17), 9403–50.

Vilà-Guerau de Arellano, J., van Heerwaarden, C. C., van Stratum, B. J. J., and van den Dries, K. 2015. *Atmospheric Boundary Layer: Integrating Air Chemistry and Land Interactions.* Cambridge University Press.

Voigt, Aiko, Stevens, Bjorn, Bader, Jürgen, and Mauritsen, Thorsten. 2013. The observed hemispheric symmetry in reflected shortwave irradiance. *Journal of Climate,* **26**(2), 468–77.

Wakimoto, R. M., and Srivastava, R. 2003. *Radar and Atmospheric Science: A Collection of Essays in Honor of David Atlas.* Meteorological Monographs. Vol. 30. American Meteorological Society.

Walker, Christopher C., and Schneider, Tapio. 2006. Eddy influences on Hadley circulations: simulations with an idealized GCM. *Journal of the Atmospheric Sciences,* **63**(12), 3333–50.

Webb, M., Senior, C., Bony, S., and Morcrette, J.-J. 2001. Combining ERBE and ISCCP data to assess clouds in the Hadley Centre, ECMWF and LMD atmospheric climate models. *Climate Dynamics,* **17**, 905–22.

Weitkamp, Claus. 2005. *Lidar: Range-Resolved Optical Remote Sensing of the Atmosphere.* Springer.

Wendisch, M., and Brenguier, J.-l. (eds). 2013. *Airborne Measurements for Environmental Research: Methods and Instruments.* Wiley-VCH Verlag GmbH & Co, Weinheim.

Westbrook, C. D., Hogan, R. J, and Illingworth, A. J. 2008. The capacitance of pristine ice crystals and aggregate snowflakes. *Journal of Atmospheric Sciences*, **65**, 206–19.

Wexler, H. 1936. Cooling in the lower atmosphere and the structure of polar continental air. *Monthly Weather Review*, **64**(4), 122–36.

Wheeler, Matthew C., and Hendon, Harry H. 2004. An all-season real-time multivariate MJO index: development of an index for monitoring and prediction. *Monthly Weather Review*, **132**(8), 1917–32.

Wild, Martin, Folini, Doris, Hakuba, Maria Z., et al. 2015. The energy balance over land and oceans: an assessment based on direct observations and CMIP5 climate models. *Climate Dynamics*, **44**(11), 3393–429.

Wild, Martin, Gilgen, H., Roesch, A., et al. 2005. From dimming to brightening: decadal changes in solar radiation at Earth's surface. *Science*, **308**, 847–50.

Wilks, D. S. 2011. *Statistical Methods in the Atmospheric Sciences*. Elsevier.

Williams, K. D., Bodas-Salcedo, A., Déqué, M., et al. 2013. The Transpose-AMIP II Experiment and its application to the understanding of Southern Ocean cloud biases in climate models. *Journal of Climate*, **26**(10), 3258–74.

Wilson, D. K. 2001. An alternative function for the wind and temperature gradients in unstable surface layers. *Boundary-Layer Meteorology*, **99**, 151–8.

Wing, A. A., Emanuel, K., Holloway, C. E., and Muller, C. 2017. Convective self-aggregation in numerical simulations: a review. *Surveys in Geophysics*, **38**, 1173–97.

WMO. 2017. International Cloud Atlas: Manual on the Observation of Clouds and Other Meteors. https://cloudatlas.wmo.int/home.html. Accessed 13/11/2017.

Wood, R. 2007. Cancellation of aerosol indirect effects in marine stratocumulus through cloud thinning. *Journal of the Atmospheric Sciences*, **64**, 2657–69.

Wood, R. 2012. Stratocumulus clouds. *Monthly Weather Review*, **140**, 2373–423.

Wood, R., and Bretherton, C. S. 2004. Boundary layer depth, entrainment, and decoupling in the cloud-capped subtropical and tropical marine boundary layer. *Journal of Climate*, **17**, 3576–88.

Wood, R., Field, Paul R., and Cotton, W. R. 2002. Autoconversion rate bias in stratiform boundary layer cloud parameterizations. *Atmospheric Research*, **65**(1), 109–28.

Wood, R., and Field, Paul R. 2011. The distribution of cloud horizontal sizes. *Journal of Climate*, **24**(18), 4800–16.

Wyant, Matthew C., Bretherton, Christopher S., Chlond, Andreas, et al. 2007. A single-column model intercomparison of a heavily drizzling stratocumulus-topped boundary layer. *Journal of Geophysical Research*, **112**(D24), D24204 -n/a.

Wyngaard, John C. 2010. *Turbulence in the Atmosphere*. Cambridge University Press.

Yanai, Michia, Esbensen, Steven, and Chu, Jan-Hwa. 1973. Determination of bulk properties of tropical cloud clusters from large-scale heat and moisture budgets. *Journal of the Atmospheric Sciences*, **30**, 611–27.

Zelinka, Mark D., and Hartmann, Dennis L. 2010. Why is longwave cloud feedback positive? *Journal of Geophysical Research*, **115**(D16).

Zelinka, Mark D., Klein, Stephen A., and Hartmann, Dennis L. 2012. Computing and Partitioning cloud feedbacks using cloud property histograms. Part II: attribution to changes in cloud amount, altitude, and optical depth. *Journal of Climate*, **25**(11), 3736–54.

Zelinka, Mark D., Klein, Stephen A., Taylor, Karl E., et al. 2013. Contributions of different cloud types to feedbacks and rapid adjustments in CMIP5. *Journal of Climate*, **26**(14), 5007–27.

Zhang, Chidong. 2005. Madden-Julian oscillation. *Reviews of Geophysics*, **43**(2).

Zhang, Guang J. 2002. Convective quasi-equilibrium in midlatitude continental environment and its effect on convective parameterization. *Journal of Geophysical Research: Atmospheres*, **107**(D14), ACL 12-1–ACL 12 -16.

Zhang, Guang J. 2003. Convective quasi-equilibrium in the tropical western Pacific: comparison with midlatitude continental environment. *Journal of Geophysical Research: Atmospheres*, **108**(D19).

Zhang, M., Bretherton, C. S., Blossey, P. N., et al. 2013. CGILS: results from the first phase of an international project to understand the physical mechanisms of low cloud feedbacks in single column models. *Journal of Advances in Modeling Earth Systems*, **5**(4), 826–42.

Zipser, Edward. 1969. The role of organized unsaturated downdrafts in the structure and rapid decay of an equatorial disturbance. *Journal of Applied Meteorology*, **8**, 799–814.

Zygmuntowska, M., Mauritsen, T., Quaas, J., and Kaleschke, L. 2012. Arctic Clouds and Surface Radiation – a critical comparison of satellite retrievals and the ERA-Interim reanalysis. *Atmospheric Chemistry and Physics*, **12**(14), 6667–77.

Index

Printed in the United States
by Baker & Taylor Publisher Services